P9-AFV-838

QUANTUM MECHANICS

Classical Results, Modern Systems,
and Visualized Examples

QUANTUM MECHANICS

Classical Results, Modern Systems, and Visualized Examples

Richard W. Robinett

Department of Physics
Pennsylvania State University

New York Oxford
OXFORD UNIVERSITY PRESS
1997

Oxford University Press

Oxford New York
Athens Auckland Bangkok Bombay
Calcutta Cape Town Dar es Salaam Delhi
Florence Hong Kong Istanbul Karachi
Kuala Lumpur Madras Madrid Melbourne
Mexico City Nairobi Paris Singapore
Taipei Tokyo Toronto

and associated companies in

Berlin Ibadan

Published by Oxford University Press, Inc.,
198 Madison Avenue, New York, New York, 10016

Library of Congress Cataloging-in-Publication Data

Robinett, Richard W. (Richard Wallace)
Quantum mechanics : classical results, modern systems, and visualized examples / Richard W. Robinett.
p. cm.
Includes bibliographical references and index.
ISBN 0-19-509202-3
1. Quantum theory. I. Title.
QC174.12.R6 1997
530.1′ 2--dc20 95-47217
CIP

Printing (last digit): 9 8 7 6 5 4 3 2 1

Printed in the United States of America on acid-free paper

We shall not cease from exploration
And the end of all our exploring
Will be to arrive where we started
And know the place for the first time.

T. S. Eliot

Contents

README.1ST (A Preface)

When I access the computerized library information system at my university, I find 39 volumes with the exact title "Quantum Mechanics" and several dozen more with quite similar names. This fact is perhaps indicative of both the importance of this subject in the physics curriculum and of the continuing need for updated presentations of the material. Many of the most familiar aspects of quantum theory that are standardly taught, after all, were already well understood almost three-quarters of a century ago. On the other hand, the central role played by quantum physics (in the guise of both wave mechanics and the spin statistics connection) in the properties of matter on many length scales in the universe guarantees that new examples and novel applications of the basic principles will continue to appear, providing new motivation to master old principles. As indicated by the title, these two themes form one important aspect of this textbook dealing with nonrelativistic quantum mechanics at the undergraduate level; it helps justify the "classical results and modern systems" part.

The other feature stressed in this text is the visual display of quantitative information and qualitative concepts, hence the "visualized examples" of the title. The inclusion of such material is motivated by several factors:

- The physical phenomena described by quantum theory are often quite remote from our daily existence, so that our intuition about the nature of quantum systems must be built up from sources of information other than direct experience. The incredibly successful application of the mathematical models and formalism of quantum theory ("the equations") is one such input, but other more complementary approaches are often useful.[1] While the "equation" that purports to relate one picture with 10^3 words is certainly a cliche, that doesn't necessarily make it wrong; a carefully selected figure, diagram, graph, or collection of plotted data can indeed often convey the desired information far more quickly and with more comprehension than many equations.
- One of the goals of physical science (as practiced by both researchers and teachers of the subject) is to discover, understand (hopefully quantitatively), and explain the connections that exist in the physical world. The meeting ground of the sometimes arcane formalism of a theorist, the observations of an experimentalist (for example, the data obtained from a bafflingly complex apparatus), and the rest of the scientific community is often the figures in a preprint, report, or published paper. These often consist of experimental data compared with theoretical predictions, usually plotted in an especially appropriate way to emphasize

[1] In this context, it is perhaps appropriate to note that many undergraduate students who have taken an introductory mechanics course, even those who have "done well" as measured by their ability to manipulate equations, often do quite poorly on standardized exams that test their conceptual understanding. Our daily experience is often not enough to appreciate fully even the notions of classical mechanics.

the most important relations between the observable quantities. It can be argued that the ability to decipher and understand the presentation of information in this manner should form an important part of the education of any scientifically literate person (whether a specialist or an interested "civilian").

- The increased use of computers can not only dramatically increase one's computational power for solving "traditional" problems, but it can also (and perhaps should) let us think in new ways and ask questions that we might not otherwise consider if we relied solely on analytic solutions. The change in emphasis from closed-form solutions to numerical results brings with it a corresponding change in the way we approach the "output" of a calculation. The visual presentation of numerical solutions via graphs or other techniques is a natural adjunct to the use of computer technology, and is an increasingly common and central feature in many fields. Much of the progress in the study of chaos and chaotic behavior in dynamical systems, for example, is connected with the reliance on visual imagery. Many of the famous images of chaotic trajectories, phase space diagrams, and fractal structures have had as much impact as the corresponding mathematical formalism.

For all of these reasons, the use of visualization is heavily emphasized in this text. Part I (The Quantum Paradigm), where the machinery of quantum mechanics is introduced, consists, to a large extent, of plots of wavefunctions that are designed to help visualize the physical content of solutions of the Schrödinger equation, both in position and momentum space. In Part II (The Quantum World), where the emphasis switches to more physical systems, an attempt is made to introduce the reader to experimental data that make contact with the more theoretical discussions.

In addition, there are other, less-standard, elements to the presentation in this text:

- The connections to classical mechanics are stressed throughout the book. Some aspects of this program include:
  - The comparison of classical and quantum probability distributions, not only for familiar bound state systems such as the harmonic oscillator (Chap. 10) and infinite well (Chap. 6) but also for unbound systems such as the accelerating particle (Sec. 5.6.2) and particle near unstable equilibrium (Sec. 10.6).
  - Wavepacket solutions for several model systems are introduced and studied.
  - Solutions in momentum space are emphasized in an attempt to appreciate better the connection to the classical velocity variable; this is complementary to the discussions regarding the uncertainty principle found in many texts. Physical observables directly related to momentum distributions are now important in many areas of experimental physics; examples range from momentum distributions obtained from neutron scattering in solids to parton distributions of the proton. The ability to extract physical information from a system in both representations is increasingly useful.
  - An expanded discussion of quantum scattering includes a more-concrete foundation in classical scattering ideas. There is also an emphasis on electromagnetic scattering as an exactly calculable example, in much the same way that the hydrogen atom is discussed for bound states. New effects beyond the contributions of Coulomb's law (such as the symmetry requirements arising from indistinguishability or the existence of the strong nuclear force) can then be introduced as deviations from the results for a simple $1/r^2$ force law.

·A self-contained appendix describing classical Hamiltonian dynamics is included so that the reader can make contact with the standard formulation of quantum mechanics based on the Schrödinger equation more easily.

·A brief discussion of the connections between classical and quantum chaos is included (Sec. 16.4) in the context of two-dimensional infinite well problems (the quantum analog of classical "billiards") illustrating some of the different methods used to characterize the "chaoticness" of a quantum system. This also complements some of the more standard discussions in the text on the important role of symmetry in both classical and quantum problems.

The emphasis on classical connections might be described as being somewhat "retro" as related discussions appear in much older textbooks. It is stressed here both for pedagogical purposes[2] and because of a new experimental emphasis on the classical-quantum interface.

- A self-contained chapter (Chap. 4) on probability and statistics is included; this is motivated not only by the statistical interpretation of quantum theory but also by the desire to be able (eventually) to apply statistical ideas to laboratory data and thereby confront theory with experiment. A short discussion of χ^2 fits is also included in Chap. 20 as an example of this connection.
- A separate chapter on two-dimensional quantum mechanics is motivated partly by the rapid growth in the importance of surface and interface science in condensed-matter physics. This also provides opportunities for a less notationally daunting introduction to the quantum version of angular momentum and a more concrete context in which to discuss symmetries. It also provides the "last chance" to completely visualize quantum wavefunctions before going to 3D.
- An extensive list of problems, most of which are directly related to the material in the text, is included; the reader is often "pointed" to the corresponding problems in the text (e.g., "... see P3.15 ..."). A fairly consistent attempt has been made to discuss only material for which at least one problem is provided. Some questions are also included which do not require detailed numerical or algebraic solutions, but which are designed to encourage the reader to examine further the concepts discussed in the chapter.
- A series of appendices is included, each of which has its own (short) set of problems. The topics include complex numbers, useful results from calculus (integrals and the like), matrix and vector algebra, and others.
- Finally, throughout the text, examples have been taken from both classic and more modern experiments in atomic, nuclear, and elementary particle physics. The reader is occasionally asked to "massage" real published data via graphing and subsequent analysis.

Because of the additional topics and new emphasis on visualization, this presentation is somewhat longer than many of the standard textbooks on the subject. Much of the material in a standard text is included, and many of the familiar "landmarks" are obviously visible throughout the book. The reader, however, is encouraged to explore the less-familiar areas presented here and beyond (by using the references).

[2]Increasingly, some students take a course in quantum theory at this level before the corresponding course in classical mechanics so that a certain amount of overlap is desirable.

The organization and inspiration of this text grew out of the several opportunities I have had in the last 6 years to teach a one-semester course in undergraduate quantum mechanics at Penn State University. I wish to express my appreciation to the many students who have taken that course and who have shared opinions on what methods work best for them in learning this subject. The contribution by the Sun Corporation of state-of-the-art workstations to the Physical Science Computing Laboratory in the College of Science at Penn State was a direct motivation to me to help develop a course with an increased emphasis on visualization methods; some of the initial attempts at the images in this text were made in that context.

On a more personal level, I feel that I have been very lucky to have received an excellent undergraduate and graduate education (at the University of Minnesota) and postdoctoral training (at both the University of Wisconsin and the University of Massachusetts-Amherst). I wish to thank the faculty of those institutions who displayed the very best attributes of scholarship, combining a true curiosity and excitement about the physical world with an enthusiastic dedication to the transmission of knowledge (both old and new) to new generations of eager students. It is also a great pleasure to acknowledge the hospitality of Argonne National Laboratory where I spent a sabbatical year. I also wish to acknowledge very useful conversations with J. Annett, J. Anderson, J. Rosner, and P. Shaw on some of the topics covered here, and thank N. Malone for careful proofreading as well as instruction on the proper use of one of the most powerful scientific tools available (namely, the English language). I also extend thanks to my wife, Sarah Malone, for her contributions to our family during this project.

It is perhaps a uniquely human trait to seek order in the universe and to search for new tools and techniques with which to "see" the connections that exist around us, to "make sense" of it all. For various peoples at various times, that search may have been in the context of mythology, religion, psychology, political theory, or science. Since the birth of my two children, their presence in my life has led to a rich structure in my personal universe which I have come to appreciate more deeply than I would have ever thought possible. I have been treated to new worlds of joy, sadness, wonder, and affection through their ever-changing attempts at learning and living. I have discovered many new things, both about the world and about myself, through their eyes. It is to my children, James and Katherine, that this book is lovingly dedicated.

Richard W. Robinett

QUANTUM MECHANICS

Classical Results, Modern Systems, and Visualized Examples

PART I

The Quantum Paradigm

CHAPTER 1

A First Look at Quantum Physics

1.1 ICONS OF QUANTUM MECHANICS

For many people, the term *icon* has changed its meaning from "a representation of some sacred personage or being" to something more like "a symbol that appears on a computer monitor screen and is used to represent a command, file, or application." However, it can also mean "a sign or representation that stands for its object by virtue of a resemblance or analogy to it," and in this more general sense human beings have been creating icons of one kind or another for millennia. Such symbols are a shorthand for our understanding of the physical world and culture in which we live, and they often help represent the content of complex systems of organized scientific knowledge, social behavior, or religious beliefs. Whatever their origin, they can facilitate communication and the sharing of information and ideas (not only between peoples but also between people and machines). Some examples of icons are shown in Fig. 1.1, and the reader can doubtless immediately think of dozens more.

In science, especially in many branches of physics, the phenomena one studies are often remote from one's everyday experience; new ways of organizing and processing information, some perhaps even suggested by the problem under study itself, are useful in working toward the goal of "understanding" the universe. Physics, predominantly but not uniquely among the sciences, has been blessed by the often wonderful congruence between the physical world and the mathematical models used to describe it, a connection that occasionally seems to border on the miraculous. It is not surprising, perhaps, that we often focus on "the equations."

Other methods of conveying the same information content in a different way—such as lecture demonstrations, computer simulations (interactive or otherwise), or even an apt analogy—are all sometimes useful in reinforcing the subject matter in a completely complementary way. Since one of the aims of this book is to present some of the standard material of quantum mechanics in a more visual format (within the context of a printed textbook), we begin our discussion with a brief "tour," using selected images of the quantum world which we take as icons representing part of what we intend to discuss.

FIGURE 1.1. *Icons. Do you have an image (so to speak) in your mind of what they all mean? If not, look at the end of the problems section in this chapter. (McDonald's logo courtesy of the McDonald's Corporation.)*

The first image we present, in Fig. 1.2, shows an electron diffraction pattern obtained with a commercial electron microscope which illustrates not only the wave properties of matter (which is, of course, one of the touchstones of quantum theory) but also the statistical nature of the measurement process in quantum mechanics. These important aspects of quantum physics help to justify the background material on classical waves (Chap. 2) and probability and statistics (Chap. 4), in addition to the more standard discussions of the Schrödinger wave equation in Chaps. 3 and 5.

Wave phenomena are perhaps most directly familiar to us from two-dimensional systems such as ripples on a pond or ocean waves, and Figs. 1.3(a) and (b) show two images of electron "standing-wave patterns" on a surface. In the case of Fig. 1.3(a), the electrons are confined to a circular "corral" formed by heavier atoms that have been adsorbed onto the surface, and are a testament to the rapid progress being made in surface science and its increasing technological importance. We devote Chap. 16 to a discussion of two-dimensional quantum systems. In addition, the images were obtained by using the technique of scanning tunneling microscopy, which relies on quantum tunneling phenomena discussed in Chap. 12.

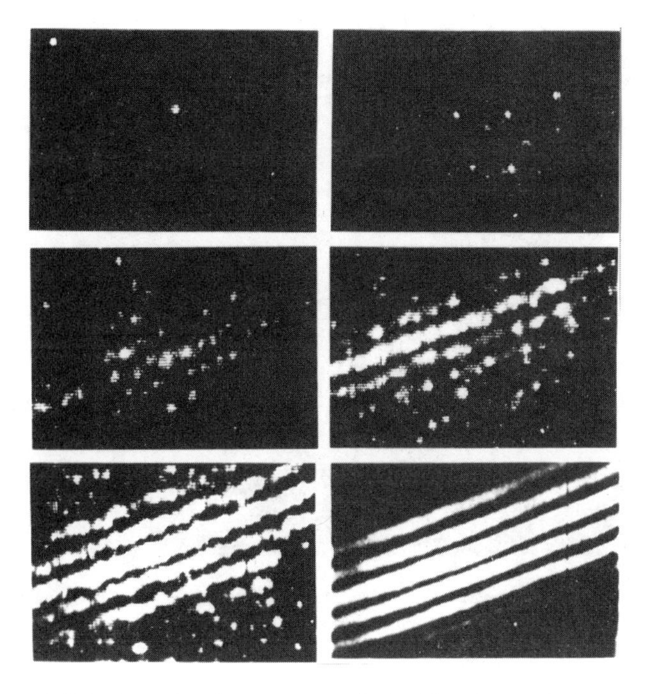

FIGURE 1.2. Interference pattern obtained by using an electron microscope showing the fringes being "built up" from an increasingly large number of measurements of individual electron events. From Merli, Missiroli, and Pozzi (1976). (Photo courtesy of G. Pozzi. Reproduced by permission of the American Institute of Physics.)

FIGURE 1.3. "Standing electron wave patterns" obtained using scanning tunneling microscopy: (a) the central portion of a circular "corral" roughly 150 Å in diameter; (continued)

FIGURE 1.3. (continued) *"Standing electron wave patterns" obtained using scanning tunneling microscopy: (b) the standing-wave patterns resulting from scattering from "steps" on a copper surface. (Photos courtesy of IBM Almaden.)*

The *quantum* in quantum mechanics is often associated with the discrete energy levels observed in bound state systems. In Fig. 1.4 we illustrate a stellar spectrum that exhibits many of the same lines of the Balmer series as the laboratory spectrum of hydrogen that brackets it above and below. This also serves to remind us of one of our most basic

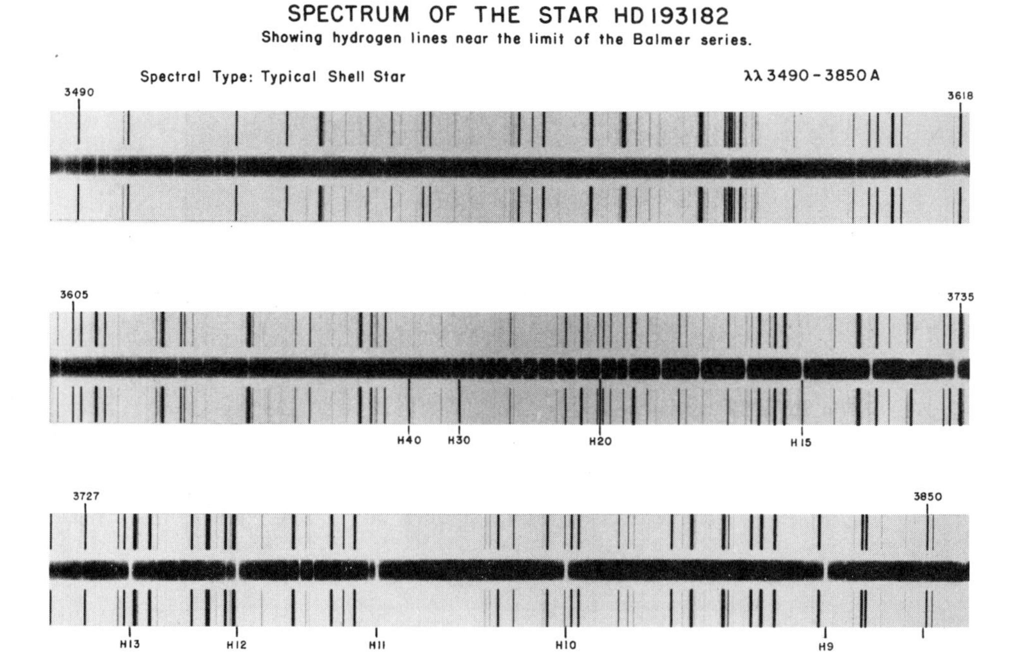

FIGURE 1.4. *Spectrum of the star HD 193182 bracketed by laboratory spectrum of hydrogen, both showing lines of the Balmer series. (Photo courtesy of the Observatories of the Carnegie Institution of Washington.)*

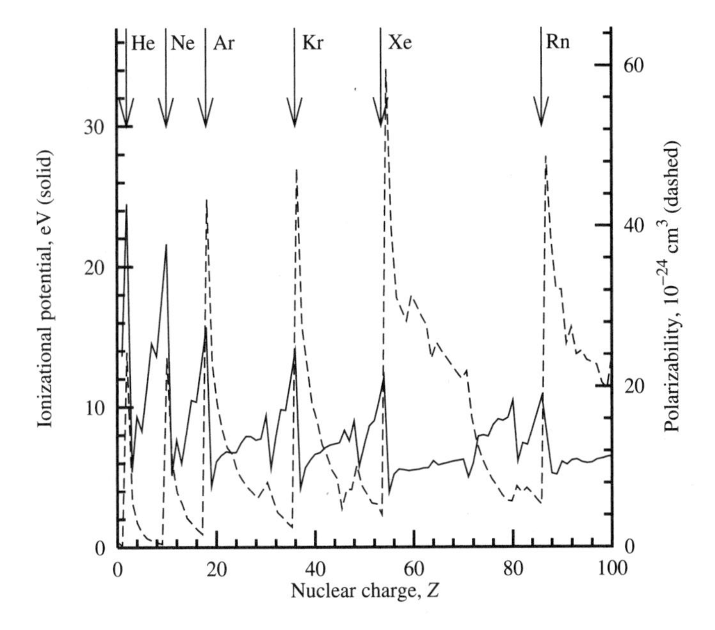

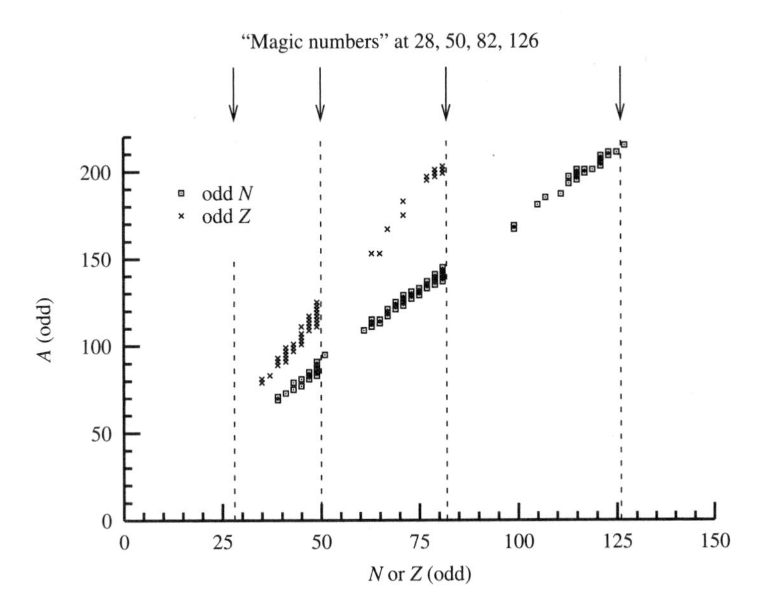

FIGURE 1.5. *(a) Plot of ionization potential (solid) and atomic polarizability (dashed) versus nuclear charge showing shell structure characterized by the noble gas atoms. (b) A compilation of the nuclei that have long-lived excited states (so-called* isomers*); nuclei with odd numbers of protons + neutrons are shown versus the number of neutrons (squares) and protons (diamonds). Numbers of protons or neutrons near the so-called* magic numbers *of 28, 58, and 82 are singled out and provide evidence for nuclear shell structure; evidence for shells corresponding to Z or N = 2, 8, or 20 are not shown.*

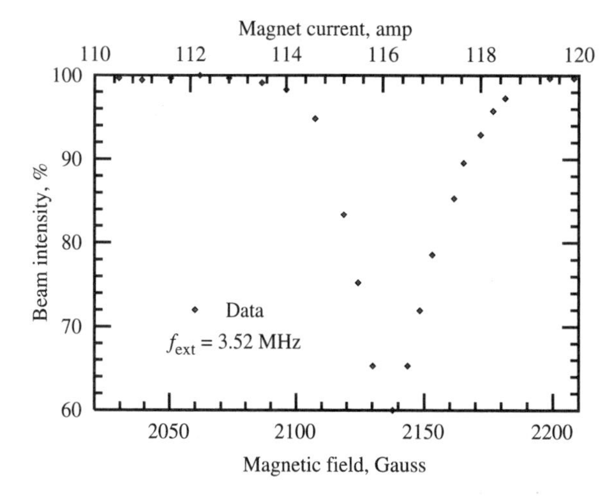

FIGURE 1.6. Atomic beam resonance absorption curve. Data taken from Rabi et al. (1938) using LiCl molecules. The resonance dip corresponding to the Li^7 nucleus is shown.

discoveries[1] in science, namely, that the physical phenomena we study, rationalize, and codify are presumably the same ones being studied everywhere else in the universe.

The collections of seemingly indistinguishable particles (such as electrons, protons, and neutrons) that combine to form nuclei, atoms, and molecules are necessarily described by wave mechanics. Much of the observable structure of aggregates of such particles, however, is equally directly attributable to the intimate connection between their intrinsic spin and the stringent requirements imposed on the quantum wavefunction of the system by the Pauli principle. Figures 1.5(a) and (b) exhibit evidence for the shell structure present in both atomic (a) and nuclear systems (b) where special configurations of particles are exceptionally stable. The role of spin in determining the "structure of matter" in quantum systems is discussed in Chaps. 8, 15, and 18.

The intrinsic spin of electrons and protons also can play a dynamic role in quantum systems. Figure 1.6 shows data from an early atomic beam magnetic resonance experiment. The "dip" signals that the nuclear spins have absorbed a microwave photon, and are evidence that the "little bar magnets" associated with the intrinsic angular momentum have "flipped" in response to the application of an external magnetic field. This basic physical phenomenon, when implemented with clever instrumentation, is the basis for magnetic resonance imaging (or MRI), and an MRI image is shown in Fig. 1.7. The dynamics of spin is explored in Chaps. 17 and 19.

Another type of "spin flip" arises when the electron spin in a hydrogen atom changes orientation relative to the proton; the two configurations differ slightly in energy due to magnetic dipole-dipole interactions (discussed in Chap. 19). The radiation emitted during such a flip is an efficient tracer of neutral atomic hydrogen atoms in the cosmos because of its unique wavelength signature. Figure 1.8 illustrates a mapping of an interstellar gas cloud (a so-called *HI region*) using this signal, the so-called *21-cm line*.

Just as bound state systems exhibit observable phenomena that have characteristics of both particle and wave behavior, so do the collisions of particles. Particle-scattering experiments have been extensively used for the study of both the nature of fundamental constituents and the basic forces of nature. Scattering is discussed at some length in both

---

[1]Feynman (1963) has said that "the most remarkable discovery in all of astronomy is that *the stars are made of atoms of the same kind as those on the earth.*"

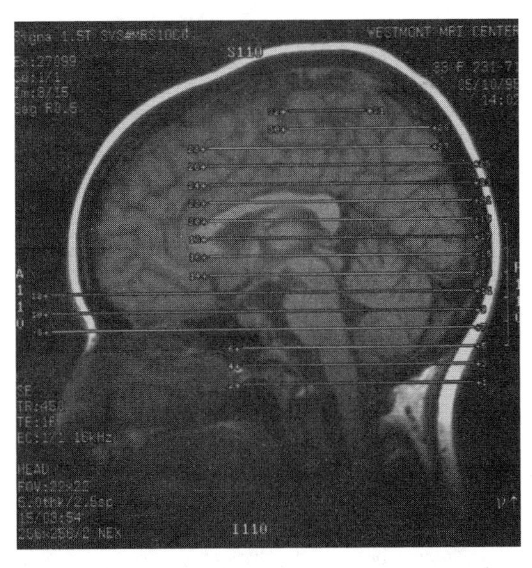

FIGURE 1.7. *Image of the human brain obtained using magnetic resonance imaging (MRI). (Photo courtesy of the Westmont MRI Center.)*

its classical and quantum versions in Chap. 20. As an example, we show in Fig. 1.9 the scattering cross section (which related to the probability of an interaction) for the collisions of α particles (indistinguishable helium nuclei). The data for experiments at two different energies are shown compared to theoretical predictions. The agreement at lower energies is evidence for the quantum description of identical particles interacting via purely electromagnetic interactions (i.e., Coulomb's law), while the disagreement of the lower curve with "theory" was one of the first pieces of evidence for the strong nuclear force that becomes important at short range. It also reminds us that experiments that "push the envelope" of the existing boundaries in available energy or ability to probe new distance scales (either the very large or very small) have often discovered radically new physical phenomena.

FIGURE 1.8. *Map of interstellar gas using the 21-cm line from the hyperfine splitting of hydrogen. (Photo courtesy of Photo Researchers.)*

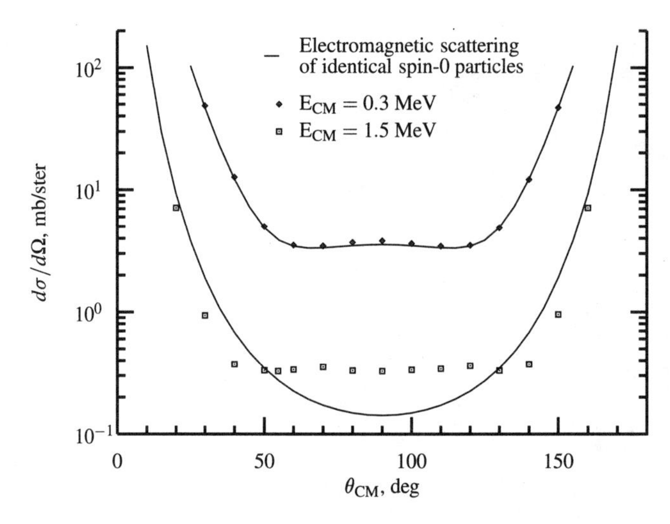

FIGURE 1.9. Cross-section for α-α scattering. Squares are experimental data; the solid curve is the prediction assuming that only the Coulomb potential is responsible for the α-α interaction.

TABLE 1.1 *Length Scales of the Observable (and Hypothesized) Universe on a Logarithmic Scale with Selected "Landmarks"*

$\log_{10}$, size *m*	System	$\log_{10}$, size *m*	System
25	Most distant galaxy	−6	1 μm
24		−7	
23		−8	Polio virus
22		−9	
21		−10	Hydrogen atom
20	Diameter of our galaxy	−11	
19		−12	
18		−13	"Muonic hydrogen"
17		−14	Heaviest nucleus
16		−15	Proton radius
15	Nearest star (~ 1 pc)	−16	
14		−17	
13		−18	Range of weak interactions
12		−19	Smallest distance probed
11	Edge of solar system	−20	(by accelerators)
10		−21	
9		−22	
8		−23	
7	Earth (white dwarf) radius	−24	
6		−25	
5		−26	
4	Neutron star radius	−27	
3	1 km	−28	
2		−29	
1		−30	
0	Esteemed reader (you!)	−31	Grand unification?
−1		−32	
−2		−33	
−3	1 mm	−34	
−4	Thickness of piece of paper	−35	Planck length (ultimate length?)
−5			

On this note, we close with a view of the length scales of the universe as one "zooms in and out" at various magnifications as shown in Table 1.1. Distances out to the "edge of the universe" have been explored (albeit remotely, via electromagnetic radiation) to length scales of the order of 10^{25} m. The structure of matter has been probed in collisions of elementary particles in terrestrial accelerators down to scales of order 10^{-19}–10^{-20} m, so that the "observed" universe might be said to have a "dynamic range" of something like 40 orders of magnitude. It is hypothesized (P1.15) that gravity and the other forces of nature (including electromagnetism) might be unified at distances of order 10^{-35} m (the so-called *Planck length*), so the universe may well be 60 orders of magnitude in scale or larger. Quantum physics in some guise (if one includes relativity and the like, as in quantum field theory) is seemingly the appropriate description of nature on length scales of atomic dimensions and below in size, perhaps all the way down to the Planck length. Quantum processes, however, are also important on macroscopic scales as well, as they are required, for example, for a complete understanding of stellar fusion and the ultimate fate of stars.

1.2 ESSENTIAL RELATIVITY

While we will consider nonrelativistic quantum mechanics almost exclusively, it is useful to review briefly some of the rudiments of relativity and the fundamental role played by the speed of light, c.

For a free particle of rest mass m moving at speed v, the total energy, momentum, and kinetic energy can be written in the relativistically correct form:

$$E = \gamma mc^2 \qquad p = \gamma mv \qquad T \equiv E - mc^2 = (\gamma - 1)mc^2 \tag{1.1}$$

where

$$\gamma = \frac{1}{\sqrt{1 - v^2/c^2}} = \left(1 - \frac{v^2}{c^2}\right)^{-1/2} \tag{1.2}$$

The nonrelativistic limit corresponds to $v/c \ll 1$, in which case we can use the series expansion

$$(1 + x)^n = 1 + nx + \frac{n(n-1)}{2}x^3 + \frac{n(n-1)(n-2)}{3!}x^3 + \cdots \tag{1.3}$$

for v^2/c^2 small to show that

$$p \approx mv \qquad \text{and} \qquad T \approx \left(1 + \frac{1}{2}\frac{v^2}{c^2} + \cdots - 1\right)mc^2 \approx \frac{1}{2}mv^2 \tag{1.4}$$

which are the familiar results for motion at speeds slow compared to the speed of light.

In quantum mechanics, the momentum is a more natural variable than v, and a useful relation can be obtained from Eqn. (1.1), namely,

$$E^2 = (pc)^2 + (mc^2)^2 \tag{1.5}$$

This form stresses the fact that E, pc, and mc^2 have the same dimensions (namely, energy), and we will often use these forms when convenient. As an example, the rest energies of various atomic particles will often be quoted in energy units; for the electron and proton, we have

$$m_e c^2 = 0.511 \text{ MeV} \qquad m_p c^2 = 938.3 \text{ MeV} \tag{1.6}$$

Recall that

$$\begin{aligned} 1 \text{ eV} &= \text{the energy gained by a fundamental charge } e \\ &\quad \text{that has been accelerated through 1 volt} \\ &= (1.6 \times 10^{-19}\ C)(1\text{ V}) = 1.6 \times 10^{-19}\text{ J} \end{aligned} \tag{1.7}$$

Atomic "masses" are often quoted in *unified atomic mass units* (formerly amu) which are given by $1\,u = 931.5$ MeV.

The nonrelativistic limit of Eqn. (1.5) is easily seen to be

$$E = mc^2\left(1 + \left(\frac{pc}{mc^2}\right)^2\right)^{1/2} = mc^2 + \frac{p^2}{2m} - \frac{p^4}{8m^3c^2} + \cdots \tag{1.8}$$

Since the rest energy is "just along for the ride" in most of the problems we consider, we will ignore its contribution to the total energy. Thus a phrase such as "... a 2 eV electron ..." should be taken to mean that the electron has $T = E - mc^2 \approx p^2/2m \approx 2$ eV. We will often write $pc = \sqrt{2(mc^2)T}$ in this limit.

At the other extreme, in the ultrarelativistic limit when $E \gg mc^2$ (or $v \lesssim c$), we can write

$$E = pc\left(1 + \left(\frac{mc^2}{pc}\right)^2\right)^{1/2} \approx pc + \frac{1}{2}\frac{m^2c^4}{pc} + \cdots \tag{1.9}$$

which is also seen to be consistent with the energy momentum relation of massless particles (such as photons), namely, $E = pc$.

We list below several typical quantum mechanical systems and the order of magnitudes of the energies involved:

- **Electrons in Atoms:** For the inner-shell electrons of an atom with nuclear charge $+Ze$, the kinetic energy is of order $T \approx Z^2\ 13.6$ eV. We can say, somewhat arbitrarily, that relativistic effects become nonnegligible when $T \gtrsim 0.05\ mc^2$ (i.e., a 5% effect). This condition is satisfied when $Z \gtrsim 43$, implying that the effects of relativity must certainly be considered for very heavy atoms.

- **Deuteron:** The simplest nuclear system is the bound state of a proton and neutron where the typical kinetic energies are $T \approx 2$ MeV. This is to be compared with $m_pc^2 \approx m_nc^2 \approx 939$ MeV so that the deuteron can be considered as a nonrelativistic system to first approximation.

- **Quarks in the Proton and Pion:** The constituent quark model of elementary particles postulates that three quarks of effective mass, roughly $m_qc^2 \approx 350$ MeV, form the proton. This implies binding energies and kinetic energies of the order of 1–10 MeV, which is consistent with "nonrelativity." The pion, on the other hand, is considered a bound state of two such quarks, but has rest energy $m_\pi c^2 \approx 140$ MeV, so that binding energies (and hence kinetic energies) of order several hundred MeV are required, and relativistic effects dominate.[2]

[2] The pion is really a quark-antiquark system. Bound states of heavier quarks and antiquarks, which are more slowly moving, can be successfully described using nonrelativistic quantum mechanics; see Sec. 18.3.5.

- **Compact Objects in Astrophysics:** The electrons in white dwarf stars and neutrons in neutron stars have kinetic energies $T_e \approx 0.08$ MeV and $T_n \approx 140$ MeV, respectively, so these objects are "barely" nonrelativistic.

1.3 QUANTUM PHYSICS: $\hbar$ AS A FUNDAMENTAL CONSTANT

Just as the speed of light, c, sets the scale for when relativistic effects are important, quantum physics also has an associated fundamental, dimensionful parameter, namely, *Planck's constant.* Its first applications came in the understanding of some of the quantum aspects of the electromagnetic field and the particle nature of EM radiation:

- In his investigations of the black-body spectrum emitted from heated objects (so-called *cavity radiation*), Planck found that he could only fit the observed intensity distribution if he made the (then radical) assumption that the electromagnetic energy of a given frequency f was quantized and given by

$$E_n(f) = nhf \qquad \text{where} \qquad n = 0, 1, 2, 3 \ldots \tag{1.10}$$

The constant of proportionality, h, was derived from a "fit" to the experimental data, and has been found to be

$$h = 6.626 \times 10^{-34} \text{ J} \cdot \text{s} \tag{1.11}$$

and is called *Planck's constant;* we will more often use the related form

$$\hbar = \frac{h}{2\pi} = 1.055 \times 10^{-34} \text{ J} \cdot \text{s} \tag{1.12}$$

which is to be read as "h-bar."

- Einstein assumed that the energy quantization of Eqn. (1.10) was a more general characteristic of light, and proposed that electromagnetic radiation was composed of photons[3] or "bundles" of discrete energy $E_\gamma = hf$. He used the photon concept to explain the *photoelectric effect,* and predicted that the kinetic energy of electrons emitted from the surface of metals after being irradiated should be given by

$$\frac{1}{2}mv_{\text{max}}^2 = E_\gamma - W = hf - W \tag{1.13}$$

where W is called the *work function* of the metal in question. Subsequent experiments were able to confirm this relation, as well as provide another, complementary measurement of h that agreed with the value obtained by Planck.

- The relativistic connection between energy and momentum for a massless particle such as the photon could be used to show that it had a momentum given by

$$p_\gamma c = E_\gamma = hf = \frac{hc}{\lambda} \qquad \text{or} \qquad p_\gamma = \frac{h}{\lambda} \tag{1.14}$$

[3]We use the notation γ (for gamma ray) to indicate a property corresponding to a photon of any energy or frequency.

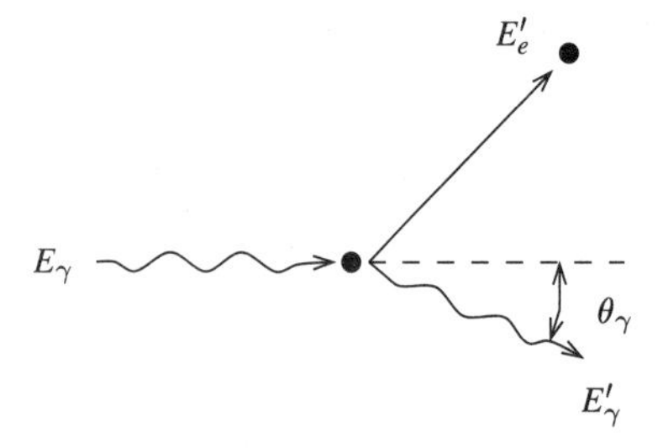

FIGURE 1.10. *Geometry for Compton scattering. The incident photon scatters from an electron, initially at rest.*

where λ is the wavelength. Arthur Compton noted that the scattering of X rays by free electrons at rest could be considered as a collision process where the incident photon has an energy and momentum given by Eqn. (1.14), as in Fig. 1.10. Conservation of energy and momentum (P1.3) can then be applied to show that the wavelength of the scattered photon, λ', is given by the Compton scattering formula

$$\lambda' - \lambda = \frac{h}{m_e c}(1 - \cos(\theta_\gamma)) \tag{1.15}$$

where θ_γ is the angle between the incident and scattered photon directions; X-ray scattering experiments confirmed the validity of Eqn. (1.15).

The connection of Planck's constant to the properties of material particles, such as electrons, came later:

■ Using yet another experimental fit to spectroscopic data, in this case the Balmer-Ritz formula for the frequencies in the spectrum for hydrogen, Bohr used semiclassical arguments to deduce that the angular momentum of the electron was quantized as

$$L = n\frac{h}{2\pi} = n\hbar \qquad \text{with} \qquad n = 1, 2, 3\ldots \tag{1.16}$$

■ Motivated by the dual wave particle nature of light exhibited, for example, in Compton scattering, de Broglie suggested that matter, specifically electrons, would exhibit wave properties. He postulated that the relation

$$\lambda = \frac{h}{p} \tag{1.17}$$

applies to material particles as well as to photons, thereby defining the *de Broglie wavelength.* He could show that Eqn. (1.17) reproduced the Bohr condition of Eqn. (1.16), and thus explain the hydrogen atom spectrum; we will discuss this in Ex. 1.1 below. The wave nature of electrons was more directly demonstrated by Davisson and Germer, who observed the diffraction of electron beams from crystals, with a wavelength consistent with Eqn. (1.17).

Example 1.1 Semiclassical Model of the Hydrogen Atom An essentially classical approach to the bound state dynamics of the electron-proton system can be extended by using the wave mechanics idea of de Broglie to derive the most important features of the hydrogen atom spectrum. The *Coulomb force* between the two particles can be written in the form

$$\mathbf{F}(\mathbf{r}) = -\frac{1}{4\pi\epsilon_0}\frac{e^2}{r^2}\hat{\mathbf{r}} = -\frac{Ke^2}{r^2}\hat{\mathbf{r}} \tag{1.18}$$

where we will conventionally write the fundamental constant of electrostatics in the form $K = 1/4\pi\epsilon_0$. This force can be derived from the *Coulomb potential,* namely,

$$V(r) = -\frac{Ke^2}{r} \tag{1.19}$$

■ Before proceeding, let us pause and make a few comments about the dimensionful constants that appear in this and other atomic and nuclear physics systems involving electromagnetism, a sort of example within an example. The combination of constants that determines the electrostatic force between two fundamental charges can be written in the form

$$Ke^2 = \left(\frac{Ke^2}{\hbar c}\right)\hbar c = \alpha\hbar c \qquad \text{where} \qquad \alpha \equiv \frac{Ke^2}{\hbar c} \approx \frac{1}{137} \tag{1.20}$$

and α is dimensionless and is called the *fine-structure constant.* The combination $\hbar c$ has dimensions and the numerical values

$$\hbar c \approx 1973 \text{ eV Å} \approx 197.3 \text{ MeV F} \approx 0.1973 \text{ GeV F} \tag{1.21}$$

which are useful for atomic/molecular, nuclear, and elementary particle physics problems, respectively. Together, these give

$$Ke^2 \approx 14.4 \text{ eV Å} \approx 1.44 \text{ MeV F} \tag{1.22}$$

Despite focusing on nonrelativistic systems, we will often manipulate factors of c to make use of these combinations. Now back to the hydrogen atom.

For circular orbits in which the electron (mass m) is assumed to orbit around the (stationary and infinitely heavy) proton, Newton's law implies that

$$m\frac{v^2}{r} = ma_C = F(r) = \frac{Ke^2}{r^2} \tag{1.23}$$

where we have used the appropriate centripetal acceleration; this relation is consistent with any value of r.

If we wish to incorporate the wave properties of the electron via the de Broglie relation

$$\lambda = \frac{h}{p} = \frac{2\pi\hbar}{p} \tag{1.24}$$

then we must presumably insist that the appropriate number of de Broglie wavelengths "fit" into the circular orbit as in Fig. 1.11, i.e., that

$$n\lambda = 2\pi R \qquad n = 1, 2, 3\ldots \tag{1.25}$$

When combined with Eqn. (1.24), this implies that

$$n\hbar = pR = mvR = L \tag{1.26}$$

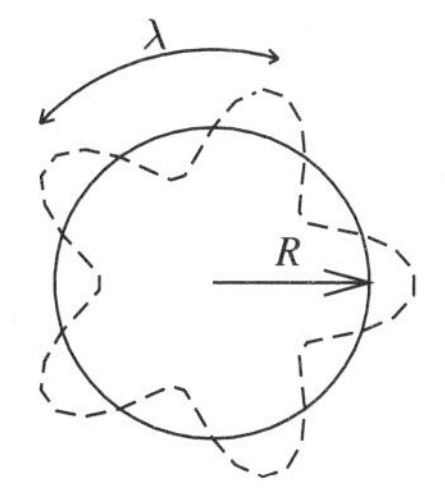

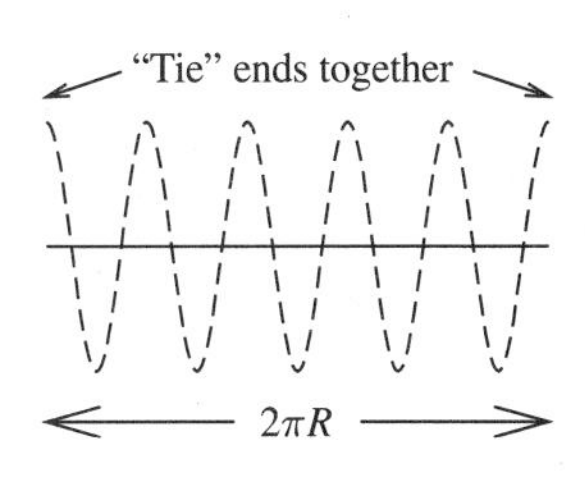

FIGURE 1.11. *Standing-wave pattern in circular orbits for deBroglie waves.*

a la Bohr, and the orbital angular momentum must be quantized. This additional constraint, along with Eqn. (1.23), gives

$$r_n = \left(\frac{\hbar^2}{mKe^2}\right)n^2 = a_0 n^2 \tag{1.27}$$

where we have defined the *Bohr radius* as

$$a_0 = \frac{\hbar^2}{mKe^2} = \frac{\hbar c}{mc^2\alpha} \approx \frac{(197.3 \text{ eV Å})(137)}{(0.511 \times 10^6 \text{ eV})} \approx 0.53 \text{ Å} \tag{1.28}$$

The corresponding speeds are also quantized and given by

$$v_n = \frac{n\hbar}{mr_n} = \left(\frac{Ke^2}{\hbar}\right)\frac{1}{n} = \frac{\alpha c}{n} \ll c \tag{1.29}$$

which reminds us that the electron is nonrelativistic. The period (τ) of the classical orbit and the corresponding frequency (f) are given by

$$\frac{1}{\tau_n} = f_n = \frac{v_n}{2\pi r_n} = \frac{1}{2\pi}\left(\frac{m(Ke^2)^2}{\hbar^3}\right)\frac{1}{n^3} = \frac{1}{2\pi}\left(\frac{mc^2\alpha^2}{\hbar}\right)\frac{1}{n^3} \tag{1.30}$$

Even more importantly, the bound state energies are also quantized since

$$\begin{aligned} E_n &= \frac{1}{2}mv_n^2 - \frac{Ke^2}{r_n} = -\frac{1}{2}\frac{m(Ke^2)^2}{\hbar^2}\frac{1}{n^2} \\ &= -\left[\frac{1}{2}mc^2\alpha^2\right]\frac{1}{n^2} \\ &\approx -\frac{1}{2}(0.51 \times 10^6 \text{eV})\left(\frac{1}{137}\right)^2\frac{1}{n^2} \\ &\approx \frac{-13.6 \text{ eV}}{n^2} \end{aligned} \tag{1.31}$$

While this has been derived assuming circular orbits, it can be shown that elliptical orbits, when properly quantized, are also described by this relation.

An important limit is suggested by Eqn. (1.26) where we note that $n \gg 1$ is required to obtain macroscopically large values of the angular momentum; the fact that quantum systems approach (in an average sense that we will discuss in later chapters) their classical counterparts in this limit is called the *correspondence principle,* and was used heavily by Bohr in his analyses. For example, he noted that the photons emitted in transitions between the quantized energy levels in Eqn. (1.31) satisfy the Balmer formula, written here in the form

$$2\pi\hbar f = E_\gamma = E_n - E_m = \frac{m(Ke^2)^2}{2\hbar^2}\left(\frac{1}{m^2} - \frac{1}{n^2}\right) \tag{1.32}$$

For transitions between neighboring states, i.e., of the form $n \to n-1$, in the large n limit the emitted radiation is of frequency

$$f = \frac{1}{2\pi\hbar}(E_n - E_{n-1}) \stackrel{n\gg 1}{\longrightarrow} \frac{m(Ke^2)^2}{2\pi\hbar^3}\frac{1}{n^3} \tag{1.33}$$

A classical particle undergoing circular acceleration would emit radiation at its orbital frequency, f, which from Eqn. (1.30) is given by exactly the limit above.

It is interesting to note in this context the role that wave mechanics and Coulomb's law (via $\hbar$ and e) play in determining the densities of "ordinary" solid matter.[4] The mass of atoms is due mostly to their nuclear constituents (the protons and neutrons), while their size is determined by the quantum properties of their electrons. For example, an order-of-magnitude estimate of the density of atomic hydrogen can be obtained by assuming that there is one proton mass in a cube of size $2a_0 \approx 1$ Å on a side; this gives a density of roughly

$$\rho \sim \frac{m_p}{(2a_0)^3} \approx 1.6 \times 10^3 \text{ kg/m}^3 \sim 1.6 \text{ gr/cm}^3 \tag{1.34}$$

which is in the right "ballpark."

The de Broglie relation contains the seeds of the *position-momentum uncertainty principle,* namely,

$$\Delta x \Delta p \geq \frac{\hbar}{2} \tag{1.35}$$

where Δx and Δp are the uncertainties in a measurement of x and p, respectively. Equation (1.35) puts fundamental limitations on one's ability to measure simultaneously the position and momentum of a particle; it also leads to the notion of zero point energy, a minimum unavoidable energy of a particle confined to a localized region of space.

The example of a particle in a one-dimensional box illustrates this most simply. A particle of mass m confined to a 1D box of length L will satisfy the "standing-wave" condition for de Broglie waves if $n(\lambda/2) = L$ (compare this to Eqn. (1.25) and explain any differences!) with $n = 1, 2, 3 \ldots$. This corresponds to quantized energies given by

$$E_n = \frac{p_n^2}{2m} = \frac{n^2\hbar^2\pi^2}{2mL^2} \qquad n = 1, 2, 3 \ldots \tag{1.36}$$

In contrast to the classical case, the particle cannot just "sit quietly in the box," but has a minimum energy. More generally, a particle localized to a region of spatial extent $\Delta x \sim L$ will have a corresponding uncertainty in momentum of order $p_{\text{min}} \sim \Delta p \sim \hbar/L$ or minimum kinetic energy

$$E_{\text{min}} \propto \frac{p_{\text{min}}^2}{2m} \propto \frac{\hbar^2}{mL^2} \tag{1.37}$$

Such "back-of-the-envelope" calculations should be used with care, but can often provide insight into the ground state of a quantum system.

▪ **Electrons in Atoms?** The electrons in atoms and molecules are confined to a region of size $\Delta x \sim a_0 \sim 1$ Å. The corresponding unavoidable spread in momentum is roughly

$$\Delta pc \sim pc \sim \frac{\hbar c}{a_0} \approx \frac{2000 \text{ eV Å}}{1 \text{ Å}} \approx 2 \text{ keV} \tag{1.38}$$

Since this is much smaller than the electron rest energy, $mc^2 \approx 0.5$ MeV, the zero point

[4]The seemingly commonplace observation that "matter held together by Coulomb forces is stable" is a remarkable and rather subtle consequence of many aspects of quantum theory; for a nice discussion, see Lieb (1976).

energy is roughly

$$E_0^{(e)} \approx \frac{p^2}{2m} = \frac{(pc)^2}{2mc^2} \approx \frac{(2000\text{ eV})^2}{2(0.5 \times 10^6\text{ eV})} \approx 4\text{ eV} \tag{1.39}$$

which is, of course, exactly the order of magnitude in Ex. 1.1.

■ **Photons in Atoms?** On the other hand, photons emitted in radiative decays cannot have been "stored" in the atom beforehand. To see this, we note that a photon "bouncing around" in an atomic-sized box will have the same $\Delta p \sim p$ as in Eqn. (1.38). Because massless photons are necessarily relativistic, the corresponding kinetic energy is then given by

$$E_0^{(\gamma)} \approx pc \approx 2000\text{ eV} \tag{1.40}$$

which is much larger than the 1–10 eV observed in typical transitions.

■ **Alpha Particles in Nuclei?** Radioactive nuclei emit α particles ($m_\alpha \approx 4m_p$) with kinetic energies of a few MeV. The minimum momentum in a heavy nucleus of radius $R \approx 5$ F is roughly

$$pc \approx \frac{\hbar c}{R} \approx \frac{200\text{ MeV F}}{5\text{ F}} \approx 40\text{ MeV} \tag{1.41}$$

which corresponds to a (nonrelativistic) zero point energy of

$$E_0^{(\alpha)} \approx \frac{(pc)^2}{2m_\alpha c^2} \approx \frac{(40\text{ MeV})^2}{8 \times 940\text{ MeV}} \approx 0.2\text{ MeV} \tag{1.42}$$

which is consistent with observations.

■ **Electrons from Neutron β Decay?** Neutrons decay via the process $n \rightarrow pe\overline{\nu}_e$, where the electron (anti-)neutrino is often not directly detected but the electron of a few MeV is easily observed. A "preexisting" electron in the neutron, "waiting around" to decay, would have pc roughly 5 times *larger* than in Eqn. (1.41) (since the neutron is roughly 5 times *smaller* than a heavy nucleus), implying relativistic electrons with kinetic energies of order $E_0^{(e)} \approx pc \approx 200$ MeV; the decay electrons must therefore be created *ex nihilo* at the time of decay. Arguments such as these were among the first pieces of evidence used to predict that a new force beyond those known classically, the so-called *weak interaction,* was required.

For single particles, it is often clear when wave mechanical effects are important. For example, in electron scattering from crystal planes, the Bragg condition for constructive interference can be written in the form $n\lambda = D\sin(\phi)$, where D is the interatomic spacing and ϕ is the scattering angle. Clearly λ must be comparable to the other spatial dimensions in the problem for the wave properties of matter to be visible. Many problems have some other natural length scale against which to compare the de Broglie wavelength.

Example 1.2 Systems of Particles: Classical or Quantum Mechanics? At high temperatures and/or low densities, the behavior of a gas can be described by classical statistical mechanics; the atoms, to a good approximation, move along classical trajectories. At low temperatures and/or high densities, quantum effects become important. The classical approximation will break down when the de Broglie wavelength of a typical particle becomes comparable to the average interparticle distance; if the number density is n, this distance is roughly $d \sim n^{-1/3}$, which can then be compared to $\lambda = h/p$. For a system in thermal equilibrium, the thermal energy is $E = p^2/2m = k_B T/2$, where k_B is Boltzmann's constant and T is the temperature. This gives

$$\lambda = \frac{h}{p} = \frac{2\pi\hbar}{\sqrt{Mk_BT}} \tag{1.43}$$

For air at STP conditions, one can estimate that $M \sim 28u$ (for diatomic nitrogen, N_2) and $T = 300°$ to find $\lambda \sim 0.45$ Å. At one atmosphere of pressure, one has $P_{\text{atm}} \sim 10^5$ N/m^2, giving[5] $n \sim 2.4 \times 10^{25}$ m^{-3} or $n^{-1/3} \sim 35$ Å. Since $d = n^{-1/3} \gg \lambda$, the system can be considered classically.

On the other hand, the conduction electrons in a metal (which for many purposes can be considered as a gas) have a de Broglie wavelength that is $\sqrt{M/m_e} \approx 225$ times *smaller* than for gas atoms, while the densities of solid matter are much larger. With a few conduction electrons per atom, electron densities of $n_e \approx (1 - 10) \times 10^{28}$ m^{-3} are typical, so that $n_e^{-1/3} \approx 2$–5 Å while $\lambda_e \sim 100$ Å and $\lambda \gg d$.

The speed of light provides a benchmark value against which the magnitudes of other velocities in a given problem can be compared. Familiar, nonrelativistic Newtonian mechanics is valid in the limit when $v/c \ll 1$; we don't need relativity in everyday situations because c is so large. In a similar way, the effects of wave mechanics on macroscopic systems often need not be considered because $\hbar$ is, in some sense, so small; but small compared to what? Equation (1.16) suggests that $\hbar$ could be compared to the typical angular momentum in a problem, but this is clearly not a fundamental connection, as quantum mechanics can be applied equally well to one-dimensional systems for which the notion of angular momentum does not arise.

The quantity in classical mechanics whose magnitude can most naturally be compared to $\hbar$ is the *classical action,*[6] S, which is used in the *Lagrangian formulation*[7] of classical mechanics.

The action takes as its argument any possible path, $x(t)$, connecting the initial (i) and final (f) points on a classical trajectory, i.e., which satisfies $x(t_i) = x_i$ and $x(t_f) = x_f$. It returns a numerical value given by the integral

$$\begin{aligned} S[x] &= \int_{t_i}^{t_f} dt\,(T - V) \\ &= \int_{t_i}^{t_f} dt\left(\frac{1}{2}mv^2(t) - V(x(t))\right) \end{aligned} \tag{1.44}$$

Hamilton's principle states that

- The path, $x(t)$, which minimizes the classical action, $S[x]$, is the trajectory as given by Newton's equations of motion.

---

[5]One uses the ideal gas law, $P = nk_BT$.

[6]Our extremely abbreviated discussion does not do justice to the notion of action, in either classical or quantum mechanics. The most elegant and seamless connection between the two descriptions of nature is provided by the path-integral approach to wave mechanics originated by Feynman. It is discussed in several undergraduate textbooks—see, e.g., Park (1992) and Sakurai (1994)—but the standard (and very readable) reference is Feynman and Hibbs (1965).

[7]See, e.g., Marion and Thornton (1988).

and provides an alternative formulation of classical mechanics. We immediately note that the action integral has the dimensions of *energy* · *time* as does $\hbar$.

Example 1.3 $\hbar$ as the "Quantum of Action" As our single example of this connection, we calculate the classical action for an electron in the hydrogen atom in a circular orbit of radius r over a single period. It is easy to show that

$$S = \int_0^{\tau} dt(T - V) = \frac{3}{2}|E|\tau = 3\pi\sqrt{mKe^2r} \tag{1.45}$$

If we now make use of the Bohr result, Eqn. (1.27), for the quantized radii,

$$r_n = a_0 n^2 = \left(\frac{\hbar^2}{m(Ke^2)}\right)n^2 \tag{1.46}$$

we find that that S is given by $S = 3\pi n\hbar$; for this reason, $\hbar$ has sometimes been called the *quantum of action.*

In classical mechanics, only the single path with the least action is realized in nature as the unique trajectory. Quantum mechanics, however, allows more leeway, in that other paths that have an action larger than the minimum, but by less than one unit of $\hbar$ (more or less), i.e.,

$$\mathcal{O}(\hbar) \gtrsim \Delta S = S[x_{\text{path}}^{\text{alternative}}] - S_{\text{min}}[x_{\text{path}}^{\text{classical}}] > 0 \tag{1.47}$$

are also possible. To gauge better the uncertainties this engenders in the notion of classical trajectory, we imagine an alternative, noncircular orbit as in Fig. 1.12 (the dashed curve). Using Eqn. (1.45) as a guide, we estimate that the change in action corresponding to a small change in radius, Δr, is given by

$$\Delta S = (3\pi)\frac{1}{2}\sqrt{\frac{m(Ke^2)}{r}}\Delta r = \frac{1}{2}\frac{\Delta r}{r}S \tag{1.48}$$

Only such changes for which $\Delta S \lesssim \hbar$ will contribute, so that we estimate that the uncertainty in a circular orbit due to quantum effects will scale roughly as

$$\frac{\Delta r}{r} \lesssim \frac{\Delta S}{S} \sim \frac{\hbar}{n\hbar} = \frac{1}{n} \tag{1.49}$$

We note that:

- Low-lying and obviously quantum mechanical states (where n is small) have actions that are already close to $\hbar$ in magnitude, so that fluctuations of $\mathcal{O}(\hbar)$ around the classical path are dramatically different. Many, quite different, paths are associated with such quantum states, and the notion of trajectory is not useful (or even well defined).
- For highly excited, quasi-classical states (with $n \gg 1$), an $\mathcal{O}(\hbar)$ change in the action makes little change in the physical dimensions of the path (P1.12, P1.13), and one approaches the classical limit of predictable trajectories.

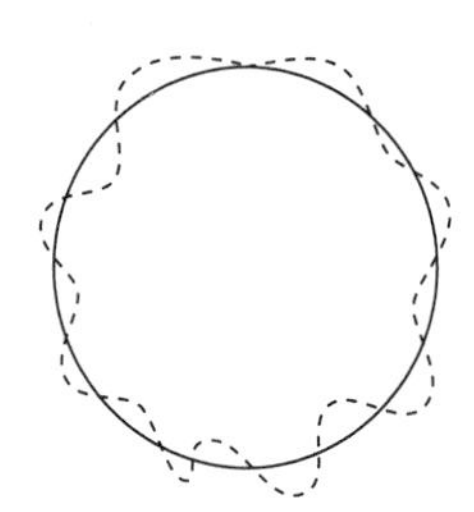

FIGURE 1.12. *Schematic picture of a circular orbit representing classical trajectory (solid) and fluctuation around it (dashed) with a larger value of the classical action.*

1.4 DIMENSIONAL ANALYSIS

Most problems in physics are ultimately to be related to measurable quantities in the "real world" and therefore have answers that carry dimensions. For purely mechanical problems, any constant or variable representing a physical property will have dimensions that can be constructed from fundamental units of mass (M), length (L), and time (T). We can formalize this statement by using the notation

$$[\chi] \equiv \text{dimensions of } \chi = M^a L^b T^c \tag{1.50}$$

where a, b, c are real, possibly fractional, exponents. Familiar examples such as force (F), pressure (P), and density (ρ) correspond to

$$[F] = MLT^{-2} \qquad [P] = ML^{-1}T^{-2} \qquad [\rho] = ML^{-3} \tag{1.51}$$

Specific conventions giving the *units* of physical observables (such as the MKS, or meter-kilogram-second system) rely on this observation, but it is more general. It can often be used to solve for the dependence of the physical quantity in question on the dimensionful parameters of the problem.

Example 1.4 The Dimensions of the Harmonic Oscillator The only dimensional inputs to the classic problem of a mass and spring system are the mass, m, and spring constant, K, which have dimensions

$$[m] = M \qquad [K] = MT^{-2} \tag{1.52}$$

The period of the oscillatory motion, τ, should presumably depend on these parameters, plus additional dimensionless constants; we therefore expect that $\tau \propto m^\alpha K^\beta$, so we write

$$T = [\tau] = [m^\alpha K^\beta] = M^\alpha (MT^{-2})^\beta \tag{1.53}$$

Comparing the powers of M, L, T, we find

$$\begin{aligned} M&: \quad 0 = \alpha + \beta \\ L&: \quad 0 = 0 + 0 \\ T&: \quad 1 = 0 - 2\beta \end{aligned} \tag{1.54}$$

which gives $\alpha = 1/2$ and $\beta = -1/2$ or

$$\tau \propto \sqrt{\frac{m}{K}} \tag{1.55}$$

The "exact" answer obtained from the solution of the equations of motion is, of course,

$$\frac{2\pi}{\omega} = \tau = 2\pi\sqrt{\frac{m}{K}} \tag{1.56}$$

This result is not atypical in that the dimensionless constant that is left unspecified by dimensional analysis is often within one to two orders of magnitude (either bigger or smaller) of unity.

In the quantum version of this problem, we have another dimensionful constant at our disposal, namely $\hbar$, which has dimensions

$$[\hbar] = ML^2T^{-1} \tag{1.57}$$

In contrast to the classical case, we can now also construct a fundamental length or amplitude by writing $A \propto m^\alpha K^\beta \hbar^\gamma$ and find that $\gamma = 1/2$ and $\alpha = \beta = -1/4$ or

$$A \propto \left(\frac{\hbar^2}{mK}\right)^{1/4} \tag{1.58}$$

The energy of the oscillator should also have typical dimensions given by

$$[E] = [KA^2] \propto \hbar\sqrt{\frac{K}{m}} \propto \hbar\omega \tag{1.59}$$

For problems involving electricity and magnetism, an additional fundamental dimensionful quantity is introduced, often the dimension of charge,[8] while thermodynamics problems require a temperature dimension (standardly taken as the degree Kelvin when used as a unit.)

1.5 QUESTIONS AND PROBLEMS

Q1.1 Everyone comes to any text with some idea of what one wants to get out of it. What particular question about or aspect of quantum mechanics interests you the most? Look in the index, and see if that topic is covered in this book. If it is, find the reference, and see what you will have to learn in order to understand it. If it is not, do a library search (or ask someone) until you learn what you really want to know.

Q1.2 Think of at least one other image that you would consider as an "icon" of quantum mechanics. Add at least one other "landmark" to Table 1.1 as an important length scale. How has the size of the observed universe in Table 1.1 changed over the course of human history? For example, how much of it was known to prehistoric society? Medieval society? Renaissance society? The discoverers of quantum mechanics in the mid-1920's?

Q1.3 Try to imagine a world[9] in which the fundamental constants of relativity and quantum mechanics were on a more human scale, namely, $c = 10$ m/s and $\hbar = 0.1$ J · s. For example, what would the zero point energy of the keys in your pocket be? How about the probability that they could quantum mechanically tunnel out of your pocket? Could you walk through a doorway and be sure of which way you were going afterwards (i.e., would you diffract?)

Q1.4 Discuss to what extent other areas of physics have a "fundamental dimensionful constant"; consider, for example, gravitation (G?), statistical mechanics (k_B?), electricity (e and/or ϵ_0?), and magnetism (e and/or μ_0?).

Q1.5 Would you expect particles with internal degrees of freedom (such as atoms or molecules) to still exhibit the same quantum wave behavior (like interference) of "point particles" such as electrons? Should quantum mechanics still work if the de Broglie wavelength of the particle is smaller than its physical size? For tests of these ideas using atoms and molecules, see Keith et al. (1991) and Chapman et al. (1995).

Q1.6 Compton scattering experiments were first carried out by scattering X rays from the valence electrons of carbon, which are bound. Why is the assumption of an initially free electron at rest still used?

---

[8]Recall that the MKSA system of units actually uses the Ampere, i.e., current, as the defining unit for EM.

[9]For an entertaining version of this "what-if exercise," see *Mr. Tompkins in Wonderland* by George Gamow (1946).

Q1.7 How would the densities of "ordinary" solid matter change if the value of $\hbar$ were somehow suddenly doubled in magnitude? How about if the electron were 200 times heavier?

P1.1 **Relativistic decays** Consider the decay of a particle with mass M (at rest) to a lighter particle of mass m and a massless photon, i.e., $M \rightarrow m + \gamma$.

(a) Using conservation of energy and momentum, find the magnitude of the momentum and the energy for both final state particles. *Hint:* Remember to use the relativistic formulae for E and pc.

(b) Show that m moves off with a speed given by

$$\frac{v}{c} = \frac{M^2 - m^2}{M^2 + m^2}$$

and comment on any obvious limiting cases.

(c) For the decay of an excited state of an atom, one has $Mc^2 = mc^2 + \Delta E$, where $\Delta E = E_\gamma$ is the energy of the photon emitted in the transition. Estimate the recoil velocity of the hydrogen atom when it emits a photon in the $2 \rightarrow 1$ transition.

P1.2 Experiments measuring the photoelectric effect often determine the *stopping potential,* V_0, required to bring the emitted photoelectrons to rest; this implies the relation

$$eV_0 = hf - W = \frac{hc}{\lambda} - W$$

The following data, taken from Millikan (1916), give the stopping potential for various wavelengths of incident light for a sodium-copper interface:

λ (Å)	**V_0(Volts)**
5461	−2.05
4339	−1.48
4047	−1.30
3650	−0.92
3216	−0.38

(Note that the values of V_0 are negative because of the effect of the additional metal interface.) Use these values to test the Einstein relation, Eqn. (1.13), and to extract a value for h. Try to estimate the errors for your value if you can.

P1.3 **Compton scattering:**

(a) Derive Eqn. (1.15) by using conservation of energy and momentum for the photon-electron collision in Fig. 1.10. *Hint:* Conservation of energy and momentum in this case look like

$$\text{momentum:} \quad \mathbf{p}_\gamma = \mathbf{p}'_\gamma + \mathbf{p}'_e$$

$$\text{energy:} \quad p_\gamma c + m_e c^2 = p'_\gamma c + \sqrt{(p'_e c)^2 + (m_e c^2)^2}$$

since the electron is initially at rest.

(b) Evaluate the fractional change in wavelength, $\Delta\lambda/\lambda$, for incident blue light ($\lambda = 4500$ Å) and X rays ($\lambda = 0.7$ Å).

P1.4 Since the derivation in P1.3 exists in most modern physics texts, here's a different version of a photon-electron collision problem. At the Stanford Linear Accelerator Center (SLAC), laser light (i.e., with visible wavelengths) has been backscattered ($\theta_\gamma = 180°$) from $E_e \approx 50$ GeV (and hence ultrarelativistic) electrons to obtain very high-energy photons as in

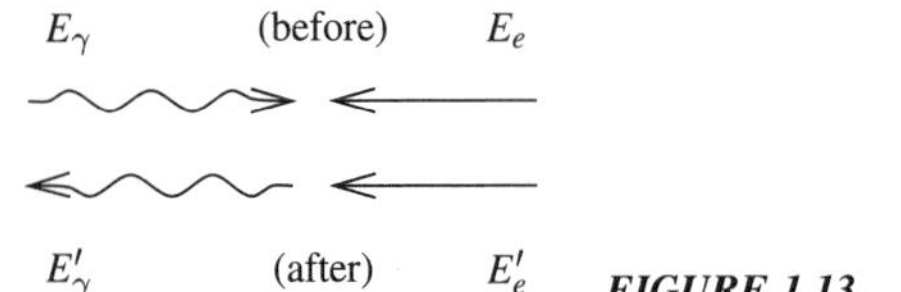

FIGURE 1.13. *Geometry for light backscattered from electrons.*

Fig. 1.13. Using any approximations suggested by this description of the problem, show that the energy of the backscattered photons is given by

$$E'_\gamma \approx E_e \left(1 + \frac{(m_e c^2)^2}{E_e E_\gamma}\right)^{-1}$$

Evaluate this energy for a "blue" laser and the electron energy above.

P1.5 **(a)** The interference patterns in Fig. 1.2 were obtained using electrons that had been accelerated through a potential difference of 80,000 V and an "effective" slit width of $D \sim 6\mu$m. Calculate the de Broglie wavelength of the electrons, and estimate the lateral size of the diffraction pattern on a screen 10 cm from the slit.

(b) Consider the diffraction pattern, due to wave mechanics effects, of a ball thrown through an open window. Using everyday values for all quantities, show *very roughly* that the first minimum of the diffraction pattern would be at a distance of one atomic radius from the central peak on a screen located at the edge of the universe.

P1.6 **Simple-minded scaling laws:** Rederive all of the results in Ex. 1.1, but for the general power law radial potential and corresponding force

$$V(r) = Ar^s \qquad \text{giving} \qquad F(r) \propto Ar^{s-1}$$

For $s = -1$ and $A = -Ke^2$, this gives the Coulomb potential, while for $s = +2$ and $A = K/2$, it represents a 3D spring (or harmonic oscillator) potential. You should also be able to show that for $A = V_0/L^s$ and $s \to \infty$, one obtains the 3D infinite well of radius L. You can use these limits to check your general results against familiar special cases.

(a) Show that the period scales as

$$\tau \propto \left(\frac{m}{n\hbar}\right)\left(\frac{n^2\hbar^2}{Am}\right)^{2/(s+2)}$$

(b) Show that the energy scales as

$$E \propto A\left(\frac{n^2\hbar^2}{Am}\right)^{s/(s+2)}$$

(c) Show that the action (when integrated over one period of a circular orbit as in Ex. 1.4) scales as

$$S \propto n\hbar$$

(d) Repeat the correspondence principle limit argument to show that the frequency of the photons emitted in the $n \to n-1$ transition scales in the same way as the classical rotation frequency when $n \gg 1$.

P1.7 **Quantum numbers for macroscopic systems:**

(a) Estimate the angular momentum quantum number n for a Ferris wheel using the Bohr condition $L = n\hbar$.

(b) The energies of a harmonic oscillator (i.e., mass and spring) are quantized via $E_n = (n + 1/2)\hbar\omega$ where $\omega = \sqrt{K/m}$ is the natural frequency of oscillation. If you push down on the back end of a car, it springs back. Estimate the quantum number associated with this classical motion.

(c) The earth (M_e)-moon (M_m) system is similar to a hydrogen atom with the electrostatic force (i.e., Ke^2) replaced by the gravitational force (GM_eM_m) and with the moon playing the role of the electron. The earth-moon system is then quantized as in Ex. 1.1, and you should first rederive all of the relevant formulae for the energy, period, and orbital radius of the system. Using, for example, the length of the month, estimate the value of n. Using the result for the classical action in Ex. 1.3, estimate the unavoidable spread in radius of a circular orbit with this quantum number.

P1.8 **Quantum or classical systems:**

(a) At one point, the "world's coldest gas" was a sample of ^{133}Cs atoms with a number density $N \sim 10^{10}\ \text{cm}^{-3}$ at a temperature of $T \sim 700\ n\text{K}$. Is this a quantum or classical system? There are hopes to go to $T \sim 200\ n\text{K}$ and $n \sim 10^{14}\ \text{cm}^{-3}$. How about that system? *Hint:* Use the ideas of Ex. 1.2.

(b) The electrons in a white dwarf star can be considered as a gas. There might be roughly 2×10^{57} of them in a volume about the size of the earth, with internal temperatures of order 10^{6-7} K. Is this a quantum or classical system?

P1.9 Use the uncertainty principle to estimate the ground state energy (zero point energy) of a mass (m) and spring (K) system. *Hint:* Write the energy in the form

$$E = \frac{p^2}{2m} + \frac{1}{2}Kx^2 \sim \frac{\hbar^2}{2mx^2} + \frac{1}{2}Kx^2$$

where one approximates $\Delta x \Delta p \sim xp \sim \hbar$; then find the minimum value of E as a function of x. What is the smallest amplitude of motion that a mass and spring system can have? Do your answers agree with the dimensional analysis result of Ex. 1.4?

P1.10 Repeat P1.9, but for a particle of mass m in a linear confining potential given by $V(x) = C|x|$. Does your result agree with the general scaling expression in P1.6, and with dimensional analysis arguments?

P1.11 **Quantum aiming errors:** In classical mechanics, it is possible, in principle, to drop a point object so precisely that it lands exactly on a point target directly below it. This is because one can (again, in principle) determine both its horizontal position and velocity exactly. In quantum mechanics, because of the uncertainty principle, one cannot do this, and there will be an unavoidable "quantum error" in the drop.

(a) Assume that one drops an object of mass m from a height L onto a target below. Assume also that there is both an initial horizontal x_0 and v_x^0 (where $v_x = p_x/m$), and determine the distance, δx, by which one misses the target in terms of x_0 and p_x^0.

(b) Use $\Delta x \Delta p_x \sim x_0 p_x^0 \sim \hbar$ to eliminate one variable, and find the minimum horizontal "error," δx, in terms of L, g, m, and $\hbar$. Does your result depend on all of the dimensionful quantities as you might expect? What is the typical error for a macroscopic object dropped 1 m? Could you have obtained a unique answer to this problem by dimensional analysis?

P1.12 **Free particle action:** In this problem, we gain some experience with the classical action and "quantum trajectories." A free particle of mass m moves in one dimension starting at the origin and arriving at position L in a time T. The path must then satisfy $x(0) = 0$ and $x(T) = L$. Such a free particle will, of course, travel in a straight-line path at constant velocity so that $x(t) = Lt/T$ is the classical trajectory. Since the potential energy vanishes, the classical action is simply

$$S = \int_0^T dt \left(\frac{1}{2}mv^2\right) = \frac{1}{2}m\left(\frac{L}{T}\right)^2 T = \frac{mL^2}{2T}$$

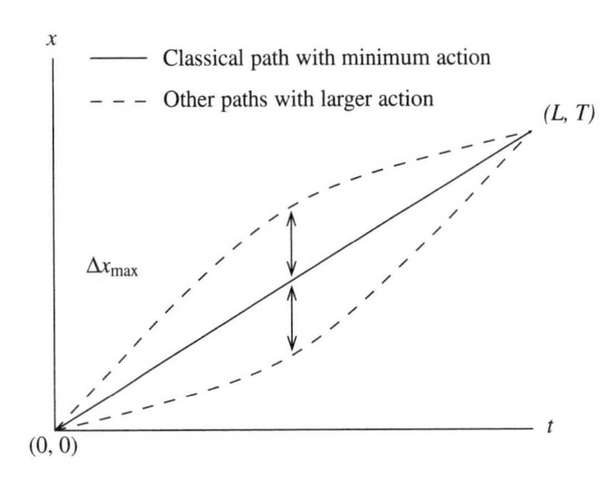

FIGURE 1.14. Classical straight-line trajectory (solid) of a free particle with the minimum classical action and other trajectories (dashed) with larger action.

(a) Assume a more general path of the form $x(t) = a + bt + ct^2$, as in Fig. 1.14, and show that the boundary conditions imply that $a = 0$ and $c = (L - bT)/T^2$. Calculate the action, and show that

$$S(b) = S = \frac{m}{2}\left(\frac{4L^2}{3T} - \frac{2bL}{3} + \frac{b^2T}{3}\right)$$

Show that this has a minimum when $b = L/T$ and $c = 0$ and therefore reproduces the classical trajectory as it should.

(b) Write $b = L/T \pm \delta$, and show that the action is bigger by an amount

$$\Delta S = S\left(b = \frac{L}{T} \pm \delta\right) - S\left(b = \frac{L}{T}\right) = \frac{mT\delta^2}{6}$$

If paths with $\Delta S \lesssim \hbar$ are allowed, show that the quantum trajectories differ at most from the classical straight-line path by

$$|\Delta x|_{\max} = \sqrt{\frac{3\hbar T}{8m}}$$

and evaluate this "shift" for a typical macroscopic particle.

(c) Repeat as much of the problem as possible by using the even more general path $x(t) = a + bt + ct^2 + dt^3$ so that there are now two parameters in the minimization problem.

P1.13 Repeat P1.12, but for a particle undergoing a uniform acceleration, $+g$, for a time T starting from rest. In this case one has $x(t) = gt^2/2$ as the "real" trajectory. Use the same trial path as in P1.12(a), but remember that the potential energy function is now $V(x) = -mgx$.

P1.14 For charged particles, the combination Ke^2 appears frequently in problems involving electromagnetic interactions and provides another natural dimensionful constant in addition to its mass, m.

(a) Using these, as well as $\hbar$ and c, show that the most general combination of these fundamental constants which gives a length is

$$L = \alpha^n\left(\frac{\hbar}{mc}\right)$$

where $\alpha = Ke^2/\hbar c$ is the fine-structure constant. *Hint:* Write

$$L = (Ke^2)^{n_1} m^{n_2} \hbar^{n_3} c^{n_4}$$

and solve for n_2, n_3, n_4 in terms of $n = n_1$.

(b) Show that $n = 0$ corresponds to the Compton wavelength of the particle, while $n = -1$ gives the Bohr radius. Do the cases with $n = +1$ and $n = -2$ look familiar?

P1.15 **(a)** The natural length scale at which gravity, quantum mechanics, and relativity are all simultaneously important is called the *Planck length, L_P*. Using dimensional analysis, find the combination of powers of G (Newton's constant of gravitation), $\hbar$, and c, which make a length.

(b) Evaluate L_P numerically, and compare to a typical scale for nuclear physics, namely, $1\text{ F} = 10^{-15}$ m.

(c) Repeat to find the Planck mass, M_P, evaluate it numerically, and compare to a typical mass of a nuclear constituent, like the proton mass.

(d) Repeat to find the Planck time, T_P, evaluate it numerically, and compare it to the light travel time across a nucleus, which is also a typical nuclear reaction time.

P1.16 Practitioners of elementary particle physics often become so accustomed to the various factors of $\hbar$ and c in their calculations that they use a shorthand notation in which they simply don't bother to write them down; this is sometimes denoted by saying that "$\hbar = c = 1$". For example, the decay rate (λ) and lifetime (τ) of the muon, calculated theoretically, are sometimes written in the form

$$\lambda_\mu = \frac{1}{\tau_\mu} = \frac{G_F^2 m_\mu^5}{192\pi^3}$$

where $G_F = 1.166 \times 10^{-5}\text{ GeV}^{-2}$ (Fermi's constant) and $m_\mu = 105.6\text{ MeV}/c^2$; this is obviously dimensionally wrong as it stands. Supply enough factors of $\hbar$ and c to make it dimensionally correct, evaluate τ_μ numerically, and compare it to the experimental value of $\tau_\mu = 2.2\,\mu$s.

The symbols in Fig. 1.1 from upper left to lower right by rows correspond to (1) Manito (the Native American Great Spirit), (2) caduceus (the life-giving wand), (3) Quauhtli or eagle (one of 20 Mexican day names), (4) ankh (Egyptian symbol for life), (5) trident (symbol of the planet Neptune), (6) Shri-Yantra (Hindu-Vedic cosmic diagram), (7) the alchemist's symbol for brimstone (sulfur), (8) Kikumon (chrysanthemum crest of the Japanese imperial family), (9) G or treble clef in musical notation, (10) symbol for British pound sterling (from Latin libra), (11) cattle brand of Don Bartolome Tapia (California, 1804), (12) postilion horn (symbol for postal services), (13) meteorological symbol for thunderstorm, (14) bio-hazard warning, (15) no-smoking sign, (16) logo for a famous chain of restaurants.

CHAPTER 2

Classical Waves

2.1 THE CLASSICAL WAVE EQUATION

As we initially approach quantum physics through the introduction of wave mechanics, it is useful to begin by recalling some of the standard results obtained from the study of the classical wave equation. For example, using Newton's laws, one can derive the equation of motion for the transverse displacement of a small piece of a stretched one-dimensional string. We obtain the equation

$$\rho \frac{\partial^2 A(x,t)}{\partial t^2} = T \frac{\partial^2 A(x,t)}{\partial x^2} \tag{2.1}$$

where T, ρ are the tension and linear mass density of the string, and $A(x, t)$ is the amplitude of the string at position x and time t. This can be written in the form

$$\frac{\partial^2 A(x,t)}{\partial t^2} = \frac{T}{\rho} \frac{\partial^2 A(x,t)}{\partial x^2} = v^2 \frac{\partial^2 A(x,t)}{\partial x^2} \tag{2.2}$$

where $v = \sqrt{T/\rho}$ is the wave velocity; this is one version of the *classical wave equation.*

In a similar way, Maxwell's equations (in vacuum) can be combined to yield the wave equation, this time in three dimensions, for the components of the electric (or magnetic) field, giving, for example,

$$\frac{\partial^2 \mathbf{E}(\mathbf{r},t)}{\partial t^2} = \frac{1}{\sqrt{\epsilon_0 \mu_0}} \nabla^2 \mathbf{E}(\mathbf{r},t) = c^2 \nabla^2 \mathbf{E}(\mathbf{r},t) \tag{2.3}$$

where the fundamental constants of electricity (ϵ_0) and magnetism (μ_0) combine to yield the speed of light. (Because of the quantum connection between classical electromagnetic waves and their corresponding particle-like quanta, namely, photons, we will be especially interested in this version.)

Both of these cases yield the classical wave equation, which we will study in the form

$$\frac{\partial^2 \phi(x,t)}{\partial t^2} = v^2 \frac{\partial^2 \phi(x,t)}{\partial x^2} \tag{2.4}$$

The most familiar solutions of Eqn. (2.4) are functions of the form

$$\sin(kx \pm \omega t) \qquad \text{or} \qquad \cos(kx \pm \omega t) \tag{2.5}$$

where the wavenumber k and angular frequency ω are related to the familiar wavelength and frequency/period via

$$k \equiv \frac{2\pi}{\lambda} \qquad \text{and} \qquad \omega \equiv 2\pi f = \frac{2\pi}{\tau} \tag{2.6}$$

In order to solve Eqn. (2.4), ω and k must be related via

$$\omega = vk \tag{2.7}$$

Any such relation between the oscillation rates in space (k) and time (ω) is called a *dispersion relation.* The two choices of sign correspond respectively to waves traveling at constant speed to the right $(-)$ or left $(+)$. This can be seen by examining a point of constant phase, $\theta = kx \pm \omega t = \theta_0$, implying that the same point on the wave satisfies $x(t) = \mp vt + \theta_0/k$. A linear relationship (dispersion relation) between ω and k is then a signal of constant speed wave motion.

While perhaps not as familiar, these traveling wave solutions can also be written compactly in complex notation; for right-moving waves, one has

$$e^{i(kx-\omega t)} \qquad \text{and} \qquad e^{-i(kx-\omega t)} \tag{2.8}$$

where we have used the fact that

$$e^{\pm iz} = \cos(z) \pm i\sin(z) \tag{2.9}$$

(See App. A for a review of complex numbers and functions.)

One of the most important features of Eqn. (2.4) is that it is a *linear differential equation,* defined by the property that if $\phi_1(x,t)$ and $\phi_2(x,t)$ are both solutions, then the linear combination

$$\Phi(x,t) = a_1\phi_1(x,t) + a_2\phi_2(x,t) \tag{2.10}$$

is also. This notion can be generalized to show that infinite (discrete) sums

$$\Phi(x,t) = \sum_{n=0}^{+\infty} a_n\phi_n(x,t) \tag{2.11}$$

and infinite (continuous) sums

$$\Phi(x,t) = \int_{-\infty}^{+\infty} a(k)\phi_k(x,t)\, dk \tag{2.12}$$

of solutions are also solutions. This is the *principle of superposition.*

For wave problems with boundaries (and hence boundary conditions that must be satisfied), it is often more useful to consider linear combinations of plane waves traveling in opposite directions, e.g.,

$$A\sin(kx - \omega t) + A\sin(kx + \omega t) = 2A\sin(kx)\cos(\omega t) \tag{2.13}$$

which gives rise to standing waves. For problems with no boundaries, say, infinitely long strings or regions of space with no conductors present (on which the electric field would have to vanish), any values of k (and hence ω) are allowed, and one has a continuous "spectrum."

If, however, one has to impose boundary conditions, there are additional constraints on the allowed solutions of the wave equations; these often require that the wave amplitude must vanish at the "edges." For a finite length string, fixed at both ends, say, at $x = 0$ and $x = L$, with a standing-wave solution of the form $A(x, t) = A\sin(kx)\cos(\omega t)$, we require that

$$A(0, t) = 0 = A(L, t) \qquad \text{for all } t \tag{2.14}$$

This implies that $\sin(k_n L) = 0$ or, equivalently, $k_n = n\pi/L$ for $n = 1, 2, \ldots$. Thus, the imposition of boundary conditions on solutions of the wave equation can give rise to quantized values of the wavenumber k and hence for the frequency ω. These quantization effects are a property of solutions to the wave equation and will necessarily appear in quantum mechanics as well, where the origin of quantized quantities can always be traced to the imposition of boundary conditions.

2.2 WAVEPACKETS AND PERIODIC SOLUTIONS

2.2.1 General Wavepacket Solutions

Plane wave solutions, characterized by a well-defined wavenumber and frequency, are useful constructs for analyzing the wave properties of a system. They are, however, far from the most general solution of the classical wave equation, so we will have to extend our analysis for several reasons:

- Because plane wave solutions imply a nonzero amplitude over all space, they can only be idealizations corresponding to wave pulses or wave trains of very long but finite duration. As we will see, this implies that they will have a finite spread in wavelength or wavenumber.
- The most general solution of the wave equation (see, e.g., P2.1) can be shown to be given by any (suitably differentiable) function of the form $\phi(x, t) = f(x \pm vt)$ since it satisfies

$$(\pm v)^2 f''(x \pm vt) = \frac{\partial^2 \phi(x, t)}{\partial t^2} = v^2 \frac{\partial^2 \phi(x, t)}{\partial x^2} = v^2 f''(x \pm vt) \tag{2.15}$$

  This implies that any appropriate initial waveform, $f(x)$, can be turned into a solution of Eqn. (2.4), $f(x \pm vt)$, which propagates to the right $(-)$ or left $(+)$ with no change in shape as in Fig. 2.1. Such a solution can be called a *wavepacket.*
- While such wavepackets are useful for describing the space-time evolution of localized wave phenomena (wave pulses traveling down a string, thunderclaps, laser pulses, etc.), they do not make the "wave content" (k or w dependence) obvious.

To understand how to extract the wavenumber dependence of a given wavepacket solution of the wave equation, it is instructive to ask the question in reverse and see how

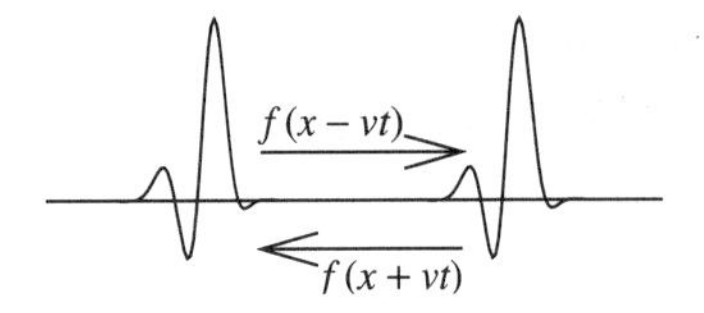

FIGURE 2.1. *Left- and right-moving wavepacket solutions to the classical wave equation.*

to construct localized wavepackets using familiar plane wave solutions. Three effects will make this possible:

1. The linearity of the wave equation ensures that one can add as many solutions as desired and still have a solution.
2. The possibility of constructive and destructive interference allows us to imagine building up a localized solution.
3. The fact that all plane wave components have the same common velocity guarantees that the entire wavepacket will not disperse as it travels, consistent with the general solution of Fig. 2.1.

It is sufficient to study the problem of obtaining localized waveforms, $f(x)$, as the complete time-dependent solution will then simply be given by $f(x - vt)$.

The simplest example that demonstrates the use of linearity and interference ideas is the phenomenon of *beats* (familiar from acoustics). We write the sum of two plane wave solutions as

$$\begin{aligned} f(x) &= A\cos(k_1 x) + A\cos(k_2 x) \\ &= 2A\cos\left[\frac{(k_1 + k_2)x}{2}\right]\cos\left[\frac{(k_1 - k_2)x}{2}\right] \end{aligned} \tag{2.16}$$

which is illustrated in Fig. 2.2. For $k_1 \approx k_2 \approx k$, the resulting waveform is a plane wave, $\cos(kx)$, modulated by the "beat envelope," $2A\cos(\Delta kx)$, where $\Delta k \equiv (k_1 - k_2)/2$. This factor gives complete destructive interference when $x = (2n + 1)\pi/2\Delta k$, implying that successive interference (beat) minima will be separated by $\Delta x = \pi/\Delta k$ or $\Delta x\Delta k \approx \pi \geq \mathcal{O}(1)$. This is our first example of a quite general feature, namely, that

- The degree of localization of a wavepacket in space making use of interference effects is inversely correlated with the spread in available k values.

2.2.2 Fourier Series

A more complex waveform, generated by selective constructive and destructive interference effects, can be obtained by the use of a *Fourier series expansion.*[1] Any (appropriately

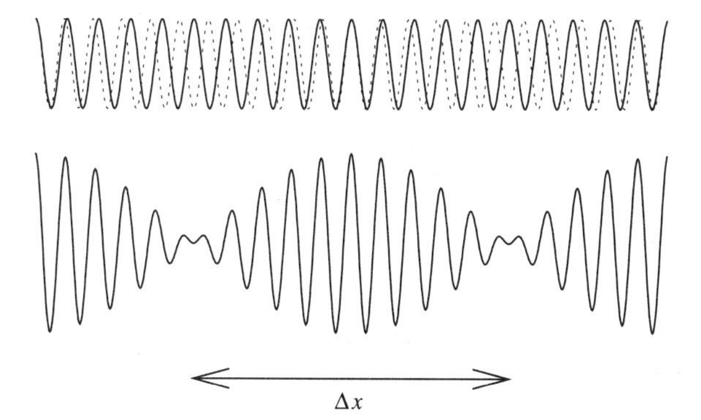

FIGURE 2.2. Interference of sinusoidal solutions giving "beats."

[1]See any standard reference on mathematical physics for more details, e.g., Butkov (1968) or Mathews and Walker (1970).

smooth) periodic function satisfying $f(x) = f(x + 2L)$ or, equivalently, $f(x - L) = f(x + L)$ can be expanded in a linear combination of wave solutions via

$$f(x) = \frac{a_0}{2} + \sum_{n=1}^{\infty} \left[a_n \cos\left(\frac{n\pi x}{L}\right) + b_n \sin\left(\frac{n\pi x}{L}\right) \right] \tag{2.17}$$

This solution will then contain varying contributions (given by the a_n, b_n) from plane wave solutions with wavenumbers $k_n = n\pi/L$. The expansion coefficients a_n, b_n can be obtained by multiplying both sides of Eqn. (2.17) by $\cos(m\pi x/L)$, $\sin(n\pi x/L)$, respectively, and integrating over one cycle. Making use of the relations

$$\int_{-L}^{+L} \cos\left(\frac{n\pi x}{L}\right) \cos\left(\frac{m\pi x}{L}\right) dx = L\delta_{n,m}$$

$$\int_{-L}^{+L} \sin\left(\frac{n\pi x}{L}\right) \sin\left(\frac{m\pi x}{L}\right) dx = L\delta_{n,m}$$

$$\int_{-L}^{+L} \cos\left(\frac{n\pi x}{L}\right) \sin\left(\frac{m\pi x}{L}\right) dx = 0 \tag{2.18}$$

we find

$$a_n = \frac{1}{L}\int_0^{2L} f(x)\cos\left(\frac{n\pi x}{L}\right) dx = \frac{1}{L}\int_{-L}^{L} f(x)\cos\left(\frac{n\pi x}{L}\right) dx$$

$$b_n = \frac{1}{L}\int_0^{2L} f(x)\sin\left(\frac{n\pi x}{L}\right) dx = \frac{1}{L}\int_{-L}^{L} f(x)\sin\left(\frac{n\pi x}{L}\right) dx \tag{2.19}$$

The $a_n(b_n)$ thus measure the "overlap" of the desired waveform with one of the basic cosine or sine solutions.

Example 2.1 Square Wave As an example, consider the square waveform defined, for one cycle $(-L, L)$, by

$$\text{(square wave)} \quad f(x) = \begin{cases} A & \text{for } |x| < L/2 \\ -A & \text{for } |x| > L/2 \end{cases} \tag{2.20}$$

The Fourier coefficients, a_n and b_n, can be easily obtained, and we find

$$\text{(square wave)} \qquad b_n = 0 \qquad a_0 = 0 \qquad a_n = 4A\frac{\sin(n\pi/2)}{n\pi} \tag{2.21}$$

where $n = 1, 2, \ldots$. In this case, the odd $\sin(n\pi x/L)$ terms cannot contribute to an even function, so that the b_n vanish for symmetry reasons. We plot the *partial* Fourier sums

$$F_N(x) = \sum_{n=1}^{N} a_n \cos\left(\frac{n\pi x}{L}\right) \tag{2.22}$$

for increasing values of N in Fig. 2.3; in Fig. 2.4 we show how the component waves contribute to a particular partial sum ($N = 5$). The Fourier series does seem to converge pointwise to the corresponding function where it should, i.e., at points where the function is appropriately smooth.[2] One useful measure of the overall convergence can be defined via

$$\Delta_N = \frac{\int_{-L}^{L} [f(x) - F_N(x)]^2\, dx}{\int_{-L}^{L} [f(x)]^2\, dx} \tag{2.23}$$

[2]The "overshoots" at the discontinuities persist even in the complete sum. These "Gibbs peaks" are discussed in many textbooks on mathematical physics, e.g., Mathews and Walker (1970).

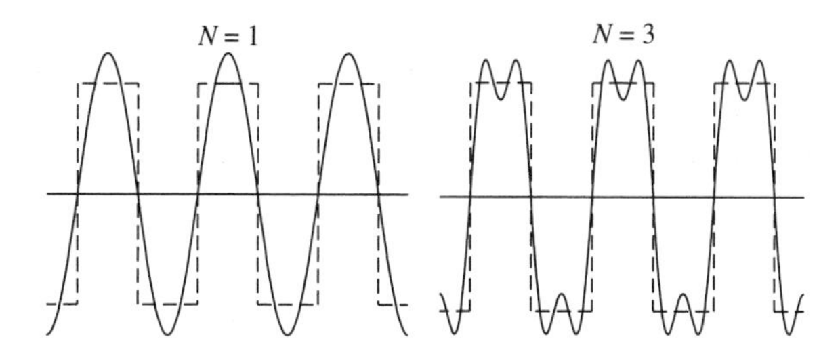

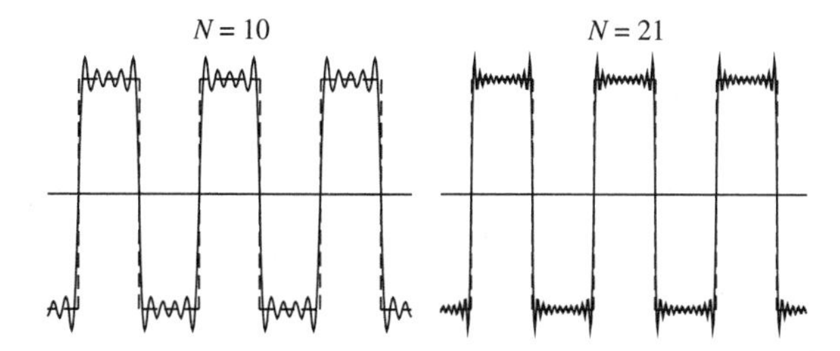

FIGURE 2.3. *Fourier series approximations, $F_N(x)$, to a square wave pulse for N = 1, 3, 5, 21.*

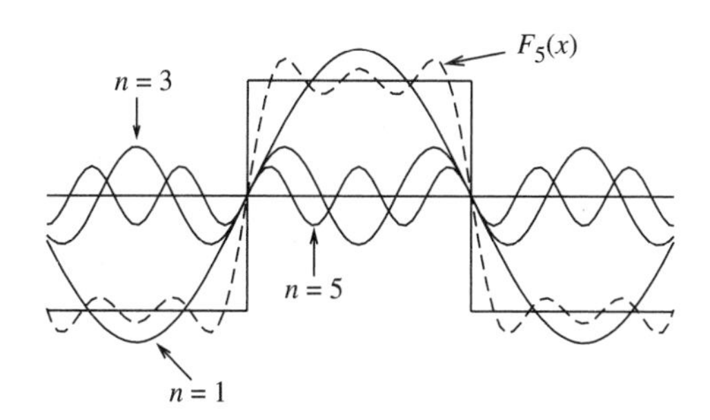

FIGURE 2.4. *Partial Fourier sum, $F_5(x)$, for square wave and component terms.*

where $F_N(x)$ is the Nth partial sum; this measures the overall deviation of the Nth partial sum from the function. For the square wave, one can show that it is given by

$$\text{(square wave)} \qquad \Delta_N = 1 - \frac{8}{\pi^2}\sum_{n=1}^{N} \frac{\sin(n\pi/2)^2}{n^2} \tag{2.24}$$

If we use the fact that

$$1 + \frac{1}{3^2} + \frac{1}{5^2} + \cdots = \sum_{k=1}^{\infty} \frac{1}{(2k-1)^2} = \frac{\pi^2}{8} \tag{2.25}$$

(discussed in App. B.2 under the topic "Zeta functions"), we see that $\Delta_N \to 0$ as $N \to \infty$; we can more and more closely approach the desired waveform, except at isolated points.

Example 2.2 Triangle Wave For comparison, we also consider a (slightly) better-behaved function, namely, a triangle waveform, defined via

$$\text{(triangle wave)} \qquad f(x) = A(L - |x|) \qquad \text{for} \qquad -L < x < L \tag{2.26}$$

This function is at least continuous at all points. We leave it as an exercise (P2.4) to show that the nonvanishing Fourier coefficients are given by

$$\text{(triangle wave)} \qquad a_n = (1 - \cos(n\pi))\frac{2AL}{n^2\pi^2} = 4AL\left(\frac{\sin(n\pi/2)}{n\pi}\right)^2 \tag{2.27}$$

while the corresponding measure of the overall convergence is given by

$$\text{(triangle wave)} \qquad \Delta_N = 1 - \frac{96}{\pi^4}\sum_{n=1}^{N}\frac{(1-\cos(n\pi))^2}{n^4} \tag{2.28}$$

This example implies that the convergence is faster (as a function of N) for smoother functions. This is clear from the form of the Fourier coefficients, since $a_n \to 1/n^2\,(1/n)$ for the triangle (square) wave.

We see that with Fourier series, we can produce any desired *periodic* waveform and extract its wavenumber content (via the a_n and b_n). But even though we have used an infinite number of plane wave components, we still evidently do not have enough "degrees of freedom" to produce a truly localized wavepacket. The combination of large numbers of different wave numbers can, however, be accomplished in other ways.

For example, consider the linear combination of cosine solutions with wavenumbers,

$$k_n = nK/N \qquad \text{where} \qquad n = -N, -(N-1), \ldots, (N-1), N \tag{2.29}$$

We can consider these to be wavenumbers sampled uniformly (i.e., each with identical weight $1/N$) from the interval $(-K, K)$; we will eventually consider the limiting case with increasingly fine spacing, i.e., letting $N \to \infty$.

We then have the solution

$$\begin{aligned}\psi_N(x) &= \frac{1}{N}\sum_{n=-N}^{n=N}\cos\left(\frac{nKx}{N}\right)\\ &= \frac{1}{N}\left(1 + 2\sum_{n=1}^{N}\cos\left(\frac{nKx}{N}\right)\right)\end{aligned} \tag{2.30}$$

This summation can actually be obtained in closed form (P2.7), giving

$$\psi_N(x) = \frac{1}{N}\left(1 + \frac{2\cos(Kx/2(1+1/N))\sin(Kx/2)}{\sin(Kx/2N)}\right) \tag{2.31}$$

This function is also periodic in x, but now with period $2\pi N/K$ (see P2.10); this implies that it can be localized by letting $N \to \infty$. We plot this summation for increasing N in Fig. 2.5, and we note that:

- The periodic "recurrences" are pushed further away from the origin as N increases, giving a truly localized waveform.

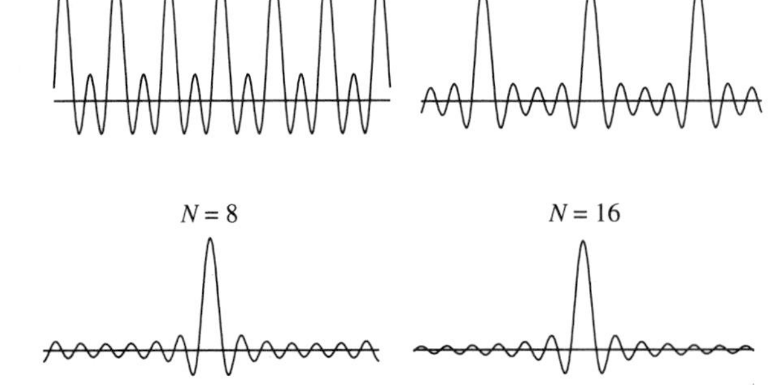

FIGURE 2.5. Linear superposition solution, $\psi_N(x)$, from Eqn. (2.31) for $N = 2, 4, 8, 16$ showing increasing localization as $N \to \infty$.

- The wavepacket still has an intrinsic width, even when $N \to \infty$, due to the finite range of k values used. This can also be seen analytically by noting that

$$\psi(x) = \lim_{N\to\infty} \psi_N(x) = \lim_{N\to\infty} \frac{2\cos(Kx/2(1+1/N))\sin(Kx/2)}{N\sin(Kx/2N)} = \frac{2\sin(Kx)}{Kx} \tag{2.32}$$

This limiting form is instructive as it clearly shows that $\psi(x) \to 0$ as $|x| \to \infty$.

- In order to accomplish the subtle destructive interference between plane wave components for arbitrarily large values of $|x|$, it seems that we must use an uncountably (continuous) large set of wavenumbers.

2.3 FOURIER TRANSFORMS

This limit of a continuous summation over wavenumbers, as well as the extension to include more general plane wave solutions, is formalized in the so-called *Fourier integral* or *Fourier transform*

$$f(x) = \frac{1}{\sqrt{2\pi}} \int_{-\infty}^{+\infty} dk\, A(k) e^{ikx} \tag{2.33}$$

The $A(k)$ gives the amplitude of each plane wave contribution to the resulting wavepacket and is the continuous analog of the discrete Fourier coefficients, a_n, b_n. The seemingly arbitrary normalization factor $\left(1/\sqrt{2\pi}\right)$ will be discussed in Sec. 2.4. As always, the final solution of the wave equation is obtained by letting $f(x) \to f(x \pm vt)$.

Example 2.3 Fourier Transform with "Flat" k Values The Fourier integral representation corresponding to the example above can be written by considering

$$A(k) = \begin{cases} 0 & \text{for } k > |K| \\ 1/\sqrt{2K} & \text{for } k < |K| \end{cases} \tag{2.34}$$

which has the resulting waveform

$$f(x) = \frac{1}{\sqrt{4\pi K}} \int_{-K}^{+K} e^{ikx}\, dk = \frac{1}{\sqrt{4\pi K}} \left. \frac{e^{ikx}}{ix} \right|_{-K}^{+K} = \sqrt{\frac{K}{\pi}} \frac{\sin(Kx)}{Kx} \tag{2.35}$$

We plot both $A(k)$ and $f(x)$ in Fig. 2.6 for two different values of K and note that the widths of the k and x distributions are inversely correlated. This arises because an increasingly large sample of wavenumbers (larger Δk) can be more efficient in the destructive interference necessary to produce a smaller, more localized (smaller Δx) wavepacket. In fact, if we make the identification of $\Delta k \approx 2K$ and estimate the spread in position by the location of the first set of

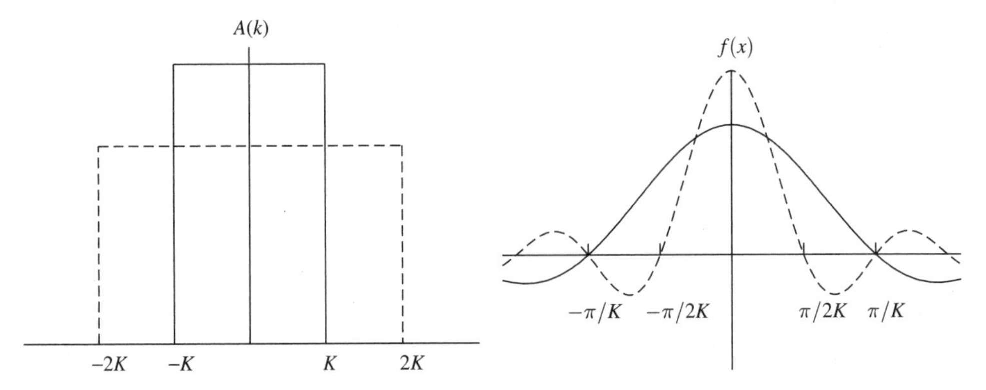

FIGURE 2.6. *"Square" A(k) and its Fourier transform f(x) from Ex. 2.1 for two values of K.*

nodes, i.e., $\Delta x \approx 2\pi/K$, we find that $\Delta k \Delta x \approx 4\pi$ or once again that

$$\Delta k\, \Delta x > \mathcal{O}(1) \tag{2.36}$$

independent of K.

Example 2.4 Fourier Transform of Exponential Another example of a Fourier transform pair is obtained by considering

$$A(k) = \frac{1}{\sqrt{K}} e^{-|k|/K} \tag{2.37}$$

which gives

$$\begin{aligned} f(x) &= \frac{1}{\sqrt{2\pi K}} \int_{-\infty}^{+\infty} e^{-|k|/K} e^{ikx}\, dk \\ &= \frac{1}{\sqrt{2\pi K}} \left(\int_{-\infty}^{0} dk\, e^{k(1/K + ix)} + \int_{0}^{+\infty} dk\, e^{-k(1/K - ix)} \right) \\ &= \sqrt{\frac{2K}{\pi}} \left(\frac{1}{1 + (Kx)^2} \right) \end{aligned} \tag{2.38}$$

which we plot in Fig. 2.7; we see that the widths also satisfy Eqn. (2.36).

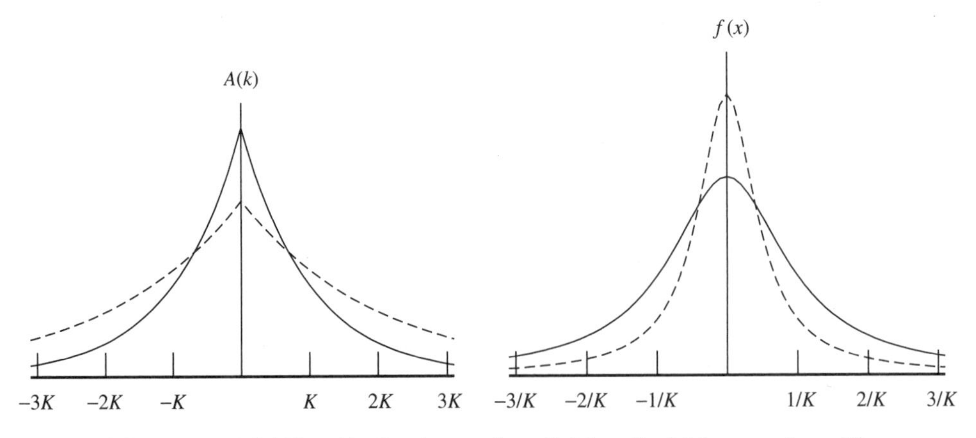

FIGURE 2.7. *Exponential A(k) and its Fourier transform f(x) from Ex. 2.2 for two values of K.*

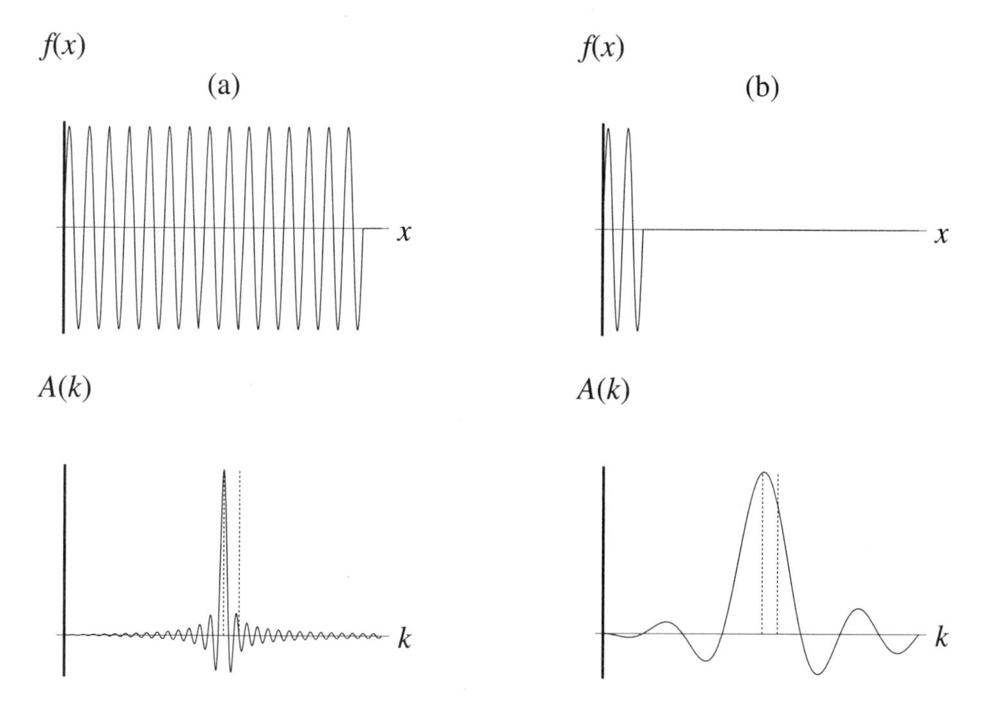

FIGURE 2.8. Schematic plot of wave amplitude for laser pulse, $f(x)$ versus **x**, *for (a) long and (b) short pulses. The corresponding wavenumber amplitudes, $A(k)$ versus k, are shown below and the dashed lines indicate two possible spectral lines.*

This unavoidable constraint on the spatial extent and wavenumber content of a localized wavepacket, $\Delta k\,\Delta x \geq 1$, is a fundamental limitation on physical systems. It restricts one's ability to make measurements or to produce physical phenomena in the same way as, for example, do the laws of thermodynamics or the limiting velocity of the speed of light.

Consider, for example, a laser pulse of finite duration (in both space and time) with the wavenumber distribution shown in Fig. 2.8(a). A long wavetrain with a correspondingly narrow k distribution has a relatively well-defined wavelength and so could resolve the two emission/absorption lines shown. If one wished to gain more real-time information on the system by exciting it with laser pulses of very short duration, Fig. 2.8(b), the corresponding wavenumber distribution would accordingly *broaden,* making it no longer possible to resolve various spectral features. (See P2.17 for an example from laser chemistry.)

Recognizing the importance of this result, we can make an immediate connection to quantum mechanics by using the de Broglie relation $p = h/\lambda = \hbar k$ to argue that any wave description of matter will necessarily satisfy

$$\Delta x\,\Delta p \approx \Delta x(\hbar\,\Delta k) \gtrsim \hbar \tag{2.39}$$

which is the content of the Heisenberg uncertainty principle. This limit on the ability to measure simultaneously the position and momentum of a quantum mechanical particle can thus be traced (in this language at least) to a wave description of mechanics.

2.4 INVERTING THE FOURIER TRANSFORM: THE DIRAC δ FUNCTION

We have seen that a truly localized wavepacket can be constructed from plane wave solutions via the Fourier transform, i.e.,

$$f(x) = \frac{1}{\sqrt{2\pi}} \int_{-\infty}^{+\infty} A(k) e^{ikx}\, dk \tag{2.40}$$

If, however, we are given a spatial waveform, $f(x)$, we would also like to be able to extract the wavenumber or wavelength "components" by somehow inverting Eqn. (2.40) to obtain $A(k)$. Knowledge of $A(k)$ allows one to determine the behavior of a wavepacket in, say, a diffraction or interference experiment, where one has simple rules for the behavior of each individual wavelength component. We devote this section to the question of how such an inversion is obtained, developing at the same time the mathematical properties of a new function that will be of continuing use, the so-called *Dirac δ function.*

The final result we will obtain is quite simple, namely that

$$A(k) = \frac{1}{\sqrt{2\pi}} \int_{-\infty}^{+\infty} f(x)\, e^{-ikx}\, dx \tag{2.41}$$

This can be taken to mean that $A(k)$ and $f(x)$ are, in a sense, inverses of each other under the Fourier transform, and the similarity in form argues for the conventional use of the common $1/\sqrt{2\pi}$ factors.

In order to derive Eqn. (2.41) we multiply both sides of Eqn. (2.40) by $\exp(-ik'x)/\sqrt{2\pi}$ and integrate over x. Doing this, we obtain

$$\begin{aligned} \frac{1}{\sqrt{2\pi}} \int_{-\infty}^{+\infty} f(x)\, e^{-ik'x}\, dx &= \frac{1}{2\pi} \int_{-\infty}^{+\infty} dx \int_{-\infty}^{+\infty} A(k) e^{ikx} e^{-ik'x}\, dk \\ &= \int_{-\infty}^{+\infty} dk\, A(k) \left[\frac{1}{2\pi} \int_{-\infty}^{+\infty} e^{i(k-k')x}\, dx \right] \\ &= \int_{-\infty}^{+\infty} dk\, A(k)\, \delta(k-k') \\ &\stackrel{?}{=} A(k') \end{aligned} \tag{2.42}$$

where we have implicitly defined a new function called the Dirac δ function via

$$\delta(k-k') \equiv \frac{1}{2\pi} \int_{-\infty}^{+\infty} e^{i(k-k')x}\, dx \tag{2.43}$$

Thus, if Eqn. (2.41) is to be true, we must have

$$\int_{-\infty}^{+\infty} A(k)\, \delta(k-k')\, dk = A(k') \tag{2.44}$$

i.e., $\delta(k-k')$ must "pick" out the value of $A(k)$ only at $k = k'$ from the continuous integral. In this regard, it is similar in function (and name) to the *discrete* or *Kronecker δ function* defined as

$$\delta_{n,m} = \begin{cases} 0 & \text{if } n \neq m \\ 1 & \text{if } n = m \end{cases} \tag{2.45}$$

which has the property that

$$\sum_{k=-\infty}^{\infty} A_k \, \delta_{n,k} = A_n \tag{2.46}$$

in a discrete summation.

To study the properties of the Dirac δ function, it suffices to consider the special case where $k' = 0$, i.e.,

$$\delta(k) = \frac{1}{2\pi} \int_{-\infty}^{+\infty} e^{ikx} \, dx \tag{2.47}$$

We can then argue (not necessarily prove!) that

$$\delta(k) = \begin{cases} \dfrac{1}{2\pi} \displaystyle\int_{-\infty}^{+\infty} 1 \, dx = \infty & \text{for } k = 0 \\ \dfrac{1}{2\pi} \displaystyle\int_{-\infty}^{+\infty} (\cos(kx) + i \sin(kx)) \, dx = 0 & \text{for } k \neq 0 \end{cases} \tag{2.48}$$

where the vanishing of the integrals of the $\sin(kx)$ and $\cos(kx)$ functions occurs because of their oscillatory behavior, leading to cancellations. This heuristic definition of $\delta(k)$, namely, that it vanishes everywhere except at $k = 0$, where it is infinite, shows that it is an extremely poorly behaved function[3] and has to be handled carefully.

We can study it more rigorously by considering the family of auxiliary functions

$$\delta_\epsilon(k) \equiv \frac{1}{2\pi} \int_{-\infty}^{+\infty} e^{-\epsilon x^2} e^{ikx} \, dx = \frac{1}{2\pi} \sqrt{\frac{\pi}{\epsilon}} e^{-k^2/4\epsilon} \tag{2.49}$$

so that

$$\lim_{\epsilon \to 0} \delta_\epsilon(k) = \delta(k) \tag{2.50}$$

(We have simply included a convergence factor so that the integral can be performed in closed form using results in App. B.1.) Using this limiting representation, it is easier to argue that

$$\delta_\epsilon(k) \propto \begin{cases} 1/\sqrt{\epsilon} \quad \longrightarrow \infty & \text{for } k = 0 \\ e^{-k^2/2\epsilon}/\sqrt{\epsilon} \longrightarrow 0 & \text{for } k \neq 0 \end{cases} \tag{2.51}$$

as $\epsilon \to 0$; we can also visualize the approach to the singular limit in Fig. 2.9.

---

[3]It is in the class of mathematical objects called *distributions* or *generalized functions* for which the standardly cited reference is Lighthill (1958) in which he discusses, for example, "Good functions and fairly good functions."

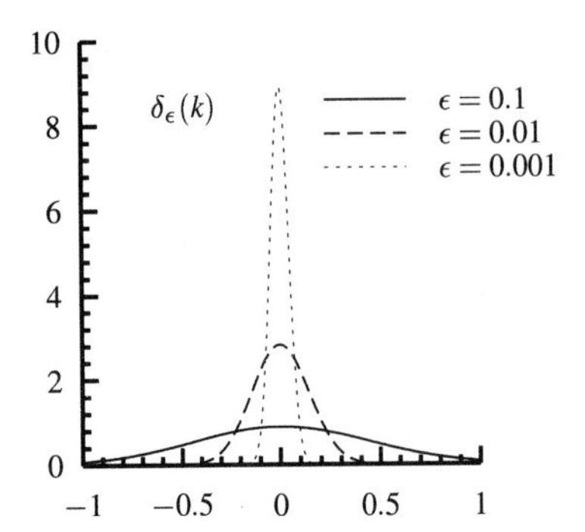

***FIGURE 2.9.** Limiting behavior of $\delta_\epsilon(k)$ in Eqn. (2.49); the dashed (solid, dotted) curves correspond to $\epsilon = 0.1\,(0.01, 0.001)$, respectively.*

This form also allows us to investigate the degree of "infiniteness" at $k = 0$ by considering

$$\int_{-\infty}^{+\infty} \delta_\epsilon(k)\,dk = \frac{1}{2\pi}\sqrt{\frac{\pi}{\epsilon}}\int_{-\infty}^{+\infty} e^{-k^2/4\epsilon}\,dk = \frac{1}{\sqrt{\pi}}\int_{-\infty}^{+\infty} e^{-q^2}\,dq = 1 \tag{2.52}$$

so that the total area under the δ function family of curves is always normalized to unity. Thus we also take

$$\int_{-\infty}^{+\infty} \delta(k)\,dk = \lim_{\epsilon\to 0}\int_{-\infty}^{+\infty} \delta_\epsilon(k)\,dk = 1 \tag{2.53}$$

in much the same way that

$$\sum_{i=-\infty}^{+\infty} \delta_{i,j} = 1 \tag{2.54}$$

(One can—very loosely!—say that $\delta(k = 0) = 1/dk$ where dk is the infinitesimal unit of measure.) We can also derive results such as

$$\int_{-\infty}^{+\infty} k\,\delta(k)\,dk = \lim_{\epsilon\to 0}\int_{-\infty}^{+\infty} k\,\delta_\epsilon(k)\,dk = 0 \tag{2.55}$$

and related ones. Using these results we can now argue that

$$\begin{aligned}
\int_{-\infty}^{+\infty} A(k)\,\delta(k-k')\,dk &= \int_{-\infty}^{+\infty} A(q+k')\,\delta(q)\,dq \\
&= \int_{-\infty}^{+\infty}\left(A(k') + qA'(k') + \frac{q^2}{2}A''(k') + \cdots\right)\delta(q)\,dq \\
&= A(k')\int_{-\infty}^{+\infty}\delta(q)\,dq + A'(k)\int_{-\infty}^{+\infty} q\,\delta(q)\,dq + \cdots \\
&= A(k')
\end{aligned} \tag{2.56}$$

where we have changed variables to $q = k - k'$ and expanded $A(k)$ in a Taylor expansion around $k = k'$. This is the desired property of the Dirac δ function and shows that $A(k)$ and $f(x)$ are indeed related by Eqns. 2.40 and 2.41.

The similarity of a spatial waveform and its Fourier transform in terms of their information content can be seen in other ways. Since we often consider complex waveforms, we will find useful the fact that

$$
\begin{aligned}
\int_{-\infty}^{+\infty} |f(x)|^2 &= \int_{-\infty}^{+\infty} f^*(x) f(x)\, dx \\
&= \int_{-\infty}^{+\infty} dx f^*(x) \left(\frac{1}{\sqrt{2\pi}} \int_{-\infty}^{+\infty} A(k) e^{ikx}\, dk \right) \\
&= \int_{-\infty}^{+\infty} A(k)\, dk \left(\frac{1}{\sqrt{2\pi}} \int_{-\infty}^{+\infty} f^*(x) e^{ikx}\, dx \right) \\
&= \int_{-\infty}^{+\infty} A(k)\, dk \left(\frac{1}{\sqrt{2\pi}} \int_{-\infty}^{+\infty} f(x) e^{-ikx}\, dx \right)^* \\
&= \int_{-\infty}^{+\infty} A(k) A^*(k)\, dk
\end{aligned}
$$

$$
\int_{-\infty}^{+\infty} |f(x)|^2 = \int_{-\infty}^{+\infty} |A(k)|^2\, dk \tag{2.57}
$$

This result is sometimes called *Parseval's theorem.* We note that both examples considered in Sec. 2.3 were chosen so that they satisfied $\int_{-\infty}^{+\infty} |A(k)|^2\, dk = 1$ (check this in P2.15) so that the corresponding spatial waveforms are guaranteed to yield $\int_{-\infty}^{+\infty} |f(x)|^2\, dx = 1$ as well. A similar relation can also be proved for the *overlap* of two different waveforms, namely,

$$
\int_{-\infty}^{+\infty} f_1^*(x) f_2(x)\, dx = \int_{-\infty}^{+\infty} A_1^*(k) A_2(k)\, dk \tag{2.58}
$$

While the Dirac δ function is used here as a tool in proving the important physical connection between $A(k)$ and $f(x)$, its usefulness in mathematical physics will become more obvious, and we make some additional comments here. Some of its other properties are discussed in App. C.8, to which we will refer the interested reader from time to time.

- The arguments of $\delta(k - k')$ are arbitrary, as we could equally well have considered the integral definition

$$
\delta(x - x') = \frac{1}{2\pi} \int_{-\infty}^{+\infty} e^{ik(x-x')}\, dk \tag{2.59}
$$

- The *dimensions* of the δ function thus depend on its argument: namely, $[\delta(z)] = 1/[z]$ where $[z]$ gives the dimensions of z. This can be inferred from the dimensionlessness of the integral $\int_{-\infty}^{\infty} \delta(z)\, dz = 1$.
- The definite integral of the δ function, defined via

$$
\theta(x) \equiv \int_{-\infty}^{x} \delta(x')\, dx' \tag{2.60}
$$

is easily seen to satisfy

$$
\theta(x) = \begin{cases} 0 & \text{for } x < 0 \\ 1 & \text{for } x > 0 \end{cases} \tag{2.61}
$$

depending on whether the integration includes the singular point or not. This function (often called the *step function* or *Heaviside function*) can describe the

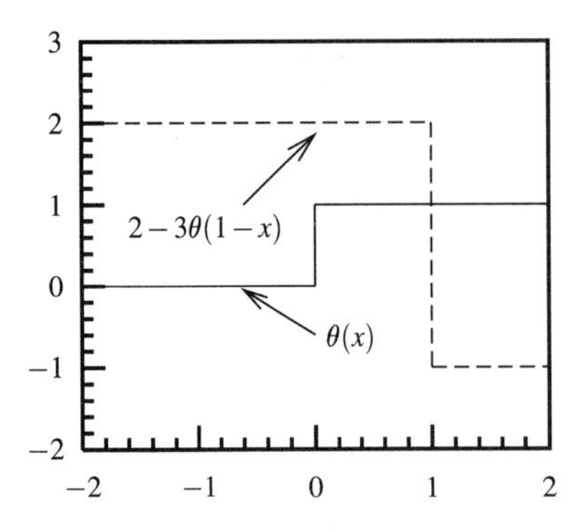

FIGURE 2.10. The Heaviside step function $\theta(x)$ (solid line) and $2 - 3\theta(1 - x)$ (dashed line) versus x.

instantaneous (and hence idealized) "turn-on"—or "turn-off" if one uses $1 - \theta(x)$—of some phenomenon or function, as shown in Fig. 2.10.

2.5 DISPERSION AND TUNNELING

2.5.1 Velocities for Wavepackets

We have seen that wavepackets constructed from plane waves satisfying the simple dispersion relation

$$\omega = \omega(k) = kv \tag{2.62}$$

propagate with no change in shape due to the constant speed, v, of each component. Many classical physical systems are characterized by dispersion relations for which this is not true and which exhibit a variety of new phenomena that have analogs in quantum mechanics, specifically *dispersion* and *tunneling;* we review these aspects of classical wave physics in this section using one model system as an example.

An important system in which both of these phenomena occur arises in the study of the propagation of electromagnetic (hereafter EM) radiation in a region of ionized gas (i.e., a plasma). The self-consistent application of Maxwell's equations (for the EM waves) and Newton's laws (for the motion of the charged particles, in this case the electrons) implies[4] that the dispersion relation is

$$\omega^2 = (kc)^2 + \omega_p^2 \tag{2.63}$$

Here ω_p is the plasma frequency,

$$\omega_p^2 \equiv \frac{n_e e^2}{\epsilon_0 m_e} \tag{2.64}$$

and n_e is the number density of electrons. The plasma frequency is the natural frequency of oscillation of the plasma system, arising, for example, when the positive ions and negative electrons are separated slightly and oscillate around their equilibrium neutral configuration.

[4]See, for example, Kittel (1971) for a derivation.

This dispersion relation can be rewritten in the form

$$\omega(k) = \frac{c}{\sqrt{1 - \omega_p^2/\omega^2}} k = v_\phi k \tag{2.65}$$

where we have defined a *phase velocity*, $v_\phi \equiv \omega(k)/k$. If we restrict ourselves to the case where $\omega > \omega_p$, it is clear that $v_\phi > c$, and this velocity exceeds the speed of light in vacuum! This is not, however, in conflict with the tenets of special relativity, as the phase velocity measures a property of one single plane wave component, which, by definition, has a trivial, sinusoidal space and time dependence. We have already argued that such a waveform is an abstraction; it is, on average, uniform over all space and time and so can lead to no transfer of information. Any *changes* in the wavefield, which might carry a signal, will be described by *modulations* in the plane wave signals, and such disturbances will propagate at speeds less than c.

To see this, consider again a linear combination of two traveling waves (as done for beats) of differing wavenumbers and frequencies, governed by a general dispersion relation, $\omega = \omega(k)$. In this case,

$$\begin{aligned}\phi(x, t) &= A\cos(k_1 x - \omega_1 t) + A\cos(k_2 x - \omega_2 t) \\ &= 2A\cos\left(\frac{(k_1 - k_2)x}{2} - \frac{(\omega_1 - \omega_2)t}{2}\right)\cos\left(\frac{(k_1 + k_2)x}{2} - \frac{(\omega_1 + \omega_2)t}{2}\right) \\ &= A_{\text{eff}}(x, t)\cos(\overline{k}x - \overline{\omega}t)\end{aligned} \tag{2.66}$$

where

$$\overline{k} \equiv \frac{k_1 + k_2}{2} \quad \text{and} \quad \overline{\omega} \equiv \frac{\omega_1 + \omega_2}{2} \tag{2.67}$$

are something like the "average" wavenumbers and frequencies. We have defined an effective amplitude

$$A_{\text{eff}}(x, t) = 2A\cos(\Delta k\, x - \Delta\omega t) \tag{2.68}$$

with

$$\Delta k \equiv \frac{k_1 - k_2}{2} \quad \text{and} \quad \Delta\omega \equiv \frac{\omega_1 - \omega_2}{2} \tag{2.69}$$

If $k_1 \approx k_2$, we can consider Eqn. (2.66) to be a plane wave, $\cos(\overline{k}x - \overline{\omega}t)$, modulated by the amplitude $A_{\text{eff}}(x, t)$; information will be carried along the modulation wave crest, at a rate given by the condition

$$\theta_{\text{eff}} = \Delta k\, x - \Delta\omega\, t = \text{constant} \tag{2.70}$$

i.e., examining a point of constant phase. This implies that

$$\left(\frac{dx}{dt}\right)_{\text{information}} = \frac{\Delta\omega}{\Delta k} \longrightarrow \frac{d\omega}{dk} \tag{2.71}$$

We are thus led to define the *group velocity*,

$$v_g(\omega) = \frac{d\omega(k)}{dk} \tag{2.72}$$

and note that information contained in modulated waveforms or wavepackets, as well as the rate of flow of energy density in a wave, are all governed by the group velocity. As an example, for EM waves in vacuum we have

$$\omega(k) = kc \qquad \text{so that} \qquad v_g = \frac{d\omega}{dk} = c \tag{2.73}$$

as expected. For the case of propagation in plasma, however, we find

$$v_g = \frac{d\omega}{dk}(k) = c\sqrt{1 - \omega_p^2/\omega^2} \leq c \tag{2.74}$$

consistent with relativity. Note that if $\omega/\omega_p \gg 1$, the background plasma cannot "keep up" with the wave, and the radiation propagates at the same rate as in vacuum; as $\omega/\omega_p \to 1$ the effective speed can become arbitrarily small and even become imaginary when the ratio is less than unity.

2.5.2 Dispersion

In this case, v_g clearly depends on frequency; this reflects the fact that different plane wave components of a wavepacket will travel at different speeds due to different *phase velocities.* Any initially localized wavepacket will necessarily contain a range of wavenumbers k (and hence frequencies ω); since the faster components will outpace the slower ones, the wave will necessarily spread or disperse as it propagates. Using Eqn. (2.74) as an example, a spread in frequencies, $\Delta\omega$, implies a range in group velocities

$$\Delta v_g = c(1 - \omega_p^2/\omega^2)^{-1/2}\frac{\omega_p^2}{\omega^3}\Delta\omega \approx c\frac{\omega_p^2}{\omega^3}\Delta\omega \tag{2.75}$$

if $\omega \gg \omega_p$; higher frequencies travel faster. If an initial pulse travels a fixed distance D, one should observe a difference in arrival times due to this effect, the two being related by $\Delta v_g = -c^2\Delta t/D$ (since higher velocity components arrive earlier). This implies that the higher-frequency components of the pulse will arrive earlier, satisfying the relation

$$\Delta\omega = -\frac{c}{D}\frac{\omega^3}{\omega_p^2}\Delta t \tag{2.76}$$

One of the most famous examples of this phenomenon is evident in the observed dispersion of pulses of microwave radiation emitted by the Crab Nebula, which must travel through a region of ionized space. Figure 2.11 illustrates one experimental realization of

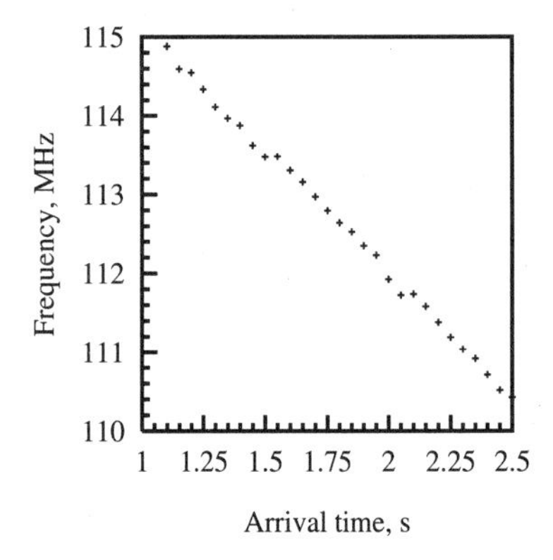

FIGURE 2.11. Plot of the frequency (MHz) versus arrival time in seconds (s) for radio pulses from the Crab Nebula showing dispersion; the highest frequencies arrive earliest. The figure uses the original data of Staelin and Reifenstein (1968).

this effect; clearly the higher frequencies do arrive earliest. From these data, the plasma frequency (and average electron density) can be easily estimated (P2.22).

This phenomenon of dispersion can be examined in more mathematical detail by constructing wavepackets consisting of a superposition of plane wave solutions, each satisfying Eqn. (2.63); these would be the analogs of the Fourier integral solutions of Eqn. (2.40). For example, we can write

$$f(x,t) = \frac{1}{\sqrt{2\pi}} \int_{-\infty}^{+\infty} dk\, A(k)\, e^{i(kx-\omega(k)t)} \tag{2.77}$$

Even if we can calculate the Fourier transform explicitly at $t = 0$ to obtain $f(x, 0)$, we can no longer assume that $f(x,t) = f(x-ct)$ unless $\omega = ck$. For a general dispersion relation, the integral must be done numerically at each desired time. We show in Fig. 2.12 the results of such a calculation. The dashed curve corresponds to a wavepacket propagating with $\omega_p \neq 0$, and we compare it to a wavepacket with the same initial shape but which propagates in vacuum (where $\omega_p = 0$); the increasing spread and the slower speed ($v_g < c$) are both evident.

If we choose $A(k)$ to be peaked around $k = k_0$, then the wavepacket

$$f(x,t) = \frac{1}{\sqrt{2\pi}} \int_{-\infty}^{+\infty} dk\, A(k-k_0)\, e^{i(kx-\omega(k)t)} \tag{2.78}$$

will contain wavenumbers centered around k_0, with some spread Δk. If we change variables to $q = k - k_0$, it is natural to expand the exponent around k_0 to obtain

$$\begin{aligned} kx - \omega t &= (k_0 x - \omega(k_0)t) + q\left(x - \frac{d\omega}{dk}(k_0)t\right) + \frac{1}{2}q^2 \frac{d^2\omega}{dk^2}(k_0)t + \cdots \\ &= (k_0 x - \omega_0 t) + q(x - v_g t) + q^2 \beta t + \cdots \end{aligned} \tag{2.79}$$

where $\beta \equiv (d^2\omega/dk^2)/2$. For EM waves in vacuum where $\omega = kc$, $\beta = 0$, the wavepacket can be written as

$$\begin{aligned} f(x,t) &= e^{i(k_0 x - \omega_0 t)} \frac{1}{\sqrt{2\pi}} \int_{-\infty}^{+\infty} dq\, A(q) e^{iq(x-v_g t)} \\ &= e^{i(k_0 x - \omega(k_0)t)} f(x - v_g t) \end{aligned}$$

so that $|f(x,t)|^2 = |f(x - v_g t)|^2$ as expected.

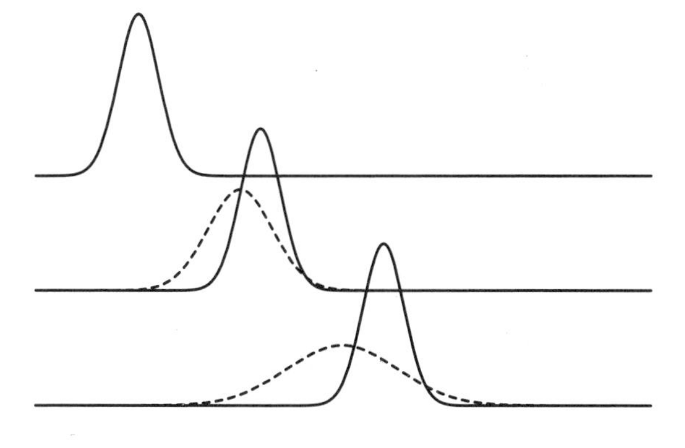

FIGURE 2.12. Electromagnetic wave packet in vacuum (solid) and in plasma (dashed) showing decrease in speed and spreading due to the dispersion relation in Eqn. (2.63).

For a general dispersion relation, $\beta \neq 0$, and β measures the spread in propagation speeds; we can see this by writing

$$q(x - v_g t) + q^2 \beta t + \cdots = q(x - v_g t - q\beta t) + \cdots = q(x - (v_g - q\beta)t) + \cdots \quad (2.80)$$

We know that the integral over q will be dominated by values in the range $(-\Delta k, \Delta k)$, so we associate

$$q\beta \approx \pm \Delta k \frac{d^2\omega}{dk^2} \approx \pm \Delta v_g \quad (2.81)$$

as the spread in velocities, implying a dispersive wave.

The rate at which the wavepacket spreads can be studied by noting that the peak of the packet is dominated by parts of the Fourier integral where $x - v_g t \approx 0$, so that as long as the next term in the expansion satisfies

$$q^2 \beta t \approx \Delta k^2 \beta t \ll 1 \quad (2.82)$$

the wavepacket will not spread significantly. This condition can be rephrased in terms of a spreading time, defined by

$$t_0 = \frac{1}{\beta \Delta k^2} \quad (2.83)$$

and noting that when $t \gtrsim t_0$, the spreading will become important.

2.5.3 Tunneling

So far we have assumed that $\omega > \omega_p$, but we can see that when $\omega < \omega_p$ we can write

$$k = \pm \frac{1}{c}\sqrt{\omega^2 - \omega_p^2} \longrightarrow \kappa = \pm \frac{i}{c}\sqrt{\omega_p^2 - \omega^2} \quad (2.84)$$

and the plane wave solutions have the forms

$$\phi_k(x, t) = e^{\pm \kappa x} e^{-i\omega t} \quad (2.85)$$

Instead of oscillatory behavior, these solutions exhibit exponential decay (or growth).

If, for example, an EM wave with $\omega < \omega_p$ propagating to the right encountered a semi-infinite wall of plasma at $x = 0$, the possible solutions for $x > 0$ are shown in

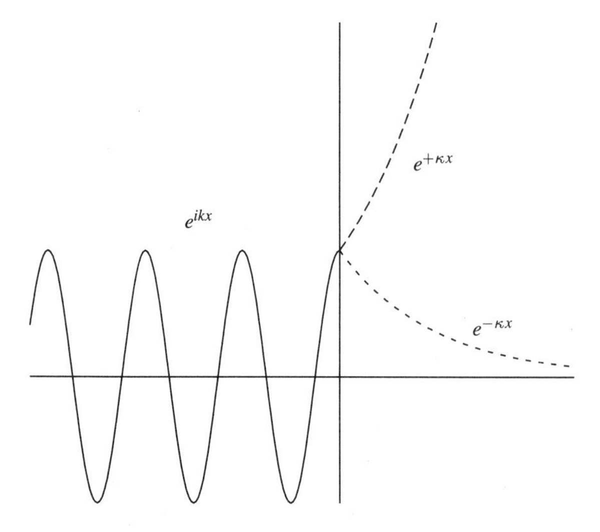

FIGURE 2.13. *Electromagnetic wave in vacuum (on the left) impinging on a region of plasma (on the right); exponentially growing (dash) and decaying solutions (dotted) are allowed in the plasma.*

Fig. 2.13. Clearly the exponentially growing solution is unacceptable if the plasma extends forever to the right and must be discarded; this leaves only the solution $e^{-\kappa x}e^{-i\omega t}$. In this case we do not have propagation but rather exponential damping or attenuation of the wave amplitude, $\phi(x, t)$. The corresponding energy density, which goes like $|\phi(x, t)|^2$, will penetrate into the plasma with a spatial dependence $e^{-2\kappa x} = e^{-x/d}$.

For static/time-independent ($\omega = 0$ or DC) or very slowly varying fields (for which we can assume that $\omega \ll \omega_p$), the penetration depth is set by $d = c/2\omega_p$ alone. An example is the plasma in fusion reactors where $n_e \approx 10^{12} - 10^{16}$ electrons/cm^3, giving $\omega_p \approx 6 \times 10^{10} - 6 \times 10^{12}\,\text{s}^{-1}$ or $d \approx 2\,\text{mm} - 20\,\mu\text{m}$; electric and magnetic fields are effectively expelled from such plasmas. Similar exponential damping occurs for EM waves in conductors (P2.24) where the fields in the nonpropagating region are sometimes called *evanescent* waves.

In this context, it is interesting to note that such waves, although exponentially attenuated, can yield a finite wave amplitude if allowed to propagate through a *finite* thickness of plasma or conductor. This is the classical wave analog of quantum mechanical tunneling discussed in Chaps. 9 and 12.

2.6 QUESTIONS AND PROBLEMS

Q2.1 Suppose you have a function, $f(x)$, which has a Fourier *series* as in Eqn. (2.17). What would the Fourier *transform* of $f(x)$ look like?

Q2.2 In Ex. 2.1 and 2.2 , one can see that $a_n = 0$ for *even* values of n (except possibly for $n = 0$). Can you sketch (either literally or figuratively) a proof of why this should be so?

Q2.3 Is there any kind of "uncertainty principle" for *Fourier series*?

Q2.4 Show that the time and frequency variables t and $\omega = 2\pi f$ form a Fourier transform pair, i.e., $f(t) \longleftrightarrow A(\omega)$. What is the corresponding classical uncertainty condition? How about its quantum analog? In this context, why would the Fourier transform pairs of Ex. 2.3 have anything to do with "ringing"? Why would the a_0 component of a Fourier series for $f(t)$ be called the DC component?

Q2.5 Why do functions with "sharp edges" have Fourier transforms that have "lots of wiggles"?

Q2.6 How would the experimental data shown in Fig. 2.11 change if the plasma density between us and the Crab Nebula were doubled? If the distance between us were doubled?

Q2.7 Assume an EM wave in vacuum is incident from the left on a semi-infinite region of plasma at $x = 0$. What is the appropriate amplitude for $x < 0$? If the wave is incident from the right on a finite thickness of plasma, what is the most general solution in that region?

P2.1 **Factoring the wave equation:** Show that the wave equation in one dimension, Eqn. (2.4), can be written in the factored form

$$0 = \left(\frac{\partial^2}{\partial t^2} - v^2\frac{\partial^2}{\partial x^2}\right)\phi(x, t) = \hat{D}_+\hat{D}_-\phi(x, t) = \hat{D}_-\hat{D}_+\phi(x, t)$$

where $\hat{D}_\pm \equiv \partial/\partial t \pm v\partial/\partial x$. Use this to show that $\hat{D}_\pm\phi(x, t) = 0$ gives the most general solutions in Sec. 2.2.1. Factorization methods are discussed in more detail in Chap. 14.

P2.2 **Reflection and transmission on a string:** Consider an infinitely long string with constant tension T that has linear mass density ρ_1 for $x < 0$ and ρ_2 for $x > 0$.

(a) Assume a solution of the form

$$A(x, t) = \begin{cases} Ie^{i(k_1 x - \omega_1 t)} + Re^{i(-k_1 x - \omega_1 t)} & \text{for } x < 0 \\ Te^{i(k_2 x - \omega_2 t)} & \text{for } x > 0 \end{cases}$$

which represents an incident and reflected wave for $x < 0$ and a transmitted wave for $x > 0$. Use the fact that the string amplitude should be continuous at $x = 0$ to show that $\omega_1 = \omega_2$ and that $I + R = T$.

(b) If the tension in both halves of the string is the same, there will be no "kink" at $x = 0$, i.e., the slope $\partial A(x, t)/\partial x$ will also be continuous. Show that this implies that $k_1(I - R) = k_2 T$.

(c) Solve for R and T in terms of I and $\rho_{1,2}$. Does your solution make sense for the cases $\rho_1 = \rho_2$? For $\rho_2 \gg \rho_1$? For $\rho_2 \ll \rho_1$?

P2.3 Verify Eqns. (2.21) and (2.24) for the Fourier series of the square waveform in Ex. 1.1.

P2.4 Verify Eqns. (2.27) and (2.28) for the Fourier series of the triangular waveform in Ex. 1.2.

P2.5 **(a)** Find the Fourier series for the waveform defined in the interval $(-L, L)$ by

$$f(x) = \begin{cases} -A & \text{for } -L < x < 0 \\ A & \text{for } 0 < x < L \end{cases}$$

(b) Compare the coefficients to those obtained for Ex. 2.1.

(c) Calculate the Δ_N for this series, and show that $\Delta_N \to 1$ as $N \to \infty$.

P2.6 Find the Fourier series for the waveform defined in the interval $(-L, L)$ by

$$f(x) = A(L^2 - x^2)^2$$

This function is continuous at $x = \pm L$ and has a continuous derivative there as well. Compare the convergence (perhaps by calculating the Δ_N) for this case, and compare to that found in Ex. 2.1 and 2.2.

P2.7 Derive an expression for the Δ_N of a Fourier series in terms of its expansion coefficients, a_n and b_n.

P2.8 **Derivatives of Fourier series:**

(a) If one periodic function is the derivative of another, how are their Fourier coefficients related?

(b) Using the definitions of the two functions in Ex. 2.1 and 2.2, show that they satisfy

$$f'_{\text{triangle}}(x) = f_{\text{square}}\left(x + \frac{L}{2}\right)$$

Differentiate the Fourier series for $f_{\text{triangle}}(x)$, and show that it also satisfies this relation.

P2.9 **(a)** Consider the summation in Eqn. (2.31) and show that

$$S_N \equiv \sum_{n=1}^{N} \cos\left(\frac{nKx}{N}\right) = \text{Re}\left(\sum_{n=1}^{N} e^{inKx/N}\right) = \text{Re}\left(\sum_{n=1}^{N} (e^{iKx/N})^n\right) \equiv \text{Re}(T_N)$$

where $\text{Re}(z)$ denotes the real part of z.

(b) Show that

$$\sum_{n=1}^{N} (e^z)^n = \frac{(e^{(N+1)z} - e^z)}{(e^z - 1)}$$

and use this to show that

$$T_N = e^{iKx(1+1/N)/2} \frac{(e^{iKx/2} - e^{-iKx/2})}{(e^{iKx/2N} - e^{-iKx/2N})}$$

so that taking the real part gives

$$S_N = \frac{\cos(Kx(1 + 1/N)/2)\sin(Kx/2)}{\sin(Kx/2N)}$$

P2.10 Show that $\psi_N(x)$ in Eqn. (2.31) is periodic with period $2\pi K/N$, and verify the limit in Eqn. (2.32).

P2.11 Show that the waveform defined by

$$\tilde{\psi}_N(x) \equiv \frac{1}{N^2} \sum_{n=-N}^{n=N} n \sin\left(\frac{nKx}{N}\right)$$

can also be evaluated in closed form. *Hint:* Show that

$$\tilde{\psi}_N(x) = -\frac{1}{K}\frac{\partial \psi_N(x)}{\partial x}$$

Find and then sketch the limiting waveform as $N \to \infty$ and verify that it is also localized. Is there an uncertainty relation of the form $\Delta k\, \Delta x > \mathbb{O}(1)$?

P2.12 Consider a Gaussian wavenumber distribution given by

$$A(k) = \frac{1}{\sqrt{K\sqrt{\pi}}} e^{-k^2/2K^2}$$

(a) Show that

$$\int_{-\infty}^{+\infty} |A(k)|^2\, dk = 1$$

(b) Find the Fourier transform $f(x)$, and confirm that it is normalized in the same way, i.e.,

$$\int_{-\infty}^{\infty} |f(x)|^2\, dx = 1$$

(c) *Estimate* the widths of the two distributions, Δk and Δx, and show that they obey the $x - k$ uncertainty principle.

P2.13 Consider the localized triangle waveform defined by

$$f(x) = \begin{cases} N(a - |x|) & \text{for } |x| < a \\ 0 & \text{for } |x| > a \end{cases}$$

(a) Find N such that $\int_{-\infty}^{+\infty} |f(x)|^2\, dx = 1$.
(b) Find the Fourier transform $A(k)$, and check that it is normalized as well.
(c) Make estimates of Δx and Δk, and show that the $x - k$ uncertainty principle holds.

P2.14 The integral in Ex. 2.3 was performed using complex exponential notation. Obtain the same result by writing e^{ikx} in terms of cosines and sines, and perform two "standard" integrals.

P2.15 Verify that the Fourier transform pairs in Ex. 2.3 and 2.4 satisfy Parseval's theorem, namely,

$$\int_{-\infty}^{+\infty} dx\, |f(x)|^2 = \int_{-\infty}^{+\infty} dk\, |A(k)|^2$$

P2.16 Verify the following properties of Fourier transform pairs, $A(k)$ and $f(x)$. Note that we must generally assume that $f(x)$ is complex.
(a) If $A(k)$ is real, then $f(-x) = f^*(x)$. Show that this implies that $|f(x)|$ is an even function.
(b) If $A(k)$ is imaginary, then $f(-x) = -f^*(-x)$.
(c) If $A(k)$ is even, then $f(-x) = f(x)$ i.e., $f(x)$ is even.
(d) If $A(k)$ is odd, then $f(-x) = -f(x)$ i.e., $f(x)$ is odd.
(e) If $A(k)$ and $f(x)$ are transform pairs, then so are $A(\alpha k)$ and $1/|\alpha| f(x/\alpha)$; this "scaling" relation is sometimes useful.

(f) If $f(x)$ and $A(k)$ are transform pairs, then so are $f(x+a)$ and $e^{ika}A(k)$.
(g) What is the Fourier transform of $f'(x)$ in terms of $A(k)$?

P2.17 **Femtosecond chemistry:** Using ultrashort laser pulses, it is possible to probe the reaction dynamics of many chemical reactions, such as $A + BC \to (ABC)^* \to AB + C$. This problem estimates some of the parameters necessary to do this.[5]

(a) If the energy available to each final state species is roughly $E \sim 0.1\,eV$, find the typical final velocities of the end products. Assume, for example, a molecule with atomic weight of 40. Show that these velocities are typically of the order $v \sim 0.01$ Å/fs where 1 fs $= 10^{-15}$s.
(b) If reactions cease to occur when the species are more than a few bond lengths away, say, 2 to 4 Å, estimate the total time for such a chemical reaction in femtoseconds.
(c) Laser pulses lasting ~ 50 fs can be generated to study such reactions. What is the spatial extent, Δx, of such a pulse?
(d) Using the $k - x$ uncertainty principle, estimate the spread in wavenumber for such a pulse, Δk.
(e) Using the fact that $k = 2\pi/\lambda$, derive an uncertainty relation between $\Delta\lambda$ and Δx.
(f) If the laser pulses have $\lambda = 300$ nm, estimate the unavoidable spread in wavelengths, $\Delta\lambda$, of such a beam.
(g) Would such a pulse be able to resolve two atomic or molecular emission or absorption lines separated by 1 Å?

P2.18 Consider the class of functions defined via

$$\delta_\epsilon(x) = \begin{cases} 0 & \text{for } |x| > \epsilon \\ 1/2\epsilon & \text{for } |x| < \epsilon \end{cases}$$

(a) Show that this is an appropriate family of functions whose limit is a Dirac δ function.
(b) Calculate $\theta_\epsilon(x) = \int_{-\infty}^{x} \delta_\epsilon(x)\,dx$, and show that it approaches a step function.

P2.19 **Overlap integrals:** If we define

$$A(k;K) = \frac{1}{\sqrt{K}} e^{-|k|/K}$$

so that its Fourier partner is

$$f(x;K) = \sqrt{\frac{2K}{\pi}} \frac{1}{(1+(Kx)^2)}$$

show that the overlap integrals satisfy

$$\int_{-\infty}^{+\infty} dk\, A(k;K_1)A(k;K_2) = \int_{-\infty}^{+\infty} dx\, f(x;K_1)f(x;K_2)$$

consistent with Eqn. (2.58). Confirm that the overlap is maximized at the value 1 when $K_1 = K_2$ as it should.

P2.20 It is sometimes useful to use a *finite well potential,* defined as

$$V(x) = \begin{cases} 0 & \text{for } |x| > a \\ -V_0 & \text{for } |x| < a \end{cases}$$

(a) Write $V(x)$ in terms of θ functions.
(b) Calculate the corresponding force via $F(x) = -dV(x)/dx$, and discuss its physical significance.

---

[5] A nice review of femtosecond laser chemistry is given by Zewail (1988, 1990).

P2.21 The dispersion relation for gravitational water waves (ignoring surface tension effects) in a fluid of depth h is

$$\omega^2 = gk\frac{(1 - e^{-2kh})}{(1 + e^{-2kh})}$$

Calculate the phase and group velocities for both deep water (defined via $h \gg \lambda$) and shallow water ($h \ll \lambda$) waves. In which limit are the waves nondispersive?

P2.22 Dispersion data from the Crab Nebula:

(a) Show that the data of Fig. 2.11 only give information on the combination $n_e D$, and use the figure to evaluate this quantity.

(b) If the distance to the Crab Pulsar is roughly 2 kpc, evaluate the *average* electron density in interstellar space; if most of the ionized gas is near the source, the interstellar electron density is less than this.

(c) Calculate the corresponding plasma frequency, ω_p, and compare that to the frequencies of the radio pulses detected in the experiment to check that $\omega \gg \omega_p$.

P2.23 **(a)** Commercial radio stations broadcast signals in both the AM band ($f = 500 - 1500$ kHz) and in the FM band ($88 - 108$ MHz). Compare the corresponding frequencies, ω, to the plasma frequency of the ionosphere, using the fact that daytime plasma densities in the ionosphere are of the order $n_e \approx 10^{12} - 10^{13}$ electrons/m^3. Will either band of frequencies propagate through the ionosphere or reflect from it?

(b) At night, one can often hear radio stations that one doesn't hear during the day. Why? (There are both physics and nonphysics reasons!)

(c) How far would DC ($f = 0$) electric and magnetic fields penetrate into the ionosphere?

P2.24 The dispersion relation for the propagation of electromagnetic waves in a conducting medium is

$$\omega^2 = (kc)^2 - i\frac{\omega g}{\epsilon}$$

where g is the electrical conductivity.

(a) Find the wavenumber k as a function of ω, and show that it corresponds to exponentially damped waves.

(b) In the limit of good conduction, i.e., $g \gg \epsilon\omega$, show that the attenuation factor (for the field amplitude) reduces to the form

$$e^{-\beta x} = e^{-x/\delta} \qquad \text{where} \qquad \beta = \sqrt{\frac{\omega g \mu}{2}} = \frac{1}{\delta}$$

The value of $\delta = 1/\beta$ is often called the *skin-depth.*

(c) Calculate the frequency and wavelength of an EM wave in water for which the skin-depth is 1 meter. Use the values

$$\mu = \mu_0 = 4\pi \times 10^{-7}\frac{N \cdot s}{C^2} \qquad \text{and} \qquad g = 4.3\,\frac{1}{\Omega \cdot \text{m}}$$

This problem has some relevance to the problem of communicating with submerged submarines.[6]

P2.25 Mass limits from dispersion

(a) For material particles (i.e., those with $m > 0$ as opposed to massless particles), the energy relation

[6]See, e.g., Reitz, Milford, and Christy (1993).

$$E = \gamma mc^2 = \frac{mc^2}{\sqrt{1 - v^2/c^2}}$$

implies that the velocity depends on energy via

$$v = c\sqrt{1 - (mc^2/E)^2}$$

Show that this implies that two particles traveling a distance D with a small difference in energy ΔE will have arrival times that differ by

$$\Delta t = -\frac{D}{c}\left(\frac{mc^2}{E}\right)^2\left(\frac{\Delta E}{E}\right)$$

when $E \gg mc^2$. Compare to Eqn. (2.76).

(b) If photons had a small mass, m_γ, we could use the Einstein relation, $E = \hbar\omega$, to find the relation

$$\Delta\omega = \frac{c}{D}\frac{\omega^3}{(m_\gamma c^2/\hbar)^2}\Delta t$$

Use this relation and the Crab Nebula data in Fig. 2.11 to obtain an upper limit on a possible photon mass of order $m_\gamma \lesssim 10^{-47}$ kg or $m_\gamma c^2 \lesssim 6 \times 10^{-12}$ eV; compare this to the mass of the electron. (An even more stringent limit is obtained in P3.9.)

(c) Supernovae are the result of a violent implosion of a star when it no longer has fuel to burn and undergoes a catastrophic gravitational collapse. During this process, neutrinos are emitted, and a burst of such particles was detected in association with the observation of SN1987A.[7] The neutrinos were emitted with energies roughly in the range $E \pm \Delta E \approx 12 \pm 3$ MeV. The collapse takes place with a time scale of order 10 s, and the neutrino pulse was detected with a time spread of the same magnitude so that no appreciable spreading took place. The distance to the precursor star is roughly $D = 50$ kpc. (Note that 1 pc = 1 parsec = 3.26 lightyear = 3.1×10^{16} m.) Use these observations to set an upper limit on the neutrino rest energy of roughly

$$mc^2(\nu_e) \lesssim 20 - 30\,\text{eV}$$

and compare to the electron rest energy. *Note:* This observation ushered in the field of neutrino astronomy, and showed once again the fruitful interactions possible between very different fields of physics.

---

[7]For a discussion of SN1987A and neutrino physics, see Burrows (1990).

CHAPTER 3

The Schrödinger Wave Equation

3.1 THE SCHRÖDINGER EQUATION

Just as it is impossible to *derive* Newton's equations of motion in classical mechanics or Maxwell's equations for electricity and magnetism (EM) from first principles, neither can we demonstrate the validity of the Schrödinger equation approach (or any other equivalent one) to quantum mechanics a priori. We can, however, make use of early quantum ideas and the connection between the classical wave equation and the photon concept to help in understanding its structure.

Let us consider one component (in one space dimension) of the classical EM wave equation, namely,

$$\frac{\partial^2 \phi(x, t)}{\partial t^2} = c^2 \frac{\partial \phi(x, t)}{\partial x^2} \tag{3.1}$$

with a plane wave solution of the form

$$\phi(x, t) = A\, e^{i(kx - \omega t)} \tag{3.2}$$

We can use the photon concept by identifying the photon energy with its frequency (a la Einstein), namely,

$$E = h\nu = \hbar\omega \tag{3.3}$$

and its momentum with its wavelength (as de Broglie did for material particles)

$$p = \frac{h}{\lambda} = \frac{\hbar 2\pi}{\lambda} = \hbar k \tag{3.4}$$

The plane wave solutions then take the form

$$\phi(x, t) = A\, e^{i(px - Et)/\hbar} \tag{3.5}$$

which satisfies the original wave equation, Eqn. (3.1), provided

$$E^2\phi(x,t) = (pc)^2\phi(x,t) \qquad \text{or} \qquad E^2 = (pc)^2 \tag{3.6}$$

We recognize this condition as the energy-momentum relation appropriate for a massless particle (namely, the photon). We can formalize this "wave-particle connection," not only for the solutions but also at the deeper level of the wave equation itself, provided we make the identifications

$$p \to \hat{p} = \frac{\hbar}{i}\frac{\partial}{\partial x} \qquad \text{and} \qquad E \to \hat{E} = i\hbar\frac{\partial}{\partial t} \tag{3.7}$$

In doing so, we have introduced two new *differential operators* representing the classical momentum and energy observables. (We will often distinguish quantum mechanical operators, $\hat{O}$, from their classical counterparts, O, by the use of this notation.) When acting on plane wave solutions, these operators have a simple effect:

$$\hat{p}\,\phi(x,t) = \frac{\hbar}{i}\frac{\partial}{\partial x}\left[e^{i(px-Et)/\hbar}\right] = p\,\phi(x,t)$$

$$\hat{E}\,\phi(x,t) = i\hbar\frac{\partial}{\partial t}\left[e^{i(px-Et)/\hbar}\right] = E\,\phi(x,t) \tag{3.8}$$

namely, they return the classical numerical values. Using this formalism, we write

$$\frac{\partial^2\phi(x,t)}{\partial t^2} = c^2\frac{\partial^2\phi(x,t)}{\partial x^2} \quad \Longrightarrow \quad \hat{E}^2\phi(x,t) = (\hat{p}c)^2\phi(x,t) \tag{3.9}$$

which is a new *operator* version of the classical wave equation.

We might then be tempted to generalize this result to material particles (i.e., particles with mass) by associating

$$E^2 = (pc)^2 + (mc^2)^2 \Longrightarrow \hat{E}^2\phi(x,t) = (\hat{p}c)^2\phi(x,t) + (mc^2)^2\phi(x,t) \tag{3.10}$$

or

$$\frac{\partial^2\phi(x,t)}{\partial t^2} = c^2\frac{\partial\phi(x,t)}{\partial x^2} - \left(\frac{mc^2}{\hbar}\right)^2\phi(x,t) \tag{3.11}$$

to obtain a (relativistically correct) wave equation for massive particles. This procedure yields the so-called *Klein-Gordon equation*, which is, in fact, a useful dynamical equation for a certain class of particles (see P3.9.). A problem arises, however, in the probabilistic interpretation of its solutions as representing a *single* particle. Very loosely speaking, the difficulty comes from the two possible signs of the square-root operation when we write $E = \pm\sqrt{(pc)^2 + (mc^2)^2}$; this can be shown to give rise to antiparticles that must be included for self-consistency. (See Chap. 5 and, especially, P5.5 for more details.)

As we are more interested in generating a wave equation based on the *nonrelativistic* connection between E and p for particles, we are led instead to write (for free particles)

$$E = \frac{p^2}{2m} \Longrightarrow \hat{E}\psi(x,t) = \frac{\hat{p}^2}{2m}\psi(x,t) \tag{3.12}$$

or

$$i\hbar\frac{\partial\psi(x,t)}{\partial t} = -\frac{\hbar^2}{2m}\frac{\partial^2\psi(x,t)}{\partial x^2} \tag{3.13}$$

This is the *time-dependent Schrödinger equation for a free particle.* We note that:

- As we will see, the free-particle Schrödinger equation, along with its solutions, gives the quantum mechanical analog of Newton's first law of motion.
- It is a linear wave equation and so supports superposition and interference effects. This will allow us to construct localized wavepacket solutions.

To allow for the possibility of interactions, we assume that any force, $F(x, t)$, felt by the particle is derivable from a potential energy function, $V(x, t)$, given by

$$F(x, t) = -\frac{\partial V(x, t)}{\partial x} \tag{3.14}$$

in which case we write

$$i\hbar\frac{\partial \psi(x, t)}{\partial t} = -\frac{\hbar^2}{2m}\frac{\partial^2 \psi(x, t)}{\partial x^2} + V(x, t)\psi(x, t) \tag{3.15}$$

Eqn. (3.15) is the *time-dependent Schrödinger equation for an interacting particle.* It is the basic dynamical equation of quantum mechanics that generalizes Newton's second law, $F = ma$. Unlike Newton's law, which has two time derivatives in the acceleration term, Eqn. (3.15) is linear in the time derivative. This implies that knowledge of a solution at $t = 0$, i.e., $\psi(x, 0)$, is sufficient to determine the wavefunction $\psi(x, t)$ at all later times.

3.2 PLANE WAVES AND WAVEPACKET SOLUTIONS

3.2.1 Plane Waves and Wavepackets

It is easy to find plane wave solutions of the free particle Schrödinger equation, as one can show that

$$\psi_p(x, t) = e^{i(px - p^2t/2m)/\hbar} \tag{3.16}$$

satisfies Eqn. (3.13). If we want solutions that can represent particle-like states, we can construct localized wavepackets using linearity, superposition, and interference ideas as before.

Because the momentum label p is more natural for a particle state, we choose to write the general linear combination solution in the form

$$\begin{aligned}\psi(x, t) &= \frac{1}{\sqrt{2\pi\hbar}}\int_{-\infty}^{+\infty} dp\, \phi(p)\, \psi_p(x, t) \\ &= \frac{1}{\sqrt{2\pi\hbar}}\int_{-\infty}^{+\infty} dp\, \phi(p)\, e^{i(px - p^2t/m)/\hbar}\end{aligned} \tag{3.17}$$

where the normalization factor $(1/\sqrt{2\pi\hbar})$ is discussed below.

Before studying an explicit example of such a wavepacket, we make the following observations:

- Each component wave, $\psi_p(x, t)$, is weighted by a different "momentum amplitude," $\phi(p)$. This function will eventually provide us with information on the momentum content of the solution, just as $A(k)$ encoded knowledge about the wavenumber dependence of a wavepacket.

- Because each plane wave solution labeled by p now corresponds to a different classical velocity, $v = p/m$, the wavepacket will be dispersive and will necessarily spread as it propagates. If these packets are meant to represent particle-like solutions in the macroscopic limit, we will have to make sure this spreading is consistent with our classical intuition and observations.
- A free, classical particle undergoes uniform, constant velocity motion, and a wavepacket description should presumably yield this in the macroscopic limit.

For the case of a free-particle wavepacket, we can also write

$$\begin{aligned}\psi(x,t) &= \frac{1}{\sqrt{2\pi\hbar}}\int_{-\infty}^{+\infty} dp\, \phi(p)\, e^{i(px-p^2t/2m)/\hbar} \\ &= \frac{1}{\sqrt{2\pi\hbar}}\int_{-\infty}^{+\infty} dp \left[\phi(p)e^{-ip^2t/2m\hbar}\right] e^{ipx/\hbar} \\ \psi(x,t) &= \frac{1}{\sqrt{2\pi\hbar}}\int_{-\infty}^{+\infty} dp\, \phi(p,t)\, e^{ipx/\hbar} \qquad (3.18)\end{aligned}$$

where

$$\phi(p,t) \equiv \phi(p)e^{-ip^2t/2m\hbar} \qquad (3.19)$$

Noting the analogy with Fourier transforms, we can invert this to obtain $\phi(p,t)$ as usual by writing

$$\begin{aligned}\frac{1}{\sqrt{2\pi\hbar}}\int_{-\infty}^{+\infty} dx\, \psi(x,t)e^{-ip'x/\hbar} &= \int_{-\infty}^{+\infty} dp\, \phi(p,t)\left[\frac{1}{2\pi\hbar}\int_{-\infty}^{+\infty} e^{i(p-p')x/\hbar}\, dx\right] \\ &= \int_{-\infty}^{+\infty} dp\, \phi(p,t)\, \delta(p-p') \\ &= \phi(p',t) \qquad (3.20)\end{aligned}$$

where we have extended the definition of the Dirac δ function. This important relation implies that

- $\psi(x,t)$ and $\phi(p,t)$ are Fourier transforms of each other

and motivates the $1/\sqrt{2\pi\hbar}$ normalization factor.

If an initial wavepacket, $\psi(x,0)$, is given, one can obtain the corresponding momentum distribution, $\phi(p) \equiv \phi(p,0)$, required to produce it via Eqn. (3.20). The subsequent time dependence of the spatial wavefunction, $\psi(x,t)$, is then given through Eqn. (3.18). This is one method of solving the *initial value problem.*

On the other hand, if $\psi(x,t)$ is somehow known, one can translate this information into knowledge of the momentum amplitude at any time t via Eqn. (3.20). For the case of a free particle only, the momentum amplitude has a trivial time dependence that implies that

$$|\phi(p,t)|^2 = |\phi(p)e^{-ip^2t/2m\hbar}|^2 = |\phi(p,0)|^2 \qquad (3.21)$$

and the momentum distribution does not change in time. This is consistent with Newtonian mechanics where a particle feeling no force would have a constant momentum (since $F = dp/dt = 0$).

This Fourier transform connection (and Eqns. 2.57 and 2.58) immediately implies that many integrals involving $\psi(x,t)$ and $\phi(p,t)$ are related, i.e.,

$$\int_{-\infty}^{+\infty} dx\,|\psi(x,t)|^2 = \int_{-\infty}^{+\infty} dp\,|\phi(p,t)|^2 \tag{3.22}$$

$$\int_{-\infty}^{+\infty} dx\,\psi_1^*(x,t)\psi_2(x,t) = \int_{-\infty}^{+\infty} dp\,\phi_1^*(p,t)\phi_2(p,t) \tag{3.23}$$

which will prove useful later.

3.2.2 The Gaussian Wavepacket

For most forms of $\phi(p)$, the integral in Eqn. (3.18) must be done numerically, and we will present some examples of such calculations in Sec. 3.3.1. In the case of a Gaussian p distribution, however, defined here by

$$\phi(p) = \sqrt{\frac{\alpha}{\sqrt{\pi}}}\,e^{-\alpha^2(p-p_0)^2/2} \tag{3.24}$$

the integrals can be done in closed form, giving an easy-to-analyze, analytic result. This distribution selects positive momentum components with values centered around p_0. This would correspond to a classical particle with speed $v_0 = p_0/m$, but there is a spread in momentum components of roughly $\Delta p \approx 1/\alpha$. We can then write

$$\begin{aligned}\psi(x,t) &= \sqrt{\frac{\alpha}{2\pi\hbar\sqrt{\pi}}}\int_{-\infty}^{+\infty} dp\,e^{-\alpha^2(p-p_0)^2/2}\,e^{i(px-p^2t/2m)/\hbar}\\ &= \sqrt{\frac{\alpha}{2\pi\hbar\sqrt{\pi}}}\,e^{i(p_0x-p_0^2t/2m)/\hbar} \qquad (\text{using } p-p_0=q)\\ &\quad\times\int_{-\infty}^{+\infty} dq\,e^{-q^2(\alpha^2/2+it/2m\hbar)}\,e^{iq(x-p_0t/m)/\hbar}\end{aligned} \tag{3.25}$$

The integral is of the form

$$\int_{-\infty}^{+\infty} e^{-ax^2-bx} = \sqrt{\frac{\pi}{a}}\,e^{b^2/4a} \tag{3.26}$$

where

$$a \equiv \frac{\alpha^2}{2} + \frac{it}{2m\hbar} \qquad \text{and} \qquad b \equiv -i(x-p_0t/m)/\hbar \tag{3.27}$$

which can be evaluated using the results in App. B.1. We find that

$$\psi(x,t) = \frac{1}{\sqrt{\alpha\hbar F\sqrt{\pi}}}\,e^{i(p_0x-p_0^2t/2m)/\hbar}\,e^{-(x-p_0t/m)^2/2\alpha^2\hbar^2F} \tag{3.28}$$

where we define

$$F \equiv 1 + \frac{it}{t_0} \qquad \text{with} \qquad t_0 = m\hbar\alpha^2 \tag{3.29}$$

To analyze this complex waveform, we evaluate its modulus squared and find

$$|\psi(x,t)|^2 = \psi^*(x,t)\psi(x,t)$$
$$= \frac{1}{\sqrt{\pi}\alpha\hbar}\frac{1}{\sqrt{(1+it/t_0)(1-it/t_0)}}e^{-(x-p_0t/m)^2/2\alpha^2\hbar^2\left(\frac{1}{1+it/t_0}+\frac{1}{1-it/t_0}\right)}$$
$$= \frac{1}{\alpha\hbar\sqrt{\pi}\sqrt{1+t^2/t_0^2}}e^{-(x-p_0t/m)^2/\alpha^2\hbar^2(1+t^2/t_0^2)}$$
$$|\psi(x,t)|^2 = \frac{1}{\beta_t\sqrt{\pi}}e^{-(x-p_0t/m)^2/\beta_t^2} \tag{3.30}$$

where

$$\beta_t = \alpha\hbar\sqrt{1+t^2/t_0^2} \tag{3.31}$$

This result illustrates several notable features:

- The central value of the wavepacket (interpreted by looking at $|\psi|^2$) is located at $x - p_0t/m = 0$, so that the peak moves at constant speed p_0/m, consistent with a particle of fixed momentum p_0 and mass m. One can show (P3.2) that if we let

$$\phi(p) \rightarrow \phi(p)e^{-ipx_0/\hbar} \tag{3.32}$$

  we can change the initial central position so that the peak location is given by $x = x_0 + p_0t/m$. This is the equivalent of fixing all of the initial conditions for a free particle.
- The width of the wavepacket, which is roughly given by β_t, does increase with time due to the variation in speed of the component plane waves. The time scale for dispersion is set by the "spreading time" defined as $t_0 = m\hbar\alpha^2$. For long times (defined as $t \gg t_0$), the spread in position goes as

$$\Delta x_t \sim \beta_t \longrightarrow \alpha\hbar\left(\frac{t}{t_0}\right) = \frac{1}{m\alpha}t \tag{3.33}$$

  Using $m\Delta v = \Delta p \approx 1/\alpha$ as the spread in velocity components in the wavepacket, we see that

$$\Delta x_t \longrightarrow \Delta v\, t \tag{3.34}$$

  which provides a very intuitive explanation for the spreading behavior.
- The initial spread of the position wavepacket is given by

$$\Delta x_0 \approx \beta_0 = \alpha\hbar \tag{3.35}$$

  so the spreading time can be estimated as

$$t_0 = m\hbar\alpha^2 \approx \frac{m\hbar}{(\Delta p)^2} \approx \frac{m(\Delta x_0)^2}{\hbar} \tag{3.36}$$

  These relations make clear the correlations between the:

  small (or large) initial spatial extent of the wavepacket

  large (or small) spread in momentum required to achieve the necessary destructive interference to form the packet

  large (or small) spread in speed of the component plane waves

  short (or long) time to exhibit the dispersion

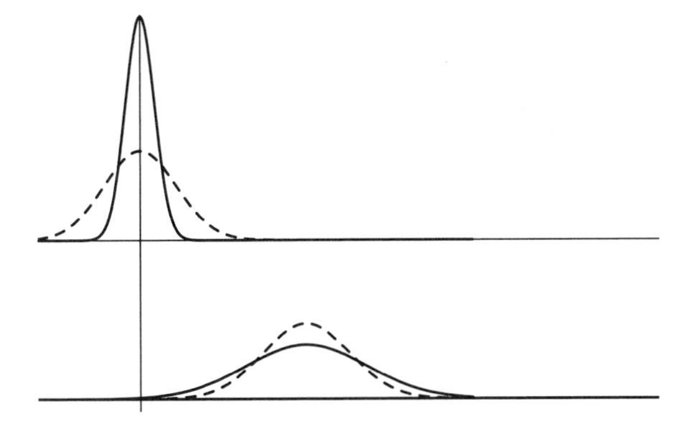

FIGURE 3.1. Spreading Gaussian wavepackets illustrated by $|\psi(x,t)|^2$ versus x at two different times. Note that the initially narrower packet (the solid curve) spreads faster.

- We see that spatially small wavepackets spread faster than those with larger initial widths, and we visualize this effect in Fig. 3.1.
- We note the subtle time dependence of both the real and imaginary parts of $\psi(x, t)$, which is required to maintain the Gaussian shape of the wavepacket (if not its width) as it moves. This is shown in Fig. 3.2 where we plot $\mathrm{Re}(\psi(x,t))$, $\mathrm{Im}(\psi(x,t))$, as well as $|\psi(x,t)|$. We stress that only for the Gaussian wavepacket does the wavepacket remain in the same "family" of shapes for later times.
- We will come to associate the "wiggliness" (or rate of spatial variation) of the wavefunction with the local kinetic energy (see Sec. 5.2.3). Figure 3.2 then shows that the "leading edge" of the wavepacket has a larger "local kinetic energy" than the "trailing edge." Compare this behavior to the spreading wavepulse discussed in Sec. 2.5.2 where the faster components outpace the slower ones.

To verify that this wavepacket spreading arises from a (quantum) wave mechanics description, we note that:

- In the spirit of Sec. 1.3, we find for a fixed initial spread or uncertainty in position, Δx_0, that $t_0 \to \infty$ as $\hbar \to 0$, so that spreading effects would be unobservable, consistent with its being a purely quantum effect.
- The spreads (or uncertainties) in x and p satisfy

$$\Delta x \Delta p \approx \left(\alpha\hbar\sqrt{1 + t^2/t_0^2}\right) \cdot \left(\frac{1}{\alpha}\right) \geq \hbar \tag{3.37}$$

as expected from the uncertainty principle.

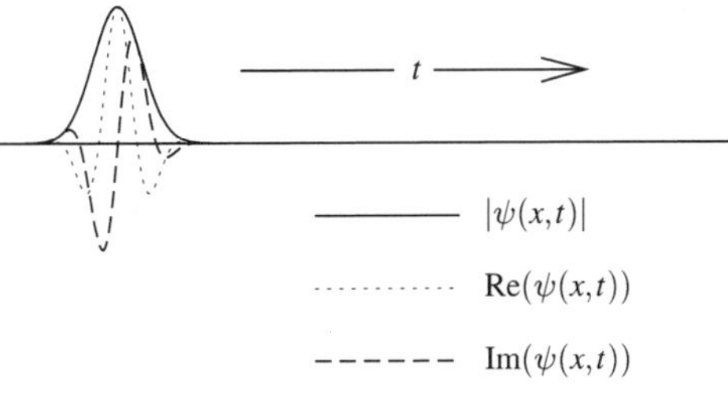

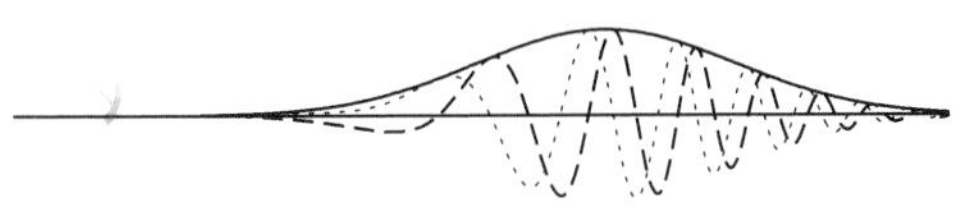

FIGURE 3.2. The real part (dotted), imaginary part (dashed), and modulus (solid) of a Gaussian wavepacket; note how the "wiggliness" is concentrated in the leading edge.

Example 3.1 Spreading Wavepackets To see if the spreading of such wavepackets has any observable consequences, we can make some numerical estimates of the spreading time in various cases. For a typical macroscopic particle with $m = 1$ kg and an initial uncertainty in position of, say, $\Delta x_0 \approx 0.1$ mm using Eqn. (3.36) we find that $t_0 \approx 10^{26}$s $\approx 3 \times 10^{16}$ years which is roughly a million times the age of the universe. Once again, the smallness of $\hbar$ on a macroscopic scale can make quantum effects unobservably small.

On the other hand, we can consider an electron in a classical circular orbit in a hydrogen atom. (This is not a perfectly valid comparison as we have dealt only with straight-line motion, but it makes the point.) In that case we have $m_e = 0.91 \times 10^{-31}$ kg, and we might assume that the electron wavepacket is initially localized to within 10% of its circumference, i.e., $\Delta x_0 \approx 1$ Å $= 10^{-10}$ m; in that case we obtain a spreading time $t_0 \approx 10^{-18}$ s. Before we can say this is a "short time" we need to estimate a typical time scale for the classical problem. For an electron with $v \approx c/100$, the typical rotation period is

$$\tau = \frac{2\pi R}{v} \approx 3 \times 10^{-16}\,\text{s} \tag{3.38}$$

so that before the electron wavepacket completes even 1/100 of a revolution, it has spread so much as to be unrecognizable as a localized particle.[1]

3.3 NUMERICAL CALCULATION OF WAVEPACKETS

3.3.1 Spreading Wavepackets

While it is possible to evaluate the wavepacket integral of Eqn. (3.18) in closed form for only a few special cases, one can always perform the p integral numerically at each point of x, t. A conceptually simple method is to approximate

$$\begin{aligned}\psi(x,t) &= \frac{1}{\sqrt{2\pi\hbar}}\int_{-\infty}^{+\infty} dp\,\phi(p)\,e^{i(px-p^2t/2m)/\hbar} \\ &\approx \frac{1}{\sqrt{2\pi\hbar}}\sum_{n=-N}^{n=N} \Delta p \phi(p_n)\,e^{i(p_nx-p_n^2t/2m)/\hbar}\end{aligned} \tag{3.39}$$

where $p_n = n\Delta p$ is evaluated on a set of discretized values ranging from $p_{\max} = N\Delta p$ to $p_{\min} = -N\Delta p$. (We make no claims that this method is the most computationally efficient way of calculating the time dependence of a wavepacket.[2])

In order to examine the generality of the results inferred from the analytic Gaussian example, we choose two initial wavepackets given by

$$(a) \qquad \psi_1(x,0) = \begin{cases} 0 & \text{for } |x| > a \\ 1/\sqrt{2a} & \text{for } |x| < a \end{cases} \tag{3.40}$$

i.e., a square wave pulse and

$$(b) \qquad \psi_2(x,0) = e^{-|x|/a}/\sqrt{a} \tag{3.41}$$

[1]See Sec. 18.2 for a discussion of the classical limit of orbits in a Coulomb potential.

[2]See, e.g., Press et al. (1986) for a comprehensive discussion of numerical Fourier transform methods.

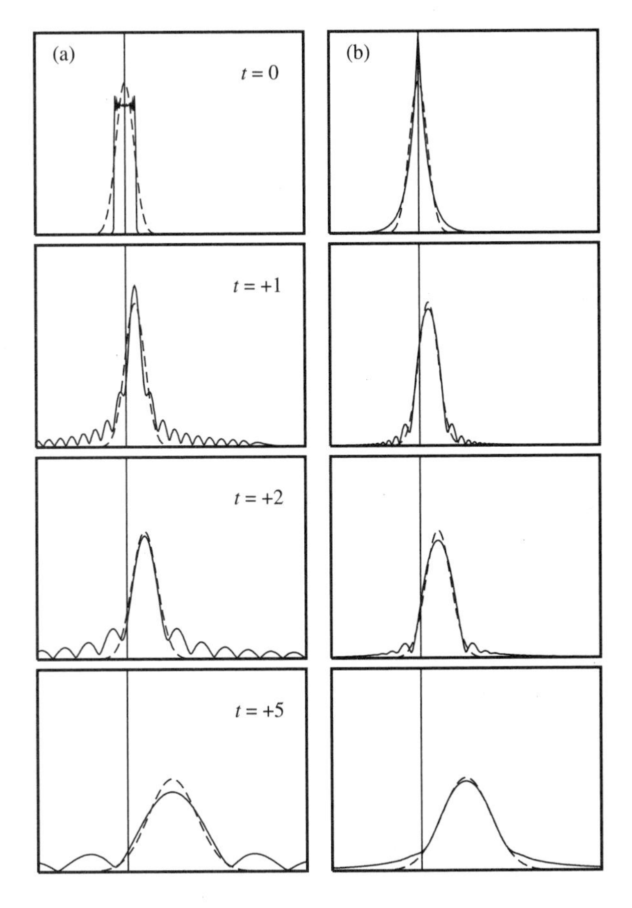

FIGURE 3.3. Spreading wavepackets, given by $|\psi(x,t)|^2$ versus x at increasing times, for non-Gaussian momentum amplitudes; (a) is for an initial "square" wavepulse, $\psi_1(x,0)$, and (b) is for an initial "exponential" wavepulse, $\psi_2(x,0)$. In each case, the dashed curve is a Gaussian wavepacket of the same initial width.

These waveforms can be used in Eqn. (3.20) to calculate the initial momentum amplitude, $\phi(p)$, required to reproduce them, namely,

$$\phi_1(p) = -\sqrt{\frac{2a}{\pi\hbar}}\left[\frac{\sin(ap/\hbar)}{ap/\hbar}\right] \tag{3.42}$$

and

$$\phi_2(p) = \sqrt{\frac{a}{\pi\hbar}}\left[\frac{1}{1+(ap/\hbar)^2}\right] \tag{3.43}$$

Once the initial momentum amplitudes are known, the resulting time-dependent wavefunctions can be obtained through Eqn. (3.18). The corresponding values of $|\psi(x,t)|^2$ are plotted in Figs. 3.3(a) and (b) for illustration. In each case, we use $\phi(p - p_0)$ so that the packets translate to the right. In addition, a Gaussian wavepacket is also plotted for comparison, and the values of a in each case are chosen to give the same initial "spread." We note that:

- Each wavepacket spreads in a way similar to the analytically obtained Gaussian wavepacket.
- The subtle interplay (constructive and destructive interference) between the component plane waves required to form the particular initial waveforms (e.g., the square bump) are rapidly altered due to the dispersive propagation; this results in rapid oscillations and changes in shape with time.

- The relative phases between the various components tend to "randomize" in some sense so that the long-term shape of the waveform seemingly approaches a Gaussian shape.

3.3.2 "Bouncing" Wavepackets

Another case of a quasi-free particle that requires a numerical solution is that corresponding to a free particle, incident from the left, which hits an impenetrable wall at $x = 0$. We can argue that the quantum mechanical particle is subject to a potential of the form

$$V(x) = \begin{cases} 0 & \text{for } x < 0 \\ \infty & \text{for } x > 0 \end{cases} \tag{3.44}$$

To solve this problem, we can look for plane wave solutions of the Schrödinger equation for $x < 0$ in the usual way, but then also enforce the boundary condition that

$$\psi(x, t) = 0 \qquad \text{for} \qquad x \geq 0 \qquad \text{for all } t \tag{3.45}$$

The complex exponential solutions $\exp(\pm ipx/\hbar)$ no longer satisfy the boundary condition at the origin, but the linear combination

$$\frac{1}{2i}(e^{ipx/\hbar} - e^{-ipx/\hbar}) = \sin(px/\hbar) \tag{3.46}$$

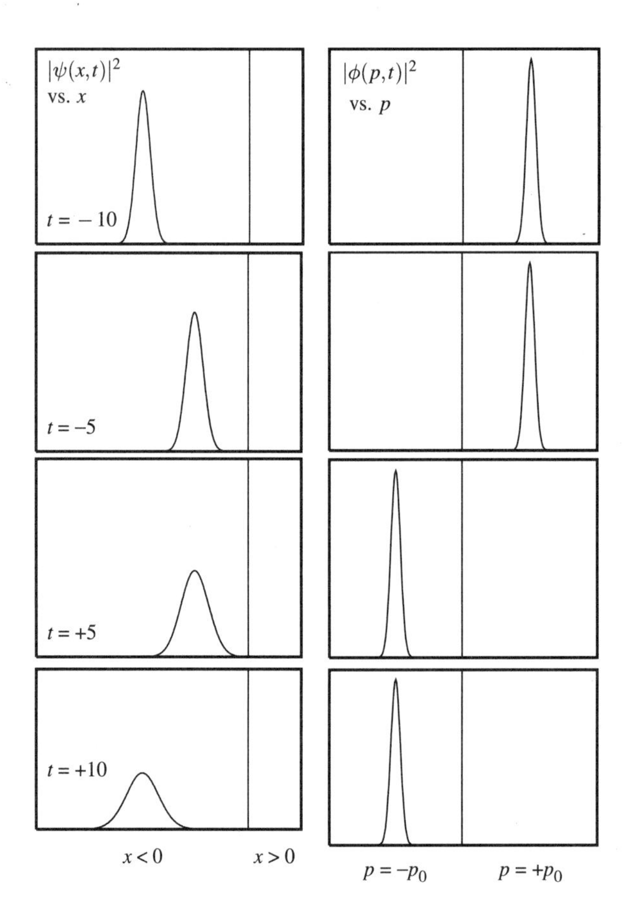

FIGURE 3.4(a). *"Bouncing" wavepacket before and after a collision with an impenetrable wall at $x = 0$. The left side shows $|\psi(x, t)|^2$ versus x, while the right side illustrates $|\phi(p, t)|^2$ versus p; (a) shows "before and after collision" pictures, (continued)*

does. The relevant plane wave solutions are now

$$\tilde{\psi}_p(x,t) = \begin{cases} \sin(px/\hbar)e^{-ip^2t/2m\hbar} & \text{for } x \leq 0 \\ 0 & \text{for } x \geq 0 \end{cases} \tag{3.47}$$

We therefore have to evaluate the integral

$$\psi(x,t) = \frac{1}{\sqrt{2\pi\hbar}} \int_{-\infty}^{+\infty} dp\, \phi(p)\, \tilde{\psi}_p(x,t) \tag{3.48}$$

to obtain the resulting wavepacket.

Because of the lack of symmetry imposed by the boundary conditions, the integrals cannot be performed analytically using standard Gaussian integral tricks, and we must evaluate Eqn. (3.48) numerically. The resulting time-dependent waveform is shown in Figs. 3.4(a) and (b), where we have again used an initial Gaussian wavepacket, as in Eqn. (3.24), and an appropriate initial value of x_0 so that the wavepacket is initially localized at $x_0 < 0$. The dispersive behavior is familiar, but the rebound, Fig. 3.4(a), and "bounce" at the wall, Fig. 3.4(b), are also obvious, consistent with a classical elastic collision.

Even more interestingly, if we use the numerical results for $\psi(x,t)$ at each value of (x,t), we can perform the Fourier inversion, using Eqn. (3.20), to obtain $\phi(p,t)$ as well; this quantity (actually $|\phi(p,t)|^2$) is shown alongside the position space plots. Because there is now an effective force on the particle, the momentum distributions are no longer

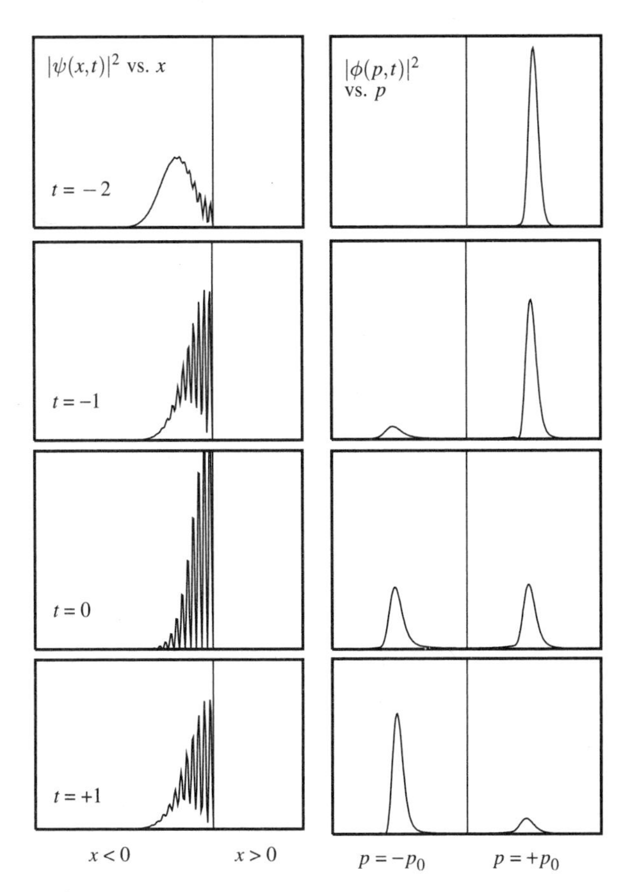

FIGURE 3.4(b). (continued) *while (b) focuses on the "impact"; note that the momentum distribution changes only over the duration of the impulsive collision.*

time-independent, and the classical impulsive change in momentum during the collision is clearly present.

It seems that the Schrödinger wave equation prescription discussed in this chapter incorporates wave mechanical effects into a dynamical equation consistent with Newton's second law. Information on the desired "wavelike" and "particlelike" phenomena appear to be contained in both $\psi(x, t)$ and $\phi(p, t)$. We have not discussed, however, the precise physical interpretation of either wave amplitude. Because both $|\psi(x, t)|^2$ and $|\phi(p, t)|^2$ will be interpreted as probability densities, we devote the next chapter to a review of the ideas of probability and statistics.

3.4 QUESTIONS AND PROBLEMS

Q3.1 Discuss why the de Broglie relation, Eqn. (3.4), is consistent with the statement in Sec. 3.2.2 that the spatial "wiggliness" of the Schrödinger wavefunction provides information on the kinetic energy of the associated particle.

Q3.2 How can the wavefunction $\tilde{\psi}_p(x, t)$ in Eqn. (3.47) contain information on the particle first moving to the right and then to the left after the bounce? What would you expect the "bouncing" wavepacket solutions to look like for the case of a particle confined between *two* infinite walls? What would the plane wave solutions look like?

P3.1 **(a)** Verify explicitly that the wavepacket solution of Eqn. (3.28) satisfies the free-particle Schrödinger equation.

(b) The "phase" piece of the solution disappears when we evaluate $|\psi(x, t)|^2$; show that $\psi(x, t)$ *without* this term is *not* a solution of the Schrödinger equation.

P3.2 **(a)** Show generally that if one lets

$$\phi(p) \rightarrow \phi(p)e^{-ipx_0/\hbar}$$

in Eqn. (3.18), then

$$\psi(x, t) \rightarrow \psi(x - x_0, t)$$

(b) Analyze the effect on $\psi(x, t)$ of letting

$$\phi(p) \rightarrow \phi(p)e^{iAp^2}$$

where A is a constant.

P3.3 Show that

$$\lim_{\beta_t \to 0} \frac{1}{\beta_t\sqrt{\pi}} e^{-(x-p_0t/m)^2/\beta_t^2} = \delta(x - p_0t/m)$$

$$\lim_{1/\alpha \to 0} \frac{\alpha}{\sqrt{\pi}} e^{-\alpha^2(p-p_0)^2} = \delta(p - p_0)$$

Discuss how your results might bear on a classical limit of quantum mechanics.

P3.4 **Other Gaussian wavepackets:** One can construct other examples of closed-form wavepackets using Eqn. (3.18) in certain special cases. Evaluate $\psi(x, t)$ for the momentum amplitude

$$\phi(p) = \sqrt{\frac{2\alpha^3}{\sqrt{\pi}}}(p - p_0)e^{-\alpha^2(p-p_0)^2/2}$$

and find the resulting expression for $|\psi(x, t)|^2$. Does it spread in a manner consistent with Eqn. (3.31)?

P3.5 Overlap integrals:

(a) Show that the Gaussian wavepacket

$$\psi(x,t;p_0) = \frac{1}{\sqrt{\alpha\hbar F\sqrt{\pi}}} e^{i(p_0 x - p_0^2 t/2m)/\hbar} e^{-(x-p_0 t/m)^2/2\alpha^2\hbar^2 F}$$

satisfies

$$\int_{-\infty}^{+\infty} dx\, |\psi(x,t;p_0)|^2 = 1$$

for all values of t and p_0. Show also that the Gaussian momentum amplitude Eqn. (3.24) satisfies

$$\int_{-\infty}^{+\infty} dp\, |\phi(p)|^2 = 1$$

for all values of p_0.

(b) Show that at $t = 0$, the values of $|\psi(x,t;p_0)|^2$ for a wavepacket with $p_0 = 0$ and $p_0 \neq 0$ are identical. We might then guess that the "overlap," defined by Eqn. (3.23), of two such solutions might be large at $t = 0$ and decrease with time as one wavepacket moves away from the other. Evaluate

$$\int_{-\infty}^{+\infty} dx\, \psi^*(x,t;0)\psi(x,t;p_0)$$

and show that the result is, in fact, independent of time and depends only on p_0 and α. (The algebra is somewhat lengthy.)

(c) Derive the same result for the overlap integral with far less work by using the corresponding momentum amplitudes. The results are guaranteed to be equal from Eqn. (3.23).

P3.6 If two wavepackets (representing the same mass particle) have different initial widths (determined by $\alpha_1 \neq \alpha_2$) and hence different spreading times, $t_0^{(1)}$ and $t_0^{(2)}$, show that the narrower one will spread faster so that after a time $t = \sqrt{t_0^{(1)} t_0^{(2)}}$ the two packets will (temporarily) have the same width.

P3.7 **(a)** Assume that the electrons in a TV set are accelerated to 30 kV and travel 10 cm before they hit the back of the screen. If each electron is associated with a wavepacket of initial size ~ 0.1 mm, estimate the amount of spreading that occurs before the electron hits the screen. How far would the electron have to travel before the size of its wavepacket would have grown to roughly 10 times its initial size?

(b) Electrons at the Stanford Linear Collider are accelerated to energies of ~ 50 GeV. If their initial wavepackets are localized to say 10 μm, what is their spatial extent at the time of their collisions when they have traveled 1 km? *Hint:* Are these particles nonrelativistic or not? If they are, what do you do?

P3.8 Show that the momentum amplitude of Eqns. (3.42) and (3.43) is indeed the Fourier transform of the initial spatial wavefunction in Eqns. (3.40) and (3.41).

P3.9 **The Klein-Gordon equation:** Consider the Klein-Gordon equation in three dimensions:

$$\nabla^2\phi(\mathbf{r},t) = \frac{1}{c^2}\frac{\partial\phi(\mathbf{r},t)}{\partial t^2} + \mu^2\phi(\mathbf{r},t)$$

where $\mu \equiv mc/\hbar$.

(a) Show that this supports plane wave solutions of the form $\phi(\mathbf{r},t) = e^{i(\mathbf{k}\cdot\mathbf{r}-\omega t)}$. *Hint:* It is easiest to use Cartesian coordinates.

(b) Consider the static limit of the Klein-Gordon equation with $m = 0$ (i.e., the classical wave equation) in three dimensions, namely,

$$\nabla^2 \phi(\mathbf{r}) = 0$$

which is just Laplace's equation of electrostatics. Show that it is satisfied by the Coulomb potential, $\phi(\mathbf{r}) = K/r$ (except perhaps at $r = 0$). Here it is easier to use spherical coordinates and App. B.3.

(c) If the photon had a small but finite mass, it would satisfy the Klein-Gordon equation, so that in the static limit, the electric potential would have to satisfy

$$\nabla^2 \phi(\mathbf{r}) = \mu^2 \phi(\mathbf{r})$$

Show that the *Yukawa potential,*

$$\phi(\mathbf{r}) = K \frac{e^{-\mu r}}{r}$$

satisfies this equation.

(d) Use the uncertainty principle to argue that the exchange of a particle of mass m would have a range given by $L = 1/\mu$, and discuss the relation with the Yukawa potential.

(e) Find the range of a force mediated by the exchange of the following particles:
1. The nuclear force mediated by π mesons of mass $\sim$ 140 MeV/c^2
2. The weak interaction mediated by W and Z bosons of mass $\sim$ 90 GeV/c^2
In each case, compare the ranges to the size of a proton, roughly 1 F.

(f) A finite photon mass will change not only electrostatic but magnetic properties as well. The most stringent limit on μ_γ (and hence on a possible photon mass, m_γ) comes from the observations of planetary magnetic fields. For simplicity, assume that the dipole field of a planet is changed (very roughly) by the presence of a photon mass via

$$|\mathbf{B}(\mathbf{r}; \mu_\gamma)| \sim |\mathbf{B}(\mathbf{r}; 0)| e^{-\mu_\gamma r}$$

Measurements of the magnetic field of Jupiter by the Pioneer 10 spacecraft[3] imply that the magnetic field at five Jupiter radii ($R = 5R_J$) is at least 30% of its expected value based on a vanishing photon mass. Given that the radius of Jupiter is $R_J \approx 7 \times 10^7$ km, find a lower limit on $L = 1/\mu_\gamma$ in meters. Use this to derive an upper bound on a possible photon mass rest energy, $m_\gamma c^2$, and compare to the rest energy of the electron.[4]

---

[3]See, e.g., Goldhaber and Nieto (1968).

[4]It is perhaps appropriate to close this chapter by noting that this method of bounding a possible photon mass was evidently first suggested by Schrödinger (1943); the proposal arose in the context of investigations of a possible unification of gravity and electromagnetism.

CHAPTER 4

Concepts of Probability

4.1 PROBABILITY AND STATISTICS

Because probability ideas play such an important role in the "orthodox" interpretation of the Schrödinger wavefunction, it is useful to review briefly some of the ideas of probability theory and statistics.[1] We can very loosely describe the difference between probability and statistics as mirroring the relationship between theoretical predictions and experimental measurements; the former we call *a priori* and the latter *a posteriori* probabilities.[2]

In the first instance, we might assume that we can enumerate beforehand the possible outcomes of some experiment or trial, say, the throw of one die. If we are interested in the likelihood of obtaining a specific outcome in any particular trial, we can define the a priori probability as the ratio of the number of "successful" outcomes (those leading to the desired result), say, n_S, to the total number of possible results, n_T, i.e,

$$P \equiv \frac{n_S}{n_T} \tag{4.1}$$

Using this definition, the probability of obtaining, say, a 3 on any one roll of a single die would be $P = 1/6$. In doing this, we tacitly assume that all of the n_T outcomes are equally likely (i.e., the die is not "loaded").

We could, of course, actually perform the experiment a large number of times, N_T, keeping track of the number of successes, N_S, and define the experimental or a posteriori probability as

$$P(N_T) \equiv \frac{N_S}{N_T} \tag{4.2}$$

[1]For a no-nonsense review of probability and statistics, see Lyons (1986).

[2]We discuss methods of comparing experimental data with theoretical predictions in Sec. 20.2.

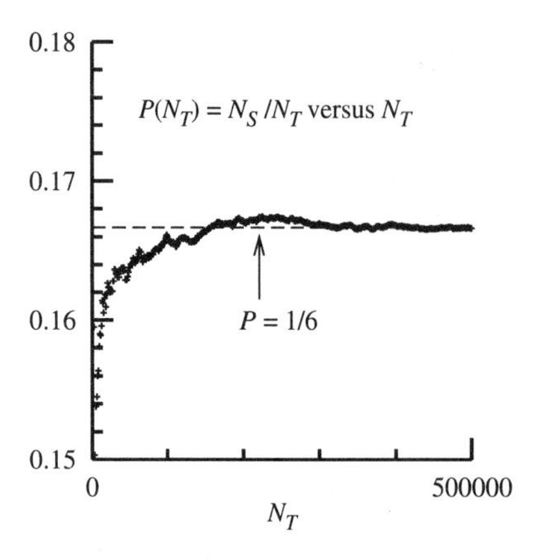

FIGURE 4.1. The experimentally determined probability of throwing a 3, P(3), versus the number of throws, N_T; the convergence to 1/6 is clear.

and our intuition would lead us to believe that the limiting value (if it exists) of this procedure would also yield the a priori probability. This is illustrated in Fig. 4.1 where we plot the probability of finding a 3 obtained for an increasingly large number of (computer-generated) throws of a die and see the approach to the limiting value.

At any stage of measurement, we do expect some deviation from the "exact" a priori result; for example, in five throws we would never measure, even in principle, a probability of 1/6 exactly. After N_T throws or measurements, we expect $N_S = PN_T$, and a simple estimate[3] of the expected error or uncertainty in this number of successes is $\Delta N_S = \sqrt{P(1-P)N_T}$. We will often approximate this as $\Delta N_S \approx \sqrt{PN_T} \approx \sqrt{N_S}$ when P is not too close to unity.

The experimental probability including its expected error is then given by

$$P \pm \Delta P = \frac{N_S \pm \Delta N_S}{N_T} = \frac{N_S}{N_T} \pm \sqrt{\frac{PN_T}{N_T^2}} \Longrightarrow P \pm \sqrt{\frac{P}{N_T}} \tag{4.3}$$

This expression for ΔP is consistent with our bias that the probability should become increasingly precisely known as the number of trials is increased. To illustrate this, we plot in Fig. 4.2 the magnitude of the deviation in the experimental probability from its

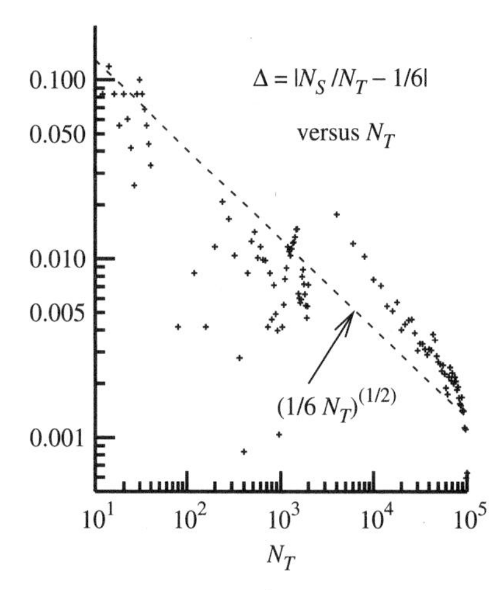

FIGURE 4.2. The deviation of the experimental probability for throwing a 3 from its a priori value of 1/6 for increasing numbers of throws; $\Delta = |N_S/N_T - 1/6|$ versus N_T is plotted. The data are compatible with a "slow" $\sqrt{P/N_T}$ falloff.

[3]This is based on an argument using the so-called *binomial probability distribution* and is discussed in P4.4.

expected value, $\Delta P = |P(N_T) - 1/6|$, after N_T throws of one die showing its slow decrease. Superimposed on the data is the value $\Delta P = \sqrt{1/6N_T}$ (which plots as a straight line on the log-log scale used) which illustrates the general trend towards increased confidence in the experimental confrontation with the theoretical predictions.

4.2 DISCRETE PROBABILITY DISTRIBUTIONS

For a given experimental situation for which the outcomes are determined by random chance, the collection of the probabilities of all the possible outcomes defines a *probability distribution.* Such a collection of probabilities for which the outcomes form a discrete set, and hence can be classified by an integer label, is called a *discrete probability distribution.* We can therefore label the possible outcomes as x_i, $i = 1, 2, \ldots, N$, and the upper limit, N, can be either finite or infinity; the corresponding probabilities are labeled $P(x_i)$.

Trivial examples that are obviously labeled by integers are the number on the ith face in one roll of a die, or more prosaically, the number of raisins found in a single slice of raisin bread; in both cases the upper limit is finite. An example in the context of quantum mechanics is the quantized energy levels of a bound system, such as the allowed energy states of a hydrogen atom, $E_n = -E_0/n^2$. The x_i in this case are the energies and are not themselves integers (or even dimensionless numbers), but a measurement of the energy of an individual atom will yield only those discrete values; the upper limit for the integer label in this case is infinity.

A simple example that exhibits many of the most general features of a discrete probability distribution is the set of probabilities corresponding to the throw of a pair of dice. In this case, we can define $x_1 = 2, \ldots, x_{11} = 12$ as the possible outcomes, with probabilities given by

$$P(x_1 = 2) = \frac{1}{36} \quad \cdots \quad P(x_6 = 7) = \frac{6}{36} \quad \cdots \quad P(x_{11} = 12) = \frac{1}{36} \tag{4.4}$$

These a priori probabilities and the experimentally measured values after 2000 "hits" are shown in Fig. 4.3, along with the error bars for each probability given by Eqn. (4.3).

Because in each measurement, *something* will be measured, the probabilities $P(x_i)$ must satisfy the obvious constraint

$$\sum_{i=1}^{N} P(x_i) = 1 \tag{4.5}$$

which is often called a *normalization* condition. While the complete set of the $P(x_i)$ encodes all of the available information concerning the system, specific combinations of the

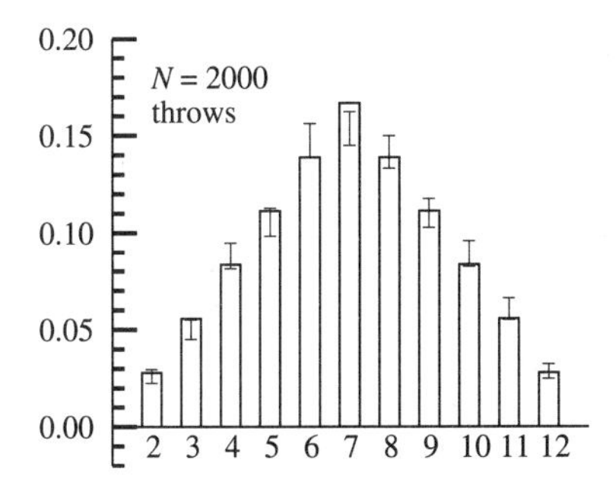

FIGURE 4.3. *The a priori probabilities for the throw of two dice. Also shown are the experimentally determined probabilities (with errors) after 2000 throws.*

$P(x_i)$ are often useful. The *average value* or *expectation value* of the discrete variable x is defined by

$$\langle x \rangle \equiv \sum_{i=1}^{N} x_i \, P(x_i) \tag{4.6}$$

as is the average value of any function of x,

$$\langle f(x) \rangle = \sum_{i=1}^{N} f(x_i) \, P(x_i) \tag{4.7}$$

Both give information on the expected results of many measurements of quantities depending on the variable. These a priori predictions can, of course, be compared to the experimental values given by repeated measurements of x or $f(x)$. If the values of a set of N_T *measurements* of the variable x are labeled $\overline{x}_s$, $s = 1, \ldots, N_T$, we define

$$\langle x \rangle_{\text{exp}} \equiv \frac{\sum_{s=1}^{N_T} \overline{x}_s}{N_T} \qquad \text{or} \qquad \langle f(x) \rangle_{\text{exp}} \equiv \frac{\sum_{s=1}^{N_T} f(\overline{x}_s)}{N_T} \tag{4.8}$$

While the values of the variable x can (but not always) cluster around the average value, a given measurement will often find x within a rather well-defined range about $\langle x \rangle$. The variable $y_i \equiv x_i - \langle x \rangle$ naturally measures the deviation of an individual value from the average, but it satisfies

$$\langle y \rangle = \langle x - \langle x \rangle \rangle = \langle x \rangle - \langle x \rangle = 0 \tag{4.9}$$

because the average deviation away from the mean vanishes. A better estimator of the *spread* or *uncertainty* in x that arises naturally is the so-called *root mean-square deviation* or *RMS deviation* or *standard deviation* defined by

$$\Delta x_{\text{RMS}} = \Delta x \equiv \left(\langle (x - \langle x \rangle)^2 \rangle \right)^{1/2} = \left(\sum_{i=1}^{N} (x_i - \langle x \rangle)^2 \, P(x_i) \right)^{1/2} \tag{4.10}$$

The name itself is a useful mnemonic for this formula, as it requires one to

1. First take the *deviation* (of each x_i from the average value $\langle x \rangle$).
2. *Square* it.
3. Then evaluate the *mean* (or average).
4. And finally take the (square) *root.*

A more calculationally useful form of this relation is obtained by noting that

$$\begin{aligned} (\Delta x)^2 &= \langle (x - \langle x \rangle)^2 \rangle \\ &= \langle x^2 - 2x\langle x \rangle + \langle x \rangle^2 \rangle \\ &= \langle x^2 \rangle - \langle x \rangle^2 \end{aligned} \tag{4.11}$$

Example 4.1 Expectation Values for Two Dice For the case of the throw of two dice, we easily find

$$\langle x \rangle = 7 \qquad \text{and} \qquad \langle x^2 \rangle = \frac{1974}{36} = 54.833 \qquad \text{so that} \qquad \Delta x = 2.415 \tag{4.12}$$

If we sum the probabilities corresponding to results within one standard deviation of the mean, i.e., in the interval $(\langle x \rangle - \Delta x, \langle x \rangle + \Delta x)$, we find $24/36 = 2/3 \sim 0.67$. We similarly find that $34/36 \sim 0.94$ of the total (unit of) probability is found within two standard deviations.

This pattern is rather typical of probability distributions that are roughly centered and somewhat peaked around their average values (see also P4.1, P4.3 to P4.5). Thus, a knowledge of $\langle x \rangle$ and Δx allows one to make reasonably reliable predictions for the likely outcome of a given measurement, as well as the reliability or uncertainty of the result.

4.3 CONTINUOUS PROBABILITY DISTRIBUTIONS

If the random variable x can take on continuous values (e.g., heights of a population, location of a particle along a line segment, etc.) we can generalize the discrete probability distribution above by noting that the $P(x_i)$ can be trivially rewritten as $P(x_i)\,\Delta x_i$ if we define the "distance" between integrally labeled measurements to be $\Delta x_i = 1$, i.e., a unit-sized bin centered at x_i. The continuous case can then be obtained by the analogy

$$P(x_i) \sim P(x_i)\,\Delta x_i \Longrightarrow P(x)\,dx \tag{4.13}$$

where dx is an infinitesimally small unit of measure of the variable x. With this identification, we are led to the interpretation of $P(x)$:

- $P(x)\,dx$ is the probability that a measurement of the variable x will find it in the interval $(x, x + dx)$, implying that
- The probability that a measurement of x will be in the finite interval (a, b) is given by

$$\text{Prob}(a < x < b) = \int_a^b P(x)\,dx \tag{4.14}$$

which we illustrate in Fig. 4.4.

Assuming for generality that the variable can take on values anywhere in the interval $(-\infty, +\infty)$, the normalization condition on the probability distribution $P(x)$ becomes

$$\sum_{i=1}^{N} P(x_i) = 1 \qquad \Longrightarrow \qquad \int_{-\infty}^{+\infty} P(x)\,dx = 1 \tag{4.15}$$

The definitions of average values and RMS deviations are easily generalized to

$$\langle x \rangle = \int_{-\infty}^{+\infty} x\,P(x)\,dx \qquad \text{or} \qquad \langle f(x) \rangle = \int_{-\infty}^{+\infty} f(x)\,P(x)\,dx \tag{4.16}$$

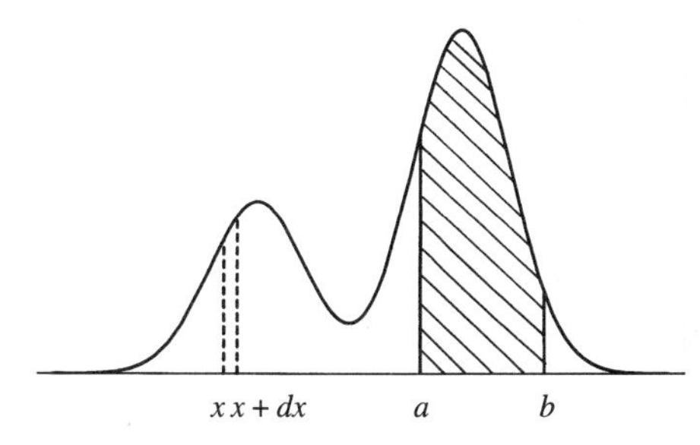

FIGURE 4.4. A generic continuous probability distribution.

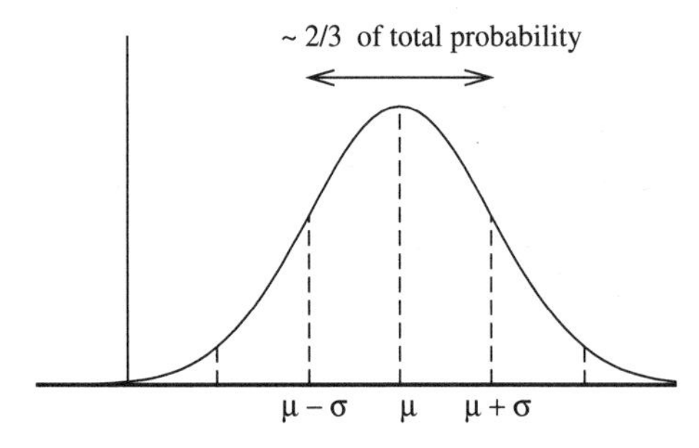

FIGURE 4.5. *Gaussian probability distribution; the mean value μ and the $\pm\sigma$ values are shown.*

and

$$(\Delta x)^2 = \int_{-\infty}^{+\infty} (x - \langle x \rangle)^2 \, P(x) \, dx = \langle x^2 \rangle - \langle x \rangle^2 \tag{4.17}$$

Example 4.2 The Gaussian Distribution One of the most frequently occurring probability distributions is the *Gaussian* (or *normal* or "bell-shaped") distribution, given by

$$P(x; \mu, \sigma) = \frac{1}{\sigma\sqrt{2\pi}} e^{-(x-\mu)^2/2\sigma^2} \tag{4.18}$$

which is shown in Fig. 4.5. This distribution arises in a profound way in the theory of statistics (via the so-called *Central Limit Theorem*[4]) but also appears in an important and natural way in quantum theory as well. Using the (appropriately named) Gaussian integrals discussed in App. B.1, it is easy to show that

$$\langle x \rangle = \mu \qquad \text{and} \qquad \Delta x = \sigma \tag{4.19}$$

so that the parameters μ and σ characterize the average values and RMS deviations directly. The probability in any finite interval (i.e., the area under $P(x)$) can only be calculated numerically (see App. F.3), and one finds

$$\begin{aligned} \text{Prob}(|x - \mu| \leq \sigma) &\approx 0.6826 \approx 68.3\% \\ \text{Prob}(|x - \mu| \leq 2\sigma) &\approx 0.9544 \approx 95.5\% \\ \text{Prob}(|x - \mu| \leq 3\sigma) &\approx 0.9974 \approx 99.7\% \end{aligned} \tag{4.20}$$

Thus, the measurement of a quantity described by a Gaussian distribution found to have a value more than three standard deviations away from its average is rare (less than 0.3% of the time) but not impossible, while we expect roughly 1/3 of the measured events to be more than 1 σ away from the mean. (This is already noticeable in Fig. 4.3, where we note that 4 to 5 out of the possible 12 measurements of probability "miss" (or just barely "hit") the predicted value by more than one error bar.)

Several points concerning $P(x)$ should be kept in mind:

- $P(x)$ is, by itself, not a probability but rather a *probability density* as it can be crudely defined as

$$P(x) \equiv \frac{d\text{Prob}(x)}{dx} \tag{4.21}$$

[4]See, e.g., Mathews and Walker (1970).

or the probability per unit x interval. The need for the infinitesimal unit of measure should always be kept in mind.

- Because of this, $P(x)$ has nontrivial dimensions, namely,

$$[P(x)] = \left[\frac{1}{dx}\right] \tag{4.22}$$

whatever the units of x happen to be (height, etc.). (Recall that $[z]$ denotes the dimensions of the variable z.) This is clear from the specific example of the Gaussian distribution in Eqn. (4.18), where $[x] = [\sigma]$ and $[P(x)] = [1/\sigma]$.

The experimental measurement of a continuous probability distribution can be accomplished in a limiting procedure similar to that for discrete distributions. We can "discretize" the problem by defining "bins" of finite width in x, of some reasonable size, δx, centered at the desired value of x. The ratio of the number of successes to total trials in that small bin, $\delta\text{Prob} = N_S/N_T$, divided by the bin width estimates the local probability density

$$P(x) = \frac{N_S}{N_T}\frac{1}{\delta x} \longrightarrow \frac{\delta\text{Prob}}{\delta x} \tag{4.23}$$

One can decrease the bin width appropriately as the number of events collected increases. This is shown in Fig. 4.6 where a Gaussian probability is "built up" by repeated measurements. The distribution of measured points is shown below each figure; the gradual accumulation of data points leading to the emergence of the probabilistic pattern should be reminiscent of Fig. 1.2. In Fig. 4.7(a) and (b) we also show one case with error bars included in each bin, based on the simple $\sqrt{N}$ estimate in Eqn. (4.3), using both linear and logarithmic scales. The second plot shows more clearly that the *fractional uncertainties* in the values are largest where the probability is smallest and the events collected are the fewest.

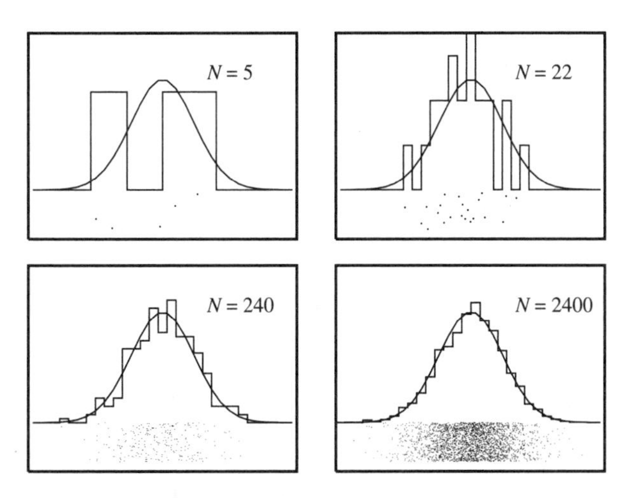

FIGURE 4.6. *"Experimental" (i.e., computer) determination of a Gaussian probability distribution for an increasingly large number of measurements.*

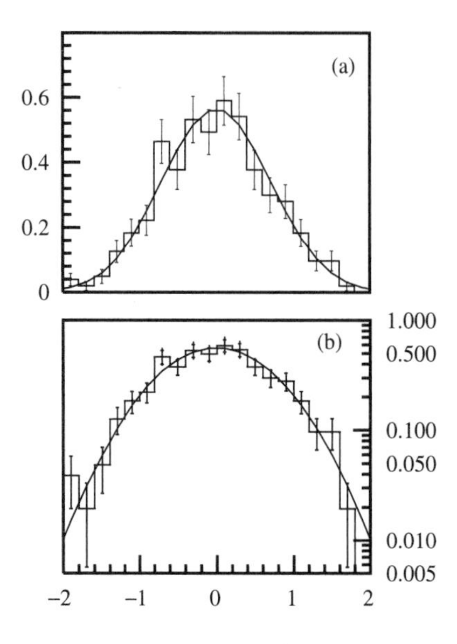

FIGURE 4.7. Experimental determination of Gaussian probability distribution including errors on a (a) linear scale and (b) logarithmic scale. A Gaussian distribution with $\mu = 0$ and $\sigma = 1/\sqrt{2}$ is used.

4.4 MULTIVARIABLE PROBABILITY DISTRIBUTIONS

Often the measurement of a physical quantity requires or yields the values of several random variables at once. An example is the throwing of a dart at a dartboard, which yields both the x and y coordinates of the "hit." In such a case, the notion of probability distributions can be easily extended to include the case of two or more variables. We can define a two-variable probability density $P(x, y)$ such that

- $P(x, y)\,dx\,dy$ is the probability that a measurement of *both* x and y will find their values to be in the regions $(x, x + dx)$ *and* $(y, y + dy)$, respectively, with the normalization condition

$$\int_{-\infty}^{+\infty} dx \int_{-\infty}^{+\infty} dy\, P(x, y) = 1 \tag{4.24}$$

Average values and deviations are obtained as before with, for example,

$$\langle f(x, y) \rangle = \int_{-\infty}^{+\infty} dx \int_{-\infty}^{+\infty} dy\, f(x, y)\, P(x, y) \tag{4.25}$$

An especially familiar case occurs when the two measured quantities are independent variables or uncorrelated with each other, so that the probability density is simply the product of two separate functions, namely, $P(x, y) = P_1(x)\,P_2(y)$. In that case we find

$$d\text{Prob} = P(x, y)\,dx\,dy = [P_1(x)\,dx][P_2(y)dy] = [d\text{Prob}_x][d\text{Prob}_y] \tag{4.26}$$

which is consistent with the basic laws of probability for independently occurring events.

Several new features of multivariable probability distributions can be motivated using various special cases of generalized Gaussian distributions. We only discuss the case of *changes of variables* and *correlated variables.*

Example 4.3 Changes of Variables We might be interested, for example, in the joint distribution of two position coordinates in the plane described by two independent Gaussian variables with common values of $\mu(=0)$ and σ; this would give

$$P(x, y) = \frac{1}{2\pi\sigma^2} e^{-(x^2+y^2)/2\sigma^2} \tag{4.27}$$

If one asks for the probability that a measurement of the position coordinates will yield values for the polar coordinates (r, θ) in a certain range, it is easy to change variables and write

$$P(x, y)\, dx\, dy \quad \longrightarrow \quad P(r, \theta)\, r\, dr\, d\theta \tag{4.28}$$

where

$$P(r, \theta) = \frac{1}{2\pi\sigma^2} e^{-r^2/2\sigma^2} \tag{4.29}$$

We now have $P(r, \theta)\, r\, dr\, d\theta$ being the probability that a measure of the position of the object will find its polar coordinates in the ranges $(r, r + dr)$ and $(\theta, \theta + d\theta)$. We must remember to make the appropriate change in the measure, $dx\, dy \to r\, dr\, d\theta$, as well as in the probability density itself.

Such a change in variables will have no effect on the total probability as it is easy to check that

$$\begin{aligned} \text{Prob}(a < r < b; 0 < \theta < 2\pi) &= \frac{1}{2\pi\sigma^2} \int_0^{2\pi} d\theta \int_a^b r\, dr\, e^{-r^2/2\sigma^2} \\ &= \frac{1}{\sigma^2} \int_a^b r\, dr\, e^{-r^2/2\sigma^2} \\ &= e^{-a^2/2\sigma^2} - e^{-b^2/2\sigma^2} \end{aligned} \tag{4.30}$$

If we let $a \to 0$ and $b \to \infty$, we find the probability of finding the coordinates *somewhere* in the plane still is unity.

Such changes of variables in multivariable probability distributions are often useful (even necessary) and easy to implement, but one must always be careful to incorporate the same change in the appropriate measure.

Variables for which the measurement of one provides nontrivial information on the other can be described as *correlated;* the heights and weights of a population are presumably correlated since *in general* taller people are heavier than shorter people.

Example 4.4 Correlated Variables An example of such a case where two variables are *not* independent and can be shown to be correlated to one another is given by

$$P(x, y) = \frac{1}{2\pi\sigma_x\sigma_y} \exp\left\{ -\frac{1}{2}\frac{1}{(1-\overline{\rho}^2)} \left(\frac{x^2}{\sigma_x^2} + \frac{y^2}{\sigma_y^2} - \frac{2\overline{\rho}xy}{\sigma_x\sigma_y} \right) \right\} \tag{4.31}$$

where $-1 \leq \overline{\rho} \leq +1$. It is not hard to show that

$$\langle x \rangle = \langle y \rangle = 0 \qquad \text{while} \qquad \Delta x = \sigma_x \qquad \Delta y = \sigma_y \tag{4.32}$$

A useful measure of the degree of correlation of x and y is the so-called *covariance,* given by

$$\text{cov}(x, y) \equiv \big\langle (x - \langle x \rangle)(y - \langle y \rangle) \big\rangle = \langle xy \rangle - \langle x \rangle \langle y \rangle \tag{4.33}$$

which reduces to the RMS deviation squared, Δx^2, if $x = y$. Recalling the definition of averaging,

$$\text{cov}(x, y) = \langle (x - \langle x \rangle)(y - \langle y \rangle) \rangle$$
$$= \int_{-\infty}^{+\infty} dx \int_{-\infty}^{+\infty} dy\, (x - \langle x \rangle)(y - \langle y \rangle)\, P(x, y) \tag{4.34}$$

we see that if $\text{cov}(x, y) > 0$ (i.e., the integral is positive), then there must be a tendency for

$$y > \langle y \rangle \text{ when } x > \langle x \rangle \qquad \text{or} \qquad y < \langle y \rangle \text{ when } x < \langle x \rangle$$

and vice versa.

For independent variables, the covariance reduces to

$$\text{cov}(x, y) = \langle (x - \langle x \rangle) \rangle \langle (y - \langle y \rangle) \rangle = 0 \tag{4.35}$$

because of the product form of the probability densities. For the correlated Gaussian variables in Eqn. (4.31), we find that

$$\text{cov}(x, y) = \overline{\rho} \sigma_x \sigma_y \tag{4.36}$$

One can generally define a *correlation coefficient,* $\rho(x, y)$, of two random variables (using a notation generalized from the Gaussian case) via

$$\rho(x, y) = \frac{\text{cov}(x, y)}{\Delta x \Delta y} \tag{4.37}$$

The correlation coefficient is obviously dimensionless and measures the degree to which the measurement of one variable provides information on the other. A two-parameter correlated Gaussian distribution is shown in Fig. 4.8(a); we also indicate by solid (dotted) vertical bands the measurement of x in an interval about $+1$ (-1). The corresponding probability distributions for a subsequent measurement of y are shown in Fig. 4.8(b) indicating the correlation. A measurement of x that is positive (negative) implies that a subsequent measurement of y will more likely be positive (negative) as well.

It can be shown that $-1 \leq \rho(x, y) \leq +1$ (P4.14) with the upper or lower limits being obtained only when the variables are linearly related, i.e.,

$$y - \langle y \rangle = A(x - \langle x \rangle) + B \tag{4.38}$$

In this case, a measurement of x implies an immediate determination of y. Clearly, a positive (negative) correlation coefficient indicates that an increase in one variable generally corresponds to an increase (decrease) in the other; if $\rho(x, y)$ vanishes, there is no such correlation.

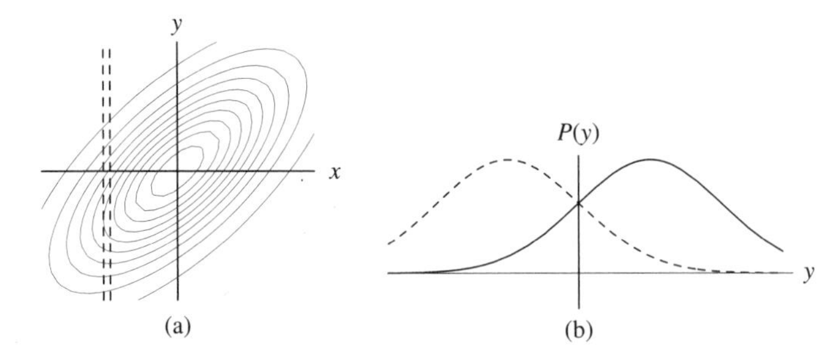

FIGURE 4.8. *(a) Contour plot of a correlated, two-parameter Gaussian distribution $P(x, y)$ with correlation coefficient $\rho(x, y) = 0.7$. (b) The single parameter distribution $P(y)$ resulting from a measurement of x. The solid (dotted) curve corresponds to x measured to be roughly $+1$ (-1).*

The applications of the notions of probability distributions, average values, and RMS deviations (or uncertainties) will be discussed in the next chapter; they form the basis for the standard interpretation of quantum theory. The importance of the idea of correlations among the observables in a multivariable probability distribution will be discussed more fully in the context of systems of indistinguishable particles in Chap. 15; the Pauli principle (and its generalization) demands that just such correlations be present in real physical quantum systems.

4.5 QUESTIONS AND PROBLEMS

Q4.1 Three experiments are performed to test the value of some quantity that has been previously measured to be 10.0. The results of the experiments (including errors) are 10.3 ± 0.4, 9.98 ± 0.004, and 13 ± 7. How would you summarize the results of these three experiments.

Q4.2 Would $\Delta x \equiv \langle |x - \langle x \rangle| \rangle$ be a good choice for the spread in x?

Q4.3 Can you show that $\langle x^2 \rangle \geq \langle x \rangle^2$ quite generally? Under what conditions is the equality possible? Can you find a probability distribution, $P(x)$, for which $\Delta x = 0$?

Q4.4 Can you think of a random quantity that has a probability distribution that is not symmetric about its mean value? Can you think of one that has more than one "peak," i.e, which could be called *bimodal?*

Q4.5 Do the variables in a multivariable probability distribution necessarily have to have the same dimensions?

Q4.6 Of what relevance is the Jacobian when making a change of variables in a multivariable probability distribution?

Q4.7 If the correlation coefficient of two variables is nonzero, does it imply some causal connection between the two variables?

P4.1 Consider the throw of a single die.

(a) Enumerate all of the possible outcomes, and evaluate the a priori probabilities, $P(x_i)$.

(b) Evaluate $\langle x \rangle$, $\langle x^2 \rangle$, and Δx.

(c) How much of the probability is found in the interval $(\langle x \rangle + \Delta x, \langle x \rangle - \Delta x)$? Within $2\Delta x$ of the average?

(d) Toss a single die (or many at once to save time, or have everyone in your class toss many), and record the numbers of 4's for increasingly large numbers of tosses, say $N_T = 10, 100, 1000$. Does the resulting probability compare (within experimental errors) to expectations?

P4.2 **(a)** Two dice are thrown 10,000 times, and the number of 7's recorded. Would you think the dice were "fair" if you obtained 1703 7's? How about if you found 1509 7's?

(b) How many tosses of the dice would you need if you wanted to check whether the probability of obtaining a 2 were really 1/36 to 10%? How about to 1%? *Hint:* This means that you want $\Delta P/P$ to be less than 10%.

P4.3 **Poisson probability distribution:** The Poisson probability distribution is defined by

$$P(n; \lambda) \equiv \frac{\lambda^n}{n!} e^{-\lambda}$$

where λ is a constant. This distribution can correspond, for example, to the probability of observing n independent events in a time interval t when the counting rate is r so that the *expected* number of events is $rt = \lambda$; this is especially relevant when $\lambda = rt$ is not too large.

(a) Show that the $P(n;\lambda)$ are properly normalized, namely, that

$$\sum_{n=0}^{\infty} P(n;\lambda) = 1$$

(b) Evaluate $\langle n \rangle$ and Δn for this distribution in terms of λ. *Hint:* One can write

$$\langle n \rangle = \sum_{n=0}^{\infty} nP(n;\lambda) = \sum_{n=0}^{\infty} n\frac{e^{-\lambda}\lambda^n}{n!} = e^{-\lambda}\left(\lambda\frac{\partial}{\partial\lambda}\right)\sum_{n=0}^{\infty}\frac{\lambda^n}{n!}$$

(c) Plot $P(n;\lambda)$ versus n for several values of λ, indicating the average value and the 1σ limits. Is the distribution symmetric around the average?

(d) Assume that this book is 600 pages long and has a total of 1200 typographical errors (which the editor assures me is not the case!). What is the average number of errors per page? What is the probability that a single page has no errors? How many pages would be expected to have less than three errors? What is the probability that a single page has four or more errors?

P4.4 Binomial probability distribution: Consider a trial that only has two outcomes, the desired one with probability p, and anything else with probability $q = 1 - p$. In N independent trials, the probability of n successful outcomes is

$$P(n;N) = \frac{N!}{n!(N-n)!}p^n q^{N-n}$$

(This result depends on two factors; the last two terms correspond to the probability that the first n trials are successful while the last $N - n$ ones are not. The first term simply counts the number of different distinguishable ways in which one can obtain the n possible outcomes.)

(a) Show that

$$\sum_{n=0}^{N} P(n;N) = (p+q)^N = 1$$

Hint: Use the binomial theorem!

(b) Evaluate $\langle n \rangle$, $\langle n^2 \rangle$, and show that $\Delta n = \sqrt{Np(1-p)}$. *Hint:* Use a differentiation trick as in P4.3

(c) Why does Δn behave as it does when $p \to 0$ or 1?

(d) Discuss the connection of this result to our estimate of the predicted error in a discrete probability measurement in Sec. 4.1.

P4.5 Consider a system consisting of differing numbers of photons, each of energy $\hbar\omega$, so that the total energy of the system can take on the values $E_n = n\hbar\omega$ where $n = 0, 1, 2 \ldots$. The probability of the system having energy E_n is $P(n,\beta) = Ne^{-E_n\beta}$, where $\beta \equiv 1/k_B T$, with k_B being Boltzmann's constant and T the temperature.

(a) Use the fact that this discrete probability distribution must be normalized to find N.

(b) Evaluate $\langle E \rangle$ and ΔE. *Hint:* It might be useful to use

$$\sum_{n=0}^{+\infty} n^K e^{-nz} = \left(-\frac{\partial}{\partial z}\right)^K \left(\sum_{n=0}^{+\infty} e^{-nz}\right)$$

(c) Discuss the limiting values of these quantities when $\hbar\omega \ll k_B T$ and $\hbar\omega \gg k_B T$.

(d) Discuss the connection of your result to Planck's expression for the black-body spectrum in which $\hbar$ first appeared.

P4.6 **(a)** Perform the necessary Gaussian integrals analytically to confirm the results in Eqn. (4.19).
(b) Numerically evaluate the integrals required to confirm the results in Eqn. (4.20).

P4.7 A continuous probability distribution has the form $P(x) = N\exp(-|x|/a)$.
(a) Find N so that $P(x)$ is properly normalized.
(b) Evaluate $\langle x \rangle$, $\langle x^2 \rangle$ and Δx.
(c) What is the probability that a measurement would find x in the interval $(0, +\infty)$? $(-a, 2a)$? $(0, 0.0001a)$?
(d) What is the probability that a measurement would find $x = 1.5$ exactly?
(e) What is the probability that $|x - \langle x \rangle| \leq \Delta x$? $\leq 2\Delta x$? $\leq 3\Delta x$? Compare these values to the Gaussian distribution.
(f) Plot the distribution above for a given value of a along with a Gaussian distribution with its parameters (μ and σ) chosen to give the same average value of x and spread Δx. Show how your plot illustrates the results of part (e).
(g) If 100,000 measurements are made of this distribution, what is the expected number of results in the interval $(a, 2a)$? What is the expected error in that number?

P4.8 Consider a particle that is located with equal probability anywhere along a one-dimensional segment of length L.
(a) Find $P(x; L)$, i.e., the normalized probability of finding the particle at position x.
(b) Evaluate $\langle x \rangle$ and Δx.
(c) What are the probabilities that the particle would be found within one standard deviation of its mean value? Why is this value lower than for the Gaussian distribution?
(d) What are the dimensions of $P(x; L)$?
(e) This is an excellent problem for which you could try to generate "measurements" of the probability distribution by using a random number generator on a computer. Write a short program that generates a random number in the range $(0, 1)$ (i.e., let $L = 1$), and put it into bins of your own choosing, counting how many fall into each bin. Collect lots of statistics, decreasing the bin size as the number of measurements increases; monitor the $\sqrt{N}$ error in each bin and see if the errors are consistent with the discussion above.

P4.9 Consider a two-dimensional probability distribution given by

$$P(x, y) = Nx^2 e^{-x^2/a^2} e^{-y^2/a^2}$$

where x, y are coordinates in a plane.
(a) Find N such that the total probability is 1. *Hint:* Use polar coordinates.
(b) What is the probability that a measurement would find $a < r < 2a$?
(c) What is the probability that a measurement would find $-\pi/4 < \theta < +\pi/4$?

P4.10 Assume that the probability density of finding a particle in three-dimensional space is given by

$$P(x, y, z) = N(a)e^{-\sqrt{x^2+y^2+z^2}/a}$$

(a) Change variables to spherical coordinates, (r, θ, ϕ), and find $N(a)$ by insisting that the distribution be properly normalized. *Hint:* Make sure you use the right measure.
(b) What are the dimensions of $P(x, y, z)$?
(c) If we define $\mathbf{r} = (x, y, z)$, find $\langle \mathbf{r} \rangle$ and $\langle |\mathbf{r}|^2 \rangle$.
(d) What is the probability that a measurement will find $|\mathbf{r}| > a$?

P4.11 Consider the measurement of a particle in a two-dimensional square region given by $x, y \in (-L, L)$. Assume a probability distribution of the form $P(x, y) = N(L^2 - x^2)(L^2 - y^2)$.
(a) Find N such that $P(x, y)$ is properly normalized.
(b) Find $\langle x \rangle$, $\langle y \rangle$, Δx, and Δy.
(c) Why is $\mathrm{cov}(x, y)$ obviously zero?

(d) What is the probability that a measurement of the position of the particle will find it in the region $x, y \in (-L/2, L/2)$?
(e) What is the probability of finding x in the interval $(-L/4, L/4)$?
(f) What are the dimensions of $P(x, y)$?

P4.12 Consider the same situation as in P4.11, but with the probability density

$$P(x, y; a) = N(a)(L^2 - x^2)(L^2 - y^2)\left(1 + \frac{a}{4}\frac{(x+y)^2}{L^2}\right)$$

(a) For what values of a is this a well-defined probability distribution?
(b) Find $N(a)$.
(c) Qualitatively, how do you expect the covariance, $\text{cov}(x, y)$, to depend on a, both its sign and magnitude?
(d) Find $\langle x \rangle$, $\langle y \rangle$, and $\text{cov}(x, y)$. Are your results for the covariance consistent with your expectations from (c)?

P4.13 Consider the normalized probability distribution given by

$$P(x, y) = \frac{9}{2L^6}(L - x)(L - y)(L^2 - xy)$$

in the interval $x, y \in (0, L)$.
(a) Find $\langle x \rangle$, $\langle y \rangle$, $\langle x^2 \rangle$, $\langle y^2 \rangle$, and $\langle xy \rangle$.
(b) Evaluate Δx, Δy, and $\rho(x, y)$.
(c) What is the probability that a measurement of y will find it to be greater than $L/2$?
(d) Suppose that one has already measured x to be $L/2$. Answer part (c) again. Are your answers different and, if so, why?

P4.14 Prove that $-1 \leq \rho(x, y) \leq +1$. *Hint:* You may wish to follow the following steps:

1. Define an auxiliary variable $z = (y - \langle y \rangle) + \lambda(x - \langle x \rangle)$ where λ is an arbitrary real number, and consider the average value

$$I(\lambda) = \langle z^2 \rangle = \int_{-\infty}^{+\infty} dx \int_{-\infty}^{+\infty} dy\, z^2\, P(x, y)$$

and find $I(\lambda)$ in terms of Δx, Δy, and $\text{cov}(x, y)$.
2. Show that $I(\lambda) \geq 0$.
3. Find the value of λ that minimizes $I(\lambda)$, and use the inequality in item 2 to complete the proof.
4. Show that the two variables must be linearly related if $\rho(x, y) = \pm 1$. *Hint:* How can the inequalities be "saturated" at either ± 1.

CHAPTER 5

Interpreting the Schrödinger Equation

5.1 PROBABILITY INTERPRETATION OF THE WAVEFUNCTION

One of our stated goals in understanding quantum physics has been to incorporate the experimentally observed wave properties of matter in a self-consistent manner with the basic (nonrelativistic) dynamical laws of motion for particles; this has led us to the time-dependent Schrödinger equation

$$i\hbar\frac{\partial\psi(x,t)}{\partial t} = -\frac{\hbar^2}{2m}\frac{\partial^2\psi(x,t)}{\partial x^2} + V(x,t)\psi(x,t) \tag{5.1}$$

The derivations leading to other wave equations make it clear that the solutions represent various physical observables (displacements of a string, an electric or magnetic field, or more generally the amplitude for some wave phenomenon), but the arguments leading to Eqn. (5.1) do not make obvious the appropriate interpretation of $\psi(x,t)$. Aside from the fact that we know that $\psi(x,t)$—or rather $|\psi(x,t)|$ since ψ is complex—for wavepackets is correlated with the position of the particle, the Schrödinger equation itself provides no obvious guidance. At the same time, we also wish to incorporate into our description of microscopic phenomena the statistical nature of the measurement process mentioned in Sec. 1.1.

These two ideas come together in a natural way in the standard interpretation of the wavefunction solutions of the Schrödinger equation, namely, that we are to interpret $|\psi(x,t)|^2$ as a *probability density* for position measurements. Specifically, if we define

$$P(x,t) = |\psi(x,t)|^2 = \psi^*(x,t)\psi(x,t) \tag{5.2}$$

then the so-called *Born interpretation*[1] states that

[1] Named after M. Born (1928).

- $P(x, t)\,dx$ is the probability that a measurement of the position of the particle described by $\psi(x, t)$, at time t, will find it in the region $(x, x + dx)$.

In this view, it seems at this point that the wavefunction itself does not make a direct confrontation with experimental measurements of position, but does so only through $|\psi(x, t)|^2$. This type of identification is not unique to quantum mechanics; the squared amplitude of a field is often used in the description of other wave phenomena where it appears, for example, in expressions for the energy density stored in the electric field given by $u(\mathbf{r}, t) = \epsilon_0|\mathbf{E}(\mathbf{r}, t)|^2/2$.

A similar and related "ensemble" interpretation of $\psi(x, t)$, which is perhaps more closely related to experimental test, is obtained by considering a (presumably large) number, N_0, of identically prepared particles, all described by the same wavefunction $\psi(x, t)$. Then

- $dN(x, t) = N_0 P(x, t)\,dx = N_0|\psi(x, t)|^2\,dx$ is the number of particles found with position in the interval $(x, x + dx)$ at time t.

In this way, one can imagine measuring the probability density by binning measurements of the position in small increments dx, each bin having $dN(x, t)$ values, so that

$$\frac{dN(x, t)}{dx} = N_0|\psi(x, t)|^2 \tag{5.3}$$

gives $|\psi(x, t)|^2$.

Just as with any other continuous probability distribution, we then argue that

$$\text{Prob}[x \in (a, b)] = \int_a^b P(x, t)\,dx = \int_a^b |\psi(x, t)|^2\,dx \tag{5.4}$$

is the probability of finding the particle in the *finite* interval (a, b); this in turn implies that

$$\int_{-\infty}^{+\infty} P(x, t)\,dx = \int_{-\infty}^{+\infty} |\psi(x, t)|^2\,dx = 1 \qquad \text{for all } t \tag{5.5}$$

since the probability of finding the particle *somewhere* in its one-dimensional universe (i.e., the real line) must be unity.

This last constraint, that the wavefunction be properly normalized, is very important, as not all solutions of the Schrödinger equation will automatically satisfy this requirement; for example, the plane wave solutions for a free particle give

$$\int_{-\infty}^{+\infty} |\psi_p(x, t)|^2\,dx = \int_{-\infty}^{+\infty} \left|e^{i(px-p^2t/2m)/\hbar}\right|^2\,dx = \int_{-\infty}^{+\infty} 1\,dx = \infty \tag{5.6}$$

As long as the solutions are at least *square-integrable,* namely,

$$\int_{-\infty}^{+\infty} |\psi(x, t)|^2\,dx = C = \text{constant} < \infty \tag{5.7}$$

then we can use the linearity of the Schrödinger equation to write

$$\tilde{\psi}(x, t) = \frac{1}{\sqrt{C}}\psi(x, t) \tag{5.8}$$

This will still be a solution with the same physical properties, but it now satisfies Eqn. (5.5). This probability constraint is the motivation for the normalization factors in all of our

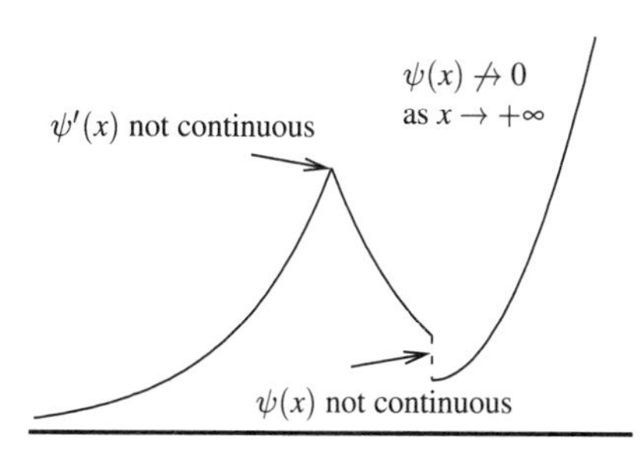

FIGURE 5.1. Unacceptable position-space wavefunction. $\psi(x)$ is assumed real, so that it can be plotted easily.

examples throughout Chaps. 3 and 5. We stress that the two steps of solving the Schrödinger equation (implementing the wave physics) and initially normalizing the solutions (ensuring a consistent probability interpretation) are independent of one another.

The requirement of square integrability implies strong constraints on $\psi(x, t)$:

- $|\psi(x, t)|$ must tend to zero sufficiently rapidly as $x \to \pm\infty$, so that the integral of $|\psi(x, t)|^2$ will converge.
- We will usually assume that ψ is well enough behaved that various spatial derivatives of ψ also vanish at infinity.
- For a probability interpretation to be valid, we must also require that $\psi(x, t)$ be continuous in x, as a discontinuous ψ would lead to ambiguous predictions for probabilities near the "jump," as in Fig. 5.1.
- Unless the potential energy function is extremely poorly behaved (see Sec. 9.1.1), we can also assume that $\psi'(x, t)$ and higher spatial derivatives are everywhere continuous.

We have argued that the requirement that $\psi(x, t)$ be properly normalized at, say, $t = 0$ is independent of the fact that it is a solution of the wave equation. It is not obvious, however, that once normalized it will continue to be so at future times; we are thus naturally led to ask whether

$$\int_{-\infty}^{+\infty} |\psi(x, t_0)|^2 \, dx = 1 \quad \Longrightarrow \quad \int_{-\infty}^{+\infty} |\psi(x, t)|^2 \, dx = 1 \tag{5.9}$$

$$\text{at } t_0 = 0 \qquad\qquad \text{for all later times } t$$

To confirm this, we assume that $\psi(x, t)$ is a solution of Eqn. (5.1), so that $\psi^*(x, t)$ satisfies the complex conjugated version of Eqn. (5.1), and we note that

$$\begin{aligned}
\frac{\partial P(x,t)}{\partial t} &= \left(\frac{\partial \psi^*}{\partial t}\psi + \psi^* \frac{\partial \psi}{\partial t}\right) \\
&= \left(\frac{i}{\hbar}\left(-\frac{\hbar^2}{2m}\frac{\partial \psi^*}{\partial x^2} + V^*\psi^*\right)\right)\psi \\
&\quad + \psi^*\left(\frac{-i}{\hbar}\left(-\frac{\hbar^2}{2m}\frac{\partial^2 \psi}{\partial x^2} + V\psi\right)\right) \\
&= -\frac{i\hbar}{2m}\left(\frac{\partial \psi^*}{\partial x^2}\psi - \psi^* \frac{\partial \psi}{\partial x^2}\right) + \frac{i}{\hbar}\left(\psi^*\psi\right)(V^* - V)
\end{aligned} \tag{5.10}$$

Classically, any potential energy function with which we are familiar is real, so that $V^*(x,t) - V(x,t) = 0$, and we can then write

$$\frac{\partial P(x,t)}{\partial t} = -\frac{\partial}{\partial x}\left[\frac{\hbar}{2mi}\left(\psi^*\frac{\partial \psi}{\partial x} - \frac{\partial \psi^*}{\partial x}\psi\right)\right] = -\frac{\partial}{\partial x}j(x,t) \tag{5.11}$$

We have defined

$$j(x,t) \equiv \frac{\hbar}{2mi}\left(\psi^*(x,t)\frac{\partial \psi(x,t)}{\partial x} - \frac{\partial \psi^*(x,t)}{\partial x}\psi(x,t)\right) \tag{5.12}$$

which can be interpreted as a *probability current* or *flux*. This relation, relating the time rate of change of a probability density to the spatial change in a flux,

$$\frac{\partial P(x,t)}{\partial t} = -\frac{\partial j(x,t)}{\partial x} \tag{5.13}$$

is called the *equation of continuity*. It is based on conservation of probability in the same way that the similarly named equation for the flow of incompressible fluids, namely,

$$\frac{\partial \rho(x,t)}{\partial t} = -\frac{\partial}{\partial x}\left[\rho(x,t)v_x(x,t)\right] \tag{5.14}$$

or in three dimensions

$$\frac{\partial \rho(\mathbf{r},t)}{\partial t} = -\nabla \cdot \left[\rho(\mathbf{r},t)\mathbf{v}(\mathbf{r},t)\right] \tag{5.15}$$

(where $\rho(x,t)$ is the fluid density), is based on the conservation of fluid mass. Integrating Eqn. (5.13) over a finite region of space we find

$$\frac{d}{dt}\left[\int_a^b P(x,t)\,dx\right] = -\int_a^b \frac{\partial j(x,t)}{\partial x}\,dx = j(a,t) - j(b,t) \tag{5.16}$$

which can be interpreted as saying

- The time rate of change of the probability in a finite interval (a, b) at any given time t is given by the difference in the rates of probability "flow" into and out of that interval.

We illustrate this connection using Gaussian wavepackets in Fig. 5.2.

Most importantly, if we specialize to $(a, b) = (-\infty, +\infty)$, i.e., the entire one-dimensional universe, we have

$$\frac{d\mathcal{P}(t)}{dt} = \frac{d}{dt}\left[\int_{-\infty}^{+\infty} P(x,t)\,dx\right] = j(-\infty,t) - j(+\infty,t) \to 0 \tag{5.17}$$

because

$$\lim_{x\to\pm\infty}(\psi(x,t)) = 0 \tag{5.18}$$

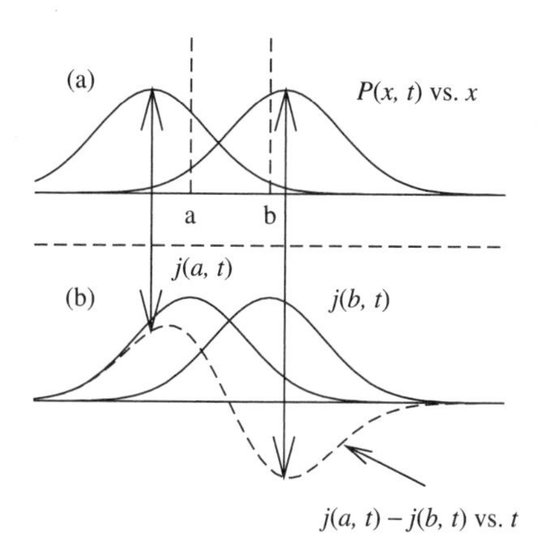

FIGURE 5.2. Position-space wavepackets and flux measurements for Gaussian wavepackets. In (a) $P(x, t)$ versus x is plotted, and two wavepackets at different times are shown, one entering and one leaving the region (a, b); in (b) the rate of probability change, $\partial Prob/\partial t = j(a, t) - j(b, t)$ is shown (dashed line) versus t. For the "entering packet," $j(a, t) > j(b, t)$, so that the probability in the region is increasing.

if the wavefunction is to be normalizable. Here

$$\mathcal{P}(t) = \int_{-\infty}^{+\infty} P(x, t)\, dx \tag{5.19}$$

is the probability of finding the particle *somewhere* as a function of time.

This shows that the total probability is indeed constant in time, provided that the:

1. Solutions satisfy the Schrödinger equation.
2. Solutions are localized so that the wavefunction can be initially normalized.
3. Potential energy function is real.

Under these conditions, the initial normalization factor is preserved by the subsequent time evolution dictated by the Schrödinger equation; we can "set it and forget it."

While the normalization of the wavefunction necessary for a probability interpretation is initially separate from the implementation of the wave equation, the Schrödinger equation guarantees that such an identification is preserved. This connection is a further assurance that the association of $|\psi(x, t)|^2$ with a probability density is a very natural one.

If we relax the constraint that the potential be real and allow it to have, for example, a constant, negative imaginary part $V(x) = V_R(x) - iV_I$, we can repeat the analysis above (P5.3). We find that

$$\frac{d\mathcal{P}(t)}{dt} = -\frac{2V_I}{\hbar}\mathcal{P}(t) = -\lambda\mathcal{P}(t) = -\frac{1}{\tau}\mathcal{P}(t) \tag{5.20}$$

where $\lambda \equiv 2V_I/\hbar \equiv 1/\tau$; this has the trivial solution

$$\mathcal{P}(t) = \mathcal{P}(0)e^{-\lambda t} = \mathcal{P}(0)e^{-t/\tau} \tag{5.21}$$

Viewed in the "ensemble" interpretation where

$$N(t) = N(0)e^{-\lambda t} = N_0 e^{-t/\tau} \tag{5.22}$$

it becomes clear that this represents the loss of probability or particles with an exponential decay law familiar from radioactivity. The decay rate, λ, and mean lifetime, τ, of an unstable particle (or ensemble thereof) can be described in quantum mechanical language at the cost of a nonintuitive (or at least nonclassical) complex potential energy.

5.2 AVERAGE VALUES

5.2.1 Average Values of Position

Given a probability density for position, we can immediately evaluate the expectation values of any x-related quantity; e.g., the average value resulting from many position measurements is

$$\langle x \rangle_t = \int_{-\infty}^{+\infty} x P(x,t)\, dx = \int_{-\infty}^{+\infty} x |\psi(x,t)|^2\, dx \tag{5.23}$$

where we stress that

- The time dependence of $\langle x \rangle_t$ comes solely from the information contained in the wavefunction, $\psi(x,t)$.

This average value is as close as one can come in quantum mechanics to the concept of a classical trajectory, $x(t)$. In a similar way, one can evaluate

$$\langle x^n \rangle_t = \int_{-\infty}^{+\infty} x^n P(x,t)\, dx \tag{5.24}$$

as well as

$$\langle f(x) \rangle_t = \int_{-\infty}^{+\infty} f(x) P(x,t)\, dx \tag{5.25}$$

for an arbitrary function of position. Using these, we can return to the example of a spreading Gaussian wavepacket considered in Sec. 3.2.2 where we studied

$$|\psi(x,t)|^2 = \frac{1}{\beta_t \sqrt{\pi}} e^{-(x-p_0 t/m)^2/\beta_t^2} \tag{5.26}$$

with $\beta_t = \alpha\hbar\sqrt{1 + t^2/t_0^2}$. Using App. B.1, we find that

$$\begin{aligned}
\langle x \rangle_t &= \int_{-\infty}^{+\infty} dx\, x |\psi(x,t)|^2 \\
&= \int_{-\infty}^{+\infty} dx\, (x - p_0 t/m + p_0 t/m) \left[\frac{1}{\beta_t \sqrt{\pi}} e^{-(x-p_0 t/m)^2/\beta_t^2} \right] \\
&= \int_{-\infty}^{+\infty} dx\, (x - p_0 t/m) \left[\frac{1}{\beta_t \sqrt{\pi}} e^{-(x-p_0 t/m)^2/\beta_t^2} \right] \\
&\quad + p_0 t/m \int_{-\infty}^{+\infty} dx \left[\frac{1}{\beta_t \sqrt{\pi}} e^{-(x-p_0 t/m)^2/\beta_t^2} \right] \\
&= (p_0 t/m) \int_{-\infty}^{+\infty} |\psi(x,t)|^2\, dx \\
\langle x \rangle_t &= \frac{p_0 t}{m}
\end{aligned} \tag{5.27}$$

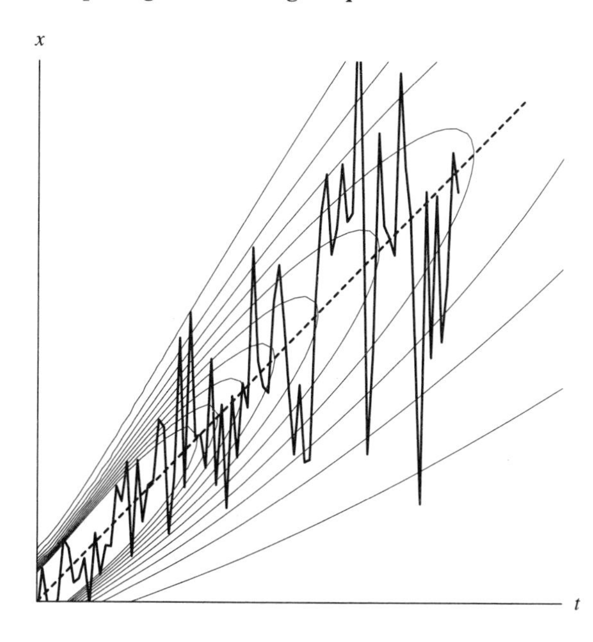

FIGURE 5.3. Contour plot of $|\psi(x,t)|^2$ versus (x,t) for a Gaussian free-particle wavepacket. Superimposed is the classical straight-line trajectory $x(t) = v_0 t$ (dashed line), as well as random measurements of the particle position for increasing times (connected by the "erratic" solid lines).

as expected. We similarly find

$$\langle x^2 \rangle_t = \left(\frac{p_0 t}{m}\right)^2 + \frac{\beta_t^2}{2} \tag{5.28}$$

so that the spread in position does indeed increase with time via

$$\Delta x_t = \sqrt{\langle x^2 \rangle_t - \langle x \rangle_t^2} = \frac{\alpha\hbar}{\sqrt{2}}\sqrt{1 + t^2/t_0^2} = \frac{\beta_t}{\sqrt{2}} \tag{5.29}$$

We can "sample" this wavefunction by randomly measuring the position of a particle at various times and comparing these measurements to the classical, straight-line trajectory in Fig. 5.3. We note that the measurements tend to cluster around the classical path, but with increasingly large excursions due to the spreading.

5.2.2 Average Values of Momentum

How to extract further information from $\psi(x,t)$ on other physically observable quantities, such as momentum and energy, which are now represented by operators, is far from obvious. For example, in Sec. 3.1 we identified the momentum with an operator, $\hat{p} = \hbar/i(\partial/\partial x)$. Does one then define

$$\langle \hat{p} \rangle_t \stackrel{?}{=} \int_{-\infty}^{+\infty} dx\, \hat{p}|\psi(x,t)|^2 \stackrel{?}{=} \int_{-\infty}^{+\infty} dx\, \psi^*(x,t)\hat{p}\psi(x,t) \tag{5.30}$$

or in some other way? No such ambiguity arises, of course, for the position variable itself as

$$\langle x \rangle_t = \int_{-\infty}^{+\infty} dx\, x|\psi(x,t)|^2 = \int_{-\infty}^{+\infty} dx\, \psi^*(x,t)x\psi(x,t) \tag{5.31}$$

To gain some guidance, we note that classically the trajectory $x(t)$ would satisfy

$$\frac{dx(t)}{dt} = \frac{p(t)}{m} \tag{5.32}$$

so we examine the time dependence of its quantum analog, $\langle x \rangle_t$, more fully. We can write

$$\begin{aligned}
\frac{d}{dt}\langle x \rangle_t &= \frac{d}{dt}\left[\int_{-\infty}^{+\infty} x|\psi(x,t)|^2 dx\right] \\
&= \int_{-\infty}^{+\infty} dx\, x \left(\frac{\partial \psi^*}{\partial t}\psi + \psi^* \frac{\partial \psi}{\partial t}\right) \\
&= \frac{\hbar}{2mi}\int_{-\infty}^{+\infty} dx \left(\frac{\partial^2 \psi^*}{\partial x^2} x\psi - \psi^* x \frac{\partial \psi}{\partial x^2}\right)
\end{aligned} \tag{5.33}$$

where we have once again used the fact that $\psi(x,t)$ satisfies the Schrödinger equation (and ψ^* its conjugate) and have assumed that $V(x,t)$ is real. To simplify this, consider

$$\begin{aligned}
\int_{-\infty}^{+\infty} dx \frac{\partial^2 \psi^*}{\partial x^2} x\psi &\overset{\text{IBP}}{=} \left(\frac{\partial \psi^*}{\partial x}(x\psi)\right)\Bigg|_{-\infty}^{+\infty} - \int_{-\infty}^{+\infty} dx \frac{\partial \psi^*}{\partial x}\frac{\partial}{\partial x}(x\psi) \\
&= -\int_{-\infty}^{+\infty} dx \frac{\partial \psi^*}{\partial x}\left(\psi + x\frac{\partial \psi}{\partial x}\right)
\end{aligned} \tag{5.34}$$

where we have used an integration by parts (IBP) and dropped the "surface" term, using the fact that the wavefunction vanishes sufficiently rapidly at infinity. We can repeat this trick once more to obtain

$$\int_{-\infty}^{+\infty} dx \frac{\partial^2 \psi^*}{\partial x^2} x\psi = 2\int_{-\infty}^{+\infty} dx\, \psi^* \frac{\partial \psi}{\partial x} + \int_{-\infty}^{+\infty} dx\, \psi^* x \frac{\partial^2 \psi}{\partial x^2} \tag{5.35}$$

so that substitution back in Eqn. (5.34) gives

$$\begin{aligned}
\frac{d\langle x \rangle_t}{dt} &= \frac{\hbar}{2mi}\int_{-\infty}^{+\infty} dx \left(2\psi^* \frac{\partial \psi}{\partial x} + \psi^* x \frac{\partial^2 \psi}{\partial x^2} - \psi^* x \frac{\partial^2 \psi}{\partial x^2}\right) \\
&= \frac{1}{m}\int_{-\infty}^{+\infty} dx\, \psi^*(x,t)\left(\frac{\hbar}{i}\frac{\partial}{\partial x}\right)\psi(x,t) \\
&= \frac{1}{m}\int_{-\infty}^{+\infty} dx\, \psi^*(x,t)\hat{p}\psi(x,t) \\
\frac{d\langle x \rangle_t}{dt} &\equiv \frac{\langle \hat{p}_t \rangle}{m}
\end{aligned} \tag{5.36}$$

This identification of the average values, which is similar to the classical one for the trajectory variables, is valid provided we adopt the following general definition:

- The average value resulting from a large number of measurements of a physically observable quantity, O, corresponding to some quantum mechanical operator, $\hat{O}$, in a state described by a wavefunction $\psi(x,t)$ is

$$\langle \hat{O} \rangle_t \equiv \int_{-\infty}^{+\infty} dx\, \psi^*(x,t)\hat{O}\psi(x,t) \tag{5.37}$$

so that the operator is "sandwiched" between ψ^* and ψ but acts "only to the right." We will speak of $\langle \hat{O} \rangle_t$ as the *average* or *expectation value* of that operator in the state $\psi(x, t)$. Clearly, this definition reduces to the standard one for any function of the position "operator" x.

Given this general result, we can now extract some (but not all) of the information contained in $\psi(x, t)$ about any other physical observable O for which we have an associated quantum operator, $\hat{O}$. The expectation value of any power of an arbitrary operator, $\hat{O}^n$, is defined in a similar way, and we can generalize this further to any function of an operator, $f(\hat{O})$, provided we have a well-defined series representation for the function $f(y)$. Thus, if

$$f(y) = f(0) + f'(0)y + \frac{f''(0)y^2}{2} + \cdots = \sum_{n=0}^{\infty} \frac{f^{(n)}(0)}{n!} y^n \tag{5.38}$$

then

$$\langle f(\hat{O}) \rangle_t = \sum_{n=0}^{\infty} \frac{f^{(n)}(0)}{n!} \langle \hat{O}^n \rangle_t \tag{5.39}$$

which is sometimes useful. The need for well-defined values of moments of the momentum operator helps justify our continued assumptions that spatial derivatives of $\psi(x, t)$ exist and are well behaved at infinity.

5.2.3 Average Values of Other Operators

We can generalize the definition of $\langle \hat{p} \rangle_t$ to include any power of the momentum operator via

$$\langle \hat{p} \rangle_t = \int_{-\infty}^{+\infty} dx\, \psi^*(x, t) \hat{p}^n \psi(x, t) = \int_{-\infty}^{+\infty} dx\, \psi^*(x, t) \left(\frac{\hbar}{i} \frac{\partial}{\partial x} \right)^n \psi(x, t) \tag{5.40}$$

so, for example, the RMS spread in momentum measurements will be given by

$$\Delta p_t = \sqrt{\langle \hat{p}^2 \rangle_t - \langle \hat{p} \rangle_t^2} \tag{5.41}$$

Returning to the canonical Gaussian wavepacket of Eqn. (3.28), we can evaluate $\langle \hat{p} \rangle_t$ (P5.8) and find

$$\begin{aligned}
\langle \hat{p} \rangle_t &= \int_{-\infty}^{+\infty} dx\, \psi^*(x, t) \left(\frac{\hbar}{i} \frac{\partial}{\partial x} \right) \psi(x, t) \\
&= \int_{-\infty}^{+\infty} dx \left(p_0 - \frac{2(x - p_0 t/m)}{2\hbar i \alpha F} \right) |\psi(x, t)|^2 \\
&= p_0 \int_{-\infty}^{+\infty} dx\, |\psi(x, t)|^2 \\
\langle \hat{p} \rangle_t &= p_0
\end{aligned} \tag{5.42}$$

where the two terms arise from differentiating the "phase" and the "Gaussian" term, respectively. In the case of position measurements, the phase term played no significant role, since only $|\psi(x, t)|^2$ appeared. For average value calculations where the operator must act on ψ *before* one squares, its effects can obviously be important. In this sense, $\psi(x, t)$, with all its phase information intact, is actually more fundamental than $|\psi(x, t)|^2$.

A similar (but more tedious) calculation shows that

$$\langle \hat{p}^2 \rangle_t = p_0^2 + \frac{1}{2\alpha^2} \tag{5.43}$$

so that $\Delta p = 1/\sqrt{2}\alpha$. Combining this result and that of Eqn. (5.29), we find that

$$\Delta x_t \Delta p = \frac{\hbar}{2}\sqrt{1 + t^2/t_0^2} \geq \frac{\hbar}{2} \tag{5.44}$$

and at $t = 0$ this wavepacket actually attains the minimum product of spreads in x and p allowed by the uncertainty principle.

The evaluation of $\langle \hat{p}^2 \rangle$ is especially interesting as it is related to the kinetic energy operator $\hat{T}$, and we can write rather generally

$$\langle \hat{T} \rangle = \frac{1}{2m}\langle \hat{p}^2 \rangle = -\frac{\hbar^2}{2m}\int_{-\infty}^{+\infty} dx\, \psi^* \frac{\partial^2 \psi(x,t)}{\partial x^2}$$

$$\overset{\text{IBP}}{=} -\frac{\hbar^2}{2m}\left(\psi^*(x,t)\frac{\partial \psi(x,t)}{\partial x}\right)\Bigg|_{-\infty}^{+\infty} + \frac{\hbar^2}{2m}\int_{-\infty}^{+\infty} dx\, \frac{\partial \psi^*}{\partial x}\frac{\partial \psi}{\partial x}$$

$$\langle \hat{T} \rangle = \frac{\hbar^2}{2m}\int_{-\infty}^{+\infty} dx \left| \frac{\partial \psi(x,t)}{\partial x} \right|^2 \tag{5.45}$$

This not only simplifies the calculation of $\langle \hat{p}^2 \rangle$ somewhat, but it also shows that the kinetic energy associated with a quantum mechanical wavefunction is related to its spatial variation, i.e., its "wiggliness." This justifies the statement made in discussing Fig. 3.2. The quantity

$$\mathcal{T}(x,t) \equiv \frac{\hbar^2}{2m}\left| \frac{\partial \psi(x,t)}{\partial x} \right|^2 \quad \text{where} \quad \int_{-\infty}^{+\infty} \mathcal{T}(x,t)\, dx = \langle \hat{T} \rangle \tag{5.46}$$

can be associated with a *kinetic energy distribution.* While it is not as fundamentally important as $|\psi(x,t)|^2$, $\mathcal{T}(x,t)$ is sometimes of use in visualizing the distribution of kinetic energy in quantum wavefunctions. A similar quantity is the *potential energy distribution,* which we can define to be

$$\mathcal{V}(x,t) \equiv |\psi(x,t)|^2 V(x,t) \tag{5.47}$$

whose integral gives the expectation value of $V(x,t)$.

Finally, the expectation value of the total energy, represented by the average value of the operator $\hat{E} = i\hbar\partial/\partial t$, can be evaluated for any state via

$$\langle \hat{E} \rangle = \int_{-\infty}^{+\infty} \psi^*(x,t)\left(i\hbar\frac{\partial}{\partial t}\right)\psi(x,t) \tag{5.48}$$

and for the Gaussian wavepacket an explicit calculation gives

$$\langle \hat{E} \rangle = \frac{1}{2m}\left(p_0^2 + \frac{1}{2\alpha^2}\right) = \frac{\langle \hat{p}^2 \rangle}{2m} \tag{5.49}$$

consistent with Eqn. (5.1) and the fact that the wavepacket is a solution of the free particle SE.

5.3 REAL AVERAGE VALUES AND HERMITIAN OPERATORS

While we have introduced a well-defined operational procedure for the calculation of expectation values of quantum mechanical operators, many questions about the connection between classical observable quantities and their quantum mechanical operator counterparts remain to be answered. For example, for $\langle x \rangle_t$ and related position averages, it is clear from Eqns. 5.23 and 5.25 that we will always find real values, as we should if we are to confront the results of measurements of observable quantities. In contrast, the explicit forms for the momentum and energy operators, for example,

$$\hat{p} = \frac{\hbar}{i}\frac{\partial}{\partial x} \qquad \text{and} \qquad \hat{E} = i\hbar\frac{\partial}{\partial t}$$

with their explicit factors of i, make it far from obvious that their expectation values will not be complex, and hence have no connection with real measurements. We presumably wish to restrict ourselves to operators, $\hat{O}$, for which we can guarantee that

$$\langle \hat{O} \rangle = \langle \hat{O} \rangle^* \tag{5.50}$$

or

$$\int_{-\infty}^{+\infty} dx\, \psi^*(x,t)\hat{O}\psi(x,t) = \left[\int_{-\infty}^{+\infty} dx\, \psi^*(x,t)\hat{O}\psi(x,t)\right]^* \tag{5.51}$$

for any physically admissible wavefunction, $\psi(x,t)$. Operators that satisfy Eqn. (5.51) are called *Hermitian,* which we can take to be an extension of the notion of "realness" to operators. The similar statement for general complex numbers would, of course, be that a complex number, z, is real provided $z = z^*$. Thus, a first test of any identification of an operator with a classical observable will be to check whether Eqn. (5.51) is satisfied.

We note that this definition can be easily extended (P5.13) to show that a Hermitian operator, $\hat{O}$, will actually satisfy

$$\int_{-\infty}^{+\infty} dx\, \psi^*(x,t)\hat{O}\phi(x,t) = \left[\int_{-\infty}^{+\infty} dx\, \psi^*(x,t)\hat{O}\phi(x,t)\right]^* \tag{5.52}$$

for any two wavefunctions $\psi(x,t)$, $\phi(x,t)$.

While Eqn. (5.37) shows implicitly that $\langle \hat{p} \rangle_t$ is real (because $\langle x \rangle_t$ is manifestly real), it is instructive to demonstrate this in a more explicit way. We can write

$$\begin{aligned}
\langle \hat{p} \rangle^* &= \left[\int_{-\infty}^{+\infty} dx\, \psi^* \left(\frac{\hbar}{i}\frac{\partial}{\partial x}\right)\psi\right]^* \\
&= \int_{-\infty}^{+\infty} dx\, \psi \left(-\frac{\hbar}{i}\frac{\partial \psi^*}{\partial x}\right) \\
&\overset{\text{IBP}}{=} -\frac{\hbar}{i}\left[\left(\psi^*\psi\right)\Big|_{-\infty}^{+\infty} - \int_{-\infty}^{+\infty} dx\, \psi^* \frac{\partial \psi}{\partial x}\right] \\
&= \int_{-\infty}^{+\infty} dx\, \psi^* \left(\frac{\hbar}{i}\frac{\partial}{\partial x}\right)\psi \\
\langle p \rangle^* &= \langle \hat{p} \rangle
\end{aligned} \tag{5.53}$$

One explicit factor of -1 arising from the complex conjugation has been canceled by a similar one from an integration by parts. This trick can be extended to show that $\langle \hat{p}^n \rangle$ is real for any power of the momentum operator if one assumes that all the relevant surface terms generated by the various integrations by parts vanish due to the behavior of ψ and its spatial derivatives at $x = \pm\infty$. Using Eqn. (5.39) then shows that an arbitrary function of $\hat{p}$ will also have real expectation values. Thus, we have found that x and $f(x)$ (trivially) and $\hat{p}$ and $f(\hat{p})$ are all Hermitian operators.

The proof that the energy operator is Hermitian is somewhat different, as we examine

$$\begin{aligned}
\langle \hat{E} \rangle - \langle \hat{E} \rangle^* &= \int_{-\infty}^{+\infty} dx\, \psi^* \left(i\hbar \frac{\partial}{\partial t} \right) \psi - \left[\int_{-\infty}^{+\infty} dx\, \psi^* \left(i\hbar \frac{\partial}{\partial t} \right) \psi \right]^* \\
&= i\hbar \int_{-\infty}^{+\infty} dx \left(\psi^* \frac{\partial \psi}{\partial t} + \frac{\partial \psi^*}{\partial t} \psi \right) \\
&= i\hbar \frac{d}{dt} \left[\int_{-\infty}^{+\infty} dx\, \psi^*(x,t) \psi(x,t) \right] \\
&= i\hbar \frac{d}{dt} \left[\int_{-\infty}^{+\infty} dx\, P(x,t) \right] \\
&= i\hbar \frac{d}{dt} \left[\mathcal{P}(t) \right] \\
&= 0
\end{aligned} \tag{5.54}$$

so that $\hat{E}$ is a Hermitian operator *provided* that the total probability, $\mathcal{P}(t)$, is constant in time. We have seen that the only situation in which this is not true is when the potential energy function, $V(x, t)$, does, in fact, have an imaginary part, in which case we might expect the classical energy to not be well defined.

5.4 THE PHYSICAL INTERPRETATION OF $\phi(p)$

We are now able to calculate average values (and higher moments) of the momentum operator, but even more detailed information on the "momentum content" of the wavefunction $\psi(x, t)$ is available. In order to extract it in a simple way, we will analyze the role played by $\phi(p)$ more carefully. So far, we have constructed localized wavepackets from plane wave solutions by using

$$\psi(x) = \frac{1}{\sqrt{2\pi\hbar}} \int_{-\infty}^{+\infty} dp\, \phi(p) e^{ipx/\hbar} \tag{5.55}$$

where $\phi(p)$ simply played the role of a weighting function, the amplitude associated with each plane wave component of definite momentum, p. Given a solution of the Schrödinger equation, $\psi(x, t)$, we can invert this to obtain

$$\phi(p, t) = \frac{1}{\sqrt{2\pi\hbar}} \int_{-\infty}^{+\infty} dx\, \psi(x, t) e^{-ipx/\hbar} \tag{5.56}$$

so that this momentum amplitude can be obtained from any solution of the Schrödinger equation. Recall that the normalizations of $\psi(x, t)$ and $\phi(p, t)$ are completely correlated as

$$\int_{-\infty}^{+\infty} dp\, |\phi(p, t)|^2 = \int_{-\infty}^{+\infty} dx\, |\psi(x, t)|^2 = 1 \tag{5.57}$$

This is true provided that we have enforced the normalization for $\psi(x, t)$ as a consequence of its interpretation as giving a probability density for position measurements. This fact strongly suggests that we make the association that

- $|\phi(p, t)|^2\, dp$ is the probability that a measurement of the *momentum* of a particle described by $\phi(p, t)$—obtained possibly via the Fourier transform of $\psi(x, t)$—will find a value in the interval $(p, p + dp)$ at time t.

We can call $\phi(p, t)$ the *momentum space wavefunction* by analogy with $\psi(x, t)$ which is the *position* or *configuration space wavefunction.* This then allows one to make more detailed predictions about the distribution of momentum values using, for example,

$$\text{Prob}[x \in (p_a, p_b)] = \int_{p_a}^{p_b} dp\, |\phi(p, t)|^2 \tag{5.58}$$

being the probability of measuring the momentum to be in the finite interval (p_a, p_b). This identification is made more compelling by the observation that

$$\begin{aligned}
\langle \hat{p} \rangle_t &= \int_{-\infty}^{+\infty} dx\, \psi^*(x, t) \left(\frac{\hbar}{i} \frac{\partial}{\partial x} \right) \psi(x, t) \\
&= \int_{-\infty}^{+\infty} \psi^*(x, t) \left(\frac{\hbar}{i} \frac{\partial}{\partial x} \right) \left[\frac{1}{\sqrt{2\pi\hbar}} \int_{-\infty}^{+\infty} dp\, \phi(p, t) e^{ipx/\hbar} \right] \\
&= \left[\frac{1}{\sqrt{2\pi\hbar}} \int_{-\infty}^{+\infty} dx\, \psi^*(x, t) \int_{-\infty}^{+\infty} dp\, p\phi(p, t) e^{ipx/\hbar} \right] \\
&= \int_{-\infty}^{+\infty} dp\, p\phi(p, t) \left[\frac{1}{\sqrt{2\pi\hbar}} \int_{-\infty}^{+\infty} dx\, \psi(x, t) e^{-ipx/\hbar} \right]^* \\
&= \int_{-\infty}^{+\infty} dp\, p\phi(p, t)\phi^*(p, t) \\
&= \int_{-\infty}^{+\infty} dp\, p|\phi(p, t)|^2 \\
&\equiv \langle p \rangle_t
\end{aligned} \tag{5.59}$$

where we have written the average value now as $\langle p \rangle_t$ (that is, without the "hat") when evaluated using the momentum space wavefunction. In this representation of the quantum mechanical solution, the momentum observable is represented by a "trivial" operator, p. Position information is now less directly obtainable, as we can write

$$\begin{aligned}
\langle x \rangle_t &= \int_{-\infty}^{+\infty} dx\, \psi^*(x,t) x \psi(x,t) \\
&= \int_{-\infty}^{+\infty} \left[\frac{1}{\sqrt{2\pi\hbar}} \int_{-\infty}^{+\infty} dp\, \phi^*(p,t) e^{-ipx/\hbar} \right] x\psi(x,t) dx \\
&= \int_{-\infty}^{+\infty} dp\, \phi^*(p,t) \left[\frac{1}{\sqrt{2\pi\hbar}} \int_{-\infty}^{+\infty} dx\, x\psi(x) e^{-ipx/\hbar} \right] \\
&= \int_{-\infty}^{+\infty} dp\, \phi^*(p,t) \left(i\hbar \frac{\partial}{\partial p} \right) \left[\frac{1}{\sqrt{2\pi\hbar}} \int_{-\infty}^{+\infty} dx \psi(x) e^{-ipx/\hbar} \right] \\
&= \int_{-\infty}^{+\infty} \phi^*(p,t) \hat{x} \phi(p,t) \\
&= \langle \hat{x} \rangle_t
\end{aligned} \tag{5.60}$$

where we now identify the *position operator* in the momentum space representation as

$$\hat{x} = i\hbar \frac{\partial}{\partial p} \tag{5.61}$$

The probability interpretation of $\phi(p,t)$ also implies that:

- $\phi(p,t)$ should be square-integrable (as a function of p) and continuous in p.
- $\phi(p,t)$ (and its spatial derivatives) should be continuous and must vanish sufficiently rapidly as $p \to \pm\infty$, so that it is square-integrable.

5.5 ENERGY EIGENSTATES, STATIONARY STATES, AND THE HAMILTONIAN OPERATOR

In much the same way that Newton's second law relates the time dependence of a particle's trajectory to the external force, the time-dependent Schrödinger equation

$$i\hbar \frac{\partial}{\partial t} \psi(x,t) = -\frac{\hbar^2}{2m} \frac{\partial^2 \psi(x,t)}{\partial x^2} + V(x,t)\psi(x,t) \tag{5.62}$$

dictates the time development of the wavefunction of a particle in the presence of an external potential. If $V(x,t)$ is truly time-dependent, the resulting partial differential equation can be difficult to solve, but we can often consider the special, but very important, case of a time-independent potential, i.e., one for which

$$V(x,t) = V(x) \tag{5.63}$$

only. In this case, the Schrödinger equation can be separated in the form

$$i\hbar \frac{\partial}{\partial t} \psi(x,t) = \left[-\frac{\hbar^2}{2m} \frac{\partial^2}{\partial x^2} + V(x) \right] \psi(x,t) = \hat{H} \psi(x,t) \tag{5.64}$$

where we have introduced the *Hamiltonian operator* that can be written as

$$\hat{H} \equiv -\frac{\hbar^2}{2m}\frac{\partial^2}{\partial x^2} + V(x) = \frac{\hat{p}^2}{2m} + V(x) \tag{5.65}$$

and is seen to be a function of position coordinates only. This operator is the quantum mechanical version of the corresponding classical Hamiltonian.[2]

Equation. (5.64) is now a *separable differential equation,* and a standard method of solution is to assume a product wavefunction of the form

$$\psi(x,t) = \psi(x)T(t) \tag{5.66}$$

If we substitute this form into Eqn. (5.64) and divide by $\psi(x,t)$, we find

$$\left[i\hbar\frac{dT(t)}{dt}\right] \Big/ T(t) = \left[\hat{H}\psi(x)\right] \Big/ \psi(x) \tag{5.67}$$

which must be true, of course, for all values of x and t. The constraint that two different functions of independent variables be identical, i.e.,

$$F(t) = G(x) \qquad \text{for all } x \text{ and } t \tag{5.68}$$

can be satisfied only if both functions are equal to a constant. This is easily shown by noting that

$$\begin{aligned} 0 = \frac{\partial F(t)}{\partial x} &= \frac{\partial G(x)}{\partial x} && \Longrightarrow && G(x) = \text{constant} \\ \frac{\partial F(t)}{\partial t} &= \frac{\partial G(x)}{\partial t} = 0 && \Longrightarrow && F(t) = \text{same constant} \end{aligned} \tag{5.69}$$

Noting the dimensions, we can then write the common constant as E, giving

$$\frac{\left[i\hbar\frac{dT(t)}{dt}\right]}{T(t)} = \frac{\left[\hat{H}\psi(x)\right]}{\psi(x)} = E \tag{5.70}$$

so that

$$T(t) = e^{-iEt/\hbar} \tag{5.71}$$

The complete wavefunction is then

$$\psi(x,t) = \psi_E(x)e^{-iEt/\hbar} \tag{5.72}$$

where $\psi_E(x)$ now satisfies the *time-independent Schrödinger equation*

$$\hat{H}\psi_E(x) = \left(-\frac{\hbar^2}{2m}\frac{d^2}{dx^2} + V(x)\right)\psi_E(x) = E\psi_E(x) \tag{5.73}$$

We will devote most of the rest of the book to examining the mathematical properties and physical meaning of solutions of the time-independent Schrödinger equation. We note that:

---

[2]See many undergraduate texts, e.g., Marion and Thornton (1988), for an introduction to the Hamiltonian formulation of classical mechanics; a brief review is also contained in App. E.

- The *number E* can certainly be identified as the uniquely defined energy of the state, since application of the energy *operator,* $\hat{E}$, gives

$$\hat{E}\psi_E(x,t) = i\hbar\frac{\partial}{\partial t}\left(\psi_E(x)e^{-iEt/\hbar}\right) = E\psi_E(x,t) \tag{5.74}$$

Moreover, calculations of expectation values of the energy operator give

$$\begin{aligned}\langle \hat{E}^n \rangle &= \int_{-\infty}^{+\infty} dx\, \psi^*(x,t)\hat{E}^n\psi(x,t) \\ &= \int_{-\infty}^{+\infty} dx \left(\psi_E^*(x)e^{iEt/\hbar}\right)\left(i\hbar\frac{\partial}{\partial t}\right)^n \left(\psi_E(x)e^{-iEt/\hbar}\right) \\ &= E^n \int_{-\infty}^{+\infty} dx\, |\psi_E(x)|^2 \\ &= E^n \end{aligned} \tag{5.75}$$

This implies that the uncertainty in energy of this state is

$$\Delta E = \sqrt{\langle \hat{E}^2 \rangle - \langle \hat{E} \rangle^2} = 0 \tag{5.76}$$

Such a state, with a precisely defined value of energy, can be called an *energy eigenstate* with *energy eigenvalue* given by *E*. (The use of the German *eigen* meaning characteristic or "belonging to" or, colloquially, "own" is appropriate here.)

- The energy eigenvalue appears as a parameter in the time-independent Schrödinger equation, so that solutions must be found for each different value of E, hence the label $\psi_E(x)$. For the case of a free particle, $V(x) = 0$, for example, we solve

$$\hat{H}\psi_E(x) = -\frac{\hbar^2}{2m}\frac{d^2\psi_E(x)}{dx^2} = E\psi_E(x) \tag{5.77}$$

to obtain

$$\psi_E(x) = e^{\pm i\sqrt{2mE}x/\hbar} \tag{5.78}$$

or

$$\psi_E(x,t) = e^{\pm i\sqrt{2mE}x/\hbar}e^{-iEt/\hbar} = e^{i(px-p^2t/2m)/\hbar} = \psi_p(x,t) \tag{5.79}$$

which are the standard plane wave solutions with the identification $E = p^2/2m$ for the *parameters E, p*.

- An equation of the form such as $\hat{H}\psi_E(x) = E\psi_E(x)$ of the form

$$\text{Operator acting on function} = \text{number times function}$$

is called an *eigenvalue problem.* Similar problems arise in matrix algebra and elsewhere; in that context, they are often of the form

$$\mathbf{M} \cdot \mathbf{v} = \lambda \cdot \mathbf{v} \tag{5.80}$$

where **M** is a matrix and **v** is a vector.

- The trivial time dependence of such states implies that the corresponding probability densities are independent of time, since

$$P(x,t) = |\psi_E(x,t)|^2 = |\psi_E(x)|^2 e^{-iEt/\hbar}e^{iEt/\hbar} = |\psi_E(x)|^2 \tag{5.81}$$

Therefore, the expectation values of most operators, $\hat{O}$, will satisfy

$$\begin{aligned}\langle \hat{O} \rangle_t &= \int_{-\infty}^{+\infty} dx \left(\psi_E^*(x) e^{iEt/\hbar} \right) \hat{O} \left(\psi_E(x) e^{-iEt/\hbar} \right) \\ &= \int_{-\infty}^{+\infty} dx \, \psi_E(x) \hat{O} \psi_E(x) \\ &= \langle \hat{O} \rangle_{t=0} \end{aligned} \tag{5.82}$$

Such states for which the probability density and other observables are "frozen in time" are called *stationary states.*

- Because of the linearity of the Schrödinger equation, the most general solution will consist of linear combinations of such stationary state or energy eigenstate solutions, i.e.,

$$\psi(x, t) = \left(\sum_E + \int dE \right) \psi_E(x) e^{-iEt/\hbar} \tag{5.83}$$

where the sum is over all possible discrete and continuous values of E. Because this function contains solutions of different energies, it will no longer be an energy eigenstate and will, in general, have $\Delta E \neq 0$. In addition, because of the possibility of "cross-terms" in $|\psi(x, t)|^2$, the probability density (and physical observables) can have nontrivial time dependence, and so it is not a stationary state. Examples of this type include the Gaussian wavepacket constructed from free-particle solutions in Sec. 3.2.2 or the accelerating wavepacket considered in the next section.

An especially simple case of such time dependence is a solution consisting of a linear combination of two eigenstates:

$$\psi(x, t) = \frac{1}{\sqrt{2}} \left(\psi_{E_1}(x) e^{-iE_1 t/\hbar} + \psi_{E_2}(x) E^{-iE_2 t/\hbar} \right) \tag{5.84}$$

where we assume that $\psi_{E_1}(x)$ and $\psi_{E_2}(x)$ are real for simplicity. In this case we can show that (P5.16)

$$\Delta E = \sqrt{\langle \hat{E}^2 \rangle - \langle \hat{E} \rangle^2} = \frac{|E_1 - E_2|}{2} \tag{5.85}$$

while

$$\begin{aligned} P(x, t) = |\psi(x, t)|^2 &= \psi_{E_1}^2(x) + \psi_{E_2}^2(x) \\ &\quad + 2\psi_{E_1}(x)\psi_{E_2}(x) \cos(|E_1 - E_2| t/\hbar) \end{aligned} \tag{5.86}$$

- This example reminds us that the actual value of E itself is not important, as for individual eigenstates the effect of the time-dependent phase, $e^{-iEt/\hbar}$, is irrelevant. In mixed states, only *energy differences* appear, and this fact is a reflection of the arbitrariness inherent in choosing the potential energy function, $V(x)$. In classical mechanics, letting $V(x) \to V(x) + V_0$ makes no change in the applied force (and hence the physics), and the choice of V_0 can change the overall energy scale, but not energy differences (P7.5).
- Energy eigenstates, characterized by $\Delta E = 0$, in order to be consistent with the energy-time uncertainty principle, $\Delta E \Delta t \geq \hbar/2$, require that $\Delta t = \infty$ in some sense; this is plausible given the static character of stationary states. For the two-

state system above, the characteristic period of the system from Eqn. (5.86) is

$$\tau = 2\pi \frac{\hbar}{|E_1 - E_2|} \tag{5.87}$$

which is perfectly consistent with

$$\Delta E \Delta t \approx \frac{|E_1 - E_2|}{2} \tau \approx \pi \hbar \tag{5.88}$$

5.6 THE SCHRÖDINGER EQUATION IN MOMENTUM SPACE

5.6.1 Transforming the Schrödinger Equation into Momentum Space

We will often concentrate on the solution of the time-dependent Schrödinger equation in position or configuration space, namely, solving

$$-\frac{\hbar^2}{2m} \frac{\partial^2 \psi(x,t)}{\partial x^2} + V(x)\psi(x,t) = i\hbar \frac{\partial \psi(x,t)}{\partial t} \tag{5.89}$$

for $\psi(x, t)$ and then obtaining $\phi(p, t)$, if desired, by the appropriate Fourier transform

$$\phi(p,t) = \frac{1}{\sqrt{2\pi\hbar}} \int_{-\infty}^{+\infty} \psi(x,t) e^{-ipx/\hbar} \, dx \tag{5.90}$$

It is possible, however, to transform the Schrödinger equation itself into momentum space and solve for $\phi(p, t)$ directly, and this strategy sometimes yields a simpler and more directly interpretable solution. (For another variation on this approach, see P5.15.) To this end, we multiply both sides of Eqn. (5.89) by $\exp(-ipx/\hbar)/\sqrt{2\pi\hbar}$ and integrate over dx to obtain

$$\begin{aligned} \frac{1}{\sqrt{2\pi\hbar}} \int_{-\infty}^{+\infty} i\hbar \frac{\partial \psi(x,t)}{\partial t} e^{-ipx/\hbar} \, dx &= \frac{1}{\sqrt{2\pi\hbar}} \int_{-\infty}^{+\infty} \left(-\frac{\hbar^2}{2m} \frac{\partial^2 \psi(x,t)}{\partial x^2} \right) \left(e^{-ipx/\hbar} \right) dx \\ &\quad + \frac{1}{\sqrt{2\pi\hbar}} \int_{-\infty}^{+\infty} V(x)\psi(x,t) e^{-ipx/\hbar} \, dx \end{aligned} \tag{5.91}$$

The order of integration and differentiation with respect to time on the right-hand side can be exchanged so that

$$\begin{aligned} \frac{1}{\sqrt{2\pi\hbar}} \int_{-\infty}^{+\infty} i\hbar \frac{\partial \psi(x,t)}{\partial t} e^{-ipx/\hbar} dx &= i\hbar \frac{\partial}{\partial t} \left(\frac{1}{\sqrt{2\pi\hbar}} \int_{-\infty}^{+\infty} \psi(x,t) e^{-ipx/\hbar} dx \right) \\ &= i\hbar \frac{\partial \phi(p,t)}{\partial t} \end{aligned} \tag{5.92}$$

The spatial derivative term can be rewritten as

$$\frac{1}{\sqrt{2\pi\hbar}}\int_{-\infty}^{+\infty}\left(-\frac{\hbar^2}{2m}\frac{\partial^2\psi(x,t)}{\partial x^2}\right)e^{-ipx/\hbar}dx \overset{\text{IBP}}{=} -\frac{\hbar^2}{2m}\frac{1}{\sqrt{2\pi\hbar}}\int_{-\infty}^{+\infty}\psi(x,t)\left(\frac{d^2}{dx^2}\left[e^{-ipx/\hbar}\right]\right)dx$$
$$= \frac{p^2}{2m}\left(\frac{1}{\sqrt{2\pi\hbar}}\int_{-\infty}^{+\infty}\psi(x,t)e^{-ipx/\hbar}dx\right)$$
$$= \frac{p^2}{2m}\phi(p,t) \tag{5.93}$$

where we have used two integrations by parts to move the derivatives onto the exponential term. Finally, the potential energy function can formally be expanded in a Taylor series to yield

$$\frac{1}{\sqrt{2\pi\hbar}}\int_{-\infty}^{+\infty}V(x)\psi(x,t)e^{-ipx/\hbar}dx = \sum_{n=0}^{+\infty}\frac{V^{(n)}(0)}{n!}\frac{1}{\sqrt{2\pi\hbar}}\int_{-\infty}^{+\infty}x^n\psi(x,t)e^{-ipx/\hbar}dx \tag{5.94}$$

and we can use

$$\frac{1}{\sqrt{2\pi\hbar}}\int_{-\infty}^{+\infty}\psi(x,t)x^n e^{-ipx/\hbar}dx = \frac{1}{\sqrt{2\pi\hbar}}\int_{-\infty}^{+\infty}\psi(x,t)\left(i\hbar\frac{\partial}{\partial p}\right)^n\left(e^{-ipx/\hbar}\right)dx$$
$$= \left(i\hbar\frac{\partial}{\partial p}\right)^n\phi(p,t) \tag{5.95}$$

in which case the potential term becomes

$$\left[\sum_{n=0}^{+\infty}\frac{V^{(n)}(0)}{n!}\left(i\hbar\frac{\partial}{\partial p}\right)^n\right]\phi(p,t) = V\left(i\hbar\frac{\partial}{\partial p}\right)\phi(p,t) \tag{5.96}$$

where we once again have identified $\hat{x} = i\hbar\partial/\partial p$. Thus, the *time-dependent Schrödinger equation in momentum space* can be written as

$$\frac{p^2}{2m}\phi(p,t) + V\left(i\hbar\frac{\partial}{\partial p}\right)\phi(p,t) = i\hbar\frac{\partial\phi(p,t)}{\partial t} \tag{5.97}$$

If the potential does not depend explicitly on time, we can write

$$\phi(p,t) = \phi(p)\exp(-iEt/\hbar) \tag{5.98}$$

and obtain the time-independent Schrödinger equation

$$\frac{p^2}{2m}\phi(p) + V\left(i\hbar\frac{\partial}{\partial p}\right)\phi(p) = E\phi(p) \tag{5.99}$$

in the usual way.

For the simple case of a free particle ($V(x) = 0$), we have

$$\frac{\partial\phi(p,t)}{\partial t} = -i\hbar\frac{p^2}{2m}\phi(p,t) \tag{5.100}$$

which is easily integrated to yield

$$\phi(p,t) = \phi_0(p)e^{-ip^2t/2m\hbar} \tag{5.101}$$

where $\phi_0(p)$ is an arbitrary initial momentum distribution. This yields the standard position space wavepacket via the Fourier transform since

$$\begin{aligned} \psi(x,t) &= \int_{-\infty}^{+\infty} dp\, \phi(p,t) e^{ipx/\hbar} \\ &= \frac{1}{\sqrt{2\pi\hbar}} \int_{-\infty}^{+\infty} \left(\phi_0(p) e^{-ip^2t/2m\hbar}\right) e^{ipx/\hbar}\, dp \\ &= \frac{1}{\sqrt{2\pi\hbar}} \int_{-\infty}^{+\infty} \phi_0(p) e^{i(px-p^2t/2m)\hbar}\, dp \end{aligned} \tag{5.102}$$

as in Sec. 3.2.1.

5.6.2 Uniformly Accelerating Particle

A more interesting case arises when one considers the quantum mechanical problem of a particle under the influence of a constant (nonvanishing) *force.* The problem of a free particle moving at constant speed is one of the simplest in introductory mechanics, and we have now discussed its quantum analog in both x and p space. A uniformly accelerating particle is the next most familiar problem of Newtonian dynamics, and we will see that its quantum solution is actually more easily obtained and understood using the momentum space version of the Schrödinger equation.

We begin in position space by assuming $F(x) = F$ so that the potential energy is $V(x) = -Fx$. We will assume, for definiteness, that $F > 0$, corresponding to uniform acceleration to the right. The time-independent Schrödinger equation in position space then reads

$$-\frac{\hbar^2}{2m}\frac{d^2\psi(x)}{dx^2} - Fx\psi(x) = E\psi(x) \tag{5.103}$$

for an arbitrary energy eigenvalue. We can simplify this by assuming a change of variables of the form

$$x = \rho y + \sigma \tag{5.104}$$

where y will be dimensionless. Substitution into Eqn. (5.103) gives

$$\frac{d^2\psi(y)}{dy^2} + \frac{2m\rho^2}{\hbar^2}\left[F(\rho y + \sigma) + E\right]\psi(y) = 0 \tag{5.105}$$

which is considerably simplified if we let

$$\rho^3 = -\frac{\hbar^2}{2mF} \qquad \text{and} \qquad \sigma = -\frac{E}{F} \tag{5.106}$$

giving

$$\frac{d^2\psi(y)}{dy^2} = y\psi(y) \tag{5.107}$$

This is the so-called *Airy differential equation,* which is discussed in some detail in App. C.2 (see also P5.21). It has two linearly independent, oscillatory solutions, $Ai(y)$ and $Bi(y)$, similar to the independent $\sin(px/\hbar)$ and $\cos(px/\hbar)$ solutions of the free-particle equation. The $Bi(y)$ solution has, however, a physically unacceptable behavior for large y and

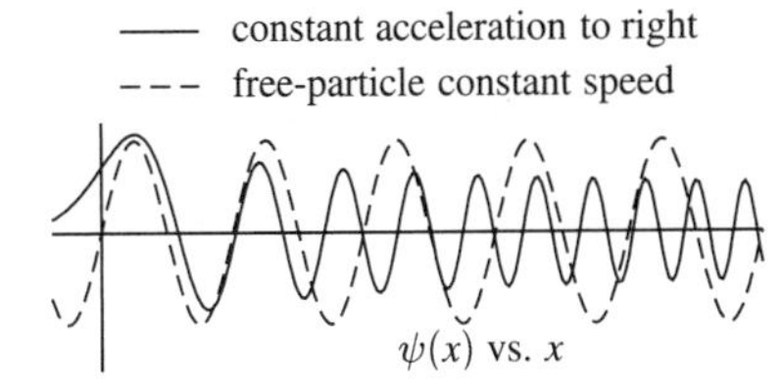

FIGURE 5.4. Wavefunctions (the real part of an energy eigenstate) for accelerating (solid) and constant speed (dashed) motion versus x. Note the increasing "wiggliness" and decreasing amplitude for the accelerating case for increasing x.

so must be excluded (i.e., its coefficient must be chosen to vanish in order to satisfy the appropriate boundary conditions at infinity). The solution is then up to an arbitrary multiplicative constant,

$$\psi(x;E) = Ai\left(\frac{x-\sigma}{\rho}\right) \tag{5.108}$$

where the dependence on E comes through σ. We plot in Fig. 5.4 the behavior of this solution for some arbitrary parameter values and compare it with a plane wave solution (in this case a cosine solution) of the free-particle case. One can see the qualitative differences, which can be understood in classical terms:

- A free particle moving at constant speed would be equally likely to be found in any region of space, implying a wavefunction that does not, on average, change over space. A uniformly accelerating particle would have increasing speed (in this case to the right) and would spend decreasing amounts of time in a given region of space as x gets large. Since $|\psi(x)|^2$ measures the probability of measuring the particle, we would expect to see $|\psi(x)|$ decrease in amplitude for larger x. The cosine solution does indeed oscillate with a constant amplitude, while the Airy function solution decreases in magnitude as $x \to +\infty$.
- The local wiggliness of the wavefunction is qualitatively related to the kinetic energy density via the spatial derivative

$$\langle \hat{T} \rangle = \frac{\hbar^2}{2m}\int_{-\infty}^{+\infty} \left|\frac{d\psi(x)}{dx}\right|^2 dx \tag{5.109}$$

  A constant speed (uniformly accelerating) particle would have constant (increasing) kinetic energy as $x \to +\infty$, implying a wavefunction that had constant (increasing) wiggliness. This is also obvious from the explicit position space solutions.
- The kinetic and potential energy densities $\mathcal{T}(x)$ and $\mathcal{V}(x)$ can be constructed from these solutions as well and are shown in Fig. 5.5 using $V(x) = -Fx$. The increasing (decreasing) kinetic energy (potential energy) is apparent.

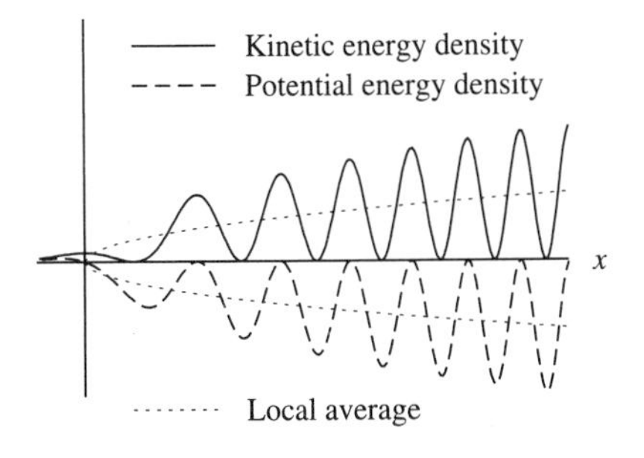

FIGURE 5.5. Kinetic energy distribution, $\mathcal{T}(x)$ (solid), and potential energy distribution, $\mathcal{V}(x)$ (dashed), versus x for an accelerating particle (energy eigenstate). The dotted lines indicate the local averages of $\mathcal{T}$ and $\mathcal{V}$, showing the increase of each with x.

Using the Airy function solutions, one can then, in principle, construct smooth, localized wavepackets[3] that demonstrate uniform acceleration, but in practice this is extremely difficult. The individual E-dependent solutions of Eqn. (5.108) must be weighted by a function of E and integrated; this process can be numerically intensive and is not terribly instructive.

In this case, however, wavepacket solutions are actually more easily obtained and analyzed using the time-dependent equation in momentum space where

$$\frac{p^2}{2m}\phi(p,t) - F \cdot \left[i\hbar\frac{\partial}{\partial p}\right]\phi(p,t) = i\hbar\frac{\partial\phi(p,t)}{\partial t} \tag{5.110}$$

or

$$i\hbar\left(F\frac{\partial\phi(p,t)}{\partial p} + \frac{\partial\phi(p,t)}{\partial t}\right) = \frac{p^2}{2m}\phi(p,t) \tag{5.111}$$

We note that the simple combination of derivatives guarantees that a function of the form $\Phi(p - Ft)$ will make the left-hand side vanish, so we assume a solution of the form $\phi(p,t) = \Phi(p - Ft)\tilde{\phi}(p)$, with $\Phi(p)$ arbitrary and $\tilde{\phi}(p)$ to be determined. Using this form, Eqn. (5.111) reduces to

$$\frac{\partial\tilde{\phi}(p)}{\partial p} = -\frac{i\hbar p^2}{2m\hbar F}\tilde{\phi}(p) \tag{5.112}$$

with the solution

$$\tilde{\phi}(p) = e^{-ip^3/6mF\hbar} \tag{5.113}$$

We can then write the general solution as

$$\phi(p,t) = \Phi(p - Ft)e^{-ip^3/6mF\hbar} \tag{5.114}$$

or, using the arbitrariness of $\Phi(p)$, as

$$\phi(p,t) = \phi_0(p - Ft)e^{i((p-Ft)^3 - p^3)/6mF\hbar} \tag{5.115}$$

where now $\phi_0(p)$ is some initial momentum distribution since $\phi(p,0) = \phi_0(p)$. If the initial momentum distribution is characterized by $\langle p\rangle_0 = p_0$, then (using an obvious change of variables)

$$\begin{aligned}
\langle p\rangle_t &= \int_{-\infty}^{+\infty} p|\phi(p,t)|^2\,dp \\
&= \int_{-\infty}^{+\infty} p|\phi_0(p - Ft)|^2\,dp \qquad \text{(and using } q = p - Ft\text{)} \\
&= \int_{-\infty}^{+\infty} q|\phi_0(q)|^2 dq + Ft\int_{-\infty}^{+\infty} |\phi_0(q)|^2 dq \\
\langle p\rangle_t &= \langle p\rangle_0 + Ft
\end{aligned} \tag{5.116}$$

Thus the average momentum value increases linearly with time, consistent with the classical result for a constant force, $F = dp/dt$. The momentum distribution "translates" uniformly to the right with no change in shape since

$$|\phi(p,t)|^2 = |\phi_0(p - Ft)|^2 \tag{5.117}$$

[3]See Churchill (1978).

The corresponding position space wavefunction can be written as

$$\psi(x,t) = \frac{1}{\sqrt{2\pi\hbar}} \int_{-\infty}^{+\infty} \phi_0(p - Ft) e^{i((p-Ft)^3 - p^3)/6mF\hbar} e^{ipx/\hbar}\, dp \tag{5.118}$$

and, because the p^3 terms cancel in the exponent, this transform can be done analytically (P5.17) for a Gaussian momentum distribution. In that case, we have

$$\phi_0(p) = \sqrt{\frac{\alpha}{\sqrt{\pi}}} e^{-\alpha^2 p^2/2} \tag{5.119}$$

so that

$$\psi(x,t) = \frac{1}{\sqrt{\alpha\hbar\sqrt{\pi}(1 + it/t_0)}} e^{iFt(x - Ft^2/6m)/\hbar} e^{-(x-Ft^2/2m)^2/2\hbar^2\alpha^2(1+it/t_0)} \tag{5.120}$$

where the spreading time is defined by $t_0 \equiv m\hbar\alpha^2$, just as in the free-particle wavepacket case of Sec. 3.2.2. The corresponding probability density is then

$$|\psi(x,t)|^2 = \frac{1}{\beta_t\sqrt{\pi}} e^{-(x-Ft^2/2m)^2/\beta_t^2} \tag{5.121}$$

where $\beta_t = \hbar\alpha\sqrt{1 + t^2/t_0^2}$, also as before. It is easy to see (P5.18) that

$$\langle x \rangle_t = Ft^2/2m \qquad \langle p^2 \rangle_t = (Ft)^2 + \frac{1}{2\alpha^2} \tag{5.122}$$

and that the uncertainty principle product is given by

$$\Delta x\, \Delta p = \frac{\hbar}{2}\sqrt{1 + t^2/t_0^2} \tag{5.123}$$

as before. A position space wavepacket with arbitrary initial position (x_0) and momentum (p_0) can then be obtained by letting

$$\phi_0(p) \to e^{ipx_0/\hbar} \phi_0(p - p_0) \tag{5.124}$$

Just as in Sec. 3.3.1, the integral in Eqn. (5.118) can be performed numerically for other initial momentum distributions. For $\phi(p, 0)$ other than a Gaussian, the position space wavepacket will change shape, but Eqn. (5.117) guarantees that the momentum distribution will not disperse.

5.7 COMMUTATORS

In the analytic wavepacket examples considered so far, the free-particle and accelerating Gaussian packets, we have explicitly demonstrated the validity of position momentum uncertainty principle

$$\Delta x\, \Delta p \geq \frac{\hbar}{2} \tag{5.125}$$

We have previously understood this as arising from a fundamental limitation imposed on wavepackets formed by constructive/destructive interference by the relation $\Delta x \, \Delta k > \mathbb{O}(1)$ from Sec. 2.3 and the identification (via de Broglie) $p = \hbar k$, i.e., as a basic constraint arising from a wave mechanics description of particle dynamics. We can approach this relation in a more formal way, using the notion of quantum mechanical operators, as a preview of the more rigorous proof of the uncertainty principle in Chap. 14.

If we choose to work in a position space representation, we note that because $\hat{p}$ is associated with a nontrivial operator, the result of the application of p and x to a wavefunction will depend on their ordering, specifically

$$x\hat{p}\psi(x) \neq \hat{p}x\psi(x) \tag{5.126}$$

We can show quite generally that

$$\begin{aligned}
\left(x\hat{p} - \hat{p}x\right)\psi(x) &= x\left(\frac{\hbar}{i}\frac{d\psi(x)}{dx}\right) - \left(\frac{\hbar}{i}\frac{d}{dx}\right)\left(x\psi(x)\right) \\
&= x\left(\frac{\hbar}{i}\frac{d\psi(x)}{dx}\right) - x\left(\frac{\hbar}{i}\frac{d\psi(x)}{dx}\right) - \frac{\hbar}{i}\psi(x) \\
&= i\hbar\psi(x) \neq 0
\end{aligned} \tag{5.127}$$

Since this is true for an arbitrary $\psi(x)$, we can write an *operator* identity, namely,

$$[x, \hat{p}] = x\hat{p} - \hat{p}x = i\hbar \tag{5.128}$$

where we have introduced the *commutator* of two operators, defined by

$$[\hat{A}, \hat{B}] \equiv \hat{A}\hat{B} - \hat{B}\hat{A} \tag{5.129}$$

It is easy to show (P5.22) that this same result is also obtained in momentum space where $\hat{x} = -i\hbar\partial/\partial p$ is now the "nontrivial" operator, so that this relation is not dependent on a specific representation. This is a more formal statement that these two operators do not commute.

To preview the connection between this lack of commutativity (as measured by the nonvanishing commutator) and the uncertainty principle, we can look again at Eqn. (5.128). If measurements of x and p gave the same result done in either ordering, we could argue that the measurement procedures for these two physical quantities did not "interfere" with each other, and independent measurements of both are possible. One could then imagine making increasingly precise measurements of both quantities until the uncertainty principle product $\Delta x \, \Delta p$ was as small as desired. The content of Eqn. (5.126), however, is to say that a measurement of, say p, will necessarily alter some of the information regarding the position x of the particle, or vice versa. This formal statement is important, as it codifies the results of a large variety of *gedanken* experiments (thought experiments) designed to violate Eqn. (5.128) (for example, see P5.23).

5.8 QUESTIONS AND PROBLEMS

Q5.1 What is the statement of the equation of continuity in electromagnetism? What is the conserved quantity in that case?

Q5.2 Why do we more often solve the Schrödinger equation in position space than in momentum space?

Q5.3 For the free-particle and accelerating wavepackets, the spread in position increases with time. Is it possible that the spread in position of a wavepacket (not necessarily for a free particle) could *shrink* with time? What would that imply about the spread in momentum? How about the case of the "bouncing" wavepacket in Sec. 3.3.2?

Q5.4 In Chap. 2, it was discussed how solutions of the classical wave equation of the form $\psi(x,t) = \psi(x - vt)$ propagated with no distortion or change in shape. Show how to calculate Δx_t for such a wave, and discuss how the spread in position remains constant in time.

Q5.5 Sketch the contour plot and "trajectory" picture for $|\phi(p,t)|^2$ corresponding to a free particle; that is, what does Fig. 5.3 look like for the momentum space distributions? How about the corresponding plots for the accelerating wavepacket in Sec. 5.6.2? The "bouncing" wavepacket of Sec. 3.3.2?

Q5.6 Sketch the real and imaginary parts and the modulus of the accelerating wavepacket in Sec. 5.6.2. Is it consistent with what you expect?

Q5.7 What are the dimensions of probability flux?

P5.1 Consider a particle described by a wavefunction

$$\psi(x,t) = Ae^{-|x|/L - iEt/\hbar}$$

(a) Find A so that $\psi(x,t)$ is normalized properly.
(b) Evaluate $\langle x \rangle$, $\langle x^2 \rangle$, and Δx.
(c) What is the probability of finding a particle in the region $(-L, +L/2)$? How about in the interval $(+L, +\infty)$? in the interval $(0.99999L, L)$? What is the probability of finding the particle exactly at $x = 3L$?
(d) If 10,000 particles were described by this wavefunction, how many would you expect to find in the region $(0, L)$? If you found 4378 in this region, what would you think? How about 4021?
(e) Try to evaluate $\langle \hat{p} \rangle$ and see what you get. Now try $\langle \hat{p}^2 \rangle$; if you have problems doing this, to what do you trace your difficulties?

P5.2 Consider the same position space wavefunction in P5.1.
(a) Using the Fourier transform, find $\phi(p,t)$.
(b) Evaluate $\langle p \rangle$, $\langle p^2 \rangle$, and Δp. Why do you think you got the results you did?
(c) What fraction of the particles would be moving with momentum in the range $p \in (\hbar/L, 2\hbar/L)$?

P5.3 Assuming that the potential has a negative imaginary part, i.e., $V(x) = V_R(x) - iV_I$, derive Eqns. 5.20 and 5.21. Is the relation between τ and V_I reminiscent of an uncertainty principle?

P5.4 **Lorentzian line shape:** Consider the time dependence of a wavefunction that has an energy with both real and imaginary parts, namely,

$$T(t) = e^{-i(E_A - iV_I)t/\hbar}$$

Use this wavefunction to evaluate $\langle t \rangle$ and $\langle t^2 \rangle$ using

$$\langle f(t) \rangle = \int_0^\infty dt\, |T(t)|^2 f(t) \bigg/ \int_0^\infty dt\, |T(t)|^2$$

Find the analog of the Fourier transform of $T(t)$, namely,

$$A(E) = \int_0^\infty dt\; T(t) e^{iEt/\hbar}$$

and show that

$$|A(E)|^2 = \frac{N}{(E - E_A)^2 + (\hbar/2\tau)^2}$$

and find the normalization factor N. Evaluate $\langle E \rangle$ and estimate the value of the spread in energy. This functional form for $A(E)$ is called a *Lorentzian* line shape and is relevant for describing the spectral line shapes of unstable quantum states.

P5.5 **The Klein-Gordon equation:** Consider the relativistic wave equation

$$\frac{\partial^2 \phi(x,t)}{\partial t^2} = c^2 \frac{\partial^2 \phi(x,t)}{\partial x^2} - \left(\frac{mc^2}{\hbar}\right)^2 \phi(x,t)$$

(a) Define the probability density for position to be

$$P(x,t) \equiv \frac{\hbar}{2mc^2 i}\left(\phi^*(x,t)\frac{\partial \phi(x,t)}{\partial t} - \frac{\partial \phi^*(x,t)}{\partial t}\phi(x,t)\right)$$

and show that the corresponding equation of continuity can be written as

$$\frac{\partial P(x,t)}{\partial t} = \frac{\partial j(x,t)}{\partial x}$$

where

$$j(x,t) \equiv \frac{\hbar}{2mi}\left(\phi^*(x,t)\frac{\partial \phi(x,t)}{\partial x} - \frac{\partial \phi^*(x,t)}{\partial x}\phi(x,t)\right)$$

(b) Evaluate the probability density for a solution of the form

$$\phi(x,t) = u(x)e^{-iEt/\hbar}$$

Is the probability density positive definite no matter what the sign of E? Compare this to the probability density defined via Eqn. (5.2) for the Schrödinger equation.

P5.6 What is the probability flux for a plane wave of the form

$$\psi_p(x,t) = Ae^{i(\pm px - p^2 t/2m)/\hbar}$$

P5.7 Consider a particle moving in an imaginary potential $V(x) = -iV_I$. What are the possible plane wave solutions of the Schrödinger equation? What would happen to a wavepacket that impinged on a region with this potential? What does such a potential have to do with absorption?

P5.8 **Free-particle wavepackets in position space:** Using the explicit form of the position space free-particle Gaussian wavepacket of Eqn. (3.28), calculate

(a) $\langle x \rangle_t$ and $\langle x^2 \rangle_t$

(b) $\langle \hat{p} \rangle_t$ and $\langle \hat{p}^2 \rangle_t$

(c) Δx and Δp

(d) $\langle \hat{E} \rangle_t$, $\langle \hat{E}_t^2 \rangle$, and ΔE.

P5.9 **Free-particle wavepackets in momentum space:** Repeat P5.7, but use the momentum space wavefunction, namely,

$$\phi(p,t) = e^{-\alpha^2(p-p_0)^2/2} e^{-ip^2 t/2m\hbar} \tag{5.130}$$

Are all your answers the same?

P5.10 Using the Gaussian free-particle wavepacket, evaluate $\mathcal{T}(x,t)$, and show that it is given by

$$\mathcal{T}(x,t) = \frac{(p_0^2 + x^2/(\alpha^2\hbar)^2)}{2m}\left(\frac{1}{1+t^2/t_0^2}\right)|\psi(x,t)|^2$$

Compare the *shapes* of $\mathcal{T}(x,t)$ and $|\psi(x,t)|^2$ in a sketch and discuss. Does your picture look like Fig. 5.6?

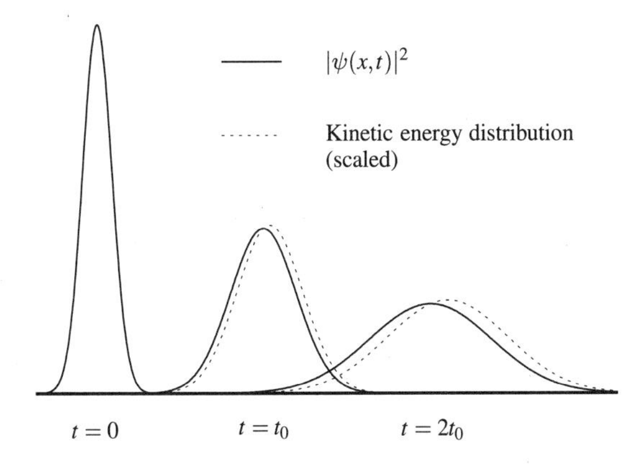

FIGURE 5.6. Kinetic energy distribution, $\mathcal{T}(x, t)$ (dashed), for the Gaussian free-particle wavepacket at several times; the solid curve indicates $|\psi(x, t)|^2$. The two curves are scaled so that they approximately coincide at $t = 0$. Note that the "leading edge" has a higher kinetic energy density than the "trailing edge."

P5.11 Which of the following operators are Hermitian and which aren't?
(a) $\partial/\partial x$
(b) $x \cdot \hat{p}$
(c) $3 - 4i$

P5.12 If the potential energy function is real, show that the Hamiltonian operator,

$$\hat{H} = \frac{\hat{p}^2}{2m} + V(x)$$

is Hermitian.

P5.13 Show that any operator, $\hat{O}$, which satisfies Eqn. (5.51) (and which is therefore Hermitian), also satisfies Eqn. (5.52). *Hint:* You might wish to proceed with the following steps.
(a) Consider a wavefunction $\zeta(x) = \psi(x) + \lambda\phi(x)$, where λ is an arbitrary *complex* number.
(b) Since ζ is an admissible wavefunction (why?), we must have

$$I(\lambda) \equiv \int_{-\infty}^{+\infty} dx\, \zeta^*(x)\hat{O}\zeta(x) = [I(\lambda)]^*$$

so that $I(\lambda) - I(\lambda)^* = 0$.
(c) Use this result and the fact that λ is arbitrary to complete the proof.

P5.14 Consider the *angular momentum operator,* defined by

$$\hat{L}_z = \frac{\hbar}{i}\frac{\partial}{\partial\phi}$$

which acts on angular wavefunctions of the form $Z(\phi)$. To be consistently defined, the wavefunctions should satisfy $Z(\phi + 2\pi) = Z(\phi)$. Explain why this is so, and then use this fact to show that the angular momentum operator is Hermitian.

P5.15 **(a)** Show that the Schrödinger equation in momentum space can be written in terms of an integral equation involving a "nonlocal" potential energy, namely,

$$\left(E - \frac{p^2}{2m}\right)\phi(p) = \int_{-\infty}^{+\infty} V(p - \overline{p})\phi(\overline{p})d\overline{p}$$

where

$$\overline{V}(q) = \frac{1}{2\pi\hbar}\int_{-\infty}^{+\infty} e^{-iqx/\hbar} V(x)\, dx$$

is essentially the Fourier transform of $V(x)$.

(b) Find the form of $\overline{V}(p - \overline{p})$ in the case of the uniformly accelerated particle.

P5.16 Calculate ΔE for the mixed-state wavefunction in Eqn. (5.84).

P5.17 Evaluate the integral in Eqn. (5.118) explicitly.

P5.18 **Accelerating wavepacket in position space:** Using the explicit form of the position-space accelerating Gaussian wavepacket of Eqn. (5.120), evaluate

(a) $\langle x \rangle_t$ and $\langle x^2 \rangle_t$

(b) $\langle \hat{p} \rangle_t$ and $\langle \hat{p}^2 \rangle_t$

(c) Δx and Δp

(d) $\langle \hat{E} \rangle_t$, $\langle \hat{E}^2 \rangle_t$, and ΔE

(e) $\langle \hat{T} \rangle$ and $\langle V(x) \rangle$. Compare their sum to $\langle \hat{E} \rangle$

P5.19 **Accelerating wavepacket in momentum-space:** Repeat P5.18, but use the momentum space wavefunction, namely,

$$\phi(p, t) = \sqrt{\frac{\alpha}{\sqrt{\pi}}}\, e^{-\alpha^2 p^2/2} e^{i((p-Ft)^3 - p^3)/6mF\hbar}$$

P5.20 Calculate the position space wavefunction for the accelerating particle using the general Gaussian momentum distribution of Eqn. (5.115), and show that the probability density is localized at the point

$$x(t) = \frac{F}{2m}t^2 + \frac{p_0}{m}t + x_0$$

P5.21 **Energy eigenstates for the accelerating particle; the Airy differential equation:**

(a) For the free particle, show that the time-independent Schrödinger equation can be written in dimensionless form as

$$\frac{d^2\psi(z)}{dz^2} = -\psi(z)$$

with solutions $\psi(z) = e^{\pm iz}$ or $\cos(z)$ and $\sin(z)$.

(b) Consider the Schrödinger equation for a uniform force F in the negative x direction (i.e., to the *left*[4]) corresponding to vanishing-energy eigenvalue ($E = 0$); the potential energy is then $V(x) = +Fx$. Show that a suitable change of variable reduces it to the form

$$-\frac{d^2\psi(z)}{dz^2} = z\psi(z) = \begin{cases} -|z|\psi(z) & \text{for } z < 0 \\ +|z|\psi(z) & \text{for } z > 0 \end{cases}$$

For $z < 0$ and large in magnitude, show that the solutions have the approximate form

$$\psi(z) = e^{\pm i2z^{3/2}/3} \quad \text{or} \quad \cos\left(\frac{2z^{3/2}}{3}\right), \sin\left(\frac{2z^{3/2}}{3}\right)$$

[4]This is in the opposite direction considered in Sec. 5.6.2.

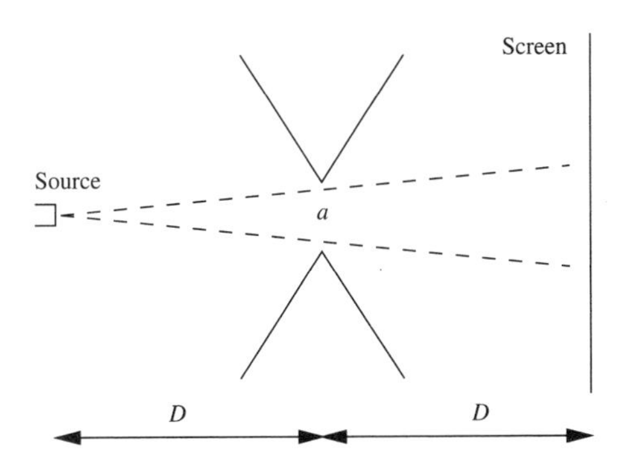

FIGURE 5.7. Schematic experimental apparatus for testing the uncertainty principle.

by direct substitution. This illustrates the increased wiggliness of the solution for large $|z|$ (see Fig. 5.4).

(c) Obtain a more realistic solution for large $|z|$ by assuming a solution of the form

$$\psi(z) = e^{i2z^{3/2}/3}(z^n + a_1 z^{n-1} + a_2 z^{n-2} + \cdots)$$

and showing that the series starts at $n = -1/4$. Thus, for large $|z|$, the probability density goes as

$$P(z) \sim \frac{\sin(2z^{3/2}/3)}{\sqrt{z}}, \frac{\cos(2z^{3/2}/3)}{\sqrt{z}}$$

which also exhibits the (slow) decrease in $|\psi(z)|$ shown in Fig. 5.4.

(d) Show that the kinetic (potential) energy density, $\mathcal{T}(z)$ ($\mathcal{V}(z)$), increases as $\sqrt{z}$ increases in this region and explain why.

(e) For $z > 0$ and $|z| >> 0$, show that the two allowed solutions correspond to exponential growth or attenuation; compare your results to those in Sec. 2.5.3.

P5.22 **(a)** Confirm that $[\hat{x}, p] = i\hbar$ in momentum space.

(b) Show that $[x, \hat{p}^2] = 2i\hbar\hat{p}$ by acting with both sides on an arbitrary function of x.

(c) Evaluate $[\hat{E}, t]$ and $[\hat{E}, x]$

P5.23 Consider the schematic experiment shown in Fig. 5.7 where electrons are emitted from a source, allowed to pass through an aperture of lateral width a, and are detected at a screen a distance D away. A naive analysis of this *gedanken* experiment might come to the conclusion that

$$\Delta p_y \sim \left(\frac{a}{D}\right)p \qquad \text{and} \qquad \Delta y \sim a$$

so that the uncertainty principle product $\Delta y\, \Delta p_y$ could be made arbitrarily small by letting $a \to 0$ and $D \to \infty$. Discuss what is wrong with such an analysis.

CHAPTER 6

The Infinite Well: Physical Aspects

6.1 THE INFINITE WELL IN CLASSICAL MECHANICS

The problem of a particle moving in an infinite well or one-dimensional box, defined by the potential

$$V(x) = \begin{cases} +\infty & \text{for } |x| > a \\ 0 & \text{for } |x| < a \end{cases} \tag{6.1}$$

is one of the most familiar and easiest to solve of all problems in introductory quantum mechanics. It is the simplest case in which to study the phenomenon of quantized energy levels in bound states, one of the "smoking guns" of wave mechanics. In this chapter and the next, we use this problem (and some related ones) as a tool in understanding the formalism used in solving the time-independent Schrödinger equation for bound states in a potential, as well as some of the formal properties of its solutions. In Chap. 8, we also use the infinite well as a model system in which to examine the role that spin and the exclusion principle play in multiparticle quantum systems.

We also wish to develop some intuition about quantum mechanical wavefunctions, both in the "quantum" limit of small quantum number and especially in the quasi-classical limit of large quantum number. To that end, we first discuss the problem of a particle moving in this potential well, treated classically, and introduce the notion of a classical probability distribution.

Classically, a particle in such a potential would move at constant speed inside the box, experiencing elastic collisions with the walls. The speed (v_0) and period (τ) of the motion are easily determined in terms of the energy to be

$$v_0 = \sqrt{2E/m} \qquad \tau = \frac{4a}{v_0} = 4a\sqrt{\frac{m}{2E}} \tag{6.2}$$

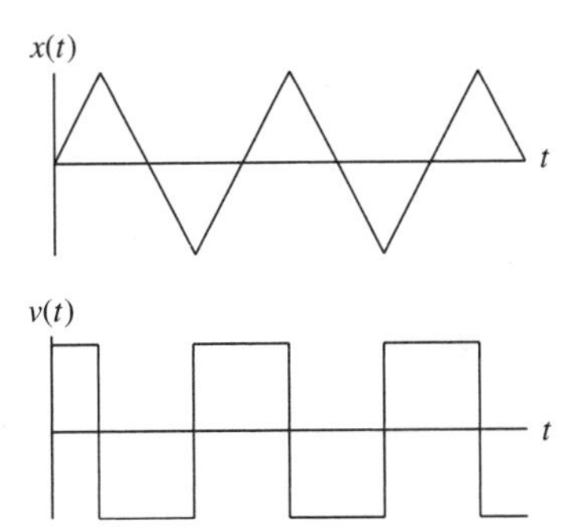

FIGURE 6.1. *Classical trajectories in the infinite well: x(t) and v(t) versus t.*

The position and velocity of a particle in such a well are shown in Fig. 6.1. We see that measurements of the velocity of the particle at any time will find only the values $\pm v_0$.

Although we know the exact trajectory, $x(t)$, for all times, we can still ask for the probability that a measurement of the position of the particle (using, for example, a large number of stroboscopic photographs of the system taken at random times) will find it in a given region inside the well. Since the particle moves at constant speed, and therefore spends equal amounts of time in all regions of the well, we must have

$$P_{CL}(x)\,dx \equiv \text{Probability}[(x,\ x+dx)] = C\,dx \tag{6.3}$$

where C is a constant, and we have introduced the notion of a *classical probability distribution, $P_{CL}(x)$*. Since the particle must be found somewhere in the box, $P_{CL}(x)$ must be normalized in the usual way so that

$$\int_{-a}^{+a} P_{CL}(x)\,dx = 1 \tag{6.4}$$

which implies that $C = 1/2a$. More generally, for motion between two classical turning points, a and b, we require

$$\int_{a}^{b} P_{CL}(x)\,dx = 1 \tag{6.5}$$

Using this classical probability distribution, we can calculate average values as usual and find, for example,

$$\langle x\rangle_{CL} = \int_{-a}^{+a} x\,P_{CL}(x)\,dx = 0 \tag{6.6}$$

$$\langle x^2\rangle_{CL} = \int_{-a}^{+a} x^2 P_{CL}(x)\,dx = \frac{1}{2a}\int_{-a}^{+a} x^2\,dx = \frac{a^2}{3} \tag{6.7}$$

so that $\Delta x_{CL} = a/\sqrt{3}$.

One can "experimentally" (via computer) generate this probability distribution by repeatedly sampling the position of the particle and binning the data, and the resulting probability densities are shown in Fig. 6.2 for increasing numbers of measurements. We also indicate the experimental value of the average position and RMS deviation obtained from the data. (We have chosen $a = 1$ for definiteness.) Clearly, the classical probability and averages are found after many such measurements. We note that the deviation of the

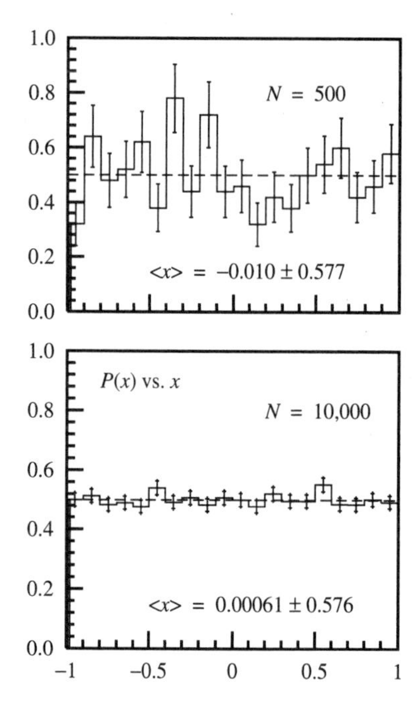

FIGURE 6.2. Statistical study of a classical particle in an infinite well. $P_{CL}(x)$ is "measured" (via computer simulation) by binning measurements of the position of the particle; the number of trials is $N = 500$ and 10,000. The average value of x and its RMS deviation are also calculated. The "data" approach the flat distribution.

estimated average (after N trials) from the real average (i.e., 0) decreases as $1/\sqrt{N}$, as expected.

One can also define a classical probability distribution for momentum (see Sec. 10.4.2 for a fuller discussion), $P_{CL}(p)$, and in this simple case where only the values $p = \pm p_0 = \pm v_0/m$ are allowed, with equal probability, such a distribution might be written as

$$P_{CL}(p) = \frac{1}{2}(\delta(p - p_0) + \delta(p + p_0)) \tag{6.8}$$

and we can easily find that $\langle p \rangle_{CL} = 0$ and $\Delta p_{CL} = p_0$.

6.2 CLASSICAL PROBABILITY DISTRIBUTIONS

The notion of a classical probability distribution[1] can be generalized to any bound state problem with a potential $V(x)$. Consider a particle of mass m and energy E moving in a general confining potential such as in Fig. 6.3 (for the case E_1 at least). The motion in such a potential is periodic, the particle bouncing back and forth between the classical turning points a and b; the time for one traversal of the well (from, say, left to right) is half the period, $\tau/2$. The amount of time, dt, the particle spends in the small region of space, dx, near the point x is given by the speed there, $v(x)$, via

$$dt = \frac{dx}{dx/dt} = \frac{dx}{v(x)} \tag{6.9}$$

[1]The comparison of classical and quantum probability distributions, in both position and momentum space, for many simple systems is discussed in Robinett (1995).

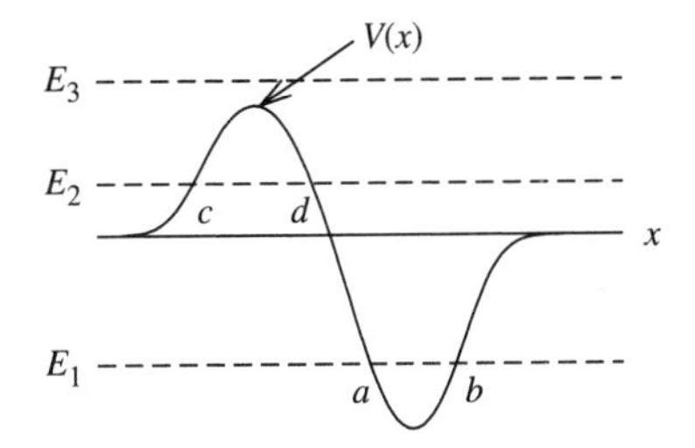

FIGURE 6.3. *Generic one-dimensional potential with bound states. For energy E_1, a and b are the classical turning points for bound motion. For energy E_2, c and d are turning points for unbound motion; particles incident from the left (right) will "rebound" at point c (d). For energy E_3, particles will slow down (speed up) as they travel over the "bump" ("well") in the potential but will not "rebound."*

The probability of finding the particle in this small region is simply the ratio of this time to the total time for one traversal, that is,

$$P_{CL}(x)\,dx \equiv \text{Probability}[(x,\ x + dx)] = \frac{dt}{\tau/2} = \frac{2}{\tau}\frac{dx}{v(x)} \tag{6.10}$$

so that

$$P_{CL}(x) = \frac{2}{\tau}\frac{1}{v(x)} \tag{6.11}$$

This definition shows that the particle will spend more time (and hence be measured more often) in regions where the classical speed is low. Since the classical speed is related to the kinetic energy by $T(x) = mv(x)^2/2$ and hence to the potential energy, we can write

$$P_{CL}(x) = \frac{2}{\tau}\sqrt{\frac{m}{2T(x)}} = \frac{2}{\tau}\sqrt{\frac{m}{2(E - V(x))}} \tag{6.12}$$

Thus the classical probability density is large (small) where the kinetic energy is small (large) or the potential energy is large (small). For the case of the infinite well, where the potential vanishes inside the well, this reduces to

$$P_{CL}(x) = \frac{2}{\tau}\sqrt{\frac{m}{2E}} = \frac{1}{2a} \tag{6.13}$$

as expected.

Example 6.1 Classical Probabilities for the Harmonic Oscillator For the case of the simple harmonic oscillator (which we will consider in its full quantum detail in Chap. 10), we have $V(x) = Kx^2/2$, $\tau = 2\pi\sqrt{m/K}$, and the amplitude A is determined from the total energy via $E = KA^2/2$; this is shown in Fig. 6.4. In this case we find

$$P_{CL}(x) = \frac{1}{\pi}\frac{1}{\sqrt{A^2 - x^2}} \tag{6.14}$$

with the classical turning points at $(-A, A)$ and one can easily check (P6.3) that this is properly normalized. This probability distribution is plotted in Fig. 6.5 along with the quantum mechanical probability distributions $P_{QM}(x; n) \equiv |\psi(x; n)|^2$ (derived in Sec. 10.2) for a relatively large value of the quantum number.

It seems that the quantum mechanical probability density is approaching the classical result in a locally averaged sense. One way to define such an averaging is to say that the limit of the quantum mechanical probability in a small, but finite, interval will approach the classical result as the quantum number gets large, i.e.,

$$\lim_{n\to\infty}\int_{x_1}^{x_2} |\psi_n(x)|^2\,dx = \int_{x_1}^{x_2} P_{CL}(x)\,dx \tag{6.15}$$

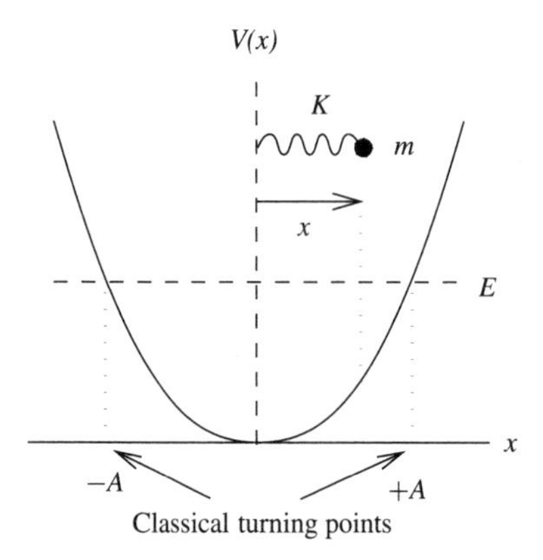

FIGURE 6.4. *Potential for the harmonic oscillator, $V(x) = Kx^2/2$; the classical turning points, $\pm A = \sqrt{2E/K}$, correspond to the maximum stretch of the spring.*

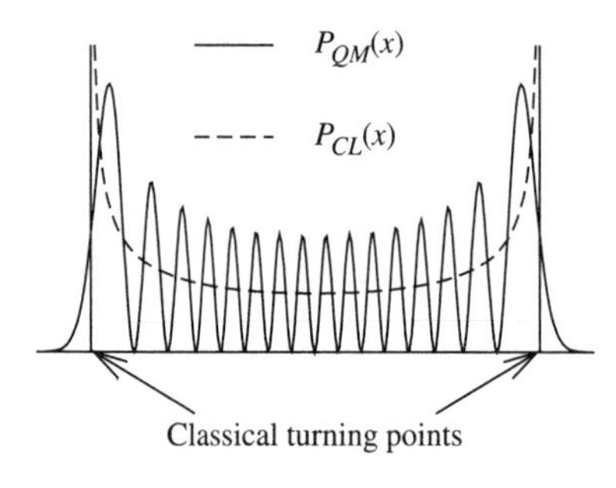

FIGURE 6.5. *Quantum (solid) and classical (dotted) probability distributions for simple harmonic oscillator; the quantum state with $n = 15$ is shown. The vertical lines represent the classical turning points.*

We also defined in Sec. 5.2.3 a kinetic energy distribution

$$\mathcal{T}(x, t) = \frac{\hbar^2}{2m}\left|\frac{\partial\psi(x, t)}{\partial x}\right|^2 \tag{6.16}$$

which implied that the spatial rate of change of the wavefunction (its wiggliness) is a measure of the kinetic energy of the particle. Crudely,

- The wavefunction $\psi(x)$ will have many (few) nodes in a region of space where the classical kinetic energy is large (small).

This feature of our growing body of "intuition" about quantum mechanical solutions[2] is also illustrated in Fig. 6.5 where the wavefunction is most (least) wiggly at the center (turning points) where the classical speed would be largest (smallest). (The classical analogs of $\mathcal{T}$ and the potential energy distribution $\mathcal{V}$ will be discussed in Sec. 10.4.1.)

Classical probability distributions, $P_{CL}(x)$, can also be easily visualized by a simple projection technique.[3] One projects bins of equal size in x on the trajectory curve (i.e., $x(t)$ versus t) onto the time axis; measurements of the time spent in each small interval can then be binned to estimate the probability density. The method is illustrated in Fig. 6.6 for the harmonic oscillator whose trajectory is a simple sinusoidal curve, $x(t) = A\cos(\omega t + \phi)$. Note again the relatively long time spent near the turning points.

[2]The use of qualitative plots to illustrate the behavior of bound state wavefunctions was pioneered by French and Taylor (1971).

[3]This technique is sometimes used in the analysis of chaotic systems.

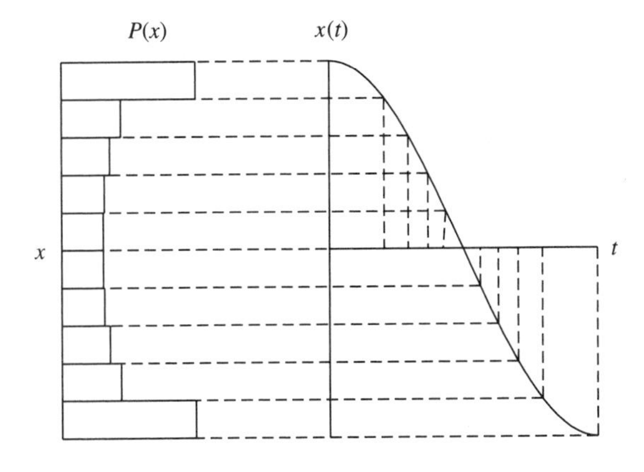

FIGURE 6.6. Projection of trajectory yielding classical position probability distribution for the simple harmonic oscillator. Equal sized bins, dx, in x are projected onto the t axis; the histogramed areas are then proportional to the time, dt, in that bin.

6.3 STATIONARY STATES FOR THE INFINITE WELL

6.3.1 The "Standard" Infinite Well

We now discuss the solution of the quantum mechanical problem of a particle in an infinite well potential. We begin by focusing on the potential well of the form

$$V(x) = \begin{cases} +\infty & \text{for } x < 0,\ x > a \\ 0 & \text{for } 0 < x < a \end{cases} \tag{6.17}$$

The time-independent Schrödinger equation becomes

$$-\frac{\hbar^2}{2m}\frac{d^2\psi(x)}{dx^2} = E\psi(x) \tag{6.18}$$

inside the well. This is of the form

$$\frac{d^2\psi(x)}{dx^2} = -k^2\psi(x) \qquad \text{where} \qquad k = \sqrt{\frac{2mE}{\hbar^2}} \tag{6.19}$$

Since the particle is not allowed outside (i.e., $\psi(x) = 0$ for $x < 0,\ x > a$), and the wavefunction should be continuous, we also must implement the boundary conditions $\psi(0) = \psi(a) = 0$. The solutions of Eqn. (6.19) which are most like standing waves are

$$\psi(x) = A\sin(kx) + B\cos(kx) \tag{6.20}$$

and application of the boundary conditions requires that

$$\begin{aligned} \psi(0) &= A\sin(0) + B\cos(0) = 0 \\ \psi(-a) &= A\sin(ka) + B\cos(ka) = 0 \end{aligned} \tag{6.21}$$

or

$$B = 0 \qquad \text{and} \qquad A\sin(ka) = 0 \tag{6.22}$$

If $A = 0$ as well, then $\psi(x)$ vanishes identically (the uninteresting case of no particle in the well), so we must have

$$\sin(ka) = 0 \qquad \text{or} \qquad k_n a = n\pi \qquad n = 1, 2, 3, \ldots \tag{6.23}$$

giving the quantized energies

$$E_n = \frac{\hbar^2 k_n^2}{2m} = \frac{\hbar^2 n^2 \pi^2}{2ma^2} \tag{6.24}$$

The corresponding normalized wavefunctions are given by

$$u_n(x) = \sqrt{\frac{2}{a}} \sin\left(\frac{n\pi x}{a}\right) \tag{6.25}$$

where the coefficients are determined by demanding that the wavefunctions be suitably normalized. We note that this normalization has the appropriate dimensionality for a one-dimensional wavefunction. The sign (or more generally the phase) of this normalization constant is purely conventional, and $-\sqrt{2}/\sqrt{a}$ or $\exp(i\theta)\sqrt{2}/\sqrt{a}$ would serve just as well.

These results exemplify a standard analytic approach to the solutions of a quantum mechanical problem of a particle in a potential well:

1. One solves the time-independent Schrödinger equation in position space, which gives the appropriate functional forms for $\psi(x)$.
2. One applies the boundary conditions that usually involve the condition, either explicitly or implicitly, that the wavefunction vanish sufficiently rapidly at infinity and gives rise to quantized energy levels. This is the analog of "fitting de Broglie waves," as in Eqn. (1.36), for a general potential.
3. Finally, one normalizes the wavefunctions to ensure that a probability interpretation is valid. Other information, such as average values of various operators or the corresponding momentum space wavefunction, $\phi(p)$, are then easily calculated.

Each step is independent of the others, and all are required to obtain the full information possible from the stationary-state solutions.

The most general solutions will be linear combinations of these energy eigenstates with their associated time dependence. The complete stationary-state solutions, including time dependence, are then given by

$$\psi_n(x,t) = u_n(x)e^{-iE_nt/\hbar} \tag{6.26}$$

The first few spatial wavefunctions, $u_n(x)$, (which we recall can be made purely real), are shown in Fig. 6.7 (along with those for the related symmetric infinite well of the next section). We note the general feature that the number of nodes increases with energy, as suggested by Eqn. (6.16), starting with a nodeless ground state. We stress that, because these are stationary-state solutions, the squared moduli of the wavefunctions, $|\psi_n(x,t)|^2$, are time-independent.

6.3.2 The Symmetric Infinite Well

A variation on this problem which we will find useful is the symmetric infinite well, defined by Eqn. (6.1). We discuss this seemingly similar case for several reasons:

1. The two potentials (and their solutions) can be obtained from each other by simple scaling arguments.
2. The symmetric infinite well introduces us to the notion of parity, which we will study further in Sec. 7.7.

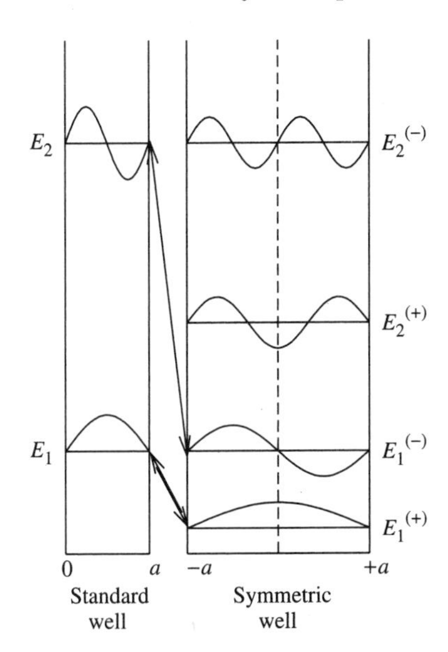

FIGURE 6.7. Energy eigenvalues and eigenfunctions for the standard and "symmetric" infinite well.

3. The momentum space solutions are somewhat more easily obtained and visualized in the symmetric case.

In this case we find the same Schrödinger equation and solutions, but the boundary conditions are implemented in a slightly different way. Now because $\psi(x) = 0$ for $|x| > a$, we impose the conditions

$$\begin{aligned} \psi(a) &= A\sin(ka) + B\cos(ka) = 0 \\ \psi(-a) &= -A\sin(ka) + B\cos(ka) = 0 \end{aligned} \tag{6.27}$$

or

$$A\sin(ka) = 0 \qquad \text{and} \qquad B\cos(ka) = 0 \tag{6.28}$$

Once again, if $A = B = 0$, then $\psi(x)$ vanishes identically, so we can consider two cases separately:

A = 0, Even Solutions In this case, only the $\cos(kx)$ (i.e., the even) solutions survive, and we set $\cos(ka) = 0$, which has solutions

$$k_n^{(+)} = (n - 1/2)\pi \qquad n = 1, 2, \ldots \tag{6.29}$$

so that the energy eigenvalues are

$$E_n^{(+)} = \frac{\hbar^2 (k_n^{(+)})^2}{2m} = \frac{\hbar^2 (2n-1)^2 \pi^2}{8ma^2} \tag{6.30}$$

where the $^{(+)}$ superscript denotes the even states. The corresponding eigenfunctions are

$$u_n^{(+)}(x) = \frac{1}{\sqrt{a}} \cos\left(\frac{(n-1/2)\pi x}{a}\right) \tag{6.31}$$

The odd states are obtained in a similar fashion.

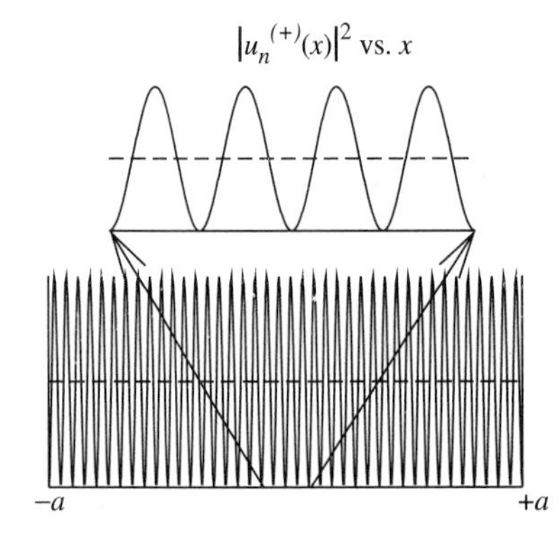

FIGURE 6.8. Probability density ($P_{QM}(x) = |\psi(x)|^2$) for the symmetric infinite well for large quantum number, $n = 40$. $P_{QM}(x)$ averages locally to the "flat" classical probability distribution.

B = 0, Odd Solutions In this case where only the $\sin(kx)$ (or the odd) solutions survive, we set $\sin(ka) = 0$, which has solutions

$$k_n^{(-)} = n\pi \qquad n = 1, 2, \ldots \tag{6.32}$$

with energy eigenvalues

$$E_n^{(-)} = \frac{\hbar^2 n^2 \pi^2}{2ma^2} \tag{6.33}$$

and eigenfunctions

$$u_n^{(-)}(x) = \frac{1}{\sqrt{a}} \sin\left(\frac{n\pi x}{a}\right) \tag{6.34}$$

We note that these solutions can be obtained from those for the standard well, in Eqn. (6.25), by first letting $a \to 2a$ (doubling the width of the well) and then letting $x \to x - a$ (shifting the center to the origin.); see Fig. 6.7 for a comparison. These solutions possess the additional property of evenness and oddness, obviously related to the symmetry of the well, and we will discuss the notion of parity in the next chapter.

To compare with the classical probability distribution discussed in Sec. 6.1, we plot the probability density corresponding to the $n = 40$ odd state in Fig. 6.8 with the classical (constant) probability distribution superimposed. The local averaging of Eqn. (6.15) can be easily implemented in this case (P6.10), and we see that the appropriate limit is reached.

To make contact with the classical concept of velocity, we can calculate the momentum space wavefunctions corresponding to the $u_n^{(\pm)}(x)$; we consider the even parity states for definiteness. We find

$$\begin{aligned}\phi_n^{(+)}(p) &= \frac{1}{\sqrt{2\pi\hbar}} \int_{-a}^{+a} dx\, u_n^{(+)}(x) e^{-ipx/\hbar} \\ &= \sqrt{\frac{a}{2\pi\hbar}} \left(\frac{\sin((n-1/2)\pi - ap/\hbar)}{((n-1/2)\pi - ap/\hbar)} + \frac{\sin((n-1/2)\pi + ap/\hbar)}{((n-1/2)\pi + ap/\hbar)} \right)\end{aligned} \tag{6.35}$$

We plot the momentum space probability distributions for even states in Fig. 6.9 for $n = 1, 3$, and 5, and note that two well-defined peaks at $p = \pm(n - 1/2)\hbar/a$ are observed, except for the ground state. This is just as expected from the classical picture of finding only $\pm v_0 = \pm p/m = \pm\sqrt{2E/m}$ as a result of a velocity measurement. One can write the

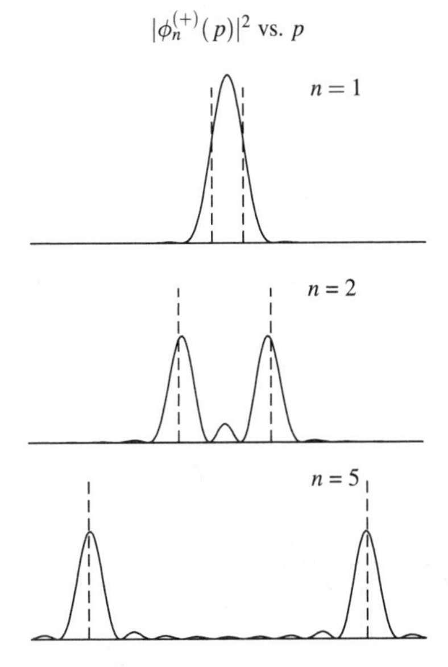

FIGURE 6.9. Momentum space probability distribution, $|\phi_n^{(+)}(p)|^2$ versus p, for the symmetric infinite well; values of n = 1, 2, 5 are shown. The dotted lines correspond to the values of p given by $p_n = \pm\sqrt{2mE_n^{(+)}}$.

momentum space wavefunction in the form

$$\phi(p) = \sqrt{\frac{\Delta p}{2\pi}}\left(\frac{\sin((p_0 - p)/\Delta p)}{(p_0 - p)} + \frac{\sin((p_0 + p)/\Delta p)}{(p_0 + p)}\right) \tag{6.36}$$

where $p_0 = (n - 1/2)\pi\hbar/a$ and $\Delta p \equiv \hbar/a$. If we take the corresponding momentum probability distribution, $|\phi(p)|^2$, in the limit that $\Delta p \to 0$ (i.e., a macroscopic limit) and use the representation of the δ function in App. C.8, we find (P6.12) that

$$\lim_{\hbar/a \to 0} |\phi(p)|^2 = \frac{1}{2}(\delta(p - p_0) + \delta(p + p_0)) = P_{CL}(p) \tag{6.37}$$

as discussed above.

The symmetry of these momentum distributions, namely, that

$$|\phi(p)|^2 = |\phi(-p)|^2 \tag{6.38}$$

suggests that the average value of momentum should vanish. If we calculate this quantity for even states in the position space representation, we indeed find that

$$\langle \hat{p} \rangle = \int_{-\infty}^{+\infty} dx\, \psi^*(x)\, \hat{p}\, \psi(x) = \frac{\hbar}{i}\int_{-a}^{+a} dx\, u_n^{(+)} \frac{du_n^{(+)}(x)}{dx} = 0 \tag{6.39}$$

with a similar result for the odd $u_n^{(-)}(x)$ states as well. This result might wrongly be attributed solely to the overall oddness of the integrand (even $\psi(x)$ times odd $\psi'(x)$) which is, in turn, related to the symmetric nature of the potential, but it is, in fact, a much more general result.

The vanishing of $\langle \hat{p} \rangle$ for generic energy eigenfunctions for *bound* states in potential wells is more directly related to the fact that the stationary-state wavefunctions in such cases can be chosen to be a purely real function (or real with multiplicative complex phases that are independent of position). It is easy to show (P6.10) that the average value of the momentum operator must vanish when evaluated with any purely real $u(x)$, even if one includes a time-dependent phase factor $\exp(-iEt/\hbar)$.

This can be thought of heuristically in classical terms as the symmetry in velocity during subsequent "back and forth" traversals of a potential during its periodic motion; the same *speed* is achieved twice during each period. The classical velocity is related to the energy and potential at a point via

$$v(x) = \pm\sqrt{\frac{2}{m}(E - V(x))} \tag{6.40}$$

so that the classical velocity distribution is necessarily symmetric. One can also show that the probability flux or current for energy eigenstates also vanishes. We stress that these results are valid only for stationary states and not for arbitrary, time-dependent solutions.

6.4 VARIATIONS ON THE INFINITE WELL

Before discussing further aspects of these solutions, we can consider two relatively straightforward generalizations of the standard infinite well problem as further illustrations of methods of solution and our "intuitive" approach to the understanding of wavefunctions.

6.4.1 The "Asymmetric" Infinite Well

The asymmetric infinite well potential, shown in Fig. 6.10, can be defined by

$$V(x) = \begin{cases} +\infty & \text{for } x < -a,\ x > b \\ 0 & \text{for } -a < x < 0 \\ V_0 & \text{for } 0 < x < b \end{cases} \tag{6.41}$$

and we will consider the case where $E > V_0$ only. This problem can be analyzed as above, with general solutions

$$\psi(x) = \begin{cases} 0 & \text{for } x < -a,\ x > b \\ A'\sin(kx) + C'\cos(kx) & \text{for } -a < x < 0 \\ B'\sin(qx) + D'\cos(qx) & \text{for } 0 < x < b \end{cases} \tag{6.42}$$

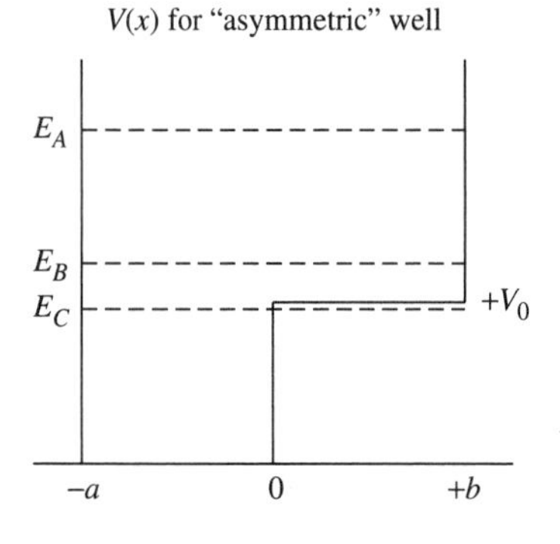

FIGURE 6.10. *Potential energy function $V(x)$ for the asymmetric well with $a = b$. The position space and momentum space probabilities for the three energies $E_{A,B,C}$ are shown in Figs. 6.11 and 6.12.*

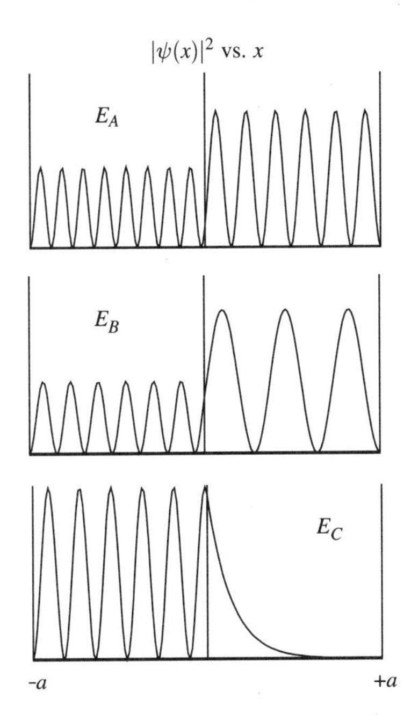

FIGURE 6.11. $|\psi(x)|^2$ *versus x for the three energies in the "asymmetric" infinite well.*

where k is defined as before, but now

$$q = \sqrt{\frac{2m(E - V_0)}{\hbar^2}} \tag{6.43}$$

where q is real, by assumption. The matching of the wavefunction at $x = -a, b$ implies a considerable simplification, giving

$$\psi(x) = \begin{cases} A\sin(k(x+a)) & \text{for } -a < x < 0 \\ B\sin(q(x-b)) & \text{for } 0 < x < b \end{cases} \tag{6.44}$$

Matching both $\psi(x)$ and $\psi'(x)$[4] at $x = 0$ then yields the energy eigenvalue condition, namely,

$$k\cos(ka)\sin(qb) = -q\sin(ka)\cos(qb) \tag{6.45}$$

which in the limit $V_0 \to 0$ and $a = b$ yields $\cos(2ka) = 0$, consistent with our earlier discussion for a well now of width $2a$.

In Fig. 6.11, we plot several different solutions corresponding to $V_0 > 0$ for $a = b$. We note that the relative magnitude of the wavefunctions squared in the various regions is consistent with our arguments above, using considerations based on classical probability distributions, as is the density of the nodes of the wavefunctions—representing the wiggliness of $\psi(x)$, i.e., $|\psi|$ is large (small)—and "not wiggly" ("wiggly") where the classical kinetic energy is small (large). We also show a solution with $0 < E \lesssim V_0$, and note that the wavefunctions and corresponding probability densities are nonvanishing outside of the classically allowed region. We will return to this new phenomenon, called *quantum tunneling,* in Chap. 9.

[4]The discontinuity in $V(x)$ at $x = 0$ is not "bad enough" to make $\psi'(x)$ discontinuous; see Sec. 9.1.1.

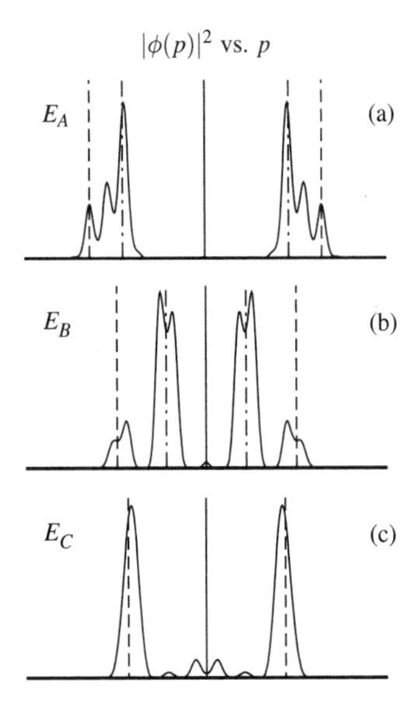

FIGURE 6.12. *$|\phi(p)|^2$ versus p for the three energies in the "asymmetric" infinite well. The vertical dashed (dotted) lines correspond to the values of $p = \pm\sqrt{2mE/\hbar^2}$ ($p = \pm\sqrt{2m(E - V_0)/\hbar^2}$), which are the allowed classical values.*

In Fig. 6.12, we also show the corresponding momentum space probability densities obtained by taking the Fourier transforms of Eqn. (6.44). For Fig. 6.12 (b) there are *four* distinct peaks corresponding recognizably to back-and-forth motion in the two different regions, and these are consistent with the classical values of p indicated by the dashed and dotted vertical lines. For the higher-energy case in Fig. 6.12(a), the two main peaks on each side have begun to coalesce as the difference in kinetic energies (or difference between k and q) becomes less important. Finally, for case (c) where the particle would classically be only in the left-hand side of the well, there are only two distinctive peaks. In each case, we also see the inherent symmetry in $|\phi(p)|^2$ under $p \to -p$ expected for stationary states.

6.4.2 The "Linear Infinite Well"

As a final example, we consider a "linear well" as shown in Fig. 6.13, defined by

$$V(x) = \begin{cases} +\infty & \text{for } x < 0 \\ +Fx & \text{for } 0 < x < a \\ +\infty & \text{for } x > a \end{cases} \tag{6.46}$$

This problem is sometimes described[5] as "the quantum bouncer in a closed court" as it can represent a ball acting under the influence of gravity, bouncing elastically from a flat floor (at $x = 0$) and rebounding from a flat ceiling (at $x = a$). In this case, the differential equation inside the well is given by

$$-\frac{\hbar^2}{2m}\frac{d^2\psi(x)}{dx^2} + Fx\psi(x) = E\psi(x) \tag{6.47}$$

The method of solution is very similar to that discussed in Sec. 5.6.2 with only the sign of F changed. The same change to dimensionless variables, $x = \rho y + \sigma$, now with

[5]For a discussion, see Aguilera-Navarro et al. (1981).

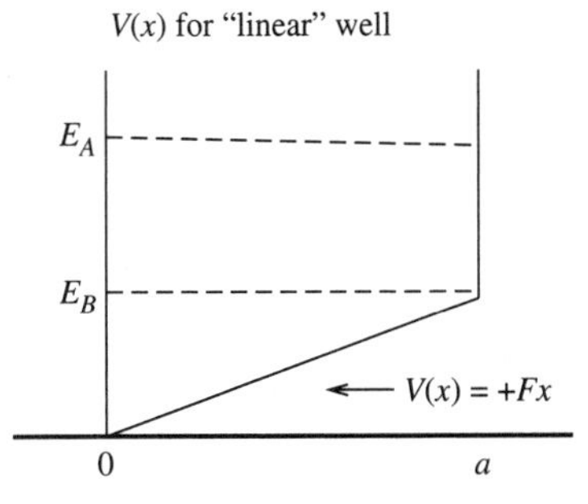

FIGURE 6.13. *Potential energy function $V(x)$ for the linear infinite well with two energy eigenvalues satisfying $E > Fa$.*

$$\rho^3 = \frac{\hbar^2}{2mF} \quad \text{and} \quad \sigma = \frac{E}{F} \tag{6.48}$$

reduces Eqn. (6.47) to the Airy equation (App. C.2):

$$\frac{d^2\psi(y)}{dy^2} = y\psi(y) \tag{6.49}$$

Because the behavior of the $Bi(y)$ at infinity is no longer a problem (as we're confined to the finite well), we must retain both linearly independent, oscillatory solutions and write

$$\psi(y) = CAi(y) + DBi(y) \tag{6.50}$$

so that

$$\psi(x) = CAi\left(\frac{x-\sigma}{\rho}\right) + DBi\left(\frac{x-\sigma}{\rho}\right) \qquad \text{for } 0 < x < a \tag{6.51}$$

Application of the boundary conditions $\psi(0) = \psi(a) = 0$ yields the eigenvalue condition

$$Ai\left(-\frac{\sigma}{\rho}\right)Bi\left(\frac{a-\sigma}{\rho}\right) - Bi\left(-\frac{\sigma}{\rho}\right)Ai\left(\frac{a-\sigma}{\rho}\right) = 0 \tag{6.52}$$

This transcendental equation can be solved numerically for the energy eigenvalues (contained, we recall, in the values of σ) and the resulting wavefunctions obtained.

Plots of $|\psi(x)|^2$ for the fourth and eighth excited states are shown in Fig. 6.14. Once again the pattern of increasing numbers of nodes and their spatial distribution, as well as the size of the wavefunction, is consistent with our discussions above. The particle has the most (least) kinetic energy on the left (right) side of the well where the wavefunction has the highest (lowest) density of nodes and where $|\psi(x)|^2$ is smallest (largest).

6.5 TIME DEPENDENCE OF GENERAL SOLUTIONS

In all of the variations of the infinite well problem we have discussed, we have focused on the stationary-state solutions and their physical interpretation. Because of the linearity of the Schrödinger equation, the most general time-dependent solution in each case will consist of a linear combination of such solutions. In the case of the standard well, for example, this implies that

$$\psi(x,t) = \sum_{n=1}^{+\infty} a_n u_n(x) e^{-iE_n t/\hbar} \tag{6.53}$$

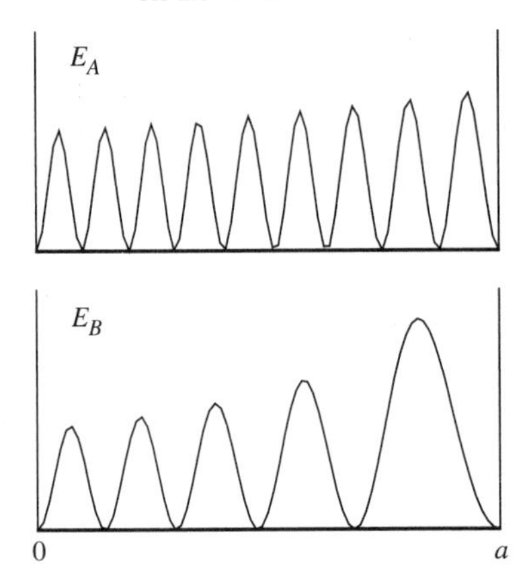

FIGURE 6.14. *Position space probability distribution, $|\psi(x)|^2$, for the two energies in Fig. 6.13.*

while for the symmetric well,

$$\psi(x,t) = \sum_{n=1}^{+\infty} \left(a_n^{(+)} u_n^{(+)}(x) e^{-iE_n^{(+)}t/\hbar} + a_n^{(-)} u_n^{(-)}(x) e^{-iE_n^{(-)}t/\hbar} \right) \tag{6.54}$$

where the a_n or $a_n^{(\pm)}$ are, at the moment, arbitrary (possibly complex) numbers. As a simple example of the more complicated time development possible with such general solutions, let us consider an equally weighted combination of the ground and first excited-state wavefunctions of the symmetric well, $\psi_C(x,t)$ (C stands for combination), i.e.,

$$\psi_C(x,t) = \frac{1}{\sqrt{2}} \left(u_1^{(+)}(x) e^{-iE_1^{(+)}t/\hbar} + u_1^{(-)}(x) e^{-iE_1^{(-)}t/\hbar} \right) \tag{6.55}$$

i.e., $a_1^{(+)} = a_1^{(-)} = 1\sqrt{2}$. (The reason for this choice of normalization will become clear shortly.) Recall that

$$E_1^{(+)} = \frac{\hbar^2\pi^2}{8ma^2} \qquad E_1^{(-)} = \frac{4\hbar^2\pi^2}{8ma^2} \tag{6.56}$$

and

$$u_1^{(+)}(x) = \frac{1}{\sqrt{a}} \cos\left(\frac{\pi x}{2a}\right) \qquad u_1^{(-)}(x) = \frac{1}{\sqrt{a}} \sin\left(\frac{\pi x}{a}\right) \tag{6.57}$$

The probability density for this solution has nontrivial time dependence, since

$$\begin{aligned} P_C(x,t) &= |\psi_C(x,t)|^2 \\ &= \frac{1}{2}\left([u_1^{(+)}(x)]^2 + [u_1^{(-)}(x)]^2 + 2[u_1^{(+)}(x)][u_1^{(-)}(x)] \cos(\Delta E t/\hbar) \right) \end{aligned} \tag{6.58}$$

where $\Delta E \equiv E_1^{(-)} - E_1^{(+)} = 3\hbar^2\pi^2/8ma^2$. A simple calculation (P6.15) shows that the average position of the particle described by this wavefunction oscillates, alternating between being more or less localized on the right and left sides of the well, specifically

$$\langle x \rangle_t = a\left(\frac{32}{9\pi^2}\right) \cos\left(\frac{\Delta E t}{\hbar}\right) \tag{6.59}$$

The corresponding average value of the momentum operator can be calculated in the standard way, and one finds

$$\langle \hat{p} \rangle_t = \int_{-\infty}^{+\infty} dx\, \psi_C^*(x,t)\hat{p}\psi_C(x,t) = -\left(\frac{4}{3}\frac{\hbar}{a}\right)\sin\left(\frac{\Delta E t}{\hbar}\right) \tag{6.60}$$

so that $\langle \hat{p} \rangle_t = md\langle x \rangle_t/dt$, as it must. The corresponding momentum space wavefunction for this state, obtained by taking the Fourier transform, shows similar structure:

$$\begin{aligned} P_C(p,t) &= |\phi_C(p,t)|^2 \\ &= \frac{1}{2}\left(\phi_1^{(+)}(p)^2 + \overline{\phi}_1^{(-)}(p)^2 + 2\phi_1^{(+)}(p)\overline{\phi}_1^{(-)}(p)\sin(\Delta E t/\hbar)\right) \end{aligned} \tag{6.61}$$

where $\phi_1^{(+)}(p)$ was defined in Eqn. (6.36), while

$$\phi_1^{(-)}(p) \equiv -i\overline{\phi}_n^{(-)}(p) \tag{6.62}$$

is the Fourier transform of the odd states, and we extract a factor of $-i$ for convenience. We have thus defined

$$\overline{\phi}_n^{(-)}(p) = \sqrt{\frac{a}{2\pi\hbar}}\left(\frac{\sin((n-1/2)\pi - ap/\hbar)}{((n-1/2)\pi - ap/\hbar)} - \frac{\sin((n-1/2)\pi + ap/\hbar)}{((n-1/2)\pi + ap/\hbar)}\right) \tag{6.63}$$

Finally, the probability flux can be calculated using Eqn. (5.12), and one finds

$$j(x,t) = F(x)\sin\left(\frac{\Delta E t}{\hbar}\right) \tag{6.64}$$

where

$$F(x) = -\frac{\hbar\pi}{2ma^2}\left(\cos(\pi x/2a)\cos(\pi x/a) + \frac{1}{2}\sin(\pi x/2a)\sin(\pi x/a)\right) \tag{6.65}$$

The resulting time-dependent position space and momentum space probability densities, as well as the flux, are shown in Fig. 6.15 for various times during one half cycle. The dotted lines indicate the position of the average position and momentum value. As the probability density "sloshes" from right to left, the flux is everywhere negative (corresponding to probability flow to the left), while the average value of momentum (the vertical dotted line) is also indicating "motion" to the left. Clearly, more dynamic behavior is allowed for general time-dependent states than for stationary states.

6.6 WAVEPACKETS IN THE INFINITE WELL

6.6.1 Gaussian Wavepacket

In earlier discussions of the free particle, we made heavy use of the concept of a wavepacket, a localized solution of the Schrödinger equation with a behavior similar to that of a classical free particle, but which inherently included the concept of spreading due to dispersion. One can construct a similar time-dependent solution in the standard infinite well[6] constructed from the solutions of Eqn. (6.25).

Motivated by the Gaussian wavepacket of Sec. 3.2.2, we can write a solution involving stationary states of the standard infinite well given by

$$\psi_{WP}(x,t) = \sum_{n=1}^{\infty} a_n u_n(x) e^{-iE_n t/\hbar} \tag{6.66}$$

[6]For a similar example, see Segre and Sullivan (1976).

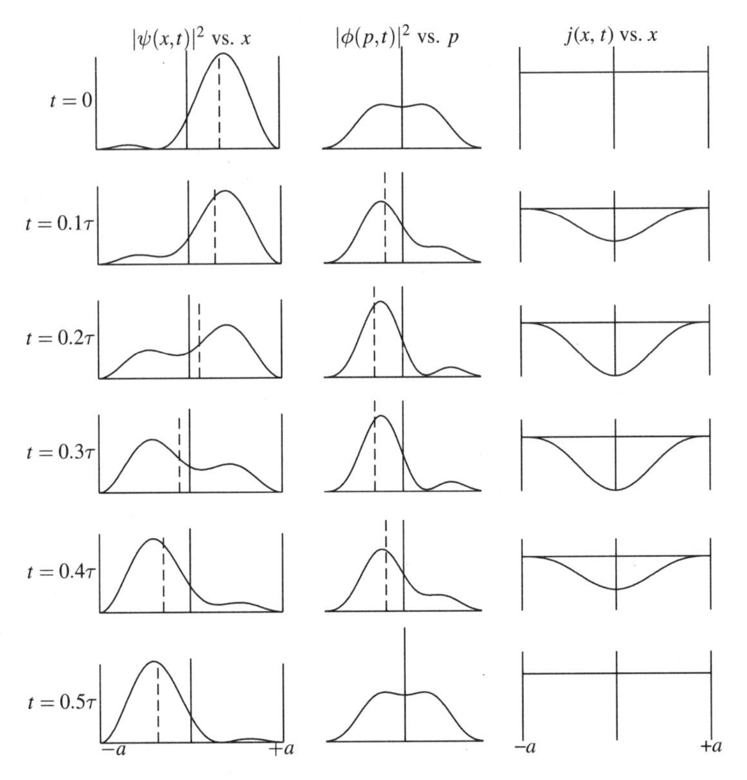

FIGURE 6.15. *Time development of two-state wavefunction. Position space probability density, $|\psi(x,t)|^2$ versus x; momentum space probability density, $|\phi(p,t)|^2$ versus p; and probability flux, $j(x,t)$ versus x are shown for various times during the first half cycle.*

where

$$a_n = e^{-((p_n - p_0)^2\alpha^2/2)} \tag{6.67}$$

Here $p_n = \hbar n/a$ is the analog of the continuous momentum variable, and $p_0 \equiv \hbar n_0/a$ defines some central value of n; this is similar to Eqn. (3.24).

This solution can be evaluated numerically (cutting off the infinite sum at some large but finite value of n) for various values of the parameters to illustrate the motion. We note that the largest contribution will come from states with $E \approx \hbar^2 n_0^2/2ma^2$, so that the classical speed of the wavepacket is roughly $v_0 \approx \hbar n_0/ma$ with period $\tau = 4ma^2/\hbar n_0$. The spreading time, by analogy with Eqn. (3.29), is $t_0 = m\hbar\alpha^2$.

We show in Fig. 6.16 the motion of this wavepacket (via its probability density); parameters have been chosen such that $t_0/\tau = 25$. The solid (dashed) curves represent motion at the start of (just after the start of) the 1st, 25th, and 50th periods, i.e., when $t \approx 0, t_0, 2t_0$; the degradation of the localized wavepacket due to the quantum mechanical spreading is clear. (Recall, however, that the spreading time for a real macroscopic particle is many times the age of the universe.) We note also that because of the special nature of the energy eigenvalues of the infinite well, namely, that they are all integral multiples of a common value, E_1, there can be "recurrences" of the wavepacket in which it regains its initial shape.

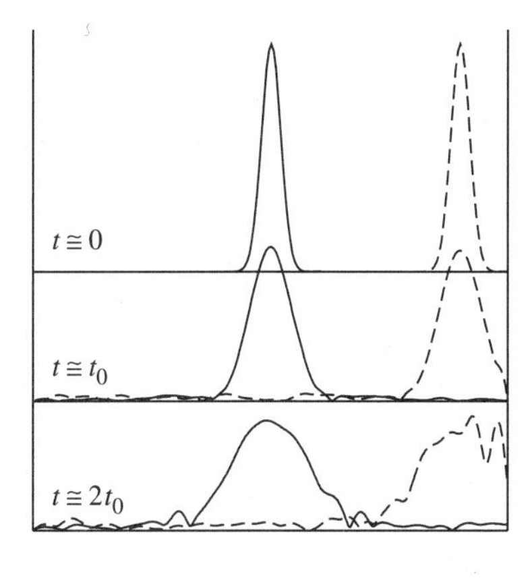

FIGURE 6.16. *Time development for a Gaussian wavepacket in the infinite well. The solid (dashed) curves correspond to times at (near) $t = 0, t_0, 2t_0$. The spreading of the wavepacket due to dispersion is clear.*

6.6.2 Wavepackets versus Stationary States

We have now been able to analyze (either analytically or numerically) examples of wavepacket solutions to the Schrödinger equation for several simple cases, namely, the free and accelerating particle and the particle in a box. We have argued that these solutions are the closest representation of something like the classical motion of a particle that we will find in quantum mechanics; they exhibit all of the expected dynamics in the classical limit.[7]

They have been constructed from energy eigenstate solutions that are, in some sense, more natural in the context of quantum physics. The $\psi_E(x)e^{-iEt/\hbar}$ give information on the quantized energies of the bound state system, one of the most important features of quantum mechanics. One might think that these solutions could contain no interesting information on the dynamics of the system as $|\psi_E(x,t)|^2$ for such states is independent of time for stationary states.

We have seen, however, that the shape of $\psi_E(x)$ does make contact with the classical motion of the particle; the local magnitude and wiggliness of the wavefunction is correlated with the local speed of the particle in a meaningful way. Furthermore the quantum probability density can be used to approach the classical distribution of probability so that information on the particle trajectory is obtained, in a time-averaged sense. Assured of these connections, we will henceforward concentrate on the physical meaning and mathematical properties of energy eigenstates.

6.7 QUESTIONS AND PROBLEMS

Q6.1 What would the projected and binned probability density look like for a particle in the infinite well; i.e., what is the analog of Fig. 6.5 for the motion in Fig. 6.1?

Q6.2 Why might the projection technique be useful in cases of chaotic motion where it might be hard to combine $x(t)$ and $v(t)$ to yield $v(x)$?

Q6.3 How would you use the projection technique to obtain a classical probability distribution for *momentum,* $P_{CL}(p)$, using the classical trajectory variables? *Hint:* What information could one gain from a plot of $v(t)$ versus t? What would $P_{CL}(p)$ look like for the particle in the infinite well? In a simple harmonic oscillator potential?

[7] See the discussion by Brown (1975) on the classical limit of quantum wavepackets.

Q6.4 How might you generalize the idea of classical probability density to the case of a bound particle in a *two-dimensional* potential?

Q6.5 Waves on a string with both ends fixed have quantized wavelengths and frequencies. Are their energies quantized as for the particle in a quantum box?

Q6.6 In the quantum version of the infinite well, the energy eigenvalues are, in principle, precisely determined. The energy is all in the form of kinetic energy, $E = p^2/2m$, and so, classically, the momentum should be exactly known as well. But $\phi(p)$ has a finite spread (as seen in Fig. 6.9). How can this happen?

Q6.7 Comment on the statement made in the text that it is usually the boundary conditions on $\psi(x)$ at $x = \pm\infty$ that determine the quantized energy levels. In what sense is this true for the infinite well examples in this chapter?

Q6.8 Referring to Fig. 6.7, why is the energy of the first excited state of the "symmetric" well identical to the ground state energy of the "standard well"? Why is $E_2 = E_2^{(-)}$? Will there be any other such "degeneracies"? What do degeneracies such as these have to do with the relative size (i.e., width) of the well?

Q6.9 What would a solution to the "asymmetric well" of the form in Eqn. (6.44) look like if k and q did not satisfy Eqn. (6.45). What would be wrong with such a solution?

Q6.10 How many eigenstates are there with energies between E_A and E_B in Fig. 6.14? What would the ground state wavefunction look like?

Q6.11 What happens to general time-dependent solutions (such as the two-state combination solution in Sec. 6.5 or the wavepackets in Sec. 6.6) when one changes the normalization constants of the individual stationary states by an arbitrary phase as one is allowed to do?

Q6.12 Can you give an argument why a wavepacket constructed from solutions of the infinite well problem might exhibit "recurrences," i.e., the initial waveform should reappear after a (possibly very) long time? Concentrate on the fact that the quantized energies of this system are all integer multiples of a common ground state value. Use the fact that Eqn. (6.66) can be written as

$$\psi_{WP}(x,t) = e^{-iE_1 t/\hbar}\left[\sum_{n=1}^{+\infty} a_n u_n(x) e^{i\Delta E t/\hbar}\right]$$

Q6.13 What would the motion of a wavepacket look like in the "asymmetric well" of Sec. 6.4.1? In the "linear well" of Sec. 6.4.2? What would the *momentum space distribution* of the wavepacket in the standard well look like as a function of time?

Q6.14 If you wanted to excite a wavepacket, would you most likely use an almost monochromatic laser beam or a broad-band (large frequency spread) laser?

P6.1 **Classical "asymmetric infinite well":** Consider a particle of mass m and energy E moving in the asymmetric infinite potential well given by

$$V(x) = \begin{cases} +\infty & \text{for } x < -a,\ x > b \\ 0 & \text{for } -a < x < 0 \\ V_0 & \text{for } 0 < x < b \end{cases}$$

where $E > V_0 > 0$.

(a) Sketch graphs of the classical trajectory variables $x(t)$ and $v(t)$ versus t.

(b) Use the expression for the classical probability distribution

$$P_{CL}(x) = \frac{2}{\tau}\sqrt{\frac{m}{2(E - V(x))}}$$

to find the classical period τ for motion in this well. Show that it reduces to the expected value when $V_0 \to 0$.

(c) What is the probability that a measurement of the position of this particle would find it with $0 < x < b$? Give a numerical answer in the case $E = 2V_0$ and $a = b$.

(d) Sketch the probability density for $a = b$ and $E = 2V_0$.

(e) Evaluate $\langle x \rangle$ for the general case, and show that it has the correct limit when $b \to a$.

(f) What would the *classical* distribution of velocities (or momenta) look like for a particle in this potential?

P6.2 Repeat P6.1—parts (a), (b), (e), and (f)—for the case of the "linear well" in Sec. 6.4.2 when $E > V_0 = Fa$.

P6.3 Consider the classical probability distribution of Eqn. (6.14) for the simple harmonic oscillator,

$$P_{CL}(x) = \frac{1}{\pi}\frac{1}{\sqrt{A^2 - x^2}}$$

with the classical turning points at $(-A, A)$ and total energy equal to $E = KA^2/2$. The potential energy is $V(x) = Kx^2/2$.

(a) Show that $P_{CL}(x)$ is properly normalized.

(b) Use $P_{CL}(x)$ to evaluate $\langle x \rangle_{CL}$, $\langle x^2 \rangle_{CL}$, and $\langle V(x) \rangle_{CL}$.

(c) Use the fact that $T(x) = E - V(x)$ to evaluate $\langle T(x) \rangle$. Show that the potential and kinetic energies satisfy

$$\langle V(x) \rangle_{CL} = \langle T \rangle_{CL} = \frac{E}{2}$$

so that, on average, the kinetic and potential energies are shared equally.

(d) Can you argue why $\langle p \rangle_{CL}$ should vanish? Use the fact that $T = p^2/2m$ to evaluate $\langle p^2 \rangle_{CL}$.

(e) Finally, evaluate Δx_{CL} and Δp_{CL}, and show that $\Delta x \cdot \Delta p$ can be made arbitrarily small in classical mechanics.

P6.4 Consider a particle of energy E moving in a linear potential given by $V(x) = C|x|$ with classical turning points at $(-A, A)$ where $E = CA$.

(a) Show that the normalized classical probability distribution is given by

$$P_{CL}(x) = \frac{1}{4\sqrt{A}}\frac{1}{\sqrt{A - |x|}}$$

(b) Evaluate the average values of the potential and kinetic energies and show that

$$\langle T(x) \rangle_{CL} = \frac{1}{2}\langle V(x) \rangle_{CL} = \frac{1}{3}E$$

so that the kinetic and potential energies are not shared equally.

P6.5 Consider a particle of mass m and energy $E > 0$ moving in the potential well

$$V(x; n) \equiv V_0 \left|\frac{x}{a}\right|^n$$

where $n \geq 1$.

(a) Show that for $n = 2$ this gives the simple harmonic oscillator potential, and identify the "spring constant."

(b) Show that this potential approaches the symmetric infinite well potential when $n \to \infty$. (This shows that results for the [ill-behaved] infinite well potential can be obtained as the limit of a general case; this is sometimes a useful result.)

(c) Use the expression for the classical probability distribution to calculate the period in this potential for arbitrary n. Show that your general answer reduces to the expected results for the cases of $n = 0, n \to \infty$. *Hint:* You might find the integrals in App. B.1 useful. You may have to make a change of variables to put them into an appropriate form.

(d) Calculate the expectation value of the potential energy, $\langle V(x) \rangle$, and compare it to the total energy for various values of n; Show that one has

$$\langle V(x) \rangle = \left(\frac{2}{2+n}\right)E$$

This shows how the energy is "shared" between potential and kinetic in potential wells of various shapes.

P6.6 Consider a uniformly accelerating particle with position given by $x(t) = at^2/2$.

(a) Show that the classical probability distribution in this case would be given by $P_{CL} \propto 1/\sqrt{x}$. Is such a distribution normalizable?

(b) Compare $P_{CL}(x)$ to the results in Fig. 5.4 and especially to the wavefunction discussion in P5.21. Does $P_{CL}(x)$ agree with the large x limit of $|\psi(x)|^2$ derived there?

P6.7 We used solutions of the form $\cos(kx)$ and $\sin(kx)$ in our discussion of the symmetric infinite well problem. A general solution of the form

$$\psi(x) = Ae^{ikx} + Be^{-ikx}$$

would be just as acceptable.

(a) Using this solution, apply the boundary conditions, determine the energy eigenvalues for the symmetric infinite well, and show that you obtain the same results as in Sec. 6.3.2.

(b) Show that the wavefunctions corresponding to these solutions are also the same as discussed previously, namely, they can be expressed in terms of sines and cosines.

P6.8 Consider a particle of mass m moving in an infinite well potential of width $2a$, centered not at the origin but rather at $x = d$.

(a) Find the solutions of the Schrödinger equation, apply the appropriate boundary conditions, and show that the resulting quantized energies are identical to those found in Sec. 6.3.2.

(b) Show how the wavefunctions in this well are related to those discussed above.

(c) Discuss why the positioning of the well at different locations should have no effect on any important physical observables.

P6.9 **(a)** Calculate the RMS deviation in position, Δx, for the odd states in the symmetric infinite well. Show that it reduces to the classical value for this quantity (which is what?) in the limit of large quantum number.

(b) For many bound systems, Δx increases without limit as the quantum number increases. Why doesn't that happen in this case?

P6.10 Consider the odd solutions to the symmetric infinite well, i.e., the $u_n^-(x)$, in the large quantum number limit.

(a) Calculate the probability that the particle will be found in an interval of length b located at some arbitrary point, $x = c$, in the well, i.e., in the interval $(c, c + b)$.

(b) Show that as $n \to \infty$, this reproduces the classical result we discussed in Sec. 6.1, i.e., that the particle is equally likely to be found anywhere in the well.

P6.11 For the odd eigenfunctions of the symmetric infinite well, $u_n^{(-)}(x)$, calculate the corresponding momentum space wavefunctions, $\phi_n^-(p)$. Sketch $|\phi_n^{(-)}(p)|^2$ for small and large values of n.

P6.12 Use the expression for $\phi(p)$ in Eqn. (6.36) to show that the corresponding momentum space probability density approaches

$$P_{CL}(p) = \frac{1}{2}(\delta(p - p_0) + \delta(p + p_0))$$

in the limit that $\Delta p \to 0$. *Hint:* Use the representation of the δ function in App. C.8.

P6.13 Show that the average value of the momentum operator, $\langle \hat{p} \rangle$, vanishes when evaluated in any state, $u(x)$, which is purely real. Do this in two ways:

(a) Use the position space representation

$$\langle \hat{p} \rangle = \int_{-\infty}^{+\infty} dx\, u^*(x)\, \hat{p}\, u(x)$$

and show that the fact that $\hat{p}$ is Hermitian requires its average in a real state, where $u^*(x) = u(x)$, to vanish.

(b) Use the momentum space representation

$$\phi(p) = \frac{1}{\sqrt{2\pi\hbar}} \int_{-\infty}^{+\infty} dx\, u(x)\, e^{-ixp/\hbar}$$

to show that $\phi(p) = f(p) + ig(p)$ where $f(p)$ and $g(p)$ are real functions that are even and odd functions of p, respectively. Show that this means that $|\phi(p)|^2$ will be symmetric under $p \to -p$.

(c) Discuss what happens to your results if $u(x)$ is multiplied by an arbitrary complex number (even a function of time such as $\exp(-iEt/\hbar)$) so long as it has no spatial dependence.

(d) Discuss how either of these proofs fail if the spatial wavefunction, $\psi(x)$, is allowed to be complex, e.g., a plane wave solution of the form $\exp(ip_0x/\hbar)$.

P6.14 Consider the asymmetric infinite well of Sec. 6.4 in the case when $0 < E < V_0$. Redo the analysis redefining q as necessary to take into account the change in sign. Show that the solutions in the right side of the well can be expressed now in terms of $\sinh(qx)$ and $\cosh(qx)$, and find the new energy eigenvalue condition. Show that the same solution is obtained if one lets $q \to iq$ in Eqn. (6.44), i.e, if one analytically continues the trigonometric functions into hyperbolic functions. *Hint:* Use the results of App. A.

P6.15 Consider a particle in the symmetric infinite well whose initial wavefunction is $\psi(x, 0) = \cos(\theta)u_1^+(x) + \sin(\theta)u_1^-(x)$, i.e., a linear combination of the ground state and first excited state.

(a) What is $\psi(x, t)$?

(b) Show that $\psi(x, t)$ is properly normalized for all t.

(c) Calculate $\langle \hat{E} \rangle_t$ for this state, and show that it is actually independent of time. What is $\langle E \rangle_t$ for $\theta = 0$? $\pi/2$?

(d) Calculate $\langle x \rangle_t$ and $\langle \hat{p} \rangle_t$ for this state, and show explicitly that they are related as they should be, namely,

$$\langle \hat{p} \rangle_t = m \frac{d}{dt} \langle x \rangle_t$$

You may find the integral

$$\int_{-1}^{+1} dy\, y \cos\left(\frac{\pi y}{2}\right) \sin(\pi y) = \frac{32}{9\pi^2}$$

useful.

(e) Verify that the expression for flux for this problem is consistent with that in Eqn. (6.65).

P6.16 Sketch your best guess for the position space ($\psi(x)$) and momentum space ($\phi(p)$) wavefunctions in the potential well shown in Fig. 6.17; the ground state (E_1) and the second and eighth excited states are shown.

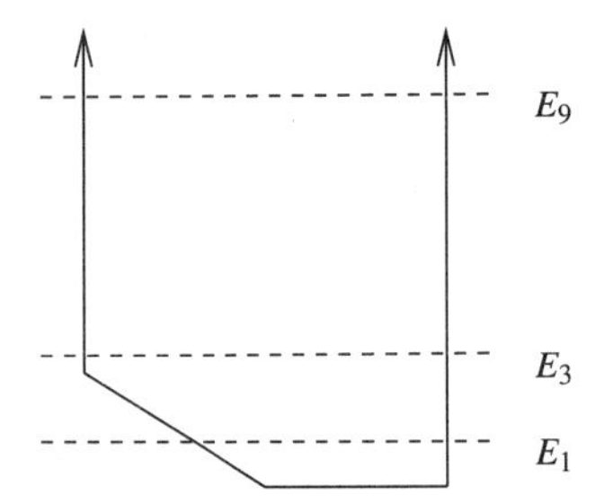

FIGURE 6.17. *What do the wavefunctions ($\psi(x)$ and $\phi(p)$) look like for the energies shown in this potential well?*

P6.17 **Classical forces:** A classical particle rattling around in a one-dimensional box would exert a force on the walls, and the same is true in quantum mechanics. To evaluate it, consider a particle in the ground state of the standard well, with walls at $(0, L)$, with energy E, and imagine moving the right-hand wall, *very slowly* (sometimes called *adiabatically*) to the right by an amount dx. It can be shown that any such slow change will leave the particle in the ground state of the evolving system.

(a) Calculate the new ground state energy, E', and the difference in energies $\Delta E = E' - E$. This difference in energies is associated with the work done *by the wall on the particle,* which we can write $dW = F_{\text{wall}} \cdot dx$; it can be used to evaluate F_{wall}.

(b) The force *of the particle on the wall* is then the same, but with opposite sign by Newton's second law. Use this force to calculate the work done on a particle by the walls if the right wall is slowly moved from L to $2L$, and show that it agrees with the change in energy.

CHAPTER 7

The Infinite Well: Formal Aspects

In the previous chapter, we concentrated on the most important physical aspects of the solutions of the Schrödinger equation in various versions of the infinite well, namely, the quantization of energy levels in a confining potential, the connection between the quantum wavefunctions (in both position and momentum space) and their classical counterparts, and wavepackets.

We now turn to an examination of the more formal properties of energy eigenstate solutions, namely, the eigenfunctions of a Hamiltonian operator. Many of these properties can be immediately generalized to include the eigenfunctions of other Hermitian operators. We begin, however, by introducing a simple but useful notation.

7.1 DIRAC BRACKET NOTATION

Motivated initially by the notation for average values used before, we define the *Dirac bracket* of a position space wavefunction via

$$\langle \psi \mid \psi \rangle \equiv \int_{-\infty}^{+\infty} dx\, \psi^*(x,t)\, \psi(x,t) \tag{7.1}$$

A properly normalized wavefunction will then have $\langle \psi \mid \psi \rangle = 1$. This is easily extended to include the overlap integrals of two different wavefunctions, namely,

$$\langle \psi_1 \mid \psi_2 \rangle \equiv \int_{-\infty}^{+\infty} dx\, \psi_1^*(x,t)\, \psi_2(x,t) \tag{7.2}$$

We note the simple connection

$$\langle \psi_1 \mid \psi_2 \rangle^* = \langle \psi_2 \mid \psi_1 \rangle \tag{7.3}$$

since such overlap integrals are not necessarily real; on the other hand, $\langle \psi \mid \psi \rangle$ is manifestly real. An equivalent definition for the momentum space representation is

$$\langle \phi_1 \mid \phi_2 \rangle \equiv \int_{-\infty}^{+\infty} dp\, \phi_1^*(p,t)\, \phi_2(p,t) \tag{7.4}$$

and Parseval's theorem (Eqn. (3.23)) guarantees that the Dirac bracket of two different wavefunctions, whether evaluated in position or momentum space, will be equal since

$$\begin{aligned} \langle \phi_1 \mid \phi_2 \rangle &= \int_{-\infty}^{+\infty} dp\, \phi_1^*(p,t)\, \phi_2(p,t) \\ &= \int_{-\infty}^{+\infty} dx\, \psi_1^*(x,t)\, \psi_2(x,t) \\ &= \langle \psi_1 \mid \psi_2 \rangle \end{aligned} \tag{7.5}$$

Expectation values of operators are written as

$$\langle \hat{O} \rangle = \langle \psi | \hat{O} | \psi \rangle \equiv \int_{-\infty}^{+\infty} dx\, \psi^*(x,t)\, \hat{O}\, \psi(x,t) \tag{7.6}$$

so that, in this simplified form, an operator is Hermitian if

$$\langle \psi | \hat{O} | \psi \rangle = \langle \psi | \hat{O} | \psi \rangle^* \tag{7.7}$$

From Eqn. (5.50), we also know that Hermitian operators will satisfy

$$\langle \psi_1 | \hat{O} | \psi_2 \rangle^* = \langle \psi_2 | \hat{O} | \psi_1 \rangle \tag{7.8}$$

Besides being of typographical convenience, the Dirac bracket notation has a geometrical significance that will be discussed in Chap. 13.

7.2 EIGENVALUES OF HERMITIAN OPERATORS

The basic equation of quantum mechanics, the time-independent Schrödinger equation,

$$\hat{H} \psi_E(x) = E \psi_E(x) \tag{7.9}$$

requires one to find the eigenvalues of a Hermitian operator, namely, the Hamiltonian (recall P5.12). The energy eigenvalues so obtained are real, which is an example of a general result:

- The eigenvalues of a Hermitian operator are real numbers.

This is easily proved by considering a Hermitian operator $\hat{A}$ satisfying

$$\hat{A} \psi_a(x) = a \psi_a(x) \tag{7.10}$$

where we work in a position space presentation for definiteness. If we multiply both sides of Eqn. (7.10) by $\psi_a^*(x)$ (on the left!) and integrate, we find that

$$\begin{aligned} \langle \hat{A} \rangle = \langle \psi_a | \hat{A} | \psi_a \rangle &= \int_{-\infty}^{+\infty} dx\, \psi_a^*(x)\, \hat{A}\, \psi_a(x) \\ &= \int_{-\infty}^{+\infty} dx\, \psi_a^*(x)\, a\, \psi_a(x) \\ &= a \langle \psi_a \mid \psi_a \rangle \end{aligned} \tag{7.11}$$

so that

$$a = \frac{\langle \psi_a | \hat{A} | \psi_a \rangle}{\langle \psi_a \mid \psi_a \rangle} \tag{7.12}$$

which is obviously real since $\hat{A}$ is Hermitian. Not surprisingly, then, the eigenvalues associated with operators corresponding to physical observables are real; familiar examples include the momentum and energy operators.

7.3 ORTHOGONALITY OF ENERGY EIGENFUNCTIONS

Starting with the standard infinite well, we found that we had to normalize the energy eigenfunctions by hand, because the relation

$$\langle u_n \mid u_n \rangle = \int_0^a |u_n(x)|^2 \, dx = \frac{2}{a} \int_0^a \sin\left(\frac{n\pi x}{a}\right)^2 dx = 1 \tag{7.13}$$

was not an automatic property of solutions of the Schrödinger equation. We now note that the overlap of two *different* solutions of this problem, i.e.,

$$\begin{aligned} \langle u_n \mid u_m \rangle &= \int_0^a u_n^*(x) u_m(x) \, dx \\ &= \frac{2}{a} \int_0^a \sin\left(\frac{n\pi x}{a}\right) \sin\left(\frac{m\pi x}{a}\right) dx \\ &= \delta_{n,m} \end{aligned} \tag{7.14}$$

vanishes if the eigenfunctions correspond to different eigenvalues, i.e., $n \neq m$. In contrast to Eqn. (7.13), this fact *is* seemingly a consequence of the fact that the eigenfunctions satisfy the Schrödinger equation. We can visualize how this occurs by examining the integrands for a pair of states with $n = 2, m = 3$ in Fig. 7.1; this figure illustrates how the "area" under the product $u_n(x)\, u_m(x)$ vanishes.

A similar result can be seen to hold for the symmetric infinite well:

$$\langle u_n^+ \mid u_m^- \rangle = 0 \quad \text{for all} \quad n, m \qquad \langle u_n^+ \mid u_m^+ \rangle = \delta_{n,m} \quad \langle u_n^+ \mid u_m^+ \rangle = \delta_{n,m} \tag{7.15}$$

where the first condition also follows from the fact that the integrand is the product of an even times an odd function.

From Eqn. (3.23), we know that the equivalent overlaps of the momentum space wavefunctions will also satisfy this orthonormality condition. For example, for the even states

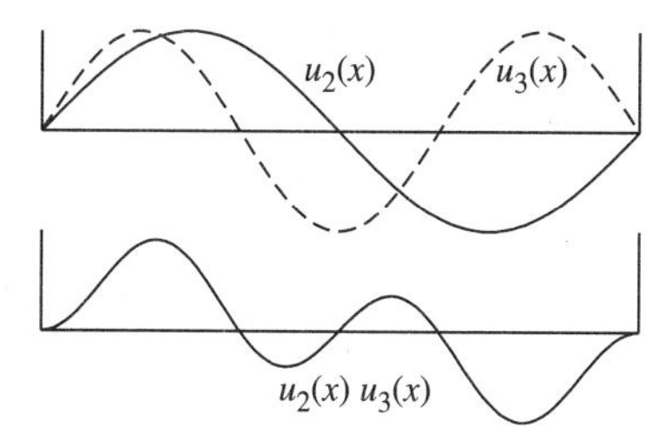

FIGURE 7.1. *$u_m(x)\, u_n(x)$ versus x for standard infinite well for $(n, m) = (2,3)$. The $u_{n,m}(x)$ are normalized, but the overlap integral vanishes because of cancelations. The "area under the curve" vanishes.*

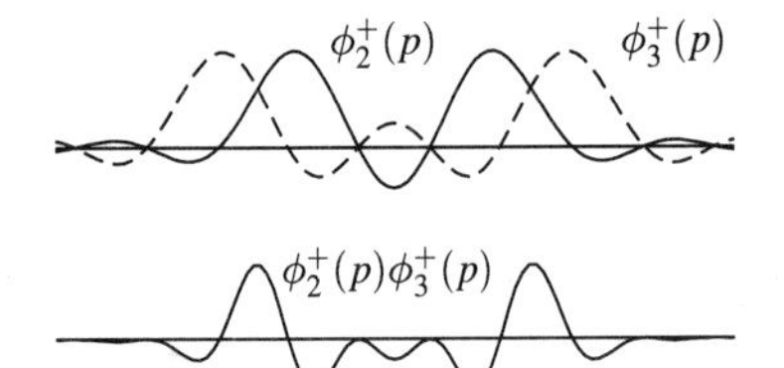

FIGURE 7.2. $\phi_n^{(+)}(p)\,\phi_m^{(+)}(p)$ *versus p for the symmetric infinite well for* $(n, m) = (2,3)$*; the product* $\phi_n(p)\,\phi_m(p)$ *has vanishing area if* $n \neq m$.

in the symmetric well we must have

$$\begin{aligned}
\langle \phi_n^{(+)} \mid \phi_m^{(+)} \rangle &= \int_{-\infty}^{\infty} (\phi_n^{(+)}(p))^* \phi_m^{(+)}(p)\, dp \\
&= \int_{-a}^{+a} u_m^{(+)}(x) u_n^{(+)}(x)\, dx \\
&= \langle u_m^{(+)} \mid u_n^{(+)} \rangle \\
&= \delta_{n,m}
\end{aligned} \tag{7.16}$$

The momentum space integrals can also be done analytically (P7.1) as a further check. We illustrate in Fig. 7.2 a case of the product of two such momentum space wavefunctions which shows how the cancellations can occur, which make the overlap integral vanish.

Finally, we also note that the infinite well is not special in this regard as we can show (P7.2), after some algebra, that the position space wavefunctions of the asymmetric well corresponding to different energies also have vanishing overlap.

We can then say that:

- Two functions are *orthogonal* if their overlap integral vanishes.

And we need not specify whether we work with position space or momentum space wavefunctions because of Eqn. (3.23). We are finding that:

- The energy eigenfunctions form an *orthonormal* set, that is, they are mutually orthogonal and normalized under a generalized form of dot or inner product (namely, the overlap integral).

This concept is very similar to the case of a set of unit vectors in an N-dimensional vector space, $\{\hat{\mathbf{e}}_i, i = 1, \ldots, N\}$, where $\hat{\mathbf{e}}_i \cdot \hat{\mathbf{e}}_j = \delta_{ij}$.

To see that this phenomenon is not restricted to solutions of the Schrödinger equation, i.e., eigenfunctions of a Hamiltonian operator, consider the eigenfunctions of the *momentum* operator

$$\hat{p}\psi_p(x) = \frac{\hbar}{i}\frac{d\psi_p(x)}{dx} = p\psi_p(x) \tag{7.17}$$

with solutions

$$\psi_p(x) = \frac{1}{\sqrt{2\pi\hbar}} e^{ipx/\hbar} \tag{7.18}$$

for any real value of the variable p. These solutions have overlap integrals given by

$$\langle \psi_p \mid \psi_{p'} \rangle = \int_{-\infty}^{+\infty} \left(\frac{1}{\sqrt{2\pi\hbar}} e^{ipx/\hbar} \right)^* \left(\frac{1}{\sqrt{2\pi\hbar}} e^{ip'x/\hbar} \right) dx$$
$$= \frac{1}{2\pi\hbar} \int_{-\infty}^{+\infty} e^{i(p'-p)x/\hbar}\, dx$$
$$= \delta(p - p') \tag{7.19}$$

which also vanish if the eigenvalues are different. In this case, the normalization condition is appropriate for a continuous label.

Thus, this generalized orthogonality is seemingly not a specific property of either the position space or momentum space representation, or even of eigenfunctions of a Hamiltonian, but a more general result. The functions above all share the property that they are eigenfunctions of some Hermitian operator, and the most general result we can derive can be stated simply as:

- The eigenfunctions of a Hermitian operator corresponding to *different* eigenvalues will be orthogonal.

We can prove that as follows. Consider two eigenfunctions of a Hermitian operator, $\hat{A}$, corresponding to two distinct eigenvalues,

$$\hat{A}\psi_a(x) = a\psi_a(x) \qquad \hat{A}\psi_b(x) = b\psi_b(x) \tag{7.20}$$

where $a \neq b$ are both real (from Sec. 7.2) since $\hat{A}$ is Hermitian; we work with position space functions for definiteness. We know from Eqn. (7.8) that any Hermitian operator will satisfy the relation

$$\langle \psi | \hat{A} | \phi \rangle = \langle \phi | \hat{A} | \psi \rangle^* \tag{7.21}$$

so we can write

$$\langle \psi_a | \hat{A} | \psi_b \rangle = \int_{-\infty}^{+\infty} dx\, \psi_a^*(x) \left[\hat{A}\psi_b(x) \right]$$
$$= \int_{-\infty}^{+\infty} dx\, \psi_a^*(x)\, b\, \psi_b(x)$$
$$= b \langle \psi_a \mid \psi_b \rangle \tag{7.22}$$

while

$$\langle \psi_b | \hat{A} | \psi_a \rangle^* = \left(\int_{-\infty}^{+\infty} dx\, \psi_b^*(x) \left[\hat{A}\psi_a(x) \right] \right)^*$$
$$= \left(\int_{-\infty}^{+\infty} \psi_b^*(x)\, a\, \psi_a(x) \right)^*$$
$$= a^* \int_{-\infty}^{+\infty} dx\, \psi_a^*(x) \psi_b(x)\, dx$$
$$= a \langle \psi_a \mid \psi_b \rangle \tag{7.23}$$

since a is real. Comparing these two quantities we see that

$$0 = \langle \psi_a | \hat{A} | \psi_b \rangle - \langle \psi_b | \hat{A} | \psi_a \rangle^* = (b - a) \langle \psi_a \mid \psi_b \rangle \tag{7.24}$$

so that $\langle \psi_a \mid \psi_b \rangle = 0$ if $b \neq a$, and the eigenfunctions are indeed orthogonal if their corresponding eigenvalues are different. *Note:* If there is more than one eigenfunction corresponding to the same eigenvector, the set of such eigenfunctions can be made orthogonal "by hand" (P7.3).

The use of Hermitian operators, which naturally correspond to physical observables, automatically induces a rich geometrical and algebraic structure on the solutions of the Schrödinger equation and other systems, and we will make extensive use of these properties.

7.4 EXPANSIONS IN EIGENSTATES

Starting with our canonical example of the standard infinite well, we saw that the general, time-dependent solution of the Schrödinger equation could be written as a linear combination of energy eigenstates via

$$\psi(x,t) = \sum_{n=1}^{\infty} a_n u_n(x) e^{-iE_n t/\hbar} \tag{7.25}$$

where the a_n are (at the moment) arbitrary and possibly complex constants. We wish to explore the physical interpretation of the "expansion" coefficients, a_n. We first note that if the solution is properly normalized, we must have

$$\begin{aligned}
1 = \langle \psi \mid \psi \rangle &= \int_0^a \psi^*(x,t)\psi(x,t)\,dx \\
&= \int_0^a \left(\sum_{n=1}^{\infty} a_n\, u_n(x) e^{-iE_n t/\hbar}\right)^* \left(\sum_{m=1}^{\infty} a_m\, u_m(x) e^{-iE_m t/\hbar}\right) dx \\
&= \sum_{n=1}^{\infty}\sum_{m=1}^{\infty} a_n^* a_m e^{-i(E_m - E_n)t/\hbar} \int_0^a u_n(x)\, u_m(x)\,dx \\
&= \sum_{n=1}^{\infty}\sum_{m=1}^{\infty} a_n^* a_m e^{-i(E_m - E_n)t/\hbar}\, \delta_{n,m} \\
1 &= \sum_{n=1}^{\infty} |a_n|^2
\end{aligned} \tag{7.26}$$

and the expansion coefficients must satisfy their own normalization condition. While a particle in any particular energy eigenstate will have a uniquely defined energy, this general solution has an expectation value for the energy given by

$$\begin{aligned}
\langle \psi | \hat{E} | \psi \rangle &= \int_0^a \psi^*(x,t) \hat{E} \psi(x,t)\,dx \\
&= \int_0^a \left(\sum_{n=1}^{\infty} a_n\, u_n(x) e^{-iE_n t/\hbar}\right)^* \left(i\hbar \frac{\partial}{\partial t}\right) \left(\sum_{m=1}^{\infty} a_m\, u_m(x) e^{-iE_m t/\hbar}\right) dx \\
&= \sum_{n=1}^{\infty}\sum_{m=1}^{\infty} a_n^* a_m E_m e^{-i(E_m - E_n)t/\hbar} \int_0^a u_n(x)\, u_m(x)\,dx \\
&= \sum_{n=1}^{\infty}\sum_{m=1}^{\infty} a_n^* a_m E_m e^{-i(E_m - E_n)t/\hbar} \delta_{n,m} \\
\langle \hat{E} \rangle &= \sum_{n=1}^{\infty} E_n |a_n|^2
\end{aligned} \tag{7.27}$$

independent of time. So, while many important properties of such a general state may vary in time, e.g., average values of x and $\hat{p}$, the average value of the energy operator does not. Similar results can be obtained for any power of the energy operator,

$$\langle \hat{E}^k \rangle = \langle \psi | \hat{E}^k | \psi \rangle = \sum_{n=1}^{\infty} (E_n)^k |a_n|^2 \tag{7.28}$$

and this allows one to calculate the energy spread, given by

$$\Delta E = \sqrt{\langle \hat{E}^2 \rangle - \langle \hat{E} \rangle^2} \tag{7.29}$$

for any state. A general state such as Eqn. (7.25) will have $\Delta E \neq 0$, unless of course $a_n = \delta_{n,\kappa}$, so that it is actually an energy eigenstate.

Taken together, these results imply that the squares of the expansion coefficients are related to the probability of "finding" the particle in one of the given energy eigenstates. Because the measurable quantity associated with the solutions of the Schrödinger equation is their energy eigenvalue, we have this more precise statement:

- In an expansion of a wavefunction $\psi(x, t)$ in terms of energy eigenstates, $|a_n|^2$ is the probability that a measurement of the energy of the particle described by $\psi(x, t)$ will yield the energy E_n as a result of such a measurement.

This definition is consistent with the expression for $\langle \hat{E} \rangle$ in Eqn. (7.27) as an average value of a discrete probability distribution. Since the $|a_n|^2$ are themselves probabilities (in contrast to, say, $|\psi(x, t)|^2$), they must be dimensionless.

Example 7.1 As an example, consider the unnormalized infinite well state given by

$$\psi(x, t) = N\left(2u_1(x)e^{-iE_1 t/\hbar} + (1 - 2i)u_2 e^{-iE_2 t/\hbar} - 3iu_3(x)e^{-iE_3 t/\hbar}\right) \tag{7.30}$$

where $E_n = \hbar^2 \pi^2 n^2 / 2ma^2$. The normalization factor N is determined by the fact that

$$1 = \sum_{n=1}^{\infty} |a_n|^2 = N^2(4 + 5 + 9) = 18N^2 \tag{7.31}$$

so $N = 1/\sqrt{18}$. Repeated measurements of the energy of an ensemble of such states would only find the values E_1, E_2, and E_3 with probabilities 2/9, 5/18, and 1/2, respectively. The average value of energy after many such measurements would be $\langle \hat{E} \rangle = (35/12)(\hbar^2 \pi^2)/ma^2$ or roughly $2.92E_1$.

This same interpretation will hold for any expansion in terms of energy eigenstates (i.e., solutions of the Schrödinger equation) but, once again, is far more general. For example, we have seen that the eigenstates of the momentum operator can be written as

$$\psi_p(x) = \frac{1}{\sqrt{2\pi\hbar}} e^{ipx/\hbar} \tag{7.32}$$

and we can write a general function as a weighted (continuous) sum of such solutions via

$$\begin{aligned} \psi(x) &= \int_{-\infty}^{+\infty} \phi(p) \psi_p(x) \, dx \\ &= \frac{1}{\sqrt{2\pi\hbar}} \int_{-\infty}^{+\infty} \phi(p) e^{ipx/\hbar} \, dx \end{aligned} \tag{7.33}$$

which is simply the Fourier transform. The expansion coefficients in this case are the $\phi(p)$ and have a continuous label, so that the probability of making a measurement of the physical observable (in this case momentum) is given by

$$\text{Prob}[(p, p + dp)] = |\phi(p)|^2 \, dp \tag{7.34}$$

as expected. We are thus led to argue:

- If we expand a function in terms of the eigenstates of a Hermitian operator, $\hat{A}$, i.e.,

$$\psi(x) = \sum_a c_a \psi_a(x) \qquad \text{where} \qquad \hat{A}\psi_a(x) = a\psi_a(x)$$

  the square of the expansion coefficients, $|c_a|^2$, gives the probability of observing that state with the value of the observable, a. Such an expansion can be either discrete or continuous.

This discussion points up again the usefulness of having many different *representations* of the same quantum mechanical system. We have already argued that the position space wavefunction, $\psi(x)$, and the momentum space wavefunction, $\phi(p)$, are equivalent in their information content, and we can now add to that list the expansion coefficients of a quantum system in terms of some set of energy eigenstates, namely, the $\{a_n\}$. These three sets of numbers give complementary information on the probabilities of measuring the position, momentum, and energy, respectively, of the particle described by them.

The different descriptions have an equivalent geometrical structure, as their norms satisfy

$$\int_{-\infty}^{+\infty} |\psi_a(x)|^2 \, dx = \int_{-\infty}^{+\infty} |\phi_a(p)|^2 \, dp = \sum_{n=1}^{\infty} |a_n|^2 = 1 \tag{7.35}$$

and generalized inner products are related by

$$\int_{-\infty}^{+\infty} \psi_a^*(x)\psi_b(x) \, dx = \int_{-\infty}^{+\infty} \phi_a^*(p)\phi_b(p) \, dp = \sum_{n=1}^{\infty} a_n^* b_n \tag{7.36}$$

if

$$\psi_a(x) = \sum_{n=1}^{\infty} a_n u_n(x) \qquad \text{and} \qquad \psi_b(x) = \sum_{n=1}^{\infty} b_n u_n(x) \tag{7.37}$$

We have explored in Chap. 5 how average values of the position and momentum operators can be obtained in the $\psi(x)$ and $\phi(p)$ representations. Section 11.5 explores how such information is obtained from the a_n.

7.5 EXPANSION POSTULATE AND TIME DEPENDENCE

We have now seen that an arbitrary linear combination of energy eigenstate solutions (each with its trivial time dependence) will also be a solution, and we have found an interpretation of the corresponding coefficients. We wish to examine whether we can invert the process—namely, if we are given an arbitrary (but physically acceptable) initial state, $\psi(x, 0)$, whether we can expand it in terms of the energy eigenstates.

This procedure, if possible, would then allow us to solve the general initial value problem, as the resulting time dependence would be simply that of Eqn. (7.25). For the infinite well, this is the quantum mechanical analog of plucking a stretched string in some initial configuration and asking about its future vibrations.

If we formally write

$$\psi(x, 0) = \sum_{n=1}^{\infty} a_n u_n(x) \tag{7.38}$$

we can try to "invert" this to find the a_n by multiplying both sides by $u_m^*(x)\,dx$ and integrating. We find

$$\begin{aligned} \int_0^a u_m(x)\,\psi(x, 0)\,dx &= \sum_{n=1}^{\infty} a_n \int_0^a u_m(x)\,u_n(x)\,dx \\ &= \sum_{n=1}^{\infty} a_n \delta_{n,m} \\ &= a_m \end{aligned} \tag{7.39}$$

We have dropped the complex conjugation in this case because the $u_m(x)$ can be made real. Thus, the contribution of the mth eigenstate to the expansion of the initial wavefunction is given by their mutual overlap, defined by the integral in Eqn. (7.39).

Another immediate similarity with a *complete set* or *basis set* of unit vectors is thus apparent. If we write a general vector as $\mathbf{A} = \sum_i A_i\,\hat{\mathbf{e}}_i$, we can extract the expansion coefficients via the inner product as

$$\mathbf{A} \cdot \hat{\mathbf{e}}_j = \left(\sum_i A_i\,\hat{\mathbf{e}}_i\right) \cdot \hat{\mathbf{e}}_j = \sum_i A_i\,\delta_{i,j} = A_j \tag{7.40}$$

With some trivial changes, this expansion also works for the symmetric well, where we write

$$\psi(x, 0) = \sum_{n=1}^{\infty} \left(a_n^{(+)} u_n^{(+)}(x) + a_n^{(-)} u_n^{(-)}\right) \tag{7.41}$$

with

$$a_m^{(\pm)} = \int_{-a}^{+a} u_m^{(\pm)}(x)\,\psi(x, 0)\,dx \tag{7.42}$$

This expansion can be seen to be formally identical to the *Fourier series* expansion discussed in Sec. 2.2.2. The only differences are that the constant term is not allowed due to the boundary conditions at the walls of the well, and that the series expansion is defined to vanish outside the well. Expressions for the average value of powers of energy can also be easily obtained.

In general, the wavefunctions can be complex, so that for an arbitrary expansion

$$\psi(x, t) = \sum_n a_n\,\psi_n(x)\,e^{-iE_n t/\hbar} \tag{7.43}$$

the overlap integrals giving the expansion coefficients must be carefully written as

$$a_n = \int_{-\infty}^{+\infty} \psi_n^*(x)\, \psi(x,0)\, dx \tag{7.44}$$

Lastly, this inversion process is not limited to sums of energy eigenstates as the expansion in momentum eigenstates

$$\psi(x) = \frac{1}{\sqrt{2\pi\hbar}} \int_{-\infty}^{+\infty} \phi(p)\, e^{ipx/\hbar}\, dp \tag{7.45}$$

has the well-known inverse

$$\phi(p) = \int_{-\infty}^{+\infty} \left(\frac{1}{\sqrt{2\pi\hbar}}\right)^* \psi(x)\, dx = \frac{1}{\sqrt{2\pi\hbar}} \int_{-\infty}^{+\infty} \psi(x)\, e^{-ipx/\hbar}\, dx \tag{7.46}$$

Taken together and generalized to include general Hermitian operators, these results are examples of the so-called *expansion postulate* that states that

- The eigenfunctions of a Hermitian operator form a *complete set* since any admissible wavefunction can be expanded in such eigenfunctions.

Example 7.2 Why Bad Wavefunctions Are Bad Let us now turn to some examples. Consider first the initial waveform, defined inside the symmetric infinite well only by

$$\psi(x) = \begin{cases} 1\sqrt{a} & \text{for } |x| < a/2 \\ 0 & \text{for } |x| > a/2 \end{cases} \tag{7.47}$$

While normalized appropriately, this is not an acceptable wavefunction due to its discontinuities, and we will use this opportunity to examine the physical consequences of such behavior. The expansion coefficients are easily calculated, giving

$$a_n^{(-)} = \int_{-a}^{+a} u_n^{(-)}(x)\, \psi(x,0)\, dx = 0 \tag{7.48}$$

because of symmetry considerations (even function times odd function integrated over symmetric interval), while

$$\begin{aligned} a_n^{(+)} &= \int_{-a}^{+a} u_n^{(+)}(x)\, \psi(x,0)\, dx \\ &= \frac{2}{a} \int_0^{+a} \cos((n-1/2)\pi x/a)\, dx \\ &= \frac{4\sin((2n-1)\pi/4)}{(2n-1)\pi} \end{aligned} \tag{7.49}$$

which is dimensionless, as it should be. The convergence is similar to that in Ex. 2.1 and is shown in Fig. 7.3. We can now use these expansion coefficients to address a more physical question, namely, to evaluate the energy of this state. We find

$$\begin{aligned} \langle\psi|\hat{E}|\psi\rangle &= \sum_{n=1}^{\infty} |a_n^+|^2 E_n^+ \\ &= \sum_{n=1}^{\infty} \left(\frac{8}{(2n-1)^2\pi^2}\right)\left(\frac{\hbar^2\pi^2(2n-1)^2}{8ma^2}\right) \\ &= \frac{\hbar^2}{ma^2} \sum_{n=1}^{\infty} 1 = \infty \end{aligned} \tag{7.50}$$

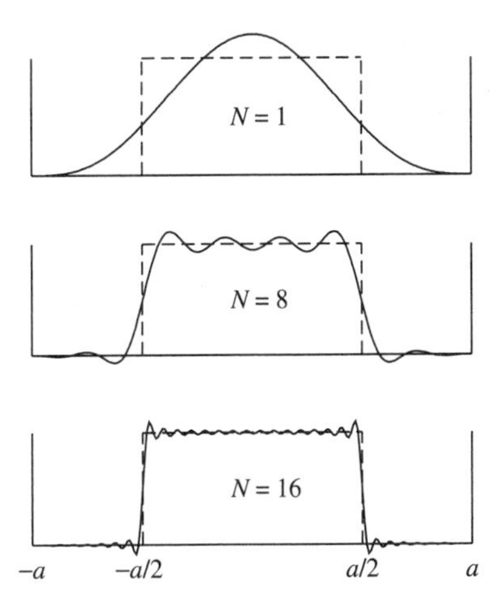

FIGURE 7.3. Generalized Fourier series expansion for square wave in symmetric infinite well for Ex. 7.2.

which is divergent! The kinetic energy corresponding to this state, measuring the wiggliness of ψ, is infinitely large due to the discontinuity. This can also be seen directly from the position and momentum space wavefunctions from a calculation of $\langle \hat{T} \rangle = \langle \hat{p}^2 \rangle / 2m$ (P7.9).

Example 7.3 The Expanding Box As a further example, let us consider the case of a particle in the symmetric well, known somehow to be in its ground state. *Very suddenly*, at $t = 0$, the walls are pulled apart symmetrically to a new width $2b$ (where $b > a$). (This is in contrast to P6.17 where slow or adiabatic changes were discussed.)

The initial wavefunction of the particle in this new well (see Fig. 7.4 at $t = 0$), defined via

$$\psi(x, 0) = \begin{cases} u_1^{(+)}(x; a) = \cos(\pi x/2a)/\sqrt{a} & \text{for } |x| < a \\ 0 & \text{for } a < |x| < b \end{cases} \tag{7.51}$$

can now be expanded in terms of the energy eigenstates of its new "universe" (the new well), i.e, in terms of the $u_n^{(\pm)}(x; b)$. The result is that

$$a_n^{(-)} = 0 \qquad a_n^{(+)} = \frac{4ab^2}{\pi\sqrt{ab}} \left(\frac{\cos((2n-1)\pi a/2b)}{(b^2 - (2n-1)^2 a^2)} \right) \tag{7.52}$$

Once again, the odd-state expansion coefficients vanish because of symmetry considerations. We plot the probabilities of finding the particle in the ground state, and first and second excited *even* states of the new well versus b/a in Fig. 7.5; as $b/a \to 0$, only the original ground state is required.

Having once calculated the expansion coefficients for the given initial waveform, the future time dependence of each term is dictated by the $\exp(-iE_n^{(\pm)}t/\hbar)$ factors. The wavefunction at later times is then given by

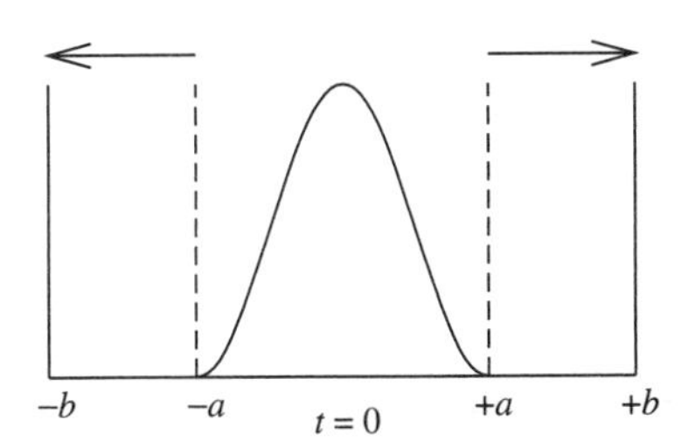

FIGURE 7.4. Initial state wavefunction of the "expanded well" state of Ex. 7.3 where the walls are suddenly moved from ±a to ±b.

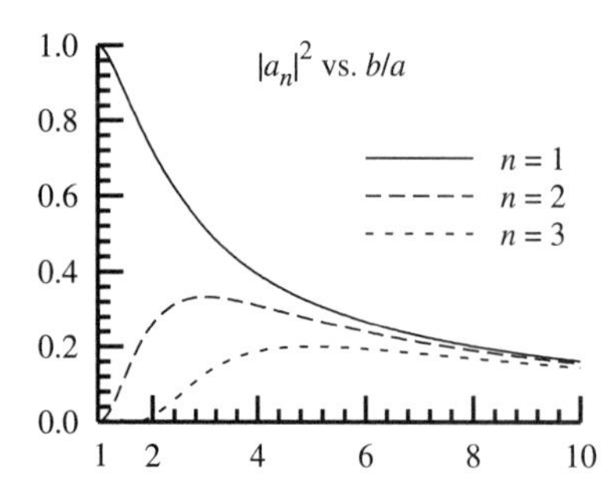

FIGURE 7.5. Expansion coefficients squared ($|a_n|^2$) for first three even levels versus b/a from Ex. 7.3; these give the probabilities of finding the state in each level.

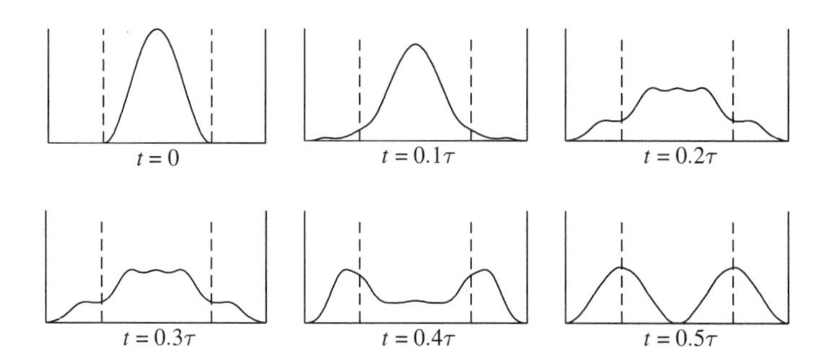

FIGURE 7.6. Time development of "expanded well" system illustrated by plots of $|\psi(x, t)|^2$ versus x for various times in the first half cycle.

$$\psi(x, t) = \sum_{n=1}^{\infty} a_n^{(+)} \frac{1}{\sqrt{b}} \cos\left(\frac{(n - 1/2)\pi x}{b}\right) e^{-iE_n^{(+)}(b)t/\hbar} \tag{7.53}$$

and we plot in Fig. 7.6 the resulting probability density (given by $|\psi(x, t)|^2$) for various future times. Because of the simplicity of the infinite well, the behavior is periodic, and we plot various times during the first half cycle only; it need not be periodic in a general potential.

If this problem seems somewhat artificial (suddenly moving the walls of an infinite well in a microscopic system is admittedly somewhat difficult to imagine realizing experimentally), it is representative of a class of interesting real-life problems in which some parameter of the potential undergoes a sudden change; we will see a more realistic example in Chap. 18 (P18.8).

7.6 ENERGY EIGENSTATES AND NORMAL MODES

Many of the properties of energy eigenstates that arise in the solutions of the time-independent Schrödinger equation in quantum mechanics are also present in the normal mode analysis of coupled systems in classical mechanics.[1] The main points that we will stress as having analogs are:

[1]For a discussion of normal modes in classical mechanics, see, e.g., Marion and Thornton (1988).

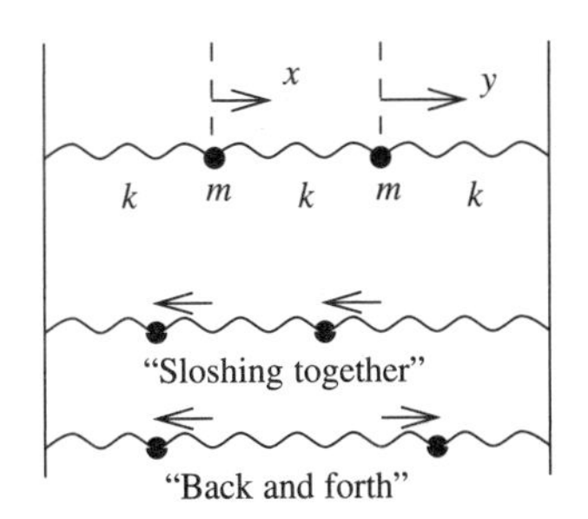

FIGURE 7.7. Mass and spring system that can exhibit normal mode vibrations.

1. A simple $e^{-iEt/\hbar}$ time dependence that turns the dynamical equation (the time-dependent SE) into an eigenvalue problem (the time-independent SE)
2. Eigenvalues that are "quantized" and correspond to special "motions" of the system
3. Eigenvectors that are automatically mutually orthogonal and that form a complete set

We will illustrate some similarities using a concrete example, namely, the system of coupled masses and springs shown in Fig. 7.7. Newton's laws for this two-degree-of-freedom problem are

$$\begin{aligned} -2kx(t) + ky(t) &= m\ddot{x}(t) \\ kx(t) - 2ky(t) &= m\ddot{y}(t) \end{aligned} \tag{7.54}$$

which can be written in the form

$$-\begin{pmatrix} 2k & -k \\ -k & 2k \end{pmatrix}\begin{pmatrix} \ddot{x}(t) \\ \ddot{y}(t) \end{pmatrix} = \begin{pmatrix} m & 0 \\ 0 & m \end{pmatrix}\begin{pmatrix} x(t) \\ y(t) \end{pmatrix} \tag{7.55}$$

or

$$-\mathbf{KX}(t) = \mathbf{M}\ddot{\mathbf{X}}(t) \tag{7.56}$$

where $\mathbf{X}(t)$ is a vector containing the position coordinates

$$\mathbf{X}(t) \equiv \begin{pmatrix} x(t) \\ y(t) \end{pmatrix} \tag{7.57}$$

while $\mathbf{M}$ and $\mathbf{K}$ are mass and spring constant matrices defined via

$$\mathbf{M} \equiv \begin{pmatrix} m & 0 \\ 0 & m \end{pmatrix} \qquad \mathbf{K} \equiv \begin{pmatrix} 2k & -k \\ -k & 2k \end{pmatrix} \tag{7.58}$$

If we assume the simple time-dependence $\mathbf{X}(t) = \mathbf{X}_0 e^{i\omega t}$, Eqn. (7.56) becomes

$$-\mathbf{KX}_0 = -\omega^2\mathbf{MX}_0 \qquad \text{or} \qquad \left(\omega^2\mathbf{M} - \mathbf{K}\right)\mathbf{X}_0 = 0 \tag{7.59}$$

or the matrix equation

$$\begin{pmatrix} m\omega^2 - 2k & k \\ k & m\omega^2 - 2k \end{pmatrix}\begin{pmatrix} x_0 \\ y_0 \end{pmatrix} = 0 \tag{7.60}$$

In order for this coupled set of equations to have a solution, the determinant of the matrix $\omega^2\mathbf{M} - \mathbf{K}$ must vanish, which implies that

$$(m\omega^2 - 2k)^2 - k^2 = 0 \qquad \text{or} \qquad \omega^2 = \frac{2k \pm k}{m} \tag{7.61}$$

This gives the two solutions $\omega_1 = \sqrt{k/m}$ and $\omega_2 = \sqrt{3k/m}$, which are the *normal mode frequencies,* i.e., the "eigenfrequencies." The eigenvectors, i.e., the $\mathbf{X}_0$ that correspond to these natural frequencies of oscillation, are

$$\mathbf{X}_0^{(1)} = \frac{1}{\sqrt{2}}\begin{pmatrix}1\\1\end{pmatrix} \qquad \text{and} \qquad \mathbf{X}_0^{(2)} = \frac{1}{\sqrt{2}}\begin{pmatrix}1\\-1\end{pmatrix} \tag{7.62}$$

They correspond, respectively, to modes of oscillation in which the two masses vibrate in phase with each other ("sloshing together") or out of phase ("back and forth"). The normal coordinates $\mathbf{X}_0^{(1),(2)}$ have been normalized by hand such that

$$\mathbf{X}_0^{(1)} \cdot \mathbf{X}_0^{(1)} = 1 \qquad \text{and} \qquad \mathbf{X}_0^{(1)} \cdot \mathbf{X}_0^{(1)} = 1 \tag{7.63}$$

but the fact that they are orthogonal is automatic, namely, that

$$\mathbf{X}_0^{(1)} \cdot \mathbf{X}_0^{(2)} = 0 \tag{7.64}$$

Note also that any point in the two-dimensional coordinate plane can be written in terms of the $\mathbf{X}_0^{(1),(2)}$, so that they form a complete set as well.

The complete solution can finally be written as

$$\mathbf{X}(t) = (A_1 e^{i\omega_1 t} + B_1 e^{-i\omega_1 t})\mathbf{X}_0^{(1)} + (A_2 e^{i\omega_1 t} + B_2 e^{-i\omega_2 t})\mathbf{X}_0^{(2)} \tag{7.65}$$

where the A_i, B_i are chosen so as to satisfy the initial conditions.

We see then that a very similar geometrical structure arises in this context. In quantum mechanics, we have discussed these properties and found them to arise because the operators associated with the eigenstates are Hermitian. We encourage the reader to find the corresponding connection in the area of normal modes.

7.7 PARITY

We have concentrated our attention so far on Hermitian operators that have familiar classical analogs; the momentum, the energy, and the Hamiltonian operators all have recognizable classical counterparts. In this section, we study a somewhat more abstract case, which arises because of a symmetry.

We have seen that the energy eigenfunctions for the symmetric well can also be characterized in terms of their evenness or oddness, i.e., their symmetry properties under reflections; these properties are obviously connected to the symmetry of the potential. To formalize this notion and to extend our experience with the properties of Hermitian operators, we are led to study the *parity operator,* defined via

$$\hat{P} f(x) \equiv f(-x) \tag{7.66}$$

which has the effect of taking the mirror reflection of a function about the origin as illustrated in Fig. 7.8.

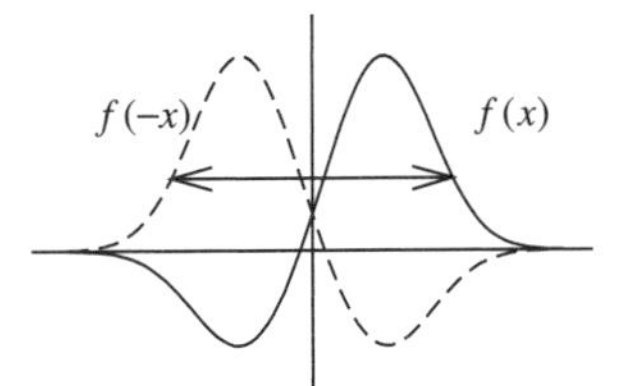

FIGURE 7.8. Parity operation applied to generic real function.

This operator is easily seen to be Hermitian since

$$\begin{aligned}
\left(\int_{-\infty}^{+\infty} dx\, \psi^*(x)\, \hat{P}\, \psi(x)\right) - \left(\int_{-\infty}^{+\infty} dx\, \psi^*(x)\, \hat{P}\, \psi(x)\right)^* \\
&= \int_{-\infty}^{+\infty} dx\, \psi^*(x)\, \psi(-x) - \int_{-\infty}^{+\infty} dx\, \psi^*(-x)\, \psi(x) \\
&= \int_{-\infty}^{+\infty} dx\, \psi^*(x)\, \psi(-x) - \int_{-\infty}^{+\infty} dy\, \psi^*(y)\, \psi(-y) \\
&= 0
\end{aligned} \tag{7.67}$$

where we have simply changed variables ($x \to -y$) in the second integral. This implies that the eigenvalues, λ_P, of the parity operator will be real (from Sec. 7.2). We can see that the definition

$$\hat{P} f(x) = \lambda_P f(x) \tag{7.68}$$

can be used twice to obtain

$$f(x) = \hat{P} f(-x) = \hat{P}^2 f(x) = \hat{P} \lambda_P f(x) = \lambda_P \hat{P} f(x) = \lambda_P^2 f(x) \tag{7.69}$$

implying that $\lambda_P = \pm 1$. Thus the eigenfunctions of the parity operator are simply even and odd functions with

$$f_E(-x) = \hat{P} f_E(x) = +f_E(x) \qquad f_O(-x) = \hat{P} f_O(x) = -f_O(x) \tag{7.70}$$

The orthogonality of eigenfunctions belonging to different eigenvalues is trivially evident as

$$\langle f_E \mid f_O \rangle = \int_{-\infty}^{+\infty} dx\, f_E^*(x)\, f_O(x) = 0 \tag{7.71}$$

for any even and odd functions. The expansion postulate can also be confirmed in this case, as we can always write a generic function in terms of even and odd solutions as

$$\begin{aligned}
f(x) &= \frac{f(x) + f(-x)}{2} + \frac{f(x) - f(-x)}{2} \\
&= f_E(x) + f_O(x) \\
&\equiv c_E \tilde{f}_E(x) + c_O \tilde{f}_O(x)
\end{aligned} \tag{7.72}$$

where $f_{E,O}(x)$ are obviously even and odd, respectively, but are not necessarily normalized properly. This expansion is illustrated in Fig. 7.9. If we write $f_{E(O)}(x) = c_{E(O)} \tilde{f}_{E(O)}(x)$ where $\tilde{f}_{E(O)}$ *is* normalized, then the expansion coefficients, namely, the $c_{E(O)}$, when squared, give the probability of finding the state with positive or negative parity (see P7.12).

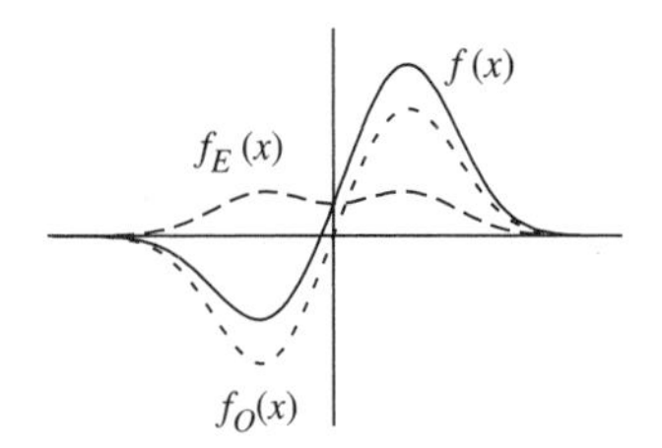

FIGURE 7.9. Generic real function ($f(x)$) with its expansion in even ($f_E(x)$) and odd ($f_O(x)$) functions.

7.8 SIMULTANEOUS EIGENFUNCTIONS

We have seen that the solutions for the symmetric infinite well problem are simultaneously eigenfunctions of both the Hamiltonian operator for that system and of the parity operator; that is, they have precisely determined values of both the energy and the parity. On the other hand, the content of the standard uncertainty principle, $\Delta x \, \Delta p \geq \hbar/2$, is to say that it's not possible to know simultaneously the values of both the position and momentum to arbitrary precision.

We are naturally led to ask under what conditions two different Hermitian operators can share eigenfunctions. The general result can be stated as:

- Two Hermitian operators, $\hat{A}$ and $\hat{B}$, can have simultaneous eigenfunctions if and only if they commute with each other, i.e.,

$$[\hat{A}, \hat{B}] \equiv \hat{A}\hat{B} - \hat{B}\hat{A} = 0$$

  or their commutator vanishes.

Before proceeding with the proof, we note that the commutator satisfies

$$[\hat{A}, \hat{B}] = -[\hat{B}, \hat{A}] \tag{7.73}$$

and

$$[\alpha\hat{A} + \beta\hat{B}, \hat{C}] = \alpha[\hat{A}, \hat{C}] + \beta[\hat{B}, \hat{C}] \tag{7.74}$$

Let us first assume that $\hat{A}$ and $\hat{B}$ commute and that we have found the eigenvalues and eigenfunctions of the operator $\hat{A}$, which we denote via $\hat{A}\psi_a = a\psi_a$. We assume for simplicity that the eigenvalue spectrum is not degenerate, i.e., that there is only one ψ_a for every eigenvalue a. (The case of degenerate eigenvalues is a straightforward extension.) We then note that

$$\begin{aligned} \hat{A}\left(\hat{B}\psi_a\right) &= \hat{B}\hat{A}\psi_a \qquad \text{(since } \hat{A} \text{ and } \hat{B} \text{ commute)} \\ &= \hat{B}\left(a\psi_a\right) \\ &= a\left(\hat{B}\psi_a\right) \end{aligned} \tag{7.75}$$

Thus, $\hat{B}\psi_a$ is a function that, when acted upon by $\hat{A}$, returns the same function multiplied by the number a. This is just the definition of ψ_a, so that $\hat{B}\psi_a$ must be, up to a multiplicative constant, the same as ψ_a, i.e., $\hat{B}\psi_a \propto \psi_a$. But this, in turn, is just the definition of an eigenfunction since

$$\hat{B}\psi_a \propto \psi_a \Longrightarrow \hat{B}\psi_a = b\psi_a \qquad \text{or} \qquad \hat{B}\psi_a^b = b\psi_a^b \tag{7.76}$$

so that ψ_a is also an eigenfunction of $\hat{B}$.

To complete the proof, we assume that we have found the simultaneous eigenfunctions of the operators $\hat{A}$ and $\hat{B}$ that satisfy

$$\hat{A}\psi_a^b = a\psi_a^b \qquad \text{and} \qquad \hat{B}\psi_a^b = b\psi_a^b \tag{7.77}$$

Clearly $\hat{A}$ and $\hat{B}$ commute on the common set of eigenfunctions since

$$[\hat{A}, \hat{B}]\psi_a^b = \left(\hat{A}\hat{B} - \hat{B}\hat{A}\right)\psi_a^b = \left(\hat{A}b - \hat{B}a\right)\psi_a^b = \left(ab - ba\right)\psi_a^b = 0 \tag{7.78}$$

Since we are dealing with Hermitian operators, we know that the ψ_a^b form a complete set, so that an arbitrary wavefunction can be written as $\psi = \sum_a c_a \psi_a^b$. This implies that

$$\left(\hat{A}\hat{B} - \hat{B}\hat{A}\right)\psi = \sum_a c_a \left(\hat{A}\hat{B} - \hat{B}\hat{A}\right)\psi_a^b = \sum_a c_a 0 = 0 \tag{7.79}$$

and the commutator of $\hat{A}$ and $\hat{B}$ vanishes when acting on an arbitrary wavefunction.

Because $[x, \hat{p}] = i\hbar \neq 0$, we now understand why we can never have simultaneous eigenfunctions of these two variables. For the case of the symmetric infinite well, the two operators are the Hamiltonian, $\hat{H} = \hat{p}^2/2m + V(x)$ and parity operator, $\hat{P}$. Their commutator, acting on a general state, can be written in two pieces:

$$\begin{aligned}
[\hat{p}^2/2m, \hat{P}]\psi(x) &= -\frac{\hbar^2}{2m}\left(\frac{d^2}{dx^2}\hat{P} + \hat{P}\frac{d^2}{dx^2}\right)\psi(x) \\
&= -\frac{\hbar^2}{2m}\left(\frac{d^2}{dx^2}\psi(-x) + \frac{d^2\psi(-x)}{dx^2}\right) \\
&= 0 \qquad \text{in general}
\end{aligned} \tag{7.80}$$

and

$$\begin{aligned}
[V(x), \hat{P}]\psi(x) &= \left(V(x)\hat{P} - \hat{P}V(x)\right)\psi(x) \\
&= \left(V(x) - V(-x)\right)\psi(-x) \\
&= 0 \qquad \text{only if } V(x) = V(-x)
\end{aligned} \tag{7.81}$$

that is, if $V(x)$ is symmetric.

Because eigenfunctions provide such detailed information on the quantum state, one strategy is to search for

- A maximally large set of *mutually commuting observable operators,* that is, the largest possible set of operators, $\hat{A}_i$, which represent observables (i.e., Hermitian operators) and which all commute with each other, namely, $[\hat{A}_i, \hat{A}_j] = 0$. Such a set is guaranteed to have simultaneous eigenfunctions.

7.9 PROPAGATORS

The expansion theorem approach to solving for the time dependence of quantum wavefunctions can be formally written in a way that is suggestive of yet another solution method familiar in classical mechanics, namely, the technique of Green's functions.[2]

[2] See, e.g., Marion and Thornton (1988).

The general solution

$$\psi(x,t) = \sum_n a_n \psi_n(x) e^{-iE_n t/\hbar} \tag{7.82}$$

with the initial conditions included by using

$$a_n = \int_{-\infty}^{+\infty} dx'\, \psi_n^*(x')\, \psi(x',0) \tag{7.83}$$

can be combined to give

$$\begin{aligned}\psi(x,t) &= \sum_n \left[\int_{-\infty}^{+\infty} dx'\, \psi_n^*(x')\, \psi(x',0)\right] \psi_n(x) e^{-iE_n t/\hbar} \\ &= \int_{-\infty}^{+\infty} dx' \left\{\sum_n \psi_n^*(x')\, \psi_n(x)\, e^{-iE_n t/\hbar}\right\} \psi(x',0) \\ &= \int_{-\infty}^{+\infty} dx'\, K(x,x';t,0)\, \psi(x',0) \end{aligned} \tag{7.84}$$

where we have defined the *propagator*

$$K(x,x';t,0) = \sum_n \psi_n^*(x')\, \psi_n(x) e^{-iE_n t/\hbar} \tag{7.85}$$

As its name implies, the propagator dictates the time dependence, or propagation in time, of the initial solution. An obvious property of $K(x',x;t,0)$ is that it must reproduce the initial state when $t \to 0$, so that one necessarily has

$$\delta(x'-x) = K(x',x;0,0) = \sum_m \psi_n^*(x')\psi_n(x) \tag{7.86}$$

which is sometimes a useful relation. We visualize this relation for the standard infinite well where we plot (in Fig. 7.10) the partial sums

$$K_N(x',x;0,0) = \sum_{n=1}^{N} u_n(x')u_n(x) \tag{7.87}$$

for two values of N; the sums increasingly cancel when $x \neq x'$ and add coherently when $x = x'$.

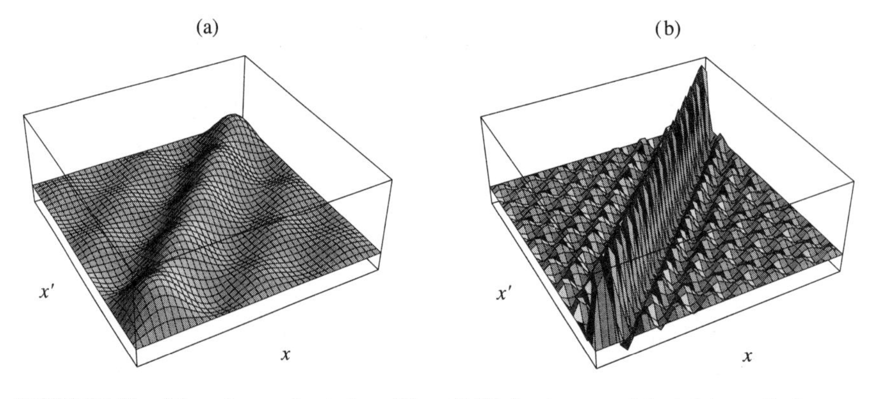

FIGURE 7.10. "Completeness" relation of Eqn. (7.87) for the case of the infinite well; the partial sums, $K_N(x',x;0,0)$, corresponding to (a) $N = 5$ and (b) $N = 15$ are plotted versus (x,x').

A generalization to states with a continuous energy eigenvalue label, such as for free particles, is straightforward. For that particular case, for example, we can write

$$K(x, x'; t, 0) = \int_{-\infty}^{+\infty} dp\, \psi_p^*(x')\, \psi_p(x) e^{-iE_p t/\hbar}$$
$$= \int_{-\infty}^{+\infty} dp \left(\frac{1}{\sqrt{2\pi\hbar}} e^{ipx'/\hbar}\right)^* \left(\frac{1}{\sqrt{2\pi\hbar}} e^{ipx/\hbar}\right) e^{-ip^2 t/2m\hbar} \tag{7.88}$$

and the integrals can be evaluated using standard Gaussian techniques (P7.19) to give

$$K(x, x'; t, 0) = \sqrt{\frac{m}{2\pi i\hbar t}}\, e^{im(x-x')^2/2\hbar t} \tag{7.89}$$

and one can check explicitly that

$$\lim_{x\to x'} K(x, x'; 0, 0) = \delta(x - x') \tag{7.90}$$

as it should. The time dependence of an initial Gaussian wavepacket,

$$\psi(x, 0) = \frac{1}{\sqrt{L\sqrt{\pi}}} e^{-(x-x_0)^2/2L^2 + ip_0(x-x_0)/\hbar} \tag{7.91}$$

is then easily evaluated using Eqn. (7.85) (P7.19), and the result agrees with our earlier derivation in Sec. 3.2.2.

A similar expansion in momentum space,

$$\phi(p, t) = \sum_n b_n \phi_n(p) e^{-iE_n t/\hbar} \tag{7.92}$$

implies a parallel formalism for propagators with

$$K_p(p', p; t, 0) = \sum_n \phi_n^*(p')\phi_n(p) e^{-iE_n t/\hbar} \tag{7.93}$$

giving

$$\phi(p, t) = \int_{-\infty}^{+\infty} dp'\, K_p(p, p'; t, 0)\phi(p', 0) \tag{7.94}$$

or suitable generalizations to continuous labels.

7.10 QUESTIONS AND PROBLEMS

Q7.1 Give an example of an operator that is *not* Hermitian, and show that its eigenvalues are not necessarily real.

Q7.2 In Ex. 7.3, what would happen to the wavefunction if the walls were suddenly *removed completely,* i.e., moved to infinity? How could you calculate the future time-dependence of the wavefunction in this case?

Q7.3 If you use the propagator formalism to calculate the future time dependence of a state that starts in an eigenstate, what happens?

P7.1 Consider the momentum space wavefunctions, $\phi_n^+(p)$, $\phi_n^-(p)$, for the symmetric infinite well given by Eqns. 6.35 and 6.63.

(a) Show that

$$\langle \phi_m^+ \mid \phi_m^+ \rangle = \int_{-\infty}^{+\infty} dp\, (\phi_n^+(p))^* \phi_m^+(p) = \delta_{n,m}$$

Hints: (1) Use partial fractions to rewrite the products found in the denominators and (2) use the integrals found in App. B.1.

(b) Show that $\langle \phi_n^+ \mid \phi_m^- \rangle = 0$ for all values of m, n.

P7.2 Consider stationary state solutions of the "asymmetric" infinite well discussed in Sec. 6.4.1. Show that two such solutions $\psi_1(x)$, $\psi_2(x)$ corresponding to *different* energy eigenvalues are mutually orthogonal, i.e, their overlap vanishes

$$\begin{aligned}
\langle \psi_1 \mid \psi_2 \rangle &= \int_{-a}^{+b} dx\, \psi_1(x)\, \psi_2(x) \\
&= \int_{-a}^{0} dx\, A_1 A_2 \sin(k_1(x+a)) \sin(k_2(x+a)) \\
&\quad + \int_{0}^{+b} dx\, B_1 B_2 \sin(q_1(x-b)) \sin(q_2(x-b)) \\
&= 0
\end{aligned}$$

Hints: Make repeated use of the boundary conditions at $x = -a, b$, as well as the definitions of k, q. You may also find useful the fact that $k_i^2 - k_2^2 = q_1^2 - q_2^2$, but you should prove it first. Do you need to have normalized the solutions first before you check whether they're orthogonal?

P7.3 **Schmidt orthogonalization procedure:** Suppose you have N wavefunctions, $\psi_i(x)$ for $i = 1, \ldots, N$, which are not necessarily normalized or orthogonal. One can generate a new set $\tilde{\psi}_i(x)$ which *are* orthogonal and normalized by taking linear combinations of the $\psi_i(x)$ using the following procedure:

1. Let $\tilde{\psi}_1(x) = \psi_1(x)$, and then normalize $\tilde{\psi}_1(x)$
2. Let $\tilde{\psi}_2(x) = \psi_2(x) - \langle \tilde{\psi}_1 \mid \psi_2 \rangle \tilde{\psi}_1(x)$. This is obviously orthogonal to $\tilde{\psi}_1(x)$ since

$$\langle \tilde{\psi}_2 \mid \tilde{\psi}_1 \rangle = \langle \tilde{\psi}_1 \mid \psi_2 \rangle - \langle \tilde{\psi}_1 \mid \tilde{\psi}_1 \rangle \langle \tilde{\psi}_1 \mid \psi_2 \rangle = 0$$

since $\tilde{\psi}_1$ is normalized. Normalize the new $\tilde{\psi}_2(x)$.

3. Let

$$\tilde{\psi}_3(x) = \psi_3(x) - \langle \tilde{\psi}_2 \mid \psi_3 \rangle \tilde{\psi}_2(x) - \langle \tilde{\psi}_1 \mid \psi_3 \rangle \tilde{\psi}_1(x)$$

and so forth, normalizing the $\tilde{\psi}_i(x)$ after each step. (If the original $\psi(x)$ are all eigenfunctions of some Hermitian operator with the same eigenvalue, the linear combinations $\tilde{\psi}_i(x)$ will be also, but will now be orthogonal.)

(a) Show that this procedure leads to an orthonormal set.

(b) Use this procedure to form an orthonormal set from the wavefunctions

$$\psi_0(x) = e^{-x^2/a^2} \qquad \psi_1(x) = xe^{-x^2/a^2} \qquad \psi_2(x) = x^2 e^{-x^2/a^2} \qquad \psi_3(x) = x^3 e^{-x^2/a^2}$$

over the interval $(-\infty, +\infty)$.

(c) Repeat for the set

$$\psi_0(x) = 1 \quad \psi_1(x) = x \quad \psi_2(x) = x^2 \quad \psi_3(x) = x^3$$

which are defined only over the interval $(-1, +1)$.

P7.4 A particle in a symmetric infinite well is described by the wavefunction

$$\psi(x,0) = N((3+2i)u_1^+(x) - 2u_1^-(x) + 3iu_2^+(x))$$

where the $u_n^{\pm}(x)$ are the infinite well wavefunctions.

(a) Find N so that $\psi(x,0)$ is properly normalized.
(b) What is $\psi(x,t)$?
(c) Evaluate $\langle\hat{E}\rangle_t$ and ΔE.
(d) What is the probability that a measurement of the energy of the particle would find $E = \hbar^2\pi^2/8ma^2$? $E = 6\hbar^2\pi^2/8ma^2$? $E = 9\hbar^2\pi^2/8ma^2$?

P7.5 Suppose that you have solved the Schrödinger equation for some arbitrary potential and found the generic eigenstates, e.g., $u_n(x)\exp(-iE_nt/\hbar)$ so that the general solution is

$$\psi(x,t) = \sum_n a_n u_n(x)\exp(-iE_nt/\hbar)$$

(a) Assume that you suddenly decide to change what you call the *origin* of coordinates along the x axis, i.e., letting $x \to x+d$, so that the new spatial solutions are $\tilde{\psi}(x,t) = \psi(x+d,t)$. Show that corresponding momentum space wavefunction is only changed by a phase factor, so that $|\phi(p,t)|^2$ is not changed at all. Discuss why this should be so.
(b) Suppose now that you decide to relabel the *zero* of your potential energy via $V(x) \to V(x) + V_0$. Find $\tilde{\psi}(x,t)$ in terms of the original solution and show that observables depending on $P(x,t) = |\psi(x,t)|^2$ are unaffected by a different choice of zero of potential. How is $|\phi(p,t)|^2$ affected?

P7.6 Consider an initial wavefunction in the symmetric infinite well defined by

$$\psi(x,0) = \begin{cases} -N & \text{for } -a < x < 0 \\ N & \text{for } 0 < x < a \end{cases}$$

(a) Find N so that $\psi(x,0)$ is properly normalized.
(b) Calculate the expansion coefficients, $a_n^{(\pm)}$, for this wavefunction.
(c) Use them to evaluate $\langle\hat{E}\rangle$. Is your answer consistent with what you know about this waveform?

P7.7 An initial wavefunction in the symmetric infinite well is given by

$$\psi(x,0) = N\left(1 - \frac{|x|}{a}\right)$$

(a) Answer the same questions as for P7.6. What is different about this waveform than that in P7.6 and Ex. 6.1.
(b) Try to evaluate ΔE in this state.

P7.8 Consider two wavefunctions in the symmetric infinite well defined by

$$\psi_1(x) = \begin{cases} 1/\sqrt{a} & \text{for } -a < x < 0 \\ 0 & \text{for } 0 < x < a \end{cases}$$

and

$$\psi_2(x) = \begin{cases} 0 & \text{for } -a < x < 0 \\ 1/\sqrt{a} & \text{for } 0 < x < a \end{cases}$$

i.e., waveforms that are completely localized within the left and right halves of the well. They are normalized properly and obviously orthogonal, i.e.,

$$\langle\psi_1 \mid \psi_2\rangle = \int_{-a}^{+a} \psi_1^*(x)\,\psi_2(x)\,dx = 0$$

(a) Calculate the corresponding momentum space wavefunctions, $\phi_{1,2}(x)$, and show explicitly that

$$\int_{-\infty}^{+\infty} dp\, \phi_1^*(p)\, \phi_2(p) = 0$$

(b) Expand both wavefunctions in energy eigenfunctions, and show that their overlap vanishes, i.e.,

$$\sum_{n=1}^{\infty} a_n^* b_n = 0.$$

Hint: Be sure to include both $a_n^{(\pm)}$ in the expansion and overlap summation.

P7.9 Consider the (unacceptable) wavefunction defined in Ex. 7.2.

(a) Write the position space wavefunction, $\psi(x)$, in terms of step functions, $\theta(x)$ (Sec. 2.4). Evaluate the average kinetic energy (which is the total energy since the potential vanishes inside the well) using the form

$$\langle \hat{T} \rangle = \frac{\langle \hat{p}^2 \rangle}{2m} = \frac{\hbar^2}{2m} \int dx \left| \frac{d\psi(x)}{dx} \right|^2$$

You may have to use some "seat-of-the-pants" mathematics to handle the Dirac δ functions you encounter.

(b) Evaluate the momentum space wavefunction, $\phi(p)$, and use

$$\langle T \rangle = \frac{\langle p^2 \rangle}{2m} = \frac{1}{2m} \int_{-\infty}^{+\infty} dp\, p^2\, |\phi(p)|^2$$

to find the kinetic energy.

P7.10 **(a)** In Ex. 7.3, show that the expansion coefficients are appropriately normalized, i.e., that

$$\sum_{n=1}^{\infty} |a_n^{(+)}|^2 = 1$$

in the case where $b = 2a$.

(b) The energy of the initial state is definitely known to be $E_1^{(+)} = \hbar^2\pi^2/8ma^2$. Evaluate $\langle \hat{E} \rangle$ after the walls have expanded for a general value of b. Show explicitly for the case $b = 2a$ that your answer agrees with the initial value. *Hint:* The summations in App. B.2 may be useful.

P7.11 Work out the normal mode frequencies and eigenvectors for the problem of three equal masses connected with equal strength springs. Check to see if the $\mathbf{X}_0^{1,2,3}$ are mutually orthogonal.

P7.12 Consider a real-valued function expanded in terms of even and odd functions as in Eqn. (7.72).

(a) Show that the probability that the wavefunction is in an even state is given by

$$c_E^2 = \frac{1}{2}\left(1 + \frac{\int_{-\infty}^{+\infty} dx\, f(x)\, f(-x)}{\int_{-\infty}^{+\infty} dx\, (f(x))^2}\right)$$

and an odd state by

$$c_O^2 = \frac{1}{2}\left(1 - \frac{\int_{-\infty}^{+\infty} dx\, f(x)\, f(-x)}{\int_{-\infty}^{+\infty} dx\, (f(x))^2}\right)$$

(b) Show that $c_E^2 = 1$ and $c_O^2 = 0$ for an even function.

P7.13 **Generalized parity operator:** If one wants to "reflect" a function in a point other than the origin, show that the generalized parity operator, defined via

$$\hat{P}_a f(x) = f(2a - x)$$

reflects the function about the point $x = a$

(a) Show that $\hat{P}_a$ is Hermitian, find its eigenvalues and eigenfunctions, and interpret them.

(b) The standard infinite well potential is symmetric about the point $x = L/2$. Show that the energy eigenfunctions for this case are eigenfunctions of $\hat{P}_{(L/2)}$, and find their eigenvalues.

P7.14 Consider the initial wavefunction in P7.4.

(a) What is the average value of the parity operator, $\hat{P}$, for this state? Does it depend on time?

(b) What is the probability of measuring the particle described by this wavefunction to have positive parity? Negative parity?

P7.15 **The complex conjugation operator:** Define the complex conjugation operator via

$$\hat{C} f(x) = f^*(x)$$

(a) Show that this operator is *not* Hermitian by finding a wavefunction for which $\langle \hat{C} \rangle = \langle \psi | \hat{C} | \psi \rangle$ is *not* real.

(b) Try to find the eigenvalues of this operator by modifying the derivation in Sec. 7.7.

(c) Show that *any* complex number $z = a + ib$ is an eigenfunction of $\hat{C}$, and find the corresponding eigenvalue.

P7.16 Which of the following pairs of operators can have simultaneous eigenfunctions?

(a) $\hat{p}$ and $\hat{T} = \hat{p}^2/2m$

(b) $\hat{p}$ and $V(x)$

(c) $\hat{E}$ and $\hat{p}$

P7.17 Consider the *translation operator*, $\hat{T}_a$, defined via

$$\hat{T}_a f(x) = f(x + a)$$

(a) Is $\hat{T}_a$ a Hermitian operator? Can you find an example where its expectation value is not real?

(b) What is $[\hat{T}_a, \hat{p}^2/2m]$?

(c) Under what conditions is $[\hat{T}_a, V(x)] = 0$?

P7.18 Show that the generalized parity operator $\hat{P}_{L/2}$ commutes with the Hamiltonian for the standard infinite well case; this explains the results found in P7.13.

P7.19 **(a)** Do the necessary Gaussian integral (assuming that it converges) to obtain Eqn. (7.89).

(b) Apply it to the initial Gaussian wavepacket in Eqn. 7.91.

(c) Show that the propagator for two arbitrary times $t_2 > t_1$ is given by

$$K(x, x'; t_2, t_1) = \sum_n \psi_n^*(x')\psi_n(x)e^{-iE_n(t_2-t_1)/\hbar}$$

(d) Show that the propagator satisfies the identity

$$K(x, x'; t_3, t_1) = \int_{-\infty}^{+\infty} dx''\, K(x, x''; t_3, t_2)\, K(x'', x', t_2, t_1)$$

and interpret the result in terms of propagating from $t_1 \to t_2 \to t_3$.

(e) Check the result of part (b) explicitly for the case of the free-particle propagator in Eqn. (7.89).

P7.20 **Position space versus momentum space propagators:**

(a) Show that the position space propagator can be derived from the one in momentum space via

$$K(x', x; t, 0) = \frac{1}{2\pi\hbar}\int_{-\infty}^{+\infty} dp'\, e^{ip'x'/\hbar}\int_{-\infty}^{+\infty} dp\, e^{-ipx/\hbar}\, K_p(p', p; t, 0)$$

Give the corresponding expression for $K_p(p', p; t, 0)$ in terms of $K(x', x; t, 0)$.

(b) Use the continuous generalization of Eqn. (7.93) to evaluate the momentum space propagator for a free particle. Apply it as in Eqn. (7.94) to an initial wavefunction $\phi(p', 0)$, and show that it reproduces the result of Eqn. 3.19.

P7.21 **Propagators for accelerating particle:** Consider the accelerating particle from Sec. 5.6.2, described by the time-independent Schrödinger equation in momentum space:

$$\frac{p^2}{2m}\phi_E(p) - i\hbar F\frac{d\phi_E(p)}{dp} = E\phi_E(p)$$

where $\phi(p, t) = \phi(p)e^{-iEt/\hbar}$.

(a) Show that an appropriate solution is

$$\phi_E(p) = Ce^{-ip^3/6m\hbar F}e^{-iEp/\hbar}$$

where C is an arbitrary constant.

(b) Use the fact that eigenstates corresponding to different eigenvalues should be properly normalized to evaluate C, and show that $C = 1/\sqrt{2\pi\hbar F}$.
Hint: Use the fact that

$$\int_{-\infty}^{+\infty} dp\, \phi_{E'}^*(p)\,\phi_E(p) = \delta(E - E')$$

(c) Construct the momentum space propagator by evaluating

$$K_p(p', p; t, 0) = \int dE\, \phi_E^*(p')\,\phi_E(p)\, e^{-iEt/\hbar}$$

where the summation is over the continuous E label. Show that the result, including the appropriate constant C, is

$$K_p(p', p; t, 0) = e^{i((p-Ft)^3-p^3)/6m\hbar F}\,\delta(p - p' - Ft)$$

and show that when applied to an initial momentum space wavefunction it reproduces Eqn. (5.112). Show also that this propagator satisfies the appropriate initial condition and that it reduces to the free-particle result when $F \to 0$.

P7.22 Using the results of P7.20 and P7.21, find the position space propagator for the accelerating particle. Show that it reduces to the free-particle case when $F \to 0$.

P7.23 **Green's functions:** The so-called *Green's function* for a quantum system can be defined via

$$G(x, x'; E) = \sum_{n=0}^{\infty}\frac{\psi_n(x)\psi_n^*(x')}{(E_n - E)}$$

The Green's function can be shown to be the transform[3] of the position space propagator, where the conjugate variables are E, t and not x, p.

(a) Show that $G(x, x'; E)$ satisfies the Schrödinger-like equation

$$\left(\hat{H} - E\right)G(x, x'; E) = -\delta(x - x')$$

(b) Show that it satisfies the "convolution" identity

$$\frac{\partial G(x, x''; E)}{\partial E} = \int dx' G(x, x'; E)\, G(x', x''; E)$$

[3]Actually the Laplace transform and not the Fourier transform.

CHAPTER 8

Many Particles in the Infinite Well: The Role of Spin and Indistinguishability

8.1 THE EXCLUSION PRINCIPLE

The self-consistent inclusion of the wave properties of matter into the dynamical equations of motion, via the Schrödinger equation, is an important aspect of any attempt to understand the structure of matter. This aspect of microscopic physics, which we might dub *ħ-physics*, cannot be underestimated, as it accounts for much of the observable phenomena we attribute to quantum mechanics. The discrete sets of spectral lines in bound state systems (atomic, molecular, nuclear, quark/antiquark), for example, arise from quantization of energy levels. Even macroscopic quantities, such as the densities of ordinary matter, can be understood using simple quantum arguments as in Ex. 1.1.

Many physical systems, however, require us to understand the organization of collections of large numbers of seemingly indistinguishable particles. The "building blocks of matter," the electrons of atomic physics, the protons and neutrons (nucleons) that form nuclei, and the quarks that, in turn, are bound together to make the nucleons and other strongly interacting particles are all in this category. These particles all share a common feature, namely, their *intrinsic angular momentum* or *spin*. The notion of "spin-up" and "spin-down" corresponding to such spin-1/2 particles will be an important additional degree of freedom that must be considered when constructing the multiparticle wavefunctions of such particles. The behavior of such particles is restricted by powerful constraints that do not arise from the machinery of wave mechanics but that, nonetheless, have a profound impact on the macroscopic properties of such systems. We will discuss this "indistinguishability physics" in much more detail in Chaps. 15 and 17, but we wish to emphasize its impact as soon as possible.

The most familiar manifestation of such effects comes from the *Pauli exclusion principle*, which states roughly that

- No two electrons (or any two indistinguishable spin-1/2 particles) can occupy the same quantum state.[1]

The fact that each atomic energy level can then accommodate only two electrons (one spin-up and one spin-down) leads to the shell structure of atomic physics and is therefore arguably responsible for much of chemistry and biology; similar shell structure occurs in nuclei as well. In this chapter, using the infinite well potential as a model, we will examine the dramatic effects this restriction can have on macroscopic numbers of particles, applying it to condensed matter and to nuclear and astrophysical systems.

8.2 ONE-DIMENSIONAL SYSTEMS

We begin by considering N_e electrons in the standard infinite well (in one dimension) with energy spectrum

$$E_n = \frac{\hbar^2 \pi^2 n^2}{2mL^2} \qquad n = 1, 2, 3, \ldots \tag{8.1}$$

For particles not required to satisfy the exclusion principle, the total energy of such a system would simply be

$$\text{(without exclusion principle)} \qquad E_0 = \frac{\hbar^2 \pi^2}{2mL^2} N_e \tag{8.2}$$

but for electrons one has to "fill up" the energy levels, two at a time, to a state characterized by $N_{\max} = N_e/2$ as in Fig. 8.1. The total energy is then

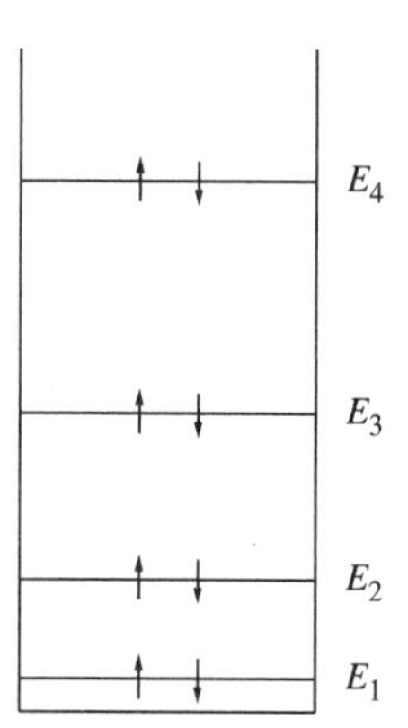

FIGURE 8.1. Filling of one-dimensional infinite well energy levels with spin-1/2 particles.

[1]Like any other physical law, the Pauli principle should be amenable to experimental verification, and it is natural to ask: How well do we know that the exclusion principle is satisfied? It turns out to be extremely difficult to construct logically self-consistent theories of quantum mechanics in which the Pauli principle is only violated by a small amount. Nonetheless, various experiments—see, e.g., Ramberg and Snow (1990)—have been taken to imply that the probability that a multi-electron system will be in a configuration that violates the exclusion principle is less than roughly 10^{-26}.

$$E_{\text{tot}} = 2\sum_{n=1}^{N_{\text{max}}} E_n = \frac{\hbar^2\pi^2}{mL^2}\sum_{n=1}^{N_{\text{max}}} n^2 \tag{8.3}$$

The summation can be done in closed form to yield

$$\sum_{n=1}^{N_{\text{max}}} n^2 = \frac{N_{\text{max}}(N_{\text{max}}+1)(2N_{\text{max}}+1)}{6} \approx \frac{N_{\text{max}}^3}{3} \tag{8.4}$$

when $N_{\text{max}} \gg 1$. In this case, the labeling of states and energy summation is trivial, but in a more realistic three-dimensional example, the enumeration of states will be more complicated, so we also do the "counting" in a slightly more formal manner. We can write

$$N_e = 2\sum_{n=1}^{N_{\text{max}}} 1 = 2\sum_{n=1}^{N_{\text{max}}} \Delta n \approx 2\int_1^{N_{\text{max}}} dn \approx 2N_{\text{max}} \tag{8.5}$$

and

$$\sum_{n=1}^{N_{\text{max}}} n^2 = \sum_{n=1}^{N_{\text{max}}} n^2\,\Delta n \approx \int_1^{N_{\text{max}}} n^2\,dn \approx \frac{N_{\text{max}}^3}{3} \tag{8.6}$$

We have identified $\Delta n = 1$ with dn and used a simple version of the more general relation between a discrete sum and continuous integral, namely, the Euler-Maclaurin formula, which is discussed in App. B.2.

In either case we find

$$\text{(with exclusion principle)} \qquad E_{\text{tot}} = \frac{\hbar^2\pi^2}{mL^2}\left(\frac{N_{\text{max}}^3}{3}\right) = \frac{\hbar^2\pi^2}{24mL^2}N_e^3 \tag{8.7}$$

The average energy per particle, $\overline{E}$, is simply

$$\overline{E} = \frac{E_{\text{tot}}}{N_e} = \frac{\hbar^2\pi^2}{24mL^2}N_e^2 \tag{8.8}$$

while the energy of the last state to be filled, the so-called *Fermi energy*, is

$$E_{\text{Fermi}} = E_F = \frac{\hbar^2\pi^2 N_{\text{max}}^2}{2mL^2} = \frac{\hbar^2\pi^2}{8mL^2}N_e^2 \tag{8.9}$$

Taken together, the electrons in this configuration consistent with the exclusion principle are sometimes colloquially said to constitute the *Fermi sea*, so that E_F is the energy of an electron at the "top of the sea." From Eqns. (8.8) and (8.9), we also note that

$$\overline{E} = \frac{1}{3}E_F \tag{8.10}$$

The role of the exclusion principle can be seen by noting that the ratio of total energies with and without this constraint is roughly

$$\frac{E_{\text{tot}}(\text{with exclusion})}{E_{\text{tot}}(\text{without exclusion})} \approx \frac{N_e^2}{12} \tag{8.11}$$

which can be an enormous difference if $N_e \gg 1$.

We are accustomed to cases in which if we get all the dimensional factors right, then the estimate of the physical observable is usually wrong by less than an order of magnitude or so, in either direction. In this case, the "$\hbar$-physics" has predicted the dimensional factors correctly, but the exclusion principle can still play just as important a role in determining the actual state of the physical system.

8.3 THREE-DIMENSIONAL INFINITE WELL

The one-dimensional example makes it clear that the exclusion principle can play a very important role. In order to make a plausible connection to a real system of spin-1/2 particles, however, we require a three-dimensional model, so we consider particles in a 3D cubical box of volume $V = L^3$, i.e., a potential given by

$$V(x) = \begin{cases} 0 & \text{for } |x|, |y|, |z| < L \\ \infty & \text{otherwise} \end{cases} \tag{8.12}$$

Either by "fitting de Broglie waves into the box" or via explicit solution of the 3D Schrödinger equation, we can easily find the quantized energies, using

$$E = \frac{\mathbf{p}^2}{2m} = \frac{p_x^2 + p_y^2 + p_z^2}{2m} \tag{8.13}$$

namely,

$$\begin{aligned} E(\mathbf{n}) &= E(n_x, n_y, n_z) \\ &= \frac{\hbar^2\pi^2}{2mL^2}\left(n_x^2 + n_y^2 + n_z^2\right) \\ &= \frac{\hbar^2\pi^2}{2mL^2}\mathbf{n}^2 \qquad n_x, n_y, n_z = 1, 2, 3 \ldots \end{aligned} \tag{8.14}$$

The quantity $\mathbf{n} \equiv (n_x, n_y, n_z)$ can be considered as a vector in the (abstract) three-dimensional number space pictured in Fig. 8.2. Two electrons can be accommodated at each discrete point of the first quadrant (since $n_x, n_y, n_z > 0$) of this $\mathbf{n}$ space. The expression for the energy levels depends on $\mathbf{n}$ in a "spherically symmetric" way as it involves only $\mathbf{n}^2 = n_x^2 + n_y^2 + n_z^2$, so we must fill up the energy levels radially outward to a "radius" in $\mathbf{n}$ space, which we call R_N. The enumeration of states is then best done using

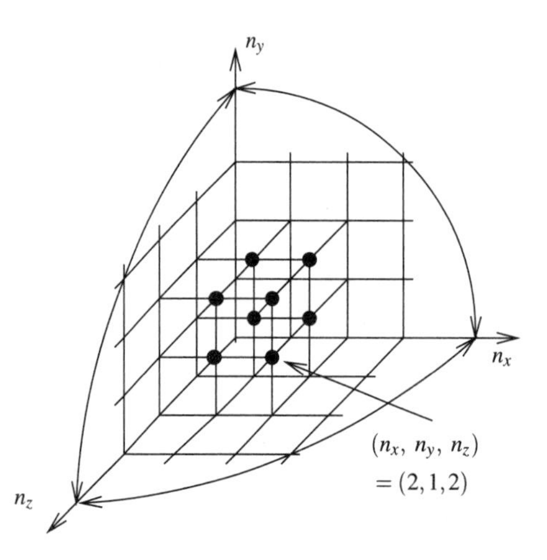

FIGURE 8.2. *Filling of three-dimensional infinite well energy levels with spin-1/2 particles.*

the integral approximations discussed above, and we find first that

$$
\begin{aligned}
N_e &= 2 \sum_{n_x,n_y,n_z=1}^{|\mathbf{n}|=R_N} 1 = 2 \sum_{n_x,n_y,n_z=1}^{|\mathbf{n}|=R_N} \Delta n_x \, \Delta n_y \, \Delta n_z \\
&\approx 2 \int dn_x \, dn_y \, dn_z \\
&= 2 \int_1^{R_N} n^2 \, dn \int d\Omega_n \\
&= 2 \left[\frac{R_N^3}{3} \right] \left(\frac{4\pi}{8} \right) \\
&\approx \frac{\pi}{3} R_N^3
\end{aligned}
\tag{8.15}
$$

where $d\Omega_n$ is the generalized solid angle in **n** space, and we integrate only over the first octant. This implies that

$$
R_N = \left(\frac{3}{\pi} N_e \right)^{1/3}
\tag{8.16}
$$

The total energy of the system is then

$$
\begin{aligned}
E_{\text{tot}} &= 2 \sum_{n_x,n_y,n_z=1}^{|\mathbf{n}|=R_N} E(\mathbf{n}) \\
&= 2 \sum_{n_x,n_y,n_z=1}^{|\mathbf{n}|=R_N} \frac{\hbar^2 \pi^2}{2mL^2} (n_x^2 + n_y^2 + n_z^2) \, \Delta n_x \, \Delta n_y \, \Delta n_z \\
&\approx \frac{\hbar^2 \pi^2}{mL^2} \int |\mathbf{n}|^2 \, d^3 n \\
&= \frac{\hbar^2 \pi^2}{mL^2} \int_1^{R_N} n^4 \, dn \left(\frac{4\pi}{8} \right) \\
&\approx \frac{\hbar^2 \pi^3}{10mL^2} R_N^5 \\
E_{\text{tot}} &= \frac{\hbar^2 \pi^3}{10mL^2} \left(\frac{3}{\pi} \right)^{5/3} N_e^{5/3}
\end{aligned}
\tag{8.17}
$$

or

$$
E_{\text{tot}} = \frac{\hbar^2 \pi^3}{10mV^{2/3}} \left(\frac{3}{\pi} \right)^{5/3} N_e^{5/3}
\tag{8.18}
$$

where we have used $L^3 = V$. The average energy and Fermi energy are obtained as before; one finds

$$
\overline{E} = \frac{\hbar^2 \pi^3}{10mL^2} \left(\frac{3}{\pi} \right)^{5/3} N_e^{2/3} = \frac{3}{5} E_F
\tag{8.19}
$$

This form is useful as it implies that

$$\overline{E} = \frac{\pi^3}{10}\left(\frac{3}{\pi}\right)^{5/3}\frac{\hbar^2}{m_e}n_e^{2/3} \tag{8.20}$$

where $n_e \equiv N_e/L^3 = N_e/V$ is the number density of electrons in the system.

Once again, because of the exclusion principle, the total energy for a system of electrons is larger than for distinguishable particles, in this case by a factor of roughly $N_e^{2/3}$. For the electrons in a white dwarf star, for example, where $N_e \approx 10^{57}$, forgetting about the constraints imposed by the exclusion principle would mean an error of roughly a factor of $(10^{57})^{2/3} \approx 10^{38}$ in an estimation of the total zero point energy in the star due to electrons. We mentioned in Chap. 1 that there is a sense in which the universe covers roughly 60 orders of magnitude in length. It is hard to imagine any other physical system in which leaving out a single physical effect, in this case the exclusion principle, could give rise to an error which is "two-thirds the size of the universe."

8.4 APPLICATIONS

In this section we will examine the role that the exclusion principle plays in some important physical systems.

8.4.1 Conduction Electrons in a Metal

Metals, with their observed large electrical and thermal conductivity, are characterized by the presence of quasi-free electrons that can move in response to electrical or thermal gradients. These *conduction electrons* can be modeled as a noninteracting gas in a three-dimensional box, just as in Sec. 8.3.

Typically, one or two electrons per atom are available so that number densities of the order $n_e \sim 3 \times 10^{22}$ cm^{-3} are appropriate. This implies (P8.6) a Fermi energy of $E_F \sim 4$ eV; this in turn gives an average electron kinetic energy due to zero point motion of roughly $\overline{E} \sim 2$ eV instead of the thermal energy of $k_B T \sim 1/40$ eV expected at room temperature on the basis of classical statistical mechanics.

This difference can be observed clearly in the behavior of the electronic contribution to the heat capacity of the metal. Classical kinetic theory would say that the thermal energy of N_e electrons would be $E \sim N_e k_B T$, so that the heat capacity due to electrons would be

$$C_{\text{el}} \sim \frac{\partial E}{\partial T} \sim N_e k_B \tag{8.21}$$

independent of temperature. This is inconsistent with experimental results that find an electronic heat capacity that varies linearly with T and which is smaller than this value by a factor of 0.01 or less.

In the quantum electron gas picture, only electrons in the Fermi sea in quantized states within roughly $k_B T$ of the Fermi energy can be thermally excited to new states (see Fig. 8.3) and so participate in the interactions required for conduction. Thus, only a fraction $k_B T/E_F$ of the electrons are available, and the heat capacity is more properly

$$C_{\text{el}} \sim \frac{\partial}{\partial T}\left(\frac{k_B T}{E_f} N_e k_b T\right) \sim N_e k_B \frac{k_B T}{E_f} \tag{8.22}$$

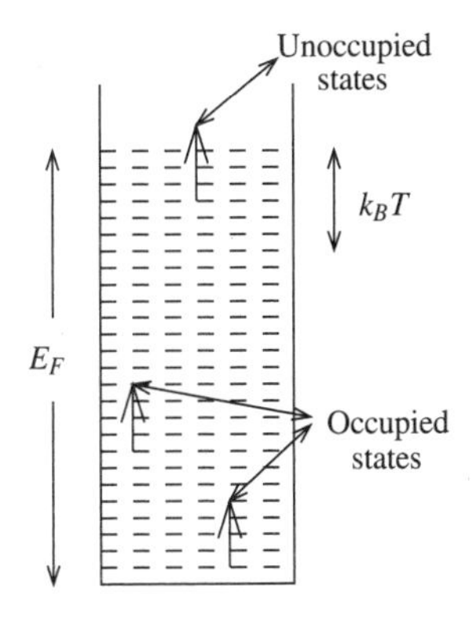

FIGURE 8.3. Filled energy levels in an electron gas picture of conduction electrons; electrons within a "distance" k_BT of the Fermi surface can be excited to unoccupied states and fully participate in thermal or electrical conduction.

The linear temperature dependence and resulting magnitude are then consistent with experiment.

Further experimental evidence for the electron gas picture comes from another macroscopic property of metals, namely, their compressibilities. The change in volume (ΔV) of a solid with applied pressure (ΔP) is often characterized by its *bulk modulus,* defined via

$$\Delta P = -B\frac{\Delta V}{V} \tag{8.23}$$

or equivalently

$$B = -V\frac{\partial P}{\partial V} \tag{8.24}$$

The appropriate pressure for an electron gas is

$$P(V) = -\frac{\partial E_{\text{tot}}}{\partial V} \tag{8.25}$$

so that from Eqn. (8.18), we find that (P8.6)

$$P_e = \frac{\pi^3}{15}\left(\frac{3}{\pi}\right)^{5/3}\frac{\hbar^2}{m_e}n_e^{5/3} \tag{8.26}$$

which implies that

$$B_e = \frac{5}{3}P_e \tag{8.27}$$

Using a value appropriate for sodium, $n_e = 2.6 \times 10^{22}\ \text{cm}^{-3}$, we find that $B_e \approx 9 \times 10^9\ N/\text{m}^2$, which is of the same order as the observed value of $6.4 \times 10^{10}\ N/\text{m}^2$. This suggests the important role that the exclusion principle plays in the compressibility of matter.

8.4.2 Neutrons and Protons in Atomic Nuclei

The nuclei of atoms are bound state systems of spin-1/2 protons and neutrons. The fact that protons are electrically charged while neutrons are not clearly makes them distinguishable from each other. The individual protons, however, are indistinguishable from each other, as are the neutrons, so the exclusion principle can play an important role in the structure of atomic nuclei.

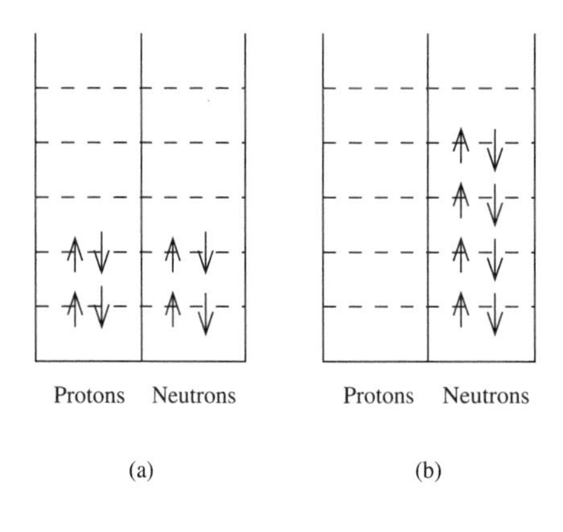

FIGURE 8.4. *Filling of nuclear energy levels with protons and neutrons; (a) corresponds to $Z = N = A/2$ and minimizes the zero point energy, while (b) corresponds to $Z = 0$ and $N = A$ to attempt to reduce the Coulomb energy.*

One of the important problems of nuclear structure is to understand which combinations of Z protons and N neutrons will be stable for a given value of the atomic number, $A = Z + N$. Given the Coulomb repulsion between electrons, it would seem that a nucleus composed solely of neutrons, i.e., $N = A, Z = 0$, would have the least energy; this is in marked contrast to the observed pattern where most light nuclei have $N \approx Z \approx A/2$.

If we model a nucleus as a 3D infinite well, the total energy of a system of Z protons and N neutrons will be obtained by filling up energy levels, consistent with the exclusion principle with a Fermi sea for each species; this is shown in Fig. 8.4(a). If, for a given value of A, we were to have all neutrons instead of protons to minimize the Coulomb potential energy, the resulting quantum zero point energy of the system would be much larger, as in Fig. 8.4(b). The energy cost to do this is called the *symmetry energy* and is discussed in P8.7. The incorporation of this effect into the so-called *semiempirical mass formula* for nuclei is discussed in many textbooks.[2]

8.4.3 White Dwarf and Neutron Stars

We are used to thinking of the gravitational force, described via classical mechanics, as playing the dominant role in determining most of the structure of large astrophysical systems, i.e., the solar system, galaxies, and beyond. The fact that stars use thermonuclear reactions as their energy source already implies that the nuclear force, necessarily described by quantum mechanics, is also important in stellar evolution. The need to understand quantum tunneling effects (Chap. 12) and scattering and reaction cross sections (Chap. 20) implies that "$\hbar$-physics" plays a key role in the structure of stars during the period when they are burning via fusion. It is perhaps not surprising then that "indistinguishability physics" can have just as important an effect in the determination of the ultimate fate of stars.

In this section, we present an extended essay on the role played by the exclusion principle on the structure of compact astrophysical objects, namely, white dwarf and neutron stars.[3] To simplify matters, we will consider a highly simplified model of a star, namely, a constant density, spherical object. We will also use the 3D infinite well as a model for the quantum zero point energy.

During its lifetime, a star is supported against gravitational collapse by the thermal pressure of the hot gas, with the energy being supplied by thermonuclear fusion reactions and radiated away in the form of electromagnetic radiation (and neutrinos). When the

[2]See, e.g., Krane (1988).

[3]For more details, see the excellent survey by Shapiro and Teukolsky (1983).

fusion reactions cease to be exothermic (no longer produce energy), the star will naturally begin to collapse. This can be seen by examining the total gravitational potential energy of a constant density, spherical object (P8.8) of mass M_*, namely,

$$V_G(R) = -\frac{3}{5}\frac{GM_*^2}{R} \tag{8.28}$$

which obviously favors smaller radii. The corresponding gravitational pressure, P_G, can be obtained from the corresponding force via

$$P_G(R) = \frac{F_G(R)}{A} = \frac{-dV_G(R)/dR}{4\pi R^2} = -\frac{dV_G(\Omega)}{d\Omega} \tag{8.29}$$

where $\Omega = 4\pi R^3/3$ is the (spherical) volume. Using Eqn. (8.28), we find that

$$P_G(\Omega) = -\frac{1}{5}\left(\frac{4\pi}{3}\right)^{1/3}\frac{GM_*^2}{\Omega^{4/3}} \tag{8.30}$$

which is radially inwards, favoring collapse. If thermal pressure is no longer sufficient to balance this, we need to look for a new source of energy or pressure that increases with decreasing radius to compensate.

One such source is the zero point energy of the (ionized and hence free) electrons in the star. We can estimate this using the infinite well model as

$$E_{\text{tot}} = \frac{\hbar^2\pi^3}{10m_e}\left(\frac{3}{\pi}\right)^{5/3}\frac{N_e^{5/3}}{\Omega^{2/3}} \tag{8.31}$$

where we have used Eqn. (8.18) and equated $V = \Omega$ with the volume of the star. (For sufficiently large numbers of particles, one can argue that the exact shape of the infinite well should be unimportant; see P8.4 for an example.)

As R (and hence Ω) decreases, this (positive) energy increases at a faster rate, proportional to $1/\Omega^{2/3}$, than the (negative) gravitational potential that goes as $1/R \propto 1/\Omega^{1/3}$. For small enough radii, it can therefore be the dominant source of energy. Equivalently, we can calculate the corresponding pressure to find

$$P_{EDP} = -\frac{dE_{\text{tot}}(\Omega)}{d\Omega} = \frac{\hbar^2\pi^3}{15m_e}\left(\frac{3}{\pi}\right)^{5/3}\frac{N_e^{5/3}}{\Omega^{5/3}} \tag{8.32}$$

We note that:

- We label this contribution as *electron degeneracy pressure* (EDP) as it arises from the zero point energy of a degenerate electron gas.
- We ignore (for the moment) the similar contributions from protons, neutrons, and other heavier nuclear species as their masses are at least $m_p/m_e \approx 2000$ times heavier and give a much smaller contribution to the zero point energy.

The total pressure, $P_G(\Omega) + P_{EDP}(\Omega)$, will vanish at some value of Ω, as shown in Fig. 8.5. The value of R at which this happens will constitute a new stable configuration as a small decrease (increase) in R will cause a net positive (negative) pressure driving the system back to equilibrium. This can also be seen by examining the total energy, $V_G(\Omega) + E_{\text{tot}}(\Omega)$ versus Ω, and noting that it has a stable minimum (positive curvature or "concave up") when the total pressure (the derivative with respect to Ω) vanishes. The dependence of P_{EDP} on $\hbar$ implies that there is a quantum effect balancing a macroscopic

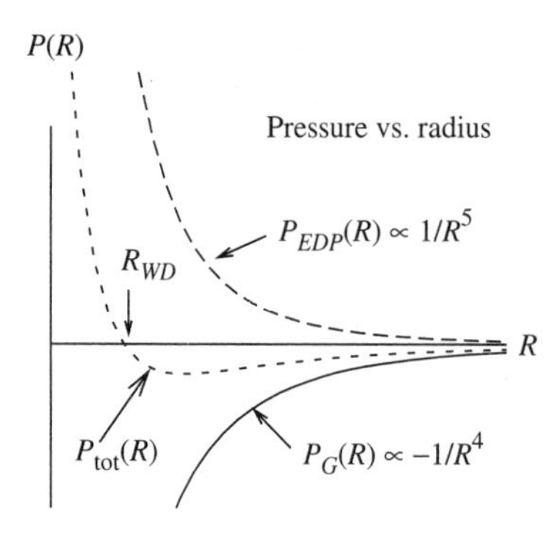

FIGURE 8.5. *Gravitational pressure ($P_G(R)$), electron degeneracy pressure ($P_{EDP}(R)$), and total pressure $P_{tot} = P_G + P_{EDP}$. The value of $P_{tot}(R = R_{WD}) = 0$ gives a stable minimum of the total energy corresponding to a white dwarf star.*

gravitational force. We might well imagine that the resulting balance could only occur at microscopically small values of R.

To find the radius of the new stable object, we simply set

$$P_G(\Omega) + P_{EDP}(\Omega) = 0 \tag{8.33}$$

or

$$\frac{1}{5}\left(\frac{4\pi}{3}\right)^{1/3} \frac{GM_*^2}{\Omega^{4/3}} = \frac{\hbar^2\pi^3}{15m_e}\left(\frac{3}{\pi}\right)^{5/3} \frac{N_e^{5/3}}{\Omega^{7/3}} \tag{8.34}$$

to find

$$R = \left(\frac{9\pi}{4}\right)^{2/3} \frac{\hbar^2 N_e^{5/3}}{m_e G M_*^2} \tag{8.35}$$

If we assume that the star was initially made of only hydrogen, i.e., equal numbers of protons (N_p) and electrons (N_e), we can take $M_* = N_p m_p$ (since $m_e \ll M_p$) or $N_e = M_*/m_p$. Then Eqn. (8.35) can also be written as

$$R = \left(\frac{9\pi}{4}\right)^{2/3} \frac{\hbar^2}{m_e m_p^2 G N_e^{1/3}} = \left(\frac{9\pi}{4}\right)^{2/3} \frac{\hbar^2}{m_e m_p^{5/3} G M_*^{1/3}} \tag{8.36}$$

showing that larger initial masses imply smaller final radii.

If we assume, for example, a two-solar-mass star, i.e., $M_* = 2M_\odot \approx 4 \times 10^{30}$ kg, we find $N_p = N_e \approx 2.5 \times 10^{57}$ and (P8.9)

$$R_{WD} = 1.8 \times 10^7 \text{ km} \approx 1.8 \times 10^4 \text{ km} \tag{8.37}$$

to be compared with the radius of the earth, namely, $R_e = 0.6 \times 10^4$ km.

This new type of compact astrophysical object is called a *white dwarf star* (hence the label WD) and is characterized by the following properties:

- Its small size and the fact that it is still hot—typical surface (interior) temperatures of roughly 8000 K–10,000 K (10^{6-7} K)—justify its name.
- Typical densities are $\rho_{WD} \sim 2 \times 10^5$ g/cm^3 compared to average stellar densities of 1–3 g/cm^3.
- Such higher densities imply that the electrons are more closely packed than in "normal" matter. One can estimate the typical electron separation to be $d \approx 0.02$ Å $\approx$ 2000 F, which is roughly halfway (on a logarithmic scale) between typical atomic and nuclear length scales.

- The average electron energy from Eqn. (8.19) is

$$\overline{E} = \frac{3}{5} E_F \approx 0.044 \text{ MeV} \tag{8.38}$$

which is smaller than the electron rest energy of 0.51 MeV. This implies that:

With this energy, the electron speeds are roughly $v/c \approx 0.4$ or $(v/c)^2 \approx 0.15$, which helps justify, a posteriori, the use of a nonrelativistic approximation. For only somewhat heavier initial stellar masses, however, the white dwarf radii will be smaller, the zero point energy larger, and the effects of relativity must be taken into account;

The classical *thermal* energy of electrons in the core is $kT \approx 0.08 - 0.8$ keV $\ll \overline{E}$, so that the electronic kinetic energy is indeed dominated by its zero point motion.

- The important role of the exclusion principle in the determination of R_{WD} is obvious, as the final numerical answer would have been a factor of $N_E^{2/3} \approx 2 \times 10^{38}$ *smaller* had we neglected its effects, giving

$$R_{WD} \approx 10^{-34} \text{ cm} \qquad \text{(without the exclusion principle)} \tag{8.39}$$

instead!

These derivations seem to imply that every star will eventually collapse and stabilize as a white dwarf star. To see under what circumstances these arguments must be refined, we first note that we can combine Eqns. (8.35) and (8.19) to see that

$$\overline{E} = \frac{1}{5}\left[\frac{2}{3\pi^2}\right]^{1/3} \frac{m_e m_p^{8/3} G^2 M_*^{4/3}}{\hbar^2} \tag{8.40}$$

This implies that a roughly sixfold increase in M_* in our example will give $\overline{E} \approx 0.51$ MeV $\approx m_e c^2$, implying that the electrons are now relativistic and the original assumptions of "nonrelativity" are invalid.

We should then use the most general energy momentum relation

$$E^2 = (\mathbf{p}c)^2 + (m_e c^2)^2 \tag{8.41}$$

but we can understand the qualitative changes by employing the ultra-relativistic limit where

$$E = |\mathbf{p}|c \tag{8.42}$$

In this case, the quantized energies are given by

$$E(\mathbf{n}) = \frac{\hbar c}{L}\sqrt{n_x^2 + n_y^2 + n_z^2} = \frac{\hbar c|\mathbf{n}|}{L} = \frac{\hbar c|\mathbf{n}|}{\Omega^{1/3}} \tag{8.43}$$

With this form, the total energy can be calculated in the same manner as Eqn. (8.10), (P8.10), and we find

$$E_{\text{tot}}^{\text{rel}}(\Omega) = \frac{\pi^2}{4}\left(\frac{3}{\pi}\right)^{4/3} \frac{\hbar c}{\Omega^{1/3}} N_e^{4/3} \tag{8.44}$$

which implies a degeneracy pressure of the form

$$P_{EDP}^{\text{rel}} = \frac{\pi^2}{12}\left(\frac{3}{\pi}\right)^{4/3} \frac{\hbar c}{\Omega^{4/3}} N_e^{4/3} \tag{8.45}$$

In this limit, the electron degeneracy pressure has the same Ω dependence as the gravitational pressure, so an equilibrium state can no longer be guaranteed by simply varying R. The fate of the star relies, instead, on the relative magnitudes of the *coefficients* of the $1/\Omega^{4/3}$ terms in Eqns. (8.30) and (8.45), namely,

$$C_{EDP} = \frac{\pi^2}{12}\left(\frac{3}{\pi}\right)^{4/3} \hbar c N_e^{4/3} \qquad \text{versus} \qquad C_G = \frac{1}{5}\left(\frac{4\pi}{3}\right)^{1/3} GM_*^2 \tag{8.46}$$

Since $C_{EDP} \propto N_e^{4/3} \propto M_*^{4/3}$ while $C_G \propto M_*^2$, the gravitational pressure will always win for sufficiently large initial stellar mass, causing continued collapse. The critical value beyond which a white dwarf star will not be stable can then be *roughly* estimated by letting $C_{EDP} = C_G$ to find

$$M_*^{\text{crit}} = \left(\frac{5}{4}\right)^{3/2} \left[\frac{9\pi}{4}\right]^{1/2} \left(\frac{\hbar c}{G}\right)^{3/2} \frac{1}{m_p^2} \approx 6.6 \times M_\odot \tag{8.47}$$

where $M_\odot$ is a solar mass.

This limit on the maximum mass of a white dwarf star is called the *Chandrasekhar limit*, and a more sophisticated analysis finds that

$$M_*^{\text{crit}} \approx 1.4\, M_\odot \tag{8.48}$$

The fact that we are within less than an order of magnitude from the "right answer" is now consistent with our belief that we have included much of the relevant physics, especially the exclusion principle, in however simplified a fashion. We note also that the Chandrasekhar mass depends only on the fundamental constants $\hbar$, c, G, and m_p in such a way that

$$M_*^{\text{crit}} \propto \frac{M_P^3}{m_p^2} \tag{8.49}$$

where M_P is the *Planck mass* discussed in P1.15:

$$M_P = \sqrt{\frac{\hbar c}{G}} \tag{8.50}$$

If $M_* > M_*^{\text{crit}}$, then the star will continue to collapse, but another stable configuration is possible. In this case, it can become energetically favorable for the electrons to interact with the protons via inverse β decay, i.e., via the reaction $e^- + p \to n + \nu_e$. The electrons and protons "disappear," leaving behind only the neutrons as the weakly interacting neutrinos escape (recall P2.25 about supernovae as the origin of a pulse of neutrinos associated with the cataclysmic collapse). This occurs when the average electron kinetic energy becomes greater than the energy cost of the reaction, namely, $Q = (m_n - m_p - m_e)c^2 \approx 0.78$ MeV, which is (coincidentally) at roughly the same point at which the nonrelativistic calculation breaks down.

This leaves a gas of indistinguishable spin-1/2 neutrons, a *neutron star*, and the same analysis as for white dwarf stars can be performed. The radius of this new configuration is given by Eqn. (8.35) with m_e replaced by m_n so that

$$R_{NS} = \frac{m_e}{m_n} R_{WD} \approx \frac{1}{1800} R_{WD} \approx 10 \text{ km} \tag{8.51}$$

Typical densities are now

$$\rho_{NS} \approx \left(\frac{m_n}{m_e}\right)^3 \rho_{WD} \approx 10^{15} \text{ g/cm}^3 \tag{8.52}$$

which is similar to the density of atomic nuclei. We leave it to the interested reader to examine the phenomenology of such interesting astrophysical objects as well as the path leading to the most intriguing final configuration, namely, black holes.

8.5 QUESTIONS AND PROBLEMS

Q8.1 We have applied a simplified picture of noninteracting spin-1/2 particles in a box to model various physical problems. In each case (conduction electrons in a metal, neutrons and protons in a nucleus, etc.), we have ignored the mutual interactions between the particles. How do the exclusion principle results and the Fermi sea picture help make this approximation more plausible? *Hint:* If two particles in the Fermi sea were to interact, thereby changing their momentum and energy, to which energy levels could they go?

Q8.2 Discuss why the exact shape of the infinite well potential should make no difference to the total zero point energy if there are many particles.

Q8.3 By how much would one side of a piece of metal 1 meter long increase in size if it were put into a vacuum so that atmospheric pressure were no longer acting on it? How does this depend on $\hbar$?

Q8.4 Discuss the following statement: There are two regions of stable nuclei:

- One with $A = 1$ to ≈ 200, with a balance between the strong attractive nuclear force and the Coulomb repulsion
- one with $A \approx 10^{57}$, with a balance between the attractive gravitational force and quantum zero point energy

Q8.5 We estimated the cutoff mass beyond which a star no longer can become a white dwarf, the so-called *Chandrasekhar limit*. Real stars are initially rotating and often have nonnegligible magnetic fields. As the star collapses, the rotation rate must increase (to conserve angular momentum) and the energy density in the magnetic field also increases (same number of field lines in a smaller volume). What qualitative changes would the inclusion of such effects make on estimates of the upper limit on masses of white dwarf stars, i.e., would the limit go up, down, or not change at all?

P8.1 A system of indistinguishable spin $S = 1/2$ particles can accommodate two states in each energy level, spin-up and spin-down, i.e., with $S_z = \pm 1/2$. A particle with spin $S = 3/2$ can have four states with the same energy, $S_z = -3/2, \ -1/2, \ 1/2,$ and $3/2$ in each level. Calculate the total, average, and Fermi energy of N such particles in a 3D infinite well. Can you generalize your results to spin $5/2, 7/2, \ldots$ and so forth where the degeneracy per level is given by $2S + 1$?

P8.2 Consider a two-dimensional system of electrons confined to a box of area $A = L^2$. Calculate the total energy of N_e such electrons, as well as the average and Fermi energies. Can you show that in d dimensions the total energy should scale as $E_{\text{tot}} \propto N_e^{(d+2)/d}$?

P8.3 **Spin and energy levels for the harmonic oscillator:**

(a) The quantized energy levels of a particle of mass m in a harmonic oscillator potential in one dimension are given by

$$E_n = (n + 1/2)\hbar\omega \quad n = 0, 1, 2, \ldots$$

where $\omega = \sqrt{K/m}$ is the classical oscillation frequency. Find the total, average, and Fermi energy of N_e (noninteracting) electrons in this potential.

(b) Can you derive the same quantities for the three-dimensional harmonic oscillator for which the energy levels are

$$E(\mathbf{n}) = (n_x + n_y + n_z + 3/2)\hbar\omega \quad n_x, n_y, n_z = 0, 1, 2, \ldots$$

P8.4 Instead of assuming a cubical 3D infinite well as in Sec. 8.3, assume that the electrons are in a box of dimensions L_x, L_y, L_z, so that $V = L_x L_y L_z$.

(a) Find the allowed energy levels as in Eqn. (8.14).

(b) Calculate the total energy of the system and compare to the expression in Eqn. (8.18). *Hint:* The volume of an ellipsoid governed by the formula

$$\frac{x^2}{a^2} + \frac{y^2}{b^2} + \frac{z^2}{c^2} = 1$$

is given by $V = 4\pi abc/3$.

P8.5 **Density of states:**

(a) Using Eqn. (8.19), show that the number of states with energy between E and $E + dE$ is given by

$$\frac{dN}{dE} \propto \sqrt{E} = C\sqrt{E}$$

which is applicable in the interval $E \in (0, E_F)$.

(b) Normalize this to find C by using the fact that

$$N_e = \int_0^{E_F} \frac{dN}{dE}\, dE$$

(c) Use this expression to evaluate $\langle E \rangle$, and show that it agrees with Eqn. (8.19).

(d) What fraction of the particles have energy less than $0.1\ E_F$? Less than $0.5\ E_F$?

P8.6 **(a)** For an electron density $n_e = 3 \times 10^{22}\ \text{cm}^{-3}$, evaluate the Fermi and average energies.

(b) Derive Eqn. (8.26).

(c) Confirm the numerical values for the bulk modulus for Na discussed in Sec. 8.4.1.

P8.7 **Spin and energy levels in nuclear physics:**

(a) Consider a heavy nucleus consisting of $Z = N_p$ protons and $N = N_n$ neutrons so that the total atomic number is $A = Z + N$. If the nucleons have volume V_0 and are bound so that they are "just touching," show that the volume of such a nucleus will go as $V = V_0 A$ so that its radius will be given by $R = R_0 A^{1/3}$. With $R_0 \approx 1.2$ to $1.4\ F$, this form works well for most nuclei.

(b) Instead of considering all the many nucleon-nucleon interactions, model this system as A nucleons in a 3D infinite well of radius R. Using Eqn. (8.18), find the total zero point energy of a system of $A = Z + N$ nucleons, i.e., $E(Z, N)$, recalling that both protons and neutrons are spin-1/2 particles and are distinguishable from each other. For simplicity, assume that the neutron and proton have the same mass, roughly $mc^2 \approx$ 940 MeV.

(c) For fixed A, minimize this energy, and show that equal numbers of neutrons and protons are favored.

(d) Assuming that $\Delta \equiv N - Z \ll A$, show that the zero point energy can be approximated by an expression of the form

$$E(Z, N) \approx E_{\text{min}}(A) + E_{\text{sym}} \frac{\Delta^2}{A} + \cdots \tag{8.53}$$

and find an expression for E_{sym}. Using the values above, show that the numerical value of E_{sym} is roughly 40 MeV. *Note:* This contribution to the total nuclear energy (or rest mass) is often called the *symmetry contribution* and is part of the well-known *semi-empirical mass formula* of nuclear physics.

P8.8 Consider a uniform sphere of mass M and radius R. Find the total gravitational potential energy stored in this configuration, namely, Eqn. (8.28). *Hint:* Consider a uniform sphere of mass $M(r)$ of some intermediate radius r and calculate the work, dW, required to add a layer of thickness dr brought in from infinitely far away; then use $W = \int dW$.

P8.9 Verify the numerical value in Eqn. (8.37).

P8.10 Repeat the calculation of the total electron energy in the relativistic case and confirm Eqn. (8.44).

P8.11 We argued that the white dwarf configuration was stable against small perturbations in radius since the total energy was a minimum at $R = R_{WD}$. Expand $E_{\text{tot}}(R)$ as a function of R around this minimum, and show that it is of the form

$$E_{\text{tot}}(R) = E_{\text{tot}}(R_{WD}) + \frac{1}{2}\frac{d^2 E_{\text{tot}}(R = R_{WD})}{dR^2}(R - R_{WD})^2 + \cdots$$

which is like a harmonic oscillator potential. Use this connection to find the frequency of oscillations around the stable minimum configuration using $\tau = 2\pi\sqrt{m/K}$, where K is the effective spring constant in the problem.

CHAPTER 9

Other 1D Potentials

Our study of the infinite well family of potentials has provided us with an array of insights into the physical meaning and mathematical structure of wave mechanics. There are other quantum mechanical properties that do not appear in such systems, and in this chapter we study several model potentials that illustrate many new aspects, both formal and intuitive, which have wide applications. We consider smoothness conditions on $\psi(x)$ and its derivatives, singular potentials and applications to models of band structure (Sec. 9.1), quantum mechanical tunneling and the large $|x|$ behavior of wavefunctions (Sec. 9.2), and the application of one-dimensional problems to three dimensions (Sec. 9.4).

9.1 SINGULAR POTENTIALS

9.1.1 Continuity of $\psi'(x)$

We have argued that the identification of $|\psi(x,t)|^2$ with an observable probability density requires that $\psi(x,t)$ be continuous in x. It is natural to ask about further smoothness conditions on the position space wavefunction and their meaning. As an example, we discuss below under what conditions we can demand that the spatial derivative of $\psi(x)$ be continuous. We note that we can rewrite the Schrödinger equation in the form

$$\frac{d^2\psi(x)}{dx^2} = \frac{2m}{\hbar^2}\big(V(x) - E\big)\psi(x) \tag{9.1}$$

and to examine the derivative at a point $x = a$ we note that we can integrate Eqn. (9.1) over the narrow interval $(a-\epsilon, a+\epsilon)$ to obtain

$$\begin{aligned}\psi'(a^+) - \psi'(a^-) &\equiv \frac{d\psi(a+\epsilon)}{dx} - \frac{d\psi(a-\epsilon)}{dx} \\ &= \int_{a-\epsilon}^{a+\epsilon} dx\, \frac{d^2\psi(x)}{dx^2} \\ &= \frac{2m}{\hbar^2}\int_{a-\epsilon}^{a+\epsilon} dx\big(V(x) - E\big)\,\psi(x)\end{aligned} \tag{9.2}$$

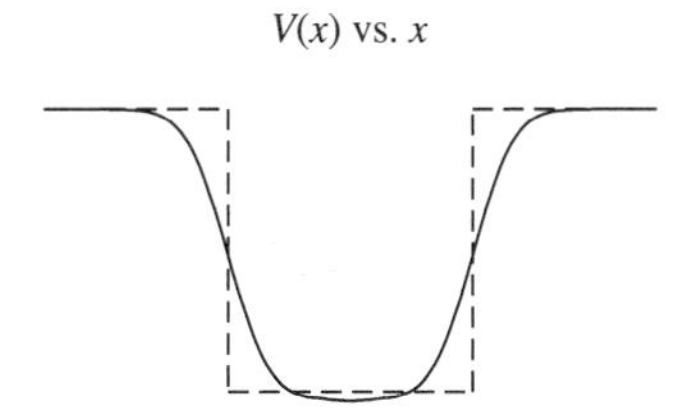

FIGURE 9.1. Simplified model of a smooth potential energy function (solid curve) using discontinuous step functions (dashed curve).

where $a^{\pm} \equiv a \pm \epsilon$ are x values that approach a infinitesimally closely from either side as $\epsilon \to 0$.

Since $E\,\psi(x)$ is everywhere finite, its contribution to the integral vanishes as $\epsilon \to 0$, and we need only worry about the behavior of $V(x)$ near a. Even if $V(x)$ is *discontinuous* at $x = a$ (say, a step or θ function as in Fig. 9.1), the integral $\int_{a-\epsilon}^{a+\epsilon} dx\, V(x)\,\psi(x)$ will still vanish as ϵ is made smaller, as the "area" under $V(x)\,\psi(x)$ still decreases; this implies that $\psi'(a^+) = \psi'(a^-)$, i.e., $\psi'(x)$ is continuous.

If, instead, the potential has a δ function singularity at $x = a$, then the area under even an infinitesimally small region around a will be finite. Specifically, if $V(x) = \pm g\delta(x - a)$ (where g is a constant with the appropriate units), we find that the appropriate boundary condition on $\psi'(x)$ is

$$\psi'(a^+) - \psi'(a^-) = \pm\frac{2mg}{\hbar^2}\psi(a) \tag{9.3}$$

We can summarize these statements by saying that

- The spatial derivative, $\psi'(x)$, is everywhere continuous except at points where the potential is *singular,* where it satisfies Eqn. (9.3).

The potential has to be this "badly behaved" in order for $\psi'(x)$ to be discontinuous. The smoothness of higher derivatives of $\psi(x)$ is explored in P9.1.

9.1.2 Single δ Function Potential

While a singular potential of the δ function type is unphysical, it can provide a model system with a highly localized attractive (or repulsive) force, and so is useful as a toy model. We examine the bound state spectrum ($E = -|E| < 0$) for a single attractive δ function potential of the form $V(x) = -g\delta(x)$ by looking for solutions of the Schrödinger equation

$$-\frac{\hbar^2}{2m}\frac{d^2\psi(x)}{dx^2} = E\psi(x) = -|E|\psi(x) \qquad \text{for } x \neq 0 \tag{9.4}$$

where $V(x) = 0$. This differential equation has the general solution

$$\psi(x) = \begin{cases} Ae^{Kx} + Be^{-Kx} & \text{for } x < 0 \\ Ce^{-Kx} + De^{Kx} & \text{for } x > 0 \end{cases} \tag{9.5}$$

where $K = \sqrt{2m|E|/\hbar^2}$. Imposing the conditions that the wavefunction be square integrable and continuous at $x = 0$ gives the physically acceptable solution

$$\psi(x) = \begin{cases} Ae^{Kx} & \text{for } x < 0 \\ Ae^{-Kx} & \text{for } x > 0 \end{cases} \tag{9.6}$$

The final boundary condition available to determine the quantized energy eigenvalues is Eqn. (9.3), which implies that

$$\psi'(0^+) - \psi'(0^-) = -AK - AK = -\frac{2mg}{\hbar^2}(A) = -\frac{2mg}{\hbar^2}\psi(0) \tag{9.7}$$

which gives

$$K = \frac{mg}{\hbar^2} \quad \text{or} \quad E = -|E| = -\frac{mg^2}{2\hbar^2} \tag{9.8}$$

We note that:

- There is a single bound state energy, and the normalized wavefunction is
$$\psi(x) = \sqrt{K}e^{-K|x|} \tag{9.9}$$
which was discussed in Sec. 3.3.1. The presence of the cusp at $x = 0$ in this wavefunction is now understood as arising from the singular potential. The fact that the expectation values of high powers of $\hat{p}$ are ill-defined (P5.1) is also plausible given the discontinuous behavior of higher derivatives of $\psi(x)$.
- The momentum space wavefunction for this solution was also derived in P5.2 by the Fourier transform and has a *Lorentzian* form
$$\phi(p) = \sqrt{\frac{2p_0}{\pi}}\left(\frac{p_0}{p^2 + p_0^2}\right) \tag{9.10}$$
where $p_0 \equiv \hbar K$; this also shows that $\langle p^4 \rangle$ (and higher powers) are not well defined.
- There is a discrete bound state spectrum for $E < 0$ (actually only one state), but there is a continuum of unbound states with $E > 0$; this is similar to the spectrum of, say, the hydrogen atom. The discrete and continuous states, *taken together*, form a complete set from which any admissible wavefunction can be constructed; the bound state solutions, by themselves, are not sufficient.

9.1.3 Single δ Function in Momentum Space

We can also solve this problem in momentum space[1] in a way that illustrates, in a manner not seen so far, how boundary conditions give rise to quantized energy levels. We can rewrite the Schrödinger equation in position space:

$$\frac{\hat{p}^2}{2m}\psi(x) - g\delta(x)\psi(x) = -|E|\psi(x) \tag{9.11}$$

in momentum space as in Sec. 5.6.1 by multiplying both sides by

$$\frac{1}{\sqrt{2\pi\hbar}}\exp(-ipx/\hbar)\,dx \tag{9.12}$$

and integrating over x to obtain

$$\frac{p^2}{2m}\phi(p) - \frac{g}{\sqrt{2\pi\hbar}}\int_{-\infty}^{+\infty} dx\; \delta(x)\,\psi(x)\,e^{-ipx/\hbar} = -|E|\phi(p) \tag{9.13}$$

[1]For a similar derivation, see Lieber (1975).

This yields

$$\left(\frac{p^2}{2m} + |E|\right)\phi(p) = \frac{g}{\sqrt{2\pi\hbar}}\psi(0) \tag{9.14}$$

or

$$\phi(p) = \frac{2mg}{\sqrt{2\pi\hbar}}\left(\frac{\psi(0)}{p^2 + (\hbar K)^2}\right) \tag{9.15}$$

which does have the Lorentzian form obtained in Eqn. (9.10), but is not yet necessarily normalized. To determine the value of quantized energy (contained in K), we recall from the definition of Fourier transform that

$$\psi(0) = \frac{1}{\sqrt{2\pi\hbar}}\int_{-\infty}^{+\infty} dp\,\phi(p) \tag{9.16}$$

so that substituting Eqn. (9.15) into this we find the condition

$$\psi(0) = \frac{1}{\sqrt{2\pi\hbar}}\left[\frac{2mg\psi(0)}{\sqrt{2\pi\hbar}}\int_{-\infty}^{+\infty}\frac{dp}{p^2 + (\hbar K)^2}\right] \tag{9.17}$$

Since this constraint is a direct application of the boundary conditions and *not* the overall normalization condition, the explicit value of $\psi(0)$ is irrelevant and drops out, and on performing the integral we find that

$$K = \frac{mg}{\hbar^2} \quad \text{or} \quad E = -\frac{mg^2}{2\hbar^2} \tag{9.18}$$

in agreement with Eqn. (9.8). The momentum space wavefunction can be normalized and the position space wavefunction obtained by an inverse Fourier transform.

9.1.4 Twin δ Function Potential

The problem defined by a twin δ function potential, namely,

$$V(x) = -g\left[\delta(x+a) + \delta(x-a)\right] \tag{9.19}$$

is a simple generalization of the problem above and can be used as a toy model for a one-dimensional diatomic ion[2]; it describes two attractive centers separated by a distance $2a$ that we can vary. As there is another natural length scale in the problem, namely, $L \equiv \hbar^2/mg$, we expect the physical results to depend on the ratio of these two quantities.

Because of the symmetry of the potential, we can restrict ourselves to the study of even and odd solutions. For example, for even solutions we can write

$$\psi^{(+)}(x) = \begin{cases} Ae^{Kx} & \text{for } x < -a \\ B\cosh(Kx) & \text{for } -a < x < a \\ Ae^{-Kx} & \text{for } x > a \end{cases} \tag{9.20}$$

where we have applied the boundary conditions at infinity and the symmetry to take the even combination of e^{-Kx} and e^{Kx}. Because of the symmetry of the wavefunction, the boundary conditions at $x = -a$ give the same constraint as those at $x = a$, which we use to write

[2]See Lapidus (1970).

$$B\cosh(Ka) = Ae^{-Ka}$$
$$\left(-AKe^{-Ka}\right) - \left(BK\sinh(Ka)\right) = -\frac{2mh}{\hbar^2}Ae^{-Ka} \quad (9.21)$$

arising from the continuity of $\psi(x)$ and "discontinuity" of $\psi'(x)$, respectively. These can be combined to yield the eigenvalue equation

$$\text{even eigenvalue condition:} \quad y\left(1 + \tanh(y)\right) = \frac{2mga}{\hbar^2} \equiv \frac{a}{a_0} \quad (9.22)$$

where we have defined

$$y \equiv Ka \quad \text{and} \quad a_0 = \frac{\hbar^2}{2mg} \quad (9.23)$$

The odd solutions, written as

$$\psi^-(x) = \begin{cases} -Ae^{Kx} & \text{for } x < -a \\ B\sinh(Kx) & \text{for } -a < x < a \\ Ae^{Kx} & \text{for } x > a \end{cases} \quad (9.24)$$

have the corresponding eigenvalue condition,

$$\text{odd eigenvalue condition:} \quad y\left(1 + \coth(y)\right) = \frac{a}{a_0} \quad (9.25)$$

The energy eigenvalues can be determined from Eqns. (9.22) and (9.25) by plotting the functions on the left-hand side versus y and looking for intersections with horizontal lines corresponding to values of a/a_0 as shown in Fig. 9.2. For comparison, in that diagram we also plot the eigenvalue condition for the *single* δ function potential, namely,

$$K = \frac{mg}{\hbar^2} \quad (9.26)$$

written in the form

$$2y = 2Ka = 2\left(\frac{mg}{\hbar^2}\right)a = \frac{a}{a_0} \quad (9.27)$$

for comparison. Plots of the position space wavefunction for three values of a/a_0 are shown in Fig. 9.3. We note several important features:

- In the limit of large separation, $a \gg a_0$, there is one even and one odd solution, with energy identical to the single δ function case; the two energy levels are *degenerate*. In this limit, the wavefunctions are simply related to those for the single δ function case (P9.3).

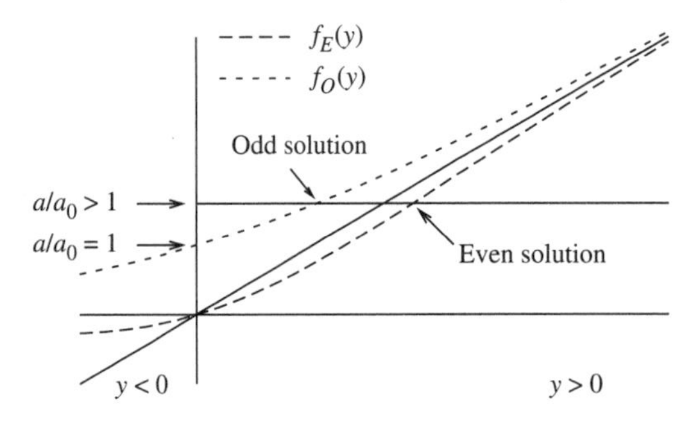

FIGURE 9.2. Energy eigenvalue condition for a twin δ function potential. Intersections of the dotted (dashed) curves with horizontal lines of constant a/a_0 give the values of y for the odd (even states). The solid line is the result for a single δ function potential. For $a < a_0$ there are no odd solutions.

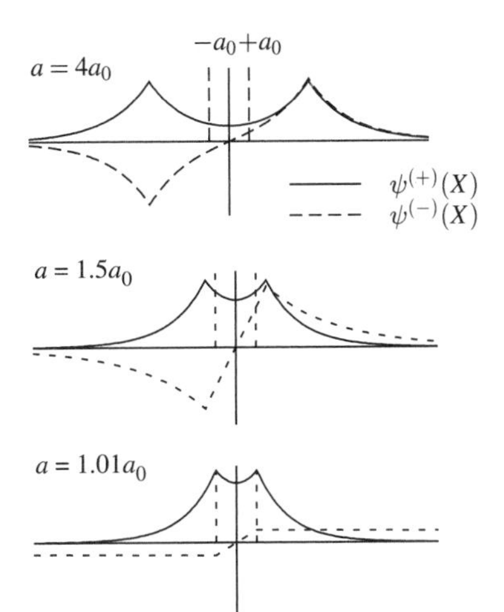

FIGURE 9.3. Position space wavefunctions for the twin δ function potential. The solid (dotted) curves correspond to the even (odd) states, respectively. Results for $a/a_0 = 4, 1.5,$ and 1.01 are shown from top to bottom. For $a/a_0 \gtrsim 1$, we can see the odd state beginning to "unbind" as it ceases to be localized.

- The odd solution (when it exists) always has a smaller value of y and hence is *less bound* than the corresponding even solution.
- For $a/a_0 < 1$, there is no odd solution, which we attribute to the increasing kinetic energy of the solution required by the node in the wavefunction at the origin.
- We can see the odd solution "unbind" for $a/a_0 \to 1$ where the position space wavefunction is becoming spatially uniform indicating no localization near the origin.
- The (unnormalized) momentum space wavefunctions can be obtained by direct Fourier transforms or by the methods of Sec. 9.1.3, and we find

$$\phi^+(p) \propto \frac{\cos(ap/\hbar)}{p^2 + (\hbar K)^2}$$
$$\phi^-(p) \propto \frac{\sin(ap/\hbar)}{p^2 + (\hbar K)^2} \tag{9.28}$$

9.1.5 Band Structure of Solids

The simple soluble model discussed above usefully illustrates some of the physics responsible for the wide variety of properties that materials can exhibit when large numbers of atoms are brought together to make solids. A typical electronic energy level diagram for an isolated atom is shown in Fig. 9.4. When N such atoms are well separated, we can think of each level shown as being N-fold degenerate where $N \gg 1$ for a macroscopic number of atoms. As the atoms are brought closer together, the degeneracy splits, with some levels becoming more or less strongly bound. When the atoms reach their equilibrium positions, the energy levels often exhibit *bands* of energy levels, regions where the energy levels are closely spaced, separated by well-defined *energy gaps* where no electronic states are allowed, as shown in Fig. 9.5. We note that overlapping energy bands are possible.

The allowed energy levels are then filled with electrons, consistent with the exclusion principle as in Chap. 8, and the position of the Fermi energy relative to the band gaps has a profound influence on the macroscopic properties of the material. If there are many unoccupied energy levels just above the Fermi surface into which electrons can easily be scattered, they are free to respond to relatively small external electric fields and thermal gradients and thus exhibit metallic behavior. If, on the other hand, there is a large energy

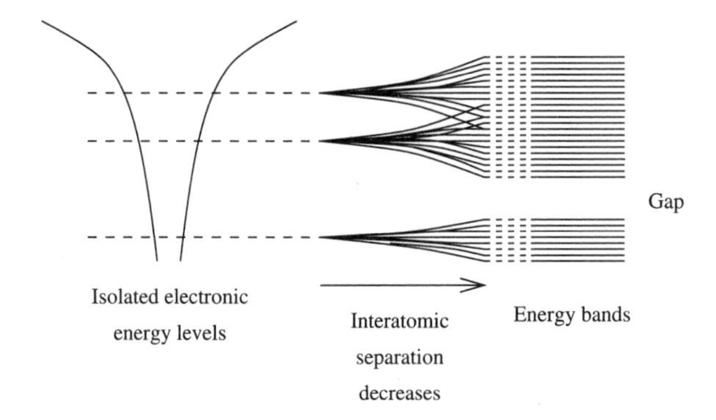

FIGURE 9.4. *Schematic description of origin of energy bands and gaps in solids. The large ($N \gg 1$) degeneracy of the energy levels of isolated atoms is split as they are brought closer together.*

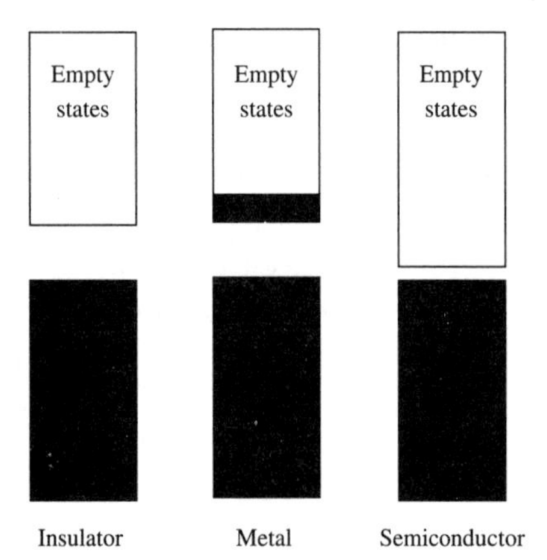

FIGURE 9.5. *Energy bands and gaps for various types of materials. The shaded regions indicate filled electron bands.*

gap that would inhibit such transitions, one has an insulator. In this case a large electric field would be required to excite the electrons across the gap; this is the equivalent of a "spark" or "breakdown." The case of a very narrow energy gap corresponds to a semiconductor for which small but finite changes in the external parameters, such as the temperature or applied field, can cause excitations.

9.2 THE FINITE WELL

9.2.1 Formal Solutions

While useful as a model system, the infinite well has at least one very unrealistic feature, namely, the fact that the particle cannot escape, i.e., be "ionized" by the addition of a sufficiently large amount of energy to the system. A more realistic version, which also illustrates several new features of quantum physics, is the finite well, here defined as

$$V(x) = \begin{cases} -V_0 & \text{for } -a < x < a \\ 0 & \text{otherwise} \end{cases} \tag{9.29}$$

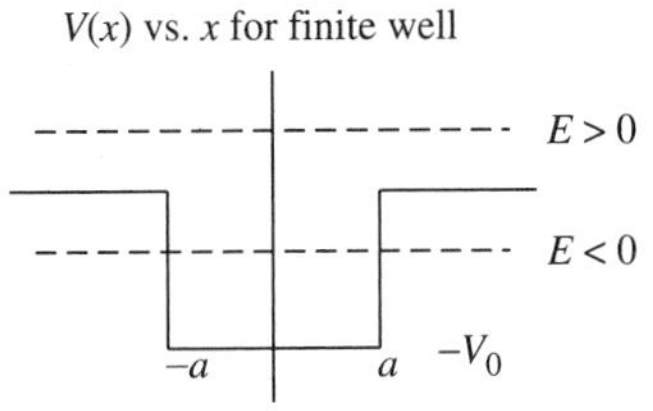

FIGURE 9.6. The finite potential well.

and we wish to examine bound states with $E \equiv -|E| < 0$ as shown in Fig. 9.6. At first glance, it might seem more appropriate to use the form

$$\tilde{V}(x) = \begin{cases} 0 & \text{for } -a < x < a \\ V_0 & \text{otherwise} \end{cases} \tag{9.30}$$

which has the symmetric infinite well as its limit when $V_0 \to \infty$, but the latter choice is trivially related to the conventional one of Eqn. (9.29) via $\tilde{V}(x) = V(x) + V_0$. This simple relation ensures (P7.5) that the two systems will have the same observable physics. The standard choice, $V(x)$, is often used, as it is conventional to pick the (arbitrary) zero of potential energy to vanish at infinite separations (e.g., the Coulomb potential $V(r) = Kq_1q_2/r$). To facilitate comparisons with the infinite well limit, we note that the quantity $V_0 - |E|$ measures the difference between the energy levels and the bottom of the well. This then should equal the usual quantized energy levels $E_n^{(\pm)}$ of the symmetric infinite well in the limit that $V_0 \to \infty$.

With this choice of potential, the Schrödinger equation in the two regions is given by

$$\begin{aligned} |x| < a: \quad & -\frac{\hbar^2}{2m}\frac{d^2\psi(x)}{dx^2} - V_0\psi(x) = -|E|\psi(x) \\ |x| > a: \quad & -\frac{\hbar^2}{2m}\frac{d^2\psi(x)}{dx^2} = -|E|\psi(x) \end{aligned} \tag{9.31}$$

or

$$\begin{aligned} |x| < a: \quad & \frac{d^2\psi(x)}{dx^2} = -q^2\psi(x) \\ |x| > a: \quad & \frac{d^2\psi(x)}{dx^2} = +k^2\psi(x) \end{aligned} \tag{9.32}$$

where

$$k = \sqrt{\frac{2m|E|}{\hbar^2}} \quad \text{and} \quad q = \sqrt{\frac{2m(V_0 - |E|)}{\hbar^2}} \tag{9.33}$$

The most general solution can be written as

$$\begin{aligned} x < -a: \quad & De^{kx} + Fe^{-kx} \\ -a < x < a: \quad & A\cos(qx) + B\sin(qx) \\ a < x: \quad & Ce^{-kx} + Ee^{kx} \end{aligned} \tag{9.34}$$

and one can immediately apply the boundary conditions at $x = \pm\infty$ (i.e., that the wavefunction be square integrable) to insist that $E = F = 0$. The fact that the potential is symmetric can be used to infer that the energy eigenfunctions will also be eigenstates of parity, i.e., even and odd functions, so we can specialize and note that for the even states we must have

$$\psi^{(+)}(x) = \begin{cases} Ce^{kx} & \text{for } x < -a \\ A\cos(qx) & \text{for } |x| < a \\ Ce^{-kx} & \text{for } a < x \end{cases} \tag{9.35}$$

while for odd states

$$\psi^{(-)}(x) = \begin{cases} -Ce^{kx} & \text{for } x < -a \\ B\sin(qx) & \text{for } |x| < a \\ Ce^{-kx} & \text{for } -a < x \end{cases} \tag{9.36}$$

Despite the discontinuous nature of the potential at $x = \pm a$, the wavefunction and its derivative are still continuous, and these conditions provide the required boundary conditions to determine the quantized energies. For the even states, for example, these continuity conditions applied at $x = a$ give

$$\begin{array}{lcl} \psi(a^-) = \psi(a^+) & \Longrightarrow & A\cos(qa) = Ce^{-ka} \\ \psi'(a^-) = \psi'(a^+) & \Longrightarrow & -qA\sin(qa) = -kCe^{-ka} \end{array} \tag{9.37}$$

and the symmetry ensures that the same conditions are obtained at $x = -a$. These can be combined to yield the condition

$$\text{even eigenvalue condition:} \qquad q\tan(qa) = k \tag{9.38}$$

which depends on the energies (through the q, k) but not on the yet-to-be-determined constants A, C. These coefficients can only be completely determined using the overall normalization condition and Eqn. (9.38). (See P9.7.) The appropriate energy eigenvalue condition for odd parity states is easily found to be

$$\text{odd eigenvalue condition:} \qquad -q\cot(qa) = k \tag{9.39}$$

The equivalent eigenvalue conditions for the symmetric infinite well were

$$\begin{array}{rl} \text{even states:} & \cos(ka) = 0 \\ \text{odd states:} & \sin(ka) = 0 \end{array} \tag{9.40}$$

respectively. These are also equations that involve transcendental functions but which can be solved analytically. In the present case, however, Eqns. (9.38) and (9.39) must be solved numerically, and a change to dimensionless variables is a useful first step. If we define

$$y \equiv qa \qquad \text{and} \qquad R = \sqrt{\frac{2mV_0a^2}{\hbar^2}} \tag{9.41}$$

we can use the defining relations for q, k to write the eigenvalue conditions as

$$\sqrt{R^2 - y^2} = y\tan(y) \tag{9.42}$$

for even states and

$$\sqrt{R^2 - y^2} = -y\cot(y) \tag{9.43}$$

for odd states. A standard method of visualizing the solution space of such equations is to plot both sides of, say, Eqn. (9.42), versus y and look for points where the two curves intersect, as such points correspond to the discrete solutions y. We illustrate this in Fig. 9.7 for both the even and odd cases for two values of the dimensionless parameter R (which in turn depends on the dimensionful parameters of the problem). We note that:

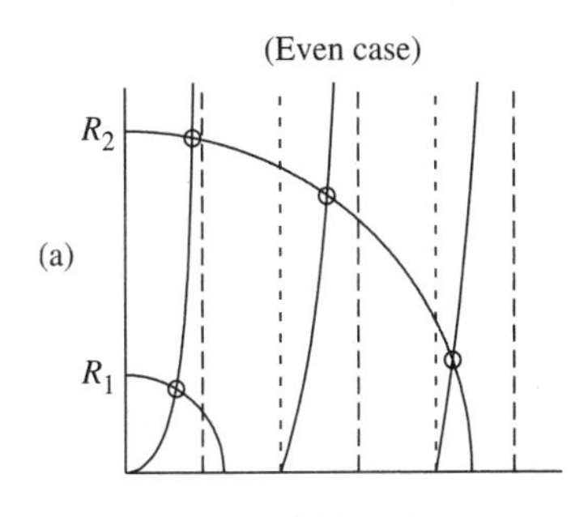

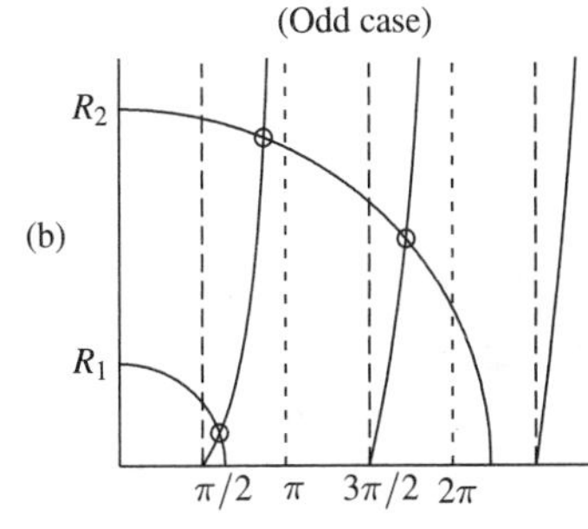

FIGURE 9.7. Energy eigenvalue conditions for the finite well for even (a) and odd (b) solutions. The vertical dashed (dotted) lines correspond to the asymptotes of tan(y) (cot(y)). Solutions corresponding to two different sets of model parameters, i.e., values of $R = \sqrt{2mV_0a^2/\hbar^2}$, are shown; for R_2, there are three even and two odd solutions.

- The number of intersections, and hence bound states, is finite but increases without bound as the values of V_0 and a increase. Thus, deeper (and hence more attractive) and wider potentials have larger numbers of bound states.
- For a fixed value of a, as $V_0 \to \infty$ the "radius" R increases without bound, and the intersection points for *even* solutions approach the *asymptotes* of tan(y), i.e., $y \to (n - 1/2)\pi$ where $n = 1, 2, 3 \ldots$. This implies that

$$V_0 - |E| \longrightarrow \frac{\hbar^2\pi^2(n - 1/2)^2}{2ma^2} \longrightarrow E_n^{(+)} \tag{9.44}$$

  in agreement with the infinite well result. The normalized wavefunctions can also be shown to have the appropriate limit as well (P9.7); the odd solutions approach the $u_n^{(-)}(x)$ in this limit.
- The number of *even* bound states can be easily determined by comparing the value of R with the various *zeroes* of tan(y) and noting that there will be $n + 1$ *even* bound states if

$$n\pi < R < (n + 1)\pi \tag{9.45}$$

  There will be n *odd* solutions provided

$$(n - 1/2)\pi < R < (n + 1/2)\pi \tag{9.46}$$

- The values of

$$V_0 - |E| = \frac{\hbar^2 q^2}{2m} = \frac{\hbar^2 y^2}{2ma^2} \tag{9.47}$$

  in the finite well are always smaller than the corresponding $E_n^{(\pm)}$ of the infinite well, and we discuss the origin of this effect below.
- No matter how shallow or narrow the well, there is always at least one intersection and hence at least one *even* bound state. It is a general feature that a purely attractive

potential (properly defined) *in one dimension* will always have at least one bound state (see P11.10), but this will not be true in higher dimensions. In contrast, there exists an odd parity bound state only if $R \geq \pi/2$.

The momentum space wavefunctions are also easily obtained (P9.10), and we will examine many of the physical interpretations of these solutions in the next section.

9.2.2 Physical Implications

A particle in the finite well described by classical mechanics would still exhibit periodic oscillatory motion between the turning points, here the two walls, provided it was in the well to begin with, i.e., if it has $|E| < 0$. Outside the classically allowed region, the kinetic energy of the particle

$$mv^2/2 = T = E - V(x) = -|E| < 0 \tag{9.48}$$

would be *negative,* so that situation would not be kinematically allowed by energy conservation. In the quantum case, however, we have found explicitly that there is a nonvanishing probability of finding the particle *outside* the classically allowed region since the wavefunction is nonzero (albeit exponentially suppressed) in that region. This phenomenon will have important implications in the area of *quantum tunneling* in Chap. 12.

A remnant of this classical discrepancy can be seen by evaluating the average value of kinetic energy for, say, an even state using the standard definition, namely,

$$\begin{aligned}\langle \hat{T} \rangle &= \int_{-\infty}^{+\infty} dx\, \psi(x)\, \frac{\hat{p}^2}{2m}\, \psi(x) \\ &= -\frac{\hbar^2}{2m}\int_{-\infty}^{+\infty} dx\, \psi(x)\, \frac{d^2\psi(x)}{dx^2} \\ &= \frac{\hbar^2 q^2 A^2}{m}\left[\int_0^a \cos^2(qx)\, dx\right] - \frac{\hbar^2 k^2 C^2}{m}\left[\int_a^{+\infty} dx\, e^{-2kx}\right]\end{aligned} \tag{9.49}$$

which shows that the contribution to $\langle \hat{T} \rangle$ from the region outside the well is indeed negative. The total value, which is what corresponds to a classical observable, is of course positive as it should be.

This behavior is similar to the tunneling wave solution found in Sec. 2.5.3 for plasma waves, and we can write the wavefunction outside the well in the form

$$\psi(x > a) = A\cos(qa)e^{k(a-x)} \propto e^{(a-x)/L} \tag{9.50}$$

where

$$L = \frac{1}{k} = \frac{\hbar}{\sqrt{2m|E|}} \tag{9.51}$$

is a penetration depth. This effect is seen to be quantum mechanical in origin in a variety of ways:

- This form immediately implies that if we formally let $\hbar \to 0$ as an effective limit of classical physics, then $L \to 0$, and the particle is constrained to stay in the well.
- The form of the penetration depth can be understood from a simple heuristic argument using the uncertainty principle. A particle with energy $E = -|E|$ can "fluctuate" to one having energy $E = 0$, which is then "free" or "outside" the well. This is plausible, provided that it does so only over a time interval consistent with the

uncertainty principle. Given this "excursion," the particle will necessarily have an uncertainty in its measured energy of $\Delta E = |E|$, so the fluctuation can last a time $\Delta t \sim \hbar/|E|$. It can then, *very roughly,* go "into" the classically disallowed region and "back" again for half this time, i.e., $\Delta t_{\text{out}} \sim \Delta t_{\text{back}} \sim \hbar/2|E|$. If we associate a classical velocity of $|E| = mv^2/2$ during this "motion," giving $v = \sqrt{2|E|/m}$, the particle could travel a distance

$$L \sim \Delta x_{\text{out}} \sim \Delta t_{\text{out}}\, v \sim \frac{\hbar}{\sqrt{2m|E|}} \tag{9.52}$$

While not to be taken as a rigorous proof in any sense, this argument (or mnemonic device) does point out one useful way of thinking about the length scale in the quantum evanescent wavefunction.

These ideas are also useful in the approximate determination of the behavior of bound state wavefunctions for large $|x|$. Imagine a particle bound in an arbitrary potential. The Schrödinger equation can be written in the form

$$\frac{d^2\psi(x)}{dx^2} = \frac{2m}{\hbar^2}(V(x) - E)\psi(x) \tag{9.53}$$

which one can integrate approximately to give

$$\psi(x) \sim \exp\left(\pm\sqrt{\frac{2m}{\hbar^2}}\int^x \sqrt{V(x) - E}\, dx\right) \tag{9.54}$$

For the finite well case, where $V(x)$ is constant, the solution is in fact exact. Of most interest is the case when $|x|$ is very large and we have $V(x) \gg E$, in which case we find

$$\psi \sim \exp\left(\pm\sqrt{\frac{2m}{\hbar^2}}\int^x \sqrt{V(x)}\, dx\right) \tag{9.55}$$

which can be a useful result.

Example 9.1 Large *x* Behavior of the Harmonic Oscillator Wavefunction As an example, consider a particle moving in a harmonic oscillator potential defined as

$$V(x) = \frac{1}{2}kx^2 = \frac{m\omega^2}{2}x^2 \tag{9.56}$$

so that

$$\int^x V(x)\, dx = \sqrt{\frac{m\omega^2}{2}}\int^x x\, dx = \sqrt{\frac{m\omega^2}{2}}\frac{x^2}{2} \tag{9.57}$$

so we expect

$$\psi_{SHO}(x) \xrightarrow{x\to\pm\infty} \exp\left(-\frac{m\omega}{\hbar}\frac{x^2}{2}\right) \tag{9.58}$$

which is indeed what is obtained from an exact solution. These arguments are also useful in that they show the rapid convergence of most realistic wavefunctions, justifying a posteriori the assumptions made previously about the "good" behavior of quantum wavefunctions at infinity.

Returning now to the finite well, we illustrate in Fig. 9.8 the energy spectrum of the infinite well and a finite well that has the same width but only three (two even and one

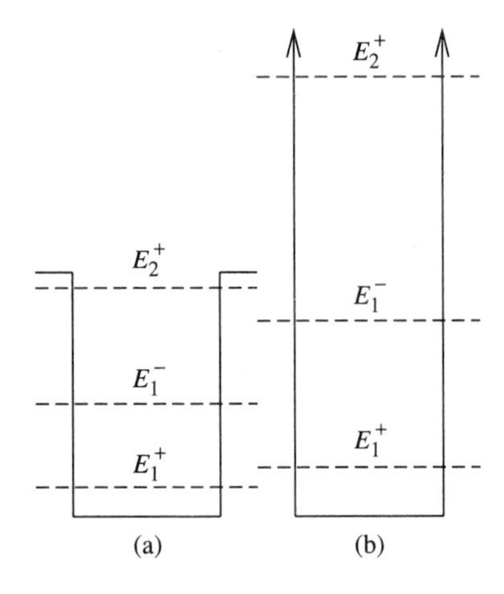

FIGURE 9.8. Comparison of energy eigenvalues for finite versus infinite well of same width.

odd) bound states. The finite well is scaled up in energy by adding V_0 for comparison, and we note again that the corresponding energy levels are *lower* in energy in the finite well. A look at the corresponding (normalized) wavefunctions in Fig. 9.9 helps illustrate the origin of this effect. The finite well solutions, while very similar in structure (i.e., numbers of nodes, parity, etc.), are allowed to tunnel into the classically disallowed region making for a "smoother" overall waveform and hence reducing its overall kinetic energy. Related effects are seen in the corresponding momentum space wavefunctions (Fig. 9.9) where one sees that:

- The distributions are somewhat narrower (Δp smaller) consistent with the fact that finite well position space wavefunctions are allowed to tunnel (giving Δx larger).
- The average values of $\langle p^2 \rangle$ for the finite well are correspondingly smaller (implied by the smaller spread in p values), consistent with less kinetic energy.

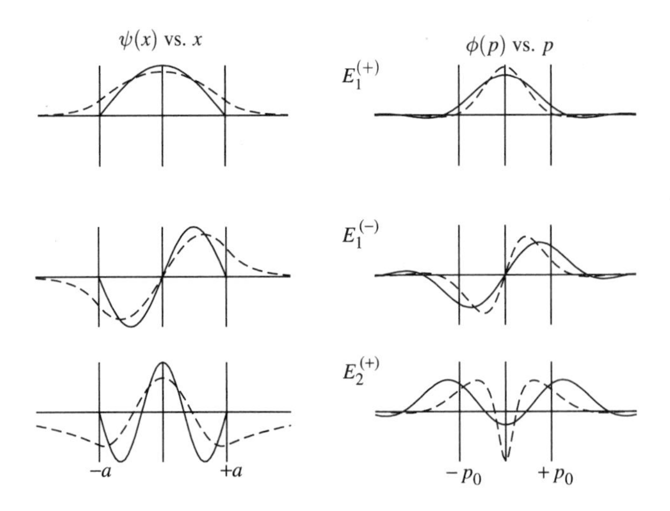

FIGURE 9.9. Position space (left) and momentum space (right) wavefunctions for the infinite well (solid) and finite well (dashed) energy levels in Fig. 9.8. On the left, the vertical solid lines indicate the edges of the well. On the right, the vertical solid lines indicate the value of p corresponding to $\pm p_0 = \pm\sqrt{2mV_0}$.

Using the fully normalized wavefunctions (P9.7), we can also calculate the probability that a position measurement would find the particle *outside* the potential, using

$$\text{Prob}\left(|x| > a\right) = \int_{|x|>a} dx\, |\psi_n^{(\pm)}(x)|^2 \tag{9.59}$$

For this example, we find that

$$\text{Prob}\left(|x| > a\right) = \begin{cases} 0.028 & \text{for } E_1^{(+)} \\ 0.127 & \text{for } E_1^{(-)} \\ 0.493 & \text{for } E_2^{(+)} \end{cases} \tag{9.60}$$

This is seen in Fig. 9.9 where the solid vertical lines indicate the position of the well boundaries; clearly the higher energy states "spend more of their time outside the well."

The corresponding calculation in momentum space can be performed to give the probability that a measurement of p would yield a value of $p^2/2m$ larger than necessary to "unbind" the particle, namely,

$$\text{Prob}\left(|p| > \sqrt{2mV_0}\right) = \int_{|p|>\sqrt{2mV_0}} dp\, |\phi_n^{(\pm)}(p)|^2 \tag{9.61}$$

The numerical values corresponding to Fig. 9.9 are

$$\text{Prob}\left(|p| > \sqrt{2mV_0}\right) = \begin{cases} 0.00134 & \text{for } E_1^{(+)} \\ 0.031 & \text{for } E_1^{(-)} \\ 0.115 & \text{for } E_2^{(+)} \end{cases} \tag{9.62}$$

In this case, the solid vertical lines indicate the values of $\pm\sqrt{2mV_0}$.

9.3 "HALF" POTENTIAL WELLS

A class of problems that at first appears rather unphysical, but which combines features of several of the one-dimensional problems considered already, is represented by potential energy functions of the form

$$\tilde{V}(x) = \begin{cases} \infty & \text{for } x < 0 \\ V(x) & \text{for } x > 0 \end{cases} \tag{9.63}$$

Trivial examples would be the potential felt by a ball bouncing elastically on a table under the influence of gravity (the "quantum bouncer"), in which case

$$\tilde{V}_g(z) = \begin{cases} \infty & \text{for } z < 0 \\ mgz & \text{for } z > 0 \end{cases} \tag{9.64}$$

or a mass connected to a spring attached to a rigid wall

$$\tilde{V}_{SHO}(x) = \begin{cases} \infty & \text{for } x < 0 \\ kx^2/2 & \text{for } x > 0 \end{cases} \tag{9.65}$$

which might represent a "quantum paddle ball."

Suppose the original function $V(x)$ is symmetric, and we have somehow found the appropriate (normalized) energy eigenstates and energy levels, namely,

$$E_n^{(\pm)} \qquad \text{and} \qquad \psi_n^{(\pm)}(x) \tag{9.66}$$

which are necessarily characterized by even and odd parities. The solutions of the "half" well described by $\tilde{V}(x)$ are then immediately obtainable, as we notice that

- The even and odd solutions for $V(x)$ solve the Schrödinger equation for $x > 0$.
- The *odd* solutions also satisfy the necessary boundary condition at $x = 0$ and can thus match smoothly onto a vanishing solution for $x < 0$.
- The energy spectrum then consists of only the *odd* energy levels of the corresponding "full"-well problem. The cases of the standard and symmetric infinite wells already provide an example of this phenomenon, and one can check that the expressions in Sec. 6.3 are consistent with these results.

Thus, the solutions for $\tilde{V}(x)$ are given by

$$\tilde{\psi}_n(x) = \begin{cases} 0 & \text{for } x < 0 \\ \sqrt{2}\psi_n^{(-)}(x) & \text{for } x \geq 0 \end{cases} \tag{9.67}$$

where the difference in normalization comes from the fact that all of the probability must now lie in the region $x > 0$.

The half-well momentum space distributions are somewhat less obviously related to their full-well counterparts, at least in the quantum limit. The odd solutions for $V(x)$ satisfy

$$\begin{aligned} \phi_n^{(-)}(p) &= \frac{1}{\sqrt{2\pi\hbar}} \int_{-\infty}^{+\infty} \psi_n^{(-)}(x) e^{-ipx/\hbar} \\ &= -2i \frac{1}{\sqrt{2\pi\hbar}} \int_0^{+\infty} \psi_n^{(-)}(x) \sin(px/\hbar)\, dx \end{aligned} \tag{9.68}$$

where we have used the fact that $\psi_n^{(-)}$ is odd. For the half-well case, we have

$$\begin{aligned} \tilde{\phi}_n(p) &= \frac{1}{\sqrt{2\pi\hbar}} \int_{-\infty}^{+\infty} \tilde{\psi}_n(x) e^{-ipx/\hbar}\, dx \\ &= \frac{1}{\sqrt{2\pi\hbar}} \sqrt{2} \int_0^{+\infty} \psi_n^{(-)}(x) \big(\cos(px/\hbar) - i \sin(px/\hbar)\big)\, dx \\ &\equiv \frac{1}{\sqrt{2}} \left(\overline{\phi}^{(-)}(p) - i\phi_n^{(-)}(p)\right) \end{aligned} \tag{9.69}$$

implicitly defining a new piece of the momentum distribution,

$$\overline{\phi}^{(-)}(p) = \int_0^{\infty} dx\, \psi_n^{(-)} \cos(px/\hbar) \tag{9.70}$$

This is now nonvanishing, since the integral is evaluated only over the interval $(0, +\infty)$. The momentum space probability distribution is then given by

$$|\tilde{\phi}_n(p)|^2 = \frac{1}{2}\left(|\overline{\phi}^{(-)}(p)|^2 + |\phi_n^{(-)}(p)|^2\right) \tag{9.71}$$

so that it is quite similar to the full-well case.

For illustration, we plot in Fig. 9.10 the position and momentum space distributions of the full- and corresponding half-harmonic oscillator potentials (using results to be derived in detail in Chap. 10) to illustrate the behavior. We note that:

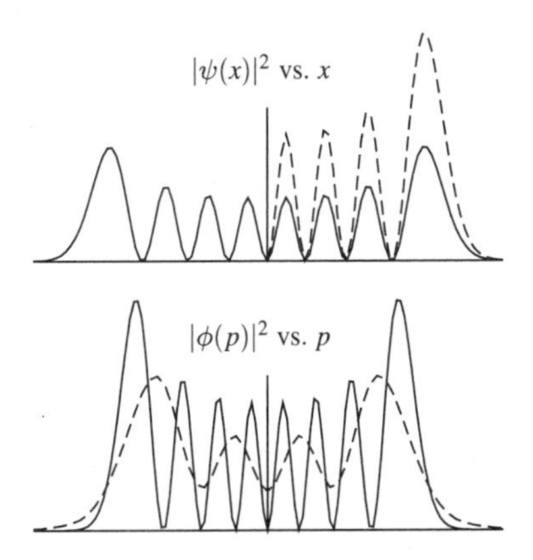

FIGURE 9.10. Position space (top) and momentum space (bottom) probability distributions for the "half" harmonic oscillator. The wavefunctions squared for the full potential (solid) and half potential (dotted) are both shown.

- Since the half-well position space wavefunctions have half the numbers of nodes, it is not implausible that the corresponding momentum space wavefunctions do also.
- The average value of momentum in both the full- and half-well cases is, of course, zero, as it is for any bound state. (Recall P6.13.) The expectation values of $\langle p^2 \rangle$ inferred from the plots of $|\phi(p)|^2$ are not obviously different, and it is indeed the case that $\langle p^2 \rangle$ is identical in the two cases. To see this, we note that for the full-well case, we have

$$\langle \hat{T} \rangle = \frac{\hbar^2}{2m} \int_{-\infty}^{+\infty} \left| \frac{d\psi_n^{(\pm)}(x)}{dx} \right|^2 dx = \frac{\hbar^2}{m} \int_0^{+\infty} \left| \frac{d\psi_n^{(\pm)}(x)}{dx} \right|^2 dx \tag{9.72}$$

where one uses the symmetry of the wavefunctions. For the half-well case one has

$$\langle \hat{T} \rangle = \frac{\hbar^2}{2m} \int_0^{+\infty} \tilde{\psi}_n(x)\, \hat{T}\, \tilde{\psi}_n(x)\, dx = \frac{\hbar^2}{m} \int_0^{+\infty} \left| \frac{d\psi_n^{(-)}(x)}{dx} \right|^2 dx \tag{9.73}$$

because of the extra factor of $\sqrt{2}$ in Eqn. (9.67). Similar statements can be made about $\langle V(x) \rangle$, so that the energy "sharing" is identical in the two cases.
- We will discuss the classical expectations for the momentum probability distributions in Chap. 10, but we can see in this case that in the large quantum number limit we seemingly have

$$|\overline{\phi}_n^{(-)}(p)|^2 \longrightarrow |\phi_n^{(-)}(p)|^2 \tag{9.74}$$

because the difference in the averaging of $\psi_n^{(-)}(x)$ over $\sin(px/\hbar)$ and $\cos(px/\hbar)$ "washes" out when the wavefunction is spread out over a large distance.

9.4 APPLICATIONS TO THREE-DIMENSIONAL PROBLEMS

9.4.1 The Schrödinger Equation in Three Dimensions

The extension of the one-particle Schrödinger equation to three dimensions is, in many ways, straightforward. A case of particular interest is when the interaction potential is a function only of the radial distance, i.e., $V(\mathbf{r}) = V(r, \theta, \phi) = V(r)$ alone, the so-called *central potentials*. In this case, it is natural to use spherical coordinates and consider $\psi(\mathbf{r}) = \psi(r, \theta, \phi)$. Specializing even further to the instance where there is no angular momentum,[3] we find (Chap. 17) that the wavefunction can be written in the form

$$\psi(r, \theta, \phi) = \psi(r) = \frac{u(r)}{r} \frac{1}{\sqrt{4\pi}} \tag{9.75}$$

where $u(r)$ satisfies a Schrödinger equation of the form

$$-\frac{\hbar^2}{2m} \frac{d^2 u(r)}{dr^2} + V(r)u(r) = Eu(r) \tag{9.76}$$

which is identical to the one-dimensional Schrödinger equation. The difference arises in that the radial coordinate is only defined for $r > 0$, and, because the wavefunction $\psi(r)$ should be well defined at the origin, one must also have $u(0) = 0$. This is then like the half-well potential problem discussed above. Furthermore, the normalization condition, derived from an integration over the full three-dimensional position space, is given by

$$\begin{aligned} 1 &= \int d^3 r \, |\psi(r, \theta, \phi)|^2 \\ &= \int_0^\infty r^2 \, dr \int d\Omega \left| \frac{u(r)}{r\sqrt{4\pi}} \right|^2 \\ &= \int_0^\infty |u(r)|^2 \, dr \end{aligned} \tag{9.77}$$

The obvious identification of $u(r)$ in three dimensions and $\tilde{\psi}(x)$ in one dimension, with their many similarities, is an example of how a relatively simple one-dimensional problem can have applications in a realistic three-dimensional system. We examine one specific example from nuclear physics in the next section.

9.4.2 Model of the Deuteron

The hydrogen atom, the two-body system consisting of an electron and proton interacting via the electromagnetic interaction, provides the simplest case in which to study realistically the effects of quantum mechanics in the atomic physics domain. The analogous system in nuclear physics, the deuteron, which consists of a neutron and proton bound by the strong nuclear force, plays somewhat the same role. It is experimentally known, however, to have only one bound state, so its spectrum is far less rich. Nonetheless, it provides some information on the strength and range of the nucleon-nucleon potential and is amenable to study using the ideas of the last section.

This two-body system in three dimensions can be described by an effective one-particle, one-dimensional Schrödinger equation provided that we:

- Use the *reduced mass* of the two-body system (to be discussed in Chap. 15)

$$\mu = \frac{m_p m_n}{m_p + m_n} \approx \frac{m_p}{2} \approx 480 \text{ MeV}/c^2 \tag{9.78}$$

[3]The special case of no angular momentum ($l = 0$) is a highly nontrivial one, as the ground state solution, which falls into this category, is important for the determination of the ultimately stable configuration of the system.

- Identify the coordinate $\mathbf{r} \equiv \mathbf{r}_1 - \mathbf{r}_2$ with the *relative coordinate* of the two bodies.

A model potential that captures some of the salient features of the nuclear force, especially its finite range, is the half-finite well, namely,

$$V(r) = V(|\mathbf{r}_1 - \mathbf{r}_2|) = \begin{cases} -V_0 & \text{for } r < a \\ 0 & \text{for } r > a \end{cases} \tag{9.79}$$

where, a priori, the depth (V_0) and range (a) of the potential are unknown. The form of the relevant *odd* energy eigenfunctions is then simply given by Eqn. (9.36), namely,

$$\psi^{(-)}(r) = \begin{cases} B\sin(qr) & \text{for } 0 < r < a \\ Ee^{-kr} & \text{for } a < r \end{cases} \tag{9.80}$$

with appropriate matching conditions at $r = a$. We must also satisfy the eigenvalue constraint for odd states, namely,

$$\sqrt{R^2 - y^2} = -y\cot(y) \tag{9.81}$$

where $R^2 = 2\mu V_0 a^2/\hbar^2$. We would then like to "fit" experimental data to determine the parameters of the potential; even in the context of a crude model, these give some indication of the properties of the true nuclear force.

Experiments find that it requires a gamma ray of energy roughly $E_\gamma = 2.23\,\text{MeV}$ to disassociate the deuteron, so that it has one bound state with $E = -|E| = -2.23\,\text{MeV}$. With only this constraint, one cannot uniquely determine V_0 and a simultaneously, but some additional information is available. Scattering experiments (Sec. 20.4) can be used to calculate the mean value of the radius squared of the system, namely,

$$\begin{aligned} \langle r^2 \rangle &\equiv \int d\mathbf{r}\, |\mathbf{r}|^2\, |\psi(r,\theta,\phi)|^2 \\ &= \int_0^\infty dr\, r^2\, |u(r)|^2 \end{aligned} \tag{9.82}$$

and the measured value is roughly $\langle r^2 \rangle \approx (4.2\,\text{F})^2$. For the appropriate quasi-odd wavefunctions, this condition reduces to

$$\langle r^2 \rangle = B^2 \int_0^R dr\, r^2\, \sin^2(kr) r + E^2 \int_R^\infty dr\, r^2\, e^{-2kr} \tag{9.83}$$

where B, E are determined by the normalization condition

$$1 = B^2 \int_0^R dr\, \sin^2(kr)\, dr + E^2 \int_R^\infty dr\, e^{-2kr} \tag{9.84}$$

and the boundary conditions at $r = a$. These constraints can be solved numerically, and one finds

$$a \approx 2.4\,\text{F} \qquad \text{and} \qquad V_0 \approx 27\,\text{MeV} \tag{9.85}$$

which do give a reasonable indication of the range of the nuclear force as determined from other experiments. We note that:

- The depth of the potential also satisfies another plausible constraint, namely, that the attractive nucleon-nucleon potential at these distances is much larger than the corresponding proton-proton repulsion due to their electrostatic interaction. Specifically, we find that

$$\begin{aligned} V_{\text{Coul}}(r = a) &= \frac{Ke^2}{a} \\ &= \frac{1.44\,\text{MeV F}}{2.4\,\text{F}} \\ &\approx 0.6\,\text{MeV} \\ &\ll V_0 = 27\,\text{MeV} \end{aligned} \tag{9.86}$$

- The fact that $|E|$ is so close to zero (compared to the value of V_0) implies that the state is extremely weakly bound. The two particles, in fact, spend roughly 60% of the time *outside* the range of the potential. This is also clear from the fact that $\sqrt{\langle r^2 \rangle} > a$!

9.5 QUESTIONS AND PROBLEMS

Q9.1 If you constructed a wavepacket for the finite well representing a bound particle, i.e., one with $\langle E \rangle < 0$, would you ever measure it with $E > 0$?

Q9.2 Why should the average values of kinetic energy and potential energy be the same in "half-well" potential as for the corresponding energy states in the "full-well" potential?

Q9.3 **What's wrong with this picture?** Figure 9.11 shows an energy level in a generic potential well. One of the position space wavefunctions below it is the real solution. Identify the wrong solution, and describe as many things wrong with the purported solution as you can. Focus on the magnitude and wiggliness of $\psi(x)$ and its behavior in the classically disallowed region.

P9.1 Show directly from Eqn. (9.1) that $\psi''(x)$ will be discontinuous if $V(x)$ is discontinuous. Can you calculate $\psi'''(x)$ and discuss the connection between its continuity and that of $V(x)$?

P9.2 For the single δ function potential, show that the average values of kinetic and potential energy satisfy

$$E = \frac{1}{2}\langle V(x) \rangle = -\langle \hat{T} \rangle$$

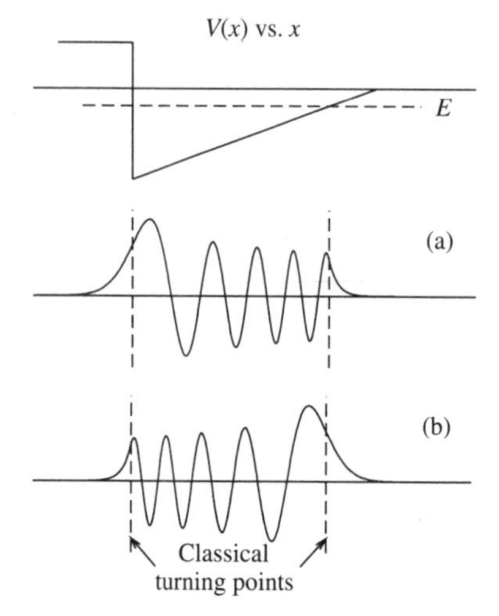

FIGURE 9.11. Energy level in a generic potential well (top) and two purported solutions (a) and (b). Which one is right?

using the position space wavefunctions. You might wish to use Eqn. (5.44) for the kinetic energy averaging to avoid problems with higher derivatives.

P9.3 **(a)** Find the normalized position space wavefunctions for the even and odd state (when it exists) for the twin δ function potential in Sec. 9.1.3. Show that for large separation (namely, $a \gg a_0$) they can be written as

$$\psi^{(\pm)}(x) \longrightarrow \frac{1}{\sqrt{2}}\big(\psi_1(x-a) \pm \psi_1(x+a)\big)$$

where $\psi_1(x)$ is the solution for the single δ function potential in Eqn. (9.9).

(b) Say one tries to localize a particle around one or the other δ function spikes by taking the initial wavefunction

$$\psi(x,0) = \frac{1}{\sqrt{2}}\left(\psi^{(+)}(x) + \psi^{(-)}(x)\right)$$

Does the particle stay localized for later times? What is the natural time scale for the problem? *Hint:* How does ΔE between the two states scale when $a/a_0 \gg 1$?

P9.4 Find the energy eigenvalue conditions for even and odd states for the three δ function potential given by

$$V(x) = -g\big(\delta(x-a) + \delta(x) + \delta(x+a)\big)$$

Discuss the energy spectrum as a gets small. Is there a critical value of a for which some states "unbind"?

P9.5 Find the form of the momentum space wave functions for the twin δ function potential by using the methods of Sec. 9.1.3. How do you obtain the energy eigenvalue condition? Would this work for a more complicated set of δ function potentials as in P9.4?

P9.6 Consider the hybrid potential[4] consisting of the sum of symmetric infinite well potential

$$V_\infty(x) = \begin{cases} 0 & \text{for } |x| < a \\ \infty & \text{for } |x| > a \end{cases}$$

and a δ function at the origin, namely,

$$V(x) = V_\infty + g\delta(x)$$

(a) Show that the potential is symmetric so that the solutions will be eigenstates of parity.

(b) Using that fact, solve the Schrödinger equation inside the well for both positive and negative parity states for *nonnegative* energies E. Apply the appropriate boundary conditions (being especially careful at $x = 0$), and determine the energy eigenvalue conditions. Show that the negative energy states are unchanged by the presence of the potential.

(c) Repeat assuming negative energy states, i.e., $E = -|E| < 0$.

(d) Plot the eigenvalue condition for even solutions, expressed as a function of the dimensionless parameter y, and discuss how the even energies vary as the value of g is varied in the range $(-\infty, +\infty)$.

(e) Find the value of g for which there is a zero energy ground state. Find and sketch its wavefunction.

(f) Show that for $g \to +\infty$, the even energies approach those of the odd states directly above them. How do their wavefunctions compare?

(g) Repeat for the case $g \to -\infty$ using the form appropriate for negative energy solutions.

P9.7 **(a)** Show that the appropriate normalization constants for the even states of the finite well are given by

[4] Some aspects of this problem are worked out in Lapidus (1982b).

$$A = \frac{1}{\sqrt{a}}\left(1 + \frac{\sin(2y)}{2y} + \frac{\cos^3(y)}{y\sin(y)}\right)^{-1/2} = \frac{1}{\sqrt{a}}\left(1 + \frac{\sin(2y)}{2y} + \frac{\cos^2(y)}{ak}\right)^{-1/2}$$

$$C = A\cos(y)e^{ka}$$

(b) Show that the even finite well wavefunctions approach those of the infinite well in the limit $V_0 \to \infty$, i.e., when $y \to (n - 1/2)\pi$.

(c) Calculate the probability that a measurement of the position of the particle will find it outside the classical disallowed region.

P9.8 **(a)** Evaluate the average value of kinetic energy for the *even* states in the finite well, and show that your result can be expressed in the form

$$\langle \hat{T} \rangle = \frac{\hbar^2 y^2}{2m}\left[\frac{1}{1 + \sin(2y)/2y + \cos^2(y)/ak}\right] \tag{9.87}$$

How does this compare to the kinetic energy in the infinite well?

(b) Show that the ratio of the contribution of the kinetic energy from *outside* the well to the total is given by

$$\langle \hat{T} \rangle_{\text{out}} \Big/ \langle \hat{T} \rangle_{\text{total}} = -\frac{\sin(2y)}{2y}$$

Show that this increases in magnitude for less bound states. Show that it has the appropriate limit for the infinite well.

P9.9 Suppose you tried to model a hydrogen atom in one dimension by assuming that the electron of rest energy $m_e c^2 = 0.51$ MeV was in a finite well of width $2a$ with $a = 0.53$ Å (which is like the Bohr radius) and of depth $V_0 = 30$ eV (which is something like the value of the Coulomb potential evaluated at $r = a$). How many bound states would you find? Would this be a very realistic model?

P9.10 Calculate the momentum space wavefunction, $\phi^{(+)}(p)$, for the finite well corresponding to the even solutions $\psi^{(+)}(x)$.

P9.11 The attractive δ function potential, $V(x) = -g\delta(x)$, can be obtained from the finite well potential by taking the limits

$$V_0 \to \infty \qquad \text{while} \qquad a \to 0 \tag{9.88}$$

in such a way that the "area" under the two potentials is kept the same, namely, $2V_0 a = g$.

(a) Show that in this limit, there is only one bound state for the finite well and that its energy reduces to that obtained in Eqn. (9.8).

(b) Show that the wavefunction approaches Eqn. (9.9) in this limit.

P9.12 Using Eqn. (9.55), find the large $|x|$ behavior of the wavefunction for a particle of mass m moving in a linear confining potential of the form $V(x) = C|x|$; compare your result to P5.21.

P9.13 How much lighter would the masses of the neutron and proton (assumed the same) have to be in order for there to be no deuteron at all, i.e., no bound state? How much heavier would they have to be to have three bound states? Assume that the parameters of the potential stay the same.

CHAPTER 10

The Harmonic Oscillator

10.1 THE IMPORTANCE OF THE SIMPLE HARMONIC OSCILLATOR

The problem of a particle moving under the influence of a linear restoring force, $F(x) = -Kx$, or equivalently a quadratic potential, $V(x) = Kx^2/2$, is a problem studied at all levels of theoretical physics, from elementary classical mechanics through quantum field theory.[1] One of its most useful features is that, at every stage of development, it is exactly soluble and so can be easily used as a closed-form, analytic example. If it had no connection to real physical systems, however, that fact would be of only academic interest. In this section, we wish to discuss two important applications of the harmonic oscillator problem to illustrate its potential wide-ranging usefulness. We will then derive its solution in nonrelativistic quantum mechanics, using a standard differential equation approach, and discuss its classical limits and experimental realizations.

In classical mechanics, a conservative system can be described by a potential energy function, $V(x)$, and states of the system that will be in equilibrium will be found at the extrema of this potential, i.e., places where

$$\left.\frac{dV(x)}{dx}\right|_{x=x_0} = -F(x_0) = 0 \tag{10.1}$$

or where the classical force vanishes, as suggested by Newton's law. The stability of the equilibrium point is governed by the sign of the second derivative with its

$$\left.\frac{d^2V(x)}{dx^2}\right|_{x=x_0} > 0 \qquad \text{or} \qquad < 0 \tag{10.2}$$

implication of stable, bounded, and periodic motion or of unstable (in fact, exponentially increasing) and unbounded motion, respectively, as in Fig. 10.1.

[1] For the most comprehensive survey of the harmonic oscillator in classical and quantum mechanics, see Pippard (1978, 1983).

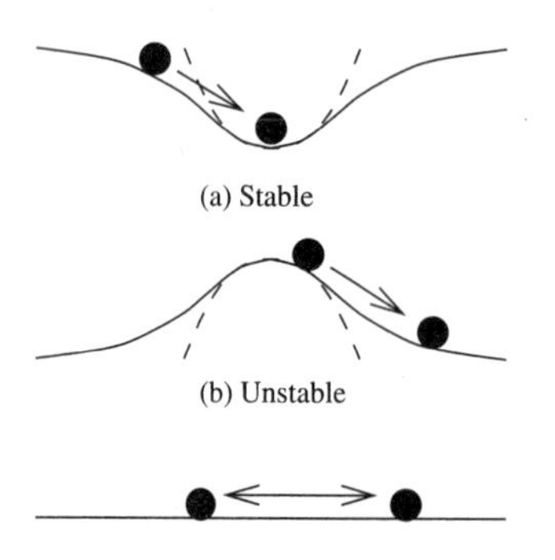

FIGURE 10.1. *Generic potential with stable, unstable, and neutral equilibrium points.*

We can expand the potential near the point of equilibrium as a power series,

$$\begin{aligned} V(x) &= V(x_0) + \left.\frac{dV(x)}{dx}\right|_{x=x_0}(x - x_0) + \left.\frac{1}{2}\frac{d^2V(x)}{dx^2}\right|_{x=x_0}(x - x_0)^2 + \cdots \\ &= V_0 + \frac{1}{2}K(x - x_0)^2 + \cdots \end{aligned} \tag{10.3}$$

where the effective "spring constant" is $K \equiv d^2V(x)/dx^2|_{x=x_0}$. Since the constant part of the potential cannot affect the physics in any meaningful way, we see that:

- The simple harmonic oscillator (hereafter SHO) potential, $V(x) = Kx^2/2$, is often the "best first guess" for the potential near a point of stable equilibrium.

If the equilibrium point happens to be at $x_0 = 0$, we have,

$$\begin{aligned} m\ddot{x}(t) &\approx -|K|x(t) \quad \text{when } K > 0 \\ m\ddot{x}(t) &\approx +|K|x(t) \quad \text{when } K < 0 \end{aligned} \tag{10.4}$$

with solutions

$$\begin{aligned} x(t) &= A\cos(\omega t) + B\sin(\omega t) \quad \text{for stable equilibrium} \\ x(t) &= Ce^{\omega t} + De^{-\omega t} \quad \text{for unstable equilibrium} \end{aligned} \tag{10.5}$$

where $\omega = \sqrt{K/m}$. The constants are determined, of course, by the initial conditions. While the first case is by far the most important, we will also briefly discuss the quantum analog of unstable motion in Sec. 10.6.

A realization of this idea comes in the study of diatomic molecules,[2] which interact via a potential of the generic type shown in Fig. 10.2. The particular form shown is often called a *Lennard-Jones* or *6-12* potential of the form

$$V(r) = 4\epsilon\left(\left(\frac{\sigma}{r}\right)^{12} - \left(\frac{\sigma}{r}\right)^{6}\right) \tag{10.6}$$

and is especially relevant for the interaction of spherical (i.e., noble gas) atoms. This functional form is not as good a representation of the potential for other types of bonding (i.e.,

[2] See any good text on modern physics for a further discussion of molecular bonding, e.g., Eisberg and Resnick (1974).

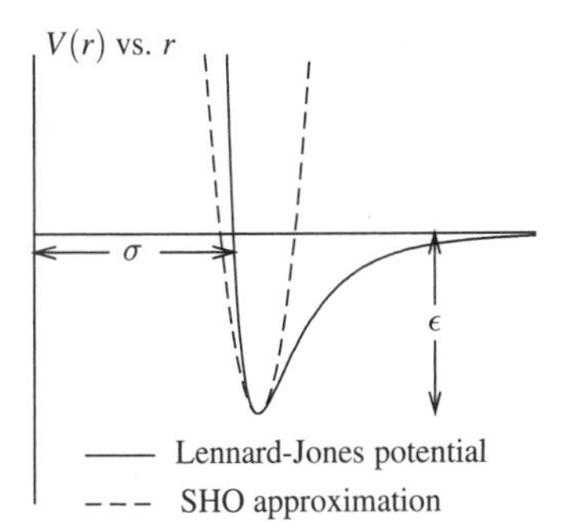

FIGURE 10.2. V(r) versus r for a Lennard-Jones potential (solid) and SHO approximation (dashed).

ionic or covalent), but there will be a stable minimum in those systems as well. In this case, the potential is a function of the distance between the two molecules, $r \equiv |\mathbf{r}_1 - \mathbf{r}_2|$, and the reduction to relative coordinates (discussed already in Sec. 9.4.2) implies that the reduced mass of the system, defined as $\mu = m_1 m_2/(m_1 + m_2)$, should be used. A harmonic oscillator potential can be used to fit the potential near the stable equilibrium point with the classical result that such diatomic molecules will oscillate around their equilibrium separation with frequency

$$\omega = \sqrt{\frac{K}{\mu}} \qquad \text{where} \qquad K = \left.\frac{d^2V(r)}{dr^2}\right|_{r=r_0}$$

with any amplitude possible. The extent to which this approximation is a good one depends on the size of the anharmonic terms, the higher-order terms in the expansion of $V(r)$, as the amplitude of the motion increases. Classically, it is always possible to have oscillations with amplitude small enough that the SHO approximation is a good one. Quantum mechanically, however, we know there will be quantized energy levels in such a confining potential. The extent to which an SHO approximation is useful now depends on the size of the energy level spacing and especially that of the lowest level given by the zero point energy. We will discuss this in more detail in Sec. 10.3.

Another example, which at first seems very remote from classical considerations of vibrating masses and springs, comes from the desire to apply quantum ideas to the oscillations of the electromagnetic (EM) field. We give a preliminary discussion here[3] and return to the subject in Chap. 19.

The total energy contained in a configuration of electromagnetic fields is given by an integral over the respective electric, $u_E(\mathbf{r}, t)$, and magnetic, $u_B(\mathbf{r}, t)$, energy densities,

$$\begin{aligned} E_{\text{tot}} &= \int d\mathbf{r} \left(u_E(\mathbf{r}, t) + u_B(\mathbf{r}, t)\right) \\ &= \int d\mathbf{r} \left(\frac{\epsilon_0}{2}|\mathbf{E}(\mathbf{r}, t)|^2 + \frac{1}{2\mu_0}|\mathbf{B}(\mathbf{r}, t)|^2\right) \end{aligned} \tag{10.7}$$

In a region of space where there is no charge density (so that the scalar potential, $\phi(\mathbf{r}, t)$, can be neglected), the electric and magnetic fields can be written in terms of the so-called *vector potential*, $\mathbf{A}(\mathbf{r}, t)$, via

$$\mathbf{E}(\mathbf{r}, t) = -\frac{\partial \mathbf{A}(\mathbf{r}, t)}{\partial t} \qquad \mathbf{B}(\mathbf{r}, t) = -\nabla \times \mathbf{A}(\mathbf{r}, t) \tag{10.8}$$

[3]For a more comprehensive discussion at this level, see Baym (1976).

We can write the vector potential in terms of its Fourier components via

$$\mathbf{A}(\mathbf{r}, t) = \frac{1}{(2\pi)^{3/2}} \int d\mathbf{k}\, \mathbf{A}(\mathbf{k}, t)\, e^{i\mathbf{k}\cdot\mathbf{r}} \tag{10.9}$$

and the total energy can be rewritten in the form

$$E_{\text{tot}} = \int d\mathbf{k} \left(\frac{\epsilon_0}{2} |\dot{\mathbf{A}}(\mathbf{k}, t)|^2 + \frac{k^2}{2\mu_0} |\mathbf{A}(\mathbf{k}, t)|^2 \right) \tag{10.10}$$

If we write the energy of a standard harmonic oscillator in terms of its amplitude, $x(t)$, as

$$E_{\text{tot}} = \frac{1}{2} m\dot{x}^2(t) + \frac{1}{2} K x^2(t) \tag{10.11}$$

we can see that there is a very real sense in which the EM field can be considered as an infinite collection (specifically an integral over $d\mathbf{k}$) of harmonic oscillators, each with amplitude $A(\mathbf{k}, t)$. Comparing the coefficients of the amplitude and derivative terms gives the appropriate frequency for both cases. For the mass and spring case we have

$$\omega = \sqrt{\frac{K/2}{m/2}} = \sqrt{\frac{K}{m}} \tag{10.12}$$

while for the EM field case we have

$$\omega_k = \sqrt{\frac{k^2/2\mu_0}{\epsilon_0/2}} = kc \qquad \text{since} \qquad c = \frac{1}{\sqrt{\epsilon_0 \mu_0}} \tag{10.13}$$

which is indeed the appropriate dispersion relation for photons. One intriguing implication of this result is that since there is a nonvanishing zero point energy for *each* $\mathbf{k}$ mode, the total vacuum energy of the EM field will be divergent. We will discuss the physical significance of this in Sec. 19.9.

10.2 SOLUTIONS FOR THE SIMPLE HARMONIC OSCILLATOR

10.2.1 Differential Equation Approach

We will now apply some standard techniques in the solution of ordinary differential equations to solve for the energy eigenvalues and eigenfunctions corresponding to the simple harmonic oscillator potential in the standard form

$$V(x) = \frac{Kx^2}{2} = \frac{m\omega^2 x^2}{2} \tag{10.14}$$

We write the time-independent Schrödinger equation in position space as

$$-\frac{\hbar^2}{2m} \frac{d^2\psi(x)}{dx^2} + \frac{m\omega^2 x^2}{2} \psi(x) = E\psi(x) \tag{10.15}$$

and we attempt a change of variables to make Eqn. (10.15) dimensionless. Specifically, we define $x = \rho y$, where ρ has the units of length and will be determined below. Substituting this above we find

$$\frac{d^2\psi(y)}{dy^2} - \frac{m^2\omega^2\rho^4}{\hbar^2}y^2\psi(y) = -\frac{2mE\rho^2}{\hbar^2}\psi(y) \tag{10.16}$$

which reduces to

$$\frac{d^2\psi(y)}{dy^2} - y^2\psi(y) = -\epsilon\psi(y) \tag{10.17}$$

provided we define

$$\rho \equiv \sqrt{\frac{\hbar}{m\omega}} \quad \text{and} \quad \epsilon = \frac{2E}{\hbar\omega} \tag{10.18}$$

where ϵ is now a dimensionless eigenvalue. We note that these are the same combinations of parameters for length and energy given simply by dimensional analysis in Sec. 1.4.

Because the behavior of $\psi(y)$ at infinity is important for the existence of normalizable solutions, we first examine Eqn. (10.17) for large y:

$$\frac{d^2\psi(y)}{dy^2} \approx y^2\psi(y) \tag{10.19}$$

which has approximate solutions, $\psi(y) = e^{\pm y^2/2}$. Choosing only the square integrable solution leads us to try to "factor out" the behavior at infinity, once and for all, by writing

$$\psi(y) \equiv h(y)e^{-y^2/2} \tag{10.20}$$

Substitution into Eqn. (10.17) then gives

$$\frac{d^2h(y)}{dy^2} - 2y\frac{dh(y)}{dy} + (\epsilon - 1)h(y) = 0 \tag{10.21}$$

Since the potential $V(x)$ is symmetric, we know in advance that the solutions will be eigenfunctions of parity, i.e., even and odd states; this implies that the $h(y)$ can be classified by their parity, and we consider even solutions first.

A standard method of solution that can be applied to this problem is to assume a *power series expansion* for $h(y)$, so that in this even case we write

$$h^{(+)}(y) = \sum_{s=0}^{\infty} a_s y^{2s} \tag{10.22}$$

which must satisfy

$$\sum_{s=0}^{\infty} 2s(2s-1)a_s y^{2s-2} + \sum_{s=0}^{\infty} (\epsilon - 1 - 4s)\, a_s y^{2s} = 0 \tag{10.23}$$

Using our freedom to relabel the dummy summation index in the first term by letting $s \to s+1$, we can combine terms to obtain

$$\sum_{s=0}^{\infty} \left[2(s+1)(2s+1)a_{s+1} + (\epsilon - 1 - 4s)a_s\right] y^{2s} = \sum_{s=0}^{\infty} B_s y^{2s} = 0 \tag{10.24}$$

which can only be true for every value of y if all of the coefficients B_s vanish separately, i.e.,

$$B_s = 0 \Longrightarrow a_{s+1} = a_s\left[\frac{4s+1-\epsilon}{2(s+1)(2s+1)}\right] \tag{10.25}$$

Eqn. (10.25) is now a *recurrence relation* that implicitly gives the solution because all of the coefficients are given in terms of the single parameter a_0 via

$$a_1 = a_0 \frac{(1-\epsilon)}{2} \tag{10.26}$$

$$a_2 = a_1 \frac{(5-\epsilon)}{12} = a_0 \frac{(5-\epsilon)(1-\epsilon)}{24} \tag{10.27}$$

$$\vdots$$

and so forth. This gives

$$\psi^{(+)}(y) = a_0 \left[1 + \frac{a_1}{a_0} y^2 + \frac{a_2}{a_0} y^4 + \cdots \right] e^{-y^2/2} \tag{10.28}$$

The fact that this solution of a second-order differential equation has only a single undetermined constant is explained by the fact that while a_0 can determine $\psi(0)$, $\psi'(0) = 0$ is already determined by the fact that $h^{(+)}(y)$ is an even function.

This solution for $h^{(+)}(y)$ is a nicely convergent series as a ratio test (App. B.2) gives

$$\frac{a_{s+1} y^{2s+2}}{a_s y^{2s}} \rightarrow \frac{y^2}{s} \rightarrow 0 \qquad \text{as } s \rightarrow \infty \text{ for fixed } y \tag{10.29}$$

We note, however, that the function $e^{y^2/2}$, which has the series expansion

$$e^{y^2} = 1 + y^2 + \frac{1}{2!}(y^2)^2 + \cdots = \sum_{s=0}^{\infty} \frac{1}{s!} \left(y^2\right)^s \tag{10.30}$$

has the identical limiting behavior, so that $h^{(+)}(y) \rightarrow e^{y^2}$ and $\psi^{(+)}(y) = h^{(+)}(y) e^{-y^2/2} \rightarrow e^{+y^2/2}$. So, while we have tried to eliminate the divergent behavior by hand, it has reemerged in the series solution.

This argument relies, however, on the comparison of ratio tests, i.e., the assumption that the series solution for $h^{(+)}(y)$ has infinitely many terms. We can avoid this problem if the series is, in fact, finite, i.e., terminates for some finite value of s. If $a_n = 0$ for some value of n, then Eqn. (10.25) guarantees that all subsequent a_{n+k} will also vanish, so that $h^{(+)}(y)$ will be simply a polynomial in y^2. This can happen only if

$$a_n = 0 \Longrightarrow \epsilon = (4n+1) \qquad \text{for } n = 0, 1, 2 \ldots \tag{10.31}$$

which implies the quantized energies

$$E_n^{(+)} = \hbar\omega(2n + 1/2) \qquad n = 0, 1, 2 \ldots \tag{10.32}$$

We plot in Fig. 10.3 the wavefunction corresponding to the first of these special even functions (i.e., $\epsilon = 1$), namely, the ground state, as well as solutions characterized by values of ϵ that are only slightly different. We note once again that the requirement of normalizable wavefunctions leads to quantized energies for bound states. The acceptable even solutions are then

$$\begin{aligned}
E_0^{(+)} &= \frac{\hbar\omega}{2} & \psi_0^{(+)}(y) &= a_0 e^{-y^2} \\
E_1^{(+)} &= \frac{5\hbar\omega}{2} & \psi_1^{(+)}(y) &= a_0(1 - 2y^2) e^{-y^2/2} \\
E_2^{(+)} &= \frac{9\hbar\omega}{2} & \psi_2^{(+)}(y) &= a_0(1 - 4y^2 + 4y^4/3) \\
&\vdots
\end{aligned} \tag{10.33}$$

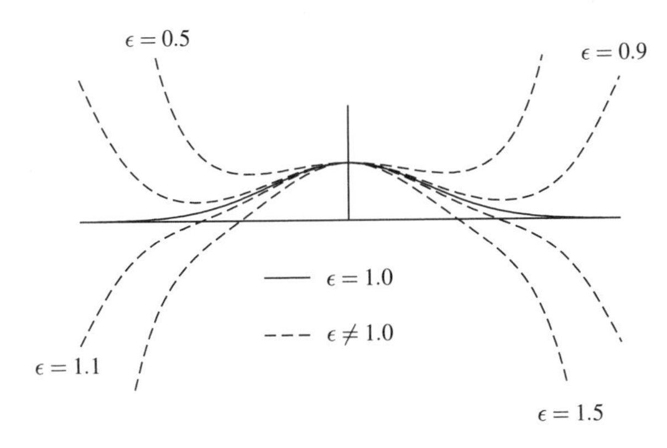

FIGURE 10.3. *Square integrable ground state wavefunction (solid curve) for the simple harmonic oscillator corresponding to $\epsilon = 1$ and divergent "near misses" (dashed curves) for various $\epsilon \neq 1$.*

The odd states can be obtained in an entirely similar way by assuming a series solution of the form

$$h^{(-)}(y) = \sum_{s=0}^{\infty} b_s y^{2s+1} \tag{10.34}$$

which yields the recursion relation

$$b_{s+1} = b_s \left[\frac{4s+3-\epsilon}{2(s+1)(2s+3)} \right] \tag{10.35}$$

and acceptable energies

$$E_n^{(-)} = (2n + 3/2)\hbar\omega \qquad n = 0, 1, 2 \ldots \tag{10.36}$$

10.2.2 Properties of the Solutions

Collecting these results, we find that the two classes of solutions can be combined with a single label to write

$$E_n = (n + 1/2)\hbar\omega \qquad n = 0, 1, 2, \ldots \tag{10.37}$$

with corresponding normalized solutions that are expressible in the form

$$\psi_n(x) = C_n h_n(y) e^{-y^2/2} \qquad \text{where} \qquad y = \sqrt{\frac{m\omega}{\hbar}}\, x \tag{10.38}$$

The $h_n(y)$ can be recognized as the *Hermite polynomials* (see App. C.3), and the normalization constants are

$$C_n = \left(\frac{\sqrt{m\omega/\hbar\pi}}{2^n n!} \right)^{1/2} = \left(\frac{1}{\rho\sqrt{\pi} 2^n n!} \right)^{1/2} \tag{10.39}$$

Several comments can be made:

- The $h_n(y)$ are polynomials in y of degree n with parity $+1\,(-1)$ depending on whether n is even (odd). The first few can easily be obtained from the recursion relations (P10.2) and are given by

$$\begin{aligned} h_0(y) &= 1 & h_1(y) &= 2y \\ h_2(y) &= 4y^2 - 2 & h_3(y) &= 8y^3 - 12y \\ h_4(y) &= 16y^4 - 48y^2 + 12 & h_5(y) &= 32y^5 - 160y^3 + 120y \end{aligned} \tag{10.40}$$

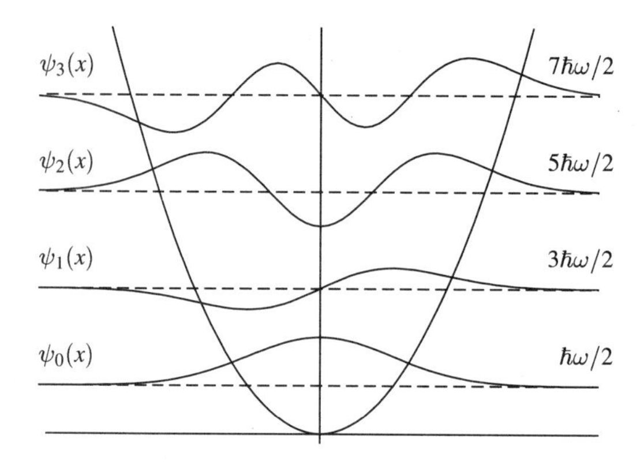

FIGURE 10.4. *SHO wavefunctions for the lowest-lying energy levels.*

- The first few energy eigenfunctions are plotted in Fig. 10.4 and show the usual pattern of increasing numbers of nodes and alternation of even and odd states.
- The nodeless ground state has a zero point energy of $E_0 = \hbar\omega/2$ consistent with the uncertainty principle discussion of P1.9. In this case, $\psi_0(x)$ is a familiar Gaussian waveform, implying that its Fourier transform, $\phi_0(p)$, is as well. It can be shown that this wavepacket has the minimum uncertainty principle product $\Delta x\,\Delta p = \hbar/2$ (Sec. 13.4).
- Because of parity, we have the expectation value

$$\langle\psi_n|x|\psi_n\rangle = \int_{-\infty}^{+\infty} dx\, x|\psi_n(x)|^2 = 0 \tag{10.41}$$

while

$$\langle\psi_n|\hat{p}|\psi_n\rangle = \int_{-\infty}^{+\infty} dx\, \psi^*(x)\hat{p}\psi(x) = 0 \tag{10.42}$$

as usual for bound states. Other average values will be more easily obtained using the operator formalism in Chap. 14. We list here for convenience the results

$$\langle\psi_n|x^2|\psi_n\rangle = \frac{\hbar}{m\omega}(n + 1/2) \tag{10.43}$$

$$\langle\psi_n|\hat{p}^2|\psi_n\rangle = \hbar m\omega(n + 1/2) \tag{10.44}$$

$$\langle\psi_n|x^4|\psi_n\rangle = \frac{3}{4}\left(\frac{\hbar}{m\omega}\right)^2(2n^2 + 2n + 1) \tag{10.45}$$

$$\langle\psi_n|\hat{p}^4|\psi_n\rangle = \frac{3}{4}(\hbar m\omega)^2(2n^2 + 2n + 1) \tag{10.46}$$

$$\langle\psi_n|x|\psi_k\rangle = \sqrt{\frac{\hbar}{2m\omega}}\left(\delta_{n,k-1}\sqrt{k} + \delta_{n,k+1}\sqrt{k+1}\right) \tag{10.47}$$

$$\langle\psi_n|\hat{p}|\psi_k\rangle = -i\sqrt{\frac{m\omega\hbar}{2}}\left(\delta_{n,k-1}\sqrt{k} - \delta_{n,k+1}\sqrt{k+1}\right) \tag{10.48}$$

- The solutions form a complete set of states, so that any acceptable wavefunction can be expanded via $\psi(x) = \sum_{n=0}^{\infty} a_n\psi_n(x)$ in the usual way.

Finally, the harmonic oscillator problem is the only one where the potential energy enters quadratically; this implies a complete symmetry between x and p representations. For example, the Schrödinger equation in momentum space reads

$$\frac{p^2}{2m}\phi(p) - \frac{\hbar^2 m\omega^2}{2}\frac{d^2\phi(p)}{dp^2} = E\phi(p) \tag{10.49}$$

which is of the same form as Eqn. (10.15) in position space. The solutions are easily obtained by the same methods, and one finds

$$\phi_n(p) = D_n h_n(q) e^{-q^2/2} \qquad \text{where} \qquad q = \frac{p}{\sqrt{m\omega\hbar}} \tag{10.50}$$

with normalization constants

$$D_n = \left(\frac{1}{n! 2^n \sqrt{m\omega\hbar\pi}}\right)^{1/2} \tag{10.51}$$

The momentum space probability densities, plotted as a function of the scaled variable q, therefore have the same form as in Fig. 10.4.

10.3 EXPERIMENTAL REALIZATION OF THE SHO

Several aspects of these quantized vibrational states are clearly evident in the study of diatomic molecules. They include:

- Approximately evenly spaced energy levels
- Vibrational contribution to the heat capacity of diatomic gases
- The variation of the zero point energy with the constituent masses

We (briefly) discuss each in turn.

The validity of a simple harmonic oscillator description of the quantized energy levels in a Lennard-Jones potential are shown in Fig. 10.5 for both relatively light (Ne-Ne) and heavy (Xe-Xe) molecules. Clearly the approximation of the potential near the minimum is much better for the heavier molecules, which sit "further down" in the potential well. In the case of NaCl, for example, where the (ionic) binding potential is much larger, there are roughly 20 such levels that can be detected spectroscopically. As the quantum number

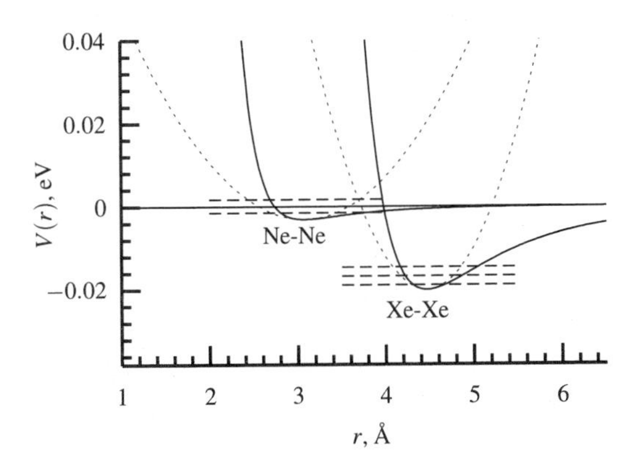

FIGURE 10.5. Lennard-Jones interaction potentials for noble gas atoms. The dashed lines indicate the first few energy levels using the SHO approximation.

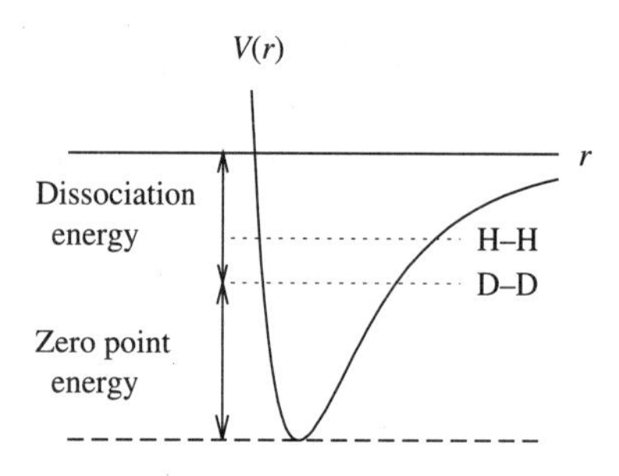

FIGURE 10.6. *Schematic representation of interatomic potential for H–H, H–D, and D–D pairs. The zero point energy (dissociation energy) for the heavier D–D pair is smaller (larger) than for the lighter H–H system due to reduced mass effects.*

is increased, however, the potential becomes more anharmonic, and the level spacings decrease somewhat. (Can you argue why?) Since diatomic molecules that vibrate can also rotate, the energy spectrum also contains rotational levels, which we discuss more fully in Chap. 17.

The vibrational motion can also contribute to the heat capacity of a system of diatomic molecules. Classically, each degree of freedom will contribute $k_B T/2$ to the thermal energy. For a monatomic gas, with only three translational degrees of freedom, this would yield a specific heat per mole given by

$$C_V = \frac{\partial E}{\partial T} = \frac{\partial}{\partial T}\left(\frac{3}{2} N_A k_B T\right) = \frac{3}{2} R \tag{10.52}$$

where $R = N_A k_B$ is the gas constant.

An additional vibrational degree of freedom would then contribute an extra $R/2$ to this value. For this mode to contribute, however, the temperature must be large enough that such energy levels can actually be excited; thus this contribution becomes important only when $k_B T \gtrsim E_0 = \hbar\omega/2$. For H_2, for example, this temperature is so high that the molecule disassociates before the vibrational states can contribute fully.

Finally, the zero point energy can be seen to scale with the reduced mass as

$$E_0 = \frac{1}{2}\hbar\omega = \frac{1}{2}\hbar\sqrt{\frac{K}{\mu}} \tag{10.53}$$

If we consider two atoms corresponding to different nuclear isotopes (i.e., the same nuclear charge but different mass), the effective interatomic potential between them should be identical (same K) since this is determined by the arrangements of the atomic electrons. The reduced mass, μ, which appears in Eqn. (10.53), however, will differ. This implies that the ground state energy (and hence the energy required to disassociate the molecule) will be different for different isotopes. This is most clearly seen using hydrogen and deuterium where the H–H, H–D, and D–D zero point energies are in ratios $1 : \sqrt{4/3} : \sqrt{2}$. The energy required to disassociate H–H is then less than for D–D (as in Fig. 10.6), and this effect is observed experimentally.

10.4 CLASSICAL LIMITS

10.4.1 Probability Distributions

The classical limit of the quantum oscillator can be approached in several ways; for example, wavepackets constructed from the solutions can be shown to oscillate with the classical oscillation period, $\tau = 2\pi/\omega$. We consider this in Sec. 10.4.3.

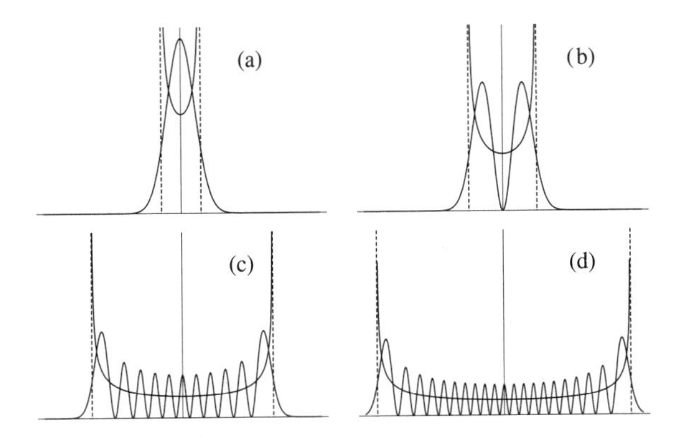

FIGURE 10.7. *Classical versus quantum probability distributions for (a) $n = 0$, (b) $n = 1$, (c) $n = 10$, and (d) $n = 20$.*

One can also examine the connection between the classical and quantum mechanical probability distributions discussed in Sec. 6.2. For the SHO, we found that

$$P_{CL}(x) = \frac{1}{\pi\sqrt{A^2 - x^2}} \tag{10.54}$$

where A is the classical turning point defined by $E_{\text{tot}} = KA^2/2$. We plot in Fig. 10.7 the quantum and classical results for the position space probability densities for the cases $n = 0, 1, 10$, and 20, and note that the quantum mechanical result, when locally averaged, does approach the classical result. The form of the classical result is easy to understand, as the particle spends more time near the turning points where it must slow down and come to rest before reversing direction, while near the origin it has its largest kinetic energy (spring is unstretched) and so spends little time.

We can also inquire about the classical distribution of both kinetic and potential energies. We consider the *potential energy density,* $V(x)$ weighted by the classical probability density, given by

$$\mathcal{V}_{CL}(x) \equiv P_{CL}(x)V(x) = P_{CL}(x)\left(\frac{1}{2}Kx^2\right) = \frac{K}{2\pi}\frac{x^2}{\sqrt{A^2 - x^2}} \tag{10.55}$$

The similar weighted value of the kinetic energy, the *kinetic energy density,* is then

$$\mathcal{T}_{CL}(x) \equiv P_{CL}(x)T(x) = P_{CL}(x)\left(\frac{1}{2}mv(x)^2\right) = \frac{K}{2\pi}\sqrt{A^2 - x^2} \tag{10.56}$$

where we have used the fact that $E = mv^2/2 + Kx^2/2$. The integrals of these quantities yield the average potential and kinetic energies, and we find

$$\begin{aligned} \langle T(x)\rangle &= \int_{-A}^{+A} P_{CL}(x)T(x)\,dx = \frac{KA^2}{4} \\ \langle V(x)\rangle &= \int_{-A}^{+A} P_{CL}(x)V(x)\,dx = \frac{KA^2}{4} \end{aligned} \tag{10.57}$$

i.e., $\langle V(x)\rangle = \langle T(x)\rangle$, so that the energy is, on average, shared equally. This result is familiar from classical mechanics where

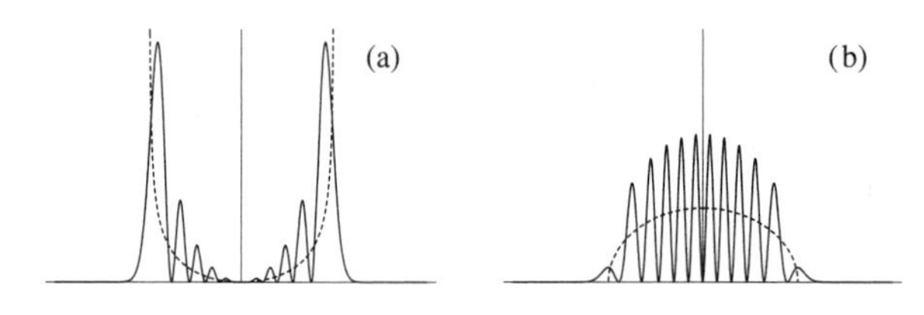

FIGURE 10.8. Classical and quantum probability distributions for the potential energy (a) and kinetic energy (b) for the simple harmonic oscillator.

$$V(x(t)) = \frac{Kx(t)^2}{2} = \frac{KA^2}{2}\cos^2(\omega t - \delta) \tag{10.58}$$

$$T(v(t)) = \frac{mv^2(t)}{2} = \frac{m\omega^2 A^2}{2}\sin^2(\omega t - \delta) \tag{10.59}$$

When these are averaged over one period (P10.9), the same result is obtained.

In a position space representation, the quantum mechanical probability density times the potential,

$$\mathcal{V}_{QM}(x) \equiv V(x)|\psi(x)|^2 \tag{10.60}$$

most closely resembles Eqn. (10.55). In addition, Eqn. (5.44) showed that

$$\mathcal{T}_{QM}(x) \equiv \frac{\hbar^2}{2m}\left|\frac{\partial\psi(x)}{\partial x}\right|^2 \tag{10.61}$$

when integrated over all space, gives the average kinetic energy, so that we can take this as a quantum mechanical equivalent of the classical kinetic energy density, the quantum version of Eqn. (10.56). In Fig. 10.8, we compare the classical and quantum versions of both the potential and kinetic energy densities. We see again that the quantum results, when locally averaged, approach the classical predictions.

The form of the quantum mechanical momentum space probability density,

$$P_{QM}(p;n) = |\phi_n(p)|^2 \tag{10.62}$$

must necessarily have the same form as $P_{QM}(x;n)$, which at first might seem somewhat surprising. One might argue classically that since the particle spends most of its time near the turning points where the classical speeds are small, $P_{CL}(p)$ should be sharply peaked for *small* values of p. Such classical behavior would then seem to be in direct conflict with the $n \to \infty$ limit of the quantum probability densities, which have the same form as Fig. 10.7. The seeming contradiction arises from the mistaken assumption that we *first* "find" the particle near the classical turning points (i.e., make a position measurement) and *then* specify its momentum. A more consistent definition of the probability distribution for measurements of momentum is clearly needed, and that is the subject of the next section.

10.4.2 Classical Momentum Distributions

In Sec. 6.2 we discussed the classical problem of a particle in a confining potential and derived an expression for the classical probability distribution for position. A very similar derivation can be carried out which gives the corresponding distribution of momenta, $P_{CL}(p)$. We begin as before by considering the probability of measuring the particle with momentum p to be

$$P_{CL}(p)\,dp \equiv \text{Probability}[(p, p+dp)] = \frac{dt}{\tau/2} = \frac{2}{\tau}\frac{dp}{|dp/dt|} \tag{10.63}$$

We then identify dp/dt with the classical force,

$$\frac{dp}{dt} = F(x) = -\frac{dV(x)}{dx} \tag{10.64}$$

so that

$$P_{CL}(p) = \frac{2}{\tau}\frac{1}{|F(x)|} \tag{10.65}$$

in a somewhat "mixed" notation. For the case of the simple harmonic oscillator, we have $F(x) = -Kx$ and $V(x) = Kx^2/2$, so

$$\begin{aligned} |F(x)| &= \sqrt{2KV(x)} = \sqrt{2K(E - mv^2/2)} \\ &= \sqrt{2K(E - p^2/2m)} = \sqrt{\frac{K}{m}(p_0^2 - p^2)} \end{aligned} \tag{10.66}$$

where $E \equiv p_0^2/2m$ defines the maximum momentum. This then gives

$$P_{CL}(p) = \frac{1}{\pi\sqrt{p_0^2 - p^2}} \tag{10.67}$$

since $\tau = 2\pi\sqrt{K/m}$. Thus the classical momentum distribution does have exactly the same form (as a function of p) as the classical distribution of positions. This is consistent with the complete symmetry between x and p for the simple harmonic oscillator.

We can also generalize the projection of trajectory technique discussed in Sec. 6.2. For the classical position space distribution, one projected the amount of time spent in equal x bins onto the vertical axis from a graph of $x(t)$ versus t, and the resulting binned values approached $P_{CL}(x)$. A knowledge of the exact trajectory, $x(t)$, implies that we can also plot $v(t)$, or equivalently and more usefully, $p(t) = v(t)/m$ versus t, and perform the same projection tricks to obtain $P_{CL}(p)$. For the harmonic oscillator, since

$$x(t) = A\cos(\omega t) \qquad \text{and} \qquad p(t) = -\frac{A\omega}{m}\sin(\omega t) \tag{10.68}$$

both have the identical sinusoidal time dependence. Time bins projected onto the x and p axes are guaranteed to yield the same distribution in either case.

Using the formulation for $P_{CL}(p)$ in Eqn. (10.65), we can now also justify the expression for the classical momentum probability distribution found for the infinite well used in Sec. 6.3, namely,

$$P_{CL}(p) = \frac{1}{2}(\delta(p - p_0) + \delta(p + p_0)) \tag{10.69}$$

You have shown (P6.5) that the infinite well can be obtained as a limiting case of the more general potential $V(x; n) = V_0|x/a|^n$, in the limit that $n \to \infty$. Using this potential, we can write

$$\begin{aligned} |F(x; n)| &= \frac{nV_0|x|^{n-1}}{a^n} = \frac{nV_0}{a}\left(\frac{V(x)}{V_0}\right)^{1-1/n} \\ &= \frac{nV_0^{1/n}}{a}\frac{(p_0^2 - p^2)^{1-1/n}}{(2m)^{1-1/n}} \end{aligned} \tag{10.70}$$

so that

$$P_{CL}(p;n) \propto \frac{1}{n(p_0^2 - p^2)^{1-1/n}} \tag{10.71}$$

where we have not included many dimensional factors for simplicity. We can examine the behavior of this solution in the $n \to \infty$ limit and find that

$$\lim_{n\to\infty} P_{CL}(p;n) = \begin{cases} 0 & \text{for } p \neq \pm p_0 \\ \infty & \text{for } p = \pm p_0 \end{cases} \tag{10.72}$$

and the normalization of Eqn. (10.65) guarantees that

$$\lim_{n\to\infty} P_{CL}(p,n) = p_0 \delta(p_0^2 - p^2) = \frac{1}{2}\big(\delta(p - p_0) + \delta(p + p_0)\big) \tag{10.73}$$

where we use results from App. C.8.

As a final example, we consider the symmetric linear potential defined by $V(x) = C|x|$. In this case, $F(x) = -Cx/|x|$, so that $|F(x)| = C$, and the momentum distribution is a constant, $P_{CL}(p) = 2/\tau C$. Using the result of P6.3 for the period, we then find

$$P_{CL}(p) = \begin{cases} 0 & \text{for } |p| > p_0 \\ \frac{2}{\tau C} = \frac{1}{2p_0} & \text{for } |p| < p_0 \end{cases} \tag{10.74}$$

which is obviously normalized properly. We will confirm this below when we solve the quantum version of the symmetric linear potential; this form can also be easily visualized by the projection technique (P10.10).

10.5 WAVEPACKETS

Wavepacket-like linear combinations of energy eigenstates can be formed for the harmonic oscillator, which also indicate even more clearly the expected quasi-classical behavior. A very convenient formulation for their construction is the propagator method introduced in Sec. 7.9. In this case we have

$$K(x, x'; t, 0) = \sum_{n=0}^{\infty} \psi_n^*(x)\psi_n(x)e^{-i(n+1/2)\omega t} \tag{10.75}$$

which gives the time development of any initial state via

$$\psi(x,t) = \int_{-\infty}^{+\infty} dx' \, \psi(x', 0)K(x, x'; t, 0) \tag{10.76}$$

The periodicity of the classical harmonic oscillator can be obtained from Eqn. (10.75) by recalling that $\tau = 2\pi/\omega$, so

$$K(x, x'; t + m\tau) = (-i)^m K(x, x'; t, 0) \tag{10.77}$$

implying that

$$|\psi(x, t + m\tau)|^2 = |(-i)^m \psi(x,t)|^2 = |\psi(x,t)|^2 \tag{10.78}$$

so that the position probability density is indeed periodic.

Using the formalism of raising and lowering operators of Chap. 15, one can actually calculate the summation in Eqn. (10.75) in closed form,[4] with the result

$$K(x, x'; t, 0) = \sqrt{\frac{m\omega}{\pi\hbar 2i \sin(\omega t)}} \exp\left(\frac{im\omega((x^2 + x'^2)\cos(\omega t) - 2xx')}{2\hbar \sin(\omega t)}\right) \quad (10.79)$$

We note that:

- This expression explicitly exhibits the periodicity in time dictated by Eqn. (10.77).
- In the limit in which the value of the spring constant, and hence ω, vanishes, the system describes a free particle; it is easy to check that $K(x, x'; t, 0)$ approaches the free-particle propagator found in Sec. 7.9 (P10.13).

Using this form and a Gaussian wavepacket representing a particle with initial position and momentum equal to x_0 and $p_0 = 0$, respectively, namely,

$$\psi(x, 0) = \frac{1}{\sqrt{L\sqrt{\pi}}} e^{-(x-x_0)^2/2L^2} \quad (10.80)$$

we can perform the time development integral in closed form and obtain for the position probability density (P10.13)

$$|\psi(x, t)|^2 = \frac{1}{\sqrt{\pi}L(t)} e^{-(x-x_0\cos(\omega t))^2/L^2(t)} \quad (10.81)$$

where

$$L(t) \equiv \sqrt{L^2\cos^2(\omega t) + (m\omega/L\hbar)^2\sin^2(\omega t)} \quad (10.82)$$

is the time-dependent width. The expectation value of position is easily evaluated, and one finds

$$\langle x \rangle_t = x_0 \cos(\omega t) \quad (10.83)$$

which is consistent with the classical solution of Newton's equations for the same initial conditions. The time evolution of this wavepacket is illustrated in Fig. 10.9. We note that the width of the distribution does not increase monotonically with time but in fact shrinks and expands. This is due to the oscillatory nature of the spread in position:

$$\Delta x_t = L(t)/\sqrt{2} \quad (10.84)$$

This is consistent with uncertainty principle arguments, as the corresponding momentum space distributions show the same behavior with the appropriate phase relation; specifically, one can show that

$$\Delta p_t = p_L(t)/\sqrt{2} \quad (10.85)$$

where

$$p_L(t) = \sqrt{(\hbar/L)^2\cos^2(\omega t) + (m\omega L)^2\sin^2(\omega t)} \quad (10.86)$$

[4]See, e.g., the excellent discussion in Saxon (1968).

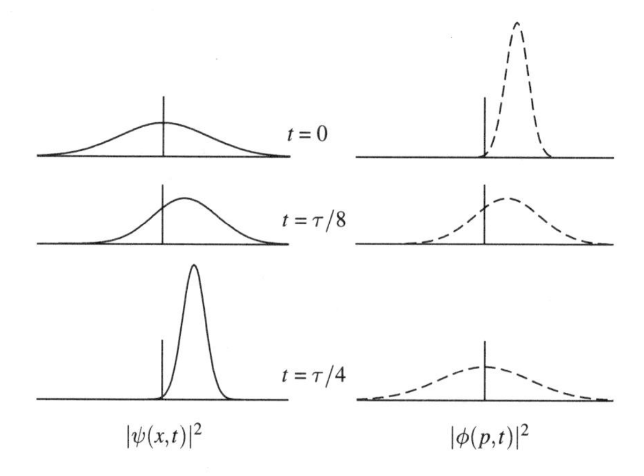

FIGURE 10.9. $|\psi(x,t)|^2$ *versus x (left) and* $|\phi(p,t)|^2$ *versus p (right) for the Gaussian SHO wavepacket during the first quarter period of motion. Note the anticorrelation between the widths of the x and p distributions.*

The propagator for the momentum space wavefunctions is obtained as easily, with the result that

$$K_p(p', p; t) = \frac{1}{\sqrt{2\pi\hbar m\omega \sin(\omega t)}} \exp\left(\frac{i((p^2 + p'^2)\cos(\omega t) - 2pp')}{2\hbar m\omega \sin(\omega t)}\right) \quad (10.87)$$

and the momentum space wavepackets can be easily constructed.

10.6 UNSTABLE EQUILIBRIUM: CLASSICAL AND QUANTUM DISTRIBUTIONS

For the case of a particle near a point of *unstable* equilibrium, we have $V(x) = -Kx^2/2$, and the dimensionless Schrödinger equation becomes

$$\frac{d^2\psi(y)}{dy^2} + y^2\psi(y) = -\epsilon\psi(y) \quad (10.88)$$

The large y dependence can be obtained as done for Eqn. (10.20), giving

$$\psi(y) \to e^{\pm iy^2/2} \quad \text{or} \quad \psi(y) \to \sin(y^2/2), \cos(y^2/2) \quad (10.89)$$

i.e., oscillatory solutions consistent with an unbound particle. If we specialize to the case of $\epsilon = 0$, we can get more information by assuming a solution of the form

$$\psi(y) = y^\alpha \sin(y^2/2) \quad (10.90)$$

and upon noting that

$$\begin{aligned}\psi''(y) + y^2\psi(y) &= (2\alpha + 1)y^\alpha \cos(y^2/2) + \alpha(\alpha - 1)y^{\alpha-2}\sin(y^2/2) \\ &\approx 0 \quad \text{up to } O(y^{\alpha-2})\end{aligned} \quad (10.91)$$

provided $\alpha = -1/2$. Thus, the next better approximation gives

$$\psi(y) \propto \frac{\sin(y^2/2)}{\sqrt{y}} \qquad \left(\text{or } \frac{\cos(y^2/2)}{\sqrt{y}}\right) \quad (10.92)$$

which gives a probability density

$$P_{QM}(y) \propto \frac{\sin^2(y^2/2)}{y} \quad (10.93)$$

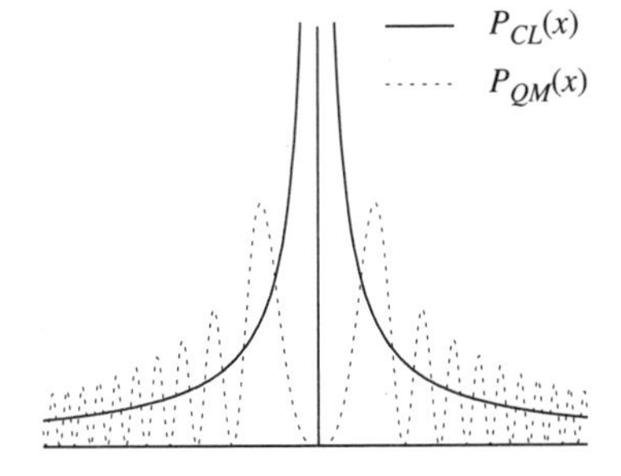

FIGURE 10.10. Classical (solid) and quantum (dotted) position space probability distributions ($|\psi(x)|^2$) for the "unstable" oscillator.

To compare to the classical probability density, we note that for large time, any unstable solution will be dominated by the increasing exponential term, giving $x(t) \rightarrow De^{\omega t}$, so that

$$v(t) = \dot{x}(t) \rightarrow D\omega e^{\omega t} \propto x(t) \tag{10.94}$$

so that

$$P_{CL}(x) \propto \frac{1}{v(x)} \propto \frac{1}{x} \tag{10.95}$$

Changing this to dimensionless variables and noting that the $\sin(y^2/2)$ term averages to 1/2 over many cycles, we can compare the quantum and classical probability distributions for an unstable particle, namely,

$$P_{CL}(y) \propto \frac{1}{2y} \quad \text{and} \quad P_{QM}(y) \propto \frac{\sin^2(y^2/2)}{y} \tag{10.96}$$

which we plot in Fig. 10.10. Once again, the increasing wiggliness as $|y|$ increases is indicative of increasing kinetic energy, while the decreasing amplitude is consistent with less and less time spent in a given y interval, i.e., larger speeds. This, then, is the quantum mechanical "picture" of "falling off a log."

10.7 SYMMETRIC LINEAR POTENTIAL

While it is not as fundamentally important as the harmonic oscillator potential, the symmetric linear potential, defined via $V(x) = F|x|$ (where $F > 0$), has several instructive features:

- The symmetry of the potential ensures that the solutions can be classified according to their parity, and the application of boundary conditions in this case relies heavily on this fact.
- The modified linear potential given by

$$\tilde{V}(z) = \begin{cases} \infty & \text{for } z < 0 \\ mgz & \text{for } 0 < z \end{cases} \tag{10.97}$$

  is the quantum equivalent of a "bouncing ball," i.e., a uniform gravitational field above a rigid flat surface.

- It is argued that the theory governing the interactions of quarks and antiquarks (called *quantum chromodynamics* or QCD) in elementary particle physics implies that the nonrelativistic limit of their interaction potential can be written as

$$V(r) = -\frac{4}{3}\frac{\alpha_s}{r} + Kr \tag{10.98}$$

so that at large separations the quark-antiquark pair "feels" a linear potential.

- Finally, it allows us a further check on the classical position and momentum distributions discussed above and for the numerical solution of the Schrödinger equation discussed in the next chapter.

Using the symmetry of the potential, we can argue that it is sufficient to solve the Schrödinger equation for $x > 0$ only, i.e.,

$$-\frac{\hbar^2}{2m}\frac{d^2\psi(x)}{dx^2} + Fx\psi(x) = E\psi(x) \tag{10.99}$$

for even, $\psi^{(+)}(x)$, and odd, $\psi^{(-)}(x)$, states and extend them to negative values of x by using

$$\psi^{(+)}(-x) = \psi^{(+)}(x) \qquad \text{and} \qquad \psi^{(-)}(-x) = -\psi^{(-)}(x) \tag{10.100}$$

As usual, we change variables to $x = \rho y + \sigma$ to simplify Eqn. (10.99) and find the result

$$\frac{d^2\psi(y)}{dy^2} = y\psi(y) \tag{10.101}$$

provided we define

$$\rho = \left(\frac{\hbar^2}{2mF}\right)^{1/3} \qquad \text{and} \qquad \sigma = \frac{E}{F} \tag{10.102}$$

This is the, by-now familiar, Airy equation with solutions $Ai(y)$, $Bi(y)$. The behavior of $Bi(y)$ for large y is unacceptable, so we have the (unnormalized) solution

$$\psi(x) = N\,Ai\left(\frac{x-\sigma}{\rho}\right) \qquad \text{for } x > 0 \tag{10.103}$$

and can extend it to $x < 0$ via Eqn. (10.100).

In this case, the boundary conditions at infinity have been easily implemented (much like the case of the finite square well) by discarding the divergent solution. The quantized energy eigenvalues are then determined by matching the solutions at the boundary between two regions where the solutions are defined differently, namely, at the origin. Because of the symmetry, we must have

$$\text{even:} \quad \psi^{(+)}(0) = \text{arbitrary} \qquad \frac{\psi^{(+)}(0)}{dx} = Ai'\left(-\frac{\sigma}{\rho}\right) = 0 \tag{10.104}$$

$$\text{odd:} \quad \frac{\psi^{(-)}(0)}{dx} = \text{arbitrary} \qquad \psi^{(-)}(0) = Ai\left(-\frac{\sigma}{\rho}\right) = 0 \tag{10.105}$$

If we label the zeroes of the $Ai'(y)$ and $Ai(y)$ functions $-y_i^{(+)}$ and $-y_i^{(-)}$, respectively, these conditions determine the quantized energy eigenvalues, since

$$E_i^{(\pm)} = y_i^{(\pm)}\left(\frac{\hbar^2F^2}{2m}\right)^{1/3} \tag{10.106}$$

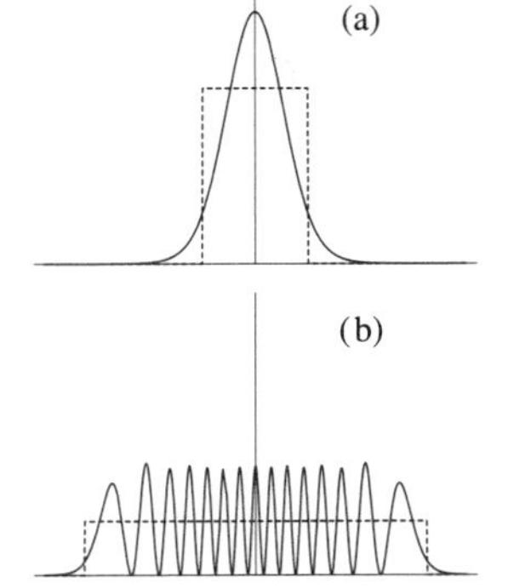

FIGURE 10.11. Classical (solid) and quantum (dashed) momentum space probability distributions ($|\phi(p)|^2$) for the symmetric linear potential. For large quantum numbers, the "flat" distribution corresponding to a classical force of constant magnitude is evident.

The first few zeros are given by

$$
\begin{aligned}
y_1^{(+)} &= 1.0188 & y_1^{(-)} &= 2.3381 \\
y_2^{(+)} &= 3.2482 & y_2^{(-)} &= 4.0879 \\
y_3^{(+)} &= 4.8201 & y_3^{(-)} &= 5.5206 \\
y_4^{(+)} &= 6.1633 & y_4^{(-)} &= 6.7867
\end{aligned}
\tag{10.107}
$$

The energies are not evenly spaced, and one can show (P10.15) that for large quantum number n, the energies go as $E_n \propto n^{2/3}$.

After normalizing the wavefunctions suitably (i.e., determining N, which must be done numerically), one can obtain the position space wavefunctions; $|\psi_n(x)|^2$ can be compared to the classical probability distribution in the large n limit. More interestingly, the momentum space wavefunctions and the corresponding $P_{QM}(p)$ can be obtained through a numerical Fourier transform and are shown in Fig. 10.11 compared with the classical result for $P_{CL}(p)$ in Eqn. (10.74). The "flat" momentum distribution is increasingly apparent.

10.8 QUESTIONS AND PROBLEMS

Q10.1 The vibrational degree of freedom in a diatomic gas does not contribute to the specific heat until $k_B T \gtrsim \hbar\omega$ because the energies are quantized. The translational degrees of freedom of a gas when confined to a box (or a room) are also quantized. At what temperatures will the translational degrees of freedom be "frozen out"?

Q10.2 Discuss the classical and quantum momentum space probability densities for the case of the unstable SHO in Sec. 10.4.4. Compare $P_{CL}(x)$ for the cases of the free particle, accelerating particle, and (exponentially) unstable particle.

Q10.3 Most solids are characterized by the fact that they expand when heated. Show that if one models the interactions of the atoms in a solid by simple harmonic forces, that any increase in energy of the individual atoms does not lead to an overall expansion. What is it about the real interactions of atoms that can lead to the observed expansion?

P10.1 Consider a particle of mass m that moves in a potential of the form

$$V(x) = V_0\left(\cosh\left(\frac{x}{a}\right) - 1\right)$$

(a) What is the "best fit" SHO approximation to this potential?
(b) What is the energy spectrum of the SHO fit?
(c) Is the approximation better for large or small a? (You should specify large or small compared to what?)
(d) For a given value of a, up to what values of n is the SHO spectrum a reasonable approximation?

P10.2 Using the recurrence relations in Eqn. (10.25), derive the first three even Hermite polynomial solutions. Do the same for the first three odd solutions using Eqn. (10.35).

P10.3 Consider a state that is a linear combination of the ground state and first excited state of the SHO, i.e.,

$$\psi(x,0) = \frac{1}{\sqrt{2}}\big(\psi_0(x) + \psi_1(x)\big)$$

(a) What is $\psi(x,t)$?
(b) Evaluate $\langle x\rangle_t$ and $\langle \hat{p}\rangle_t$, and show explicitly that $d\langle x\rangle_t/dt = \langle \hat{p}\rangle_t/m$.
(c) What is $\langle \hat{E}\rangle_t$?

P10.4 A particle of mass m is in the ground state of a simple harmonic oscillator with spring constant K. Somehow, the spring constant is suddenly made 4 times smaller. What is the probability that the particle is in the ground state of the new system? In the first excited state? In the second excited state?

P10.5 A particle of mass m in a harmonic oscillator potential is described by the wavefunction

$$\psi(x,0) = \frac{1}{\sqrt{a\sqrt{\pi}}}e^{-x^2/a^2}$$

What is the probability that a measurement of its energy would give the value $\hbar\omega/2$? $3\hbar\omega/2$?

P10.6 Consider a charged particle of mass m in a SHO potential but which is also subject to an external electric field E. The potential for this problem is now given by

$$V(x) = \frac{1}{2}m\omega^2x^2 - qEx$$

where q is the charge of the particle.
(a) Show that a simple change of variables makes this problem completely soluble in terms of the standard SHO solutions. *Hint:* Complete the square.
(b) Find the new eigenfunctions and energy eigenvalues.
(c) Show that for a particular value of E the ground state energy can be made to vanish. Does this mean that there is no zero point energy in this case?
(d) Evaluate $\langle x\rangle$ and $\langle \hat{p}\rangle$.
(e) What are the new momentum space wavefunctions? Can you evaluate $\langle \hat{x}\rangle$ and $\langle p\rangle$ in this representation?

P10.7 **Lennard-Jones potentials:** Consider the potential in Eqn. (10.6) for pairs of noble gas atoms.
(a) Show that the minimum of the potential is at $r_{\min} = 2^{1/6}\sigma$ and that the depth of the potential there is $V(r_{\min}) = -\epsilon$.
(b) "Fit" the potential near the minimum with a SHO potential, and find the effective spring constant in terms of σ and ϵ. What is the energy spectrum of the SHO potential in terms of these quantities?
(c) A measure of the "quantum-ness" of the system is given by the ratio E_0/ϵ. Calculate this in terms of the parameters of the problem?
(d) Using the numerical values on the next page, evaluate E_0/ϵ in each case and comment.

Atom	$\mu(amu)$	ϵ(eV)	σ(Å)
He	2	0.000875	2.56
Ne	10	0.00312	2.74
Ar	20	0.01	3.40
Kr	42	0.014	3.65
Xe	66	0.02	3.98

P10.8 The NaCl diatomic molecule has $\hbar\omega \approx 0.04$ eV.

(a) Use the atomic weights of Na (~ 23) and Cl (~ 35) to estimate the effective spring constant K. Express this in eV/Å and N/m.

(b) Estimate the amplitude of the vibrational motion in the ground state ($n = 1$) and in the last bound state ($n \sim 20$) by equating $E_n \sim KA^2/2$. Express your answer in Å, and compare to the equilibrium separation of the Na and Cl, namely, 2.4 Å.

(c) At roughly what temperature will the vibrational degree of freedom of NaCl contribute to the specific heat?

P10.9 Average values:

(a) The average value of a function of position, $x(t)$, in a periodic system can be defined via

$$\langle f(x(t))\rangle = \left[\int_0^\tau f(x(t))\,dt\right] \Big/ \left[\int_0^\tau dt\right] = \frac{1}{\tau}\int_0^\tau f(x(t))\,dt$$

Use the fact that $v(x) = dx/dt$ to show that this definition is equivalent to the definition of average value used for functions of $x = x(t)$ in Eqn. (6.11), namely,

$$\langle f(x)\rangle = \int_a^b P_{CL}(x)\, f(x)\,dx \qquad \text{with} \qquad P_{CL}(x) = \frac{2}{\tau}\frac{1}{v(x)}$$

(b) Use the first method to calculate the average values of potential and kinetic energies using the forms in Eqn. (10.58) and (10.59).

P10.10 Use the projection technique for the symmetric linear potential to confirm that the momentum distribution is flat.

P10.11 Consider a ball of mass m with energy E bouncing elastically on a level floor under the influence of gravity.

(a) What potential does the ball feel?

(b) Find $P_{CL}(z)$, and use it to evaluate $\langle z\rangle$.

(c) Find $P_{CL}(p_z)$, and use it to evaluate $\langle p_z\rangle$.

P10.12 Using the general potential $V(x;n) = V_0|x/a|^n$ and the result for the period, τ, in this potential from P6.4, find the properly normalized classical momentum distribution, $P_{CL}(p)$. Show explicitly that it is normalized, i.e., demonstrate that

$$\int_{-p_0}^{p_0} P_{CL}(p)\,dp = 1$$

where $p_0 = \sqrt{2mE}$ is the maximum classical value of momentum.

P10.13 **(a)** Show that the $\omega \to 0$ limit of Eqn. (10.79) gives the free-particle propagator.

(b) Use Eqn. (10.79) to find the $\psi(x,t)$ for the initial Gaussian in Eqn. (10.80).

(c) Evaluate $\langle x\rangle_t$ to confirm Eqn. (10.83).

P10.14 Wavepacket for unstable equilibrium: The unstable quadratic potential, $V(x) = -Kx^2/2$, can be trivially obtained from the stable harmonic oscillator by the identification $K \to -K$.

(a) Show that this corresponds to the identification $\omega \to i\omega$.

(b) Use this and the relations of App. B to show that a wavepacket satisfying Eqn. (10.88) with initial mean position and momentum given by $x_0 = 0$ and p_0 is given by

$$|\psi(x,t)|^2 = \frac{1}{\sqrt{\pi}L(t)} e^{-(x-p_0/m\omega \sinh(\omega t))^2/L(t)^2}$$

where

$$L^2(t) = L^2\cosh^2(\omega t) + (\hbar/m\omega L)^2 \sinh^2(\omega t).$$

(c) Evaluate $\langle x \rangle_t$, and show that it is consistent with the classical solutions of Newton's equations with the same initial conditions.

P10.15 Use the fact that the Airy function solutions, $Ai(y)$, for large negative y go like

$$\psi(y) \propto \frac{\sin\left(2y^{3/2}/3 + \pi/4\right)}{\sqrt{y}}$$

to show that the energy eigenvalues scale as ($E_n \propto n^{2/3}$) for large n. Compare this to the "simple-minded scaling" result in P1.6.

P10.16 The classical anharmonic oscillator:

(a) Consider the *anharmonic oscillator* potential given by

$$V(x) = \frac{1}{2}Kx^2 - \lambda K x^3$$

and consider λ to be small. Show that the resulting classical equation of motion is

$$\ddot{x}(t) = -\omega^2 x(t) + 3\omega^2 \lambda x^2(t)$$

(b) Attempt a solution of the form

$$x(t) = A\sin(\omega t) + B_1 + \left[C_1\cos(2\omega t) + D_1 \sin(2\omega t)\right] + \cdots$$

where A is fixed and B_1, C_1, D_1 are all of order λ, and solve for the undetermined coefficients.

(c) In contrast to the (symmetric) harmonic oscillator, show that $\langle x \rangle = 3\lambda A^2/2 \neq 0$ where the average value is defined via

$$\langle f(t) \rangle = \frac{1}{\tau}\int_0^\tau f(t)\,dt$$

as in P10.9.

(d) Discuss Q10.3 in the context of your answer.

(e) Evaluate the "anharmonic" term in the SHO approximation to the Lennard-Jones potential by expanding Eqn. (10.6) to the appropriate order.

For the quantum version of the anharmonic oscillator, see P11.24 and P11.25.

CHAPTER 11

Alternative Methods of Solution and Approximation Methods

It is a common practice to approach quantum mechanics through the study of a few, exactly soluble examples using the Schrödinger equation in position space. The number of potential energy functions for which such closed-form solutions are available is, however, quite small. Luckily, many of them actually correspond reasonably well to actual physical systems. Examples include the infinite well as a model of a free particle in a "box" (Chaps. 6 and 8), the harmonic oscillator (Chap. 10), the rigid rotator (Chap. 17), and the Coulomb potential for the hydrogen atom (Chap. 18).

Nonetheless, it is important to recognize that other methods can be used to study the properties of a quantum system. Some of them are quite different from the Schrödinger equation approach, and many are amenable for use as numerical and approximation methods in problems for which analytic solutions are not available.

In this chapter, we focus on several methods that can be used to study the spectrum of energy eigenvalues and wavefunctions for time-independent[1] systems, not only as calculational tools for possible numerical analysis but also as examples of different ways of approaching quantum mechanics. We make several general comments:

- Many (but by no means all) of the alternative approaches discussed here are most useful for the study of the ground state of the system. Because the structure of matter is ultimately determined by the lowest energy configuration, the determination of the properties of the ground state is arguably the most important; it is the "first among equals."
- Any method that is to be used as a numerical approximation technique should be capable of increased precision (usually at the cost of increased calculational difficulty), as well as providing an estimate of the errors made in the approximation. We will not focus extensively on these questions, but the reader should always keep

[1] For a discussion of numerical solutions of the time-dependent Schrödinger equation, see Press et al. (1986).

in mind how each method can be extended in precision (and the possible effort involved in doing so).

- As our ultimate goal is to understand the physics behind the equations, we may well have to rethink what it means to "solve" a problem when we approach it numerically. For example, do we need an analytic functional form for $\psi(x)$, or is an array of numbers or an interpolating function enough? How precisely do we need to know the energy eigenvalues? When are we "done"?
- Finally, the use of numerical methods is often nicely complementary to the study of analytic examples. One often looks at a problem in a much different way when one approaches it expecting to write a computer program to "solve" it, and such new insights can be valuable. For example, the study of chaotic dynamics in classical mechanics owes much of its success to the application of numerical as opposed to analytic techniques to otherwise familiar problems.

In each section, we first discuss the formalism of each method and then give an example of its possible use as a computational tool.

11.1 NUMERICAL INTEGRATION

Classical and quantum mechanics share the fact that their basic dynamical equations of motion are second-order differential equations, Newton's law for a point particle

$$m\frac{d^2x(t)}{dt^2} = F(x) \tag{11.1}$$

and the time-independent Schrödinger equation

$$-\frac{\hbar^2}{2m}\frac{d^2\psi(x)}{dx^2} + V(x)\psi(x) = E\psi(x) \tag{11.2}$$

We are used to thinking of Eqn. (11.1) as being completely deterministic,[2] in that if we are given the appropriate initial conditions, namely, $x_0 = x(0)$ and $v_0 = \dot{x}(0)$, the future time development of $x(t)$ is then predicted. To see how a particle "uses" Eqn. (11.1) to "know where it should be" at later times, we can use a conceptually simple method[3] to integrate Newton's law directly. We first approximate the acceleration (the second derivative) via

$$\begin{aligned}
\frac{d^2x(t)}{dt^2} &= \ddot{x}(t) \\
&= \lim_{\delta\to 0}\left(\frac{\dot{x}(t+\delta) - \dot{x}(t)}{\delta}\right) \\
&= \lim_{\delta\to 0}\left[\frac{\lim_{\delta\to 0}\left(\frac{x(t+2\delta)-x(t+\delta)}{\delta}\right) - \lim_{\delta\to 0}\left(\frac{(x(t+\delta)-x(t))}{\delta}\right)}{\delta}\right] \\
&= \lim_{\delta\to 0}\left(\frac{x(t+2\delta) - 2x(t+\delta) + x(t)}{\delta^2}\right) \\
&\approx \frac{x(t+2\delta) - 2x(t+\delta) + x(t)}{\delta^2}
\end{aligned} \tag{11.3}$$

[2]We ignore any complications such as the extreme sensitivity to initial conditions present in chaotic systems.

[3]Much more powerful techniques, such as the Runge-Kutta method, are discussed in all textbooks dealing with numerical methods.

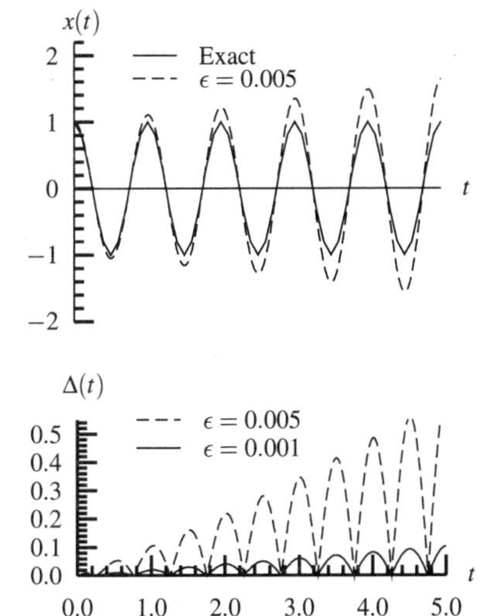

FIGURE 11.1. *The exact (solid) and numerical (solution) of the harmonic oscillator differential equation is shown in (a); the difference between the numerical and exact solutions versus time for two different step sizes, δ, are plotted in (b).*

With this approximation, Newton's law can be written as

$$x(t + 2\delta) \approx 2x(t + \delta) - x(t) + \delta^2 \frac{F(x(t))}{m} \tag{11.4}$$

which is now a *difference equation* for $x(t)$, evaluated at the discretized times $t = n\delta$. Since

$$v_0 = \dot{x}(0) \approx \frac{x(\delta) - x(0)}{\delta} \qquad \text{we have} \qquad x(\delta) \approx x(0) + \delta\dot{x}(0) \tag{11.5}$$

and the values of $x(t)$ at the first two of the discretized times, $n = 0, 1$, are fixed by the initial conditions. For later times, the $x(t = n\delta)$ with $n \geq 2$ are then determined by Eqn. (11.4).

Example 11.1 Numerical Integration of the Classical Harmonic Oscillator The classical equation for a mass and a spring is of the form

$$\ddot{x}(t) = -\omega^2 x(t) \tag{11.6}$$

where $\omega = \sqrt{K/m}$. For any numerical problem, we must specialize to definite values, so we choose

$$\omega = 2\pi \qquad x(0) = 1 \qquad v(0) = \dot{x}(0) = 0 \tag{11.7}$$

which has the exact solution $x(t) = \cos(2\pi t)$. In Fig. 11.1(a), we show the result of a numerical solution of Eqn. (11.6) (dotted curve) compared to the exact solution (solid curve); in Fig. 11.1(b) the difference between the numerical and exact solutions is seen to increase with t but also to be smaller for smaller step sizes, δ, as expected.

The same strategy can be used to solve the Schrödinger by approximating Eqn. (11.2) as

$$\psi(x + 2\delta) \approx 2\psi(x + \delta) - \psi(x) + \delta^2 \left[\frac{2m}{\hbar^2}(V(x) - E)\right]\psi(x) \tag{11.8}$$

and using

$$\psi(0) \qquad \text{and} \qquad \psi(\delta) \approx \psi(0) + \psi'(0)\delta \tag{11.9}$$

In this sense, Eqn. (11.2) is just as deterministic as Newton's laws; the chief differences are:

- The choice of $x = 0$ as the "initial" value is arbitrary.
- The differential equation can (and should) be integrated "to the left" as well to obtain $\psi(x)$ for $x < 0$.
- Most importantly, the Schrödinger equation can be integrated (solved) for any value of the energy eigenvalue, E; the solutions so obtained, however, will not necessarily be physically acceptable, i.e., square integrable.

To illustrate the usefulness of this approach to the isolation of energy eigenvalues and their corresponding eigenfunctions, we restrict ourselves to the special case of a symmetric potential for reasons that will become clear. In that case, we know that the solutions will also be eigenfunctions of parity and hence satisfy

$$\text{even solutions:} \quad \psi(0) = \text{arbitrary} \quad \text{and} \quad \psi'(0) = 0 \tag{11.10}$$

and

$$\text{odd solutions:} \quad \psi(0) = 0 \quad \text{and} \quad \psi'(0) = \text{arbitrary} \tag{11.11}$$

The arbitrariness in $\psi(0)$ or $\psi'(0)$ present at this point is eventually removed when the wavefunction is properly normalized, but that is separate from the solution of the Schrödinger equation itself.

We now focus on the behavior of the wavefunction at large $|x|$ for various values of E. For the even case, for example, we can start at $x = 0$ with an arbitrary value of $\psi(0)$, use the oddness of $\psi'(x)$ to determine $\psi(\delta) = \psi(0)$, and then use Eqn. (11.8) to integrate numerically to arbitrarily large values of $x = n\delta$. We find the generic behavior shown in Fig. 11.2(a). If we call the lowest even energy eigenvalue $E_1^{(+)}$, then for values of $E < E_1^{(+)}$, the solutions diverge as $\psi(x) \to +\infty$ as $x \to +\infty$. When $E \gtrsim E_1^{(+)}$, the solutions are still poorly behaved at infinity, but now diverge with the opposite sign. Clearly, the energy of the square-integrable ground state solution lies between E_a and E_b; this behavior is familiar from our study of the harmonic oscillator and Fig. 10.3.

Once such a pair of energy values that brackets the "acceptable" ground state solution is found, one can determine $E_1^{(+)}$ with increasing *precision* by a systematic exploration in the interval (E_a, E_b), finding values of E that bracket the "true value" with decreas-

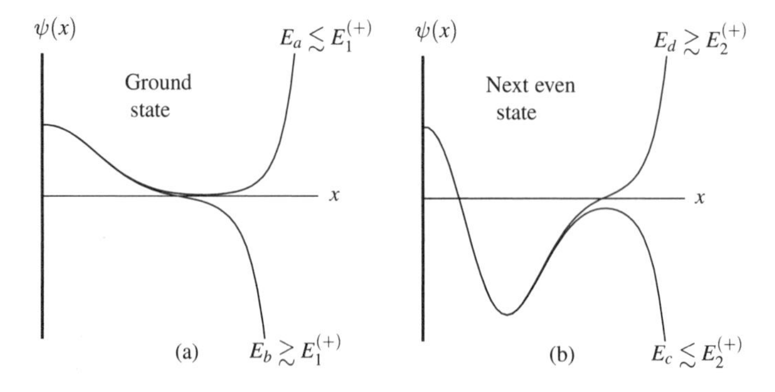

FIGURE 11.2. *Numerical solutions of the Schrödinger equation for a symmetric potential. The energy parameters $E_a < E_1^{(+)} < E_b$ bracket the true ground state energy; $E_c < E_2^{(+)} < E_d$ brackets the first excited even state.*

ing error. The resulting *accuracy* of the estimated value of E_0, however, will still depend on the integration method used (Q11.2). As the energy parameter E is increased further, additional changes in sign of the wavefunction at infinity are encountered—Fig. 11.1(b)—and the energy spectrum can be systematically mapped out by finding pairs of energy values that bracket a "sign change."

Example 11.2 Energy Eigenvalues for the Harmonic Oscillator The numerical solution of the Schrödinger equation for the harmonic oscillator potential is easy to implement using Eqn. (11.8), provided the problem is put into dimensionless form as in Sec. 10.2.1, namely,

$$\frac{d^2\psi(y)}{dy^2} = (y^2 - \epsilon)\psi(y) \tag{11.12}$$

where the dimensionless eigenvalues are $\epsilon_n = 2E_n/\hbar\omega = (2n+1)$. The *even* states have $\epsilon_n = 1, 5, 9, \ldots$ and so forth. Values of E_a, E_b that bracket the ground state ($E_1^{(+)}$) and first even excited state ($E_2^{(+)}$) energies for several values of δ are given by

δ	$E_1^{(+)}$	$E_2^{(+)}$
0.1	(1.191,1.192)	(5.510,5.511)
0.01	(1.0171,1.0172	(5.0431,5.0432)
0.001	(1.00169,1.00170)	(5.00423,5.00424)

so that the effect of decreasing the step size on the reliability of the results is clear.

It is useful to keep in mind that before applying any numerical technique to a new problem, it is best to test it on a well-understood example if at all possible.

Once an approximate energy eigenvalue is found, the wavefunction for each energy eigenvalue is obtained from the numerical integration as the collection of points $\psi(x = n\delta)$, and can be fit to a smooth function using interpolation techniques if desired. In any case, it can be normalized and used to extract further information about the quantum system. The odd states are found in a similar way (Q11.4) by making use of Eqn. (11.11).

11.2 THE VARIATIONAL OR RAYLEIGH-RITZ METHOD

Many branches of physics can be formulated in terms of a simple minimum principle using the methods of the calculus of variations. Examples include minimum surface problems (bubble problems and the like), Fermat's formulation of geometrical optics using a principle of least time and, most importantly, the principle of least-action approach to classical mechanics.

In each case, the object of study is a *functional,* so called because it takes as its argument a function and returns a number as its output. The classical action, $S[x(t)]$, is just such an example; it takes any possible classical path, $x(t)$, and returns the numerical value

$$S[x(t)] = \int_{t_a}^{t_b} dt \left(\frac{1}{2}m\dot{x}^2(t) - V(x(t))\right) \tag{11.13}$$

and the trajectory is the unique path that minimizes Eqn. (11.13).

It is perhaps not surprising that quantum mechanics can also be formulated in such a manner as well. We will first discuss just such an approach and then discuss how it can be applied as a calculational tool to approximate energy eigenvalues and wavefunctions.

Consider a Hamiltonian, $\hat{H}$, defining the bound state spectra of some system. We assume that it will have a discrete spectrum of bound state energies, E_n, with corresponding, already normalized wavefunctions $\psi_n(x)$. We can define an *energy functional* for any trial wavefunction, $\psi(x)$, via

$$\begin{aligned} E[\psi] \equiv \langle\psi|\hat{H}|\psi\rangle &= \langle\psi|\hat{T}|\psi\rangle + \langle\psi|V(x)|\psi\rangle \\ &= \int_{-\infty}^{+\infty} dx\psi^*(x)\hat{H}\psi(x) \\ &= \frac{1}{2m}\int_{-\infty}^{+\infty} dx\psi^*(x)\hat{p}^2\psi(x) + \int_{-\infty}^{+\infty} dxV(x)|\psi(x)|^2 \end{aligned} \tag{1.14}$$

This is defined whether $\psi(x)$ is an eigenfunction or not. It is often convenient to use the alternative form of the average value of kinetic energy, i.e.,

$$\langle\psi|\hat{T}|\psi\rangle = \langle\hat{T}\rangle = \frac{\hbar^2}{2m}\int_{-\infty}^{+\infty} dx\left|\frac{d\psi(x)}{dx}\right|^2 \tag{11.15}$$

If for some reason the trial wavefunction ψ is not already normalized, we can simply write

$$E[\psi] \equiv \frac{\langle\psi|\hat{H}|\psi\rangle}{\langle\psi\mid\psi\rangle} \tag{11.16}$$

It is easy to see that this functional simply returns an energy eigenvalue when its argument is an eigenstate, since

$$E[\psi_n] = \int_{-\infty}^{+\infty} dx\psi_n^*(x)\hat{H}\psi_n(x) = \int_{-\infty}^{+\infty} dx\psi_n^*(x)E_n\psi_n(x) = E_n \tag{11.17}$$

For a general wavefunction, $\psi(x)$, we assume we can use the expansion theorem and write $\psi(x) = \sum_{n=0}^{\infty} a_n\psi_n(x)$, and we find that

$$\begin{aligned} E[\psi] &= \langle\psi|\hat{H}|\psi\rangle \\ &= \sum_{n=0}^{\infty}\sum_{m=0}^{\infty} a_m^* a_n\langle\psi_m|\hat{H}|\psi_n\rangle \\ &= \sum_{n=0}^{\infty}\sum_{m=0}^{\infty} a_m^* a_n E_n\delta_{n,m} \\ &= \sum_{n=0}^{\infty}|a_n|^2 E_n \end{aligned} \tag{11.18}$$

This derivation is similar to that of Sec. 7.4 for the average value of the energy operator, $\hat{E}$, in a general state, but the quantity that appears in the energy functional here is the expectation value of the appropriate Hamiltonian for the problem, which in general acts only on spatial degrees of freedom.

We assume that the energy eigenvalues are ordered, i.e., $\cdots \geq E_2 \geq E_1 \geq E_0$ so that

$$E[\psi] = \sum_{n=0}^{\infty}|a_n|^2 E_n \geq \sum_{n=0}^{\infty}|a_n|^2 E_0 = E_0 \qquad \text{or} \qquad E[\psi] \geq E_0 \tag{11.19}$$

because the expansion coefficients, when squared, sum to unity. The lower bound is only "saturated" when $\psi(x) = \psi_0(x)$, so that $a_n = \delta_{n,0}$ and only the ground state energy term contributes.

This is then the desired minimum principle, namely, that

- The energy functional, defined via Eqn. (11.14), always gives an energy at least as large as the true ground state energy, i.e., $E[\psi] \geq E_0$ for all ψ.

To use this property as a calculational tool, we first note that if the wavefunction used in the functional has an arbitrary parameter, e.g., $\psi(x) = \psi(x; a)$, then the energy functional yields a function of one variable, namely,

$$E[\psi(x; a)] = E(a) \tag{11.20}$$

An example of this would be the family of Gaussian variational wavefunctions, $\psi(x; a) = \exp(-x^2/2a^2)/\sqrt{a\sqrt{\pi}}$, with a variable width.

Because the functional satisfies the minimum principle for each value of the parameter, one can minimize the variational function $E(a)$ and be assured that the resulting minimum is still greater than the true ground state energy. Thus, one can find the trial wavefunction, in the one parameter family considered, which has the lowest energy. The minimizing wavefunction accomplishes this by somehow "adjusting" to be as similar as possible to the exact ground state solution. This approach is similar in spirit to the zero point energy argument of P1.9, but it is more powerful because:

- The guaranteed lower bound of Eqn. (11.19) provides a method of assessing the reliability of the approximations.

  Of two variational estimates of the ground state energy, the lower one is always closer to the true value.

  In this context, "lower is always better" as we know that we can never "overshoot" E_0 on the negative side.
- It also provides an approximation to the wavefunction as well as to the energy; one can then use it to estimate expectation values and to find the approximate momentum space wavefunction. As an aside, because the argument leading to Eqn. (11.19) is not specific to a position space representation, one can also use the variational method with momentum space wavefunctions (P11.5).

For illustrative purposes, we will sometimes calculate $|a_0|^2$ as a measure of the "overlap" of the trial solution with the exact ground state wavefunction (if known); it can be used as a quantitative measure of the similarity of any two functions. We reiterate, however, that the trial wavefunction of a given class that minimizes the energy is not necessarily the one that has the largest overlap with the true ground state wavefunction; i.e., it does not necessarily maximize $|a_0|^2$ (see, for example, P11.9).

Example 11.3 Variational Estimate for the Harmonic Oscillator I As an example of the method, consider approximating the ground state energy and eigenfunction of the simple harmonic oscillator by using the family of trial wavefunctions $\psi(x; \alpha) = \exp(-x^2/2a^2)/\sqrt{a\sqrt{\pi}}$ mentioned above. Because the true ground state solution is also a Gaussian, we expect to find the exact answer. We have to evaluate Eqn. (11.14) with $V(x) = m\omega^2 x^2/2$, and we find that

$$E[\psi(x; a)] = E(a) = \langle \hat{T} \rangle + \langle V(x) \rangle = \frac{\hbar^2}{4ma^2} + \frac{1}{4} m\omega^2 a^2 \tag{11.21}$$

Minimizing this expression, we find

$$\frac{dE(a)}{da} = -\frac{\hbar^2}{2ma^3} + \frac{1}{2}m\omega^2 a = 0 \tag{11.22}$$

which yields $a_{\min} = \sqrt{\hbar/m\omega}$ and $E(a_{\min}) = \hbar\omega/2$ as expected.

Example 11.4 Variational Estimate for the Harmonic Oscillator II To illustrate the principle in the case where the form of the ground state wavefunction is not known, consider as a trial wavefunction for the SHO the wavefunction

$$\psi(x;a) = \begin{cases} 0 & \text{for } |x| > a \\ N(a^2 - x^2)^2 & \text{for } |x| < a \end{cases} \tag{11.23}$$

where the variational parameter is again a and the normalization constant is given by $N = \sqrt{315/256a^9}$. A similar calculation to the one above shows (P11.3) that the energy function is

$$E[\psi] = E(a) = \frac{3\hbar^2}{2ma^2} + \frac{m\omega^2 a^2}{22} \tag{11.24}$$

This has a minimum value at $a^2_{\min} = \sqrt{33}\hbar/m\omega$, yielding

$$E(a_{\min}) = \frac{\hbar\omega}{2}\sqrt{\frac{12}{11}} \tag{11.25}$$

which is only 4.4% greater than the exact value.

The trial wavefunctions, along with their corresponding energies for several choices of a are shown in Fig. 11.3 along with the value of $|a_0|^2$. We plot in Fig. 11.4 the fractional difference between the variational energy and the exact ground state value $(E(\text{var}) - E(\text{exact}))/E(\text{exact})$ as well as the probability that the variational wavefunction is *not* in the ground state, i.e., $1 - |a_0|^2$, versus the variational parameter a. We note that variations in a seem to have a much larger effect on the energy functional than on the wavefunction itself.

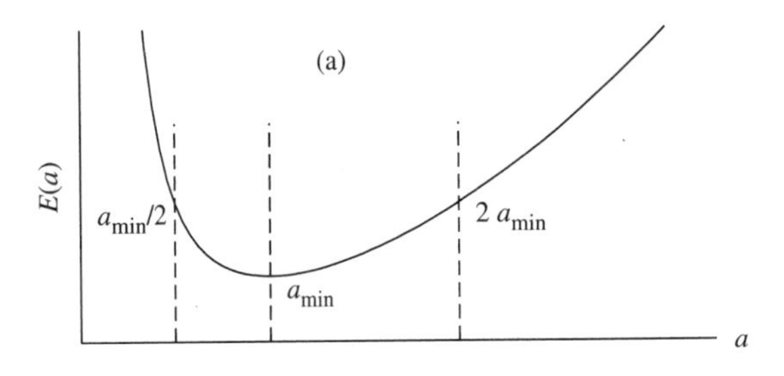

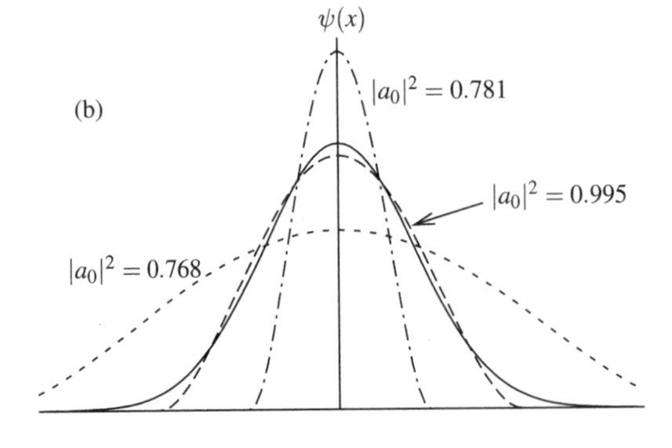

FIGURE 11.3. (a) The variational energy $E(a)$ versus a showing the value of $a_{\min}$ that minimizes the energy functional and two other values. (b) The corresponding variational wavefunctions along with the exact ground state; values of the overlap given by $|a_0|^2$ for each variational waveform are also shown.

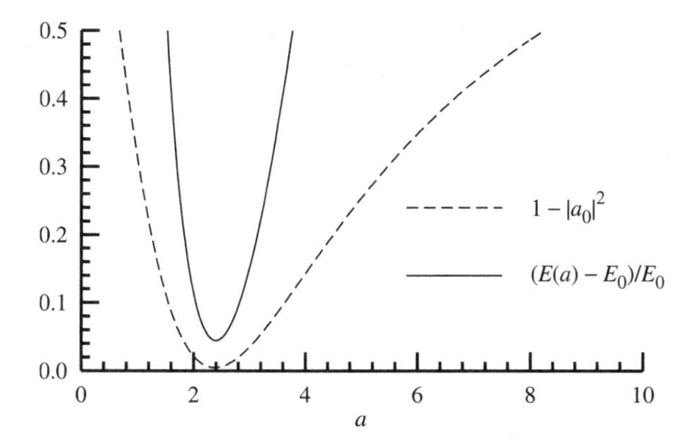

FIGURE 11.4. *The fractional energy error (solid) and the degree of "nonoverlap" (dotted) versus variational parameter a for Ex. 11.4. This illustrates that first-order changes in the wavefunction give second-order changes in the energy functional.*

To formalize this last observation further, let us imagine making small variations around the exact ground state wavefunction, $\psi_0(x)$, parameterized by $\psi_0(x) \to \psi_0(x) + \lambda\phi(x)$ so that $\phi(x)$ represents a first-order change in the wavefunction; we use λ to keep track of the expansion.

Consider then the energy functional—where we use Eqn. (11.16) since the new wavefunction is not properly normalized—and we find

$$\begin{aligned}
E[\psi_0 + \lambda\phi] &= \frac{\langle \psi_0 + \lambda\phi | \hat{H} | \psi_0 + \lambda\phi \rangle}{\langle \psi_0 + \lambda\phi \mid \psi_0 + \lambda\phi \rangle} \\
&= \frac{E_0\langle \psi_0 \mid \psi_0 \rangle + \lambda(\langle \psi_0 | \hat{H} | \phi \rangle + \langle \phi | \hat{H} | \psi_0 \rangle) + \lambda^2 \langle \phi | \hat{H} | \phi \rangle}{\langle \psi_0 \mid \psi_0 \rangle + \lambda\langle \psi_0 | \phi \rangle + \lambda\langle \phi_0 \mid \psi_0 \rangle + \lambda^2 \langle \phi \mid \phi \rangle} \\
&= E_0 \left(\frac{1 + \lambda(\langle \psi_0 \mid \phi \rangle + \langle \phi \mid \psi_0 \rangle) + \lambda^2 \langle \phi | \hat{H} | \phi \rangle / E_0}{1 + \lambda(\langle \psi_0 \mid \phi \rangle + \langle \phi \mid \psi_0 \rangle) + \lambda^2 \langle \phi \mid \phi \rangle} \right) \\
&= E_0(1 + \mathcal{O}(\lambda^2))
\end{aligned} \tag{11.26}$$

This shows that, in general,

- *First-order changes,* $\mathcal{O}(\lambda)$, in the trial wavefunction, away from the true ground state solution, give rise to *second-order changes,* $\mathcal{O}(\lambda^2)$, in the corresponding energy functional.

This fact is reflected in Fig. 11.4 as the fractional change in energy seems to vary quadratically with deviations away from the minimum value of the variational parameter, while the deviation in the wavefunction itself (as measured by $1 - |a_0|^2$) seems to vary much more weakly on a. This is a typical feature of problems involving the calculus of variations.

If the variational method is to be useful as an approximation method, there should be some possibility of further refinement of the estimation of the ground state energy. This can be accomplished by simply taking as a trial wavefunction one with a larger number of variational parameters. For example, one might consider

$$\psi(x; a, b) = e^{-x^2/2a^2}(1 + bx^2) \tag{11.27}$$

which has an addition parameter, b, but which reduces to the original choice in some limit (namely, $b = 0$). In this case, we are guaranteed to have

$$E(a_{\rm min}) \equiv E[\psi(x; a_{\rm min})] > E(a_{\rm min}, b_{\rm min}) \equiv E[\psi(x; a_{\rm min}, b_{\rm min})] \geq E_0 \quad (11.28)$$

because any variational energy must be larger than the true ground state and because the minimum with nonzero values of b will be at least as small as for $b = 0$. The new minimum value will be determined by

$$\frac{\partial E(a,b)}{\partial a} = \frac{\partial E(a,b)}{\partial b} = 0 \quad (11.29)$$

By adding more and more variational parameters, we can allow the trial wavefunction to conform as closely as possible to the exact ground state.

Example 11.5 Variational Estimate for the Harmonic Oscillator III We illustrate the improvement possible with multi-parameter trial wavefunctions by using the (unnormalized) function

$$\psi(x; a, b) = \begin{cases} 0 & \text{for } |x| > a \\ (a^2 - x^2)^2(1 + bx^2) & \text{for } |x| < a \end{cases} \quad (11.30)$$

as a trial solution for the ground state of the simple harmonic oscillator. We plot in Fig. 11.5 a contour plot of $E(a, b)$ versus a, b. The small asterisk on the dotted line indicates the minimum for the $b = 0$ case, while the small + indicates the new global minimum that does indeed have somewhat lower energy. The values of the exact, one-parameter, and two-parameter fits are shown below.

quantity	exact	$\psi(x; a)$	$\psi(x; a, b)$
$E_0/(\hbar\omega/2)$	1	1.0445	1.0198
$\|a_0\|^2$	1	0.9951	0.9977
$\langle x^2\rangle/\rho^2$	1/2	0.5222	0.5099
$\langle x^4\rangle/\rho^4$	3/4	0.6923	0.6884

(11.31)

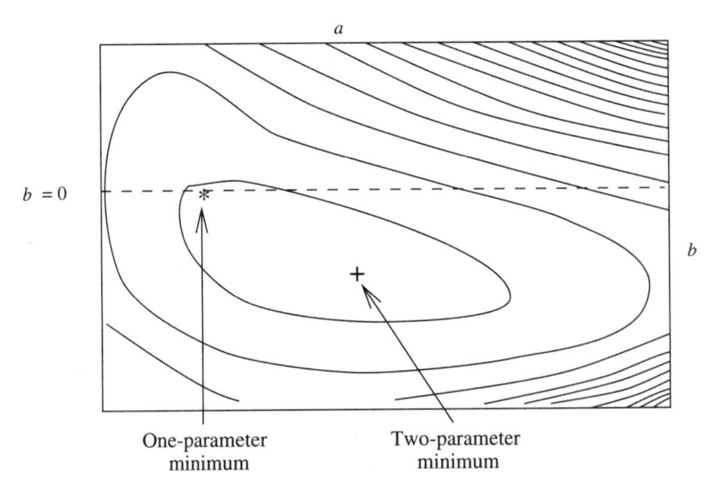

FIGURE 11.5. Contour plot of the two-parameter variational energy $E(a, b)$ versus (a, b); the dotted line corresponds to the one-parameter family.

The energy is lower and the overall fit is better ($|a_0|^2$ is closer to 1) than in the one-parameter case. It is clear, however, that various higher moments (i.e., average values of x^{2n}) never fit very well with this particular form, which is not surprising given its lack of a realistic "tail."

The process can be continued with as many variational parameters as one can handle, presumably improving the agreement with experiment, if not providing much more useful insight into the basic physics.[4] Some of the most famous tour de force calculations of this type are variational calculations of the ground state of the helium atom which use trial wavefunctions with hundreds of parameters.[5]

In some cases, it is also possible to extend the variational method to give rigorous lower bounds for excited states as well as the ground state. Suppose, for example, that one could choose a trial wavefunction that was somehow known to be orthogonal to the true ground state, i.e., $\langle\psi \mid \psi_0\rangle = 0$ so that $a_0 = 0$ in the expansion theorem. The standard argument would then give

$$\psi = \sum_{n=1}^{\infty} a_n \psi_n \quad \Longrightarrow \quad E[\psi] = \sum_{n=1}^{\infty} |a_n|^2 E_n \geq \sum_{n=1}^{\infty} |a_n|^2 E_1 = E_1 \tag{11.32}$$

Various symmetries of the problem can often be used to restrict the form of the trial wavefunction so as to satisfy this constraint. For example, in a one-dimensional problem with a symmetric potential, $V(x) = V(-x)$, we know that the ground state will be an even function; therefore any odd trial wavefunction will have $a_0 = 0$ and hence satisfy Eqn. (11.32) and give a good estimate of the first excited-state energy. Less prosaically, in three-dimensional problems with spherical symmetry, the ground state will have no angular momentum (i.e., $l = 0$), and excited states with higher values of l are automatically orthogonal to the ground state.

11.3 MONTE CARLO METHODS

There are many calculational methods in theoretical physics that rely on the use of randomness and that have generically become known as *Monte Carlo techniques*. They include algorithms for the evaluation of complicated multidimensional integrals by the use of random numbers[6] and the solution of differential and integral equations.

There is a similar array of techniques that come under the general title of quantum Monte Carlo methods.[7] We will illustrate here only one of the simplest ones, namely, the random walk method[8] of solving the Schrödinger equation, as it makes interesting connections with diffusion and heat conduction problems and classical statistical mechanics. As with the variational method, it is most specifically useful for extracting the ground state energy and wavefunction of a bound state system. We will consider only the simplest case

[4]It is said that, when confronted with the result of an impressive numerical calculation, Eugene Wigner said: "It is nice to know that the computer understands the problem. But I would like to understand it too." See Nussenzveig (1992).

[5]See Bethe and Jackiw (1968) or Park (1992) for discussions.

[6]For a nice discussion with applications to elementary particle physics, see Barger and Phillips (1987); see also Press et al. (1986).

[7]For an introduction, at a level rather more sophisticated than we consider here, see Ceperley (1984).

[8]See Anderson (1975).

of one particle in one dimension, but the method can be easily extended to many-body problems in three dimensions.

We begin by recalling that a general time-dependent solution of the Schrödinger equation for some Hamiltonian system can be written as

$$\psi(x, t) = \sum_{n=0} a_n \psi_n(x) e^{-iE_n t/\hbar} \tag{11.33}$$

where the $\psi_n(x)$, E_n are the stationary states and energy eigenvalues of the system; we also assume that $E_n > E_0$ for $n \geq 1$.

If we formally make a change of variables to "imaginary time" via $it = \tau$, we then have

$$\begin{aligned} \psi(x, \tau) &= \sum_{n=0}^{\infty} a_n \psi_n(x) e^{-E_n \tau/\hbar} \\ &= e^{-E_0 \tau/\hbar} \left(a_0 \psi_0(x) + \sum_{n=1}^{\infty} a_n \psi_n(x) e^{-(E_n - E_0)\tau/\hbar} \right) \end{aligned} \tag{11.34}$$

so that

$$\psi(x, \tau) \longrightarrow a_0 \psi_0(x) e^{-E_0 \tau/\hbar} \qquad \text{as } \tau \to \infty \tag{11.35}$$

since $E_n - E_0 > 0$ for $n > 0$.

If one starts with any general initial waveform, $\psi(x, 0)$, for which $a_0 \neq 0$, and if one can actually evaluate the time-dependent wavefunction, $\psi(x, \tau)$, then

- The *shape* of the solution for large (imaginary) times gives the (unnormalized) ground state wavefunction $\psi_0(x)$ independent of the initial conditions. Here, large times are defined as $\tau \gg \hbar/(E_1 - E_0)$, as this ensures that the other terms in the summation become negligible compared to the a_0 term.
- The simple exponential time dependence in Eqn. (11.35) then gives the ground state energy; for example, one can plot $\psi(x, \tau)$ versus τ for any x value and fit to an exponential.

We see that arbitrary solutions of the Schrödinger equation in imaginary time can be thought to "relax" to the ground state wavefunction for large τ.

To make use of this observation as a calculational tool, we must be able to evaluate the time dependence of $\psi(x, \tau)$. We first note that with this change of time coordinate, the standard time-dependent Schrödinger equation

$$i\hbar \frac{\partial \psi(x, t)}{\partial t} = -\frac{\hbar^2}{2m} \frac{\partial^2 \psi(x, t)}{\partial x^2} + V(x)\psi(x, t) \tag{11.36}$$

becomes

$$\frac{\partial \psi(x, \tau)}{\partial \tau} = \frac{\hbar}{2m} \frac{\partial^2 \psi(x, \tau)}{\partial x^2} - \frac{V(x)}{\hbar} \psi(x, \tau) \tag{11.37}$$

This has the form of a *classical diffusion equation* but also includes the possibility of exponential growth or decay, since we can write it in the form

$$\frac{\partial \psi(x, \tau)}{\partial \tau} = D \frac{\partial^2 \psi(x, \tau)}{\partial x^2} - k(x)\psi(x, \tau) \tag{11.38}$$

where $D \equiv \hbar/2m$ is the *diffusion constant* and $k(x) \equiv V(x)/\hbar$ gives the (spatially dependent) probability per unit time for growth or extinction. [Equation (11.38) is also of the form of the heat conduction equation in which case D is the conductivity.]

In this language, $\psi(x, \tau)$ might represent the concentration of some liquid in a chemical reaction or the density of plasma (ionized gas) in a region of one-dimensional space. If $k = 0$, one has a standard diffusion equation for which the concept of random walks is very useful; on the other hand, if $D = 0$, one has the trivial solution $\psi(x, \tau) = \psi(x)\exp(-k(x)\tau)$, corresponding to exponential decay (growth) if k is positive (negative).

These two limiting cases are the basis for a numerical method of solution of this equation based on random-walk processes. The conceptual algorithm involves the following steps:

1. Divide the position coordinate, i.e., x, into discrete points, separated by "steps" of size Δx, so that the wavefunction is evaluated on a finite mesh of points, i.e., $\psi(x, \tau) \to \psi(n\Delta x, \tau)$; this discretization is familiar from the numerical integration methods in Sec. 11.1
2. Select an arbitrary initial configuration, $\psi(x, 0)$, which is as close to the likely solution as possible (to make sure $a_0 \neq 0$), which for the ground state would be a smooth, nodeless function. This gives $\psi(n\Delta x, 0)$ after discretization.
3. For the next time step, choose $\Delta\tau$ to be consistent with the already defined diffusion coefficient, namely,

$$\frac{\hbar^2}{2m} = D = \frac{(\Delta x)^2}{2\Delta\tau} \tag{11.39}$$

To implement the effect of the diffusion term, let the "random walkers" (i.e., the values of $\psi(n\Delta x, \tau)$) at position $x = n\Delta x$ "diffuse" with equal probabilities to both the left and right, giving the intermediate result (ψ_{int})

$$\psi_{\text{int}}(n\Delta x, \tau + \Delta\tau) = \frac{1}{2}\psi((n-1)\Delta x, \tau) + \frac{1}{2}\psi((n+1)\Delta x, \tau) \tag{11.40}$$

4. To take into account the "growth/decay" term, multiply this new configuration by $(1 - k(x)\Delta\tau)$ where $x = n\Delta x$; then give the complete approximation for the next time step

$$\psi(n\Delta x, \tau + \Delta\tau) = (1 - k(n\Delta x)\Delta\tau)\psi_{\text{int}}(n\Delta x, \tau + \Delta\tau) \tag{11.41}$$

Steps (3) and (4) implement the time development of $\psi(x, \tau)$ dictated by the two processes on the right-hand side of Eqn. (11.38).
5. Continue this process, incrementing the imaginary time by steps $\Delta\tau$ until a large enough value of τ is obtained for which the *shape* of the waveform is not changing; the overall magnitude of the solution will, of course, be exponentially decreasing in time.
6. Extract the ground state wavefunction, $\psi_0(x)$, from the spatial dependence of the solution obtained and the ground state energy from the time dependence of $\psi(x, \tau)$ for large values of τ.

Example 11.6 Random-Walk Solution of the Particle in a Box As an example of the method, we will solve the simple problem of a particle in the standard infinite well (walls at *0* and L) for which the (unnormalized) ground state solution and energy, in this notation, are

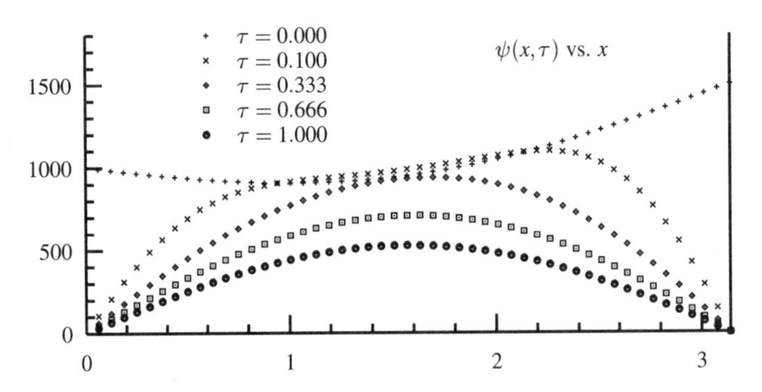

FIGURE 11.6. Random-walk solutions in imaginary time, τ, to the particle in a box problem in Ex. 11.6. For large τ, the shape of the solutions approaches the true ground state wavefunction.

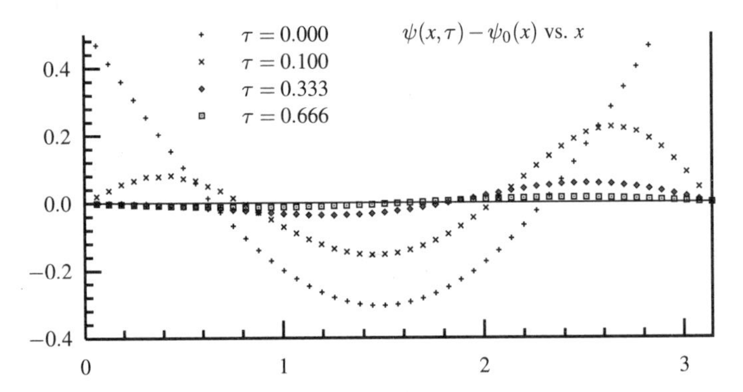

FIGURE 11.7. Deviation of the normalized random-walk solution from the true ground state solution for increasing values of imaginary time, τ.

$$\psi_0(x) = \sin\left(\frac{\pi x}{L}\right) \qquad \text{and} \qquad E_0 = \frac{\hbar^2\pi^2}{2mL^2} \tag{11.42}$$

We set $D = 1$ and $L = \pi$ for convenience so that $E_0/\hbar(E_1/\hbar) = 1(4)$ in these units. We divide the box into 50 points so that $\Delta x = \pi/50$, which gives, via Eqn. (11.39), $\Delta\tau = \pi^2/5000$. Thus, to get to the "asymptotic" regime, where $\tau \gg \hbar/(E_1 - E_0)$, we require $\tau \gg 1/3$.

We plot, in Fig. 11.6, $\psi(x, \tau)$ versus x for increasing values of τ starting with an arbitrary initial configuration. The values of $\psi(x, \tau)$ are unnormalized, but the shape for large τ does seem to be approaching the expected one. To confirm this, we take the discretized values of $\psi(x, \tau)$, normalize them, and plot in Fig. 11.7 the difference between the normalized solution and the true ground state solution of Eqn. (11.42). We note that the differences do disappear as $\tau \to \infty$. Finally, in Fig 11.8, we plot the value of the wavefunction at $x = L/2$ and $L/4$ versus τ on a logarithmic scale, and note the eventual exponential dependence. From the "slope" on the logarithmic plot, we can calculate the rate of exponential decay and indeed find that $E_0/\hbar \approx 1$.

While one can easily extend such techniques to higher dimensions and more particles, one drawback is that it is generally most easily implemented for wavefunctions that are guaranteed to be nonnegative; the nodeless ground state of many systems satisfy this criteria. This can be understood in our case by the analogy to diffusion/reaction/heat conduction

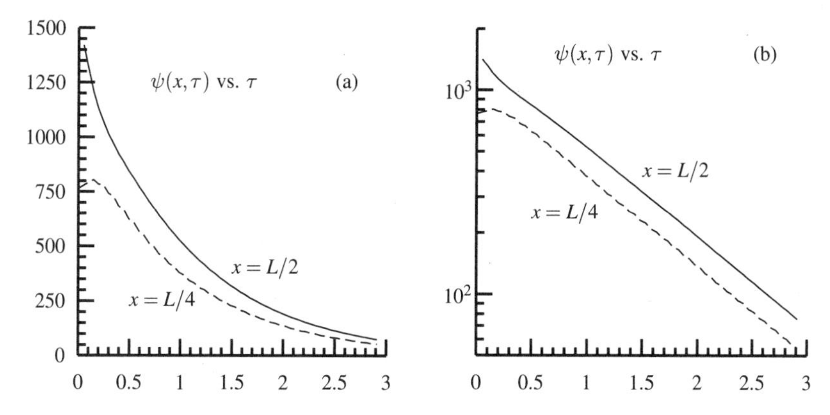

FIGURE 11.8. *(a) The value of the "random-walk" solution for two values of x versus τ on a linear (a) and a "log-log" plot (b). The value of the ground state energy can be extracted from the "slope" of the straight line in (b).*

problems where we normally assume that $\psi(x, \tau)$ is positive (no negative "concentrations" or densities or temperatures). This implies that the application to excited states (which have nodes and hence sign changes) or systems of fermions[9] often poses problems.

11.4 THE WKB METHOD

Both the variational method and the imaginary time formulation of the Schrödinger equation are best suited to evaluating the properties of the ground state solution, i.e., for $n = 0$. It is useful to have a complementary approach that is more appropriate for the quasi-classical regime where $n \gg 1$. We have argued that this limit is also attained, in some sense, when $\hbar \to 0$. Such an approach was first discussed in the context of quantum mechanics by Wentzel, Kramers, and Brillouin[10] and is therefore often called the *WKB method.*

11.4.1 WKB Wavefunctions

Motivated by the simple form for a free-particle de Broglie wave, i.e.,

$$\psi(x) = Ae^{i2\pi x/\lambda} = Ae^{ikx} = Ae^{ipx/\hbar} \tag{11.43}$$

we attempt a solution of the time-independent Schrödinger equation of the form

$$\psi(x) = A(x)e^{iF(x)/\hbar} \tag{11.44}$$

where $A(x)$ and $F(x)$ are an amplitude and phase term, respectively. We retain the explicit factor of $\hbar$ and will use it to parameterize the smallness of various terms. We also assume, for the moment, that we are in the classically allowed region, so that $E > V(x)$, as in Fig. 11.9, so that $a < x < b$.

[9]This is because the total wavefunction of such a system must be antisymmetric, and hence have nodes implying changes in sign.

[10]It was also studied independently by Jeffries; the name *WKBJ approximation* is therefore sometimes used.

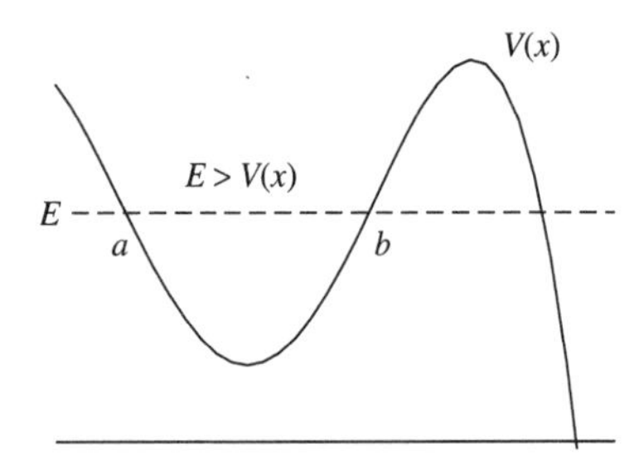

FIGURE 11.9. *Generic potential with classical turning points for the WKB approximation.*

With this *ansatz,*[11] the Schrödinger equation becomes

$$0 = A(x)\left[\frac{1}{2m}\left(\frac{dF(x)}{dx}\right)^2 - (E - V(x))\right]$$
$$- \hbar\left(\frac{i}{2m}\right)\left[2\frac{dA(x)}{dx}\frac{dF(x)}{dx} + A(x)\frac{d^2F(x)}{dx^2}\right]$$
$$- \hbar^2\left[\frac{1}{2m}\frac{d^2A(x)}{dx^2}\right] \tag{11.45}$$

At this point, we can either consider $\hbar$ as an arbitrary small parameter and set the first two terms [of order $\mathcal{O}(\hbar^0)$ and $\mathcal{O}(\hbar^1)$, respectively] separately to zero, or else we can require that both the real and imaginary parts of Eqn. (11.46) are satisfied. In either case, we neglect the last term [being of order $\mathcal{O}(\hbar^2)$] and discuss the validity of this approximation below.

The $\mathcal{O}(\hbar^0)$ equation (or real part) is easily written as

$$\frac{dF(x)}{dx} = \pm\sqrt{2m(E - V(x)} \equiv \pm p(x) \tag{11.46}$$

or

$$F(x) = \pm \int^x p(x)\,dx \tag{11.47}$$

where $p(x)$ is simply the classical momentum. The $\mathcal{O}(\hbar^1)$ (or imaginary part) then gives

$$2\frac{dA(x)}{dx}p(x) + A(x)\frac{dp(x)}{dx} = 0 \tag{11.48}$$

which can be multiplied on both sides by $A(x)$ to obtain

$$\left(2A(x)\frac{dA(x)}{dx}\right)p(x) + [A(x)]^2\frac{dp(x)}{dx} \equiv \frac{d}{dx}([A(x)]^2 p(x)) = 0 \tag{11.49}$$

or

$$[A(x)]^2 p(x) = C \tag{11.50}$$

[11]The German term *ansatz* meaning, in this context, something like "assumed form of the solution" is appropriate here.

where C is a constant. The two linearly independent solutions [corresponding to right (+) and left-moving (−) waves] are then given by

$$\psi_{\pm}(x) = \frac{C_{\pm}}{\sqrt{p(x)}} e^{\pm i \int^x p(x)\,dx/\hbar} \quad \propto \quad \frac{1}{\sqrt{v(x)}} e^{\pm i \int^x k(x)\,dx} \tag{11.51}$$

where $p(x) = \hbar k(x)$ defines the "local wavenumber" $k(x)$, and $v(x)$ is the local speed. This remarkably simple solution has several obvious features:

- The corresponding probability density, $|\psi(x)|^2$, satisfies

$$|\psi(x)|^2 \propto \frac{1}{p(x)} \propto \frac{1}{v(x)} \tag{11.52}$$

  which is exactly of the form of the *classical probability distribution* first discussed in Secs. 6.1 and 6.2. This implies (recall Fig. 10.7) that the wavefunctions for the low-lying energy levels will be *poorly* described by the WKB solutions. The quantum wavefunctions for large quantum numbers will, however, approach these semiclassical solutions when suitably locally averaged.
- The phase of the wavefunction can be written as

$$\int^x k(x)\,dx = \int^x d\phi(x) \tag{11.53}$$

  where

$$d\phi(x) = k(x)\,dx = \frac{2\pi}{\lambda(x)}dx \qquad \text{or} \qquad \frac{d\phi}{2\pi} = \frac{dx}{\lambda(x)} \tag{11.54}$$

  Thus, as the particle moves a distance dx, or a fraction of a "local" wavelength, $df = dx/\lambda(x)$, through the potential, it acquires a phase $d\phi = 2\pi df$.
- The solutions can be easily extended to the case where $E < V(x)$, i.e., in the classically disallowed regions by appropriate changes in sign giving

$$\psi_{\pm} = \frac{\tilde{C}_{\pm}}{\sqrt{p(x)}} \exp\left(\pm\sqrt{2m/\hbar^2}\int^x \sqrt{V(x) - E dx}\right) \tag{11.55}$$

  which are the exponentially suppressed solutions discussed in Sec. 9.2.2; these give rise to quantum tunneling effects. The WKB wavefunction thus has features of both the classical probability distribution, arising from averaging over the trajectory, and the quantum wavefunction.

With this form of the solution, we can examine the effect of neglecting the $\mathcal{O}(\hbar^2)$ term in Eqn. (11.46). Taking the ratio of the last term to the first, we find something of the order

$$\frac{\hbar^2}{F'(x)^2}\left(\frac{A''(x)}{A(x)}\right) \quad \propto \quad \frac{\hbar^2}{p(x)^2}\frac{1}{l^2} \quad \propto \quad \frac{1}{[lk(x)]^2} \quad \propto \quad \left(\frac{\lambda(x)}{2\pi l}\right)^2 \tag{11.56}$$

where l is a typical distance scale over which $E - V(x)$ changes. Thus, if the "local de Broglie wavelength," $\lambda(x)$, is much shorter than the distance scale over which the potential changes, the semiclassical approximation is a good one. This is obviously not the case near the classical turning points where the explicit $1/\sqrt{p(x)}$ factors in Eqn. (11.51) actually diverge, indicating that the solution is poorly behaved there.

To obtain a complete description of the wavefunction, the solutions inside and outside the well must be smoothly matched onto each other. The formalism for doing this is not beyond the level of this text, but we choose to quote only the results.[12] For example, we can take linear combinations of the complex exponential solutions near the left turning point to write

$$\psi_L(x) = \frac{A_L}{\sqrt{p(x)}} \cos\left(\int_a^x k(x)\, dx - C_L \pi\right) \tag{11.57}$$

For an infinite wall-type boundary condition, it is easy to see that $C_L = 1/2$, since we require that

$$\psi_L(a) \quad \propto \quad \cos(-C_L \pi) = 0 \tag{11.58}$$

For a smoother potential, one for which one can approximate $V(x)$ near $x = a$ by a linear function,[13] the appropriate value of C_L turns out to be 1/4. This can be interpreted as saying that the quantum wavefunction penetrates $\pi/4 = 2\pi/8$ or $\sim 1/8$ of a local wavelength into the classically disallowed region.

11.4.2 Quantized Energy Levels

One of the most useful results arising from the WKB method is a semiclassical estimate for the quantized energy levels in a potential. Matching the WKB wavefunctions at each of the two classical turning points yields two, presumably equivalent descriptions of $\psi(x)$ inside the well, namely,

$$\psi_L(x) = \frac{A_L}{\sqrt{p(x)}} \cos\left(\int_a^x k(x)\, dx - C_L \pi\right) \tag{11.59}$$

and

$$\psi_R(x) = \frac{A_R}{\sqrt{p(x)}} \cos\left(\int_x^b k(x)\, dx - C_R \pi\right) \tag{11.60}$$

If these two solutions are to agree, we must clearly have $|A_L| = |A_R|$; then comparing the arguments of the cosines we find that

$$\int_a^b k(x)\, dx - (C_L + C_R)\pi = n\pi \qquad \text{for } n = 0, 1, 2\ldots \tag{11.61}$$

This implies that

$$\int_a^b k(x)\, dx = (n + C_L + C_R)\pi \qquad \text{for } n = 0, 1, 2\ldots \tag{11.62}$$

or

$$\int_a^b \sqrt{2m(E - V(x))}\, dx = (n + C_L + C_R)\pi\hbar \tag{11.63}$$

---

[12]See, e.g., Park (1992); I also like the discussion in Migdal and Krainov (1969).

[13]In this case, the solution which interpolates between the inside and outside can be described by an Airy function (See Appendix C.2).

Recalling that $k(x) = 2\pi/\lambda(x)$, we see that Eqn. (11.62) is simply a more sophisticated version of "fitting an integral number of de Broglie half wavelengths in a box" and generalizes the Bohr-Sommerfeld quantization condition. The value of n can be seen to count the number of nodes in the quantum wavefunction.

Example 11.7 Infinite Well and Harmonic Oscillator For the standard infinite well, we have $C_L = C_R = 1/2$ and $k(x) = \sqrt{2mE/\hbar^2}$, so that the WKB quantization condition gives

$$\int_0^a \sqrt{2mE}dx = (n+1)\pi \tag{11.64}$$

or

$$E_n = \frac{\hbar^2(n+1)^2\pi^2}{2ma^2} \qquad \text{for } n = 0, 1, 2 \ldots \tag{11.65}$$

More interestingly, the WKB quantization also gives the correct answer for the harmonic oscillator. In that case we have

$$p(x) = \sqrt{2m(E - m\omega^2x^2/2)} \qquad \text{or} \qquad k(x) = \frac{m\omega}{\hbar}\sqrt{A^2 - x^2} \tag{11.66}$$

where $E = m\omega^2A^2/2$. Since $C_L = C_R = 1/4$ in this case, we find

$$\int_{-A}^{+A} k(x)\,dx = \frac{m\omega}{\hbar}\int_{-A}^{+A} \sqrt{A^2 - x^2}\,dx = (n + 1/2)\pi \tag{11.67}$$

or

$$E_n = (n + 1/2)\hbar\omega \tag{11.68}$$

We have noted that we have dropped terms of order $\mathcal{O}(\hbar^2)$ or $1/(lk)^2$. Since typically we find $k_n \propto n/l$, we expect the WKB estimates of the energies to have errors of order $\mathcal{O}(1/n^2)$. This is consistent with our keeping the C_L, C_R terms in Eqn. (11.62), which, in this language, are of order $\mathcal{O}(1/n)$.

11.5 MATRIX METHODS

The variational and imaginary time methods both rely on the expansion of a general quantum state in terms of energy eigenstates. In this section, we describe a matrix approach that also uses the algebraic structure inherent in the Schrödinger equation, but in a rather different way.

Suppose that we have solved for the energy eigenstates of some Hamiltonian operator, $\hat{H}$. We call them $\psi_n(x)$, where we let the label n start with $n = 1$ for notational convenience; with this labeling, the ground state is $\psi_1(x)$, the first excited state $\psi_2(x)$, and so on.

We know that a general wavefunction can be expanded in such eigenstates via $\psi(x) = \sum_{n=1}^{\infty} a_n\psi_n(x)$, and from Sec. 7.4 we know that the information content in $\psi(x)$ and the expansion coefficients, $\{a_n\}$, is the same. We can write the collected $\{a_n\}$ as an (infinite-dimensional) vector $\mathbf{a}$

$$\psi(x) \Longleftrightarrow \{a_n\} \Longleftrightarrow \begin{pmatrix} a_1 \\ a_2 \\ a_3 \\ \vdots \end{pmatrix} \Longleftrightarrow \mathbf{a} \tag{11.69}$$

where we demand that $\sum_n^\infty |a_n|^2 = 1$ for proper normalization. In this language, individual energy eigenstates are written as

$$\psi_1(x) \Longleftrightarrow \{a_1 = 1, a_{n>1} = 0\} \Longleftrightarrow \begin{pmatrix} 1 \\ 0 \\ 0 \\ \vdots \end{pmatrix} \Longleftrightarrow \mathbf{e}_1 \tag{11.70}$$

and so forth. The set of vectors, $\mathbf{e}_i$, corresponding to eigenfunctions are said to form a *basis* for the infinite-dimensional vector space; they are like the unit vectors of a more physical vector space. We then have $\mathbf{a} = \sum_i a_i \mathbf{e}_i$.

The Schrödinger equation $\hat{H}\psi = E\psi$ can be written in the form

$$\hat{H}\left(\sum_m a_m \psi_m(x)\right) = E\left(\sum_m a_m \psi_m(x)\right) \tag{11.71}$$

so that if we multiply both sides by $\psi_n^*(x)$ *on the left* and integrate, we find that

$$\sum_m \langle \psi_n | \hat{H} | \psi_m \rangle a_m = E \sum_m \langle \psi_n \mid \psi_m \rangle a_m = E \sum_m \delta_{n,m} a_m = E a_n \tag{11.72}$$

We then choose to identify

$$\langle \psi_n | \hat{H} | \psi_m \rangle \equiv \mathbf{H}_{nm} \tag{11.73}$$

with the n, mth element of a *matrix* $\mathbf{H}$, in which case the Schrödinger equation takes the form of a matrix eigenvalue problem (see App. D.1), namely,

$$\begin{pmatrix} \mathbf{H}_{11} & \mathbf{H}_{12} & \mathbf{H}_{13} & \cdots \\ \mathbf{H}_{21} & \mathbf{H}_{22} & \mathbf{H}_{23} & \cdots \\ \mathbf{H}_{31} & \mathbf{H}_{32} & \mathbf{H}_{33} & \cdots \\ \vdots & \vdots & \vdots & \ddots \end{pmatrix} \begin{pmatrix} a_1 \\ a_2 \\ a_3 \\ \vdots \end{pmatrix} = E \begin{pmatrix} a_1 \\ a_2 \\ a_3 \\ \vdots \end{pmatrix} \tag{11.74}$$

or

$$\mathbf{Ha} = E\mathbf{a} \tag{11.75}$$

for short. The $\mathbf{H}_{nm}$ are called the *matrix elements* of the Hamiltonian and are said to form a *matrix representation* of the operator $\hat{H}$. We note that they have been evaluated using a particular set of basis vectors, namely, the eigenfunctions of $\hat{H}$ itself. In this particular case, the matrix takes an especially simple form, namely because

$$\mathbf{H}_{nm} = \langle \psi_n | \hat{H} | \psi_m \rangle = \langle \psi_n | E_m | \psi_m \rangle = E\delta_{nm} \tag{11.76}$$

so that the matrix $\mathbf{H}$ is *diagonal.* Thus, Eqn. (11.74) takes the form

$$\begin{pmatrix} E_1 & 0 & 0 & \cdots \\ 0 & E_2 & 0 & \cdots \\ 0 & 0 & E_3 & \cdots \\ \vdots & \vdots & \vdots & \ddots \end{pmatrix} \begin{pmatrix} a_1 \\ a_2 \\ a_3 \\ \vdots \end{pmatrix} = E \begin{pmatrix} a_1 \\ a_2 \\ a_3 \\ \vdots \end{pmatrix} \tag{11.77}$$

The only way Eqn. (11.77) can be satisfied is if

$$\det(\mathbf{H} - E\mathbf{1}) = \det \left| \begin{pmatrix} E_1 - E & 0 & 0 & \cdots \\ 0 & E_2 - E & 0 & \cdots \\ 0 & 0 & E_3 - E & \cdots \\ \vdots & \vdots & \vdots & \ddots \end{pmatrix} \right| = 0 \tag{11.78}$$

where $\mathbf{1}$ is the unit matrix. This is equivalent to

$$(E_1 - E)(E_2 - E)(E_3 - E)\cdots = \prod_{n=1}^{\infty}(E_n - E) = 0 \tag{11.79}$$

so that the energy eigenvalues are simply the E_n as we knew. The corresponding eigenvectors of the matrix equation are then simply the $\mathbf{e}_i$ (why?). We then say that

- The matrix representation of a Hamiltonian, when evaluated using its eigenfunctions as a basis, is diagonal, and the diagonal entries are just its energy eigenvalues.

Matrix representations of other operators can also be generated. For example, the position and momentum operators, x and $\hat{p}$, have matrix counterparts denoted by $\mathbf{x}$ and $\mathbf{p}$ and defined via

$$\mathbf{x}_{nm} = \langle\psi_n|x|\psi_m\rangle \qquad \text{and} \qquad \mathbf{p}_{nm} = \langle\psi_n|\hat{p}|\psi_m\rangle \tag{11.80}$$

Such matrix representations satisfy the usual rules of matrix algebra, namely,

$$(\mathbf{x}^2)_{nm} = \sum_k \mathbf{x}_{nk}\mathbf{x}_{km} \tag{11.81}$$

or more explicitly

$$\begin{pmatrix} \mathbf{x}_{11}^2 & \mathbf{x}_{12}^2 & \cdots \\ \mathbf{x}_{21}^2 & \mathbf{x}_{22}^2 & \cdots \\ \vdots & \vdots & \ddots \end{pmatrix} = \begin{pmatrix} \mathbf{x}_{11} & \mathbf{x}_{12} & \cdots \\ \mathbf{x}_{21} & \mathbf{x}_{22} & \cdots \\ \vdots & \vdots & \ddots \end{pmatrix} \cdot \begin{pmatrix} \mathbf{x}_{11} & \mathbf{x}_{12} & \cdots \\ \mathbf{x}_{21} & \mathbf{x}_{22} & \cdots \\ \vdots & \vdots & \ddots \end{pmatrix} \tag{11.82}$$

The matrix representation for the kinetic energy operator is, for example, $\mathbf{T} = \mathbf{p}^2/2m$ or

$$\mathbf{T}_{nm} = \frac{1}{2m}\sum_k \mathbf{p}_{nk}\mathbf{p}_{km} \tag{11.83}$$

Example 11.8 Matrix Representation of the Harmonic Oscillator We can make use of the results of Chap. 10 to evaluate many of these matrix representations for the specific case of the harmonic oscillator. Using the standard energy eigenvalues we find that

$$\mathbf{H} = \frac{\hbar\omega}{2}\begin{pmatrix} 1 & 0 & 0 & \cdots \\ 0 & 3 & 0 & \cdots \\ 0 & 0 & 5 & \cdots \\ \vdots & \vdots & \vdots & \ddots \end{pmatrix} \tag{11.84}$$

Using the results in Sec. 10.2.2, we then find that

$$\mathbf{x}_{nm} = \sqrt{\frac{\hbar}{2m\omega}}\left(\delta_{n,m-1}\sqrt{m} + \delta_{n,m+1}\sqrt{m+1}\right) \tag{11.85}$$

and we also quote the result

$$\mathbf{x^2}_{nm} = \frac{\hbar}{2m\omega}\left(\delta_{n,m+2}\sqrt{(m+1)(m+2)} + (2n+1)\delta_{n,m} + \delta_{n,m-2}\sqrt{m(m-1)}\right) \tag{11.86}$$

We can check that the matrix equation

$$\mathbf{x^2}_{nm} = \sum_k \mathbf{x}_{nk}\mathbf{x}_{km} \tag{11.87}$$

holds explicitly by comparing

$$\frac{\hbar}{2m\omega}\begin{pmatrix} 1 & 0 & \sqrt{1\cdot 2} & 0 & \cdots \\ 0 & 3 & 0 & \sqrt{2\cdot 3} & \cdots \\ \sqrt{1\cdot 2} & 0 & 5 & 0 & \cdots \\ 0 & \sqrt{2\cdot 3} & 0 & 7 & \cdots \\ \vdots & \vdots & \vdots & \vdots & \ddots \end{pmatrix} \stackrel{?}{=}$$

$$\sqrt{\frac{\hbar}{2m\omega}}\begin{pmatrix} 0 & \sqrt{1} & 0 & 0 & \cdots \\ \sqrt{1} & 0 & \sqrt{2} & 0 & \cdots \\ 0 & \sqrt{2} & 0 & \sqrt{3} & \cdots \\ 0 & 0 & \sqrt{3} & 0 & \cdots \\ \vdots & \vdots & \vdots & \vdots & \ddots \end{pmatrix} \cdot \sqrt{\frac{\hbar}{2m\omega}}\begin{pmatrix} 0 & \sqrt{1} & 0 & 0 & \cdots \\ \sqrt{1} & 0 & \sqrt{2} & 0 & \cdots \\ 0 & \sqrt{2} & 0 & \sqrt{3} & \cdots \\ 0 & 0 & \sqrt{3} & 0 & \cdots \\ \vdots & \vdots & \vdots & \vdots & \ddots \end{pmatrix} \tag{11.88}$$

Similar results hold for $\mathbf{p}$ and $\mathbf{p}^2$, and one can show (P11.15) that

$$\mathbf{H}_{nm} = \frac{1}{2m}\mathbf{p}^2_{nm} + \frac{m\omega^2}{2}\mathbf{x}^2_{nm} \tag{11.89}$$

holds as a matrix equation.

The *average* or *expectation value* of an operator in any state can also be written in this language. For example, we have

$$\begin{aligned} \langle x\rangle = \langle\psi\,|x|\,\psi\rangle &= \left\langle \sum_n a_n\psi_n \middle| x \middle| \sum_m a_m\psi_m \right\rangle \\ &= \sum_{n,m} a_n^* \langle\psi_n|x|\psi_m\rangle a_n \\ &= \sum_{n,m} a_n^* \mathbf{x}_{nm} a_m \end{aligned} \tag{11.90}$$

with similar expressions for other operators. We can also easily include the time dependence for any state via

$$\mathbf{a}(t) \Leftarrow\Rightarrow \sum_i a_i \mathbf{e}_i e^{-iE_i t/\hbar} \Leftarrow\Rightarrow \begin{pmatrix} a_1 e^{-iE_1 t/\hbar} \\ a_2 e^{-iE_2 t/\hbar} \\ \vdots \end{pmatrix} \tag{11.91}$$

The expectation value of the energy operator can be checked to satisfy

$$\langle \hat{E}\rangle_t = \sum_n |a_n|^2 E_n \tag{11.92}$$

independent of time because the energy matrix is diagonal. Other average values have less trivial time dependence (P11.17) in agreement with earlier examples.

Thus far we have considered only the case in which we already know the energy eigenfunctions and eigenvalues of the Hamiltonian operator $\hat{H}$. In this instance, the discussion above is interesting but provides little new information; we have just provided yet another representation of the solution space. If, on the other hand, we did not know the stationary states, we could still proceed as follows:

1. Pick a convenient set of energy eigenfunctions to *some* problem, called $\zeta_n(x)$; we immediately know that they form a complete set so that the expansion theorem will work.
2. Evaluate the Hamiltonian matrix using this set of basis functions, i.e., calculate
$$\mathbf{H}_{nm} \equiv \langle \zeta_n | \hat{H} | \zeta_m \rangle = \frac{1}{2m} \langle \zeta_n | \hat{p}^2 | \zeta_m \rangle + \langle \zeta_n | V(x) | \zeta_m \rangle \tag{11.93}$$
In this case, $\mathbf{H}$ will no longer be diagonal.
3. The Schrödinger equation in matrix form is still an eigenvalue problem of the form in Eqn. (11.74); its eigenvalues are determined by the condition that $\det(\mathbf{H} - E\mathbf{1}) = 0$.
4. If the eigenvalues are labeled via E_i and the corresponding eigenvectors by $\mathbf{a}^{(i)}$, the position space wavefunctions are given by $\psi_i(x) = \sum_n^\infty a_n^{(i)} \zeta_n(x)$.

Since finding the *exact* eigenvalues and eigenvectors of an infinite-dimensional matrix is only possible in very special cases, to use this method as a real calculational tool we most often restrict ourselves to a truncated version of the problem. More specifically, we try to diagonalize the $N \times N$ submatrix in the upper left-hand corner for some finite value of N. As N is made larger, we expect to obtain an increasingly good representation of the exact result. Because there exist powerful techniques for diagonalizing large matrices, especially if they happen to have large numbers of vanishing components (so-called *sparse* matrices), this technique is well suited for numerical computations.

Example 11.9 As an example of this method, consider the potential discussed in P9.6, namely, a symmetric infinite well defined via

$$V(x) = \begin{cases} 0 & \text{for } |x| < a \\ +\infty & \text{for } |x| > a \end{cases} \tag{11.94}$$

plus a δ function potential spike at the origin,

$$V_g(x) = g\delta(x) \tag{11.95}$$

This problem can be solved exactly and hence is useful as a testing ground for various approximation techniques. [For a thorough discussion, see Lapidus (1987).]

We know that the odd states are unaffected by $V_g(x)$ since they all possess nodes at $x = 0$. We thus consider the even states only, for which the energy eigenvalue condition can be written as

$$\lambda = -2y\cot(y) \tag{11.96}$$

where

$$\lambda \equiv \frac{2mag}{\hbar^2} \qquad \text{and} \qquad E = \frac{\hbar^2 y^2}{2ma^2} \tag{11.97}$$

We naturally choose as a set of basis functions the even solutions of the symmetric well *without* the ϵ function potential, i.e.,

$$\psi_n(x) = \frac{1}{\sqrt{a}} \cos\left(\frac{(n - 1/2)\pi x}{a}\right) \tag{11.98}$$

Evaluating the Hamiltonian matrix with this basis set, we find

$$\mathbf{H} = \begin{pmatrix} \hbar^2\pi^2/8ma^2 + g/a & g/a & g/a & \cdots \\ g/a & 9\hbar^2\pi^2/8ma^2 + g/a & g/a & \cdots \\ g/a & g/a & 25\hbar^2\pi^2/8ma^2 + g/a & \cdots \\ \vdots & \vdots & \vdots & \ddots \end{pmatrix} \tag{11.99}$$

or

$$\mathbf{H} = \left(\frac{\hbar^2\pi^2}{8ma^2}\right) \begin{pmatrix} 1+\epsilon & \epsilon & \epsilon & \cdots \\ \epsilon & 9+\epsilon & \epsilon & \cdots \\ \epsilon & \epsilon & 25+\epsilon & \cdots \\ \vdots & \vdots & \vdots & \ddots \end{pmatrix} \tag{11.100}$$

where $\epsilon \equiv 4\lambda/\pi^2$. Since we rely on matrix diagonalization methods ("canned" packages exist in many programming languages that will find the eigenvalues and eigenvectors of matrices), we must choose some specific numerical values. For $\lambda = 5$, the exact even energy eigenvalues (in terms of $\hbar^2\pi^2/8ma^2$) obtained from Eqn. (11.96) are:

$$2.2969\ (1) \qquad 10.8048\ (9) \qquad 26.9303\ (25) \qquad 50.9743\ (49) \tag{11.101}$$

where the terms in parentheses are the values without the δ function term.

Using an available package (in this case *Mathematica*®), we find the eigenvalues for increasingly large $N \times N$ truncated basis sets:

1×1	3.02642	—	—	—
2×2	2.54241	11.5104	—	—
3×3	2.44832	11.1412	27.4898	—
4×4	2.40672	11.0335	27.2396	51.4298
$\vdots$	$\vdots$	$\vdots$	$\vdots$	$\vdots$
10×10	2.33857	10.8861	27.0277	51.07991
$\vdots$	$\vdots$	$\vdots$	$\vdots$	$\vdots$
50×50	2.30506	10.8204	26.9487	50.9937

(11.102)

It does seem that the eigenvalues of the truncated set approach the exact values as N is increased. Such programs give the eigenvectors as well. The components corresponding to the ground state solution are:

1×1	(1)
2×2	(0.97264, −0.23232)
3×3	(0.97451, −0.21543, −0.06258)
4×4	(0.97575, −0.20818, −0.06075, −0.02946)

(11.103)

Using these values, we illustrate in Fig. 11.10 the approximations to the ground state wavefunction for the first and fourth approximations, comparing them to the exact solution (with the cusp expected from the singular δ function; the convergence to the exact solution is not particularly rapid).

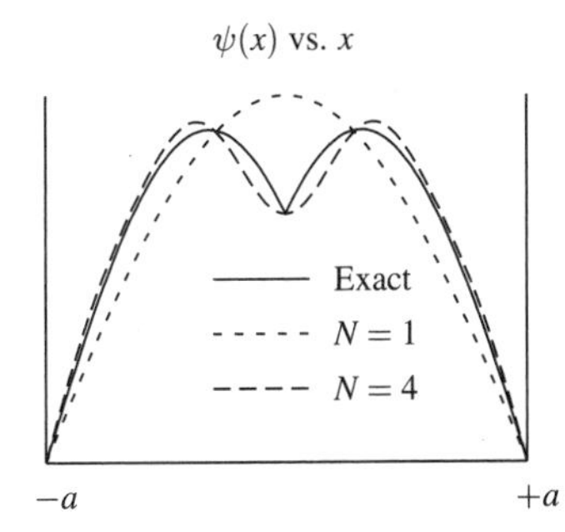

FIGURE 11.10. *The exact (solid) and approximate solutions for Ex. 11.9.*

11.6 PERTURBATION THEORY

We now turn to what is undoubtedly the most widely used approximation method we will discuss, perturbation theory.[14] We are certainly used to the notion of the systematic expansion of some quantity in terms of a small parameter; a familiar example is the series expansion of a function,

$$f(x) = f(0) + f'(0)x + \frac{1}{2!}f''(0)x^2 + \cdots \tag{11.104}$$

Such an expansion may well formally converge for all values of x [such as for the series for $\exp(x)$], but is often most useful as a calculational tool when $|x| \ll 1$.

Perturbation theory extends this notion to quantum mechanics in cases where the system under study can be described by an "unperturbed" Hamiltonian, $\hat{H}_0$, for which the energy eigenstates can be obtained exactly, i.e.,

$$\hat{H}_0\psi_n^{(0)} = E_n^{(0)}\psi_n^{(0)} \tag{11.105}$$

We began this chapter with the observation that many important systems such as the hydrogen atom or the harmonic oscillator can actually be solved exactly. One can then imagine "turning on" an additional perturbing interaction, $\hat{H}'$, which will change the spectrum and wavefunctions. Examples include the addition of an electric field acting on a charged particle (via a term $\hat{H}' = V(x) = -qEx$ in one dimension) or a magnetic field acting on a magnetic moment ($\hat{H} = -\boldsymbol{\mu} \cdot \mathbf{B}$). [While we will most often consider the case where the perturbation is a (small) additional potential energy function, other cases are possible (P11.24).] We can then write

$$\hat{H} = \hat{H}_0 + \lambda\hat{H}' \tag{11.106}$$

where we introduce a dimensionless parameter λ (which can be set equal to unity at the end of the calculation) to act as an expansion parameter. Our goal is then to solve the Schrödinger equation for the complete system,

$$\hat{H}\psi_n = E_n\psi_n \tag{11.107}$$

as a series in λ.

[14]We will focus only on time-independent or stationary-state perturbation theory; see Saxon (1968) for an excellent discussion of time-dependent methods.

11.6.1 Nondegenerate States

We will begin by making the assumption that the energy levels of the unperturbed system are all distinct, i.e., that there are no degeneracies where $E_n^{(0)} \approx E_l^{(0)}$ for some pair l, n. Then, as we imagine $\lambda \to 0$, we can unambiguously write

$$\psi_n \overset{\lambda\to 0}{\longrightarrow} \psi_n^{(0)} \qquad \text{and} \qquad E_n \underset{\lambda\to 0}{\longrightarrow} E_n^{(0)} \tag{11.108}$$

and make a unique identification of each perturbed state with its unperturbed counterpart. Motivated by these assumptions, we first write

$$E_n = E_n^{(0)} + \lambda E_n^{(1)} + \lambda^2 E_n^{(2)} + \cdots \tag{11.109}$$

as a series in λ. Then, since the unperturbed eigenstates form a complete set, we always have

$$\psi_n = \sum_{j=0}^{\infty} a_{nj}\psi_j^{(0)} = a_{nn}\psi_n^{(0)} + \sum_j{}' a_{nj}\psi_j^{(0)} \tag{11.110}$$

where $\sum_j{}'$ denotes the infinite sum with the $j = n$ term removed. The coefficients have slightly different expansions in λ:

$$\begin{aligned} a_{nn} &= a_{nn}^{(0)} + \lambda a_{nn}^{(1)} + \lambda^2 a_{nn}^{(2)} + \cdots \\ a_{nj} &= \qquad\quad \lambda a_{nj}^{(1)} + \lambda^2 a_{nj}^{(2)} + \cdots \qquad \text{for } j \neq n \end{aligned} \tag{11.111}$$

because Eqn. (11.108) implies that

$$\lim_{\lambda\to 0} a_{nj} = \delta_{nj} \tag{11.112}$$

We can constrain the expansion coefficients of Eqn. (11.110) further by noting that the normalization condition

$$\sum_{j=1}^{\infty} |a_{nj}|^2 = 1 \qquad \text{for all } n \tag{11.113}$$

implies that

$$1 = |a_{nn}|^2 + \sum_j{}'|a_{nj}|^2 = |a_{nn}|^2 + \sum_j{}'(\lambda a_{nj}^{(1)} + \cdots)^2 = |a_{nn}|^2 + \mathcal{O}(\lambda^2) \tag{11.114}$$

so that

$$a_{nn} \approx 1 \quad \text{to} \quad \mathcal{O}(\lambda^2) \qquad \text{which implies that} \quad a_{nn}^{(1)} = 0 \tag{11.115}$$

The Schrödinger equation can now be written [to $\mathcal{O}(\lambda^2)$] in the form

$$(\hat{H}_0 + \lambda\hat{H}')\left(\psi_n + \lambda\sum_j{}' a_{nj}\psi_j^{(0)} + \cdots\right) =$$

$$(E_n^{(0)} + \lambda E_n^{(1)} + \lambda^2 E_n^{(2)} + \cdots)\left(\psi_n + \lambda\sum_j{}' a_{nj}\psi_j^{(0)} + \cdots\right) \tag{11.116}$$

We first multiply Eqn. (11.116) by $(\psi_n^{(0)})^*$ on the left, and integrate and obtain

$$\left\langle \psi_n^{(0)} \middle| \hat{H}_0 + \lambda \hat{H}' \middle| \left(\psi_n + \lambda {\sum_j}' a_{nj} \psi_j^{(0)} + \cdots \right) \right\rangle =$$

$$(E_n^{(0)} + \lambda E_n^{(1)} + \lambda^2 E_n^{(2)} + \cdots) \left\langle \psi_n^{(0)} \middle| \left(\psi_n + \lambda {\sum_j}' a_{nj} \psi_j^{(0)} + \cdots \right) \right\rangle \quad (11.117)$$

Equating powers of λ and making extensive use of the orthogonality of the unperturbed wavefunctions, namely, that $\langle \psi_n^{(0)} | \psi_j^{(0)} \rangle = \delta_{nj}$, we find

$$\mathcal{O}(\lambda^0): \quad E_n^{(0)} = \langle \psi_n^{(0)} | \hat{H}_0 | \psi_n^{(0)} \rangle \quad (11.118)$$

$$\mathcal{O}(\lambda^1): \quad E_n^{(1)} = \langle \psi_n^{(0)} | \hat{H}' | \psi_n^{(0)} \rangle \equiv \mathbf{H}'_{nn} \quad (11.119)$$

$$\mathcal{O}(\lambda^2): \quad E_n^{(2)} = {\sum_j}' a_{nj}^{(1)} \langle \psi_n^{(0)} | \hat{H}' | \psi_j^{(0)} \rangle \equiv {\sum_j}' a_{nj}^{(1)} \mathbf{H}'_{nj} \quad (11.120)$$

These expressions all require the *matrix elements* of the perturbing Hamiltonian, evaluated using the unperturbed eigenfunctions, $\mathbf{H}'_{nk}$.

The $\mathcal{O}(\lambda^0)$ term simply reproduces the unperturbed energy spectrum. The equation for $E_n^{(1)}$ in Eqn. (11.119) is a very important result as it states that

- The first-order shift in the energy of level n due to a (small) perturbation is given by the diagonal matrix element of the perturbing Hamiltonian, $\mathbf{H}'_{nn}$, evaluated with the unperturbed wavefunctions, i.e.,

$$E_n^{(1)} = \langle \psi_n^{(0)} | \hat{H}' | \psi_n^{(0)} \rangle = \mathbf{H}'_{nn} \quad (11.121)$$

which we repeat because of its extreme importance.

Using Eqn. (11.121), we see that it can sometimes happen that the first-order energy shift vanishes identically because of symmetry. For example, a charged particle in the infinite symmetric well subject to a weak electric field, given by a potential of the form $V(x) = -qEx$, would have a first-order energy shift given by

$$E_n^{(1)} = \begin{cases} -qE\langle u_n^{(+)} | x | u_n^{(+)} \rangle & \text{for even states} \\ -qE\langle u_n^{(-)} | x | u_n^{(-)} \rangle & \text{for odd states} \end{cases} \quad (11.122)$$

which vanishes for all states. In such cases, the second-order term $E_n^{(2)}$ is the leading correction. Eqn. (11.120) suggests the more general result:

- The kth order correction to the energy levels requires knowledge of the $(k-1)$-th order wavefunctions

so that to determine $E^{(2)}$ we require the leading-order expansion coefficients, $a_{nj}^{(1)}$.

To obtain information on the expansion coefficients, we multiply Eqn. (11.116) by $(\psi_j^{(0)})^*$ with $j \neq n$ and integrate. The $\mathcal{O}(\lambda^0)$ terms are absent, while the $\mathcal{O}(\lambda^1)$ terms require that

$$a_{nk}^{(1)} = \frac{\langle \psi_k^{(0)} | \hat{H}' | \psi_n^{(0)} \rangle}{(E_n^{(0)} - E_k^{(0)})} = \frac{\mathbf{H}'_{nk}}{(E_n^{(0)} - E_k^{(0)})} \quad (11.123)$$

The first-order wavefunction thus receives contributions from every state for which the *off-diagonal* matrix element $\mathbf{H}'_{nk} \neq 0$. Combining Eqns. (11.123) and (11.120), we find that the second-order corrections to the energies are given by

$$\begin{aligned} E_n^{(2)} &= \sum_k{}' \frac{\langle \psi_k^{(0)} | \hat{H}' | \psi_n^{(0)} \rangle}{(E_n^{(0)} - E_k^{(0)})} \langle \psi_n^{(0)} | \hat{H}' | \psi_k^{(0)} \rangle \\ &= \sum_k{}' \frac{|\langle \psi_n^{(0)} | \hat{H}' | \psi_k^{(0)} \rangle|^2}{(E_n^{(0)} - E_k^{(0)})} \\ &= \sum_k{}' \frac{|\mathbf{H}'_{nk}|^2}{(E_n^{(0)} - E_k^{(0)})} \end{aligned} \tag{11.124}$$

In the last step, we have made use of the fact that $\hat{H}'$ is Hermitian to write $\mathbf{H}'_{nk} = (\mathbf{H}'_{kn})^*$; this form makes it clear that the second-order shift in energy is manifestly real. This result has many interesting consequences:

- The second-order shift depends on the off-diagonal matrix elements inversely weighted by the "distance in energy" to the state in question; thus, in general, states nearby in energy have a larger effect.
- This form also implies that the spacing in energy levels must be larger than the matrix elements of the perturbation for the expansion to be valid, i.e., we demand that

$$\mathbf{H}'_{nk} \ll |E_n^{(0)} - E_k^{(0)}| \tag{11.125}$$

  This shows that degenerate energy levels must be handled in a different way.
- States with energy below (above) a given level induce a second-order energy shift that is positive (negative); this effect is often referred to as *level repulsion.*
- The second-order shift in the ground state energy is clearly always negative, as all the other states lie above it. For many problems for which there are large numbers of levels, one can argue heuristically that the second-order shift for any fixed energy level will be negative due to the large number of states above it; this can be motivated on more physical grounds[15] and is often observed.

The second-order expansion coefficients (the $a_{nj}^{(2)}$) are too complicated to reproduce here, but, for reference, we state without proof that the result for the third-order shift in energies is

$$E_n^{(3)} = \sum_k{}' \sum_j{}' \frac{\mathbf{H}'_{nk} \mathbf{H}_{kj} \mathbf{H}'_{jn}}{(E_n^{(0)} - E_k^{(0)})(E_n^{(0)} - E_j^{(0)})} - \mathbf{H}'_{nn} \sum_k{}' \frac{|\mathbf{H}'_{nk}|^2}{(E_n^{(0)} - E_k^{(0)})^2} \tag{11.126}$$

We see that the work required to continue the perturbation theory expansion increases rapidly, so that often only the first- and second-order corrections are calculated. We now turn to some examples.

Example 11.10 Harmonic Oscillator with Applied Electric Field The problem of a charged oscillator in a constant electric field, described by the Hamiltonian

$$\hat{H} = \frac{1}{2} m\omega^2 x^2 - Fx \tag{11.127}$$

[15] See the nice discussion by Saxon (1968).

(where $F = qE$) was investigated in P10.6 where it was shown that it could be solved exactly; the resulting energy spectrum is

$$E_n = (n + 1/2)\hbar\omega - \frac{F^2}{2m\omega^2} \tag{11.128}$$

Let us approach this problem by considering the electric field interaction to be a small perturbation about the unperturbed oscillator, i.e., $\hat{H}' = -Fx$; we can use F as an expansion coefficient to count powers in perturbation theory. We have $E_n^{(0)} = (n + 1/2)\hbar\omega$, of course, while the first-order correction vanishes (because of symmetry) since

$$E_n^{(1)} = \langle \psi_n | \hat{H}' | \psi_n \rangle = -qE\langle \psi_n | x | \psi_n \rangle = 0 \tag{11.129}$$

The second-order correction is given by

$$\begin{aligned} E_n^{(2)} &= F^2 \sum_k{}' \frac{|\langle \psi_n | x | \psi_k \rangle|^2}{(E_n^{(0)} - E_k^{(0)})} \\ &= F^2 \sum_k{}' \frac{|\langle n|x|k\rangle|^2}{(n-k)\hbar\omega} \end{aligned} \tag{11.130}$$

Using the results in Sec. 10.2.2, we know that

$$\langle n|x|k\rangle = \sqrt{\frac{\hbar}{2m\omega}}\left(\sqrt{n}\delta_{k,n-1} + \sqrt{n-1}\delta_{k,n+1}\right) \tag{11.131}$$

Inserting this result into Eqn. (11.130), we find that

$$\begin{aligned} E_n^{(2)} &= \frac{F^2}{\hbar\omega}\left(\frac{\hbar}{2m\omega}\right)\sum_k{}' \frac{(n\delta_{k,n-1} + (n+1)\delta_{k,n+1})}{(n-k)} \\ &= \frac{F^2}{2m\omega^2}\left(\frac{n}{1} + \frac{(n+1)}{-1}\right) \\ &= -\frac{F^2}{2m\omega^2} \end{aligned} \tag{11.132}$$

which reproduces the exact answer. One would then expect that all of the higher-order corrections to the energy would vanish identically, and one can confirm explicitly (P11.21) that the third-order correction in Eqn. (11.126) is indeed zero in this case. It is also an example where the second-order corrections are, in fact, negative for all energy levels.

This does not imply, however, that the expansion coefficients have a similarly simple series behavior. To see this, we can make use of the exact ground state solution to the complete problem (see P10.6 again)

$$\psi(x; F) = \frac{1}{\sqrt{\rho\sqrt{\pi}}} e^{-(x-x_0)^2/2\rho^2} \tag{11.133}$$

where $x_0 = F/m\omega^2$ and $\rho = \sqrt{\hbar/m\omega}$. The expansion coefficient a_{00} is then given by

$$a_{00} \equiv \int (\psi(x; F = 0))^* \psi(x; F)\, dx = e^{-x_0^2/4\rho^2} = e^{-F^2/F_0^2} \tag{11.134}$$

where $F_0 \equiv 2\sqrt{\hbar m\omega}$. Expanding a_{00} in powers of F, we find that

$$a_{00} = 1 - \frac{F^2}{F_0^2} + \frac{1}{2}\frac{F^4}{F_0^4} + \cdots \tag{11.135}$$

Thus, while the perturbation series for the energies terminates at second order, the expansion coefficients require the full series to converge to the exact answer.

Example 11.11 Consider the problem, discussed in P9.6, of the symmetric infinite square well potential plus a δ function potential at the origin. In this case, let the δ function constitute the perturbation so that $\hat{H}' = g\delta(x)$.

For the odd case, the explicit application of the boundary conditions for the full problem requires that the $u_n^{(-)}(x)$ vanish at the origin and gives the same energy eigenvalue condition as for the infinite well alone. This can be confirmed to any order in perturbation theory, since all of the relevant matrix elements in Eqns. (11.121) , (11.124), and (11.126) vanish explicitly. (See P11.22.)

For the even case, the exact eigenvalue condition was given by $\lambda = -2y\cot(y)$ where $\lambda \equiv 2mag/\hbar^2$ and $E = \hbar^2 y^2/2ma^2$. Focusing only on the ground state, this eigenvalue condition can be expanded to second order to yield

$$y = \frac{\pi}{2} + \frac{\lambda}{\pi} - 2\frac{\lambda^2}{\pi^3} + \mathcal{O}(\lambda^3) = \frac{\pi}{2}\left(1 + \frac{2\lambda}{\pi^2} - \frac{4\lambda^2}{\pi^4}\right) + \cdots \tag{11.136}$$

or

$$\begin{aligned} E_1^{(+)} &\approx \frac{\hbar^2}{2ma^2}\frac{\pi^2}{4}\left(1 + \frac{2\lambda}{\pi^2} - \frac{4\lambda^2}{\pi^4} + \cdots\right)^2 \\ &\approx \frac{\hbar^2\pi^2}{8ma^2} + \lambda\left(\frac{\hbar^2}{2ma^2}\right) - \lambda^2\left(\frac{\hbar^2}{2\pi^2 ma^2}\right) + \cdots \end{aligned} \tag{11.137}$$

The first-order perturbation result for even states is simply

$$(E_n^{(+)})^{(1)} = \langle u_n^{(+)}|g\delta(x)|u_n^{(+)}\rangle = \frac{g}{a} = \lambda\left(\frac{\hbar^2}{2ma^2}\right) \tag{11.138}$$

independent of n for all even states. This obviously agrees with the explicit expansion of the eigenvalue condition for the ground state in Eqn. (11.137). The second-order correction is then

$$\begin{aligned} (E_n^{(+)})^{(2)} &= \sum_{k=2} \frac{|\langle u_n^{(+)}|g\delta(x)|u_k^{(+)}\rangle|^2}{(\hbar^2\pi^2/8ma^2)(1-(2k-1)^2)} \\ &= -\lambda^2\frac{2\hbar^2}{m\pi^2 a^2}S \qquad \text{where} \qquad S \equiv \sum_{k=2}^{\infty}\frac{1}{(2k-1)^2 - 1} = \frac{1}{4} \\ &= -\lambda^2\left(\frac{\hbar^2}{2\pi^2 ma^2}\right) \end{aligned} \tag{11.139}$$

which also agrees with the expansion of the exact result.

While the technical details of the calculation are beyond our level, it is appropriate to note here that one of the most spectacularly successful predictions in all of physics makes use of perturbation theory. The magnetic moment of both the electron and the muon can be calculated in the theory of quantum electrodynamics,[16] using more advanced perturbation theory methods. A recent theoretical result for the electron magnetic moment (expressed as a dimensionless number) is

$$g_e(\text{theory}) = 2.0023193048(8) \tag{11.140}$$

[16] See Perkins (1987) and references therein for a discussion at an undergraduate level.

where the uncertainty is indicated in the last significant digit. Amazingly, it can also be measured to a similar precision with the result

$$g_e(\text{experiment}) = 2.0023193048(4) \tag{11.141}$$

11.6.2 Degenerate Perturbation Theory

When two (or more) energy levels of the unperturbed system are degenerate, any linear combinations of the corresponding wavefunctions, $\psi_n^{(0)}(x)$ and $\psi_l^{(0)}(x)$, still give the same energy eigenvalue. (Such combinations can still be made orthogonal, of course.) This implies, however, that the unique identification of each perturbed state with an unperturbed counterpart as in Eqn. (11.108) is not possible. The breakdown of the perturbation method in this case is clearly signaled by the appearance of small energy denominators in Eqn. (11.123); states nearby in energy can play an important role and have to be considered on a more equal footing.

In this limit, it is convenient to return to the matrix formulation of the eigenvalue problem and consider the 2×2 submatrix involving the states in question, namely,

$$\begin{pmatrix} E_n^{(0)} + \lambda \mathbf{H}'_{nn} & \lambda \mathbf{H}'_{nl} \\ \lambda \mathbf{H}'_{ln} & E_l^{(0)} + \lambda \mathbf{H}'_{ll} \end{pmatrix} \begin{pmatrix} a_n \\ a_l \end{pmatrix} = E \begin{pmatrix} a_n \\ a_l \end{pmatrix} \tag{11.142}$$

where the a_n, a_l are the expansion coefficients. In general, for a case with N degenerate levels, the corresponding $N \times N$ submatrix must be considered.

This system of linear equations (for the $a_{n,l}$) will only have a nontrivial solution if the appropriate determinant vanishes, i.e.,

$$\det \begin{pmatrix} E_n^{(0)} + \lambda \mathbf{H}'_{nn} - E & \lambda \mathbf{H}'_{nl} \\ \lambda \mathbf{H}'_{ln} & E_l^{(0)} + \lambda \mathbf{H}'_{ll} - E \end{pmatrix} = 0 \tag{11.143}$$

The special case of an exact degeneracy where $E_n^{(0)} = E_l^{(0)} \equiv \mathscr{E}$ is easiest to treat; in this case, the energy eigenvalues are determined by the condition

$$(E - [\mathscr{E} + \lambda \mathbf{H}'_{nn}])(E - [\mathscr{E} + \lambda \mathbf{H}'_{ll}]) - \lambda^2 \mathbf{H}'_{ln} \mathbf{H}'_{nl} = 0 \tag{11.144}$$

or

$$E_{\pm} = \mathscr{E} + \lambda \frac{1}{2}(\mathbf{H}'_{n} + \mathbf{H}'_{ll}) \pm \lambda \frac{1}{2} \sqrt{(\mathbf{H}'_{nn} - \mathbf{H}'_{ll})^2 + 4 \mathbf{H}'_{ln} \mathbf{H}'_{ln}} \tag{11.145}$$

The first term is obviously the (common) value of the unperturbed energy, while the second is the average of the first-order energy shifts in each level, consistent with the nondegenerate case. The third term, however, splits the two levels and removes the degeneracy.

Substituting the result of Eqn. (11.145) into the matrix equation Eqn. (11.142), we find that the expansion coefficients are given by

$$\frac{a_n^{(\pm)}}{a_l^{(\pm)}} = \frac{2\mathbf{H}'_{nl}}{(\mathbf{H}'_{ll} - \mathbf{H}'_{nn}) + \sqrt{(\mathbf{H}'_{nn} - \mathbf{H}'_{ll})^2 + 4 \mathbf{H}'_{ln} \mathbf{H}'_{ln}}} \tag{11.146}$$

so that the appropriate (unnormalized) eigenfunctions are given by

$$\psi^{(\pm)}(x) \quad \propto \quad a_n^{(\pm)} \psi_n(x) + a_l^{(\pm)} \psi_l(x) \tag{11.147}$$

The actual energy splitting in Eqn. (11.145) clearly depends on λ, i.e., on the *magnitude* of the perturbation. The appropriate linear combinations, however, do not, but are determined by the *form* of the perturbation, i.e., the *relative* sizes of the matrix elements $\mathbf{H}'_{nn}$, $\mathbf{H}'_{ll}$, and $\mathbf{H}'_{ln}$ (P11.26).

Because degeneracy of energy levels is far more common in multiparticle or multidimensional systems, we postpone presenting examples until later chapters.

11.7 QUESTIONS AND PROBLEMS

Q11.1 If you are given a numerical solution of the Schrödinger equation in the form of a list of values at discrete points, i.e., $\psi(x = n\epsilon)$, how would you normalize the solution? How would you find $\langle x \rangle$? How about $\langle \hat{p} \rangle$? How would you calculate the momentum space wavefunction, $\phi(p)$?

Q11.2 Distinguish carefully between the *precision* and the *accuracy* of a measurement. For a given approximation method, you can imagine determining the range over which the solutions change their sign at infinity more and more precisely. Is this increased precision or increased accuracy? Do you think that decreasing the step size ϵ or using a better integration method results in increased accuracy or precision?

Q11.3 Assume that you have a program that numerically integrates the Schrödinger equation and that you have found two energy values, E_a and E_b, that bracket an acceptable (i.e., square integrable) solution of the SE. Describe an efficient strategy to get arbitrarily close to the "real" energy. *Hint:* If someone tells you they have a number between 1 and 1000, what is the optimal strategy to find their number using the minimum number of "Yes-No" questions?

Q11.4 If you have a program that solves the Schrödinger equation for the even solutions in a symmetric potential, what lines of code would you have to change to let it solve for the odd solutions?

P11.1 **Numerical integration; Classical:** Pick some simple technique designed to integrate numerically second-order differential equations, perhaps even the simple one used in Sec. 11.1. Write a short program (using a computer language, programmable calculator, or even a spread sheet program) to solve Newton's laws for a general potential or force law.

(a) Apply it to the differential equation

$$\frac{d^2x(t)}{dt^2} = -x(t) \qquad \text{where} \quad x(0) = 1 \quad \text{and} \quad \dot{x}(0) = 0$$

Compare your results for decreasing step size with the exact solution [which is, of course, $x(t) = \cos(t)$.] Try to reproduce Fig. 11.1.

(b) Try the same thing for the equation

$$\frac{d^2x(t)}{dt^2} = x(t) \qquad \text{where} \qquad x(0) = 1 \qquad \text{and} \qquad \dot{x}(0) = -1$$

and also for the initial conditions $x(0) = +1$ and $\dot{x}(0) = +1$. What are the exact solutions, how well does your program work in these cases, and why?

P11.2 **Numerical integration; the Schrödinger equation:** Using your experience from P11.1, modify your program to solve the Schrödinger equation for a symmetric potential.

(a) Apply it to the case of the harmonic oscillator written in dimensionless coordinates as

$$\frac{d^2\psi(y)}{dy^2} - y^2\psi(y) = -\epsilon\psi(y)$$

where ϵ are dimensionless eigenvalues. Try to reproduce the values in Ex. 11.2. Repeat for the odd case where the eigenvalues are $\epsilon = 3, 7, 11, \ldots$.

(b) Use your program to find the (dimensionless) eigenvalues for a particle in the symmetric linear potential of Sec. 10.7. Compare your results with those listed in Eqn. (10.107).

(c) Apply your program to the case of a quartic potential, i.e., $V(x) = Cx^4$. Write the Schrödinger equation in dimensionless variables, and find the first two even and odd energy eigenvalues; use your previous experience to estimate the errors in your calculation.

P11.3 Show that the trial wavefunction in Ex. 11.4 yields the energy function in Eqn. (11.24). You might also like to try this problem with the nonzero piece of the wavefunction given by $N(a^2 - x^2)^n$ with $n = 1, 3, 4$ as well, and compare your results.

P11.4 Estimate the ground state energy of the simple harmonic oscillator by using the family of trial wavefunctions

$$\psi(x; a) = \sqrt{\frac{1}{a}} e^{-|x|/a}$$

Why is your answer so much worse than that using the cutoff polynomial expression of Ex. 11.4?

P11.5 The momentum space wavefunction corresponding to P11.4 is

$$\phi(p) = \sqrt{\frac{2p_0}{\pi}} \left(\frac{p_0}{p^2 + p_0^2} \right)$$

where $p_0 \equiv \hbar/a$. Evaluate the energy functional in momentum space using this trial wavefunction for the SHO, and show that you get the same result (for the energy and trial parameter) as in position space.

P11.6 Estimate the energy of the first excited state of the SHO potential by using a trial wavefunction of the form

$$\psi(x; a) = \begin{cases} 0 & \text{for } |x| > a \\ Nx(a^2 - x^2)^2 & \text{for } |x| > a \end{cases}$$

P11.7 Use a Gaussian trial wavefunction to estimate the ground state energy for the quartic potential $V(x) = gx^4$. Show that your answer is

$$E_{\text{min}} = \left(\frac{3}{4} \right)^{4/3} \left(\frac{\hbar^4 g}{m^2} \right)^{1/3}$$

Compare this to the "exact" answer (determined by numerical integration) that has the prefactor 0.668.

P11.8 **(a)** Estimate the ground state energy of the symmetric infinite well using the family of trial wavefunctions

$$\psi(x) = \begin{cases} 0 & \text{for } |x| > a \\ N(a^\lambda - |x|^\lambda) & \text{for } |x| < a \end{cases}$$

where λ is the variational parameter and you must determine the normalization constant N.

(b) Estimate the energy of the first excited state by using the wavefunction in (a) multiplied by x to make an appropriate odd trial wavefunction. You will have to renormalize the wavefunction, of course.

P11.9 **(a)** Estimate the ground state energy of symmetric infinite well by using the wavefunction

$$\psi(x) = \begin{cases} 0 & \text{for } |x| > L \\ N(L^2 - x^2) & \text{for } |x| < L \end{cases}$$

Evaluate $E(\text{var})/E_0 - 1$ and $1 - |a_0|^2$ for this state. Note that this has no variational parameter.

(b) Now consider the family of trial functions

$$\psi(x;b) = \begin{cases} 0 & \text{for } |x| > L \\ N'(L^2 - x^2)(1 + bx^2/L^2) & \text{for } |x| < L \end{cases}$$

which does have an additional parameter. Calculate both the variational energy $E(b) = E[\psi(x;b)]$ and $1 - |a_0|^2$. Find the values of b that minimize each of these two quantities, and show that they are slightly different. Specifically, show that

$$b_{\text{min}} = \left(\frac{504 - 51\pi^2}{7\pi^2 - 72}\right) \approx -0.223216$$

for the overlap maximum, while

$$b_{\text{min}} = \frac{(-98 + 8\sqrt{133})}{26} \approx -0.22075$$

for the energy minimum. This demonstrates that while there is a strong correlation between the wavefunction that minimizes the variational energy and the one that maximizes the overlap with the ground state wavefunction, the two criteria are ultimately independent.

P11.10 Using the variational method, show that any purely attractive potential in one dimension has at least one bound state. By purely attractive, we mean that $V(x) \leq 0$ for all x. *Hint:* Show that we can find a (perhaps very shallow and narrow) finite square well potential, $V_{\square}(x)$, that satisfies $0 > V_{\square}(x) > V(x)$ for all x, and use the fact that a finite square well always has at least one bound state.

P11.11 Show that the matching of solutions leading to the WKB quantization condition implies that $A_L = A_R(-1)^n$.

P11.12 Apply the WKB quantization condition to the symmetric linear potential, $V(x) = F|x|$. Compare your results with the exact values found in Sec. 10.5. Does the agreement get better with increasing n as expected?

P11.13 Apply the WKB quantization condition to the "half harmonic oscillator" potential, namely,

$$V(x) = \begin{cases} +\infty & \text{for } x < 0 \\ m\omega^2 x^2/2 & \text{for } x > 0 \end{cases}$$

What are the appropriate values of C_L, C_R? How does the WKB quantization method "pick out" only the odd states of the full potential?

P11.14 Apply the WKB quantization condition to estimate the bound state energies of the potential

$$V(x) = -\frac{V_0}{\cosh(x/a)^2}$$

(a) Show that your results can be written in the form

$$E_n = -\left(\sqrt{V_0} - (n + 1/2)\sqrt{\frac{\hbar^2}{2ma^2}}\right)^2$$

Hint: You might use the integral

$$\int_0^A \frac{\sqrt{A^2 - u^2}}{1 + u^2}\,du = \frac{\pi}{2}\left(\sqrt{1 + A^2} - 1\right)$$

(b) If $V_0 \gg \hbar^2/2ma^2$, show that your result approximates the harmonic oscillator approximation for this potential.

(c) One might think that one could take the limit $V_0 \to \infty$ and $a \to 0$ in such a way as to reproduce an attractive δ function potential. Discuss the WKB approximation to the energy levels in this limit. If it works, does it reproduce the result of Sec. 9.1.2? If it doesn't, why not?

P11.15 (a) Evaluate $\mathbf{p}_{nm}$ for the harmonic oscillator using the methods in Ex. 11.8. Using your result, show that the commutator $[x, p] = i\hbar$ holds as a matrix equation.

(b) Evaluate $\mathbf{p^2}_{nm}$ and use your result to show that Eqn. (11.89) holds by evaluating both sides as matrices.

P11.16 (a) Evaluate the matrix elements $\mathbf{p}_{nm}$ using the "standard" infinite well energy eigenstates as a basis. How would you show that

$$\mathbf{H}_{nm} = \frac{1}{2m}\sum_k \mathbf{p}_{nk}\mathbf{p}_{km}$$

is the (diagonal) Hamiltonian matrix?

(b) Evaluate the matrix elements $\mathbf{x}_{nm}$. Can you show that $[x, p] = i\hbar$ holds as a matrix equation?

P11.17 What is the expectation value, $\langle \hat{x} \rangle_t$, for a state vector in a matrix representation for general t, i.e., how does Eqn. (11.90) generalize? What does your expression look like for a state with only two components?

P11.18 Show that the wavefunction to second order in perturbation theory (assuming no degeneracies) is given by

$$\psi_n^{(2)} = \sum_m{}' \sum_k{}' \frac{\mathbf{H}'_{mk}\mathbf{H}'_{kn}}{(E_n^{(0)} - E_k^{(0)})(E_n^{(0)} - E_m^{(0)})}\psi_m^{(0)} - \sum_m{}' \frac{\mathbf{H}'_{nn}\mathbf{H}_{mn}}{(E_n^{(0)} - E_m^{(0)})^2}\psi_m^{(0)}$$
$$- \frac{1}{2}\psi_n^{(0)} \sum_m{}' \frac{|\mathbf{H}'_{mn}|^2}{(E_n^{(0)} - E_m^{(0)})^2}$$

P11.19 We have seen in P7.5 that a constant shift in the potential energy function, i.e., $V(x) \to V(x) + V_0$ can have no effect on the observable physics. Consider such a shift as a perturbation, and evaluate (1) the first-, second-, and third-order changes in the energy of any state using Eqns. (11.121), (11.124), and (11.126); (2) the first-order shift in the wavefunction using Eqn. (11.123), and discuss your results. You can also use the results of P11.18 to check the second-order shift in the wavefunction.

P11.20 The nonrelativistic series for the kinetic energy in Eqn. (1.8) is given by

$$T = \frac{p^2}{2m} - \frac{p^4}{8m^3c^2} + \cdots$$

Using first-order perturbation theory, evaluate the effect of the second term in this expansion (with p replaced by the operator $\hat{p}$) on the nth level of a harmonic oscillator.

P11.21 Using Eqns. (11.124) and (11.126), show that the second- and third-order energy shifts in Ex. 11.10 vanish.

P11.22 Evaluate the first-, second-, and third-order shifts in energy for the odd states in Ex. 11.11 due to the $\delta(x)$ perturbation.

P11.23 A particle of mass m in a harmonic oscillator potential $V(x) = m\omega^2 x^2/2$ is subject to a small perturbing potential of the same type, namely, $V'(x) = \lambda x^2$.

(a) Show that the energy spectrum can be derived exactly with the result

$$E'_n = \left(n + \frac{1}{2}\right)\hbar\overline{\omega}$$

where $\overline{\omega} = \omega\sqrt{1 + 2\lambda/m\omega^2}$. Expand this for small λ to $\mathcal{O}(\lambda^2)$ for comparison with part (b).

(b) Evaluate the first- and second-order shifts in energy using Eqns. (11.121) and (11.124), and compare your results to the exact answer in part (a). You will find the matrix elements of $\langle n|x^2|k\rangle$ in Ex. 11.8 useful.

P11.24 Evaluate the effect of a small anharmonic term of the form

$$V'(x) = -\lambda K x^3$$

on the spectrum of the harmonic oscillator in first- and second-order perturbation theory. You may find the following matrix element useful:

$$\langle\psi_n|x^3|\psi_k\rangle = \left(\frac{\hbar}{2m\omega}\right)^{3/2}\Big(\sqrt{(n+1)(n+2)(n+3)}\,\delta_{k,n+3} + 3(n+1)^{3/2}\delta_{k,n-1} + 3n^{3/2}\delta_{k,n+1} + \sqrt{n(n-1)(n-2)}\,\delta_{k,n-3}\Big)$$

P11.25 Evaluate the effect of the anharmonic term $V'(x)$ from P11.24 on the wavefunctions of the harmonic oscillator using first-order perturbation theory, i.e., Eqn. (11.123). The expectation value of position in the unperturbed system vanishes since $\langle\psi_n^{(0)}|x|\psi_n^{(0)}\rangle = 0$ as in Eqn. (10.41). Evaluate $\langle x\rangle$ in the perturbed system using the first-order wavefunction and compare it to the results of P10.16.

P11.26 **Degenerate states in perturbation theory:**

(a) Show that the first-order shifts in energy in Eqn. (11.145) are real as they should be. Do the $a_{n,l}$ have to be real?

(b) Using Eqn. (11.146), show that the linear combinations $\psi^{(+)}(x)$ and $\psi^{(-)}(x)$ are always orthogonal.

(c) Discuss the energy levels and mixing of eigenstates in the case where $\mathbf{H}'_{nn} = \mathbf{H}'_{ll} = 0$ but $\mathbf{H}'_{ln} = (\mathbf{H}'_{nl})^* \neq 0$; show that the eigenfunctions are "completely mixed."

(d) Discuss the case where the degenerate states are not connected in the Hamiltonian to lowest order, i.e., for which $\mathbf{H}'_{ln} = (\mathbf{H}'_{nl})^* = 0$.

CHAPTER 12

Scattering in One Dimension

12.1 SCATTERING IN ONE-DIMENSIONAL SYSTEMS

12.1.1 Bound and Unbound States

Besides the bounded, periodic motion of particles in potentials as shown in Fig. 6.3 (for energy E_1), there is also the possibility of unbound states that are not localized in space and not repetitive in time. These correspond to particles incident on the potential and which subsequently "bounce" off (energy E_2 in Fig. 6.3) or temporarily change their speeds as they go over the potential (energy E_3). Classical mechanics, which solves for the exact trajectories in either case, makes little distinction between the two types of motion, aside from some technical details, such as the use of Fourier series in the study of periodic motion. In quantum mechanics, on the other hand, the experimental realizations of these two classes of classical motion are quite distinct, so that the theoretical formalisms and methods used to analyze them are necessarily somewhat different.

One main source of experimental information on microscopic systems which allows us to test the ideas of quantum mechanics is spectroscopy, the study of bound states and their radiative decays. This corresponds most closely to bound states as studied in Chaps. 6 to 11. The quantization of energy levels for particles bound in potentials gives rise to a discrete spectrum of photons (or other particles) when excited states decay into lower energy levels. A precise map of the photon energies can allow one to reconstruct the energy level diagram, which, of course, then conveys information on the nature of the bound particles and their interactions.

Quantum scattering experiments, however, make use of rather different experimental techniques. Typically, a beam of incident particles is directed towards a target, and the scattered particles are collected (detected) and counted at various angular locations. A collection of classical trajectories corresponding to unbound motions would, as a function of, say, the particle energy and impact parameter, also allow for a mapping of the scattering force. Quantum mechanically, the particle trajectories are replaced, at best, by probabilistic wavepacket motion, but variations of the probability of scattering at different angles and energies still gives information on the nature of the scattering potential.

We will discuss the formalism of scattering theory much more fully in Chap. 20, but we find it useful here to introduce some of the ideas and notation of three-dimensional scattering before specializing to one-dimensional problems.

In a three-dimensional scattering experiment, a given intensity of incident particles

$$j_{\text{inc}}^{(3)}(\mathbf{r}, t) \equiv \frac{dN_{\text{inc}}}{dt\, dA} \tag{12.1}$$

that is, the number of particles incident on a target per unit time per unit area, can be directly associated with the probability flux, defined in three dimensions via

$$\mathbf{j}(\mathbf{r}, t) = \frac{\hbar}{2mi}\left[\psi^*(\mathbf{r}, t)\nabla\psi(\mathbf{r}, t) - \nabla\psi^*(\mathbf{r}, t)\psi(\mathbf{r}, t)\right] \tag{12.2}$$

which is a generalization of the 1D result. The number of particles scattered into the given solid angle ($d\Omega$) at an angular location specified by (θ, ϕ), per unit time, as in Fig. 12.1, is described by

$$j_{\text{scatter}}^{(3)}(\theta, \phi) = \frac{dN_{\text{scatter}}}{dt\, d\Omega} \tag{12.3}$$

and will certainly depend on the incident intensity. An appropriate ratio that measures the probability of a scattering event is

$$\frac{j_{\text{scatter}}^{(3)}}{j_{\text{inc}}^{(3)}} = \frac{dN_{\text{scatter}}}{dt\, d\Omega} \Big/ \frac{dN_{inc}}{dt\, dA} \equiv \frac{d\sigma(\theta, \phi)}{d\Omega} \tag{12.4}$$

which defines a *differential cross section* for scattering, $d\sigma/d\Omega$, which has the dimensions of an effective area. This quantity can be calculated from a knowledge of the scattering potential, $V(\mathbf{r})$, and can also be directly compared to experiment. For classical "specular" (equal angle reflective) scattering, idealized by scattering small masses (e.g., BB's) from larger, heavy shapes (e.g., billiard balls), the differential cross section gives direct information on the size and shape of the scatterers, hence the notion of "cross-sectional area" or cross section. In quantum mechanics, the wave properties of the scatterers will

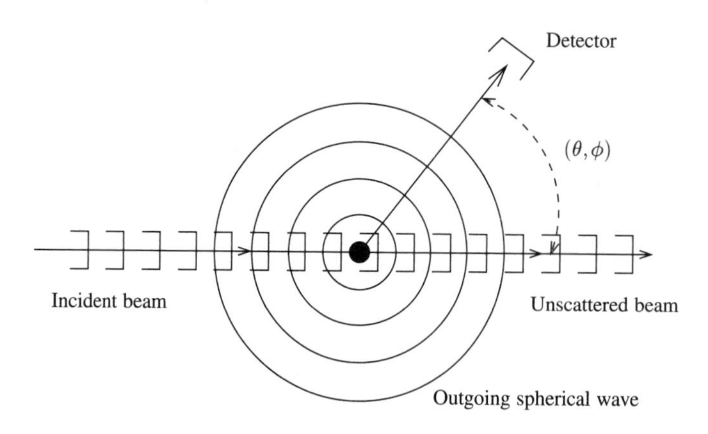

FIGURE 12.1. *Geometry of a scattering experiment in three dimensions.*

also be important, and the analogs of effects such as interference and diffraction will be apparent.

For scattering in two dimensions, the similar quantity involves ratios of particles incident per unit time per unit length and numbers of particles scattered per unit time per unit angle, i.e.,

$$\frac{j^{(2)}_{\text{scatter}}}{j^{(2)}_{\text{inc}}} = \frac{dN_{\text{scatter}}}{dt\,d\theta} \Big/ \frac{dN_{\text{inc}}}{dt\,dx} \equiv \frac{d\rho(\theta)}{d\theta} \tag{12.5}$$

This is an effective "width" or lateral size of the target for scattering through various angles θ in the plane.

In one dimension, the geometric situation is far simpler, as there are only two possible directions. In this case, incident particles that continue forward are *transmitted* while those undergoing a backscatter are called *reflected.* The incident particles will be described by the incoming number per unit time

$$j^{(1)}_{\text{inc}} = \frac{dN_{\text{inc}}}{dt} \tag{12.6}$$

with similar expressions for the scattered (i.e., reflected) and transmitted fluxes,

$$j^{(1)}_{\text{ref}} = \frac{dN_{\text{ref}}}{dt} \quad \text{and} \quad j^{(1)}_{\text{trans}} = \frac{dN_{\text{trans}}}{dt} \tag{12.7}$$

The ratio of reflected to incident flux, $j^{(1)}_{\text{ref}}/j^{(1)}_{\text{inc}}$ is then the analog of the scattering cross section or size, but is dimensionless.

A description of scattering in 1, 2, or 3 dimensions involving wavepackets (with obvious connections to classical particle trajectories) is possible, but following our discussion for free particles in Chap. 3, we will begin by considering the fluxes (and their ratios) corresponding to the scattering of plane wave solutions as it contains much of the physics involved.

12.1.2 Plane Wave Solutions

If we deal with plane wave solutions, we immediately come up against the problem of how to use nonnormalizable wavefunctions. We can avoid any such questions in a highly physical way by considering only the concept of probability flux or currents in Eqns. (12.6) and (12.7). For a right- or left-moving plane wave of the form

$$\psi(x,t) = Ae^{i(\pm px - p^2t/2m)/\hbar} = Ae^{i(\pm kx - \hbar k^2 t/2m)} \tag{12.8}$$

the probability flux is given by (P5.6)

$$j(x,t) = \pm\frac{\hbar k}{m}|A|^2 \tag{12.9}$$

independent of time. The factor $\pm\hbar k/m = \pm p/m = \pm v$ simply corresponds to the classical velocity of the particle beam. Since we expect (normalizable) one-particle wavefunctions in one dimension to have dimensions given by $[A] = 1/\sqrt{\text{length}}$, the dimensions of flux can be thought of as the number of particles per unit time; thus we take the probability flux as equivalent to Eqn. (12.6). The flux or incident intensity of a particle beam can be made larger by increasing A corresponding, say, to more intensity in the source, or by changing the speed of the particles, i.e., increasing k (i.e., having them "come at you faster").

Ratios of fluxes are then free from any ambiguities from the lack of normalizability and correspond most naturally to the real experimental situation where just such ratios are measured. The conservation of particle flux should also follow immediately from the fact that we have introduced no possible absorption processes, and this can be used as a check in any calculation. We will implement these ideas in a series of examples involving plane waves in the next few sections; we will also exhibit some wavepacket solutions for comparison.

12.2 SCATTERING FROM A STEP POTENTIAL

The simplest one-dimensional potential that can be investigated analytically corresponds to scattering from a step potential. Any physical "step" will have smooth edges as in Fig. 12.2(a), but it is easiest to treat the discontinuous potential in Fig. 12.2(b); this is defined by

$$V(x) = \begin{cases} 0 & \text{for } x < 0 \\ V_0 & \text{for } x > 0 \end{cases} \tag{12.10}$$

where we consider both $V_0 > 0$, $V_0 < 0$, but allow only $E > 0$.

If a quantum mechanical free particle in one dimension corresponds to classical waves on a string, this potential is the analog of two strings of differing mass density, and hence with different propagation velocities, "tied together" at the origin. The corresponding classical particle picture would be described by a force of the form $F(x) = -V_0\delta(x)$, implying an impulsive "kick" each time the particle crosses the origin, changing the magnitude (and/or direction) of its momentum. For example, a particle incident on the step from the left with energy $E > V_0 > 0$ ($V_0 > E > 0$), would slow down but continue over (bounce back from) such a potential step.

Considering first the case where $E > V_0$, the allowed plane wave solutions in the two regions are simply

$$\psi(x,0) = \psi(x) = \begin{cases} Ie^{ikx} + Re^{-ikx} & \text{for } x < 0 \\ Te^{iqx} + Se^{-iqx} & \text{for } x > 0 \end{cases} \tag{12.11}$$

where

$$k = \sqrt{\frac{2mE}{\hbar^2}} \quad \text{and} \quad q = \sqrt{\frac{2m(E - V_0)}{\hbar^2}} \tag{12.12}$$

The solutions for $x < 0$ are taken to correspond to an initial plane wave (Ie^{ikx}) incident from $-\infty$ (i.e., from the left) and a reflected component (Re^{-ikx}). For $x > 0$, the $T\exp(iqx)$

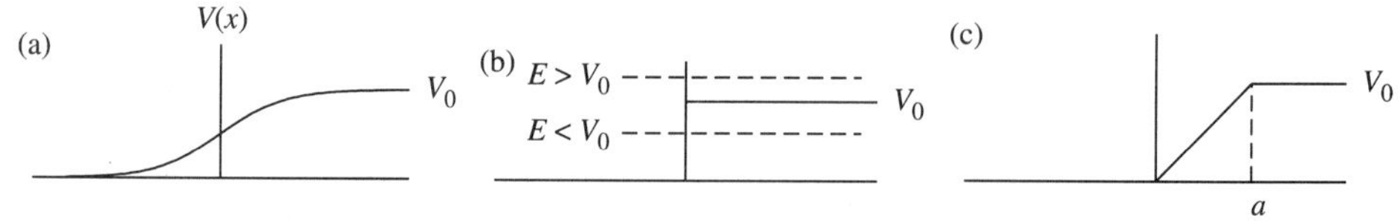

FIGURE 12.2. *Models of a step potential in one dimension; (a) a physically acceptable smooth step, (b) an idealization of (a) as a discontinuous potential, and (c) a "linear step potential" where the potential is allowed to "turn on" over a distance a.*

term certainly represents the right-moving transmitted wave; as we expect there to be no left-moving solution for $x > 0$ if particles are incident on the step from the left, we require $S = 0$. The wavefunction must still satisfy the appropriate continuity conditions on $\psi(x)$ and $\psi'(x)$, so we insist that

$$\begin{aligned} \psi(0^-) &= \psi(0^+) \Longrightarrow I + R = T \\ \psi'(0^-) &= \psi'(0^+) \Longrightarrow ikI - ikR = iqT \end{aligned} \tag{12.13}$$

which gives

$$R = I\left(\frac{k-q}{k+q}\right) \qquad \text{and} \qquad T = I\left(\frac{2k}{k+q}\right) \tag{12.14}$$

As expected, we cannot solve for R, T completely, as they depend on the arbitrary incident amplitude I, and we cannot normalize the plane wave solutions. Ratios of fluxes, however, will be well defined and independent of I.

A check on this (trivial) calculation is to confirm that the probability fluxes are also consistent with expectation. If we calculate the fluxes corresponding to $x < 0$ (j_L) and $x > 0$ (j_R) (see P12.1), we find that

$$j_L(x,t) = \frac{\hbar k}{m}|I|^2 - \frac{\hbar k}{m}|R|^2 = \frac{\hbar k}{m}|I|^2 \frac{4kq}{(k+q)^2} \tag{12.15}$$

$$j_R(x,t) = \frac{\hbar q}{m}|T|^2 = \frac{\hbar k}{m}|I|^2 \frac{4kq}{(k+q)^2} \tag{12.16}$$

are equal as expected. The fact that, classically, the particle is moving at a different speed for $x > 0$ (and so q and not k appears in the transmitted flux) is obviously important to remember.

The quantities that would be most similar to experimental observables would be the ratios of reflected and transmitted to incident fluxes, i.e.,

$$\frac{j_{\text{ref}}}{j_{\text{inc}}} = \left|\frac{R}{I}\right|^2 = \left(\frac{k-q}{k+q}\right)^2 = \left(\frac{\sqrt{E} - \sqrt{V_0 + E}}{\sqrt{E} + \sqrt{V_0 + E}}\right)^2 \tag{12.17}$$

$$\frac{j_{\text{trans}}}{j_{\text{inc}}} = \frac{q}{k}\left|\frac{T}{I}\right|^2 = \frac{2kq}{(k+q)^2} = \frac{2\sqrt{E(V_0+E)}}{(\sqrt{E} + \sqrt{V_0+E})^2} \tag{12.18}$$

Using these expressions, it is easy to check that the probability of a backscatter goes to zero when $E \gg V_0$.

We plot in Fig. 12.3(a) and (b) plots of the real part of the incident, reflected, and transmitted wavefunctions corresponding to two signs of the potential step. These figures illustrate several aspects of both the wave and particle aspects of scattering:

- The wavefunction is wigglier (less wiggly) and smaller (larger) in magnitude when the kinetic energy is larger (smaller), as expected from our earlier discussions of the intuitive connections between classical and quantum probability distributions; this shows that such ideas are not restricted to bound states.
- A classical particle with energy larger than the step height would, however, never scatter back, so the reflected flux is purely a wave phenomenon. This is not at all apparent from Eqn. (12.17) for the ratio of scattered to incident flux, as it depends

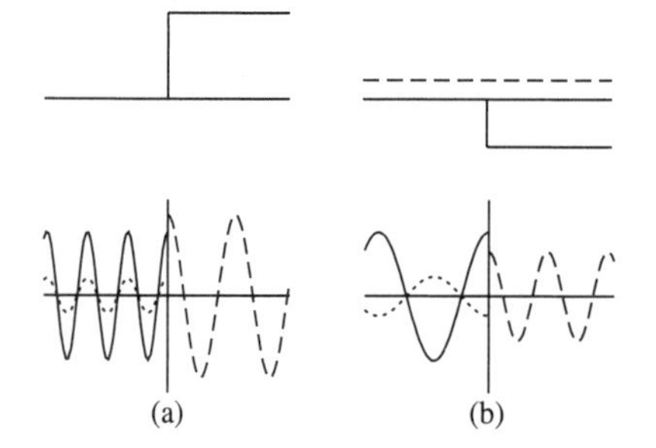

FIGURE 12.3. Plane wave scattering from a discontinuous step potential. Two cases where (a) $E > V_0 > 0$ and (b) $E > 0 > V_0$ are shown. The real part of the incident and reflected amplitude (solid and dotted curves for $x < 0$) and transmitted amplitude (dashed curve for $x > 0$) are illustrated.

only on E, V_0 and has no explicit factor of $\hbar$ as would be expected for a purely wave (and hence quantum) effect. This is evidently an artifact of the discontinuous nature of the step potential, and we will examine a more realistic "smoothed" step potential in the next section for comparison.

- The phase of the reflected wave relative to that of the incident one depends on the sign of V_0 (or equivalently the classical propagation speed) in a manner that is familiar from other wave phenomena. In going from a region of large to small speed, for example, there is a phase change on reflection.

For the case of $V_0 > E > 0$, we can either find the solutions for $x > 0$ directly or analytically continue the ones of Eqn. (12.11) by letting

$$q = \sqrt{\frac{2m(E - V_0)}{\hbar^2}} \longrightarrow \kappa = iq = \sqrt{\frac{2m(V_0 - E)}{\hbar^2}} \tag{12.19}$$

so that, for example, the new solutions satisfy

$$\psi(x) = Te^{iqx} \longrightarrow \psi(x) = Te^{-\kappa x} \qquad \text{for } x > 0 \tag{12.20}$$

Such a real function has vanishing probability flux, so the particles must be completely reflected from this step potential. The wavefunction itself, however, is nonzero in the classically disallowed region, as we expect from tunneling ideas. The reflection coefficient is then given by

$$R = I\left(\frac{k - i\kappa}{k + i\kappa}\right) \tag{12.21}$$

which can be written in terms of a simple phase

$$R = Ie^{-2i\phi} \qquad \text{where} \qquad \tan(\phi) = \kappa/k \tag{12.22}$$

Using these, we indeed find that

$$\frac{j_{\text{ref}}}{j_{\text{inc}}} = \left|\frac{R}{I}\right|^2 = 1 \qquad \text{and} \qquad \frac{j_{\text{trans}}}{j_{\text{inc}}} = 0 \tag{12.23}$$

as expected. Fig. 12.4 illustrates two such cases where we can observe the tunneling behavior as the amount of energy "cheating" increases. One obvious limit is when $V_0 \gg E$, in which case we have the "wall" scattering case considered in Sec. 3.3.2. We note that in this limit we have $\kappa \gg k$ so that $R \to -I$ and the wave solution in the allowed region ($x < 0$) is then

$$\psi_L(x) = Ie^{ikx} + Re^{-ikx} \longrightarrow \left(Ie^{ikx} - Ie^{-ikx}\right) \sim \sin(px/\hbar) \tag{12.24}$$

as discussed previously.

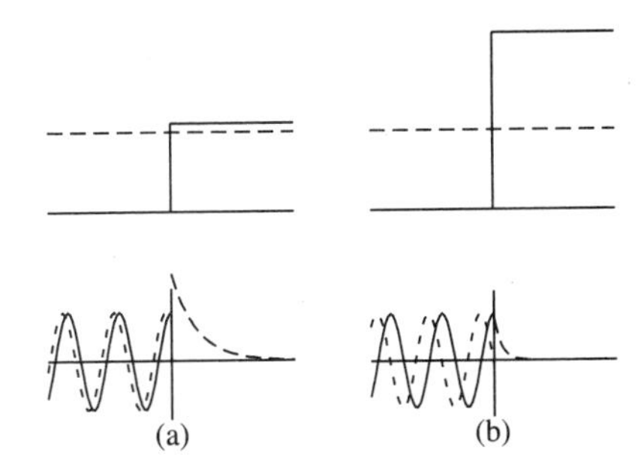

FIGURE 12.4. *Same as for Fig. 12.3, but for two cases where $V_0 > E > 0$.*

To see the connections to wavepacket scattering, we write the general time-dependent plane wave solution in the form

$$\psi_p(x,t) = \begin{cases} Ie^{i(px-Et)/\hbar} + I\left(\frac{p-p_q}{p+p_q}\right)e^{i(-px-Et)/\hbar} & \text{for } x < 0 \\ I\left(\frac{pp_q}{p+p_q}\right)e^{i(p_qx-Et)/\hbar} & \text{for } x > 0 \end{cases} \tag{12.25}$$

where $p = \hbar K$, $p_q = \hbar q$, and $E = p^2/2m$. This can then be used with a Gaussian weighting distribution as in Sec. 3.3.2 to generate the time-dependent wavepacket. We illustrate the results for two cases in Fig. 12.5 and note many similarities with the plane wave results:

- The dotted curves corresponding to unscattered (but spreading) wavepackets help illustrate the "slow-down" and "speed-up" of the classical particle when $V_0 > 0$ (a) and $V_0 < 0$ (b).
- The probability of reflection and transmission of such packets can be determined by the relative areas under the separate "bumps" of $|\psi(x,t)|^2$ calculated (numerically) for times long after the scatter; the complete wavepacket, of course, remains properly normalized at all times. One should not think, therefore, of the particle somehow "splitting" into two separate "blobs"; as with all quantum measurements, a measurement of the particle will find it in a definite position, but with probability related to the relative sizes of the bumps.

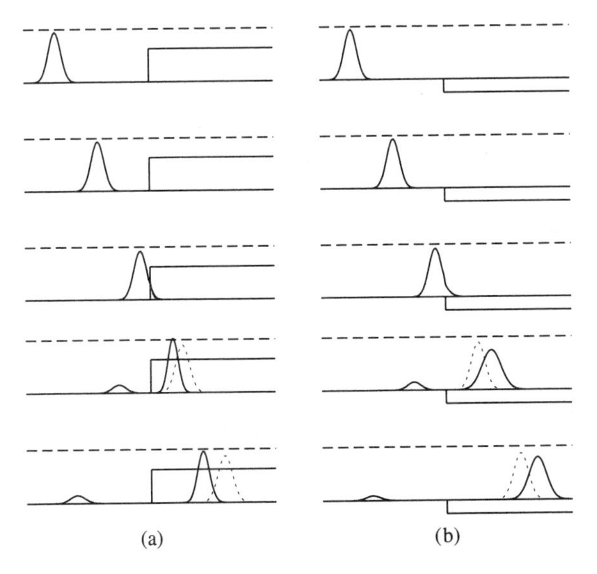

FIGURE 12.5. *Gaussian wavepacket scattering from a discontinuous step potential for (a) $E > V_0 > 0$ and (b) $E > 0 > V_0$.*

12.3 "LINEAR" STEP POTENTIAL

A somewhat more realistic version of the step potential[1] is shown in Fig. 12.2(c) where we allow the impulsive force to occur over a finite interval, or equivalently, where the potential "turns on" more or less slowly. The functional form of the "linear step" potential is then

$$V(x) = \begin{cases} 0 & \text{for } x < 0 \\ V_0 x/a & \text{for } 0 < x < a \\ V_0 & \text{for } x > a \end{cases} \tag{12.26}$$

The incident and reflected waves (for $x < 0$) and the transmitted waves (for $x > a$) all have the same form as before. Using the change of variables $x = \rho y + \lambda$ in the region $(0, a)$, we once again find the Airy differential equation, and the full solution is then given by

$$\psi(x) = \begin{cases} Ie^{ikx} + Re^{-ikx} & \text{for } x < 0 \\ EAi\left(\frac{x-\lambda}{\rho}\right) + FBi\left(\frac{x-\lambda}{\rho}\right) & \text{for } 0 < x < a \\ Te^{iqx} & \text{for } a > x \end{cases} \tag{12.27}$$

Insisting on the continuity of $\psi(x)$ and $\psi'(x)$ at both $x = 0, a$ allows one to solve for T, R in terms of I and as functions of the Airy functions and their derivatives. These must be evaluated numerically, and we illustrate in Fig. 12.6 some representative plots of the real parts of the incident, scattered, and transmitted waves for a case corresponding to $E > V_0 > 0$ and both $a = 0$ and $a \neq 0$. One immediately notes that the smoother potential gives rise to much less reflection, as the wave has more time to "accommodate" itself to the new region of potential. From the plots, one can infer that if the potential changes over a distance of roughly the size of the *largest* wavelength in the problem, the scattering is greatly diminished.

We can also see this more globally in Fig. 12.7 where we plot the ratio of reflected to incident fluxes, $|R/I|^2$ (for two fixed values of $E/V_0 > 1$), versus the dimensionless ratio

$$\zeta \equiv \frac{a}{\rho} = \left(\frac{2mV_0a^2}{\hbar^2}\right)^{1/3} \tag{12.28}$$

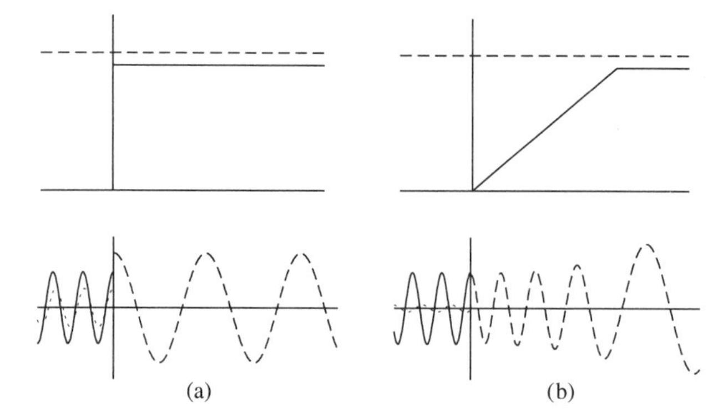

FIGURE 12.6. *Plane wave scattering from the "linear step potential" for $E > V_0 > 0$ and (a) vanishing value of a (i.e., discontinuous potential) and (b) $a > 0$. The real part of the incident and reflected amplitude (solid and dotted curves for $x < 0$) and transmitted amplitude (dashed curve for $x > 0$) are shown.*

[1] The results in this section require more tedious calculations to yield analytic and graphical results than in Sec. 12.2: we thus provide fewer details and concentrate more on the physically meaningful results. Many of the gory details are contained in Delbourgo (1977).

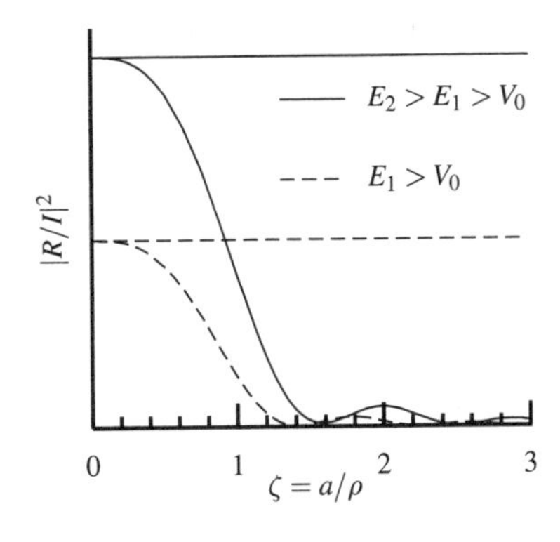

FIGURE 12.7. The ratio of reflected to incident fluxes, $|R/I|^2$, versus the dimensionless ratio $\zeta \equiv a/\rho = (2mV_0a^2/\hbar^2)^{1/3}$. The solid (dashed) curves corresponding to $E/V_0 = 3\,(4)$ are shown. The corresponding horizontal lines show the values of $|R/I|^2$ for the discontinuous step potential (i.e., for $a = 0$) as obtained from the simple analytic results of Sec. 12.2.

The variable ζ appears naturally as an argument in the Airy functions. These plots show that:

- As the distance over which the change in potential is allowed to occur, i.e., as the change becomes more gradual, the reflected amplitude decreases. This is a familiar effect from other branches of wave physics, with examples including impedance matching in signal propagation and even the shape of musical instruments in acoustics; "sharp edges" induce more wave scattering than "soft" ones.
- One can also view the limit of large ζ as keeping a, E, V_0 all fixed, but letting $\hbar \to 0$; this illustrates the fact that in the classical limit there is no scattering. Any realistic potential step will have a finite "rise" distance, so the classical limit, as obtained by letting $\hbar \to 0$, should be taken *before* letting $a \to 0$.
- Similarly in the limit of a very slow particle, i.e., with a, E, V_0 fixed but $m \to \infty$, the reflected flux vanishes. All of these limits correspond to "slow" or *adiabatic* changes.

12.4 SCATTERING FROM THE FINITE SQUARE WELL

An important, analytically calculable, example of scattering in one dimension, which again illustrates classical connections but which also introduces several new features, especially quantum tunneling, is the finite well or barrier defined by the potential

$$V(x) = \begin{cases} -V_0 & \text{for } |x| < a \\ 0 & \text{for } |x| > a \end{cases} \tag{12.29}$$

This potential satisfies the more realistic condition that $V(x) \to 0$ when $|x|$ becomes large. Please note that with this notation, the case where $V_0 > 0$ corresponds to an attractive well, while $V_0 < 0$ is a repulsive barrier.

12.4.1 Attractive Well

For the case $V_0 > 0$, we can easily solve the Schrödinger equation in the three appropriate regions. Assuming a plane wave solution incident from the left, we find

$$\psi(x) = \begin{cases} Ie^{ikx} + Re^{-ikx} & \text{for } x < -a \\ Ee^{iqx} + Fe^{-iqx} & \text{for } -a < x < a \\ Te^{ikx} & \text{for } a < x \end{cases} \tag{12.30}$$

where

$$k = \sqrt{\frac{2mE}{\hbar^2}} \qquad \text{and} \qquad q = \sqrt{\frac{2m(E + V_0)}{\hbar^2}} \tag{12.31}$$

as usual; inside the well, we expect both left- and right-moving waves due to reflections at the edges. In contrast to the bound state problem, the boundary conditions at $x = \pm a$ are not equivalent because of the asymmetric nature of the incident scattering. The four independent boundary conditions are given by

$$\begin{aligned} Ie^{-ika} + Re^{ika} &= Ee^{-ika} + Fe^{ika} \\ ik(Ie^{-ika} - Re^{ika}) &= iq(Ee^{-ika} - Fe^{ika}) \\ Ee^{iqa} + Fe^{-iqa} &= Te^{ika} \\ iq(E^{iqa} - Fe^{-iqa}) &= ikTe^{ika} \end{aligned} \tag{12.32}$$

We can then solve for R, T, E, F in terms of I; for the determination of the scattering and transmission probabilities, we require R, T, and one can obtain (P12.5)

$$R = Iie^{-2ika} \frac{(q^2 - k^2)\sin(2qa)}{2kq\cos(2qa) - i(q^2 + k^2)\sin(2qa)} \tag{12.33}$$

$$T = Ie^{-2ika} \frac{2kq}{2kq\cos(2qa) - i(q^2 + k^2)\sin(2qa)} \tag{12.34}$$

while E, F can also be determined if one wishes to see the wavefunction over the entire region.

We illustrate an example of the real parts of the various wavefunctions for a typical case in Fig. 12.8, and we note some general features:

- A classical particle speeds up as it goes over the (attractive) well, and this is consistent with the wavefunction shown in the figure.
- Another quantum remnant of this increase in velocity over the well is present in the *phase* of the transmitted wave. If we write the wavefunction for $x > a$ as

$$\psi(x) = |T|e^{i\phi}e^{ikx} \tag{12.35}$$

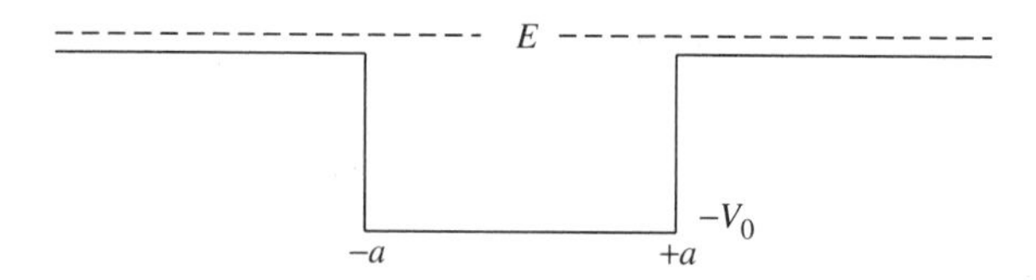

Particle speeds up
over potential well

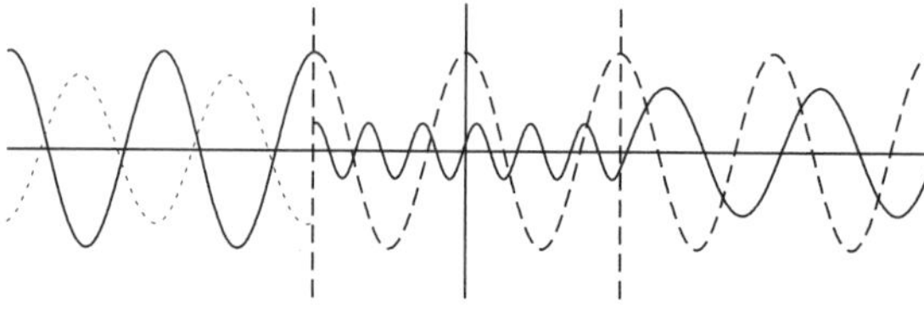

FIGURE 12.8. The real part of the incident, reflected, and transmitted plane wave solutions for a finite well. For $x < -a$, the dotted curve represents the reflected wave; for $x > -a$, the dashed curve simply continues the incident wave showing an unscattered wave for comparison.

where ϕ is the phase of the (complex) transmission coefficient, we find that

$$\phi = \tilde{\phi} - 2ka \qquad \text{where} \qquad \tan(\tilde{\phi}) = \frac{k^2 + q^2}{2kq} \tan(2qa) \tag{12.36}$$

In the limit that $E \gg V_0$, we have

$$\frac{k^2 + q^2}{2kq} = \frac{2E + V_0}{2\sqrt{E(E + V_0)}} \longrightarrow 1 + O(E^2/V_0^2) \approx 1 \tag{12.37}$$

so that $\phi \to 2a(q - k)$; in this case, the transmitted wave is of the simple form

$$I e^{i(\kappa x + 2a(q - \kappa))} \tag{12.38}$$

If we compare this form to an incident plane wave that does not interact (i.e., $V_0 = 0$), we find that at a *given time*, the position of a point *of the same phase* on the two waves satisfies

$$\begin{array}{ccc} \text{scattered} & \Leftarrow\!\Rightarrow & \text{unscattered} \\ kx' + 2a(q - k) & = & kx \end{array}$$

so that the difference in positions between the two cases is given by

$$\Delta x \equiv x' - x = \frac{2a}{k}(k - q) \tag{12.39}$$

If we write these quantities in terms of the classical speeds in the two regions,

$$\hbar k = m v_0 \qquad \text{and} \qquad \hbar q = m v \tag{12.40}$$

we find that

$$\Delta x = \frac{2a}{v_0}(v_0 - v) = t_{\text{across}}\, \Delta v \tag{12.41}$$

where $\Delta v < 0$, since the particle speeds up ($v > v_0$) over the well. Here t_{across} is just the time it would take the (unscattered) particle to cross the well. This is shown in Fig. 12.8 where the transmitted waves do indeed "lead" a wave (the dashed curve) that does not interact. This appearance of the classical speed-up in the phase of the transmitted wave is related to the so-called *phase-shift* of three-dimensional scattering.

We next plot the probabilities of both reflection and transmission versus incident energy in Fig. 12.9 (for a particular choice of V_0 and a) where we see that:

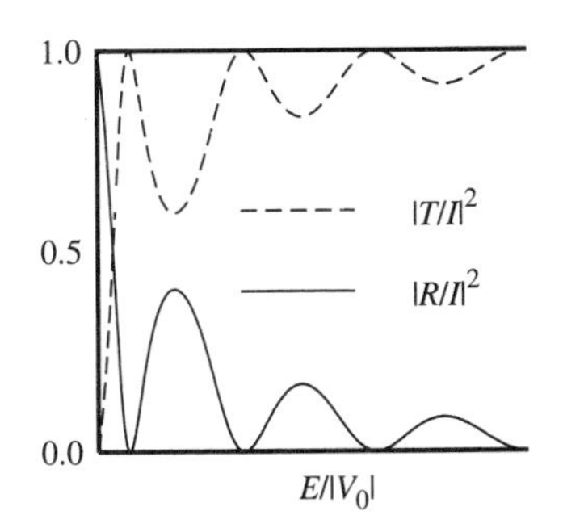

FIGURE 12.9. *The ratio of reflected (solid) and transmitted (dashed) to incident flux versus energy for the attractive square well. The zeroes in* $|R/I|^2$ *correspond to transmission resonances.*

- In the limit $E \gg V_0$, one has $q^2 - k^2 \ll 2kq, q^2 + k^2$, so that there is little reflection as expected.
- A completely nonclassical phenomenon is readily apparent, as there are special values of E for which there is no reflection, corresponding to so-called *transmission resonances*. These points can be traced to the vanishing of $\sin(2qa)$ for $2qa = n\pi$, i.e., for energies satisfying

$$E + V_0 = \frac{n^2\hbar^2\pi^2}{8ma^2} \qquad \text{for} \qquad n = 1, 2, 3, \ldots \tag{12.42}$$

This effect is easily understood in wave terms as due to the complete destructive interference between waves scattered at the first "step" (for which there is a phase change on reflection) and the second (for which there is no phase change). In that case one requires that

$$\text{Back-and-forth distance across the well} = 4a = n\lambda \tag{12.43}$$

so that the difference in path lengths between the two waves is an integral number of wavelengths, while one additional phase change from a reflection from an edge guarantees a minimum instead of a maximum. This effect is familiar from geometrical optics and is often used to minimize the amount of reflected light for optical instruments. A similar effect is seen in atomic systems (hence in three dimensions) where low-energy electrons are scattered from atoms of certain noble gases (such as neon or argon). In this case, called the *Ramsauer-Townsend effect,* the scattering cross-section exhibits dramatic dips as a function of energy, indicating near perfect transmission.

12.4.2 Repulsive Barrier

All of the formulae for the attractive well can be taken over to the case of the repulsive barrier by simply letting $V_0 \to -V_0$, provided that the incident energy is larger than the step size, i.e., $E > V_0$; for example,

$$q = \sqrt{\frac{2m}{\hbar^2}(E - V_0)} \tag{12.44}$$

and the particle now slows down classically over the barrier.

A more interesting case arises when the energy is less than the height of the barrier, $E < V_0$. The form of the solutions for $|x| > a$ are unchanged, but in the barrier region the Schrödinger equation now takes the form

$$\frac{d^2\psi(x)}{dx^2} = \kappa^2\psi(x) \tag{12.45}$$

where $\kappa = \sqrt{2m(V_0 - E)/\hbar^2}$. We can write the most general solution in the region $|x| < a$ as

$$\psi(x) = Ae^{-\kappa x} + Be^{+\kappa x} \tag{12.46}$$

but we can save considerable work by noting that the case under consideration can be easily obtained from an analytic continuation of earlier results, since we have

$$q = \sqrt{\frac{2m}{\hbar^2}(E - V_0)} \longrightarrow i\kappa = i\sqrt{\frac{2m}{\hbar^2}(V_0 - E)} \tag{12.47}$$

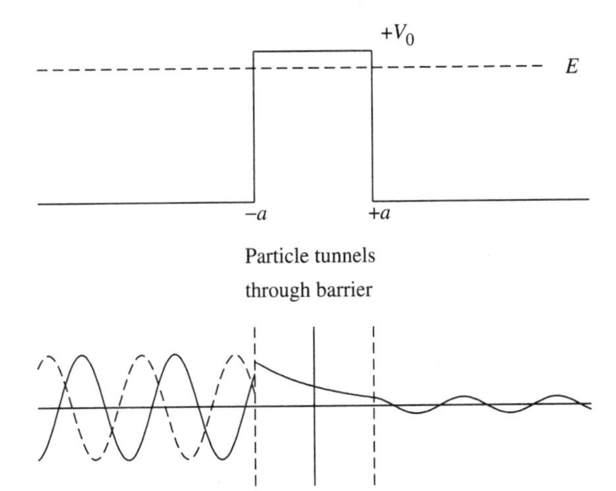

FIGURE 12.10. Representation of a plane wave incident on a square barrier showing tunneling wavefunction.

The results of Eqns. (12.33) and (12.34) can then be taken over by using the relations

$$\sin(iz) = i\sinh(z) \qquad \cos(iz) = \cosh(z) \tag{12.48}$$

so that, for example, the transmission coefficient is given by

$$T = Ie^{-2ika} \frac{2k\kappa}{2k\kappa\cosh(2qa) - i(k^2 - \kappa^2)\sinh(2qa)} \tag{12.49}$$

We illustrate the corresponding wavefunctions in Fig. 12.10 for a generic case.

A particle incident on such a rectangular barrier would have a probability of penetration given by

$$\frac{|T|^2}{|I|^2} = \frac{(2\kappa k)^2}{(k^2+\kappa^2)\sinh^2(2\kappa a) + (2k\kappa)^2} \approx \left(\frac{4k\kappa}{k^2+\kappa^2}\right)^2 e^{-4\kappa a} \tag{12.50}$$

We note that the dominant exponential piece of the tunneling probability in Eqn. (12.50) is just that given by the WKB formula of Sec. 11.4.1. We can make a useful connection between this square-well result and the WKB formulae.

We derive this by noting that in order for a particle to tunnel through the general potential of Fig. 12.11, it must successfully tunnel through each of the thin rectangular barriers in turn, with probability $P_i(x_i)$ for each barrier. We make the following identifications

$$\begin{aligned} \text{width, } 2a &\longrightarrow \Delta x \\ \kappa &\longrightarrow \kappa(x_i) \\ P = e^{-2(\kappa)2a} &\longrightarrow P_i(x_i) = e^{-2(\kappa(x_i))\Delta x} \end{aligned} \tag{12.51}$$

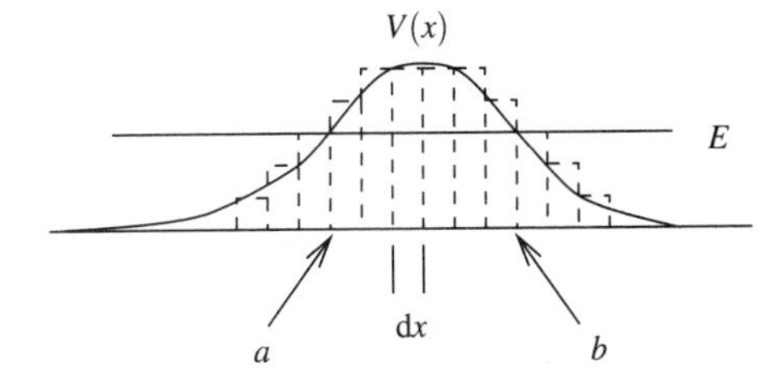

FIGURE 12.11. Generic potential barrier, $V(x)$, approximated as a series of square barriers making connection to the WKB approximation for the tunneling wavefunction.

Because each of the "successes" is an independent event, the total probability is the product of the $P_i(x_i)$. We thus have

$$\begin{aligned}
P_T &= \prod_i P_i(x_i) \\
&\approx \prod_i e^{-2\kappa(x_i)\Delta x} \qquad \text{where } \kappa(x) \equiv \sqrt{2m/\hbar^2(V(x)-E)} \\
&= e^{-2\sum_i \kappa(x_i)\Delta x} \\
&\approx e^{-2\int_a^b \kappa(x)\,dx}
\end{aligned}$$

$$P_T = \exp\left(-2\sqrt{\frac{2m}{\hbar^2}}\int_a^b \sqrt{V(x)-E}\,dx\right) \tag{12.52}$$

where a, b are the classical turning points. We will use this approximate expression for the transmission probability for barrier penetration in the examples discussed in the next section.

12.5 APPLICATIONS OF QUANTUM TUNNELING

12.5.1 Field Emission

A subject often discussed in modern physics courses is the *photoelectric effect* in which electrons are emitted from a metal by the absorption of sufficiently energetic photons. The effect is illustrated in Fig. 12.12(a) where the Fermi energy of the filled electron sea is still an energy W below the threshold for a free particle; W is often called the *work function* of the metal. A photon of energy E_γ can extract an electron from the sea, provided $E_\gamma \geq W$, with any remaining energy transferred to the electron as kinetic energy. Such experiments provide evidence for the photon concept and the quantization of the electromagnetic field.

A completely different form of electron emission that relies instead on a purely classical electric field, but which makes use of quantum tunneling, is *field emission*. In this case, shown in Fig. 12.12(b), an external electric field $\mathcal{E}$ is applied to the sample; electrons at the top of the sea can now tunnel through the triangular-shaped potential barrier.

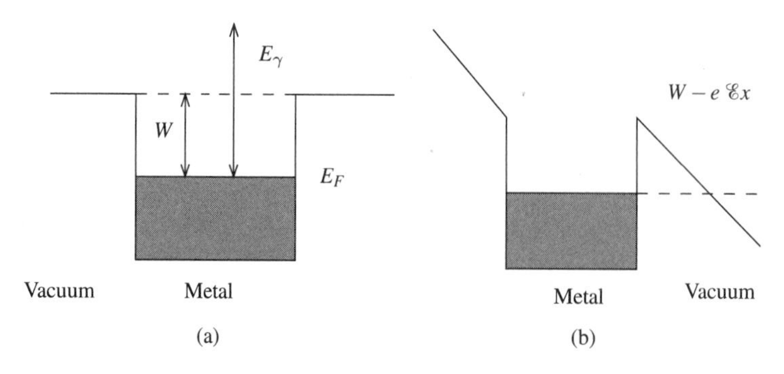

FIGURE 12.12. (a) Allowed electron states for a metal showing the filled Fermi sea and the photoelectric effect. (b) An external electric field is applied illustrating the triangular barrier giving rise to field emission.

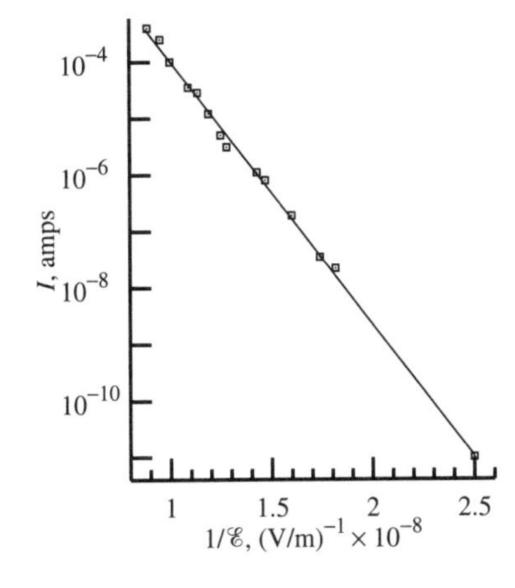

FIGURE 12.13. Semilog plot of tunneling current, I, versus 1/ℰ, where ℰ is the applied electric field. The data are taken from Millikan and Eyring (1926).

In this simple approximation, the probability of tunneling corresponding to Eqn. (12.52) is (P12.8)

$$P_T = \exp\left(-\frac{4}{3}\sqrt{\frac{2mW^3}{\hbar^2}}\frac{1}{e\mathcal{E}}\right) = \exp\left(-\frac{\mathcal{E}_0}{\mathcal{E}}\right) \tag{12.53}$$

This expression shows the strong dependence on the local value of the work function W at the surface. The resulting electron current due to quantum tunneling should be directly proportional to this probability, namely,

$$I = I_0 e^{-\mathcal{E}_0/\mathcal{E}} \quad \text{or} \quad \log(I) = \log(I_0) - \frac{\mathcal{E}_0}{\mathcal{E}} \tag{12.54}$$

We compare this prediction with some of the data from one of the original experiments[2] in Fig. 12.13.

This effect is also used as the basis of an imaging device called the *field ion microscope*[3] (*FIM*), which was the first microscope to achieve atomic resolutions enabling one to "see" individual atoms. The device works roughly as follows:

- A sharp, metallic tip with radius of curvature in the range 100 Å to 200 Å is placed in a vacuum and charged to a large voltage, typically 1–20 keV. This process itself helps to smooth the surface by selective field ionization of the metal atoms to what can be called *atomic smoothness*.
- A very dilute gas of noble gas atoms (often helium) is introduced; this is used as the imaging gas. These are adsorbed onto the surface of the probe due to dipole-dipole attractions to the tip atoms (remember that both atomic species are initially neutral).
- The image gas atoms, once attached to a tip atom, can be ionized via field emission, losing an electron to the tip. The resulting positively charged *ions* are then accelerated by the electric field toward a phosphorescent screen some tens of centimeters away forming the image. A schematic representation of the process is shown in Fig. 12.14; an FIM image is shown in Fig. 12.15.
- In this device, the electron tunneling serves only to initiate ion formation, and the electrons themselves do not participate in the image formation. In the original *field*

[2]We replot the data of Millikan and Eyring (1926) in the way suggested by Eqn. (12.54).
[3]For many technical details, see the excellent books by Tsong (1990) and Müller and Tsong (1969).

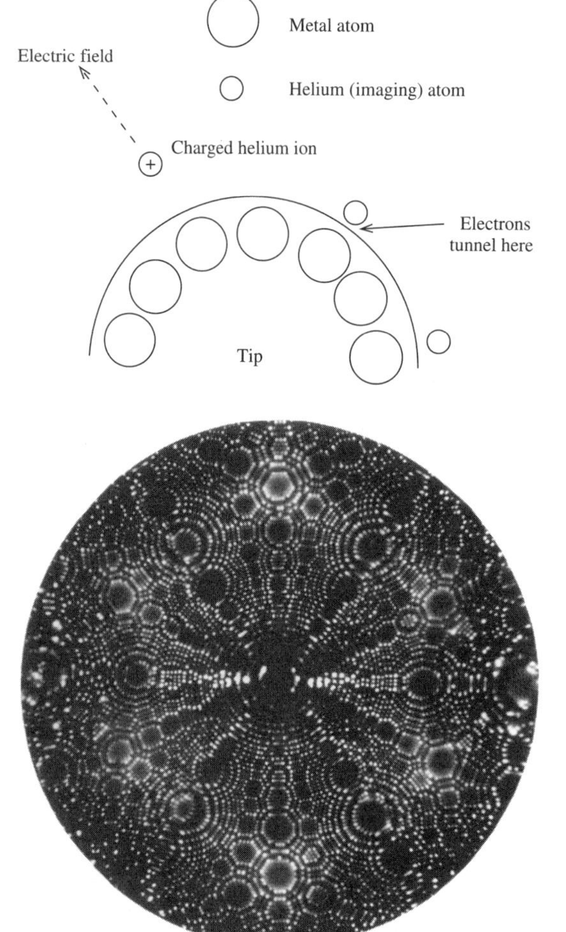

FIGURE 12.14. Schematic representation of field ion microscope. The metal atoms (large circles) forming the surface of the smooth probe tip attract (via dipole-dipole forces) the atoms of the imaging gas (small circles are helium). Once bound, electrons from the He can tunnel into the tip via field emission. The resulting charged ions travel along the electric field lines (the tip is held at a large electric potential) and form an image on a screen.

FIGURE 12.15. Field ion microscope image of a tungsten tip of radius ~ 400 Å. The original image had a magnification of 3 million. (Photo courtesy of T. Tsong.)

emission microscope, the electrons emitted via field ionization were used to image the surface. This process relied on the local variations of the work function W on the surface, which give rise to large variations in the tunneling probability, and in turn the electron current. The large lateral velocity spread of the emitted electrons, as well as de Broglie wave diffraction effects, limited this technique to resolutions of the order 20 to 25 Å.

12.5.2 Scanning Tunneling Microscopy (STM)

A newer technique that has had great success in obtaining images of atomic structures on (typically graphite or silicon) surfaces is *scanning tunneling microscopy.*[4] A schematic representation of the physics involved is shown in Fig. 12.16:

1. Two metal electrodes are placed close together (often only Angstroms apart), one being the sample, while the other is the tip.

[4] See the recent books by Stroscio and Kaiser (1993) and Chen (1993) for many details. The images in Fig. 1.3 were obtained using this technique.

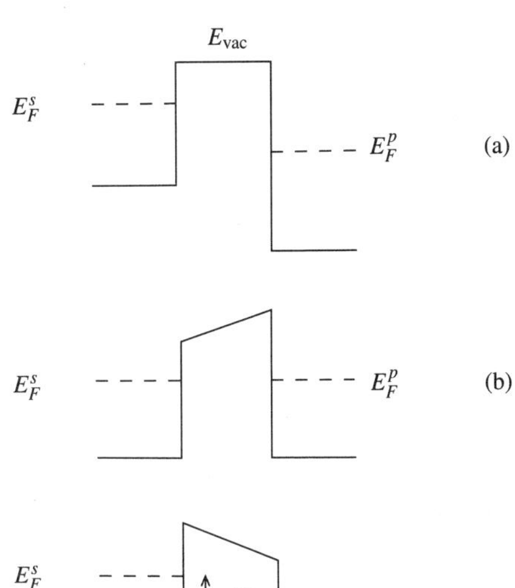

FIGURE 12.16. Schematic representation of the energy levels relevant for a scanning tunneling microscope (STM).

2. Their Fermi surfaces differ, and electrical equilibrium is reached only when enough electrons have tunneled through the junction (from left to right in this case). The resulting charge separation results in an electric field in the vacuum region between the electrodes.
3. An external voltage difference is applied to the tip shifting the Fermi energies again and allowing electron tunneling to occur.

As the tip is scanned over a plane surface, feedback circuits monitor the tunneling current, adjusting the tip height to maintain it at a constant value. The resulting height profile provides a map of the surface.

An estimate of the lateral resolution possible with this instrument can be made by assuming a simple shape for the STM tip probe, as in Fig. 12.17. Assuming a parabolic probe with radius of curvature R of say, 1000 Å, one finds a current profile as a function of distance away from the closest point d given by

$$I(x) \propto e^{-2(d+x^2/2R)\kappa} \propto e^{-x^2\kappa/R} \propto e^{-x^2/\rho^2} \tag{12.55}$$

where $\rho \equiv \sqrt{R/\kappa}$; typically $\kappa \approx \sqrt{2mW/\hbar^2} \approx 1\,\text{Å}^{-1}$. This familiar Gaussian distribution has a spread in lateral position given by $\Delta x = \rho/\sqrt{2} \approx 20\,\text{Å}$. As mentioned above, even smaller tip sizes are possible, but because of the exponential sensitivity of the tunneling current to the tip-to-surface distance d, it can well be the case that the best images arise from the tunneling from a very few surface atoms forming an atomic-scale "dimple" closest to the surface.

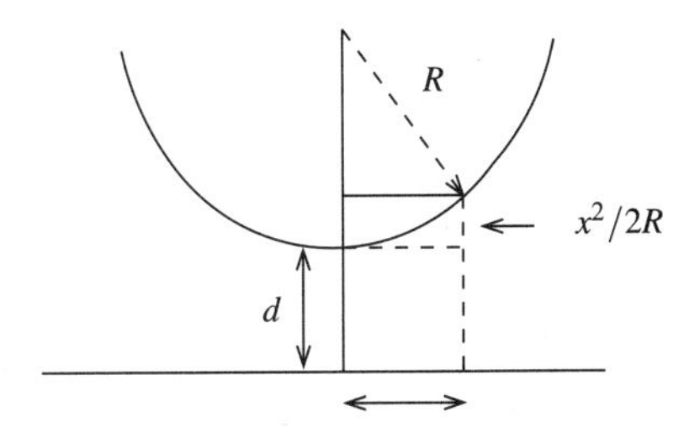

FIGURE 12.17. *Typical geometry of the tip of an STM probe.*

12.5.3 Alpha Particle Decay of Nuclei

One of the most famous early (semiquantitative) successes of quantum tunneling theory in nuclear physics was the understanding of the process of α-particle tunneling. In this process, a heavy nucleus decays to a lighter one by the emission of an α particle, i.e., the nucleus He^4. Using a compact notation, the process can be written as

$$ {}^{Z}_{N}X^{A} \longrightarrow {}^{(Z-2)}_{(N-2)}Y^{(A-4)} + {}^{2}_{2}\text{He}^{4} \tag{12.56}$$

where Z, N, A are, respectively, the numbers of protons, neutrons, and total nucleons in the nuclear species denoted by X (the "parent") or Y (historically the "daughter" nucleus).

Because this is a two-body decay, the energy of the emitted α is determined uniquely from conservation of energy and momentum, and can be calculated from a knowledge of the masses of the parent and daughter nuclear species. For the nuclei for which α decay is an important decay mechanism, the range in numerical values for the appropriate dimensionful parameters in the problem is not very large:

$$R \sim 2-4F \qquad E_\alpha \sim 2-8\,\text{MeV} \qquad Z \sim 50-100 \tag{12.57}$$

while the observed lifetimes have been measured over an incredibly large range

$$\tau \sim 10^{17}\,\text{sec}–10^{-12}\,\text{sec}$$

A simple model for this process assumes that the α particle moves in the potential of the *daughter* nucleus, modeled by a combination of an attractive square well (as in Sec. 9.4.2), along with the mutual Coulomb repulsion. This can be written as

$$V(r) = \begin{cases} -V_0 & \text{for } r < R \\ Z_1 Z_2 K e^2/r & \text{for } r > R \end{cases} \tag{12.58}$$

We would then take $Z_1 = Z_\alpha = 2$ and $Z_2 = Z - 2$ where Z is the charge of the *parent* nucleus. This potential is illustrated in Fig. 12.18, and the α particle is assumed to have positive energy E_α equal to its observed final kinetic energy. The model pictures the α particle as "rattling around" inside the nucleus with a small quantum tunneling probability of escaping each time it "hits" the Coulomb barrier. The tunneling probability for this process is then given from Eqn. (12.52) by

$$P_T = \exp\left[-2\sqrt{\frac{2\mu}{\hbar^2}}\int_a^b dr\sqrt{\frac{Z_1 Z_2 K e^2}{r} - E}\right] = e^{-2G} \tag{12.59}$$

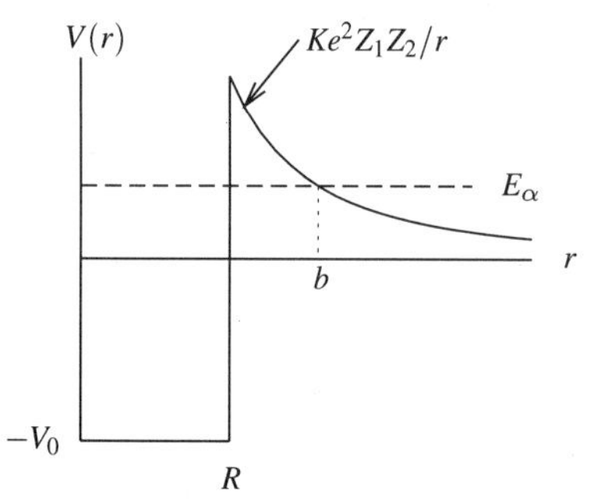

FIGURE 12.18. *Simple model for the effective potential seen by an α particle undergoing tunneling.*

where the factor in the exponential (G) is known as the *Gamow factor.* The classical turning points are taken to be

$$a = R \qquad \text{and} \qquad b = \frac{Z_1 Z_2 K e^2}{E_\alpha} \tag{12.60}$$

and we have used the reduced mass μ as is appropriate for a two-body problem (see Sec. 15.3). Since the daughter nucleus is much heavier than the α particle, one has $\mu \approx m_\alpha$. The Gamow factor can be written in the form

$$G = Z_1 Z_2 \alpha \sqrt{\frac{2\mu c^2}{E_\alpha}} \int_{w^2}^{1} d\eta \sqrt{\frac{1}{\eta} - 1} \tag{12.61}$$

where $\alpha = Ke^2/\hbar c$ as always and $w^2 \equiv R/b$. The integral can be done in closed form giving

$$\int_{w^2}^{1} d\eta \sqrt{\frac{1}{\eta} - 1} = \frac{\pi}{2} - \sin^{-1}(w) - \sqrt{w^2(1-w^2)}$$

$$\downarrow$$

$$\frac{\pi}{2} \quad \text{for} \quad w = \sqrt{b/R} \ll 1 \tag{12.62}$$

One then has very roughly that

$$2G \approx 4\frac{(Z-2)}{\sqrt{E_\alpha(\text{MeV})}} \tag{12.63}$$

The decay lifetime may be estimated by noting that there is roughly an e^{-2G} probability of a tunneling "escape" every time the α "hits" the electrostatic barrier. The time between such "escape attempts" can be estimated as

$$T_0 \approx \frac{2R}{v_\alpha} \qquad \text{where} \qquad v_\alpha \approx \sqrt{\frac{2E_\alpha}{m_\alpha}} \approx \frac{1}{20}c \tag{12.64}$$

and $R \approx 5-8$ F is a typical (heavy) nuclear radius. This gives $T_0 \approx 10^{-21}$ sec, which is a typical nuclear reaction time. The lifetimes in this simple picture then scale as

$$\tau = T_0 e^{+2G} \qquad \text{or} \qquad \log(\tau) \approx \log(T_0) + 4\frac{(Z-2)}{\sqrt{E_\alpha(\text{MeV})}} \tag{12.65}$$

This behavior is most easily studied by examining the α decay lifetimes of different isotopes of the same element (so that the value of Z is fixed and only E_α varies). We plot the lifetimes for several such series (on a log scale) versus $1/\sqrt{E_\alpha}$ in Fig. 12.19 (a so-called *Geiger-Nuttall* plot) and note the reasonable straight-line fits. The simple approximations made here can be refined,[5] but they provide convincing evidence for the importance of quantum tunneling effects in nuclear decay processes.

[5]See, e.g., Park (1992) for a more extensive discussion.

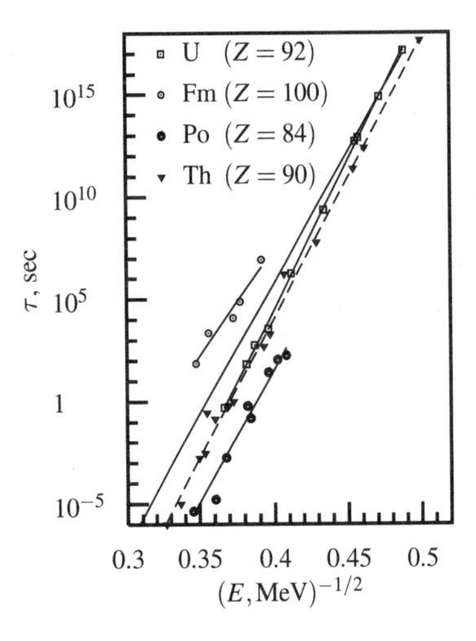

FIGURE 12.19. Semilog plot of α decay lifetime [τ (sec)] versus $1/\sqrt{E_\alpha}$ (MeV) *for four different isotope series, the so-called Geiger-Nuttall plot. The data are taken from a recent edition of the Chart of the Nuclides [Walker (1983)].*

12.5.4 Nuclear Fusion Reactions

For α decay of heavy nuclei, the repulsive Coulomb potential forms a barrier through which α particles must tunnel to get out of the nucleus. For *fusion reactions*, we must consider the inverse process in which light nuclei must penetrate their mutual Coulomb barrier in order to get close enough to participate in strong nuclear or weak interactions, which are short-ranged forces.

For example, in the production of thermonuclear energy in stars, protons (^{1}H) are eventually converted to helium nuclei via the so-called *pp cycle*. The overall reaction can be written as

$$\text{pp cycle:} \qquad 4\,^1\text{H} \to {}^4\text{He} + 2e^+ + 2\nu_e \qquad (Q = 26.7\,\text{MeV}) \tag{12.66}$$

where Q is the energy released per reaction. The net reaction is the result of a series of two-body interactions given by

$$\begin{aligned} {}^1\text{H} + {}^1\text{H} &\to {}^2\text{H} + e^+ + \nu && (Q = 1.44\,\text{MeV}) \\ {}^2\text{H} + {}^1\text{H} &\to {}^3\text{He} + \gamma && (Q = 5.49\,\text{MeV}) \\ {}^3\text{He} + {}^3\text{He} &\to {}^4\text{He} + 2\,^1\text{H} + \gamma && (Q = 12.86\,\text{MeV}) \end{aligned} \tag{12.67}$$

The reaction rate for such two-body processes is determined not only by the probability for the quantum mechanical tunneling event (given by the interaction cross section) but also by the available number of initial particle pairs, and each of these factors has a very different dependence on the center-of-mass energy in the collision. The cross section must include the Gamow factor, e^{-2G}, while the number density of particles in the hot gas is proportional to the Boltzmann factor, e^{-E/k_BT}. This means that the interaction rate is proportional to

$$\text{Reaction rate} \propto e^{-2G-E/k_BT} = e^{-f(E)} \tag{12.68}$$

where

$$f(E) = Z_1 Z_2 K e^2 \pi \sqrt{\frac{2\mu c^2}{E}} + \frac{E}{k_B T} \tag{12.69}$$

and the *maximum* event rate corresponds to the *minimum* value of $f(E)$. The value of energy at this point is given (P12.10) by

$$\left(\frac{E_{\max}}{k_B T}\right)^{3/2} = Z_1 Z_2 \pi \alpha \sqrt{\frac{\mu c^2}{2 k_B T}} \tag{12.70}$$

This implies that if one wishes to study the dynamics of the nuclear reactions responsible for stellar fusion at a given temperature, one should do nuclear scattering experiments in terrestrial accelerators at energies very near $E_{\max}$ as given by Eqn. (12.70). The interaction rate for values not too different from $E_{\max}$ will be hugely suppressed due to the exponential sensitivity in Eqn. (12.68).

12.6 QUESTIONS AND PROBLEMS

Q12.1 If you wanted to construct a wavepacket for the attractive finite well, what would be the complete set of states you would have to use?

Q12.2 Consider a wavepacket of mean energy $\overline{E}$ incident on a step potential of height $V_0 > 0$. What would $\langle x \rangle_t$, Δx_t, $\langle \hat{p} \rangle_t$, and Δp_t all look like as functions of time for the cases (1) $\overline{E} \gg V_0$, (2) $\overline{E} \gtrsim V_0$, and (3) $V_0 \gg \overline{E} > 0$?

Q12.3 To what temperature would you have to heat a typical metal to remove electrons by thermal excitations? Why is field emission sometimes referred to as *cold emission*?

Q12.4 Why are neutrons most often used in initiating fission processes in heavy nuclei? Would protons work as well? How about antiprotons?

P12.1 Show that the probability flux corresponding to the wavefunction

$$\psi(x,t) = I e^{i(kx - \hbar k^2 t/2m)} + R e^{i(-kx - \hbar k^2 t/2m)}$$

is given by

$$j(x,t) = \frac{\hbar k}{m}|I|^2 - \frac{\hbar k}{m}|R|^2$$

P12.2 Consider plane waves of amplitude I incident on a single attractive δ function potential, $V(x) = -g\delta(x)$.

(a) Match boundary conditions to calculate R and T in terms of I and the other parameters of the problem.

(b) Calculate the ratios of the reflected and transmitted flux to the incident flux, and check conservation of flux.

(c) Express the reflected and transmitted fluxes in terms of E, and sketch them versus E.

(d) For what value of E will 99% of the incident flux be reflected? 1%?

(e) What changes if we choose a repulsive potential instead?

P12.3 **Absorptive scattering:** Consider a potential similar to that of Prob. 12.2 except that the scattering center now has a negative imaginary part, i.e.,

$$V(x) = (g - ig_0)\delta(x).$$

(a) Repeat the analysis above, and show that flux is *not* conserved in this case. This result is expected from the discussion of Sec. 4.3, where we noted that this type of potential corresponds to absorption.

(b) Calculate the rate of loss of particles "into" the absorption potential and discuss its energy dependence. Also discuss the reflected and transmitted fluxes for a *purely* absorptive potential, i.e., in the limit when $g = 0$ but $g_0 > 0$. What would $g_0 < 0$ mean?

P12.4 Consider scattering from a twin repulsive δ function potential defined by $V(x) = g(\delta(x-a) + \delta(x+a))$.

(a) Calculate R and T.

(b) Show that flux is conserved.

(c) Show that this potential exhibits transmission resonances in the same way as the square barrier, with the same physical condition being required.

[For a discussion of this problem, see Lapidus (1982c).]

P12.5 Derive the expressions for R and T in Eqns. (12.33) and (12.34). Also solve for the E and F coefficients. Show that flux is conserved in this reaction.

P12.6 The attractive δ function potential, $-g\delta(x)$, can be thought of as the limiting case of a deep and narrow attractive well where the width and depth satisfy $2aV_0 = g$ as $a \to 0$ and $V_0 \to \infty$. Show that your results for $|T/I|^2$ in the last problem in this limit gives the same result as in P12.2.

P12.7 **(a)** Show that the smooth potential function defined via

$$V(x) = V_0\left(\frac{1}{1 + e^{-x/L}}\right)$$

approaches the discontinuous step potential for $L \to 0$.

(b) One can show[6] that the ratio of reflected to incident flux is given by

$$\frac{|R|^2}{|I|^2} = \left(\frac{\sinh(\pi(k-q)L)}{\sinh(\pi(k+q)L)}\right)^2$$

Show that the $L \to 0$ limit reduces to the result for the step potential.

(c) Discuss the $L \to \infty$, $\hbar \to 0$, and $m \to \infty$ limits as in Sec. 12.3.

P12.8 Derive Eqn. (12.53).

P12.9 **Geiger-Nuttall plot for thorium:** The lifetimes (in various units) and observed α particle energies (in MeV) for decays of various isotopes of thorium ($Z = 90$) are given below:

Mass number	τ	E_α**(MeV)**
232	1.4×10^{10} yr	4.01
230	7.7×10^{4} yr	4.69
229	7.34×10^{3} yr	4.85
228	1.91 yr	5.42
227	18.7 day	6.04
226	31 min	6.34
225	8 min	6.48
224	1.04 s	7.17
223	0.66 s	7.29
222	2.9 ms	7.98
221	1.68 ms	8.15
220	10 μs	8.79
219	1.05 μs	9.34
218	0.11 μs	9.67

Plot $\log(\tau)$ versus $1/\sqrt{E_\alpha}$, and try to "fit" a straight line through the data points (either by eye or via some fitting routine if you know how). Also plot the theoretical expression in Eqn. (12.65) on your graph, and compare the "slopes" and "intercepts" of your lines. This exercise gives some feel for the reliability of this simplest estimate of tunneling effects.

P12.10 Evaluate the optimal energy, $E_{\max}$, for stellar fusion for the two situations

$$\begin{array}{lllll} \text{H–H reactions} & Z_1 = Z_2 = 1 & \mu \approx m_p/2 & T = 10^7\,\text{K} \\ \text{C–C reactions} & Z_1 = Z_2 = 6 & \mu \approx 6m_p & T = 10^9\,\text{K} \end{array}$$

What is the ratio of interaction rates as one varies $E = rE_{\max}$ over the range $r = (0.5, 2)$? Over what range of energies around $E_{\max}$ does the interaction rate drop to half its peak value?

[6]See Landau and Lifschitz (1965).

CHAPTER 13

More Formal Topics

In this chapter, we investigate in more depth many of the formal properties of quantum mechanics introduced in Chaps. 5 and 7. We extend the discussions of Hermitian operators (Sec. 13.1), the vector structure of quantum wavefunctions (13.2), and commutators (13.3). We use these techniques to prove important new results involving uncertainty principles (13.4) and the time development of quantum states and its connection to conservation laws (13.5).

13.1 HERMITIAN OPERATORS

We have stressed that for a quantum operator, $\hat{O}$, to correspond to a classical observable quantity, O, it must be Hermitian. We have defined this by insisting that its expectation value in any quantum state be real, i.e.,

$$\langle\hat{O}\rangle^* = \left[\int \psi^* \hat{O}\, \psi\right]^* = \int \left(\hat{O}\psi\right)^* \psi \stackrel{?}{=} \int \psi^* \hat{O}\, \psi = \langle\hat{O}\rangle \tag{13.1}$$

where any proof consists in verifying the equality in question. We also have, even more compactly, in bracket notation

$$\langle\psi|\hat{O}|\psi\rangle^* = \langle\psi|\hat{O}\psi\rangle^* = \langle\hat{O}\psi|\psi\rangle \stackrel{?}{=} \langle\psi|\hat{O}|\psi\rangle \tag{13.2}$$

and we will often use whichever notation is more convenient. From Sec. 7.1, we know that Hermitian operators will thus satisfy the more general requirement

$$\langle\chi|\hat{O}|\psi\rangle^* = \langle\hat{O}\chi|\psi\rangle = \langle\psi|\hat{O}|\chi\rangle \tag{13.3}$$

It will turn out to be extremely useful to generalize the notion of complex conjugation used here for average values to the operators themselves. To this end, and motivated by Eqn. (13.1), we define the *Hermitian conjugate* of the operator $\hat{O}$, denoted by $\hat{O}^\dagger$, by the relation

$$\int \left(\hat{O}\psi\right)^* \psi \equiv \int \psi^* \hat{O}^\dagger\, \psi \tag{13.4}$$

or equivalently

$$\langle \hat{O}\psi|\psi\rangle \equiv \langle\psi|\hat{O}^\dagger|\psi\rangle \tag{13.5}$$

Using our established definition of Hermitian-ness, we see that:

- An operator is Hermitian provided
$$\langle\psi|\hat{O}|\psi\rangle^* = \langle\psi|\hat{O}\psi\rangle^* = \langle\hat{O}\psi|\psi\rangle \equiv \langle\psi|\hat{O}^\dagger|\psi\rangle = \langle\psi|\hat{O}|\psi\rangle \tag{13.6}$$
for all ψ or, more simply, if
$$\hat{O}^\dagger = \hat{O} \tag{13.7}$$
This is a test performed directly on the operator.

The analogous condition for a complex number, c, to be real is simply

$$c = c^* \tag{13.8}$$

and it is easy to show for any complex constant

$$\int \psi^* c^\dagger \psi \equiv \int (c\psi)^* \psi = \int \psi^* c^* \psi \qquad \text{so that} \qquad c^\dagger = c^* \tag{13.9}$$

This is then our first example of Hermitian conjugation and shows that this operation reduces to complex conjugation when applied to complex numbers. It is similarly easy to show that $x^\dagger = x$ in this language so that x is Hermitian, as is any real function of x.

A less trivial example is the calculation of $(d/dx)^\dagger$, which gives

$$\begin{aligned}\int_{-\infty}^{+\infty} \psi^* \left(\frac{d}{dx}\right)^\dagger \psi &\equiv \int_{-\infty}^{+\infty} \left(\frac{d}{dx}\psi\right)^* \psi \\ &\overset{\text{IBP}}{=} -\int_{-\infty}^{+\infty} dx\, \psi^* \frac{d\psi}{dx} \\ &= \int_{-\infty}^{+\infty} \psi^* \left(-\frac{d}{dx}\right)\psi\end{aligned} \tag{13.10}$$

whence

$$\left(\frac{d}{dx}\right)^\dagger = -\frac{d}{dx} \tag{13.11}$$

Combining these results, we presumably then have (see below)

$$\hat{p}^\dagger = \left(\frac{\hbar}{i}\frac{d}{dx}\right)^\dagger = \left(\frac{\hbar}{-i}\right)\left(-\frac{d}{dx}\right) = \frac{\hbar}{i}\frac{d}{dx} = \hat{p} \tag{13.12}$$

and $\hat{p}$ is Hermitian. This example demonstrates some of the utility of a definition of Hermitian-ness involving operators themselves. If one establishes, via direct calculation, a small class of known Hermitian operators, the Hermitian conjugate operation and its properties can be used to test new combinations of operators in a straightforward way.

Some of the most useful properties (whose proofs will be left to the problems) are:

- The Hermitian conjugate satisfies the more general relation
$$\langle\hat{A}\chi|\psi\rangle = \langle\chi|\hat{A}^\dagger|\psi\rangle \tag{13.13}$$
for all states ψ, χ.

- This can be used to show that

$$(\hat{A}^\dagger)^\dagger = \hat{A} \tag{13.14}$$

in much the same way that $(c^*)^* = c$ for complex conjugation.

- The Hermitian conjugate reverses the order of a product of operators, i.e.,

$$(\hat{A}\hat{B})^\dagger = \hat{B}^\dagger \hat{A}^\dagger \tag{13.15}$$

This can be shown by considering

$$\begin{aligned}
\langle\psi|(\hat{A}\hat{B})^\dagger|\psi\rangle &\equiv \langle(\hat{A}\hat{B})\psi|\psi\rangle \\
&= \langle\hat{A}(\hat{B}\psi)|\psi\rangle \\
&= \langle\hat{B}\psi|\hat{A}^\dagger|\psi\rangle \\
&= \langle\psi|\hat{B}^\dagger|\hat{A}^\dagger\psi\rangle \\
&= \langle\psi|\hat{B}^\dagger\hat{A}^\dagger|\psi\rangle
\end{aligned} \tag{13.16}$$

- This implies that:

  If both $\hat{A}$ and $\hat{B}$ are separately Hermitian, their product is Hermitian provided that $\hat{A}$ and $\hat{B}$ also commute, i.e., $[\hat{A}, \hat{B}] = 0$

  This helps justify the derivation in Eqn. (13.12).

- The combinations

$$\hat{A}\hat{A}^\dagger \qquad (\hat{A} + \hat{A}^\dagger)/2 \qquad -i(\hat{A} - \hat{A}^\dagger)/2 \tag{13.17}$$

are all Hermitian, even if $\hat{A}$ itself is not; these correspond, respectively, to

$$cc^* = |c|^2 \qquad \mathrm{Re}(c) \qquad \mathrm{Im}(c) \tag{13.18}$$

which are the modulus squared, and the real and imaginary parts of an ordinary complex number, c.

- The combination

$$i[\hat{A}, \hat{B}] \tag{13.19}$$

is Hermitian provided both $\hat{A}$ and $\hat{B}$ are also. This will prove useful in our derivations of generalized uncertainty principles.

- For real functions of operators, we have

$$f(\hat{O})^\dagger = f(\hat{O}^\dagger) \tag{13.20}$$

which can be seen by using a series expansion for $f(x)$.

Finally, we can make a connection with the matrix representation of operators in Sec. 11.5. If we define the matrix corresponding to some operator using a complete set of states as

$$\mathrm{O}_{nm} \equiv \langle u_n|\hat{O}|u_m\rangle \tag{13.21}$$

then the definition of Hermitian conjugate gives

$$\langle u_n|\hat{O}^\dagger|u_m\rangle \equiv \langle\hat{O}u_n|u_m\rangle^* = \langle u_m|\hat{O}|u_n\rangle^* \tag{13.22}$$

This can be written in the form

$$\left(\mathrm{O}^\dagger\right)_{nm} = \left(\mathrm{O}^*\right)_{mn} = \left(\mathrm{O}^{T*}\right)_{nm} \tag{13.23}$$

where O^T defines the *transpose* of the matrix, i.e., the matrix resulting from interchanging rows and columns (see App. D.1); thus

- The *Hermitian conjugate of a matrix* is obtained by taking its transpose and the complex conjugate of all its elements, i.e., $O^\dagger = O^{T*}$. A matrix is then described as Hermitian if $O^\dagger = O$.

13.2 QUANTUM MECHANICS, LINEAR ALGEBRA, AND VECTOR SPACES

We have made occasional reference to the algebraic structure of solutions of the Schrödinger equation, especially in position space, noting similarities to the vectors of a linear vector space. In this section, we wish to extend and amplify this identification by collecting more examples of the apparent similarities. We know that the different representations of the Schrödinger wavefunction

$$\begin{aligned}\psi(x) \longleftrightarrow \phi(p) &= \frac{1}{\sqrt{2\pi\hbar}}\int_{-\infty}^{+\infty} dx\, e^{-ipx/\hbar}\,\psi \\ \longleftrightarrow \mathbf{a} &= \left\{a_n = \int_{-\infty}^{+\infty} (u_n(x))^*\psi(x)\,dx;\; n = 1, 2\ldots\right\}\end{aligned} \tag{13.24}$$

all have the same information content and satisfy the same normalization condition

$$\int_{-\infty}^{+\infty} |\psi(x)|^2\,dx = \int_{-\infty}^{+\infty} |\phi(p)|^2\,dp = \sum_{n=1}^{\infty} |a_n|^2 = \mathbf{a}^*\cdot\mathbf{a} \tag{13.25}$$

This connection of $\psi(x)$ and $\phi(p)$ with the discrete, but infinite dimensional, complex vector $\mathbf{a}$ motivates us to consider all three as simply different representations of the same *quantum state vector.* Extending the Dirac bracket notation, we often denote

$$\begin{aligned}&\text{“ket” vector} \longleftrightarrow |\psi_a\rangle \longleftrightarrow \psi_a(x), \phi_a(p), \mathbf{a} \\ &\text{“bra” vector} \longleftrightarrow \langle\psi_b| \longleftrightarrow \psi_b^*(x), \phi_b^*(p), \mathbf{b}^*\end{aligned} \tag{13.26}$$

The various overlap integrals (and sums) can now be described as an *inner product* or *generalized dot product* of two such vectors; their common value is then given by

$$\int_{-\infty}^{+\infty} \psi_b^*(x)\psi_a(x)\,dx = \int_{-\infty}^{+\infty} \phi_b^*(p)\,\phi_a(p)\,dp = \mathbf{b}^*\cdot\mathbf{a} = \langle\psi_b|\psi_a\rangle \tag{13.27}$$

where a bra vector “dotted into” a ket vector forms the familiar Dirac bra · ket or bracket.

We can generalize the notion of a *linear vector space* to include the following features:

- In position space, for example, the vectors are associated with square integrable functions, $\psi(x)$, where the continuous label x generalizes the discrete label $\mathbf{a} = \{a_n; n = 1, 2, 3\ldots\}$. Such an infinite dimensional vector space, conventionally denoted as $\mathcal{H}$, is often called a *Hilbert space*.
- We assume that $\alpha\psi_a(x) + \beta\psi_b(x) \in \mathcal{H}$ for complex constants α, β, provided that ψ_a, ψ_b are also; the space $\mathcal{H}$ is therefore closed.
- The eigenfunctions of a Hermitian operator, which we know form a complete set, can be thought of as a set of orthonormal *unit vectors* since $\langle u_n|u_m\rangle = \delta_{nm}$.

- We will concentrate on *linear operators* on this space, namely, ones that satisfy

$$\hat{O}\left(\alpha\phi_a + \beta\psi_b\right) = \alpha\left(\hat{O}\phi_a\right) + \beta\left(\hat{O}\psi_b\right) \tag{13.28}$$

A very important set of operators are those that preserve the inner product among vectors; these can be thought of as generalized rotations that keep the "length" of a function fixed. We will denote them by $\hat{U}$, with the notation $|\psi'\rangle = \hat{U}|\psi\rangle$. If the dot product is to remain unchanged under such transformations, we must have

$$\langle\psi_a|\psi_b\rangle = \langle\psi'_a|\psi'_b\rangle = \langle\hat{U}\psi_a|\hat{U}\psi_a\rangle \equiv \langle\psi_b|\hat{U}^\dagger\hat{U}|\psi_a\rangle \tag{13.29}$$

or

$$\hat{U}^\dagger\hat{U} = \hat{1} \tag{13.30}$$

where $\hat{1}$ is the unit operator and we have used the definition of Hermitian conjugate. Operators that satisfy Eqn. (13.30) are called *unitary*. Just as ordinary rotations preserve the length of standard vectors, unitary transformations preserve the norm of a vector in Hilbert space ($\sqrt{\langle\psi|\psi\rangle}$) and hence conserve probability.

In a matrix notation (P13.5), where the operator is defined via

$$\mathbf{a}' = \mathbf{U}\mathbf{a} \qquad \text{or} \qquad a'_i = \sum_j \mathbf{U}_{ij}a_j \tag{13.31}$$

we have the equivalent statement

$$\mathbf{U}^{T*}\mathbf{U} = \mathbf{U}^\dagger\mathbf{U} = 1 \qquad \text{or} \qquad \left(\mathbf{U}^{T*}\mathbf{U}\right)_{ij} = \sum_k \left(\mathbf{U}^{T*}\right)_{ik}\mathbf{U}_{kj} = \delta_{ij} \tag{13.32}$$

Example 13.1 A simple example involving the rotation matrix for two-dimensional real vectors is a useful reminder. In this case we have

$$\mathbf{x}' = \mathbf{R}x \qquad \text{or} \qquad \begin{pmatrix} x' \\ y' \end{pmatrix} = \begin{pmatrix} \cos(\theta) & \sin(\theta) \\ -\sin(\theta) & \cos(\theta) \end{pmatrix}\begin{pmatrix} x \\ y \end{pmatrix} \tag{13.33}$$

which rotates the vector **x** through θ radians. It is then easy to check that

$$\mathbf{R}^\dagger\mathbf{R} = \begin{pmatrix} \cos(\theta) & -\sin(\theta) \\ \sin(\theta) & \cos(\theta) \end{pmatrix}\begin{pmatrix} \cos(\theta) & \sin(\theta) \\ -\sin(\theta) & \cos(\theta) \end{pmatrix} = \begin{pmatrix} 1 & 0 \\ 0 & 1 \end{pmatrix} = 1 \tag{13.34}$$

In this case $\mathbf{R}^\dagger = \mathbf{R}^T$ since **R** is real.

Example 13.2 One example of such a norm-preserving or unitary transformation of more relevance to quantum mechanics is the Fourier transform

$$\phi(p) = \frac{1}{\sqrt{2\pi\hbar}}\int_{-\infty}^{+\infty} dx\, \psi(x)e^{-ipx/\hbar} \tag{13.35}$$

which we already know satisfies $\langle\phi|\phi\rangle = \langle\psi|\psi\rangle$. If we write the Fourier transform in the form

$$\begin{aligned} \phi(p) &= \int_{-\infty}^{+\infty} dx \left(\frac{e^{-ipx/\hbar}}{\sqrt{2\pi\hbar}}\right)\psi(x) \\ &= \int_{-\infty}^{+\infty} dx\, R_{px}\, \psi(x) \end{aligned} \tag{13.36}$$

we can note the immediate similarity to Eqn. (13.31), with the discrete labels and summations replaced by continuous variables and integrations. The "rotation matrix elements" are

$$R_{px} \equiv \frac{1}{\sqrt{2\pi\hbar}} e^{-ipx/\hbar} \tag{13.37}$$

and satisfy their own (continuous) version of Eqn. (13.32), namely,

$$\int_{-\infty}^{+\infty} dx\,(R_{p'x})^* R_{xp} = \frac{1}{2\pi\hbar}\int_{-\infty}^{+\infty} dx\, e^{i(p-p')x/\hbar} = \delta(p-p') \tag{13.38}$$

This identification with norm-preserving transformations makes it even clearer that $\psi(x)$ and $\phi(p)$ share the same information content. A very similar set of results for the expansion coefficients for a complete set of states can also be shown; they follow from the definition

$$\psi(x) = \sum_n a_n u_n(x) \equiv \sum_n U_{xn} a_n \tag{13.39}$$

where

$$a_n = \int dx\,(u_n(x))^*\,\psi(x) = \int dx\left(U^*_{nx}\right)\psi(x) \tag{13.40}$$

In this case, the norm-preserving condition of Eqn. (13.32) reads

$$\int_{-\infty}^{+\infty} dx\,(U_{nx})^* U_{xm} = \int_{-\infty}^{+\infty} dx\, u_n^*(x) u_m(x) = \delta_{nm} \tag{13.41}$$

which is familiar from the orthogonality of eigenfunctions belonging to different energy eigenvalues; similarly we find

$$\sum_n R^*_{x'n} R_{n,x} = \sum_n u_n^*(x') u_n(x) = \delta(x-x') \tag{13.42}$$

which was first inferred from our discussion of the propagator in Sec. 7.9.

The powerful analogy between the Hilbert space of quantum mechanics and ordinary vector spaces can be used to great benefit to prove new results that have an obvious geometrical analogy in the more familiar setting of ordinary dot products. An example is the so-called *Schwartz* or *triangle inequality* for vectors, which reads

$$\mathbf{A}^2\mathbf{B}^2 \geq \left(\mathbf{A}\cdot\mathbf{B}\right)^2 \tag{13.43}$$

which in two and three dimensions can be easily translated into the statement that

$$A^2B^2 \geq A^2B^2\cos^2(\theta) \qquad \text{or} \qquad 1 \geq \cos^2(\theta) \tag{13.44}$$

where θ is the angle between **A** and **B**; the equality is only achieved when **A**, **B** are parallel or antiparallel, i.e., $\cos(\theta) = \pm 1$.

A completely analogous result can be proved for complex wavefunctions, namely,

$$\left[\int_{-\infty}^{+\infty} dx\,|\psi(x)|^2\right]\left[\int_{-\infty}^{+\infty} dx\,|\chi(x)|^2\right] \geq \left|\int_{-\infty}^{+\infty} dx\,\psi^*(x)\chi(x)\right|^2 \tag{13.45}$$

or more generally in terms of quantum states,

$$\langle\psi|\psi\rangle\langle\chi|\chi\rangle \geq |\langle\psi|\chi\rangle|^2 \tag{13.46}$$

We will prove Eqn. (13.46) in a formal, abstract way by generalizing the standard argument for ordinary vectors. We begin by considering a general quantum state given by a linear combination of $|\psi\rangle$ and $|\chi\rangle$, whose ket vector is written as

$$|\zeta\rangle \equiv |\psi\rangle\langle\chi|\chi\rangle - |\chi\rangle\langle\chi|\psi\rangle \tag{13.47}$$

where we recall that the overlaps $\langle\chi|\chi\rangle$ and $\langle\chi|\psi\rangle$ are inner products of quantum states and hence simply (complex) numbers. The corresponding bra vector is then

$$\begin{aligned}\langle\zeta| &= \langle\chi|\chi\rangle\langle\psi| - \langle\chi|\psi\rangle^*\langle\chi| \\ &= \langle\chi|\chi\rangle\langle\psi| - \langle\psi|\chi\rangle\langle\chi|\end{aligned} \tag{13.48}$$

The overlap of any generalized vector with itself must necessarily be nonnegative so that $\langle\zeta|\zeta\rangle \geq 0$, with the lower bound of 0 only being saturated if $|\zeta\rangle$ itself vanishes. This then implies that

$$\begin{aligned}0 \leq \langle\zeta|\zeta\rangle &= \{\langle\chi|\chi\rangle\langle\psi| - \langle\psi|\chi\rangle\langle\chi|\}\{|\psi\rangle\langle\chi|\chi\rangle - |\chi\rangle\langle\chi|\psi\rangle\} \\ &= \langle\chi|\chi\rangle\langle\chi|\chi\rangle\langle\psi|\psi\rangle - \langle\psi|\chi\rangle\langle\chi|\psi\rangle\langle\chi|\chi\rangle \\ &\qquad - \langle\chi|\chi\rangle\langle\psi|\chi\rangle\langle\chi|\psi\rangle + \langle\chi|\chi\rangle\langle\psi|\chi\rangle\langle\chi|\psi\rangle \\ &= \langle\chi|\chi\rangle\langle\chi|\chi\rangle\langle\psi|\psi\rangle - \langle\psi|\chi\rangle\langle\chi|\psi\rangle\langle\chi|\chi\rangle\end{aligned} \tag{13.49}$$

or

$$\langle\psi|\psi\rangle\langle\chi|\chi\rangle \geq \langle\psi|\chi\rangle\langle\chi|\psi\rangle = \langle\psi|\chi\rangle\langle\psi|\chi\rangle^* = |\langle\psi|\chi\rangle|^2 \tag{13.50}$$

Once again, the equality is only achieved when the state $|\zeta\rangle$ vanishes identically, which implies that

$$|\psi\rangle = \left(\frac{\langle\chi|\psi\rangle}{\langle\chi|\chi\rangle}\right)|\chi\rangle \qquad \text{or} \qquad \psi(x) \propto \chi(x) \tag{13.51}$$

i.e., $|\psi\rangle$ and $|\chi\rangle$ are proportional to each other, or "parallel" in a generalized sense.

13.3 COMMUTATORS

Since it is obvious that the commutator of two operators plays such a fundamental role in the formalism of quantum mechanics, we find it useful to collect below some basic results on commutators which can make the evaluation of complicated combinations easier. We can easily show by direct manipulation (P13.7) that:

- The commutator of a nontrivial operator with any (complex) number vanishes trivially:

$$[c, \hat{A}] = 0 \quad \text{for any complex number } c \tag{13.52}$$

- The commutator of an operator with itself vanishes, i.e.,

$$[\hat{A}, \hat{A}] = \hat{A}\hat{A} - \hat{A}\hat{A} = 0 \tag{13.53}$$

  This statement is not as trivial as it may at first seem.
- The commutator "distributes" in that

$$[\hat{A} + \hat{B}, \hat{C}] = [\hat{A}, \hat{C}] + [\hat{A}, \hat{C}] \tag{13.54}$$

- The commutator is an antisymmetric function, namely,

$$[\hat{A}, \hat{B}] = -[\hat{B}, \hat{A}] \tag{13.55}$$

so that the ordering of the operators is important.

- More involved commutators can often be simplified by repeated use of the relation

$$[\hat{A}\hat{B}, \hat{C}] = \hat{A}[\hat{B}, \hat{C}] + [\hat{A}, \hat{C}]\hat{B} \tag{13.56}$$

or its equivalent

$$[\hat{A}, \hat{B}\hat{C}] = \hat{B}[\hat{A}, \hat{C}] + [\hat{A}, \hat{B}]\hat{C} \tag{13.57}$$

We note that the last five relations [Eqns. (13.53)–(13.57)] have a structure similar to the vector- or cross-product of two vectors (Q13.1).

- Commutators can be applied repeatedly, so that, for example,

$$[\hat{A}, [\hat{B}, \hat{C}]] = \hat{A}[\hat{B}, \hat{C}] - [\hat{B}, \hat{C}]\hat{A} \tag{13.58}$$

which is to be distinguished from Eqn. (13.57).

- A sometimes useful relation among double commutators is the so-called *Jacobi identity* that states

$$[\hat{A}, [\hat{B}, \hat{C}]] + [\hat{A}, [\hat{B}, \hat{C}]] + [\hat{A}, [\hat{B}, \hat{C}]] = 0 \tag{13.59}$$

As an example of the use of these relations, we can make use of the relation $[x, \hat{p}] = i\hbar$ to show that

$$[x, \hat{p}^2] = \hat{p}[x, \hat{p}] + [x, \hat{p}]\hat{p} = 2i\hbar\hat{p} \tag{13.60}$$

instead of evaluating it directly as in Sec. 5.7 or P5.22.

13.4 UNCERTAINTY PRINCIPLES

We have already seen several cases where it is possible to have simultaneous eigenfunctions of two different Hermitian operators, $\hat{A}$ and $\hat{B}$, for example, the energy and parity eigenfunctions for Hamiltonians with a symmetric potential.

In this section, we can combine several of our formal results to derive a general form for the minimum uncertainty principle product for any two observable quantities, A, B represented by quantum operators, $\hat{A}$, $\hat{B}$. To this end, we note that

- The spread or RMS deviation in the measurement of any observable is given by

$$(\Delta A)^2 = \langle\psi|(\hat{A} - \langle\hat{A}\rangle)^2|\psi\rangle \tag{13.61}$$

where $\langle\hat{A}\rangle = \langle\psi|\hat{A}|\psi\rangle$, with a similar expression for ΔB.

- Since we are dealing with observable quantities, we will assume that both $\hat{A}$ and $\hat{B}$ are Hermitian.

Using the same trick as in the proof of the Schwartz inequality, we define a state given by

$$|\zeta\rangle \equiv (\hat{A} - \langle\hat{A}\rangle)\,|\psi\rangle + i\lambda(\hat{B} - \langle\hat{B}\rangle)\,|\psi\rangle \tag{13.62}$$

where λ is an arbitrary real number. The corresponding bra vector is then

$$\langle\zeta| = \left\langle\left(\hat{A} - \langle\hat{A}\rangle\right)\psi\right| - i\lambda\left\langle\left(\hat{B} - \langle\hat{B}\rangle\right)\psi\right| \tag{13.63}$$

The inner product $\langle\zeta|\zeta\rangle$ is then a function of λ and, of course, is nonnegative, so that $\langle\zeta|\zeta\rangle \geq 0$. It can then be written as:

$$\begin{aligned} 0 \leq I(\lambda) = \langle\zeta|\zeta\rangle &= \left\langle(\hat{A} - \langle\hat{A}\rangle)\psi\,\middle|(\hat{A} - \langle\hat{A}\rangle)\psi\right\rangle \\ &+ \lambda^2\left\langle(\hat{B} - \langle\hat{B}\rangle)\psi\,\middle|(\hat{B} - \langle\hat{B}\rangle)\psi\right\rangle \\ &+ i\lambda\left[\left\langle(\hat{A} - \langle\hat{A}\rangle)\psi\,\middle|(\hat{B} - \langle\hat{B}\rangle)\psi\right\rangle\right. \\ &\left.-\lambda\left\langle(\hat{B} - \langle\hat{B}\rangle)\psi\,\middle|(\hat{A} - \langle\hat{A}\rangle)\psi\right\rangle\right] \end{aligned} \tag{13.64}$$

We can make repeated use of the fact that $\hat{A}$, $\hat{B}$ are Hermitian, and that $\langle\hat{A}\rangle$, $\langle\hat{B}\rangle$ are therefore real, to move all the terms "out from under" the bra vector. We then write this as

$$\begin{aligned} I(\lambda) &= \left\langle\psi\,\middle|(\hat{A} - \langle\hat{A}\rangle)^2\middle|\,\psi\right\rangle + \lambda^2\left\langle\psi\,\middle|(\hat{B} - \langle\hat{B}\rangle)^2\middle|\,\psi\right\rangle \\ &+ i\lambda\left\langle\psi\,\middle|(\hat{A} - \langle\hat{A}\rangle)(\hat{B} - \langle\hat{B}\rangle) - (\hat{B} - \langle\hat{B}\rangle)(\hat{A} - \langle\hat{A}\rangle)\middle|\,\psi\right\rangle \end{aligned} \tag{13.65}$$

We note that the first two terms can be written as

$$(\Delta A)^2 + \lambda^2(\Delta B)^2 \tag{13.66}$$

while the third term can be written in terms of a commutator, namely,

$$(\hat{A} - \langle\hat{A}\rangle)(\hat{B} - \langle\hat{B}\rangle) - (\hat{B} - \langle\hat{B}\rangle)(\hat{A} - \langle\hat{A}\rangle) = \hat{A}\hat{B} - \hat{B}\hat{A} = [\hat{A}, \hat{B}] \tag{13.67}$$

with the simplification due to the fact that the commutator of an operator with any (complex) number vanishes. The resulting term has the form

$$i\lambda\left\langle\psi\,\middle|[\hat{A}, \hat{B}]\middle|\,\psi\right\rangle = \lambda\langle\psi|\hat{F}|\psi\rangle \tag{13.68}$$

where we know that the combination $\hat{F} \equiv i[\hat{A}, \hat{B}]$ is a Hermitian operator (P13.3) that must necessarily have real expectation values. We can thus write $I(\lambda)$ in terms of manifestly real quantities, namely,

$$I(\lambda) = (\Delta A)^2 + \lambda^2(\Delta B)^2 + \lambda\langle\psi|\hat{F}|\psi\rangle \geq 0 \tag{13.69}$$

Since this is, by construction, nonnegative for all values of λ, it will be so at the minimum, namely, for λ determined by

$$0 = \frac{dI(\lambda)}{d\lambda} = 2\lambda_{\text{min}}(\Delta B)^2 + \langle\psi|\hat{F}|\psi\rangle \tag{13.70}$$

or

$$\lambda_{\text{min}} = -\frac{\langle\psi|\hat{F}|\psi\rangle}{2(\Delta B)^2} \tag{13.71}$$

The inequality for this value of λ can then be written in the form of an uncertainty product bound, namely,

$$(\Delta A)^2(\Delta B)^2 \geq \frac{\langle\psi|\hat{F}|\psi\rangle^2}{4} = \frac{\langle\psi|i[\hat{A},\hat{B}]|\psi\rangle^2}{4} \tag{13.72}$$

where the right-hand side is obviously real and positive (despite the explicit i) since $\langle\psi|\hat{F}|\psi\rangle$ is real. This general result shows that:

- It is, in principle, possible to have simultaneous eigenvalues, i.e., states for which *both* ΔA and ΔB are vanishing provided $[\hat{A},\hat{B}] = 0$. The uncertainty principle product can be written as

$$\Delta A \Delta B \geq \frac{\left|\langle\psi\left|[\hat{A},\hat{B}]\right|\psi\rangle\right|}{2} \tag{13.73}$$

The famous Heisenberg uncertainty principle is now simply a special case with

$$\hat{A} = \hat{p} \qquad \text{and} \qquad \hat{B} = x \qquad \text{so that} \qquad \hat{F} = i[\hat{A},\hat{B}] = \hbar \tag{13.74}$$

so that

$$\Delta x \Delta p \geq \frac{\hbar}{2}\langle\psi|\psi\rangle = \frac{\hbar}{2} \tag{13.75}$$

This derivation also allows us to calculate the minimum uncertainty product waveform for the pair x, p, as we know that the lower bound $\Delta x \Delta p = \hbar/2$ will only be saturated when $|\zeta\rangle$ vanishes identically. Using explicit position-space forms, this occurs when

$$0 = \zeta(x) = \left(\frac{\hbar}{i}\frac{d}{dx} - \langle\hat{p}\rangle\right)\psi(x) + i\lambda(x - \langle x\rangle)\psi(x) \tag{13.76}$$

or using

$$\lambda_{\text{min}} = -\frac{\hbar}{2(\Delta x)^2} \tag{13.77}$$

we have

$$\frac{d\psi(x)}{dx} = \frac{ip_0}{\hbar}x - \frac{(x - x_0)}{2(\Delta x)^2}\psi(x) \tag{13.78}$$

where $\langle x\rangle \equiv x_0$, $\langle\hat{p}\rangle = p_0$. This has the simple solution

$$\psi(x) \propto e^{-(x-x_0)^2/4(\Delta x)^2} e^{ip_0 x/\hbar} \tag{13.79}$$

which is the familiar Gaussian wavefunction with arbitrary initial position and "speed."

13.5 TIME DEPENDENCE AND CONSERVATION LAWS IN QUANTUM MECHANICS

In classical mechanics, Newton's laws yield not only particle trajectories, $x(t)$ but through them also the time behavior of all other observables. For example, the kinetic energy varies in time via

$$T(t) = \frac{1}{2}mv^2(t) = \frac{1}{2}m\left(\frac{dx(t)}{dt}\right)^2 \tag{13.80}$$

In a similar way, we have seen that the time dependence of the expectation values of the quantum operator analogs of classical observables for most quantities arises solely through the time dependence of the wavefunctions satisfying the Schrödinger equation, i.e.,

$$\langle \hat{O} \rangle_t = \int_{-\infty}^{+\infty} dx\, \psi^*(x,t)\, \hat{O}\, \psi(x,t) \tag{13.81}$$

Thus, for example,

$$\langle \hat{T} \rangle_t = \frac{\hbar^2}{2m}\int_{-\infty}^{+\infty} dx \left|\frac{\partial \psi(x,t)}{\partial x}\right|^2 \tag{13.82}$$

An important special case occurs in classical mechanics when the time development of an observable is trivial; that is, it is constant in time. This results in so-called *conservation laws;* examples include the conservation of energy, momentum, and angular momentum under the appropriate circumstances. The use of such conservation laws and the study of the conditions under which they are valid is an important topic in classical mechanics, and in this section we extend these notions to the quantum arena by examining in detail the time dependence of expectation values of quantum operators.

A deceptively simple result arises when one considers energy eigenstates or stationary states. In that case the time dependence is trivial:

$$\psi(x,t) = \psi_n(x)e^{-iE_nt/\hbar} \tag{13.83}$$

so that the average values of operators satisfy

$$\begin{aligned}\langle \hat{O} \rangle_t &= \int_{-\infty}^{+\infty} dx\, \psi^*(x)e^{-Et/\hbar}\hat{O}\, \psi(x)e^{-iEt/\hbar} \\ &= \int_{-\infty}^{+\infty} dx\, \psi^*(x)\hat{O}\, \psi(x) = \langle \hat{O} \rangle_0 \end{aligned} \tag{13.84}$$

which is trivially time-independent and further justifies the name *stationary state*. For a more general state, $\psi(x,t) = \sum_n a_n\psi_n(x)e^{-iE_nt/\hbar}$, no such simple cancellation of time factors occurs, and in general one has $d\langle \hat{O} \rangle_t/dt \neq 0$.

Specializing, for familiarity, to a position-space representation, we can calculate the time dependence of the expectation value of a generic operator $\hat{O}$ in some quantum state $\psi(x,t)$, by writing

$$\begin{aligned}\frac{d\langle \hat{O} \rangle_t}{dt} = \frac{d}{dt}\langle \hat{O} \rangle_t &= \frac{d}{dt}\int_{-\infty}^{+\infty} dx\, \psi^*(x,t)\hat{O}\, \psi(x,t) \\ &= \int_{-\infty}^{+\infty} dx\, \psi^* \left(\frac{\partial \hat{O}}{\partial t}\right)\psi \\ &\quad + \int_{-\infty}^{+\infty} dx \left[\frac{\partial \psi^*}{\partial t}\hat{O}\psi + \psi^*\, \hat{O}\frac{\partial \psi}{\partial t}\right] \end{aligned} \tag{13.85}$$

where we have assumed that $\hat{O}$ itself can have an *explicit* time dependence. Since $\psi(x,t)$ satisfies the Schrödinger equation, we can write

$$i\hbar\frac{\partial \psi}{\partial t} = \hat{H}\psi \qquad \text{and hence} \qquad -i\hbar\frac{\partial \psi^*}{\partial t} = (\hat{H}\psi)^* \tag{13.86}$$

and using these relations in Eqn. (13.85) and switching to the more compact bracket notation, we have

$$\begin{aligned}\frac{d\langle\hat{O}\rangle_t}{dt} &= \left\langle\psi\left|\frac{\partial\hat{O}}{\partial t}\right|\psi\right\rangle + \frac{i}{\hbar}\langle\hat{H}\psi|\hat{O}|\psi\rangle - \frac{i}{\hbar}\langle\psi|\hat{O}|\hat{H}\psi\rangle \\ &= \left\langle\psi\left|\frac{\partial\hat{O}}{\partial t}\right|\psi\right\rangle + \frac{i}{\hbar}\left(\langle\psi|\hat{H}\hat{O}|\psi\rangle - \langle\psi|\hat{O}\hat{H}|\psi\rangle\right) \\ &= \left\langle\frac{\partial\hat{O}}{\partial t}\right\rangle + \frac{i}{\hbar}\left\langle[\hat{H},\hat{O}]\right\rangle_t \end{aligned} \quad (13.87)$$

where we have used the fact that $\hat{H}$ is Hermitian to move it from the bra to the ket vector.

This is an extremely important result as it states that

- If the quantum operator analog $\hat{O}$ of the classical observable O
  (a) is not an explicit function of time
  (b) it commutes with the Hamiltonian, i.e., $[\hat{H},\hat{O}] = 0$

  then the expectation value of $\hat{O}$ in any quantum state is *independent of time*. This implies that the classical observable O is *conserved!*

Example 13.3 Before discussing the meaning of this result in more detail, let us consider several examples. The quantum momentum operator $\hat{p}$ obviously has no explicit time dependence, and its commutator with a generic Hamiltonian is given by

$$\begin{aligned}[\hat{H},\hat{p}] &= [\hat{p}^2/2m + V(x),\hat{p}] \\ &= \frac{1}{2m}[\hat{p}^2,\hat{p}] + [V(x),\hat{p}] \\ &= 0 - \frac{\hbar}{i}\left(\frac{dV(x)}{dx}\right) \\ &= \frac{\hbar}{i}F(x)\end{aligned} \quad (13.88)$$

so that

$$\frac{d}{dt}\langle\hat{p}\rangle_t = \langle F(x)\rangle \quad (13.89)$$

where $F(x)$ is the classical force. We thus have the quantum version of the familiar classical condition on conservation of momentum, namely, that $\langle\hat{p}\rangle_t$ will be constant in time provided the classical force vanishes. Similar conditions relating to the conservation of energy can be discussed (Q13.4).

Eqn. (13.87) can also be used to derive already familiar relations in a more compact form, such as

$$\begin{aligned}\frac{d\langle x\rangle_t}{dt} &= 0 + \frac{i}{\hbar}\langle[\hat{H},x]\rangle \\ &= \frac{i}{\hbar}\left\langle\frac{1}{2m}[\hat{p}^2,x] + [V(x),x]\right\rangle \\ &= \frac{i}{2m\hbar}\langle\hat{p}[\hat{p},x] + [\hat{p},x],\hat{p}\rangle \\ &= \frac{\langle\hat{p}\rangle_t}{m}\end{aligned} \quad (13.90)$$

This can be combined with the result of Eqn. (13.89) to give

$$m\frac{d^2\langle x\rangle_t}{dt^2} = m\frac{d}{dt}\left(\frac{\langle\hat{p}\rangle_t}{m}\right) = \langle F(x)\rangle = -\left\langle\frac{dV(x)}{dx}\right\rangle \tag{13.91}$$

This quantum version of Newton's second law is called *Ehrenfest's theorem.*

Returning to Eqn. (13.87), it is easy to understand that a quantum operator must have no explicit time dependence in order to be conserved. To appreciate why the commutator with the $\hat{H}$ operator arises, let us examine more fully the role that the Hamiltonian plays in quantum mechanics. The time-dependent Schrödinger equation can be written as

$$i\hbar\frac{\partial\psi(x,t)}{\partial t} = \hat{H}\psi(x,t) \qquad \text{or} \qquad \frac{\partial\psi(x,t)}{\partial t} = -\frac{i\hat{H}}{\hbar}\psi(x,t) \tag{13.92}$$

which we can *formally* integrate to yield

$$\psi(x,t) = e^{-i\hat{H}t/\hbar}\psi(x,0) \tag{13.93}$$

thus solving the initial value problem. Recall that the exponential of an operator is only defined via the series expansion of e^x, i.e.,

$$e^{\hat{O}} = 1 + \hat{O} + \frac{1}{2!}\hat{O}^2 + \cdots = \sum_{n=0}^{\infty}\frac{\hat{O}^n}{n!} \tag{13.94}$$

While this relation is of somewhat limited use as a calculational tool,[1] it does show that

- The Hamiltonian acts as the *time-development operator* in quantum mechanics.

For example, for small enough times, the approximation

$$\psi(x,t+dt) \approx \psi(x,t) - \frac{i}{\hbar}(\hat{H}\psi(x,t))\,dt \tag{13.95}$$

will be a good one, so that the operation of $-i\hat{H}\,dt/\hbar$ "translates" the wavefunction ahead in time by an infinitesimal amount dt.

Example 13.4 Time Development of Free-Particle Gaussian Wavepacket One case in which the time development operation in Eqn. (13.93) can be performed in closed form is for the free-particle propagation of the Gaussian wavepacket.[2] To be consistent with the notation in Sec. 3.2, we write the initial wavefunction in the form

$$\psi(x,0) = \frac{1}{\sqrt{\alpha\hbar\sqrt{\pi}}}e^{-x^2/2\alpha^2\hbar^2} = \sqrt{\frac{\alpha\hbar}{2\sqrt{\pi}}}\left[\frac{1}{\sqrt{w}}e^{-x^2/4w}\right] \tag{13.96}$$

where we have written $\alpha^2\hbar^2 = 2w$ for use in the "trick" we will employ below; this is the same as in Sec. 2.2, but we choose $p_0 = 0$ for simplicity.

The free-particle Hamiltonian is simply

$$\hat{H} = -\frac{\hbar^2}{2m}\frac{\partial^2}{\partial x^2} \tag{13.97}$$

[1] See, however, Press et al. (1986).
[2] See Blinder (1968).

so the time development operator is

$$e^{-i\hat{H}t/\hbar} = \sum_{n}^{\infty} \left(\frac{it\hbar}{2m} \frac{\partial^2}{\partial x^2} \right)^n \tag{13.98}$$

and we can use the identity

$$\frac{\partial^2}{\partial x^2} \left[\frac{1}{\sqrt{w}} e^{-x^2/4w} \right] = \frac{\partial}{\partial w} \left[\frac{1}{\sqrt{w}} e^{-x^2/4w} \right] \tag{13.99}$$

to write Eqn. (13.98) as

$$\sum_{n}^{\infty} \left(\frac{it\hbar}{2m} \frac{\partial}{\partial w} \right)^n = e^{c\partial/\partial w} \tag{13.100}$$

where $c = it\hbar/2m$. We note that this is true only when acting on Gaussian functions. We can then use the results of P13.6 to show that

$$e^{c\partial/\partial w} f(w) = f(w + c) \tag{13.101}$$

so that the net result of the time development operator is to give

$$\begin{aligned} \psi(x, t) &= \sqrt{\frac{\alpha\hbar}{2\sqrt{\pi}}} \left[\frac{1}{\sqrt{w + it\hbar/2m}} e^{-x^2/4(w+it\hbar/2m)} \right] \\ &= \frac{1}{\sqrt{\alpha\hbar F \sqrt{\pi}}} e^{-x^2/2\alpha^2\hbar^2 F} \end{aligned} \tag{13.102}$$

where $F = 1 + it/m\hbar\alpha$, just as in Sec. 3.2.2.

Using the identification in Eqn. (13.95), we can now discuss the conditions under which a quantity might be conserved, i.e., have a constant value in time. If such a quantity O is to be conserved, we should obtain the same value from a measurement at different times. We can check this by comparing

$$\begin{gathered} \hat{H}\hat{O}\psi(x,t) \Leftrightarrow (\textit{then} \text{ evolve in time})(\text{measure } O \textit{ first}) \\ ? \quad ? \\ = \quad = \\ ? \quad ? \\ \hat{O}\hat{H}\psi(x,t) \Leftrightarrow (\textit{then} \text{ measure } O)(\text{evolve in time } \textit{first}) \end{gathered} \tag{13.103}$$

Thus, whether the operation of measuring O (via $\hat{O}$ acting on ψ) and the evolution of the wavefunction in time (via $\hat{H}$ on ψ) "interfere" with each other, as measured by whether $[\hat{H}, \hat{O}] = 0$ or not, helps determine whether O will change in time.

Defining the time development operator as $\hat{U}_t = e^{-i\hat{H}t/\hbar}$, we note that $\hat{U}_t$ is unitary since

$$\hat{U}_t^\dagger \hat{U}_t = e^{i\hat{H}t/\hbar} e^{-i\hbar H t/\hbar} = \hat{1} \tag{13.104}$$

where we have used the fact (P13.18) that $e^{i\hat{O}}$ is unitary if $\hat{O}$ is Hermitian. This is the "fancy" version of the statement that Schrödinger wavefunctions, if initially normalized, stay normalized at later times.

13.6 QUESTIONS AND PROBLEMS

Q13.1 What are the analogous expressions to Eqns. (13.53) to (13.57) for the cross-products of ordinary vectors?

Q13.2 Is the complex conjugation operator, defined in P7.15, a linear operator as defined by Eqn. (13.28)?

Q13.3 Would the operator $\hat{O} = x\hat{p}$ have an observable classical counterpart? How would you decide?

Q13.4 Using the ideas in Sec. 13.5, discuss under what conditions the average value of the energy operator will be constant in time.

P13.1 Which of the following are Hermitian operators?

(a) $x\hat{p}$?

(b) $x\hat{p} + \hat{p}x$?

(c) $\hat{E}x$?

(d) $x\hat{p}x$?

P13.2 Show that $(\hat{A}^\dagger)^\dagger = \hat{A}$.

P13.3 Show that the combinations in Eqn. (13.17) and (13.19) are Hermitian.

P13.4 If $\hat{A}$ is Hermitian, show that $\langle\psi|\hat{A}^2|\psi\rangle \geq 0$ for any $|\psi\rangle$. Why do we expect this kind of result?

P13.5 **(a)** Consider the matrix operator relation defined by Eqn. (13.31). Show that if $\mathbf{a}' \cdot \mathbf{a}' = \mathbf{a} \cdot \mathbf{a}$, then the matrix U satisfies Eqn (13.32).

(b) Show that the rows and columns of a Hermitian matrix can be thought of as orthonormal vectors.

(c) Show that the matrix

$$\mathsf{U} = \frac{1}{5\sqrt{2}}\begin{pmatrix} 4-4i & 3-3i \\ -3\sqrt{2}i & 4\sqrt{2}i \end{pmatrix}$$

is unitary.

(d) Find values of b and c for which the matrix

$$\mathsf{U} = \frac{1}{5}\begin{pmatrix} 3+i & \sqrt{15}i \\ b & ci \end{pmatrix}$$

is unitary. Are your choices unique?

P13.6 Consider a *translation operator* defined by

$$\hat{T}_a \equiv e^{ia\hat{p}/\hbar} = e^{a\partial/\partial x}$$

where a is a constant.

(a) Show that $\hat{T}_a f(x) = f(x+a)$. *Hint:* Expand both the operator and the function in a series expansion and compare.

(b) Show that $\hat{T}_a$ is a unitary operator, and interpret this statement.

P13.7 Show that the set of translation operators, $\hat{T}_a$, where a is any real number, form a group. *Hint:* Show that this set satisfies all of the group requirements in App. D.2.

P13.8 Verify the commutator relations in Eqns. (13.52) to (13.59).

P13.9 Since we know that an operator commutes with itself, we would expect that

$$e^{a\hat{O}}e^{b\hat{O}} = e^{(a+b)\hat{O}}$$

since there would be no problems with operator ordering. Show this explicitly by comparing series expansions for both sides; i.e., show that

$$\left(\sum_{n=0}^{\infty}\frac{(a\hat{O})^n}{n!}\right)\left(\sum_{m=0}^{\infty}\frac{(b\hat{O})^m}{m!}\right)=\sum_{J=0}^{\infty}\frac{[(a+b)\hat{O}]^J}{J!}$$

Hint: Relabeling the sums and using the binomial expansion helps. A general identity, valid for any two operators $\hat{A}$ and $\hat{B}$, is given by the so-called *Campbell-Baker-Hausdorff formula,* namely,

$$e^{\hat{A}}e^{\hat{B}} = e^{\hat{O}}$$

where

$$\hat{O} = \hat{A} + \hat{B} + \frac{1}{2}[\hat{A},\hat{B}] + \frac{1}{12}\left([[\hat{A},\hat{B}],\hat{B}] + [\hat{A},[\hat{A},\hat{B}]]\right) + \frac{1}{24}[[[\hat{B},\hat{A}],\hat{A}],\hat{B}] + \cdots$$

is an infinite series of nested commutators.

P13.10 Using explicit representations of operators in momentum space, find the minimum-uncertainty wavefunction corresponding to Eqn. (13.79).

P13.11 Consider a general quantum state that has been expanded in energy eigenfunctions, $\psi(x,0) = \sum_n a_n u_n(x)$. What is the effect of operating on $\psi(x,0)$ with the time development operator, $\hat{U}_t = e^{-i\hat{H}t/\hbar}$?

P13.12 **The Virial theorem:**

(a) Use the fact that average values of operators in energy eigenstates are time-independent to show that

$$\langle\hat{T}\rangle = \frac{1}{2}\left\langle x\frac{dV(x)}{dx}\right\rangle$$

when evaluated using energy eigenstates! Do this by considering the time dependence of the expectation value of the operator $x\hat{p}$, i.e., calculate

$$\frac{d}{dt}\langle x\hat{p}\rangle$$

(b) This relation often can give information on how the potential and kinetic energy in a quantum system are "shared." Show this by considering power law potentials of the form $V(x) = Kx^n$ and finding the fraction of the total energy, which, on average, is in the form of potential and kinetic energy. Specifically, show that

$$\frac{\langle V(x)\rangle}{E} = \frac{2}{n+2} \quad \text{and} \quad \frac{\langle\hat{T}\rangle}{E} = \frac{n}{n+2}$$

(c) Try to confirm your results in (b) by considering both the harmonic oscillator potential ($n = 2$) and infinite well potential ($n = \infty$).

(d) Compare this problem to the classical results of P6.3 to P6.5.

P13.13 For a system described by a Hamiltonian, $\hat{H} = \hat{p}^2/2m + V(x)$, obtain an expression for $d\langle\hat{T}\rangle/dt$ where $\hat{T}$ is the kinetic energy operator. Discuss the relation of your result to the classical work energy theorem.

P13.14 If $\hat{A}$ and $\hat{B}$ are arbitrary, perhaps time-dependent operators, show that

$$\frac{d}{dt}\langle\hat{A}\hat{B}\rangle = \left\langle\frac{\partial\hat{A}}{\partial t}\hat{B}\right\rangle + \left\langle\hat{A}\frac{\partial\hat{B}}{\partial t}\right\rangle + \frac{i}{\hbar}\langle[\hat{H},\hat{A}]\hat{B}\rangle + \frac{i}{\hbar}\langle\hat{A}[\hat{H},\hat{B}]\rangle$$

P13.15 Sum rules:
(a) Derive the so-called *Thomas-Reiche-Kuhn sum rule,*

$$\sum_k (E_k - E_j)|\langle k|x|j\rangle|^2 = \frac{\hbar^2}{2m}$$

where the $|j\rangle$ are energy eigenstates. The following steps might be useful:

1. Use the uncertainty principle and a complete set of states to write

$$\frac{\hbar}{i} = \langle j|\hat{p}x - x\hat{p}|j\rangle = \sum_k \left[\langle j|\hat{p}|k\rangle\langle k|x|j\rangle - \langle j|x|k\rangle\langle k|\hat{p}|j\rangle\right]$$

2. Use the relation

$$\langle j|\hat{p}|k\rangle = m\frac{d}{dt}\langle j|x|k\rangle$$

and generalize Eqn. (13.87).
(b) Check the sum rule explicitly for the harmonic oscillator using Eqn. (10.46).
(c) Derive the sum rule

$$\sum_k (E_j - E_k)^2|\langle k|x|j\rangle|^2 = \frac{2\hbar^2}{m}\langle j|\hat{T}|j\rangle$$

and check it with the harmonic oscillator results.

Such sum rules are important, as they can be tested (using experimental data) even when the complete mathematical problem is not solved.

P13.16 **(a)** Generalize the result of Ex. 13.4 to include the case where the Gaussian wavepacket has a nontrivial initial position (x_0) and momentum (p_0).
(b) The use of Eqn. (13.93) is not restricted to position space wavefunctions. What is the time development operator for free-particle momentum space wavefunctions?
(c) Can you generalize Ex. 13.4 to find the explicit time dependence of the initial wavefunction

$$\psi(x;0) = \sqrt{\frac{2}{\sqrt{\pi}\beta^3}}\, x e^{-x^2/2\beta^2}$$

Hint: You will need to generate a new "tricky identity" to replace Eqn. (13.99).

P13.17 Use the Hamiltonian for a particle subject to a uniform force, corresponding to $V(x) = -Fx$, to calculate the effect of the time development operator on both position and momentum space wavefunctions. *Hint:* Use the Hausdorff formula in P13.9, or see Robinett (1996b).

P13.18 Prove that $e^{i\hat{O}}$ is unitary if $\hat{O}$ is Hermitian.

P13.19 **(a)** If we let

$$|\psi\rangle_t = e^{-i\hat{H}t/\hbar}|\psi\rangle_0$$

show that the time-dependent expectation value of any operator can be written as

$$\langle\hat{O}\rangle_t = \langle\psi_0|e^{i\hbar Ht/\hbar}\hat{O}e^{-i\hat{H}t/\hbar}\rangle = \langle\psi_0|\hat{O}(t)|\psi_0\rangle$$

where $\hat{O}(t) = e^{\hat{H}t/\hbar}\hat{O}e^{\hat{H}t/\hbar}$. In this view, the quantum state vectors, $|\psi_0\rangle$, are fixed once and for all, but the operators evolve in time. This is often referred to as the *Heisenberg picture* of quantum mechanics as opposed to the *Schrödinger picture* that we have adopted.

(b) Use these results to show that the time derivative of a Heisenberg operator can be written as

$$\frac{d}{dt}\hat{O} = \frac{i}{\hbar}[\hat{H}, \hat{O}]$$

in the case where $\hat{O}$ itself has no explicit time dependence. *Hint:* Simply recall the definition of derivative as

$$\frac{d}{dt}\hat{O} = \lim_{dt \to 0}\left(\frac{\hat{O}(t+dt) - \hat{O}(t)}{dt}\right)$$

This result generalizes Eqn. (13.87) for the time development of matrix elements of operators.

P13.20 **(a)** Using the results of the last problem, find the time-dependent operators, $\hat{x}(t)$ and $\hat{p}(t)$, in the Heisenberg picture in the free-particle case where the Hamiltonian is $\hat{H} = \hat{p}^2/2m$. Comment on any classical analogs to your results. *Hint:* Use the operator identity

$$e^{\hat{A}}\hat{B}e^{-\hat{A}} = \hat{B} + [\hat{A}, \hat{B}] + \frac{1}{2}[\hat{A}, [\hat{A}, \hat{B}]] + \frac{1}{3!}[\hat{A}, [\hat{A}, [\hat{A}, \hat{B}]]] + \cdots$$

(b) Repeat part (a), but use the Hamiltonian corresponding to the case of a constant force, F, in the $+x$ direction, namely,

$$\hat{H}_F = \frac{\hat{p}^2}{2m} - Fx$$

CHAPTER 14

Operator and Factorization Methods for the Schrödinger Equation

14.1 FACTORIZATION METHODS

The special and recurring role that the simple harmonic oscillator plays in quantum mechanics can be attributed both to its physical relevance and to its simple solutions. The fact that the ground state solution is the minimum uncertainty wavepacket (Sec. 13.4) and the highly constrained connection between the position space and momentum space wavefunctions (Sec. 10.2.2) also indicate that this problem occupies a special niche. The symmetry between x and p present in the solutions is obviously a reflection of the fact that only for the SHO is the potential energy function quadratic in x. It is often the case in physics that systems with a high degree of symmetry are amenable to solution in a variety of ways, sometimes quite unexpected. In this chapter, we discuss a powerful method of solving the harmonic oscillator problem involving the factorization of the differential equation using differential operators; we then discuss extensions and applications of operator methods[1] to other physical problems.

Factorization methods are often used in the solution of linear differential equations with constant coefficients, i.e., equations of the form

$$a_n \frac{d^n}{dx^n} y(x) + a_{n-1} \frac{d^{n-1}}{dx^{n-1}} y(x) + \cdots + a_1 \frac{d}{dx} y(x) + a_0 y(x) = 0 \qquad (14.1)$$

One standard approach is to define a differential operator, $\hat{D} = d/dx$, and write Eqn. (14.1) in operator form as

$$\left(a_n \hat{D}^n + a_{n-1} \hat{D}^{n-1} + \cdots + a_1 \hat{D} + a_0\right) y(x) = 0 \qquad (14.2)$$

[1]The most comprehensive treatment of operator methods in quantum mechanics is DeLange and Raab (1991).

which is similar to a polynomial equation in $\hat{D}$. If we can factor the associated polynomial, finding its real and imaginary roots, we can write this equation as

$$a_n[(\hat{D} - r_n)(\hat{D} - r_{n-1})\cdots(\hat{D} - r_1)]y(x) = 0 \tag{14.3}$$

where the r_i are the n roots of the polynomial equation $a_n x^n + \cdots + a_1 x + a_0 = 0$. The n independent solutions of Eqn. (14.1) are then obtained from

$$(\hat{D}_i - r_i)y_i(x) = 0 \qquad \text{for } i = 1, \ldots, n \tag{14.4}$$

which have exponential solutions of the form

$$y_i(x) = C_i e^{-r_i x} \tag{14.5}$$

so that the general solution is

$$y(x) = \sum_{i=1}^{n} C_i e^{-r_i x} \tag{14.6}$$

Special care is necessary when there are multiple roots, but the extension to that case is conceptually easy (P14.1). This method then allows us to solve an nth-order differential equation in terms of n first-order ones. A trivial example is the differential equation in Sec. 6.3 for the infinite well:

$$\frac{d^2\psi(x)}{dx^2} + k^2\psi(x) = 0 \tag{14.7}$$

which we can write as

$$(\hat{D}^2 + k^2)\psi(x) = (\hat{D} + ik)(\hat{D} - ik)\psi(x) = 0 \tag{14.8}$$

which has the solutions $\psi(x) = \exp(\pm ikx)$ or $\sin(kx)$ and $\cos(kx)$ as expected.

14.2 FACTORIZATION OF THE HARMONIC OSCILLATOR

The form of the Hamiltonian for the harmonic oscillator problem

$$\hat{H}\psi(x) = \left(\frac{\hat{p}^2}{2m} + \frac{1}{2}m\omega^2 x^2\right)\psi(x) = E\psi(x) \tag{14.9}$$

gives rise to a differential equation that does not have constant coefficients. Nonetheless, the form of $\hat{H}$ suggests that a similar factorization might be possible as classically

$$\frac{p^2}{2m} + \frac{1}{2}m\omega^2 x^2 = \left(\frac{p}{\sqrt{2m}} + i\sqrt{\frac{m\omega^2}{2}}x\right)\left(\frac{p}{\sqrt{2m}} - i\sqrt{\frac{m\omega^2}{2}}x\right) \tag{14.10}$$

but the lack of commutativity between the operators x and p in quantum mechanics will make things less simple. We are therefore motivated to consider such a possible factorization, and to that end we first write

$$\hat{H} = \hbar\omega\left(\frac{\hat{p}^2}{2m\hbar\omega} + \frac{m\omega}{2\hbar}x^2\right) \tag{14.11}$$

as we know, if only on grounds of dimensional analysis, that the energies will be given in terms of $\hbar\omega$. We then define

$$\hat{A} \equiv \sqrt{\frac{m\omega}{2\hbar}}x + i\frac{\hat{p}}{\sqrt{2m\omega\hbar}} \tag{14.12}$$

from which one can immediately obtain

$$\hat{A}^\dagger \equiv \sqrt{\frac{m\omega}{2\hbar}}x - i\frac{\hat{p}}{\sqrt{2m\omega\hbar}} \tag{14.13}$$

because x, $\hat{p}$ are Hermitian. We first note that $\hat{A}$, $\hat{A}^\dagger$ are not Hermitian operators themselves (as $\hat{A}^\dagger \neq \hat{A}$) and so cannot represent physical observables. They do not commute either as

$$\begin{aligned}[\hat{A}, \hat{A}^\dagger] &= \left[\sqrt{\frac{m\omega}{2\hbar}}x + i\frac{\hat{p}}{\sqrt{2m\omega\hbar}}, \sqrt{\frac{m\omega}{2\hbar}}x + i\frac{\hat{p}}{\sqrt{2m\omega\hbar}}\right] \\ &= \frac{i}{2\hbar}\left\{-[x, \hat{p}] + [\hat{p}, x]\right\} = 1 \end{aligned} \tag{14.14}$$

since $[x, \hat{p}] = -[\hat{p}, x] = i\hbar$. Thus, we have

$$\hat{A}\hat{A}^\dagger = \hat{A}^\dagger\hat{A} + 1 \qquad \text{and} \qquad \hat{A}^\dagger\hat{A} = \hat{A}\hat{A}^\dagger - 1 \tag{14.15}$$

which will prove useful.

To see if these operators factorize Eqn. (14.9), we can invert Eqns. (14.12) and (14.13) and write

$$x = \sqrt{\frac{\hbar}{2m\omega}}(\hat{A} + \hat{A}^\dagger) \qquad \hat{p} = i\sqrt{\frac{m\omega\hbar}{2}}(\hat{A}^\dagger - \hat{A}) \tag{14.16}$$

If we substitute these into the Hamiltonian, we obtain

$$\begin{aligned}\hat{H} &= \frac{1}{2m}\left(-\frac{m\hbar\omega}{2}\right)\left(\hat{A}^\dagger - \hat{A}\right)^2 + \frac{m\omega^2}{2}\left(\frac{\hbar}{2m\omega}\right)\left(\hat{A} + \hat{A}^\dagger\right)^2 \\ &= \frac{\hbar\omega}{2}\left(\hat{A}^\dagger\hat{A} + \hat{A}\hat{A}^\dagger\right) \\ &= \hbar\omega\left(\hat{A}^\dagger\hat{A} + \frac{1}{2}\right) \qquad \text{(using Eqn. (14.15))} \\ &\equiv \hbar\omega\left(\hat{N} + \frac{1}{2}\right) \end{aligned} \tag{14.17}$$

where we have defined a new *number operator* via $\hat{N} \equiv \hat{A}^\dagger\hat{A}$.

Because of the lack of commutativity of x, p, we have not achieved a complete factorization, but we have shown that the Hamiltonian can be put into a form that is highly suggestive of the known result for the energy eigenvalues themselves. Thus, instead of considering the energy eigenvalue problem using the Hamiltonian, written as

$$\hat{H}\psi_n(x) = E_n\psi_n(x) \qquad \text{in position space} \tag{14.18}$$

$$\hat{H}\phi_n(p) = E_n\phi_n(p) \qquad \text{in momentum space} \tag{14.19}$$

we abstractly examine the eigenvalues of the number operator, i.e.,

$$\hat{N}|n\rangle = n|n\rangle \tag{14.20}$$

While we label the number eigenstates by n, we cannot assume (at least initially) that they are nonnegative integers; however, since $\hat{N}$ is Hermitian,

$$\hat{N}^\dagger = \left(\hat{A}^\dagger \hat{A}\right)^\dagger = \hat{A}^\dagger \hat{A} = \hat{N} \tag{14.21}$$

we do know that the eigenvalues of $\hat{N}$ are real.

To examine the effects of $\hat{A}$ and $\hat{A}^\dagger$, and to provide constraints on the "number spectrum," we first note that

$$\begin{aligned}
\hat{N}\left(\hat{A}|n\rangle\right) &= \left(\hat{A}^\dagger \hat{A}\right)\left(\hat{A}|n\rangle\right) \\
&= \left(\hat{A}\hat{A}^\dagger - 1\right)\left(\hat{A}|n\rangle\right) \\
&= \hat{A}\left(\hat{N}|n\rangle\right) - \hat{A}|n\rangle \\
&= (n-1)\left(\hat{A}|n\rangle\right)
\end{aligned} \tag{14.22}$$

where we have used the commutation relations between $\hat{A}$ and $\hat{A}^\dagger$. This can be interpreted as saying that the number operator, $\hat{N}$, acting on the state $\hat{A}|n\rangle$ yields the same eigenvalue, namely $n-1$, as when it acts on the state $|n-1\rangle$. A similar derivation shows that

$$\hat{N}\left(\hat{A}^\dagger|n\rangle\right) = (n+1)\left(\hat{A}^\dagger|n\rangle\right) \tag{14.23}$$

Taken together, these imply that:

- The operator $\hat{A}$ $(\hat{A}^\dagger)$ acting on $|n\rangle$ must give, up to an arbitrary constant, the state $|n-1\rangle(|n+1\rangle)$, i.e.,

$$\hat{A}|n\rangle \propto |n-1\rangle \qquad \hat{A}^\dagger|n\rangle \propto |n+1\rangle \tag{14.24}$$

- The eigenvalues of the number operator are thus separated by integer differences, and the $\hat{A}$ and $\hat{A}^\dagger$ operators act to move one up and down the ladder of possible eigenvalues, as shown in Fig. 14.1. For this reason, $\hat{A}$ and $\hat{A}^\dagger$ can be called *lowering* and *raising* operators, respectively; collectively they are called *ladder operators.*

It is still not clear (from this algebraic derivation) that the number eigenvalues are given by integers (and not say, π, $\pi+1$, $\pi+2\ldots$) or even if the spectrum is bounded from below. To address the second question, using the explicit form of $\hat{N}$ and the definition of Hermitian conjugate, we can easily show (using P13.4) that the expectation value of the number operator in the state $|n\rangle$ satisfies

$$n = n\langle n|n\rangle = \langle n|\hat{N}|n\rangle = \langle n|\hat{A}^\dagger \hat{A}|n\rangle = \langle \hat{A}n|\hat{A}n\rangle \geq 0 \tag{14.25}$$

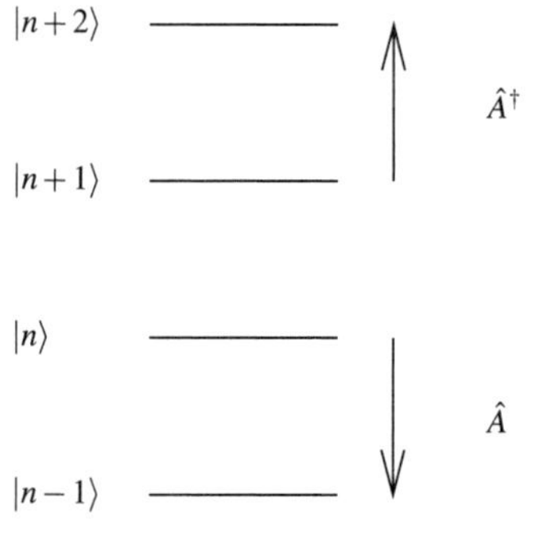

$|n+2\rangle$

$|n+1\rangle$

$|n\rangle$

$|n-1\rangle$

$\hat{A}^\dagger$

$\hat{A}$

FIGURE 14.1. *The operation of the raising (or creation) operator $\hat{A}^\dagger$ and lowering (or annihilation) operator $\hat{A}$ on number eigenstates.*

Thus, the n are nonnegative numbers and therefore there must be a smallest one, which we label $n_{\min}$. By assumption, there can be no state with a lower value of n, so we must have

$$\hat{A}|n_{\min}\rangle \propto |n_{\min} - 1\rangle = 0 \tag{14.26}$$

since $|n_{\min} - 1\rangle$ does not exist. The number operator when acting on this lowest state then gives

$$n_{\min}|n_{\min}\rangle = \hat{N}|n_{\min}\rangle = \left(\hat{A}^{\dagger}\hat{A}\right)|n\rangle = \hat{A}^{\dagger}\left(\hat{A}|n\rangle\right) = 0 \tag{14.27}$$

so that $n_{\min} = 0$; we sometimes say that $\hat{A}$ "annihilates the vacuum." Thus, the number and energy spectrum are given by

$$\begin{aligned} \hat{N}|n\rangle &= n|n\rangle & n &= 0, 1, 2, \ldots \\ \hat{H}|n\rangle &= \hbar\omega\left(n + \frac{1}{2}\right)|n\rangle & n &= 0, 1, 2, \ldots \end{aligned} \tag{14.28}$$

and these results have been obtained in a purely algebraic way.

While these techniques have yielded the energy spectrum in an elegant fashion, we might also wish to extract information from the position space or momentum space wavefunctions. The first step toward obtaining the wavefunctions is to note that any desired state can be obtained from the ground state by successive applications of the raising operator, namely,

$$|n\rangle \propto \left(\hat{A}^{\dagger}\right)^{n}|0\rangle \tag{14.29}$$

provided we have an explicit representation of the ground state. This can be obtained by using the fact that the lowering operator annihilates the lowest state, i.e., $\hat{A}|0\rangle = 0$. Using an explicit position space representation, we write

$$\begin{aligned} 0 = \hat{A}\psi_0(x) &= \left(\sqrt{\frac{m\omega}{2\hbar}}x + i\frac{\hat{p}}{\sqrt{2m\hbar\omega}}\right)\psi_0(x) \\ &= \sqrt{\frac{\hbar}{2m\hbar\omega}}\left(\frac{m\omega}{\hbar}x + \frac{d}{dx}\right)\psi_0(x) \end{aligned} \tag{14.30}$$

giving

$$\frac{d\psi_0(x)}{dx} = -x\left(\frac{m\omega}{\hbar}\psi_0(x)\right) = -\frac{x}{\rho^2}\psi_0(x) \tag{14.31}$$

or

$$\psi_0(x) = \frac{1}{\sqrt{\rho\sqrt{\pi}}}e^{-x^2/2\rho^2} \tag{14.32}$$

when finally normalized. The momentum space wavefunctions can be obtained from $(\hat{A}^{\dagger})^n\phi_0(p)$ by solving $\hat{A}\phi_0(p) = 0$.

The nth eigenstate is obtained by repeated applications of the raising operator on the ground state as in Eqn. (14.29), and we can easily determine the precise normalization by considering

$$|n\rangle = c_n(\hat{A}^{\dagger})^n|0\rangle \qquad \text{or equivalently} \qquad \langle n| = \langle 0|\hat{A}^n c_n \tag{14.33}$$

Assuming that all states are to be properly normalized, we note that

$$
\begin{aligned}
n = n\langle n|n\rangle &= \langle n|\hat{N}|n\rangle \\
&= \langle n|\hat{A}^\dagger \hat{A}|n\rangle \\
&= \langle n|\left(\hat{A}\hat{A}^\dagger - 1\right)|n\rangle \\
&= \langle n|\hat{A}\hat{A}^\dagger|n\rangle - 1
\end{aligned} \tag{14.34}
$$

so that

$$
\begin{aligned}
n + 1 &= \langle n|\hat{A}\hat{A}^\dagger|n\rangle \\
&= c_n^2\langle 0|(\hat{A})^n\left(\hat{A}\hat{A}^\dagger\right)(\hat{A}^\dagger)^n|0\rangle \\
&= c_n^2\langle 0|\hat{A}^{n+1}(\hat{A}^\dagger)^{n+1}|0\rangle \\
&= c_n^2\frac{\langle n+1|n+1\rangle}{c_{n+1}^2} \\
&= \frac{c_n^2}{c_{n+1}^2}
\end{aligned} \tag{14.35}
$$

Thus, the normalization constants are related via

$$
c_{n+1} = c_n/\sqrt{n+1} \qquad \text{which implies that} \qquad c_n = 1/\sqrt{n!} \tag{14.36}
$$

since $c_0 \equiv 1$. Thus

$$
|n\rangle = \frac{1}{\sqrt{n!}}\left(\hat{A}^\dagger\right)^n|0\rangle \tag{14.37}
$$

and this can be used (P14.2) to show that

$$
\hat{A}|n\rangle = \sqrt{n}|n-1\rangle \qquad \text{and} \qquad \hat{A}^\dagger|n\rangle = \sqrt{n+1}|n+1\rangle \tag{14.38}
$$

These same techniques can be used to show explicitly that eigenfunctions corresponding to different eigenvalues are orthogonal (P14.3).

The power of these general algebraic methods evident in generating the properly normalized wavefunctions can also be used in the efficient evaluation of expectation values of many operators.

Example 14.1 Expectation Values for the Harmonic Oscillator For example, the vanishing of $\langle n|x|n\rangle$ is physically obvious from the parity properties of the wavefunctions. In operator language, this relation arises because

$$
\begin{aligned}
\langle n|x|n\rangle &= \left\langle n\left|\sqrt{\frac{\hbar}{2m\omega}}(\hat{A} + \hat{A}^\dagger)\right|n\right\rangle \\
&= \sqrt{\frac{\hbar}{2m\omega}}\left\{\langle n|\hat{A}|n\rangle + \langle n|\hat{A}^\dagger|n\rangle\right\} \\
&\propto \sqrt{n}\langle n|n-1\rangle + \sqrt{n+1}\langle n|n+1\rangle = 0
\end{aligned} \tag{14.39}
$$

because these states are orthogonal; this proof gives no new information but illustrates the method. The expectation value of the potential energy, $V(x) = m\omega^2x^2/2$, requires

$$
\begin{aligned}
\langle n|x^2|n\rangle &= \frac{\hbar}{2m\omega}\left\langle n\left|\hat{A}^2 + \hat{A}\hat{A}^\dagger + \hat{A}^\dagger\hat{A} + \hat{A}^{\dagger 2}\right|n\right\rangle \\
&= \frac{\hbar}{2m\omega}\langle n|(2\hat{N} + 1)|n\rangle \\
&= \frac{\hbar}{m\omega}\left(n + \frac{1}{2}\right)
\end{aligned} \tag{14.40}
$$

so that

$$\langle n|V(x)|n\rangle = \frac{\hbar\omega}{2}\left(n + \frac{1}{2}\right) = \frac{1}{2}E_n \tag{14.41}$$

with a similar result for $\langle n|\hat{T}|n\rangle = \langle n|\hat{p}^2|n\rangle/2m$.

14.3 CREATION AND ANNIHILATION OPERATORS

The importance of raising and lowering operators is not limited to the study of the quantum version of the classical oscillating particle. We have seen in Sec. 10.1 that the vibrations of the electromagnetic field can be represented by an ensemble of effective harmonic oscillators where the "amplitude" is essentially the vector potential $\mathbf{A}(\mathbf{k}, t)$. This similarity can be pushed even further if we define generalized ladder operators $\hat{a}_{\mathbf{k}}$ and $\hat{a}^\dagger_{\mathbf{k}}$ for each wavenumber[2] mode $\mathbf{k}$.

To proceed formally, we need only specify the appropriate commutation relations, i.e.,

$$[\hat{a}_{\mathbf{k}}, \hat{a}^\dagger_{\mathbf{k}}] = \delta_{\mathbf{k}\mathbf{k}'} \tag{14.42}$$

while we also note that the $\hat{a}_{\mathbf{k}}$, $\hat{a}^\dagger_{\mathbf{k}}$ always commute with themselves, i.e.,

$$[\hat{a}_{\mathbf{k}}, \hat{a}_{\mathbf{k}}] = 0 \qquad \text{and} \qquad [\hat{a}^\dagger_{\mathbf{k}}, \hat{a}^\dagger_{\mathbf{k}}] = 0 \tag{14.43}$$

The number operator for each mode is simply $\hat{N}_{\mathbf{k}} = \hat{a}^\dagger_{\mathbf{k}}\hat{a}_{\mathbf{k}}$ with corresponding integer eigenvalues $n_{\mathbf{k}}$; the Hamiltonian is simply $\hat{H}_{\mathbf{k}} = (\hat{N}_{\mathbf{k}} + 1/2)\hbar\omega_{\mathbf{k}}$. We then formally have relations such as

$$\hat{a}^\dagger_{\mathbf{k}}|n_{\mathbf{k}}\rangle = \sqrt{n_{\mathbf{k}} + 1}|n_{\mathbf{k}}\rangle \qquad \text{and} \qquad \hat{a}|n_{\mathbf{k}}\rangle = \sqrt{n_{\mathbf{k}}}|n_{\mathbf{k}} - 1\rangle \tag{14.44}$$

corresponding to increasing or decreasing the energy of the system by one unit of $\hbar\omega_{\mathbf{k}}$ where $\omega_{\mathbf{k}} = |\mathbf{k}|c$. These results are consistent with the following interpretation:

- The radiation field, quantized using such ladder operators, consists of an ensemble of photons, each of quantized energy $\hbar\omega_{\mathbf{k}}$
- The number of photons of wavenumber $\mathbf{k}$ is given by $n_{\mathbf{k}}$
- The shift operator $\hat{a}^\dagger_{\mathbf{k}}$ ($\hat{a}_{\mathbf{k}}$) increases (decreases) the number of photons by one; this justifies the use of the term *creation (annihilation)* operator for $\hat{a}_{\mathbf{k}}$ ($\hat{a}^\dagger_{\mathbf{k}}$).

Such operators describing the particle-like quanta of the electromagnetic field are useful in many aspects of quantum optics, laser physics, and beyond. It turns out that they are also the prototypes of the general class of creation and annihilation operators for an entire class of particles. Particles can be classified by the value of their intrinsic angular momentum or spin, J, with particles of integral spin, $J = 0, 1, 2, \ldots$, called *bosons,* while those with half-integral spin $J = 1/2, 3/2, 5/2, \ldots$ called *fermions*. This last class includes the most familiar components of matter, the spin-1/2 electrons, protons, and neutrons that are known to satisfy the exclusion principle.

[2]For simplicity, we neglect the additional labels that describe the photon states of polarization; see Baym (1969) for more complete details.

The creation and annihilation operators for bosons are obviously inappropriate for fermions as they allow arbitrarily many particles to be in the same quantum state. Remarkably, the corresponding operators for fermionic degrees of freedom satisfy a set of commutation relations that are very similar in appearance to Eqns. (14.42) and (14.43), namely,

$$\{\hat{b}_{\mathbf{k}}, \hat{b}^{\dagger}_{\mathbf{k}}\} = \delta_{\mathbf{k}\mathbf{k}'} \tag{14.45}$$

and

$$\{\hat{b}_{\mathbf{k}}, \hat{b}_{\mathbf{k}}\} = 0 \qquad \text{and} \qquad \{\hat{b}^{\dagger}_{\mathbf{k}}, \hat{b}^{\dagger}_{\mathbf{k}}\} = 0 \tag{14.46}$$

The new symbol, { , }, denotes the so-called *anticommutator* of two operators and is defined by

$$\{\hat{A}, \hat{B}\} \equiv \hat{A}\hat{B} + \hat{B}\hat{A} \tag{14.47}$$

It is also sometimes written $\{\hat{A}, \hat{B}\} = [\hat{A}, \hat{B}]_{+}$.

While formally quite similar, the physical implications of these anticommutation relations are entirely different:

- The anticommutation relation of $\hat{b}^{\dagger}$ with itself implies that
$$\hat{b}^{\dagger}_{\mathbf{k}}\hat{b}^{\dagger}_{\mathbf{k}} = 0 \tag{14.48}$$
which may at first appear to be somewhat puzzling. This is simply the restatement of the exclusion principle that no two identical fermions can be put into the same quantum state. The state $|0\rangle$ with no quanta (the ground state or "vacuum") and $|\mathbf{k}\rangle = \hat{b}^{\dagger}_{\mathbf{k}}|0\rangle$ with one quantum are both allowed, but operating twice with $\hat{b}^{\dagger}_{\mathbf{k}}$ gives a vanishing result.
- The *number operator*, $\hat{N}_{\mathbf{k}} = \hat{b}^{\dagger}_{\mathbf{k}}\hat{b}_{\mathbf{k}}$, satisfies (with the $\mathbf{k}$ label suppressed)
$$\begin{aligned} \hat{N}^2 &= \left(\hat{b}^{\dagger}\hat{b}\right)\left(\hat{b}^{\dagger}\hat{b}\right) \\ &= \hat{b}^{\dagger}\left(\hat{b}\hat{b}^{\dagger}\right)\hat{b} \\ &= \hat{b}^{\dagger}\left(1 - \hat{b}^{\dagger}\hat{b}\right)\hat{b} \\ \hat{N}^2 &= \hat{b}^{\dagger}\hat{b} = \hat{N} \end{aligned} \tag{14.49}$$
This implies that the corresponding *number eigenvalues,* given by $\hat{N}|n\rangle = n|n\rangle$, satisfy $n^2 = n$, which has only the trivial (but appropriate) solutions $n = 0, 1$. Again, *at most one* fermion is allowed per quantum state.
- Ordinary numbers or operators cannot satisfy the anticommutation relations of Eqns. (14.45) and (14.46). One can check that an appropriate *matrix representation* of the $\hat{b}$ and $\hat{b}^{\dagger}$ is given by (P14.8):
$$\hat{b} = \begin{pmatrix} 0 & 1 \\ 0 & 0 \end{pmatrix} \qquad \text{and} \qquad \hat{b}^{\dagger} = \begin{pmatrix} 0 & 0 \\ 1 & 0 \end{pmatrix} \tag{14.50}$$

One of the most important implications of this formalism is that the wavefunctions for multiparticle quantum states are highly correlated. For example, if we label a state of two particles of similar type—say, both photons or both electrons, as $|1; 2\rangle$—we can write

$$|1; 2\rangle = \hat{a}^{\dagger}_1\hat{a}^{\dagger}_2|0\rangle = +\hat{a}^{\dagger}_2\hat{a}^{\dagger}_1|0\rangle = +|2; 1\rangle \tag{14.51}$$

for bosons while

$$|1;2\rangle = \hat{b}_1^\dagger \hat{b}_2^\dagger |0\rangle = -\hat{b}_2^\dagger \hat{b}_1^\dagger |0\rangle = -|2;1\rangle \tag{14.52}$$

for fermions. Such wavefunctions are classified, respectively, as being symmetric or antisymmetric under the interchange of the two particles. This constraint on systems of similar particles imposes important new constraints on the Schrödinger wavefunction. This will be explored further in the next chapter.

We see that far from being only a clever way to solve a special problem in quantum mechanics, the method of factorization, with its resulting ladder operators, plays a central role in the description of systems of particles of all possible spins.

14.4 FACTORIZATION METHODS AND SUPERSYMMETRIC QUANTUM MECHANICS

Given the elegance and power of the factorization method as applied to the harmonic oscillator, it is natural to ask whether there are other model systems to which this attractive method can be applied. To study this, let us begin by assuming that we know the exact ground state solution, $\psi_0(x)$, for some potential $V(x)$. We require knowledge of only the ground state, and we label its energy E_0. We next define a shifted potential energy function via $V_-(x) \equiv V(x) - E_0$, so that ψ_0 satisfies the Schrödinger equation

$$\hat{H}_- \psi_0 \equiv \left(-\frac{\hbar^2}{2m}\frac{d^2}{dx^2} + V_-(x) \right) \psi_0(x) = 0 \tag{14.53}$$

and $V_-(x)$ has a zero energy ground state. Since $\psi_0(x)$ is assumed known, using Eqn. (14.53), we can write $\hat{H}_-$ in the form

$$\hat{H}_- = \frac{\hbar^2}{2m}\left(-\frac{d^2}{dx^2} + \frac{\psi_0''(x)}{\psi_0(x)} \right) \tag{14.54}$$

A little algebra (P14.9) shows that if we define the ladder operators

$$\hat{A} \equiv \frac{\hbar}{\sqrt{2m}}\left(\frac{d}{dx} - \frac{\psi_0'}{\psi_0} \right) \quad \text{so that} \quad \hat{A}^\dagger \equiv \frac{\hbar}{\sqrt{2m}}\left(-\frac{d}{dx} - \frac{\psi_0'}{\psi_0} \right) \tag{14.55}$$

we then have $\hat{A}^\dagger \hat{A} = \hat{H}_-$; i.e., $\hat{H}_-$ is indeed factorizable. We can also use the short-hand notation

$$\hat{A} = \frac{\hbar}{\sqrt{2m}} + W(x) \quad \text{and} \quad \hat{A}^\dagger = -\frac{\hbar}{\sqrt{2m}} + W(x) \tag{14.56}$$

where

$$W(x) \equiv -\frac{\hbar}{\sqrt{2m}}\left(\frac{\psi_0'(x)}{\psi_0(x)} \right) \tag{14.57}$$

where $W(x)$ is usually called the *superpotential* for the problem.

The operators $\hat{A}$ and $\hat{A}^\dagger$, which generalize the raising and lowering operators of the harmonic oscillator problem, have a more complicated commutation relation than for that problem, since

$$[\hat{A}, \hat{A}^\dagger] = \hat{A}\hat{A}^\dagger - \hat{A}^\dagger\hat{A} = \frac{2\hbar}{\sqrt{2m}} W'(x) \tag{14.58}$$

Using the harmonic oscillator ground state wavefunction, one finds that $W(x) = \sqrt{\hbar\omega^2/2}x$ so that $[\hat{A}, \hat{A}^\dagger] = \hbar\omega$ in this notation, and the normalization of the $\hat{A}$, $\hat{A}^\dagger$ is slightly different from that in Sec. 14.2. Most importantly, however, the commutator is still a constant in that case, and the same general results hold.

While $\hat{A}^\dagger\hat{A} = \hat{H}_-$ factorizes the original Hamiltonian (up to an additive constant), the related combination, $\hat{A}\hat{A}^\dagger$, can be seen to define an entirely new potential since

$$\hat{A}\hat{A}^\dagger \equiv \hat{H}_+ = -\frac{\hbar^2}{2m}\frac{d^2}{dx^2} + V_+(x) \tag{14.59}$$

where

$$\begin{aligned} V_+(x) &= V_-(x) - \frac{\hbar^2}{m}\frac{d}{dx}\left(\frac{\psi_0'(x)}{\psi_0(x)}\right) \\ &= -V_-(x) + \frac{\hbar^2}{m}\left(\frac{\psi_0''(x)}{\psi_0(x)}\right) \\ &= -V_-(x) + 2W^2(x) \end{aligned} \tag{14.60}$$

The potentials $V_-(x)$ and $V_+(x)$ can be written in the form

$$V_\pm(x) = W^2(x) \pm \frac{\hbar}{\sqrt{2m}}W'(x) \tag{14.61}$$

and are called *supersymmetric partner potentials*. This is motivated by the fact, which we prove below, that

- $V_-(x)$ and $V_+(x)$ have the same energy level spectrum, $E_n^{(-)}$ and $E_n^{(+)}$, except that the zero energy ground state of $V_-(x)$, $E_0^{(-)}$, has no counterpart for $V_+(x)$.

  This connection can be trivially checked for the harmonic oscillator where the partner potentials and energy spectrum are given by

$$V_-(x) = \frac{1}{2}m\omega^2x^2 - \frac{\hbar\omega}{2} \Leftrightarrow E_n^{(-)} = n\hbar\omega, n = 0, 1, 2, 3, \ldots \tag{14.62}$$

$$V_+(x) = \frac{1}{2}m\omega^2x^2 + \frac{\hbar\omega}{2} \Leftrightarrow E_n^{(+)} = (n+1)\hbar\omega, n = 0, 1, 2, 3, \ldots \tag{14.63}$$

  To prove this connection for general superpartner potentials, we first denote the eigenfunctions of $\hat{H}_-$ and $\hat{H}_+$ by $\psi_n^{(-)}(x)$ and $\psi_n^{(+)}(x)$ with energy eigenvalues $E_n^{(-)}$ and $E_n^{(+)}$, respectively. The integer label $n = 0, 1, 2\ldots$ counts the number of nodes in the wavefunctions. Following the methods in Sec. 14.1, we can show that:

- If $\psi_n^{(-)}$ is any eigenfunction of $\hat{H}_-$ with eigenvalue $E_n^{(-)}$, then $\hat{A}\psi_n^{(-)}$ is an eigenfunction of $\hat{H}_+$ with the same eigenvalue.

This is easily seen since

$$\begin{aligned} \hat{H}_+\left(\hat{A}\psi_n^{(-)}\right) &= \hat{A}\hat{A}^\dagger\left(\hat{A}\psi_n^{(-)}\right) \\ &= \hat{A}\left(\hat{H}_-\psi_n^{(-)}\right) \\ &= E_n^{(-)}\left(\hat{A}\psi_n^{(-)}\right) \end{aligned} \tag{14.64}$$

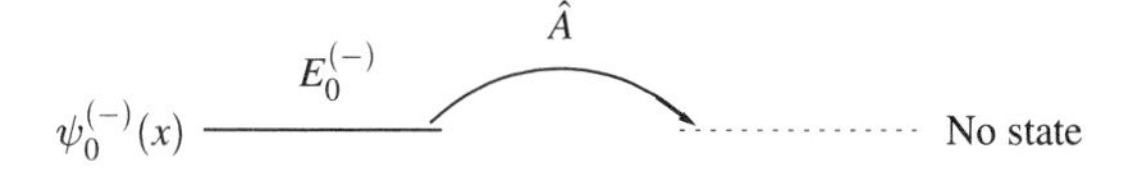

FIGURE 14.2. *The generalized ladder operators of supersymmetry that connect states of the same energy.*

One can similarly show that if $\psi_n^{(+)}$ is an eigenfunction of $\hat{H}_+$ with eigenvalue $E_n^{(+)}$, then $\hat{A}^\dagger \psi_n^{(+)}$ is an eigenfunction of $\hat{H}_-$ with the same eigenvalue. Taken together, these relations can be shown to imply (P14.10)

$$E_n^{(+)} = E_{n+1}^{(-)} \tag{14.65}$$

$$\psi_n^{(+)} = \frac{1}{\sqrt{E_{n+1}^{(-)}}} \hat{A}\, \psi_{n+1}^{(-)} \tag{14.66}$$

$$\psi_{n+1}^{(-)} = \frac{1}{\sqrt{E_n^{(+)}}} \hat{A}^\dagger\, \psi_n^{(+)} \tag{14.67}$$

These connections are illustrated schematically in Fig. 14.2, where we see that the $\hat{A}$ and $\hat{A}^\dagger$ connect states of the same energy for two, in principle, quite different potentials. For the harmonic oscillator, the superpartner potentials are actually identical up to an additive constant. In this very special case, the $\psi_n^{(-)}$ and $\psi_n^{(+)}$ are equivalent, and the $\hat{A}$ and $\hat{A}^\dagger$ can be thought of as connecting states of different energy in the same potential, thereby "solving" for the spectrum.

Example 14.2 Supersymmetric Partner of the Infinite Well A surprisingly simple example of this connection between the energy levels of two different potentials comes from the most familiar of all model systems, namely, the standard infinite well. For the unshifted problem, we have the energies and eigenfunctions

$$E_n = \frac{\hbar^2 \pi^2 (n+1)^2}{2ma^2} \quad \text{and} \quad \psi_n(x) = \sqrt{\frac{2}{a}} \sin\left(\frac{(n+1)\pi x}{a}\right) \tag{14.68}$$

where $n = 0, 1, 2 \ldots$. The notation is slightly different than that used in Sec. 6.3.1, since we want n to count the wavefunction nodes. If we now subtract the ground state energy, we have the potential

$$V_-(x) = V(x) - \frac{\hbar^2 \pi^2}{2ma^2} \tag{14.69}$$

with the ground state wavefunction

$$\psi_0^{(-)}(x) = \sqrt{\frac{2}{a}} \sin\left(\frac{\pi x}{a}\right) \tag{14.70}$$

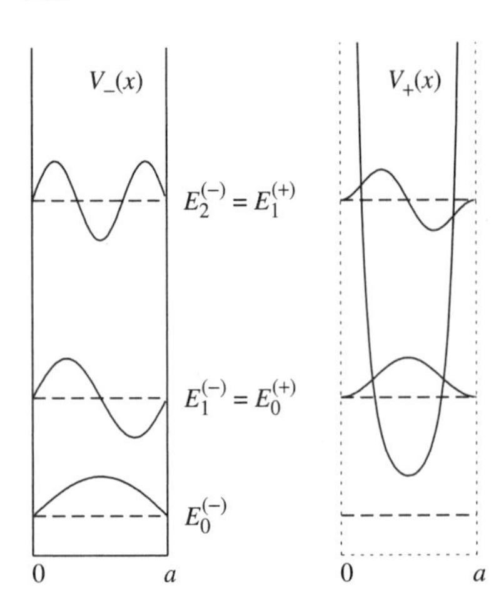

FIGURE 14.3. *Energy spectrum and wavefunctions for the infinite well, $V_-(x)$, and its supersymmetric partner potential, $V_+(x)$.*

The wavefunctions $\psi_n^{(-)}(x)$ are, of course, just those of Eqn. (14.68), while the quantized energies are

$$E_n^{(-)} = \frac{\hbar^2\pi^2 n(n+2)}{2ma^2} \qquad n = 0, 1, 2, \ldots \tag{14.71}$$

Using Eqn. (14.57), we find that the superpotential is given by

$$W(x) = -\frac{\hbar}{\sqrt{2m}}\frac{\pi}{a}\cot\left(\frac{\pi x}{a}\right) \tag{14.72}$$

This can be used to calculate the superpartner potential to the infinite well, namely,

$$V_+(x) = \frac{\hbar^2\pi^2}{2ma^2}\left[2\csc^2\left(\frac{\pi x}{a}\right) - 1\right] \tag{14.73}$$

The wavefunctions for $V_-(x)$ are then obtained from Eqn. (14.66), and one finds

$$\begin{aligned} \psi_n^{(+)}(x) &\propto \hat{A}\psi_{(n+1)}^{(-)}(x) \\ &\propto \left[\frac{d}{dx} - \frac{\pi}{a}\cot\left(\frac{\pi x}{a}\right)\right]\sin\left(\frac{(n+2)\pi x}{a}\right) \end{aligned} \tag{14.74}$$

We compare the potential energy, energy levels, and wavefunctions for the first few states in Fig.14.3, which shows the patterns of nodes.

Supersymmetric quantum mechanics,[3] which studies the relationship between pairs of superpartner potentials, can shed light on a number of interesting issues. We mention a few here and provide further references for those interested.

- The extent to which the observed energy spectrum of a system can be used to determine the potential energy function is an important one. Can one "invert" the energy levels and determine a $V(x)$ that gives them? Is such an inversion unique? The same questions arise not only for bound states but for scattering as well, hence the name

[3]For a recent review, see Cooper, Khare, and Sukhatme (1995).

inverse scattering problem. Supersymmetric quantum systems with their almost identical spectra provide new insight into this problem.

- There are a handful of potentials for which the Schrödinger equation can be solved analytically. These include the infinite well, the harmonic oscillator (in one, two, and three dimensions), the Coulomb potential, and a few others. All of these potentials share the fact that they can be factorized using operator techniques, have supersymmetric analogs, and have the property of "shape invariance,"[4] namely, that their corresponding superpartner potentials are similar in shape, differing only in the parameters in their definition.
- Finally, supersymmetric quantum mechanics can be written in a way that is suggestive of a connection between the fermion and boson degrees of freedom[5] discussed in Sec. 14.3.

14.5 QUESTIONS AND PROBLEMS

P14.1 Consider the factorized differential equation

$$y''(x) - 2ry'(x) + r^2 y(x) = (\hat{D} - r)^2 y(x) = 0$$

which has r as a double root.

(a) Show that $y(x) = e^{rx}$ is still a solution.

(b) To extract the second, linearly independent solution, imagine that there are two *different* roots, r, s. The linear combination given by

$$y(x) = \frac{e^{rx} - e^{sx}}{r - s}$$

will also be a solution in this case. Let $s \to r$ to obtain a second solution, and explicitly check that it works.

(c) Extend these results to the case where there is a root with an N-fold degeneracy.

P14.2 Confirm that Eqn. (14.38) gives the proper normalization for the action of $\hat{A}$ and $\hat{A}^\dagger$ on a general state $|n\rangle$.

P14.3 **Orthogonality of harmonic oscillator states:** We can use purely algebraic techniques to show that the harmonic oscillator states are mutually orthogonal, namely, that

$$\langle n|m\rangle = \delta_{n,m}$$

It is easiest to use a proof by induction as follows:

(a) Since $|n\rangle \propto (\hat{A}^\dagger)^n|0\rangle$, we also have $\langle n| \propto \langle 0|(\hat{A})^n$, so that

$$\langle n|0\rangle = \langle 0|(\hat{A})^n|0\rangle = \langle 0|(\hat{A})^{n-1}\hat{A}|0\rangle = 0$$

since $\hat{A}$ "annihilates the vacuum state." This gives

$$\langle n|0\rangle = 0 \quad \text{for} \quad n > 0$$

(b) *Assume* that it has been shown that

$$\langle n|m\rangle = 0$$

[4] See Dutt, Khare, and Sukhatme (1988) for a nice discussion.

[5] See Schwabl (1992).

for *all* n values for *some* value of m. Show that

$$\langle n|m+1\rangle \propto (m+1)\langle n-1|m\rangle$$

Show that these two pieces, taken together, constitute a proof by induction.

P14.4 Consider a Hamiltonian written in terms of raising and lowering operators of the form

$$\hat{H} = \epsilon_1\left(\hat{A}^\dagger\hat{A}\right) + \epsilon_2\left(\hat{A} + \hat{A}^\dagger\right)$$

where $\epsilon_{1,2}$ are real constants with units of energy and $[\hat{A}, \hat{A}^\dagger] = 1$.

(a) Show that $\hat{H}$ is Hermitian.

(b) Find the energy eigenvalues of the Hamiltonian *in a purely algebraic way,* i.e., do not transform back to x, $\hat{p}$ operators. *Hint:* Introduce *new* operators $\hat{D} = \alpha\hat{A} + \beta$ (which defines $\hat{D}^\dagger$), and choose the α, β so as to make the resulting problem look like the harmonic oscillator.

(c) Repeat for the Hamiltonian

$$\hat{H} = \epsilon_3\left(\hat{A}^\dagger\hat{A}\right) + \epsilon_4 i\left(\hat{A} - \hat{A}^\dagger\right)$$

(d) Transform these Hamiltonians into position or momentum space differential operators, solve the appropriate differential equations, and show that the spectra are the same as given using purely operator methods.

P14.5 Use raising and lowering operators to show that

$$\langle n|x|k\rangle = \sqrt{\frac{\hbar}{2m\omega}}\left(\sqrt{n+1}\,\delta_{n+1,k} + \sqrt{n}\,\delta_{n-1,k}\right)$$

and

$$\langle n|x^2|k\rangle = \left(\frac{\hbar}{2m\omega}\right)\left[\sqrt{(n+1)(n+2)}\,\delta_{n+2,k} + (2n+1)\delta_{n,k} + \sqrt{n(n-1)}\,\delta_{n-2,k}\right]$$

Find the corresponding relations for the momentum operator, $\hat{p}$. These relations are useful both for the matrix formulation of the harmonic oscillator in Ex. 11.8 and for perturbation theory.

P14.6 Raising and lowering operators:

(a) Show that the raising and lowering operators satisfy

$$[\hat{H}, \hat{A}] = -\hbar\omega\hat{A} \qquad \text{and} \qquad [\hat{H}, \hat{A}^\dagger] = \hbar\omega\hat{A}^\dagger$$

(b) In the Heisenberg picture (see P13.19), where operators have a nontrivial time dependence, we can think of the harmonic oscillator states, $|n\rangle$, as being time-independent, while the raising and lowering operators are given by

$$\hat{A}(t) = e^{i\hat{H}t/\hbar}\,\hat{A}\,e^{-i\hat{H}t/\hbar}$$

and similarly for $\hat{A}^\dagger(t)$. Show that $[\hat{A}(t), \hat{A}^\dagger(t)] = 1$, independent of t. Show that the commutation relations of part (a) are also valid for the $\hat{A}(t)$.

(c) Use these facts, and the time development equation for operators, $d\hat{O}(t)/dt = (i/\hbar)[\hat{H}, \hat{O}]$, to show that $\hat{A}(t) = \hat{A}(0)e^{-i\omega t}$ and $\hat{A}^\dagger(t) = \hat{A}^\dagger(0)e^{i\omega t}$. Use the defining relations for the $\hat{A}(0)$, $\hat{A}^\dagger(0)$ to find expressions for the *operators* $x(t)$, and $\hat{p}(t)$.

P14.7 Coherent states: It is possible to construct wavepacket-like quantum states for which the expectation value of $x(t)$ is given by the classical, trajectory-like, functional form using raising and lowering operators.

(a) Consider (somewhat abstractly at first) an eigenstate of the lowering operator $\hat{A}$, namely,

$$\hat{A}|z\rangle = z|z\rangle$$

Such a state $|z\rangle$ is called a *coherent state.* Since $\hat{A}$ is not Hermitian, the eigenvalue z is not necessarily real; we thus have the corresponding "bra" relation

$$\langle z|\hat{A}^{\dagger} = z^{*}\langle z|$$

Use the results of P14.6(c) to show that the expectation value of x in this state is given by

$$\begin{aligned}\langle z|x(t)|z\rangle &= \sqrt{\frac{\hbar}{2m\omega}}\langle z|\hat{A}(t) + \hat{A}^{\dagger}(t)|z\rangle \\ &= B\cos(\omega t - \phi)\end{aligned}$$

where we have used $z = |z|e^{i\phi}$ and defined $B = |z|\sqrt{2\hbar/m\omega}$. Find the corresponding expression for $\langle z|p(t)|z\rangle$. Such states have an obvious classical analog, and the amplitude and phase of the sinusoidal behavior is directly related to the amplitude and phase of the eigenvalue z.

(b) Using $\hat{H} = \hbar\omega(\hat{A}^{\dagger}\hat{A} + 1/2)$, evaluate

$$\langle z|\hat{H}|z\rangle \qquad \text{and} \qquad \langle z|\hat{H}^2|z\rangle$$

and show that $\Delta H/\langle H\rangle \to 1/|z|$ for large z.

(c) We can find a representation of such a coherent state in terms of number eigenstates, $|n\rangle$, by writing

$$|z\rangle = \sum_{n=0}^{\infty} a_n|n\rangle \qquad \text{where} \qquad a_n = \langle n|z\rangle \quad \text{is the expansion coefficient.}$$

To do this, multiply both sides (on the left) first by $\hat{A}$ and then the bra vector $|j-1\rangle$ to obtain the recursion relation

$$a_j = \left(\frac{z}{\sqrt{j}}\right)a_{j-1}$$

and show that this implies that

$$a_n = \frac{z^n}{\sqrt{n!}}a_0$$

The coherent state can then be written as

$$|z\rangle = a_0\sum_{n=0}^{\infty}\frac{z^n}{\sqrt{n!}}|n\rangle$$

Finally, evaluate a_0 by insisting that $|z\rangle$ be normalized; show that $a_0 = \exp(-|z|^2/2)$.

(d) Show that the probability that a measurement of the energy will yield the value $E_n = (n+1/2)\hbar\omega$ is given by a Poisson distribution with mean value $\lambda = |z|^2$, i.e.,

$$\text{Prob}(E_n = (n+1/2)\hbar\omega) = P_n = \frac{\lambda^n}{n!}e^{-\lambda} \qquad \text{where } \lambda = |z|^2$$

(e) Use the P_n to show that

$$\begin{aligned}\langle E\rangle &= \sum_{n=0}^{\infty} E_n P_n = \left(|z|^2 + 1/2\right)\hbar\omega \\ \langle E^2\rangle &= \sum_{n=0}^{\infty} E_n^2 P_n = \left(|z|^4 + 2|z|^2 + 1/4\right)(\hbar\omega)^2\end{aligned}$$

so that $\Delta E = |z|\hbar\omega$. Discuss the fractional uncertainty in energy $\Delta E/E$ in the classical limit where $|z| \gg 1$.

(f) Evaluate Δx and Δp, and show that their product is the minimum value allowed by the uncertainty principle.

(g) Show that

$$\langle z_1|z_2\rangle = \exp\left(-|z_1|^2/2 - |z_2|^2/2 + z_1^* z_2\right)$$

so that coherent states are not orthogonal.

P14.8 Show that the matrices in Eqn. (14.50) satisfy the anticommutation relations in Eqns. (14.45) and (14.46).

P14.9 Show that the ladder operators in Eqn. (14.55) factorize $\hat{H}_-$, i.e., $\hat{A}^\dagger\hat{A} = \hat{H}_-$.

P14.10 Confirm the relations in Eqns. (14.65) to (14.67).

P14.11 Find the first three $\psi_n^{(+)}(x)$ wavefunctions in the supersymmetric partner potential of the infinite well in Ex. 14.2. Confirm that they satisfy the Schrödinger equation for the $V_+(x)$ given in Eqn. (14.73).

P14.12 **(a)** Find the supersymmetric partner potentials, $V_-(x)$ and $V_+(x)$, corresponding to the single δ function potential in Sec. 9.1.2. Show that $V_+(x)$ doesn't admit any bound states and explain why.

(b) Analyze the supersymmetric partner potential for the twin δ function potential in Sec. 9.1.4. See Goldstein et al. (1994) for details.

CHAPTER 15

Multiparticle Systems

15.1 GENERALITIES

In classical mechanics, Newton's laws for a multiparticle system have the form

$$m_i \frac{d^2 x_i(t)}{dt^2} = \mathcal{F}_i(x_i) + \sum_{j}^{j \neq i} F_{ij}(x_i - x_j) \qquad \text{for } i = 1, 2, \ldots, N \tag{15.1}$$

where we have specialized to the case of external forces $\mathcal{F}_i(x_i)$ that act on each particle separately and mutual two-body interactions $F_{ij} = F_{ij}(x_i - x_j)$. The functional form of the F_{ij} is consistent with Newton's third law, which requires that $F_{ij} = -F_{ji}$. When solved self-consistently, these equations predict the time dependence of the coordinates, $x_i(t)$, once the initial conditions are specified.

The time development of the corresponding quantum system is dictated by the *multi-particle Hamiltonian* operator

$$\hat{H} = \sum_i \frac{\hat{p}_i^2}{2m_i} + \sum_i \mathcal{V}_i(x_i) + \sum_{i,j}^{i>j} V_{ij}(x_i - x_j) \tag{15.2}$$

Here the external and two-body potentials give the corresponding forces via

$$\mathcal{F}_i(x_i) = -\frac{\partial \mathcal{V}_i(x_i)}{\partial x_i} \qquad \text{and} \qquad F_{ij}(x_i - x_j) = -\frac{\partial V_{ij}(x_i - x_j)}{\partial x_i} \tag{15.3}$$

and the restriction $i > j$ on the double sum is to avoid double counting. The momentum operator corresponding to each coordinate is $\hat{p}_i = (\hbar/i)\partial/\partial x_i$, and one has $[x_j, \hat{p}_k] = i\hbar\delta_{jk}$.

This acts on a multiparticle wavefunction $\psi(x_1, x_2, \ldots, x_n; t)$ and generalizes the Schrödinger equation to

$$\hat{H}\psi(x_1, x_2, \ldots, x_N; t) = i\hbar \frac{\partial}{\partial t}\psi(x_1, x_2, \ldots, x_n; t) \tag{15.4}$$

If none of the potentials depends on time, the usual exponential time dependence is found so that $\psi(x_1, x_2, \ldots, x_N; t) = \psi_E(x_1, x_2, \ldots, x_n)e^{-iEt/\hbar}$ where ψ_E satisfies the time-independent Schrödinger equation

$$\hat{H}\psi_E(x_1, x_2, \ldots, x_N) = E\psi_E(x_1, x_2, \ldots, x_N) \tag{15.5}$$

The multiparticle wavefunction is then associated with a probability amplitude so that

$$P(x_1, x_2, \ldots, x_N; t) = |\psi(x_1, x_2, \ldots, x_N; t)|^2 \tag{15.6}$$

is a *multivariable probability density.* A more concrete definition is:

- $|\psi(x_1, x_2, \ldots, x_n; t)|^2\, dx_1\, dx_2, \ldots, dx_n$ is the probability that a measurement of the positions of the N particles, at time t, would find

  particle 1 in the interval $(x_1, x_1 + dx_1)$ and
  particle 2 in the interval $(x_2, x_2 + dx_2)$ and
  ⋮
  and
  particle N in the interval $(x_N, x_N + dx_N)$

Just as with any multivariable probability distribution, this implies that the wavefunction must be normalized so that

$$\int_{-\infty}^{+\infty} dx_1 \int_{-\infty}^{+\infty} dx_2 \cdots \int_{-\infty}^{+\infty} dx_n |\psi(x_1, x_2, \ldots, x_n; t)|^2 = 1 \tag{15.7}$$

and the total probability of measuring "something" is unity and not, for example, N. We emphasize that $|\psi|^2$ does not "count the number of particles" but rather specifies the probability of the system of particles all being in a particular state.

All of the usual conditions on the smoothness and convergence of one-dimensional wavefunctions can be easily generalized, as can the expressions for expectation values; for example, one has

$$\langle \hat{O} \rangle_t = \int_{-\infty}^{+\infty} dx_1 \int_{-\infty}^{+\infty} dx_2 \cdots \int_{-\infty}^{+\infty} dx_n\, \psi^*(x_1, x_2, \ldots, x_N; t) \hat{O} \psi(x_1, x_2, \ldots, x_N; t) \tag{15.8}$$

for the average value of any operator $\hat{O}$.

Example 15.1 Correlations in Two-particle Wavefunctions Consider a two-particle wavefunction

$$\psi(x_1, x_2) = N e^{-(ax_1^2 + 2bx_1x_2 + cx_2^2)/2} \tag{15.9}$$

where we must have $a, c > 0$ in order for the wavefunction to be normalizable. The normalization constant is determined by the condition that

$$\begin{aligned} 1 &= \int_{-\infty}^{+\infty} dx_1 \int_{-\infty}^{+\infty} dx_2 |\psi(x_1, x_2)|^2 \\ &= N^2 \int_{-\infty}^{+\infty} dx_2 e^{-(c - b^2/a)x_2^2} \int_{-\infty}^{+\infty} dx_1 e^{-a(x_1 + bx_2/a)^2} \\ &= N^2 \frac{\pi}{\sqrt{ac - b^2}} \end{aligned} \tag{15.10}$$

This form (obtained by completing the square in the exponent) is useful in that it shows that one must also have $ac - b^2 > 0$ in order for the state to be acceptable. It is then easy to show that $\langle x_1 \rangle = \langle x_2 \rangle = 0$ while the covariance is given by

$$\big\langle (x_1 - \langle x_1 \rangle)(x_2 - \langle x_2 \rangle) \big\rangle = \langle x_1 x_2 \rangle = -\frac{b}{2(ac - b^2)} \tag{15.11}$$

This shows that the positions of the coordinates are correlated with each other, so that a measurement of one provides nontrivial information on the other. When $b \to 0$, we note that the wavefunction becomes the (uncorrelated) product of two Gaussians.

15.2 SEPARABLE SYSTEMS

A great simplification occurs if the mutual interactions of the particles can be ignored ($V_{ij} = 0$), because the Hamiltonian then takes the form

$$\hat{H} = \sum_i \frac{\hat{p}_i^2}{2m_i} + \sum_i \mathcal{V}(x_i) = \sum_i \left(\frac{\hat{p}_i^2}{2m_i} + \mathcal{V}(x_i) \right) \equiv \sum_i \hat{H}_i \tag{15.12}$$

With no mutual interactions present to give rise to dynamical correlations, it is natural to assume a product solution of the form

$$\psi_E(x_1, x_2, \ldots, x_N) = \psi_1(x_1)\psi_2(x_2)\cdots\psi_N(x_N) \tag{15.13}$$

The time-independent Schrödinger equation, Eqn. (15.5), can then be written as

$$\begin{aligned} E\left[\psi_1(x_1)\psi_2(x_2)\cdots\psi_N(x_N)\right] &= \left[\hat{H}_1\psi_1(x_1)\right]\psi_2(x_2)\cdots\psi_N(x_N) \\ &\quad + \psi_1(x_1)\left[\hat{H}_2\psi_2(x_2)\right]\cdots\psi_N(x_N) \\ &\quad + \cdots + \psi_1(x_1)\psi_2(x_2)\cdots\left[\hat{H}_N\psi_N(x_N)\right] \end{aligned} \tag{15.14}$$

Using the usual separation of variables trick, we divide both sides by Eqn. (15.13) and find

$$\frac{[\hat{H}_1\psi_1(x_1)]}{\psi_1(x_1)} + \frac{[\hat{H}_2\psi_2(x_2)]}{\psi_2(x_2)} + \cdots + \frac{[\hat{H}_N\psi_N(x_N)]}{\psi_N(x_N)} = E \tag{15.15}$$

This is only consistent if

$$\hat{H}_i\psi_i(x_i) = E_i\psi_i(x_i) \qquad \text{for } i = 1, 2, \ldots, N \tag{15.16}$$

where $E_1 + E_2 + \cdots + E_N = E$; we then have to solve N "versions" of the one-dimensional problem. Several comments can be made:

- If each component wavefunction is properly normalized, then the product solution is also, since

$$\int_{-\infty}^{+\infty} dx_1 \cdots \int_{-\infty}^{+\infty} dx_N |\psi(x_1, x_2, \ldots, x_N)|^2 = \prod_i \left[\int_{-\infty}^{+\infty} dx_i |\psi_i(x_i)|^2 \right] = 1 \tag{15.17}$$

- The overall time-dependence can also be factorized, since

$$e^{-iEt/\hbar} = e^{-iE_1t/\hbar} \ldots e^{-iE_nt/\hbar}$$

which implies that

$$\psi(x_1, x_2, \ldots, x_N; t) = \psi_1(x_1, t)\psi_2(x_2, t)\cdots\psi_N(x_N, t) \tag{15.18}$$

This is potentially useful, as wavepackets for each particle can be constructed using superposition techniques, so that products of such wavepackets will also be valid solutions for the noninteracting case.

Example 15.2 Degeneracy in Two-particle Systems Consider two particles of the same mass m confined to the standard infinite well; for the moment, we neglect any mutual interactions.[1] The general solution of this two-particle system is

$$\psi_{(n_1,n_2)}(x_1, x_2) = u_{n_2}(x_1)u_{n_2}(x_2) \tag{15.19}$$

where $u_n(x) = \sqrt{2/a}\sin(n\pi x/a)$ with the corresponding energy spectrum

$$E_{(n_1,n_2)} = \frac{\hbar^2\pi^2}{2ma^2}\left(n_1^2 + n_2^2\right) \tag{15.20}$$

The ground state energy is $\hbar^2\pi^2/ma^2$, corresponding to $(n_1, n_2) = (1, 1)$ and is unique. The first excited state, given by the two choices $(1, 2)$ and $(2, 1)$, is doubly degenerate, and the corresponding wavefunctions can be written as $\psi_\alpha = \psi_{(1,2)}(x_1, x_2)$ and $\psi_\beta = \psi_{(2,1)}(x_1, x_2)$. These choices are not unique because we can invoke the linearity of the Schrödinger equation to show that any linear combination of these two is also a solution with energy $E_{(1,2)} = 5\hbar^2\pi^2/2ma^2 = E_{(2,1)}$.

We can now use this example to illustrate the methods of degenerate perturbation theory. We add a small mutual interaction term given by $V'(x_1, x_2) = g\delta(x_1 - x_2)$ where positive (negative) g corresponds to a repulsive (attractive) interaction between the two particles. Referring to Sec. 11.6.2, we require the various matrix elements of the perturbing interaction; for example,

$$\begin{aligned}
\mathbf{H}'_{\alpha\alpha} &= \langle\psi_\alpha|V(x_1 - x_2)|\psi_\alpha\rangle \\
&= \int_0^a dx_1 \int_0^a dx_2[u_1(x_1)u_2(x_2)](g\delta(x_1 - x_2))[u_1(x_1)u_2(x_2)] \\
&= g\left(\frac{2}{a}\right)^2 \int_0^a \sin^2\left(\frac{\pi x_1}{a}\right) dx_1 \sin^2\left(\frac{2\pi x_1}{a}\right) \\
&= \frac{g}{a}
\end{aligned} \tag{15.21}$$

with identical answers for $\mathbf{H}'_{\beta\beta}$ and $\mathbf{H}'_{\alpha\beta}$. The condition determining the (split) energy eigenvalues (Eqn. (11.143)) then reads

$$\det\begin{pmatrix} \mathscr{E} + g/a - E & g/a \\ g/a & \mathscr{E} + g/a - E \end{pmatrix} = 0 \tag{15.22}$$

where $\mathscr{E} = E_{(2,1)} = E_{(1,2)}$ is the initially degenerate energy level. The resulting polynomial equation is easily solved and yields

$$E^{(\pm)} = \mathscr{E} + \frac{g}{a} \pm \frac{g}{a} \tag{15.23}$$

The two energy eigenvalues and corresponding (normalized) eigenstates are given by

$$E^{(-)} = \mathscr{E} \qquad \psi^{(-)}(x_1, x_2) = \frac{1}{\sqrt{2}}\left(\psi_{1,2}(x_1, x_2) - \psi_{2,1}(x_1, x_2)\right) \tag{15.24}$$

$$E^{(+)} = \mathscr{E} + \frac{2g}{a} \qquad \psi^{(+)}(x_1, x_2) = \frac{1}{\sqrt{2}}\left(\psi_{1,2}(x_1, x_2) + \psi_{2,1}(x_1, x_2)\right) \tag{15.25}$$

The antisymmetric combination state, corresponding to $E^{(-)}$, is unshifted in energy because the wavefunction vanishes where the perturbation has any effect, namely, for $x_1 = x_2$. The

[1] If they are both in the same 1D well, this implies that they are somewhat "ghostlike" as they must be able to "pass through" each other.

symmetric solution has a larger probability of having $x_1 = x_2$ than does either ψ_α or ψ_β individually, and it can "feel" the effect of the perturbation; the energy of the symmetric state is increased or decreased depending on the sign of g.

15.3 TWO-BODY SYSTEMS

While much of classical mechanics is concerned with the motion of single particles under the influence of external forces, many standard problems, especially in gravitation, are concerned with the motion of two bodies subject only to their mutual interaction. While general methods of solution for the N-body problem[2] (with $N \geq 3$) do not exist, a simple change of variables is often enough to transform Newton's equations for two particles into an effective one-particle problem that can then be approached using a variety of familiar techniques.

Such techniques are perhaps even more important in quantum mechanics where many of the "textbook" examples are two-body systems; examples include diatomic molecules, the hydrogen atom, the deuteron (proton-neutron bound state), and quarkonia (quark-antiquark bound state). In these cases, we are often more interested in probing the (sometimes unknown) force between the particles, so that not having to deal with the complications of many particles is extremely important. Even though it is implemented in a very different way, the same coordinate transformation "trick" works in both classical and quantum mechanics, and we begin our study by reviewing the classical case.

15.3.1 Classical Systems

The classical equations of motion for a two-particle system with no external forces and only mutual two-body interactions are

$$m_1 \ddot{x}_1(t) = F_{21}(x_1 - x_2) \qquad \text{and} \qquad m_2 \ddot{x}_2(t) = F_{12}(x_1 - x_2) \tag{15.26}$$

and we recall that $F \equiv F_{21} = -F_{12}$ from Newton's third law. Two combinations of these equations then immediately suggest themselves and naturally select out a new set of variables. If, for example, we add the two equations, we obtain

$$\begin{aligned} 0 = F_{21} + F_{12} &= m_1 \ddot{x}_1(t) + m_2 \ddot{x}_2(t) \\ &= (m_1 + m_2)\left(\frac{m_1 \ddot{x}_1(t) + m_2 \ddot{x}_2(t)}{m_1 + m_2}\right) \\ 0 &= M\ddot{X}(t) \end{aligned} \tag{15.27}$$

where we define the *total mass* $M = m_1 + m_2$ and the *center-of-mass coordinate*

$$X(t) = \frac{m_1 x_1(t) + m_2 x_2(t)}{m_1 + m_2} \tag{15.28}$$

We note that Eqn. (15.27) is the standard result, that if there are no net external forces, the center of mass of a system moves at constant speed. A related variable is the *total momentum* given by

$$P(t) = M\dot{X}(t) = m_1 v_1(t) + m_2 v_2(t) = p_1(t) + p_2(t) \tag{15.29}$$

so that Eqn. (15.27) also shows that the total momentum is conserved.

[2]See, e.g., Symon (1971).

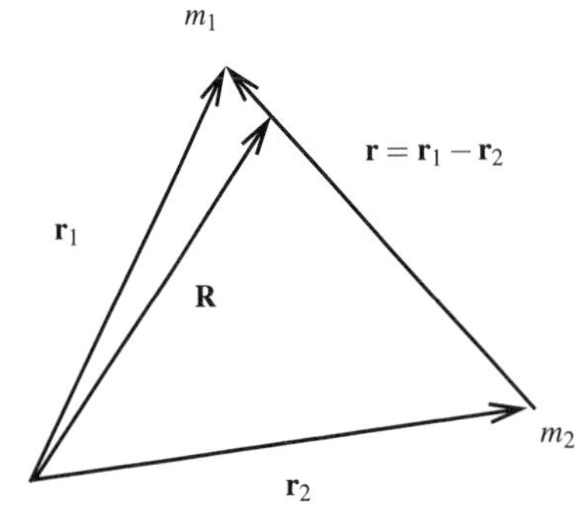

FIGURE 15.1. Center of mass (R) and relative ($r = r_1 - r_2$) coordinates for a two-body system in two-dimensions, which generalizes Eqn. (15.37). Estimate the ratio m_1/m_2 from the figure.

If we now divide both sides of Eqns. (15.26) by the respective masses and subtract, we find

$$F\left(\frac{1}{m_1} + \frac{1}{m_2}\right) = \frac{F_{21}}{m_1} - \frac{F_{12}}{m_2} = \ddot{x}_1(t) - \ddot{x}_2(t) \tag{15.30}$$

or

$$F(x) = \mu\ddot{x}(t) \tag{15.31}$$

where we have defined the *reduced mass* via

$$\frac{1}{\mu} \equiv \frac{1}{m_1} + \frac{1}{m_2} \qquad \text{or} \qquad \mu = \frac{m_1 m_2}{m_1 + m_2} \tag{15.32}$$

and the *relative coordinate* via

$$x(t) = x_1(t) - x_2(t) \tag{15.33}$$

while $F(x) = F_{21}(x_1 - x_2)$. The nontrivial dynamics of the system is then described by Eqn. (15.31); the interesting physics is all contained in the relative coordinate that describes the motion of a fictitious particle of effective mass μ. The change of variables can also be inverted to give

$$x_1 = X + \frac{m_2}{M}x \qquad \text{and} \qquad x_2 = X - \frac{m_1}{M}x \tag{15.34}$$

so that the motion of each particle can be extracted if so desired. The same variable change works in more realistic two- and three-dimensional systems, and we visualize the new coordinates in 2D in Fig. 15.1.

15.3.2 Quantum Case

The quantum version of this problem requires us to solve the two-particle Schrödinger equation given by

$$\hat{H}\psi(x_1, x_2) = E\psi(x_1, x_2) \tag{15.35}$$

where the Hamiltonian is given by

$$\begin{aligned}\hat{H} &= \frac{\hat{p}_1^2}{2m_1} + \frac{\hat{p}_2^2}{2m_2} + V(x_1 - x_2) \\ &= -\frac{\hbar^2}{2m_1}\frac{\partial^2}{\partial x_1^2} - \frac{\hbar^2}{2m_2}\frac{\partial^2}{\partial x_2^2} + V(x_1 - x_2)\end{aligned} \tag{15.36}$$

The change to center of mass and relative coordinates

$$X = \frac{m_1 x_1 + m_2 x_2}{m_1 + m_2} \qquad \text{and} \qquad x = x_1 - x_2 \tag{15.37}$$

(where we drop the t dependence of the coordinates) is trivially implemented for the potential energy term where $V(x_1 - x_2) = V(x)$. For the kinetic energy operators, we need to rewrite

$$\hat{p}_i = \frac{\hbar}{i} \frac{\partial}{\partial x_i} \qquad \text{for } i = 1, 2 \tag{15.38}$$

in terms of the spatial derivatives of the X, x coordinates, which give the momentum operators corresponding to those variables, namely,

$$\hat{P} \equiv \frac{\hbar}{i} \frac{\partial}{\partial X} \qquad \text{and} \qquad \hat{p} \equiv \frac{\hbar}{i} \frac{\partial}{\partial x} \tag{15.39}$$

This requires the chain rule relation

$$\frac{\partial}{\partial x_{1,2}} = \frac{\partial X}{\partial x_{1,2}} \frac{\partial}{\partial X} + \frac{\partial x}{\partial x_{1,2}} \frac{\partial}{\partial x} \tag{15.40}$$

which, using Eqn. (15.37), gives

$$\hat{p}_1 = \frac{m_1}{M} \hat{P} + \hat{p} \qquad \text{and} \qquad \hat{p}_2 = \frac{m_2}{M} \hat{P} - \hat{p} \tag{15.41}$$

We note that adding these two equations gives $\hat{P} = \hat{p}_1 + \hat{p}_2$, now as an operator relation. The kinetic energy operators in the Hamiltonian can now be written as

$$\begin{aligned} \frac{\hat{p}_1^2}{2m_1} + \frac{\hat{p}_2^2}{2m_2} &= \frac{1}{2m_1} \left(\frac{m_1}{M} \hat{P} + \hat{p} \right)^2 + \frac{1}{2m_2} \left(\frac{m_2}{M} \hat{P} - \hat{p} \right)^2 \\ &= \frac{1}{2m_1} \left(\frac{m_1^2}{M^2} \hat{P}^2 + \frac{m_1}{M} \left(\hat{p}\hat{P} + \hat{P}\hat{p} \right) + \hat{p}^2 \right) \\ &\quad + \frac{1}{2m_2} \left(\frac{m_2^2}{M^2} \hat{P}^2 - \frac{m_2}{M} \left(\hat{p}\hat{P} + \hat{P}\hat{p} \right) + \hat{p}^2 \right) \\ &= \frac{\hat{P}^2}{2M} + \frac{\hat{p}^2}{2\mu} \end{aligned} \tag{15.42}$$

We have been careful with the ordering of $\hat{P}$ and $\hat{p}$, but it is easy to see that $[\hat{P}, \hat{p}] = 0$. This can also be used to confirm that the total momentum of the system is a constant, since $\hat{P}$ commutes with the Hamiltonian operator, i.e.,

$$[\hat{H}, \hat{P}] = \frac{1}{2M} [\hat{P}^2, \hat{P}] + \frac{1}{2\mu} [\hat{p}^2, \hat{P}] + [V(x), \hat{P}] = 0 \tag{15.43}$$

We emphasize that this is true only under our assumption that the potential is only a function of the relative coordinate; if there are external forces, an equivalent classical result is obtained (P15.3).

Most importantly, in the new coordinates, the Hamiltonian is now separable since

$$\hat{H} = \frac{\hat{P}^2}{2M} + \left(\frac{\hat{p}^2}{2\mu} + V(x) \right) = \hat{H}_X + \hat{H}_x \tag{15.44}$$

so a product wavefunction of the form

$$\left[\Psi(X)e^{-iE_X t/\hbar}\right]\left[\psi(x)e^{-iE_x t/\hbar}\right] \tag{15.45}$$

satisfying

$$\hat{H}_X\Psi(X) = E_X\Psi(X) \qquad \text{and} \qquad \hat{H}_x\psi(x) = E_x\psi(x) \tag{15.46}$$

is a solution. The center-of-mass equation has trivial plane wave solutions of the form

$$\Psi_P(X) = e^{i(PX - P^2t/2M)/\hbar} \tag{15.47}$$

(with P a number) from which free-particle wavepackets can be constructed.

Example 15.3 Reduced Mass Effects in Two-particle Systems A simple model of a one-dimensional diatomic molecule consists of two masses m_1, m_2 interacting via the potential $V(x_1 - x_2) = K(x_1 - x_2 - l)^2/2$, where l is the equilibrium separation of the two masses. The equation for the relative coordinate

$$\left(\frac{\hat{p}^2}{2\mu} + V(x - l)\right)\psi(x) = E_x\psi(x) \tag{15.48}$$

has the standard harmonic oscillator solutions $\psi(x) = \phi_n(x - l)$ with quantized energy levels given by $E_x^{(n)} = (n + 1/2)\hbar\omega$, where $\omega = \sqrt{K/\mu}$. The dependence on μ of the zero point energy of vibrational states in diatomic molecules was mentioned in Sec. 10.3.

In systems where $m_1 \approx m_2$, the reduced mass is roughly $\mu \approx m_1/2 \approx m_2/2$, and its effect is obviously important to include. In cases such as the hydrogen atom, where one has $m_1 = m_e \ll m_p = m_2$ and $\mu = m_e/(1 + m_e/m_p) \approx m_e$, the effect is much smaller (since $m_e/m_p \approx 1/2000$) and is sometimes not stressed sufficiently. The discrete energy spectrum of a hydrogen-like atom (a single electron interacting via a Coulomb force with a nucleus of charge Z) is given by Eqn. (1.30):

$$E_n = -\frac{1}{2}\mu c^2 Z^2 \alpha^2 \frac{1}{n^2} \tag{15.49}$$

where the reduced mass μ now properly appears. The frequencies of the photons emitted in a transition are

$$\hbar\omega_{nl} = \hbar 2\pi f_{nl} = E_n - E_l = \frac{1}{2}\mu c^2 Z^2 \alpha^2 \left(\frac{1}{n^2} - \frac{1}{l^2}\right) \tag{15.50}$$

The dependence on μ implies that the corresponding lines in atoms with the same value of Z but with different nuclear masses (i.e., isotopes) will have slightly differing wavelengths. This effect was utilized in the discovery of the "hydrogen isotope of mass 2," now known as *deuterium*. (See P15.6 for details.)

15.4 SPIN WAVEFUNCTIONS

In describing the quantum state of a particle, we have concentrated on the wavefunctions corresponding to observable quantities such as position ($\psi(x)$), momentum ($\phi(p)$), or energy eigenvalues ($\{a_n; n = 0, 1, \ldots\}$). The spin-up and spin-down label necessary to describe spin-1/2 particles must also be included in the multiparticle wavefunctions for such particles. A more comprehensive discussion of spin in quantum mechanics is given in the next chapter, but we introduce here, for convenience, some of the basic formalism for spin-1/2 particles.

A convenient (matrix) representation of the spin operator (quantized along some convenient direction, often the z axis) is given by

$$\mathbf{S}_z = \frac{\hbar}{2}\begin{pmatrix} 1 & 0 \\ 0 & -1 \end{pmatrix} \tag{15.51}$$

which acts on a (complex) *spinor wavefunction*

$$\chi = \begin{pmatrix} \alpha \\ \beta \end{pmatrix} \tag{15.52}$$

In order to be normalized, such spinors must satisfy

$$(\chi)^* \chi = (\alpha^*, \beta^*) \begin{pmatrix} \alpha \\ \beta \end{pmatrix} = |\alpha|^2 + |\beta|^2 = 1 \tag{15.53}$$

The eigenvectors and eigenvalues of $\mathbf{S}_z$ are seen to be

$$\chi^{(+)} = \begin{pmatrix} 1 \\ 0 \end{pmatrix} \qquad \mathbf{S}_z \chi^{(+)} = +\frac{\hbar}{2}\chi^{(+)} \tag{15.54}$$

$$\chi^{(-)} = \begin{pmatrix} 0 \\ 1 \end{pmatrix} \qquad \mathbf{S}_z \chi^{(-)} = -\frac{\hbar}{2}\chi^{(-)} \tag{15.55}$$

Since $\mathbf{S}_z$ is a Hermitian (matrix) operator (note that $\mathbf{S}_z^\dagger = \mathbf{S}_z$), it is not surprising that its eigenvalues, $\pm\hbar/2$, are real. For the same reason, the eigenfunctions (in this case eigenvectors) also satisfy the usual orthonormality conditions, since

$$(\chi^{(+)})^* \chi^{(+)} = 1 = (\chi^{(-)})^* \chi^{(-)} \qquad \text{and} \qquad (\chi^{(+)})^* \chi^{(-)} = 0 = (\chi^{(-)})^* \chi^{(+)} \tag{15.56}$$

and the expansion theorem in the form

$$\chi = \begin{pmatrix} a^{(+)} \\ a^{(-)} \end{pmatrix} = a^{(+)} \begin{pmatrix} 1 \\ 0 \end{pmatrix} + a^{(-)} \begin{pmatrix} 0 \\ 1 \end{pmatrix} = a^{(+)}\chi^{(+)} + a^{(-)}\chi^{(-)} \tag{15.57}$$

This form makes it clear that $|a^{(+)}|^2$ ($|a^{(+)}|^2$) is the probability that a measurement of the spin (projected onto the z axis) will yield a value of $+\hbar/2$ ($-\hbar/2$); it is also consistent with the expectation value

$$\langle \chi | \mathbf{S}_z | \chi \rangle = (a^{(+)*}, a^{(-)*}) \frac{\hbar}{2} \begin{pmatrix} 1 & 0 \\ 0 & -1 \end{pmatrix} \begin{pmatrix} a^{(+)} \\ a^{(-)} \end{pmatrix} = \frac{\hbar}{2} \left(|a^{(+)}|^2 - |a^{(+)}|^2 \right) \tag{15.58}$$

A spin-1/2 particle can then carry information on its spin state in its quantum wavefunction, $\psi(x, \chi)$, and inner products between different quantum states must be generalized; for example, the overlap "integral" of $\psi_a(x)\chi_a$ and $\psi_b(x)\chi_b$ will be

$$\langle \psi_a \mid \psi_b \rangle = \left[\int_{-\infty}^{+\infty} dx \psi_a^*(x) \psi_b(x) \right] \left[(\chi_a)^* \chi_b \right] \tag{15.59}$$

and wavefunctions can be orthogonal because of different spin dependences. An expansion in energy and spin eigenstates might then have the form

$$\psi(x, \chi) = \sum_n \left(a_n^{(+)} u_n(x)\chi^{(+)} + a^{(-)} u_n(x)\chi^{(-)}\right) \tag{15.60}$$

Example 15.4 Expansion in Energy and Spin Eigenstates Consider a spin-1/2 particle in a harmonic oscillator potential described by the wavefunction

$$\psi(x, \chi) = N\left(3\psi_0(x)\chi^{(+)} - (2+i)\psi_1(x)\chi^{(-)} + \sqrt{6}\psi_1(x)\chi^{(+)}\right) \tag{15.61}$$

The normalization constant can be determined by the requirement that

$$\sum_n (|a_n^{(+)}|^2 + |a_n^{(-)}|^2) = 1 \qquad \text{so that} \qquad N = 1/\sqrt{20} \tag{15.62}$$

The average value of the energy is

$$\langle \hat{E} \rangle = \left(\frac{9}{20}\right)\left(\frac{1}{2}\hbar\omega\right) + \left(\frac{5+6}{20}\right)\left(\frac{3}{2}\hbar\omega\right) = \frac{21}{20}\hbar\omega \tag{15.63}$$

while the expectation value of $\mathbf{S}_z$ is

$$\langle \mathbf{S}_z \rangle = \frac{\hbar}{2}\left(\frac{9+6}{20}\right) - \frac{\hbar}{2}\left(\frac{5}{20}\right) = \frac{\hbar}{4} \tag{15.64}$$

The probability that a measurement will find $\mathbf{S}_z = +\hbar/2$ and $E = 3\hbar\omega/2$ is $P = 3/10$.

For multiparticle wavefunctions, we have to specify the position and spin label for each particle,[3] e.g., $\psi(x_1, \chi_1; x_2, \chi_2; \ldots; x_n, \chi_n)$. We will often denote all of the relevant labels for a given particle by simply specifying the common numerical index, i.e. $\psi(1; 2; \ldots; N)$.

For example, for two (noninteracting) electrons in an infinite well, the following wavefunctions will all turn out to be physically acceptable:

$$\psi_A(1;2) = \frac{1}{\sqrt{2}}\left(\chi_1^{(+)}\chi_2^{(-)} - \chi_1^{(-)}\chi_2^{(+)}\right) u_1(x_1)u_1(x_2) \tag{15.65}$$

$$\psi_B(1;2) = \chi_1^{(+)}\chi_2^{(+)}\frac{1}{\sqrt{2}}\left(u_1(x_1)u_2(x_2) - u_2(x_1)u_1(x_2)\right) \tag{15.66}$$

$$\psi_C(1;2) = \frac{1}{\sqrt{2}}\left(u_1(x_1)u_2(x_2)\chi_1^{(+)}\chi_2^{(-)} - u_2(x_1)u_1(x_2)\chi_1^{(-)}\chi_2^{(+)}\right) \tag{15.67}$$

The inner product for spin states for different particles is generalized to be

$$\langle \chi_1, \chi_2 | \chi_1, \chi_2 \rangle = \langle \chi_1 | \chi_1 \rangle \langle \chi_2 | \chi_2 \rangle = (|\alpha_1|^2 + |\beta_1|^2)(|\alpha_2|^2 + |\beta_2|^2) = 1 \tag{15.68}$$

You should now be able to show that the wavefunctions in Eqns. (15.65) to (15.67) are properly normalized and mutually orthogonal and be able to calculate the energy of each state.

---

[3] We do not consider other possible degrees of freedom that might be labeled, such as "isospin" for nucleons and "color" for quarks.

15.5 INDISTINGUISHABLE PARTICLES

Thus far, we have focused almost exclusively on exploring the consequences of a wave description of particles, its implications for observable phenomena, and the connection between the classical and quantum limits; we have thus concentrated on what we have termed *$\hbar$ physics*. There is another dichotomy between the extreme quantum and classical limits of particles, which we can call *indistinguishability physics;* hereafter, indistinguishable will be abbreviated IND for convenience.

Unlike a set of billiard balls that have different colors and even distinct numeric labels, each of the electrons in an atom is seemingly equivalent to every other electron. In the same way, all protons are effectively identical, all neutrons are equivalent, and so forth. No experiment has yet been able to discern any measurable differences between individual electrons, individual protons, etc. The same statement holds for particles other than the "building blocks" such as photons and all other "elementary" particles.

We can roughly define

- A set of *indistinguishable* (IND) particles to be one in which the interchange of any two particles has no observable effect on any property of the system

The notion of indistinguishability raises an interesting question when one considers the total number of wavefunctions that can have the same total energy, i.e., the degeneracy. For simplicity, say we have a product wavefunction of N IND particles of the form

$$\psi(1;2;\ldots;N) = \phi_a(1)\phi_a(2)\cdots\phi_a(N) \tag{15.69}$$

where each particle is in the same quantum state; the total energy is simply $E_{\text{tot}} = NE_a$. The assumption of indistinguishability means that any permutation of the indices will give a state with the same energy. Because of the special form of Eqn. (15.69), however, the resulting exchanges do not change the wavefunction, and there is only one distinct wavefunction with this energy.

Contrast this to the situation where all N particles are in totally different one-particle configurations:

$$\psi(1;2;\ldots;N) = \phi_{a_1}(1)\phi_{a_2}(2)\cdots\phi_{a_N}(N) \qquad \text{with} \qquad a_1 \neq a_2 \neq \cdots \neq a_N \tag{15.70}$$

with energy $E_{\text{tot}} = E_{a_1} + E_{a_2} + \cdots + E_{a_N}$. One can use N different labels for the first state ϕ_{a_1}, leaving $N-1$ for the second state ϕ_{a_2}, and so forth. There are thus $N!$ different permutations of the labels, each of which gives a state of the same total energy, and in this case all $N!$ wavefunctions are distinct. For example, with three particles in the $n = 0, 1, 2$ levels of the harmonic oscillator, we might have (ignoring spin labels) the $3! = 6$ states

$$\begin{array}{ccc} \phi_0(x_1)\phi_1(x_2)\phi_2(x_3) & \phi_0(x_1)\phi_1(x_3)\phi_2(x_2) & \phi_0(x_2)\phi_1(x_1)\phi_2(x_3) \\ \phi_0(x_2)\phi_1(x_3)\phi_2(x_1) & \phi_0(x_3)\phi_1(x_1)\phi_2(x_2) & \phi_0(x_3)\phi_1(x_2)\phi_2(x_1) \end{array} \tag{15.71}$$

For *distinguishable* particles, each of these choices corresponds to a different physical system since one can, by definition, tell the particles apart. The probability density for a hydrogen atom with an electron "here" and a proton "there" is obviously different from the exchanged system. For IND particles, however, we have the possibility of $N!$ wavefunctions that supposedly all describe the same (presumably unique) physical system. The obvious question is:

- Which one (if any) of these $N!$ choices is the appropriate wavefunction?

To help answer this question, we first formalize the notion of interchange by defining the *exchange operator,* $\hat{\mathscr{E}}_{ij}$, which has the effect of exchanging particles i and j with *all* their appropriate labels. Recalling the notation

$$\psi(x_1, \chi_1; x_2, \chi_2; \ldots; x_N, \chi_N) \equiv \psi(1; 2; \ldots; N) \tag{15.72}$$

we see that

$$\hat{\mathscr{E}}_{ij}\psi(1; 2; \ldots i; \ldots j; \ldots; N) = \psi(1; 2 \ldots j; \ldots i; \ldots; N) \tag{15.73}$$

so that $x_i \leftrightarrow x_j$ and $\chi_i \leftrightarrow \chi_j$.

If, for example, one has two IND spin-1/2 particles in the infinite well with wavefunction

$$\psi(1; 2) = u_4(x_1)u_7(x_2)\chi_1^{(+)}\chi_2^{(-)} \tag{15.74}$$

we will have

$$\hat{\mathscr{E}}_{12}\psi(1; 2) = u_4(x_2)u_7(x_1)\chi_2^{(+)}\chi_1^{(-)} \tag{15.75}$$

The two-particle wavefunctions in Eqns. (15.65) to (15.67) are easily seen to be antisymmetric under the action of $\hat{\mathscr{E}}_{12}$.

There are $N(N-1)$ distinct $\hat{\mathscr{E}}_{ij}$ that, by themselves, exchange labels pairwise. The complete set of all the $N!$ permutations of the particle indices is generated by taking products of the individual $\hat{\mathscr{E}}_{ij}$. For example, when $N = 3$, we have

$$\psi(3; 2; 1) = \hat{\mathscr{E}}_{13}\psi(1; 2; 3) \tag{15.76}$$

while

$$\psi(2; 3; 1) = \hat{\mathscr{E}}_{12}\psi(1; 3; 2) = \hat{\mathscr{E}}_{12}\hat{\mathscr{E}}_{23}\psi(1; 2; 3) \tag{15.77}$$

The set of the $\hat{\mathscr{E}}_{ij}$ and their products forms a *permutation group* (P15.7).

We can establish several important properties of the exchange operators using two-particle systems as an example for ease of notation:

- The exchange operator is Hermitian. We show this explicitly for the position degree of freedom by noting that

$$\begin{aligned}
\langle\hat{\mathscr{E}}_{12}\rangle^* &= \left[\int_{-\infty}^{+\infty} dx_1 \int_{-\infty}^{+\infty} dx_2\, \psi^*(x_1, x_2)\hat{\mathscr{E}}_{ij}\psi(x_1, x_2)\right]^* \\
&= \int_{-\infty}^{+\infty} dx_1 \int_{-\infty}^{+\infty} dx_2\, \psi^*(x_2, x_1)\psi(x_1, x_2) \\
&= \int_{-\infty}^{+\infty} dy_2 \int_{-\infty}^{+\infty} dy_1\, \psi^*(y_1, y_2)\psi(y_2, y_1) \\
&= \int_{-\infty}^{+\infty} dy_2 \int_{-\infty}^{+\infty} dy_1\, \psi^*(y_1, y_2)\hat{\mathscr{E}}_{12}\psi(y_1, y_2) \\
&= \langle\hat{\mathscr{E}}_{12}\rangle
\end{aligned} \tag{15.78}$$

  where a simple relabeling of variables is used; the similar proof for spin wavefunctions is discussed in P15.8.

- $\hat{\mathscr{E}}_{ij}$ certainly commutes with the many-body Hamiltonian because

$$\begin{aligned}
[\hat{H}, \hat{\mathscr{E}}_{12}]\psi(1; 2) &= \hat{H}\hat{\mathscr{E}}_{12}\psi(1; 2) - \hat{\mathscr{E}}_{12}\hat{H}\psi(1; 2) \\
&= \hat{H}\psi(2; 1) - \hat{\mathscr{E}}_{ij}E_{(12)}\psi(1; 2) \\
&= (E_{(21)} - E_{(12)})\psi(1; 2) = 0
\end{aligned} \tag{15.79}$$

since the energy is an observable that should be unchanged by interchange. We then know that the states of the system are simultaneous eigenfunctions of both the energy and *all* the exchange operators.

- The square of the exchange operator is just the identity, since

$$\left(\hat{\mathcal{E}}_{12}\right)^2 \psi(1;2) = \hat{\mathcal{E}}_{12}\hat{\mathcal{E}}_{12}\psi(1;2) = \hat{\mathcal{E}}_{12}\psi(2;1) = \psi(1;2) \tag{15.80}$$

Just as with the parity operator (Sec. 7.7), this fact implies that the eigenvalues of the exchange operator are ± 1 corresponding to states that are symmetric $(+1)$ and antisymmetric (-1) under interchange of any two particles, i.e.,

$$\text{Symmetric} \qquad \psi(2;1) = \hat{\mathcal{E}}_{12}\psi(1;2) = +\psi(1;2) \tag{15.81}$$

or

$$\text{Antisymmetric} \qquad \psi(2;1) = \hat{\mathcal{E}}_{12}\psi(1;2) = -\psi(1;2) \tag{15.82}$$

This is certainly consistent with the fact that the probability distribution of two exchanged wavefunctions should be the same under exchange, namely,

$$P(2;1) = |\psi(2;1)|^2 = |\pm\psi(1;2)|^2 = P(1;2) \tag{15.83}$$

- These last two points, taken together, imply a very powerful constraint on the total wavefunction (by which we mean both position and spin degrees of freedom), namely,
- The wavefunction of N indistinguishable particles must be either totally symmetric (S) or totally antisymmetric (A) under the exchange of any two of the IND particles, i.e., either

$$\text{Totally symmetric} \qquad \hat{\mathcal{E}}_{ij}\psi_S(1;2;\ldots;N) = +\psi_S(1;2;\ldots;N) \tag{15.84}$$

or

$$\text{Totally antisymmetric} \qquad \hat{\mathcal{E}}_{ij}\psi_A(1;2;\ldots;N) = -\psi_A(1;2;\ldots;N) \tag{15.85}$$

for all possible pairs (i, j).

This important result is the key ingredient in determining the correct form of the quantum wavefunction for a system of IND particles. Using the states in Eqn. (15.71) as an example, we see that none of them satisfies either (15.84) or (15.85) by itself. One can see, however, that the linear combinations

$$\begin{aligned}\psi_S(1;2;3) = C_S\big(&\phi_0(x_1)\phi_1(x_2)\phi_2(x_3) + \phi_0(x_1)\phi_1(x_3)\phi_2(x_2)\\ &+ \phi_0(x_2)\phi_1(x_1)\phi_2(x_3) + \phi_0(x_2)\phi_1(x_3)\phi_2(x_1)\\ &+ \phi_0(x_3)\phi_1(x_1)\phi_2(x_2) + \phi_0(x_3)\phi_1(x_2)\phi_2(x_1)\big)\end{aligned} \tag{15.86}$$

and

$$\begin{aligned}\psi_A(1;2;3) = C_A\big(&\phi_0(x_1)\phi_1(x_2)\phi_2(x_3) - \phi_0(x_1)\phi_1(x_3)\phi_2(x_2)\\ &- \phi_0(x_2)\phi_1(x_1)\phi_2(x_3) + \phi_0(x_2)\phi_1(x_3)\phi_2(x_1)\\ &+ \phi_0(x_3)\phi_1(x_1)\phi_2(x_2) - \phi_0(x_3)\phi_1(x_2)\phi_2(x_1)\big)\end{aligned} \tag{15.87}$$

are, respectively, symmetric and antisymmetric under the interchange of any two labels. The constants C_S, C_A are determined, of course, by the overall normalization.

A general prescription for the construction of such properly symmetrized or antisymmetrized linear combinations is easy to generate. For the symmetric combination, one can take

$$\text{Completely symmetric} \qquad \psi_S(1;2;\ldots;N) = C_S \sum_P \psi(1;2;\ldots;N) \tag{15.88}$$

where $\sum_P$ denotes the sum over all possible permutations of the N indices; this form certainly reproduces Eqn. (15.86). While there can be as many as $N!$ terms in this sum, if the IND particles are not all in different quantum states, the number of terms can be far less. For example, if the N particles are all in the same state given by Eqn. (15.69), the permutations in (15.88) are all identical, and we find

$$\psi_S(1;2;\ldots;N) = C_S N! \psi(1;2;\ldots;N) \tag{15.89}$$

In this language, the antisymmetric combination can be written schematically as

$$\text{Completely antisymmetric} \qquad \psi_A(1;2;\ldots;N) = C_A \sum_P (-1)^{n_P} \psi(1;2;\ldots;N) \tag{15.90}$$

- Here n_P is the *number of two-particle permutations or exchanges* that are required to achieve the overall permutation denoted by P starting from the canonical ordering $(1;2\ldots;N)$. This factor gives the alternating signs required by the antisymmetry.

As an example of this last case, we note that

$$\begin{aligned} \psi(2;1;3) &= \hat{\mathcal{E}}_{12}\psi(1;2;3) &\implies \quad n_P = 1,\ (-1)^{n_P} = -1 \\ \psi(3;1;2) &= \hat{\mathcal{E}}_{13}\hat{\mathcal{E}}_{23}\psi(1;2;3) &\implies \quad n_P = 2,\ (-1)^{n_P} = +1 \end{aligned} \tag{15.91}$$

as in Eqn. (15.87). This form also implies that:

- No two particles described by a totally antisymmetric wavefunction can be in the same quantum state

which we can see as follows. Suppose particles i and j were in the same quantum state, namely, $i = j$. Then since the overall wavefunction must be antisymmetric under $\hat{\mathcal{E}}_{ij}$, we have

$$\begin{aligned} \psi(1;2;\ldots j;\ldots;i;\ldots;N) &= -\psi(1;2;\ldots j;\ldots;i;\ldots;N) && \text{by antisymmetry} \\ &= -\psi(1;2;\ldots;i;\ldots;j;\ldots;N) && \text{since } i = j \\ &= 0 \end{aligned} \tag{15.92}$$

The more careful way of stating this result is that:

- The wavefunction (and hence the probability density) for two particles in a completely antisymmetric state to occupy the same quantum "niche" vanishes.

For the case of noninteracting particles where the wavefunction can be written in product form, the antisymmetric combination can be written in an especially simple form using a determinant, namely,

$$\psi_A(1;2;\ldots;N) = \frac{1}{\sqrt{N!}}\det\begin{pmatrix} \phi_{a_1}(1) & \phi_{a_2}(1) & \cdots & \phi_{a_N}(1) \\ \phi_{a_1}(2) & \phi_{a_2}(2) & \cdots & \phi_{a_N}(2) \\ \vdots & \vdots & \ddots & \vdots \\ \phi_{a_1}(N) & \phi_{a_2}(N) & \cdots & \phi_{a_N}(N) \end{pmatrix} \tag{15.93}$$

The "recipe" for constructing this matrix is to put the N (necessarily different) single-particle wavefunctions in succeeding *columns*, while the particle state labels $1, 2, \ldots, N$ are then inserted in different *rows*. The overall antisymmetry of the wavefunction is guaranteed by the linear algebra result that the exchange of any two rows (or columns) of a matrix introduces a (-1) in the determinant. The overall normalization constant is correct provided that each ϕ_{a_i} is properly normalized (P15.9). This form is called a *Slater determinant* and is useful even when the particles interact with each other, as it can be used as a trial wavefunction for a variational calculation.

A similar shorthand notation for the antisymmetric state is

$$\psi_S(1;2;\ldots;N) = C_S \det\begin{pmatrix} \phi_{a_1}(1) & \phi_{a_2}(1) & \cdots & \phi_{a_N}(1) \\ \phi_{a_1}(2) & \phi_{a_2}(2) & \cdots & \phi_{a_N}(2) \\ \vdots & \vdots & \ddots & \vdots \\ \phi_{a_1}(N) & \phi_{a_2}(N) & \cdots & \phi_{a_N}(N) \end{pmatrix}_+ \tag{15.94}$$

where the $+$ subscript indicates that the determinant should be taken with all positive signs. For symmetric states, more than one particle can be in a given state, not all of the resulting terms will necessarily be different, and the normalization must be determined case by case. An example is Eqn. (15.89) and P15.9.

The requirement that IND particles have totally symmetric or antisymmetric wavefunctions has therefore reduced the possible ambiguity in the number of quantum states describing the same physics from being as large as $N!$ possible choices to only two. The physical property of the particles in question, which determines which choice is actually realized in nature, is their intrinsic angular momentum or spin, specifically whether the particles have integral spin ($J = 0, 1, 2 \ldots$) or half-integral spin ($1/2, 3/2, 5/2, \ldots$). The former are called *bosons*, while the latter are known as *fermions*. This distinction is the content of the *spin statistics theorem*, which states that

- The total wavefunction (including both spin and position information) of a system of indistinguishable bosons (fermions) must be symmetric (antisymmetric) under the interchange of any two particles.

The exclusion principle, as stated in Sec. 7.1, that "no two electrons may be in the same quantum state" is seen to be an immediate consequence of this result from Eqn. (15.92). The same result must then hold for neutrons and protons in nuclear systems, quarks inside nucleons, and all other particles with half-integral spin. Besides yielding a "no-go" theorem for what is not allowed, the spin statistics theorem provides a prescription for the appropriate wavefunction for a system of IND particles via Eqns. (15.88) and (15.90) and, as such, is a much more powerful statement about how nature organizes itself.

Example 15.5 Two Electrons with Spin in a Box Consider two noninteracting electrons in an infinite well potential. The ground state of the system is achieved when both particles are in the lowest allowed energy state with $E_{\text{tot}} = 2E_1$. This is allowed provided that their spins are different, as in Fig. 15.2(a), in which case the Slater determinant wavefunction is

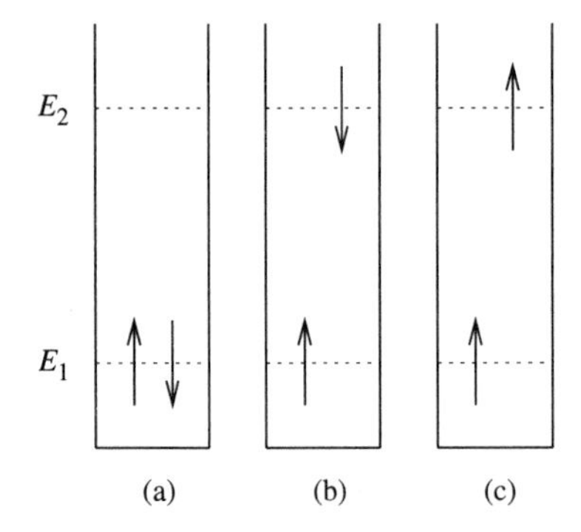

FIGURE 15.2. *Allowed states of two spin-1/2 particles in an infinite well; (a) is the ground state; (b) and (c) show two possible first excited states.*

$$\begin{aligned}\psi(1;2) &= \frac{1}{\sqrt{2!}}\det\begin{pmatrix} u_1(x_1)\chi_1^{(+)} & u_1(x_1)\chi_1^{(-)} \\ u_1(x_2)\chi_2^{(+)} & u_1(x_2)\chi_2^{(-)}\end{pmatrix} \\ &= u_1(x_1)u_2(x_2)\frac{1}{2}\left(\chi_1^{(+)}\chi_2^{(-)} - \chi_2^{(+)}\chi_1^{(-)}\right) \end{aligned} \tag{15.95}$$

The antisymmetric wavefunction with both particles in the ground state with spins aligned vanishes as in Eqn. (15.92).

For the first excited state of the system, one electron can be "elevated" to the next energy level so that $E_{\text{tot}} = E_1 + E_2$, and both spin configurations in Fig. 15.2(b) and (c) are possible; the corresponding wavefunctions are then

$$\begin{aligned}\psi(1;2) &= \frac{1}{\sqrt{2!}}\det\begin{pmatrix} u_1(x_1)\chi_1^{(+)} & u_2(x_1)\chi_1^{(-)} \\ u_1(x_2)\chi_2^{(+)} & u_2(x_2)\chi_2^{(-)}\end{pmatrix} \\ &= \frac{1}{\sqrt{2}}\left(u_1(x_1)u_2(x_2)\chi_1^{(+)}\chi_2^{(-)} - u_2(x_1)u_1(x_2)\chi_1^{(-)}\chi_2^{(+)}\right) \end{aligned} \tag{15.96}$$

and

$$\begin{aligned}\psi(1;2) &= \frac{1}{\sqrt{2!}}\det\begin{pmatrix} u_1(x_1)\chi_1^{(+)} & u_2(x_1)\chi_1^{(+)} \\ u_1(x_2)\chi_2^{(+)} & u_2(x_2)\chi_2^{(+)}\end{pmatrix} \\ &= \chi_1^{(+)}\chi_1^{(+)}\frac{1}{\sqrt{2}}\left(u_1(x_1)u_2(x_2) - u_2(x_1)u_1(x_2)\right) \end{aligned} \tag{15.97}$$

with a similar state with both spins down also possible. These are just the wavefunctions of Eqns. (15.65) to (15.67).

We know that the presence of mutual interactions between particles will induce dynamical correlations between them that are reflected in their quantum wavefunctions. What is more surprising is that IND particles exhibit such correlations even when they do not interact, simply due to the requirement of indistinguishability. These can be called an effective *Fermi repulsion* and *Bose attraction,* which we illustrate in the next example.

Example 15.6 Correlations Due to Indistinguishability Consider two particles of mass m in a harmonic oscillator potential. Assume that we somehow know that there is one particle in the ground state (ψ_0) and one in the first excited state (ψ_1). We define the three wavefunctions

$$\psi_D(x_1, x_2) = \psi_0(x_1)\psi_1(x_2) \tag{15.98}$$

$$\psi_B(x_1, x_2) = \frac{1}{\sqrt{2}}\left(\psi_0(x_1)\psi_1(x_2) + \psi_1(x_1)\psi_0(x_2)\right) \tag{15.99}$$

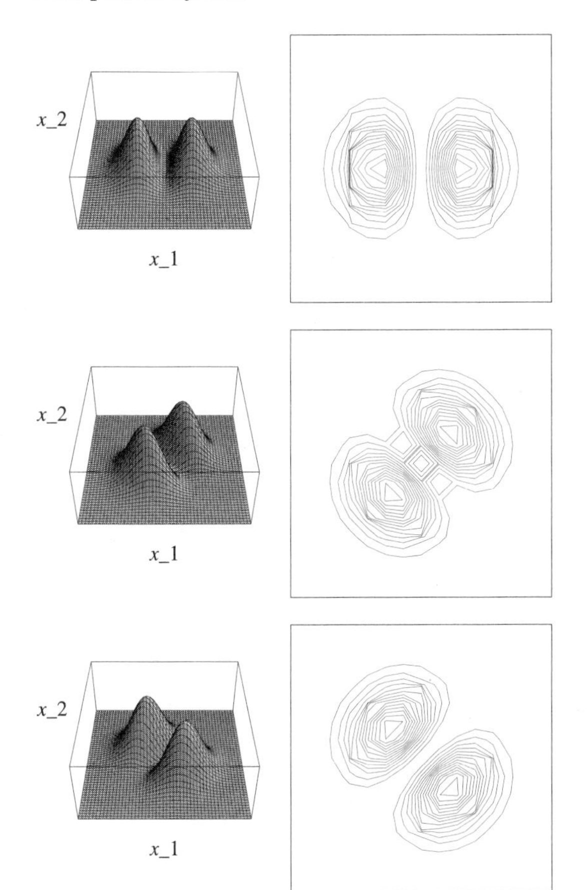

FIGURE 15.3. *Contour plot of $|\psi(x_1, x_2)|^2$ versus x_1, x_2 for the case of (top) distinguishable particles, (middle) indistinguishable bosons, and (bottom) indistinguishable fermions.*

$$\psi_F(x_1, x_2) = \frac{1}{\sqrt{2}}\big(\psi_0(x_1)\psi_1(x_2) - \psi_1(x_1)\psi_0(x_2)\big) \tag{15.100}$$

where

$$\psi_0(x) = \sqrt{\frac{1}{\rho\sqrt{\pi}}}e^{-x^2/2\rho^2} \qquad \text{and} \qquad \psi_1(x) = \sqrt{\frac{2}{\rho\sqrt{\pi}}}\left(\frac{x}{\rho}\right)e^{-x^2/2\rho^2} \tag{15.101}$$

and the labels stand for distinguishable (D), boson (B), and fermion (F), respectively. We imagine, for example, that the bosons have no spin, so that their wavefunction must be symmetric in the position coordinate, while the fermions have a symmetric spin wavefunction (which we do not exhibit), implying that their position space wavefunction must be odd under exchange. There is, of course, another distinguishable wavefunction with $1 \leftrightarrow 2$.

To see the correlations contained in these wavefunctions, we note that $\langle x_1\rangle = \langle x_2\rangle = 0$ for all cases, but:

$$\begin{array}{lllll} D: & \langle x_1^2\rangle = \rho^2/2 & \langle x_2^2\rangle = 3\rho^2/2 & \langle x_1x_2\rangle = 0 & \langle (x_1-x_2)^2\rangle = 2\rho^2 \\ B: & \langle x_1^2\rangle = \rho^2 & \langle x_2^2\rangle = \rho^2 & \langle x_1x_2\rangle = \rho^2/2 & \langle (x_1-x_2)^2\rangle = \rho^2 \\ F: & \langle x_1^2\rangle = \rho^2 & \langle x_2^2\rangle = \rho^2 & \langle x_1x_2\rangle = -\rho^2/2 & \langle (x_1-x_2)^2\rangle = 3\rho^2 \end{array} \tag{15.102}$$

The two particles described by the fermion (boson) wavefunction are, on average, farther apart (closer together) than if they were distinguishable particles. This is illustrated in Fig. 15.3 for the three cases where we plot $|\psi(x_1, x_2)|^2$ versus $x_1,\ x_2$. A "slice" through these plots along

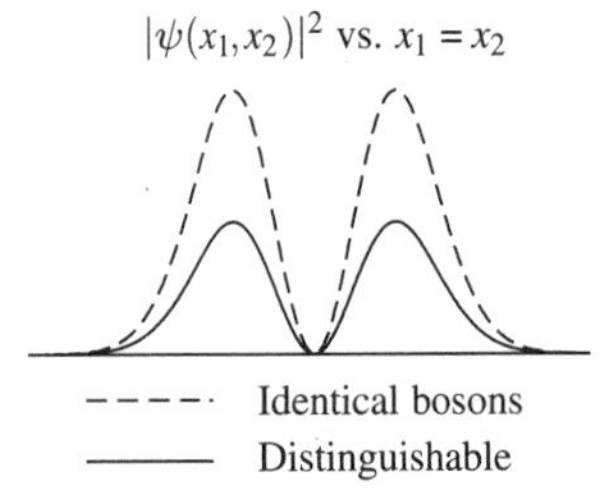

FIGURE 15.4. A "slice" through the contour plots of Fig. 15.3 for $x_1 = x_2$. The fermion probability distribution vanishes, and the identical boson configuration is twice as likely as the indistinguishable particle state.

the line $x_1 = x_2$ is shown in Fig.15.4 where we see that the totally symmetric wavefunction is twice as probable to be found with the particles in the same state; the totally antisymmetric wavefunction, of course, vanishes identically in this case.

We can further explore the implications of these correlations by "turning on" a mutual interaction between the two particles of the form

$$V'(x_2 - x_2; d) \equiv \frac{\lambda}{d\sqrt{\pi}} \exp\left(-\frac{(x_2 - x_2)^2}{2d^2}\right) \tag{15.103}$$

This function has the nice properties that

$$\lim_{d\to 0}[V(x_2 - x_2; d)] = \lambda\delta(x_2 - x_2) \tag{15.104}$$

so that the particles only interact when they're on "top of each other" in this limit. This form is also convenient, as an estimate of the effect of this interaction can be made using first-order perturbation theory, and the necessary overlap integrals can be done analytically (P15.10). The shift in energy due to this perturbation at this order can be written as

$$\begin{aligned}
E_D^{(1)} &= \Delta E_D = \frac{\lambda}{\sqrt{\pi}} \frac{\rho^2 + d^2}{(2\rho^2 + d^2)^{3/2}} = V_b\left[\frac{1 + z^2}{(2 + z^2)^{3/2}}\right] \\
E_B^{(1)} &= \Delta E_B = \frac{\lambda}{\sqrt{\pi}} \frac{2\rho^2 + d^2}{(2\rho^2 + d^2)^{3/2}} = V_b\left[\frac{2 + z^2}{(2 + z^2)^{3/2}}\right] \\
E_F^{(2)} &= \Delta E_F = \frac{\lambda}{\sqrt{\pi}} \frac{d^2}{(2\rho^2 + d^2)^{3/2}} = V_b\left[\frac{z^2}{(2 + z^2)^{3/2}}\right]
\end{aligned} \tag{15.105}$$

where $V_b \equiv \lambda/b\sqrt{\pi}$ and $z \equiv d/\rho$; we plot these results in Fig.15.5. One sees that when $d \ll \rho$ ($z \ll 1$ or "range of mutual interaction" $\ll$ "particle separation"), the perturbation "samples" the various wavefunctions in a region where the correlations are dramatic, and the resulting energy shifts are very different. When $d \gg \rho$ ($z \gg 1$), the effects of indistinguishability are less important.

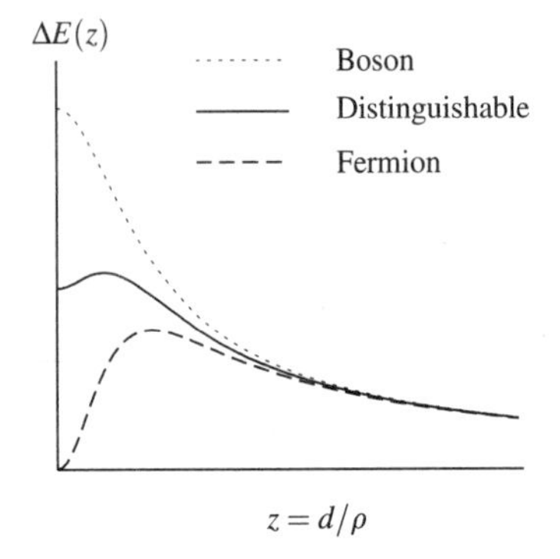

FIGURE 15.5. Energy shift, $\Delta E(z)$, versus the range of the mutual interaction, $z = d/\rho$. Results for the distinguishable (solid), indistinguishable boson (dotted) and indistinguishable fermion (dashed) state are shown.

15.6 QUESTIONS AND PROBLEMS

Q15.1 What are the *dimensions* of an N-particle wavefunction in one dimension? In three dimensions?

Q15.2 Does the notion of probability flux generalize to multiparticle wavefunctions?

Q15.3 Referring to Fig. 15.1, show how to *estimate* the ratio m_1/m_2.

P15.1 Multiparticle momentum space wavefunctions:

(a) Show that the definition

$$\phi(p_1, p_2) = \frac{1}{2\pi\hbar} \int_{-\infty}^{+\infty} dx_1 \int_{-\infty}^{+\infty} dx_2 e^{-i(p_1x_1+p_2x_2)/\hbar} \psi(x_1, x_2)$$

is appropriate for the momentum space wavefunction corresponding to a two-particle position space $\psi(x_1, x_2)$. Specifically, show that $\phi(p_1, p_2)$ is normalized if $\psi(x_1, x_2)$ is, that the appropriate inverse relation holds, and anything else you think is important.

(b) Evaluate the momentum space wavefunction corresponding to $\psi(x_1, x_2)$ in Ex. 15.1, and show that it is proportional to

$$\phi(p_1, p_2) = \exp\left(-\frac{(cp_1^2 - 2bp_1p_2 + ap_2^2)}{2\hbar^2(ac - b^2)}\right)$$

Normalize this wavefunction, show that this form has the right limit as $b \to 0$ and $ac - b^2 \to 0$, and interpret your results.

(c) Find the covariance for the variables p_1, p_2, show that it is opposite in sign to that for x_1, x_2, and interpret your result.

P15.2 Consider two distinguishable and noninteracting particles of mass m moving in the same harmonic oscillator potential.

(a) What is the ground state energy E_0 and wavefunction $\psi_0(x_1, x_2)$ of the two-particle system? Is the ground state energy degenerate?

(b) Show that the first excited state E_1 is doubly degenerate, and write down the two wavefunctions, $\psi_1^{a,b}(x_1, x_2)$. Consider a δ function interaction between the particles as in Ex. 15.2, $V'(x_1 - x_2) = g\delta(x_1 - x_2)$ as a small perturbation. Find the perturbed energies and eigenfunctions.

(c) Show that the second excited state is triply degenerate, and write down the possible wavefunctions. Show if a $g\delta(x_1 - x_2)$ mutual interaction is added that the resulting eigenstates are given by

$$\psi_2^a(x_1, x_2) = \frac{1}{\sqrt{2}}\left(\psi_0(x_1)\psi_2(x_2) - \psi_2(x_1)\psi_0(x_2)\right)$$

$$\psi^b(x_1, x_2) = \frac{1}{2}\left(\psi_0(x_1)\psi_2(x_2) - \sqrt{2}\psi_1(x_1)\psi_1(x_2) + \psi_2(x_1)\psi_0(x_1)\right)$$

$$\psi^c(x_1, x_2) = \frac{1}{2}\left(\psi_0(x_1)\psi_2(x_2) + \sqrt{2}\psi_1(x_1)\psi_1(x_2) + \psi_2(x_1)\psi_0(x_1)\right)$$

where the $\psi_n(x)$ are the one-particle SHO eigenfunctions. Show that these states are mutually orthogonal. If the mutual interaction is repulsive ($g > 0$), which state has the highest energy? The lowest? What are the energy eigenvalues? *Hint:* For this three-state system you have to diagonalize a 3×3 matrix. You presumably know one linear combination that would be unaffected by the perturbation and hence one eigenvector and eigenfunction. Extracting this one helps you solve for the other two.

P15.3 Consider the operator representing the total momentum of a multiparticle system, namely,

$$\hat{P} = \hat{p}_1 + \hat{p}_2 + \cdots + \hat{p}_N$$

The time dependence of the expectation value of this operator will be given (recall Sec. 13.5) by

$$\frac{d}{dt}\langle \hat{P} \rangle = \frac{i}{\hbar}\langle [\hat{H}, \hat{P}] \rangle$$

where $\hat{H}$ is the multiparticle Hamiltonian in Eqn. (15.2). Show that this can be written as

$$\frac{d}{dt}\langle \hat{P} \rangle = \langle \mathcal{F}_1 + \mathcal{F}_2 + \cdots + \mathcal{F}_N \rangle = \langle \mathcal{F}_{\text{tot}} \rangle$$

where $\mathcal{F}_{\text{tot}}$ corresponds to the total external force; if this vanishes, the total momentum is conserved as in the classical case. *Note:* This result implies that the effects of the two-body mutual forces cancel, and you should show that

$$\frac{\partial V_{ij}}{\partial x_i} + \frac{\partial V_{ij}}{\partial x_j} = 0$$

P15.4 In changing to center of mass and relative coordinates for a two-body system, the wavefunction in terms of the original x_1, x_2 coordinates can be recovered by using

$$\Psi(X)\phi(x) = \Psi(X(x_1, x_2))\phi(x(x_1, x_2)) \longrightarrow \psi(x_1, x_2)$$

We have stressed that a probability interpretation requires not only the value of $|\psi|^2$ but also the measure, so it is more relevant to compare

$$|\Psi(X(x_1, x_2))\phi(x(x_1, x_2))|^2 dX\, dx \overset{?}{\longleftrightarrow} |\psi(x_1, x_2)|^2 dx_1\, dx_2$$

Use the Jacobian of the transformation from x_1, x_2 to x, X to show that $dX\, dx = dx_1\, dx_2$. *Hint:* The Jacobian of a coordinate transformation from (w, v) to (x, y) is given by

$$dw\, dv = \left| \det \begin{pmatrix} \partial w/\partial x & \partial w/\partial y \\ \partial v/\partial x & \partial v/\partial y \end{pmatrix} \right| dx\, dy$$

As a test, recall the change from Cartesian to polar coordinates, and show that $dxdy = r\,dr\,d\theta$.

P15.5 Consider two particles of mass m moving in the potential

$$V(x_1, x_2) = \frac{1}{2}m\omega^2 x_1^2 + \frac{1}{2}m\omega^2 x_2^2$$

(a) Using the fact that the potential is separable in the x_1, x_2 coordinates, find the energy spectrum and ground state wavefunction. Since this is a product wavefunction, there can be no correlations between the the two particles; show that $\text{cov}(x_1, x_2)$ vanishes in any state.

(b) Show that the potential also separates when expressed in center-of-mass and relative coordinates, and find the energy spectrum and ground state wavefunction in these coordinates. Show that the degeneracy is the same in each representation and that the ground state wavefunctions agree.

(c) Add an additional mutual interaction of the form

$$V'(x_1, x_1) = V'(x_1 - x_2) = \lambda(x_1 - x_2)^2$$

and find the energy spectrum exactly using center-of-mass and relative coordinates. Are the energy levels changed in the way you expect from the form of V'?

(d) When such a mutual interaction is present, we expect the positions of the particles to be correlated. Evaluate $\text{cov}(x_1, x_2)$ for the ground state wavefunction, and show that it is proportional to λ when λ is small. Convince yourself that the correlation should be positive (negative) when $\lambda > 0\ (< 0)$. *Hint:* Use the coordinate transformations to show things like

$$\left\langle x_1 \right\rangle = \left\langle X + \frac{1}{2}x \right\rangle = 0 + 0$$

and

$$\langle x_1^2 \rangle = \langle X^2 \rangle + \langle Xx \rangle + \frac{1}{4}\langle x^2 \rangle$$

P15.6 Reduced mass effects in "heavy hydrogen":

(a) The energy levels of hydrogen-like atoms with a proton nucleus (ordinary hydrogen or H^1) will be slightly different from those with a deuteron nucleus (so-called *heavy hydrogen* or deuterium, H^2 which has roughly twice the mass of a proton) due to reduced mass effects. Show that the shift in wavelength of a given line for H^2 relative to H^1 is roughly

$$\frac{\Delta\lambda}{\lambda} \approx m_e \left(\frac{1}{M_D} - \frac{1}{m_p} \right)$$

where $M_D \approx 2m_p$. Evaluate this fractional change numerically.

(b) The original discovery of deuterium was made by looking for such shifts in the *visible,* atomic Balmer spectra of hydrogen. The original paper[4] says that

> When with ordinary hydrogen, the times of exposure required to just record the strong H^1 lines were increased 4000 times, very faint lines appeared at the calculated positions for the H^2 lines . . . on the short wavelength side and separated from them by between 1 and 2 Å.

Using the result of part (a), quantitatively explain the wavelength shifts observed. Estimate the relative abundance of H^2 and H^1 in normal hydrogen.

P15.7 Permutation groups: We have seen that permutations play an important role in the physics of IND particles, and in this problem you are asked to study some of the properties of the permutation group, using the case of three particles as an example. Consider three ($N = 3$) objects, labeled a, b, and c which can be in the three positions $1, 2, 3$; there are then $N! = 3! = 6$ different ways in which the labels can be placed. These permutations can all be obtained from one standard labeling, say (a, b, c), by the action of six permutation operators:

$$\begin{aligned}
1(a, b, c) &\longrightarrow (a, b, c) \\
(12)(a, b, c) &\longrightarrow (b, a, c) \\
(13)(a, b, c) &\longrightarrow (c, b, a) \\
(23)(a, b, c) &\longrightarrow (a, c, b) \\
(231)(a, b, c) &\longrightarrow (b, c, a) \\
(312)(a, b, c) &\longrightarrow (c, a, b)
\end{aligned}$$

The element 1 is the identity operator. The natural multiplication on these group elements is obtained by letting the permutations act in order, e.g.,

$$(g_1 \cdot g_2)(a, b, c) \Longrightarrow g_1(g_2(a, b, c))$$

[4] Urey, Brickwedde, and Murphy (1932).

so that, for example,

$$(12)\cdot(231) = (13) \qquad \text{and} \qquad (23)\cdot(13) = (312)$$

Complete the multiplication table below, and show that these elements form a group satisfying all of the requirements in the definition of App. D.2.

	1	**(12)**	**(13)**	**(23)**	**(231)**	**(312)**
1	1	(12)	(13)	(23)	(231)	(312)
(12)	(12)				(13)	
(13)	(13)					
(23)	(23)		(312)			
(231)	(231)					
(312)	(312)					

P15.8 **(a)** Generalize the proof in Eqn. (15.78) to show that the exchange operator $\hat{\mathscr{E}}_{ij}$ is Hermitian by showing that $\langle\hat{\mathscr{E}}_{ij}\rangle$ is real when evaluated with any multiparticle position space wavefunction.

(b) Show that the expectation value of $\hat{\mathscr{E}}_{12}$ is real when evaluated between spinstates; i.e., show that

$$\langle\chi_1;\chi_2|\hat{\mathscr{E}}_{12}|\chi_1;\chi_2\rangle = (\alpha_1^*, \beta_1^*)(\alpha_2^*, \beta_2^*)\hat{\mathscr{E}}_{12}\begin{pmatrix}\alpha_1\\ \beta_1\end{pmatrix}\begin{pmatrix}\alpha_2\\ \beta_2\end{pmatrix} = |\alpha_1\alpha_2^* + \beta_1\beta_2^*|^2$$

is real.

P15.9 **(a)** Show that the Slater determinant in Eqn. (15.93) is properly normalized.

(b) Four spinless particles move in the same harmonic oscillator potential; two are in the ground state and two in the first excited state. Write down the normalized wavefunction for this system.

P15.10 Confirm the results in Eqn. (15.105).

P15.11 Consider two IND spin-1/2 particles that interact via the potential $V(x_1 - x_2) = k(x_1 - x_2)^2/2$.

(a) Ignoring spin for the moment, show that the position space wavefunctions can be written as

$$\psi(x_1, x_2) \propto e^{iP(x_1+x_2)/2\hbar}\psi_n(x_1 - x_2)$$

where the ψ_n are the harmonic oscillator eigenstates.

(b) Show that under the exchange $1 \leftrightarrow 2$ that these solutions satisfy

$$\hat{\mathscr{E}}_{12}\psi(x_1, x_2) = (-1)^n\psi(x_1, x_2)$$

(c) Add the appropriate symmetric or antisymmetric spin wavefunctions, and find the allowed states of the system.

PART 2

The Quantum World

CHAPTER 16

Two-Dimensional Quantum Mechanics

One-dimensional (1D) systems provide examples of many of the most important features of quantum mechanics, but it is also instructive to consider two-dimensional (hereafter 2D or planar) systems for several reasons:

- Systems with two spatial degrees of freedom provide more opportunities to study multivariable probability concepts, separation of coordinates techniques, and new mathematical methods and special functions. It also allows for the visualization of many quantum phenomena that arise in more realistic three-dimensional systems but that are obviously difficult to plot in 3D.
- Two-dimensional systems naturally exhibit symmetries not present in 1D systems and provide a glimpse of the intimate connection between symmetries and the degeneracy of energy levels.
- Two-dimensional systems allow one to study rotational motion and its symmetries, as well as the properties of angular momentum, both quantum mechanically and in its approach to the classical limit. For example, charged particles in a uniform magnetic field classically can undergo circular planar orbits (see Sec. 19.4.1). A similar quantum system of electrons in two dimensions in a uniform **B** field has important implications for the understanding of the so-called *quantum Hall effect* in condensed-matter physics.
- Finally, while often considered of pedagogical use only, two-dimensional systems of particles are rapidly becoming of more practical importance as their realization in surface physics becomes increasingly easy. It is now possible, using modern crystal growth techniques such as molecular-beam epitaxy and other methods, to fabricate semiconductor nanostructures, artificially created patterns of atoms whose atomic composition and sizes are controllable at the nanometer scale, which is comparable to interatomic distances. At such length scales, quantum effects obviously become increasingly important. Even more dramatically, scanning tunnel microscopy (STM) techniques can now be used to manipulate individual atoms and molecules[1]

[1]See, e.g., Stroscio and Eigler (1991).

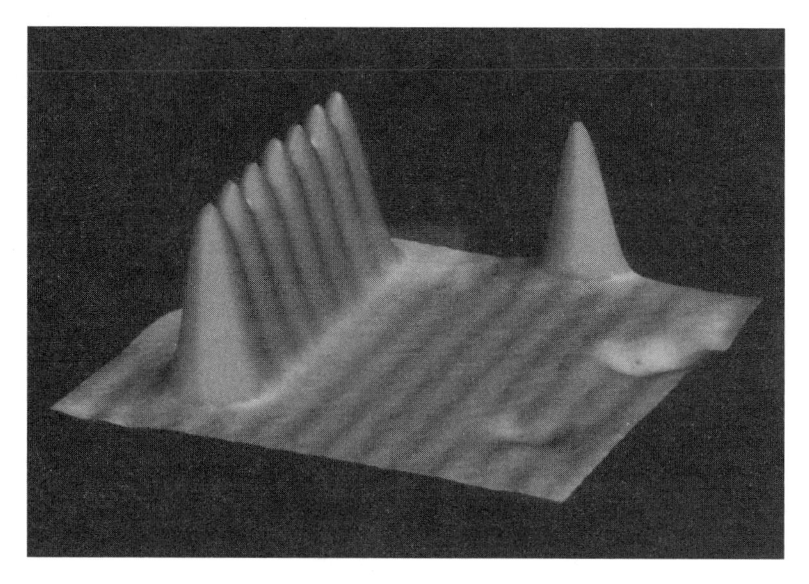

FIGURE 16.1. *The first atomic structure built "by hand" using STM techniques. Seven Xe atoms are bonded together to form a chain; the image is 50 Å on a side. (Photo courtesy of IBM Almaden.)*

with atomic scale precision. Figure 16.1 shows the first "hand-built" atomic structure using these methods.

As an example of such a system, consider a two-dimensional electron gas, bound to a surface or interface between surfaces by the potential shown in Fig. 16.2. The typical localization scale (determined, say, by the thickness of the interface layer) might be $L \sim 40$ Å. This implies quantized energies in the direction perpendicular to the surface (here the z direction) of the order of

$$E_n \sim \frac{\hbar^2 \pi^2 n^2}{2m_e L^2} \sim n^2 25 \text{ meV} \tag{16.1}$$

so that the energy required to excite the electrons in this direction would be roughly $\Delta E \sim$ 75 meV. This can be compared to other typical energy scales for the two-dimensional electron gas that might be either:

- The thermal energy, $k_B T \lesssim 25$ meV, for temperatures at, or below, room temperature (300K) or
- The Fermi energy of the system. From Prob. 8.2, we have that

$$E_F^{(2D)} = \frac{\hbar^2 \pi}{m_e} n_e^{(2D)} \tag{16.2}$$

where $n_e^{(2D)} = N_{tot}/L^2$ is the electron surface density in two dimensions. Using a typical value for $n_e^{(2D)}$, this can be written in the form

$$E_F^{(2D)} = 2.5 \text{ meV} \left(\frac{n_e^{(2D)}}{10^{12} e^- \text{cm}^{-2}} \right) \tag{16.3}$$

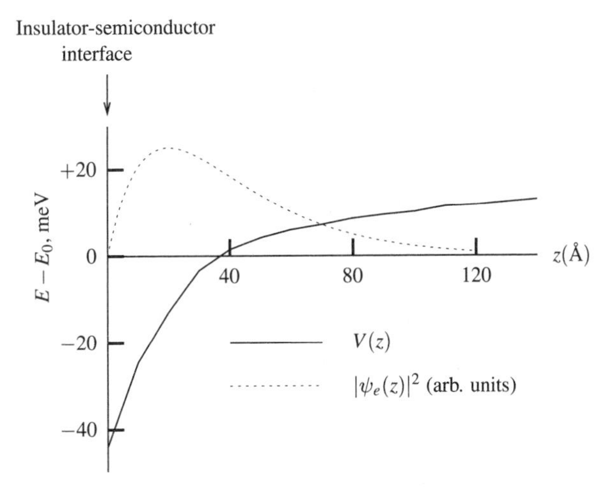

FIGURE 16.2. *Semischematic representation of the potential energy, $V(z)$, and electron wavefunction, $|\psi_e(z)|^2$, versus the distance from the surface, z, for a two-dimensional electron gas near an insulator-semiconductor interface, indicating the approximate localization and energy scales. [Adapted from von Klitzing (1987).]*

and we see that typical 2D collisions will not be able to "excite" electrons in the z direction, provided that the density is not too high. Thus the electron wavefunctions will stay effectively localized within the interface or on the surface (as in Fig. 16.2), and one can study an effectively two-dimensional problem. We note that artificial zero-dimensional systems, variously called *quantum dots*[2] or *artificial atoms*[3] can also be constructed.

16.1 CARTESIAN 2D SYSTEMS

The simplest example of a quantum system in two dimensions is one described by Cartesian coordinates with a Hamiltonian of the form

$$\hat{H} = \frac{1}{2m}\left(\hat{p}_x^2 + \hat{p}_y^2\right) + V(x, y) \quad \text{where} \quad \hat{p}_x = \frac{\hbar}{i}\frac{\partial}{\partial x} \quad \text{and} \quad \hat{p}_y = \frac{\hbar}{i}\frac{\partial}{\partial y} \tag{16.4}$$

with a wavefunction

$$\psi(x, y; t) = \psi(x, y)e^{-iEt/\hbar} \tag{16.5}$$

satisfying the time-dependent Schrödinger equation

$$\hat{H}\psi(x, y) = -\frac{\hbar^2}{2m}\left(\frac{\partial^2}{\partial x^2} + \frac{\partial^2}{\partial y^2}\right)\psi(x, y) + V(x, y)\psi(x, y) = E\psi(x, y) \tag{16.6}$$

[2] Reed (1993).
[3] Kastner (1993).

The probability density (now in two dimensions) is given by $|\psi(x, y; t)|^2$, which must satisfy

$$\int_{-\infty}^{+\infty} dx \int_{-\infty}^{+\infty} dy |\psi(x, y; t)|^2 = 1 \tag{16.7}$$

Average values are calculated in the usual way via

$$\langle \hat{O} \rangle_t \equiv \int_{-\infty}^{+\infty} dx \int_{-\infty}^{+\infty} dy\, \psi^*(x, y; t)\, \hat{O}\, \psi(x, y; t) \tag{16.8}$$

If we consider *separable* potentials of the form

$$V(x, y) = V_x(x) + V_y(y) \tag{16.9}$$

we can assume a factorized form for the wavefunction

$$\psi(x, y; t) = X(x)Y(y)e^{-i(E_x+E_y)t/\hbar} \tag{16.10}$$

where each coordinate satisfies its own, one-dimensional Schrödinger equation

$$\frac{\hat{p}_x^2}{2m} X(x) + V_x(x)X(x) = E_x X(x)$$

$$\frac{\hat{p}_y^2}{2m} Y(y) + V_y(y)Y(y) = E_y Y(y)$$

and the total energy of the system is given by $E = E_x + E_y$.

16.1.1 2D Infinite Well

A simple and instructive case is that of the two-dimensional infinite well (or square box) with walls at $x = 0, L$ and $y = 0, L$. This is of the form above, as we can define

$$V_\square(z; L) = \begin{cases} 0 & \text{for } 0 < z < L \\ \infty & \text{otherwise} \end{cases} \tag{16.11}$$

in which case the potential is of the form $V(x, y) = V_\square(x; L) + V_\square(y; L)$. The fully normalized solutions can be written in the form

$$u_{n,m}(x, y) = u_n(x)u_m(y) = \frac{2}{L} \sin\left(\frac{n\pi x}{L}\right) \sin\left(\frac{m\pi y}{L}\right) \tag{16.12}$$

with

$$E_{n,m} = E_n + E_m = \frac{\hbar^2 \pi^2}{2mL^2}\left(n^2 + m^2\right) \tag{16.13}$$

and the spectrum is illustrated in Fig. 16.3. The wavefunctions for several sets of n, m are shown in Fig. 16.4, illustrating the wave properties of the system. We note that if the one-dimensional infinite well corresponds to waves on a string with fixed ends, then this case corresponds to the vibrations of a (square) drumhead. Such wavefunctions are not only of pedagogical interest, as very similar patterns of "standing electron waves" can be measured on surfaces using STM techniques; this was shown in Fig. 1.3(b).

Particle-like, wavepacket solutions for this separable potential can be formed by using the linearity of the Schrödinger equation to write

$$\psi_{WP}(x, y; t) = \psi_{WP}(x; t)\psi_{WP}(y; t) \tag{16.14}$$

(1,5)
(3,4)
(2,4)
(3,3)
(1,4)
(2,3)
(1,3)
(2,2)
(1,2)
(1,1)

FIGURE 16.3. The spectrum of the 2D infinite square well. The values of (n, m) for the lowest-lying states are shown; states for which $n \neq m$ are doubly degenerate.

where one has

$$\psi_{WP}(x, t) = \sum_{n=1}^{\infty} a_n^{(x)} u_n(x) e^{-iE_n^{(x)}t/\hbar} \tag{16.15}$$

For simplicity, we can use Gaussian weighting factors

$$a_n^{(x)} = e^{-(p_n - p_0^{(x)})^2 \alpha^2/2} e^{-ip_n x_0} \tag{16.16}$$

with $p_n = \hbar n/L$, as in Sec. 6.6. We can localize the packet initially to be centered at (x_0, y_0) with central value of momentum $\mathbf{p} = (p_0^x, p_0^y)$. The ballistic propagation of such a wavepacket with elastic collisions from the walls is illustrated in Fig. 16.5, where we have chosen $(x_0, y_0) = (L/2, L/2)$ and $p_0^x = 2p_0^y$. Thus, the system can exhibit both wave- and particle-like behavior in a fashion similar to the one-dimensional case.

The solutions also form a complete set in that any allowable wavefunction in the 2D infinite well can be written in the form

$$\psi(x, y; t) = \sum_{n=1}^{\infty} \sum_{m=1}^{\infty} a_{n,m} u_{n,m}(x, y) e^{-iE_{n,m}t/\hbar} \tag{16.17}$$

where $|a_{n,m}|^2$ is the probability that a measurement of the energy associated with $\psi(x, y)$ will yield the value $E_{n,m}$. Equation (16.14) is then a special case of the time development of such a solution.

The most interesting new feature of this system is the fact that more than one independent energy eigenstate corresponds to the same energy level, at least for $n \neq m$, where the exchange $n \leftrightarrow m$ gives the same energy. Such a system is said to exhibit *degeneracy*, and we say that:

- The quantum value of an observable quantity is *degenerate* when two (or more) independent eigenfunctions of an operator yield the *same* eigenvalue.

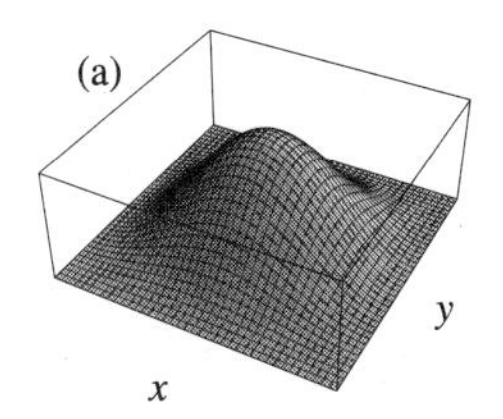

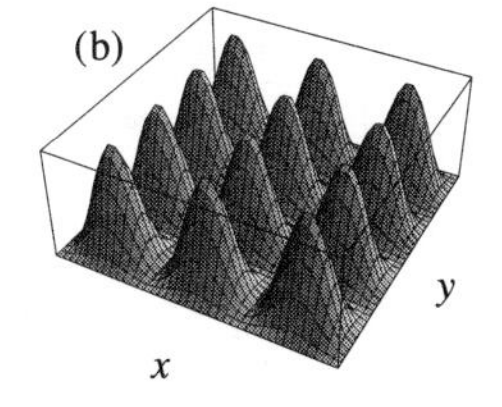

FIGURE 16.4. Plots of $|u_{n,m}(x, y)|^2$ versus (x, y) for (a) $(n, m) = (1, 1)$ and (b) $(n, m) = (3, 4)$.

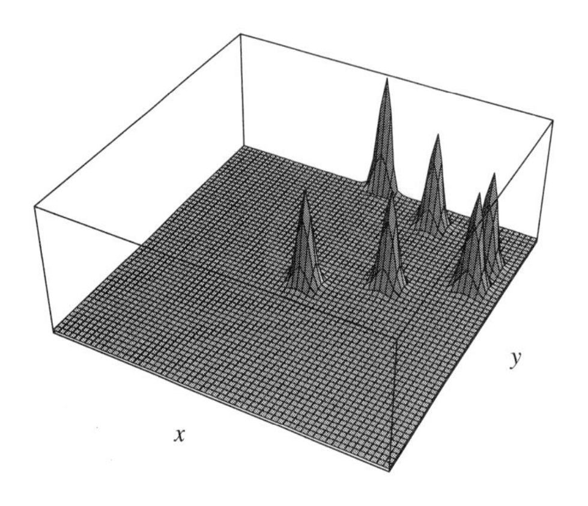

FIGURE 16.5. *Propagation of a quasi-Gaussian wavepacket in a two-dimensional square infinite well; the packet is initially localized in the center of the well with* $p_0^{(x)} = 2p_0^{(y)}$.

In this case, the degeneracy is easily traced to the symmetry of the potential (and kinetic energies) since the exchange $x \leftrightarrow y$ has no observable effect on the system, leading naturally to a doubly degenerate set of levels (when $n \neq m$). We can formalize this notion by introducing an *exchange operator,* $\hat{E}_{(x,y)}$, defined via

$$\hat{E}_{(x,y)} f(x, y) \equiv f(y, x) \tag{16.18}$$

which can be easily shown to be Hermitian, even though its classical connection is far from obvious. This operator is very similar to the multiparticle exchange operator used in Chap. 15.

Following the discussion of Sec. 13.5, we note that the statement above that the exchange $x \leftrightarrow y$ "has no effect on the system" can be associated with the fact that this operator commutes with the Hamiltonian, i.e.,

$$\left(\hat{H}\hat{E}_{(x,y)} - \hat{E}_{(x,y)}\hat{H}\right)\psi(x, y) = 0 \tag{16.19}$$

or $[\hat{H}, \hat{E}_{(x,y)}] = 0$. This implies that there will be simultaneous eigenfunctions of *both* $\hat{H}$ and $\hat{E}_{x,y}$. By invoking the same arguments used previously for the parity operator [mostly the fact that $(\hat{E}_{(x,y)})^2 = 1$], the eigenvalues (eigenfunctions) of the exchange operator can be seen to be ± 1 (even-odd functions under exchange), i.e.,

$$\hat{E}_{(x,y)} f^{(+)}(x, y) = +f^{(+)}(x, y) \qquad \hat{E}_{(x,y)} f^{(-)}(x, y) = -f^{(-)}(x, y) \tag{16.20}$$

The $u_{n,m}(x, y)$ solutions individually do not, however, immediately satisfy this requirement. We note that any linear combination of degenerate energy eigenstates will also be an energy eigenstate with the same energy eigenvalue since

$$\begin{aligned}
\hat{H}\left(\sum_E a_E \psi_E(x)\right) &= \sum_E a_E \hat{H}\psi_E(x) \\
&= \sum_E E a_E \psi_E(x) \\
&= E\left(\sum_E a_E \psi_E(x)\right)
\end{aligned} \tag{16.21}$$

We are thus free to take appropriate linear combinations of degenerate solutions provided they remain orthogonal, and it is easy to see that the required combinations are

$$u_{n,m}^{(\pm)}(x, y) \equiv \frac{1}{\sqrt{2}}\big(u_{n,m}(x, y) \pm u_{m,n}(x, y)\big) \tag{16.22}$$

which do satisfy Eqn. (16.20).

It is something of a "folk-theorem" (meaning roughly a statement that is universally accepted as being true, but difficult to state precisely and to prove in each case) that:

- Most degeneracies are necessarily a result of *some* symmetry (sometimes not obvious) of the system under consideration.

The study of symmetries in quantum mechanics (under the guise of group theory) has had profound applications in atomic, nuclear, and elementary particle physics.

16.1.2 2D Harmonic Oscillator

Another separable Cartesian system is described by an isotropic harmonic oscillator, i.e., a 2D mass and spring, defined by the potential energy

$$V(x, y) = \frac{1}{2}K(x^2 + y^2) = \frac{1}{2}m\omega^2(x^2 + y^2) \tag{16.23}$$

The product wavefunctions are given by

$$\psi_{n,m}(x, y) = \psi_n(x)\psi_m(y) \tag{16.24}$$

where the $\psi_n(x)$ are the solutions of Sec. 10.2.2. The energy spectrum is given by

$$E_{n,m} = E_n + E_m = (n + m + 1)\hbar\omega = (N + 1)\hbar\omega \tag{16.25}$$

which is illustrated in Fig. 16.6. As before, the energy levels with $n \leftrightarrow m$ are degenerate. The total degeneracy, i.e., the number of distinct states, N_s, with energy value labeled by N is $N_s = N$; this is much larger than expected solely on the basis of the $x \leftrightarrow y$ symmetry. The enlarged degeneracy is partly due to the fact that the system also exhibits a symmetry under *rotations* since the potential can also be written in the circularly symmetric way

$$V(x, y) = \frac{1}{2}K(x^2 + y^2) = \frac{1}{2}Kr^2 = V(r) \tag{16.26}$$

and the system can be separated in cylindrical coordinates as well, as we will see in Sec. 16.3.3.

Two-dimensional wavepackets, a la Eqn. (16.14), can also be constructed in this case using, for example, the special Gaussian packet of Sec. 10.5, and can be shown to undergo

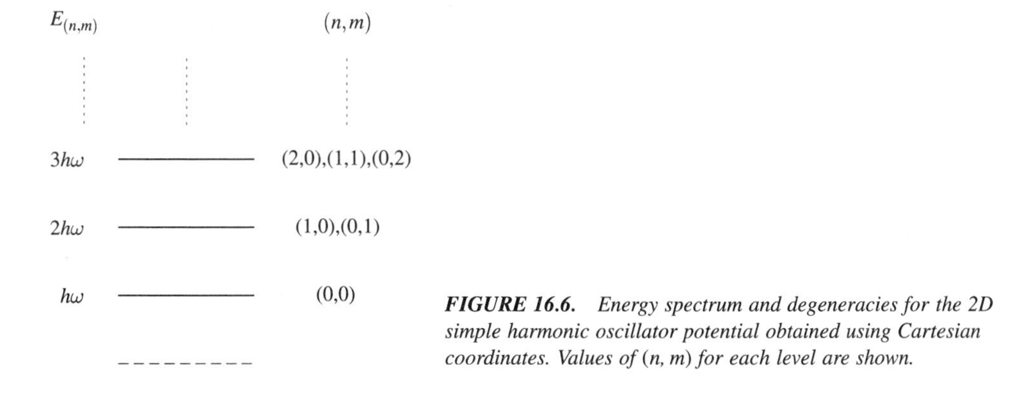

FIGURE 16.6. *Energy spectrum and degeneracies for the 2D simple harmonic oscillator potential obtained using Cartesian coordinates. Values of (n, m) for each level are shown.*

semiclassical motion (P16.11). This is especially interesting in the case of a *nonisotropic spring,* i.e., a potential of the form

$$V(x, y) = \frac{1}{2}K_x x^2 + \frac{1}{2}K_y y^2 \tag{16.27}$$

which is still separable. In this case, the natural vibration frequencies are different, $\omega_{x,y} = \sqrt{K_{x,y}/m}$, and the "trajectories" of the wavepackets will, in general, not be periodic. In the special case where the frequencies are *commensurate,* namely, rational multiples of each other, i.e.,

$$\frac{\omega_x}{\omega_y} = \frac{p}{q} \tag{16.28}$$

(with p, q integers) the classical motion *is* periodic, and the quantum wavepackets can reproduce the classical "Lissajous figures" discussed in many classical mechanics texts.[4]

16.2 CENTRAL FORCES AND ANGULAR MOMENTUM

16.2.1 Classical Case

Cartesian coordinates may not be the most natural set of variables for the study of many systems, and this is especially true for two-dimensional systems described by a cylindrically symmetric potential of the form

$$V(\mathbf{r}) = V(r, \theta) = V(r) \tag{16.29}$$

In this case, the classical force is given by

$$\begin{aligned}\mathbf{F}(\mathbf{r}) &= -\boldsymbol{\nabla} V(r, \theta) \\ &= -\frac{\partial V(r, \theta)}{\partial r}\hat{\mathbf{r}} - \frac{1}{r}\frac{\partial V(r, \theta)}{\partial \theta}\hat{\boldsymbol{\theta}} \\ &= -\frac{dV(r)}{dr}\hat{\mathbf{r}} \\ &= F(r)\hat{\mathbf{r}}\end{aligned} \tag{16.30}$$

where $\hat{\mathbf{r}}$ and $\hat{\boldsymbol{\theta}}$ are unit vectors in the radial and tangential directions, respectively. Thus, for a central potential, the force is directed radially towards (or away from) the origin.

The corresponding classical torque, $\boldsymbol{\tau} = \boldsymbol{r} \times \boldsymbol{F}$, then vanishes and the relation

$$\frac{d\mathbf{L}}{dt} = \tau = 0 \tag{16.31}$$

guarantees that the classical angular momentum, $\mathbf{L}$, is conserved, i.e., is constant in time. Because of its importance as an additional conserved quantity (along with the total energy), we will discuss the quantum version of angular momentum extensively.

The classical equations of motion for the particle in polar coordinates can be derived by those in Cartesian coordinates, namely,

$$\mathbf{F}(\mathbf{r}) = F(r)\hat{\mathbf{r}} = m\mathbf{a}(t) \tag{16.32}$$

[4]See, e.g., Marion and Thornton (1988).

giving

$$x \qquad F(r)\left(\frac{x(t)}{r(t)}\right) = m\ddot{x}(t) \tag{16.33}$$

$$y \qquad F(r)\left(\frac{y(t)}{r(t)}\right) = m\ddot{y}(t) \tag{16.34}$$

and by using the relations

$$x(t) = r(t)\cos(\theta(t)) \qquad \text{and} \qquad y(t) = r(t)\sin(\theta(t)) \tag{16.35}$$

to obtain

$$\frac{F(r)}{m}\cos(\theta) = \ddot{r}\cos(\theta) - 2\dot{r}\sin(\theta)\dot{\theta} - r\cos(\theta)\dot{\theta}^2 - r\sin(\theta)\ddot{\theta} \tag{16.36}$$

$$\frac{F(r)}{m}\sin(\theta) = \ddot{r}\sin(\theta) + 2\dot{r}\cos(\theta)\dot{\theta} - r\sin(\theta)\dot{\theta}^2 + r\cos(\theta)\ddot{\theta} \tag{16.37}$$

The combination Eqn. (16.37) $\times\ (r\cos(\theta))$ − Eqn. (16.36) $\times\ (r\sin(\theta))$ implies that

$$0 = 2\dot{r}r\dot{\theta} + r^2\ddot{\theta} = \frac{d}{dt}\left(r^2\dot{\theta}\right) = 0 \tag{16.38}$$

which is another statement of conservation of angular momentum as $L_z = rp = mrv = mr^2\dot{\theta}$ in polar coordinates. Using this identification, the other obvious combination, Eqn. (16.37) $\sin(\theta)$ + Eqn. (16.36) $\cos(\theta)$, then gives the dynamical equation of motion for the radial coordinate

$$F(r) = m\ddot{r} - mr\dot{\theta}^2 = m\ddot{r} - \frac{L_z^2}{mr^3} \tag{16.39}$$

which is the standard Newton's law result including the "centrifugal force" term. The derivation makes clear that this term is solely due to the proper accounting of the rotational motion and not to any "fictitious force."

Most importantly for connections to quantum mechanics, the total energy will be constant for a conservative potential, so we can write

$$E = \frac{1}{2}m\mathbf{v}^2(t) + V(x, y) \tag{16.40}$$

and the chain rule and Eqn. (16.35) give

$$\mathbf{v}^2 = v_x^2 + v_y^2 = \dot{x}^2 + \dot{y}^2 = \dot{r}^2 + r^2\dot{\theta}^2 \tag{16.41}$$

so that the total energy can be written as

$$E = \frac{m}{2}(\dot{r} + r^2\dot{\theta}^2) + V(r,\theta) = \frac{m\dot{r}^2}{2} + \frac{L_z^2}{2mr^2} + V(r,\theta) \tag{16.42}$$

The second term can be put in the form

$$T_{\text{rot}} = \frac{L_z^2}{2mr^2} = \frac{1}{2}mr^2\dot{\theta}^2 = \frac{1}{2}I\omega^2 \tag{16.43}$$

where $I = mr^2$ is the rotational moment of inertia for a point mass. This makes it clear that it represents the rotational kinetic energy. It is this form for the energy that can be most easily generalized to a quantum mechanical Hamiltonian.

16.2.2 Quantum Angular Momentum in 2D

To extend the notion of angular momentum to quantum mechanical operators, it is most natural to start from the classical definition

$$\mathbf{L} = \mathbf{r} \times \mathbf{p} \tag{16.44}$$

so that the relevant component for two-dimensional motion is the angular momentum about the z axis, namely,

$$L_z = xp_y - yp_x \tag{16.45}$$

Motivated by the position representation of operators, we replace the classical momentum components by their operator analogs and define

$$L_z \longrightarrow \hat{L}_z \equiv x\hat{p}_y - y\hat{p}_x = \frac{\hbar}{i}\left(x\frac{\partial}{\partial y} - y\frac{\partial}{\partial x}\right) \tag{16.46}$$

It is easy to show that $\hat{L}_z$ is Hermitian (P5.14), and that it is also the infinitesimal generator of rotations around the z axis (P16.12), just as $\hat{p}_x$ is responsible for translations along the x axis.

To express this more naturally in polar coordinates, we again use the defining relations

$$x = r\cos(\theta) \quad y = r\sin(\theta) \tag{16.47}$$

or their inverses

$$r = \sqrt{x^2 + y^2} \quad \tan(\theta) = \frac{y}{x} \tag{16.48}$$

and the chain rule and find

$$\frac{\partial}{\partial x} = \cos(\theta)\frac{\partial}{\partial r} - \frac{\sin(\theta)}{r}\frac{\partial}{\partial \theta} \tag{16.49}$$

$$\frac{\partial}{\partial y} = \sin(\theta)\frac{\partial}{\partial r} + \frac{\cos(\theta)}{r}\frac{\partial}{\partial \theta} \tag{16.50}$$

This then gives

$$\hat{L}_z = \frac{\hbar}{i}\frac{\partial}{\partial \theta} \tag{16.51}$$

and θ, $\hat{L}_z$ can be seen (P16.12) to have many similarities with the conjugate pair x, $\hat{p}_x$.

The *eigenfunctions of angular momentum* (in 2D), labeled $\Theta_m(\theta)$, are then determined by the equation

$$\frac{\hbar}{i}\frac{d\Theta_m(\theta)}{d\theta} = \hat{L}_z\Theta_m(\theta) = L_z\Theta_m(\theta) = m\hbar\Theta_m(\theta) \tag{16.52}$$

which yields

$$\Theta_m(\theta) = \frac{e^{im\theta}}{\sqrt{2\pi}} \tag{16.53}$$

where we have chosen to write the dimensionful angular momentum eigenvalue, L_z, in terms of the natural unit of $\hbar$. The normalization constant is chosen so as to satisfy

$$1 = \int_0^{2\pi} d\theta |\Theta_m(\theta)|^2 \tag{16.54}$$

which is natural for an angular wavefunction.

One difference between these solutions and momentum eigenstates arises because of the periodic nature of the variable θ; this presumably requires us to identify coordinates separated by $\theta = 2\pi$ as representing the same physical point,[5] i.e.,

$$\Theta_m(\theta + 2\pi) = \Theta_m(\theta) \Longrightarrow e^{i2\pi m} = 1 \tag{16.55}$$

or $m = 0, \pm 1, \pm 2, \ldots$ and the angular momentum is quantized

$$L_z = 0, \pm\hbar, \pm 2\hbar, \ldots \tag{16.56}$$

This quantization once again arises because of the need to impose boundary conditions. It also guarantees that eigenfunctions corresponding to different eigenvalues are orthogonal, namely,

$$\langle n \mid m \rangle = \int_0^{2\pi} \Theta_n^*(\theta)\Theta_m(\theta) d\theta = \frac{1}{2\pi}\int_0^{2\pi} \left(e^{in\theta}\right)^* e^{im\theta} d\theta = \delta_{n,m} \tag{16.57}$$

These complex angular wavefunctions correspond most closely to the plane wave solutions (traveling waves) for momentum; linear combinations can yield $\Theta(\theta) = \sin(m\theta)$ or $\cos(m\theta)$, which are more like standing waves, and which may be more appropriate for some bound state problems or for visualization purposes.

The appropriate Hamiltonian operator in Cartesian coordinates,

$$\begin{aligned} \frac{\hat{p}_x^2 + \hat{p}_y^2}{2m} + V(x, y) &= -\frac{\hbar^2}{2m}\left(\frac{\partial^2}{\partial x^2} + \frac{\partial^2}{\partial y^2}\right) + V(x, y) \\ &= -\frac{\hbar^2}{2m}\boldsymbol{\nabla}^2 + V(x, y) \end{aligned} \tag{16.58}$$

can be written in polar coordinates by expressing the two-dimensional gradient squared in terms of (r, θ) by extending the chain rule arguments used above. One finds (P16.13) that

$$\frac{\partial^2 \psi(x, y)}{\partial x^2} + \frac{\partial^2 \psi(x, y)}{\partial y^2} = \frac{1}{r}\frac{\partial}{\partial r}\left(r\frac{\partial \psi(r, \theta)}{\partial r}\right) + \frac{1}{r^2}\frac{\partial^2 \psi(r, \theta)}{\partial \theta^2} \tag{16.59}$$

The Hamiltonian operator in polar coordinates can thus be written as

$$\hat{H}_{(r,\theta)} = -\frac{\hbar^2}{2\mu}\left(\frac{\partial^2}{\partial r^2} + \frac{1}{r}\frac{\partial}{\partial r} + \frac{1}{r^2}\frac{\partial^2}{\partial \theta^2}\right) + V(r, \theta) \tag{16.60}$$

where we have expanded the radial derivative operators. We will also henceforward write the mass as μ to avoid confusion with the angular momentum quantum number m; this will also be appropriate for two-body problems where the use of the reduced mass μ is natural. The angular derivative term can be written in the form

$$\hat{T}_\theta = -\frac{\hbar^2}{2\mu r^2}\frac{\partial^2}{\partial \theta^2} = \frac{1}{2\mu r^2}\hat{L}_z^2 \tag{16.61}$$

[5] This is true for the angular momentum associated with the orbital motion of particles; for the case of intrinsic angular momentum or spin, see Sec. 17.4.

which is indeed the obvious quantum operator analog of the classical energy of rotation. The appropriate Schrödinger equation is then

$$\hat{H}_{(r,\theta)}\psi(r,\theta) = E\psi(r,\theta) \tag{16.62}$$

along with the normalization condition for the corresponding probability density,

$$1 = \int_0^\infty r\,dr \int_0^{2\pi} d\theta |\psi(r,\theta)|^2 \tag{16.63}$$

This condition is associated with the fact that the probability of finding the particle simultaneously in the small coordinate intervals $(r, r+dr)$ and $(\theta, \theta + d\theta)$ is

$$d\text{Prob}(r,\theta) = |\psi(r,\theta)|^2 r\,dr\,d\theta \tag{16.64}$$

and we will see that the additional factor of r in the "measure" is important.

The case of central-force motion for which the potential has no angular dependence, i.e., $V(\mathbf{r}) = V(r)$, is the most important, and we note that:

- In this case, $[\hat{H}, \hat{L}_z] = 0$, so that the energy eigenfunctions will also be eigenfunctions of the (planar) angular momentum; this fact also implies that the angular momentum will be a conserved quantity.
- The Schrödinger equation is separable, so we can assume solutions of the form $\psi(r,\theta) = R(r)\Theta_m(\theta)$.

Performing the separation of variables in the SE, we find that

$$\frac{r^2}{R(r)}\left\{-\left(\frac{d^2R(r)}{dr^2} + \frac{1}{r}\frac{dR(r)}{dr}\right) + \frac{2\mu}{\hbar^2}(V(r) - E)R(r)\right\} = \frac{1}{\Theta_m(\theta)}\frac{d^2\Theta_m(\theta)}{d\theta^2} = -m^2 \tag{16.65}$$

so that the Schrödinger equation for the radial wavefunction, the quantum analog of Eqn. (16.39), is

$$-\frac{\hbar^2}{2\mu}\left(\frac{d^2R(r)}{dr^2} + \frac{1}{r}\frac{dR(r)}{dr}\right) + \left(V(r) + \frac{\hbar^2 m^2}{2\mu r^2}\right)R(r) = ER(r) \tag{16.66}$$

Given the already defined normalization properties of the $\Theta_m(\theta)$, the radial wavefunction must satisfy

$$1 = \int_0^\infty r\,dr |R(r)|^2 \tag{16.67}$$

A simple example of two-dimensional motion for which this formulation is useful is the case of a mass connected to a light but rigid rod of length r_0, free to rotate around the origin; such a system is sometimes called a *rigid rotator.* In this case, the Hamiltonian is simply $\hat{H} = \hat{L}_z^2/2\mu r_0^2$, with eigenfunctions given by the $\Theta_m(\theta)$, and quantized energies given by $E_m = \hbar^2 m^2/2\mu r_0^2$. Particles corresponding to $\pm m$ have the same quantized energies corresponding to the equivalence of clockwise versus counterclockwise motion. The same result can be inferred from the complete radial Schrödinger equation, Eqn. (16.66), if we assume that there is no potential, $V(r,\theta) = 0$, and that the radius is fixed, so that spatial derivatives of $R(r)$ vanish. We turn to less trivial examples of rotational motion in the next section.

16.3 QUANTUM SYSTEMS WITH CIRCULAR SYMMETRY

16.3.1 Free Particle

The Schrödinger equation for a free particle in polar coordinates reads

$$-\frac{\hbar^2}{2\mu}\left(\frac{d^2R(r)}{dr^2} + \frac{1}{r}\frac{dR(r)}{dr}\right) + \frac{\hbar^2 m^2}{2\mu r^2}R(r) = ER(r) \tag{16.68}$$

which can be written in terms of the dimensionless variable $z = kr$ (where $k = \sqrt{2\mu E/\hbar^2}$) as

$$\frac{d^2R(z)}{dz^2} + \frac{1}{z}\frac{dR(z)}{dz} + \left(1 - \frac{m^2}{z^2}\right)R(z) = 0 \tag{16.69}$$

which can be recognized from the mathematical literature as *Bessel's equation* (see App. C.4). Similarly to the case of a free particle in one dimension, it has two linearly independent solutions for each value of m^2, the so-called *regular solution,* $J_{|m|}(z)$, standardly labeled *cylindrical Bessel functions of order* $|m|$ (or Bessel functions of the first kind), and the irregular solutions, $Y_{|m|}(z)$, (Neumann functions or Bessel functions of the second kind). We will explore the mathematical properties and physical meaning of these solutions in this section.

We can exhibit the behavior of the solutions for large z by noting that the equation becomes approximately

$$\frac{d^2R(z)}{dz^2} = -R(z) \tag{16.70}$$

in this limit, so that the behavior is oscillatory, i.e., $R(z) \to \sin(z), \cos(z)$ or $\exp(\pm iz)$. We can do better by assuming a solution of the form

$$R(z) \longrightarrow z^{\alpha}\cos(z) \tag{16.71}$$

where we assume that $\alpha < 0$, and substitution into Eqn. (16.69) implies that (P16.14) the next-order term ($z^{\alpha-1}$) also vanishes when $\alpha = -1/2$. These results help justify the well-known asymptotic expansions

$$J_{|m|}(z) \longrightarrow \sqrt{\frac{2}{\pi z}}\cos\left(z - \frac{|m|\pi}{2} - \frac{\pi}{4}\right)\left[1 + O(1/z^2)\right] \tag{16.72}$$

$$Y_{|m|}(z) \longrightarrow \sqrt{\frac{2}{\pi z}}\sin\left(z - \frac{|m|\pi}{2} - \frac{\pi}{4}\right)\left[1 + O(1/z^2)\right] \tag{16.73}$$

which are valid for $z \gg 0$. This behavior has an immediate physical interpretation, as the *probability density times measure* gives

$$|R(r)\Theta_{|m|}(\theta)|^2 r\,dr\,d\theta \propto \cos\left(kr - \frac{|m|\pi}{2} - \frac{\pi}{4}\right)^2 dr\,d\theta \tag{16.74}$$

implying that in a spatially averaged sense, there is a uniform distribution of probability corresponding to constant speed motion everywhere in the plane. Compare this to the case

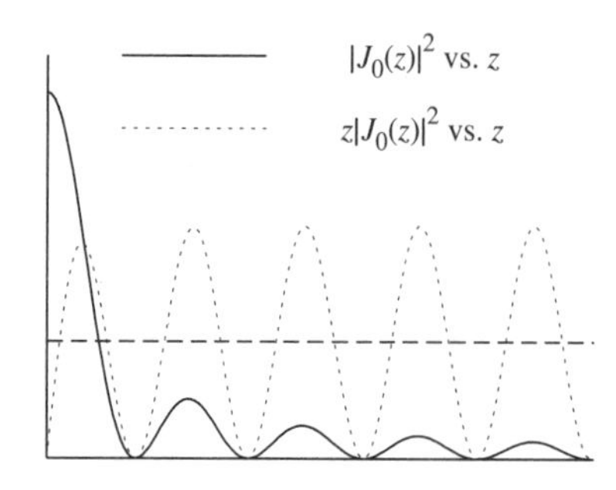

FIGURE 16.7. Plot of $|J_0(z)|^2$ (solid) and $z|J_0(z)|^2$ (dotted) versus z showing the effect of the "measure" for the free-particle wavefunction in two dimensions in polar coordinates. The horizontal dashed line corresponds to a probability distribution for constant speed motion in the plane.

of the 2D free particle in Cartesian coordinates (P16.1) and the corresponding probability distribution. This can also be contrasted with the one-dimensional case of the unstable harmonic oscillator (Sec. 10.6) where $\psi(x) \propto 1/\sqrt{x}$ as well, but which there corresponded to an accelerating particle. This is a reminder of the importance of the coordinate measure in the implementation of a probability interpretation. In Fig. 16.7, we plot $|J_0(kr)|^2$, both with and without the extra factor of r, to show the effect.

A similar analysis can be used to examine the $r \to 0$ behavior. We assume a series solution of the form

$$R(z) \to \sum_{s=\beta}^{\infty} a_s z^s = a_\beta z^\beta + a_{\beta+1} z^{\beta+1} + \cdots \tag{16.75}$$

with β to be determined. Once again, substitution into the differential equation Eqn. (16.69) yields the condition

$$\beta^2 = m^2 \qquad \text{or} \qquad \beta = \pm|m| \tag{16.76}$$

The regular, i.e., well-behaved at the origin, Bessel functions are conventionally chosen to have $\beta = +|m|$, while the ill-behaved Neumann functions are described by $\beta = -|m|$ near the origin.[6] (The behavior of the two functions is somewhat similar to the exponentially growing and decaying solutions found in tunneling problems, the rotational kinetic term $\hbar^2 m^2/2\mu r^2$ playing the role of an "angular momentum barrier" in this case.) Because of its divergence at the origin, it is often necessary to exclude this solution by hand, i.e., use the freedom to pick its coefficient in the most general solution to vanish. We plot $J_{0,1}(z)$ and $Y_1(z)$ in Fig. 16.8 for illustration.

The small r behavior of the Bessel functions solution is also intuitively physical. The probability of being "near" the origin when the particle is in a state of angular momentum

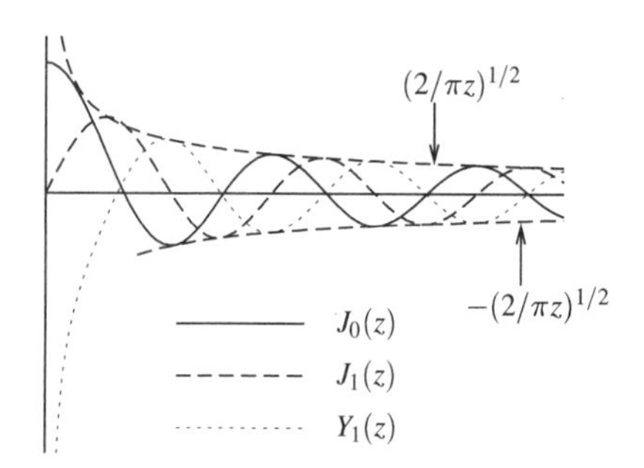

FIGURE 16.8. Plots of the regular $J_{0,1}(z)$ and irregular $Y_1(z)$ solutions of Bessel's equation showing the small and large x behavior. The behavior for large z is consistent with Eqns. (16.72) and (16.73).

[6]The behavior of $Y_0(z)$ near the origin requires special treatment, as $m = 0$ in that case corresponds to a logarithmic behavior; specifically $Y_0(z) \to 2J_0(z)\, log(\gamma z/2)/\pi$.

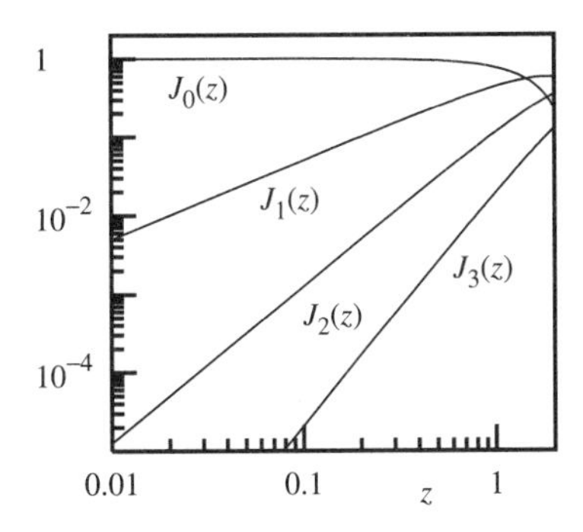

FIGURE 16.9. *Plot of $J_m(z)$ versus z with a logarithmic scale showing the power-law behavior $J_m(z) \propto z^m$ for small z.*

$\pm m\hbar$ is given by

$$d\text{Prob} \propto (kr)^{2|m|+1} \qquad \text{when} \qquad r \to 0 \tag{16.77}$$

where the additional factor of r comes from the measure. This suppression can be understood as arising from the centrifugal barrier term in Eqn. (16.61), which demands a large cost in energy to be near the origin for rotating particles. The behavior of the first few Bessel functions, $J_m(z)$, for $m = 0, 1, 2, 3$ for small argument is shown in Fig. 16.9 for illustration, and the plot on log-log paper demonstrates the increasingly large power law behavior near the origin.

16.3.2 Circular Infinite Well

A simple use of the free-particle wavefunctions arises in the study of the infinite circular well,[7] defined by the potential

$$V(r) = \begin{cases} 0 & \text{for } r < R \\ \infty & \text{for } r > R \end{cases} \tag{16.78}$$

Inside the well, where the particle is free, the solutions are

$$\psi(r, \theta) = J_{|m|}(kr)\Theta_m(\theta) \tag{16.79}$$

or, perhaps more appropriately for bound states,

$$J_{|m|}(kr)\sin(m\theta) \qquad \text{and} \qquad J_{|m|}(kr)\cos(m\theta) \tag{16.80}$$

where we have excluded the irregular $Y_{|m|}(kr)$ solutions as discussed above. The boundary conditions at the edge of the well are satisfied for all values of θ, provided that $J_m(kR) = 0$. If we label the nth zero of the mth Bessel function by $a_{(n,m)}$, we can see that the corresponding radial wavefunction will have $n_r = n - 1$ radial nodes. We see that the boundary conditions once again determine the quantized energies, in this case giving

$$E_{n,m} = \frac{\hbar^2 k^2_{(n_r,m)}}{2\mu} = \frac{\hbar^2 a^2_{(n_r,m)}}{2\mu R^2} \tag{16.81}$$

The notation n_r is motivated by the fact that it counts the number of nodes in the radial wavefunction. Some of the lowest-lying zeroes are given by

[7]For a discussion of the visualization of the solutions for this problem in both quantum and classical mechanics, see Robinett (1996a).

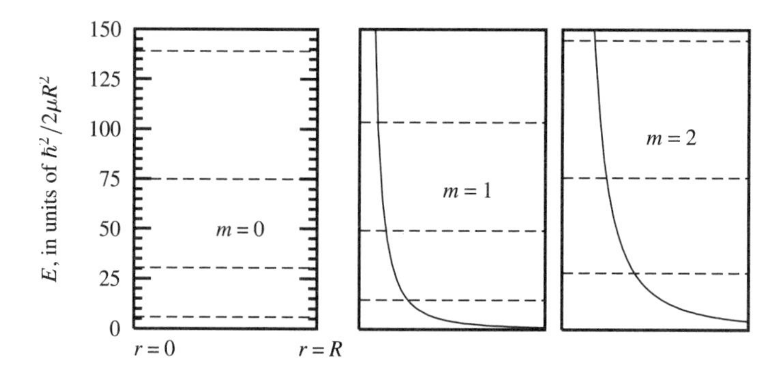

FIGURE 16.10. *Part of the energy spectrum for the infinite circular well obtained from using the Bessel function zeroes in Eqn. (16.82). Note the increase in energy of corresponding levels due to more rotational kinetic energy as m is increased.*

$$\begin{array}{llllll} m = 0: & 2.40483 & 5.52008 & 8.65373 & \cdots \\ m = 1: & 3.83171 & 7.01559 & 10.1735 & \cdots \\ m = 2: & 5.13562 & 8.41724 & \cdots & \\ m = 3: & 6.38016 & 9.76102 & \cdots & \end{array} \qquad (16.82)$$

and part of the resulting energy spectrum is shown in Fig. 16.10. Each state with $m \neq 0$ state is doubly degenerate because the two values of $\pm|m|$, corresponding to rotations in opposite senses, give the same energy.

To see connections to both wave physics and classical particle motion, we plot in Figs. 16.11 to 16.13 $|\psi(r, \theta)|^2$ for several cases:

- In Fig. 16.11, we show the ground state corresponding to $m = 0$ and the first radial zero, i.e., $a_{(0,0)} = 2.404$, with no rotational kinetic energy and the least amount of radial kinetic energy. The similarity to the shape of a circular drumhead is obvious.
- Figures 16.12(a) and (b) show $|\psi(r, \theta)|^2$ for the lowest-lying state with $|m| = 1$ ($a_{(0,1)} = 3.8318\ldots$) with both the $\cos(\theta)$ and the $\sin(\theta)$ solutions plotted to illustrate their similarity and to help to visualize their degeneracy.
- Figure 16.13(a) shows a radially excited state (large n_r), but still with no angular momentum, $(n_r, m) = (4, 0)$, with $a_{(4,0)} = 14.43$. This corresponds to a classical particle bouncing back and forth through the origin in a particle interpretation or a spherically symmetric wave reflecting from the walls. We note that "corrals" of heavy atoms can be constructed which approximate infinite circular potential wells on surfaces, and the measured electron densities in such a configuration closely match these predictions; an example was shown in Fig. 1.3(a).

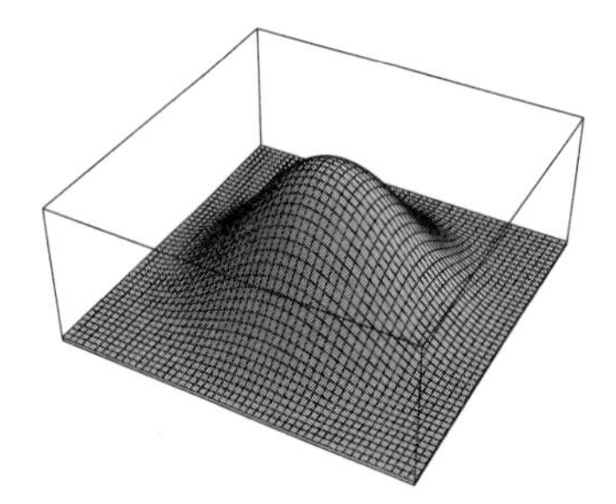

FIGURE 16.11. *Plots of $|\psi(r, \theta)|^2$ for the ground state with $(n_r, m) = (0, 0)$.*

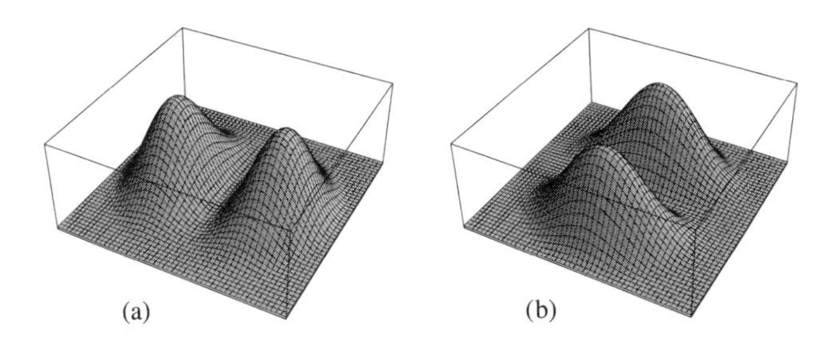

FIGURE 16.12. *Plot of* $|\psi(r,\theta)|^2$ *for the lowest-lying* $m = 1$ *state; both the* $\sin(\theta)$ *(a) and* $\cos(\theta)$ *(b) cases are plotted to help visualize the double degeneracy.*

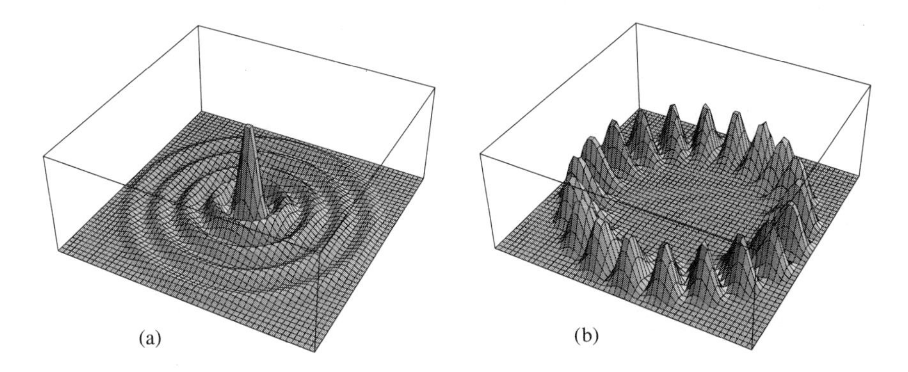

FIGURE 16.13. *Plot of* $|\psi(r,\theta)|^2$ *for "radial" and "angular" states: (a) corresponds to a radially excited state,* $(n_r, m) = (4, 0)$*, with no angular momentum, while (b) is for* $(n_r, m) = (10, 0)$*, with large angular momentum (and rotational kinetic energy) and little radial kinetic energy.*

- Finally, Fig. 16.13(b) shows a state with large angular momentum ($m = 10$) but with the smallest radial quantum number possible, specifically $a_{(0,10)}$, with $a_{(0,10)} = 14.48$. This corresponds to a particle in such a well undergoing uniform circular motion, and the expected peaking of the quantum wavefunction near the walls (classically the particle would, after all, roll around the inner edge) is apparent. This might be the quantum equivalent of a "roulette wheel." The energies of the (4, 0) and (0, 10) states differ by less than 1%, but the distribution between radial and rotational kinetic energy is very different.

It is, in principle, possible to construct localized wavepackets and track the quasi-classical ballistic motion of particles bouncing in the circular well, but we will not consider that here due to the technical complexity.

16.3.3 Isotropic Harmonic Oscillator

We return to the isotropic harmonic oscillator in two dimensions with potential

$$V(r) = \frac{1}{2}Kr^2 = \frac{1}{2}\mu\omega^2 r^2 \tag{16.83}$$

and Schrödinger equation

$$-\frac{\hbar^2}{2\mu}\left(\frac{d^2R(r)}{dr^2} + \frac{1}{r}\frac{dR(r)}{dr}\right) + \left(\frac{\hbar^2 m^2}{2\mu r^2} + \frac{\mu\omega^2}{2}r^2\right)R(r) = ER(r) \tag{16.84}$$

A standard change of variables, $r = \rho y$ with $\rho^2 = \hbar/\mu\omega$, reduces this to

$$\frac{d^2R(y)}{dy^2} + \frac{1}{y}\frac{dR(y)}{dy} + \left(\epsilon - y^2 - \frac{m^2}{y^2}\right)R(y) = 0 \tag{16.85}$$

with the dimensionless energy eigenvalue $\epsilon = 2E/\hbar\omega$. The large y dependence can be extracted as in Sec. 10.2.1, while the behavior near the origin is guaranteed to be of the form $y^{|m|}$ from Eqn. (16.76). We are thus led to write

$$R(y) = y^{|m|}e^{-y^2/2}G(y) \tag{16.86}$$

leading to

$$\frac{d^2G(y)}{dy^2} + \left(\frac{2|m|+1}{y} - 2y\right)\frac{dG(y)}{dy} + \left(\epsilon - 2 - 2|m|\right)G(y) = 0 \tag{16.87}$$

A somewhat less obvious change of variables to $z = y^2$ then yields the differential equation

$$\frac{d^2G(z)}{dz^2} + \frac{dG(z)}{dz}\left(\frac{|m|+1}{z} - 1\right) + G(z)\left(\frac{\epsilon - 2(|m|+1)}{4z}\right) = 0 \tag{16.88}$$

(which, to the mathematically sophisticated, can be recognized as *Laguerre's equation* as discussed in App. C.7). A series solution of the form $G(z) = \sum_{s=0}^{\infty} b_s z^s$ yields the recursion relation

$$\frac{b_{s+1}}{b_s} = \frac{(s + (|m|+1)/2 - \epsilon/4)}{(s+1)(s+|m|+1)} \longrightarrow \frac{1}{s} \tag{16.89}$$

in the limit of large s implying (as in Sec. 10.2) that $G(z) \sim e^z \sim e^{y^2}$, which would be inconsistent with the desired behavior in Eqn. (16.86). Once again we find the series must terminate, yielding a polynomial of finite degree in z. The quantized energies are given in terms of the maximum power of this polynomial, $s_{\max} = n_r$, by the condition

$$\epsilon = 2|m| + 2 + 4s_{\max} \equiv 2|m| + 2 + 4n_r \tag{16.90}$$

and we note that n_r also counts the number of radial nodes of the resulting polynomial. This leads to the quantized energies

$$E_{n_r,m} = \hbar\omega\left(|m| + 2n_r + 1\right) \tag{16.91}$$

and the corresponding $G(z)$ are *generalized Laguerre polynomials,* denoted as $L_{n_r}^{(|m|)}(z)$. The first few of these are given here for later use:

$$\begin{aligned} L_0^{(k)}(z) &= 1 \\ L_1^{(k)}(z) &= 1 + k - z \\ L_2^{(k)}(z) &= \frac{1}{2}\left(2 + 3k + k^2 - 2z(k+2) + z^2\right) \end{aligned} \tag{16.92}$$

The resulting constant polynomials in the case of $n_r = 0$ are especially important for the classical limit.

The complete (but unnormalized) solutions in polar coordinates are then given by

$$\psi_{n_r,m}(r,\theta) \propto r^{|m|}e^{-r^2/2\rho^2}L_{n_r}^{(|m|)}(r^2/\rho^2)e^{im\theta} \tag{16.93}$$

The energy spectrum and degeneracies thusly derived from polar coordinates are shown in Fig. 16.14, and we see that the degeneracies agree with those found using Cartesian coordinates. The wavefunctions for a given energy level in the two different schemes are necessarily linear combinations of each other, which can be shown explicitly in simple cases (P16.21).

A particularly easy classical limit to exhibit in this case is that corresponding to uniform circular motion, in which case one would use $|m| \gg 1$ and $n_r = 0$ corresponding to the minimum possible radial kinetic energy. In this case, the Laguerre polynomials, $L_0^{|m|}(z)$ are constants, so the radial probability density is proportional to

$$P(r) = |\psi_{0,m}(r,\theta)|^2 \sim r^{2|m|}e^{-r^2/\rho^2} \tag{16.94}$$

which has a maximum value when

$$0 = \frac{dP(r)}{dr} = \left(2|m|r^{2|m|-1} - \frac{2r^{2|m|+1}}{\rho^2}\right)e^{-r^2/\rho^2} \tag{16.95}$$

or $r_{max}^2 = |m|\rho^2$. Recalling the definition ρ, we find that

$$r_{max}^2 = \frac{|m|\hbar}{\mu\omega} = \frac{L_z}{\mu\omega} \tag{16.96}$$

The classical circular orbit, of constant radius r_0, is determined by Eqn. (16.39) where we take $\ddot{r}(t) = 0$ implying that

$$-\mu\omega^2 r_0 = -Kr_0 = F(r_0) = -\frac{L_z^2}{\mu r_0^3} \quad \text{or} \quad r_0^2 = \frac{L_z}{\sqrt{\mu K}} = \frac{L_z}{\mu\omega} \tag{16.97}$$

in agreement with Eqn. (16.96), and with the correspondence principle. Somewhat surprisingly, the form of this Schrödinger equation for a 2D SHO and its solutions are very similar to that for a charged particle in a uniform magnetic field (Sec. 19.5), which partly motivates our detailed study of it here.

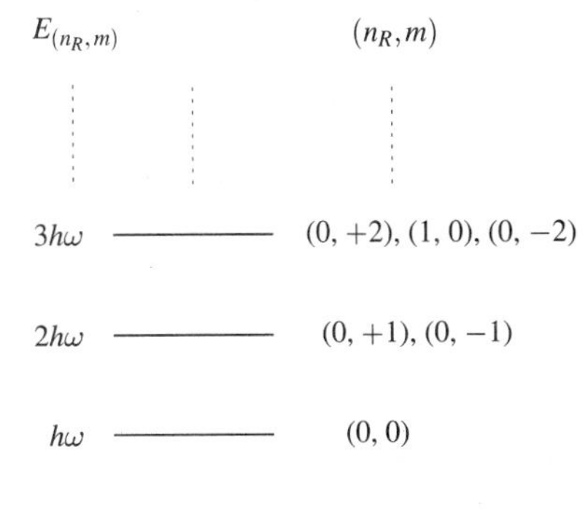

FIGURE 16.14. *Energy spectrum for 2D simple harmonic oscillator problem as obtained in polar coordinates showing the same level degeneracy as in Fig. 16.6; values of (n_r, m) for each level are shown.*

16.4 CLASSICAL AND QUANTUM CHAOS

Many texts rightfully stress the important role played by symmetry in the study of both classical and quantum mechanics. The harmonic oscillator (in any number of dimensions) and the hydrogen atom (or inverse-square law or Kepler) problem, for example, in addition to their close connection to real physical systems, are studied extensively because they are exactly soluble, in both their classical and quantum treatments. This, in turn, is due in part to the extraordinary amount of symmetry present in these systems. The classical trajectories and quantum wavefunctions both exhibit a remarkable simplicity for these famous problems.

One of the most noteworthy advances in the last two decades in the study of dynamical systems[8] has been the appreciation of the widespread and universal appearance of *chaos* and *chaotic behavior.* One manifestation of chaos in classical mechanics systems is the notion of *sensitive dependence on initial conditions.* This expression often carries the implication that the solutions of the differential equations (say, Newton's laws)—while being, in principle, completely deterministic—are so sensitive to the initial values used as inputs that the resulting motion is almost completely unpredictable in practice.[9]

One obvious question is how such classically chaotic trajectories are related to the corresponding quantum wavefunctions and energy spectra. Many of the most important connections between chaos[10] in classical and quantum systems are easily visualized in two-dimensional systems, and in this section we briefly discuss *classical and quantum billiards,* i.e., point particles confined to two-dimensional infinite wells.

The ballistic motion of a point particle in a circular enclosure (i.e, a point mass in a circular infinite well) can exhibit rather simple trajectories, as shown in Fig. 16.15. The circular symmetry guarantees conservation of angular momentum, which, in turn, is responsible for the minimum radius that is evident in the figure.

A simple generalization of the circular boundary is given by the *stadium problem* illustrated in Fig. 16.16, where the circle of radius R is "cut" at the top and bottom, and straight sections of length a are added. An example of a typical trajectory in one such stadium (with $a = R$) is shown in Fig. 16.17. There is no longer any rotational symmetry (or conserved angular momentum), and a single trajectory, if followed over a long enough time, will appear to "paint" the entire interior region. The trajectories of almost all ray trajectories[11] will be *ergodic;* that is, the particle will come arbitrarily close to any point in phase space[12] if one waits long enough.

---

[8]Crudely defined as a set of variables that are governed by differential equations that purport to predict the future time dependence of the system; examples include hydrodynamics, biology, and even economics.

[9]The classic example comes from meteorology where chaotic behavior was first recognized by the meteorologist Lorenz who was performing numerical simulations of weather patterns. He found that small changes in his input data caused large changes in the resulting predictions. It has been described as being akin to "the flap of a butterfly's wing in Texas giving rise to a hurricane in Brazil," hence the term *butterfly effect.* For a popular review of chaos, see Gleick (1988).

[10]We will not carefully define chaos anywhere in this section, but rather present results that demonstrate some of the methods of analysis that are used to distinguish between predictable (often called *integrable*) and chaotic systems. To borrow a phrase from art criticism, "I may not be able to define chaos, but I know it when I see it." The most comprehensive treatment of the relation between classical and quantum chaos is Gutzwiller (1990).

[11]Very special cases where the particle bounces back and forth between the flat sections and the like are possible.

[12]The region of position and momentum variables that are allowed to the particle as it moves along the trajectory determined by the initial conditions.

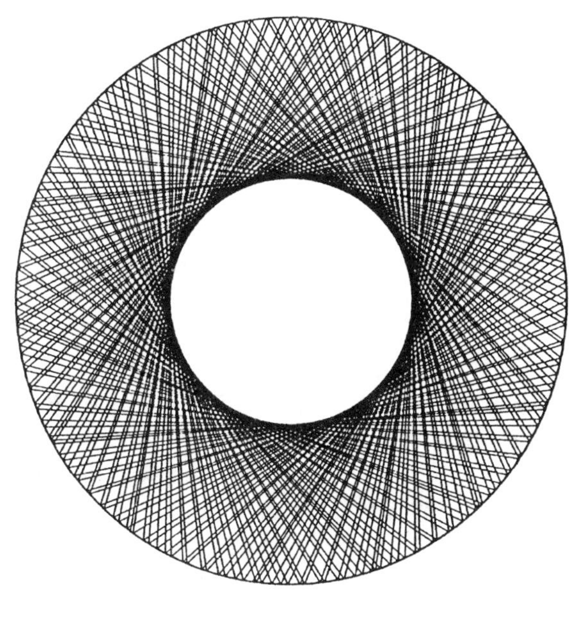

FIGURE 16.15. *Typical classical trajectory for a particle in a circular well. (Permission of Allan Kaufman.)*

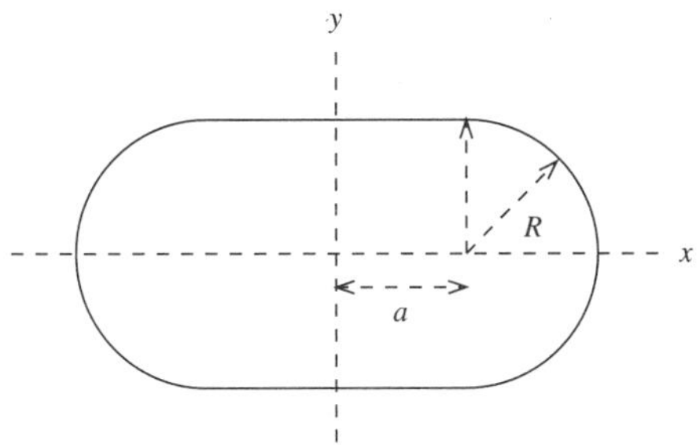

FIGURE 16.16. *Geometry for the "stadium."*

For the case of the corresponding quantum billiard, the Schrödinger equation for a free particle in a 2D region with infinite walls corresponds to the Helmholtz equation (familiar from other wave problems such as the propagation of electromagnetic radiation in waveguides[13]), i.e.,

$$\left(\nabla^2 + k_n^2\right)\psi_n(\mathbf{r}) = 0 \tag{16.98}$$

with $\psi_n(\mathbf{r}) = 0$ on the boundary. Solutions for the rectangular and circular billiard problem have been discussed in Secs. 16.1.1 and 16.3.2 and are available in closed-form solutions.

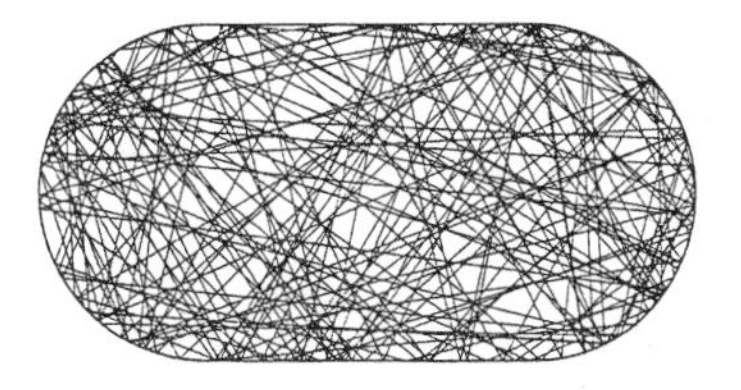

FIGURE 16.17. *Typical classical trajectory for a particle in a "stadium." (Permission of Allan Kaufman.)*

[13]In fact, the eigenfrequencies of stadium-shaped resonant cavities have been studied by microwave absorption and give the quantum mechanical eigenvalue spectrum for the same shape; see Stöckmann and Stein (1990).

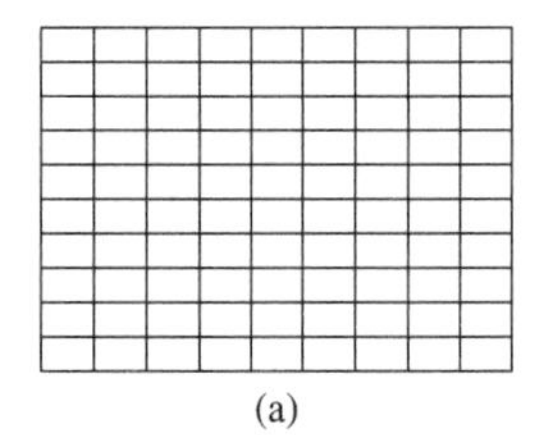
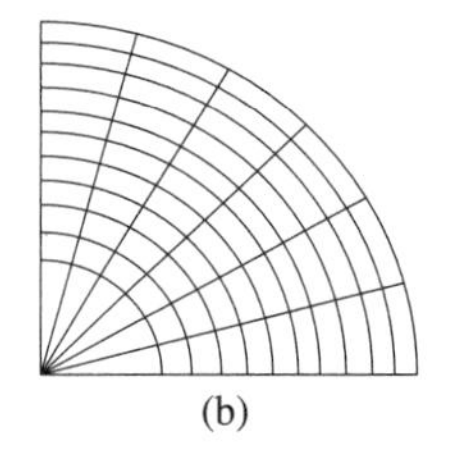

(a) (b)

FIGURE 16.18. Pattern of nodal lines (given by $|\psi(r)| = 0$) for quantum wavefunctions in (a) rectangular and (b) circular wells.

Results for the stadium problem can be obtained numerically.[14] To better visualize the differences, we plot in Figs. 16.18 and 16.19 the *nodal lines,* that is, the curves corresponding to $|\psi(\mathbf{r})|^2 = 0$ in the plane for rectangular/circular and stadium shapes, respectively. To facilitate comparison, we choose examples that satisfy the following requirements:

1. The circle, rectangle, and stadium problems still possess separate parity symmetries about the x and y axes, so that the wavefunctions will satisfy $\psi(-x, -y) = (-1)^{P_x}(-1)^{P_y}\psi(x, y)$. We choose solutions with double-odd parity which vanish along both coordinate axes, so we need only plot the first quadrant.
2. The area of "billiard table" in each case is kept constant as the shape is changed.
3. Finally, the "energy eigenvalue," namely, the value of k_n, is chosen to be as similar as possible in each case.

We note that:

- The separation of the nodal curves in all cases is approximately regular and is roughly given by the half wavelength, π/k, consistent with waves on a surface.[15]
- On the other hand, the directions of the nodal lines in the chaotic case are highly irregular, and they exhibit no crossings in the interior; this is in direct contrast to the highly symmetric (and integrable) rectangular and circular wells.

Further information on the structure of the solutions is available in the spectrum of energy eigenvalues; the distribution of spacings between one energy level and the next is characteristically different for predictable and chaotic systems. A systematic approach can be followed for any billiard system:

1. For a given geometry, find the energy eigenvalues and order them.
2. Evaluate the difference between successive energy levels, given by $\Delta E = E_n - E_{n-1}$.
3. Calculate the number of occurrences, dN, of a given value of ΔE, in a given bin size.
4. For comparison, normalize the distribution to the total number of energy levels used, and plot dN/N versus ΔE.

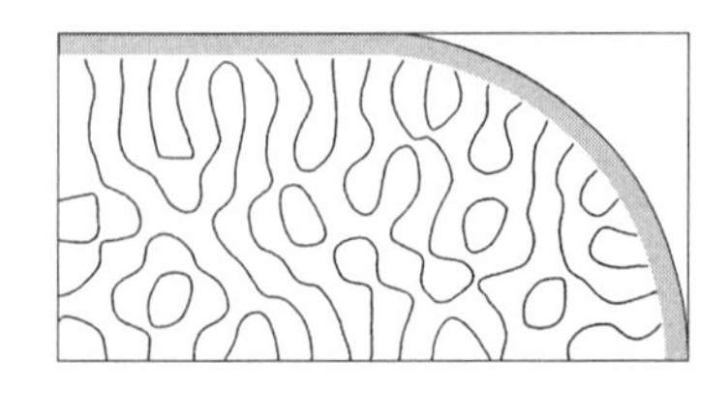

FIGURE 16.19. Pattern of nodal lines for "stadium." (Permission of Allan Kaufman.)

[14] We use the results of McDonald and Kaufman (1979, 1988).

[15] What would a stadium-shaped drumhead sound like?

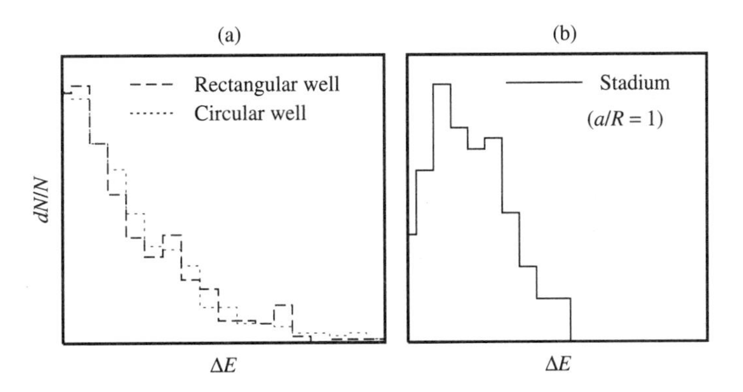

FIGURE 16.20. Distribution of differences between successive energy levels in the rectangular/circular wells (a) and stadium (b).

We plot such distributions in Fig. 16.20(a) and (b) for the rectangular/circular and stadium systems; once again, there are clear differences between the integrable and chaotic systems:

- The rectangular and circular systems are characterized by a large number of levels with close spacing as the distribution is peaked at $\Delta E \approx 0$. In contrast, the chaotic system has a maximum at a finite value of ΔE, which is often said to correspond to a "mutual repulsion" of energy levels. This form is also characteristic of the eigenvalue spectrum of large matrices with random elements, and is a "trademark" of chaotic systems. Such distributions of energy level spacings have been observed experimentally in the spectra of heavy nuclei and atoms in strong magnetic fields.

16.5 QUESTIONS AND PROBLEMS

Q16.1 Estimate the zero point energy and spread in position of an electron bound to a (horizontal) surface because of gravity. Assume for simplicity that the potential is given by

$$V(z) = \begin{cases} \infty & \text{for } z < 0 \\ m_e g z & \text{for } z > 0 \end{cases}$$

What does this potential look like superimposed on Fig. 16.2?

Q16.2 What would a plot of $|\phi(p_x, p_y)|^2$ versus (p_x, p_y) look like for the $(n, m) = (1, 1)$ state in the infinite square well? How about for $(n, m) = (10, 10)$ or $(15, 30)$?

Q16.3 What would $|\phi(p_x, p_y; t)|^2$ look like for the "bouncing 2D wavepacket" in Fig. 16.5 as a function of time?

Q16.4 Recall the wavefunctions in the circular well shown in Fig. 16.13 (a) and (b) corresponding to "radial" and "angular" motion, respectively. The momentum space distributions can be evaluated numerically, and the resulting distributions are plotted in Fig. 16.21(a) and (b). Explain why they have the form they do. No numerical calculations are required. Simply use your physical intuition, and think about what the momentum vectors in 2D would look like for the two cases of "radial" and "angular" motion.

Q16.5 How would you solve the 2D harmonic oscillator problem using raising and lowering operators? Go as far as you can in generalizing the arguments of Sec 14.2.

Q16.6 Are there any uncertainty relations in two dimensions for pairs of variables like $x, \hat{p}_y$? Can you find an example of a state that has $\Delta x \Delta p_y = 0$ or arbitrarily small?

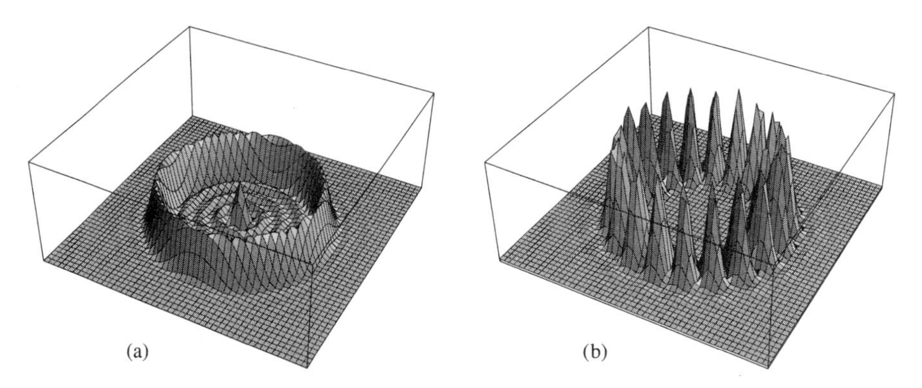

FIGURE 16.21. *Momentum space probability distributions, $|\phi(p_x, p_y)|^2$ versus p_x, p_y, corresponding to the "radial" (a) and "angular" (b) wavefunctions in the circular infinite well of Fig. 16.13. Why do they look rather similar?*

Q16.7 What are the appropriate commutation relations for θ and $\hat{L}_z$? Is there an associated uncertainty principle? Is there any problem[16] due to the fact that θ is only defined up to multiples of 2π, so that $\Delta\theta$ cannot be arbitrarily large?

P16.1 **(a)** Find the plane wave solutions of the time-dependent free-particle Schrödinger equation in two dimensions using Cartesian components. Show that they can be written in the form

$$\psi(\mathbf{r}; t) = e^{i(k_x x - k_y y - \omega t)} = e^{i(\mathbf{k}\cdot\mathbf{r} - \omega t)}$$

and find the dispersion relation relating $\mathbf{k}$ and ω.

(b) Explicitly construct a localized Gaussian wavepacket with central momentum value $\mathbf{p} = (p_0^x, p_0^y)$ with initial central position $\mathbf{r}_0 = (x_0, y_0)$. Calculate $\langle x \rangle_t$ and $\langle y \rangle_t$ for this state.

P16.2 Consider a two-dimensional potential given by

$$V(x, y) = \begin{cases} 0 & \text{for } y < 0 \\ V_0 > 0 & \text{for } y > 0 \end{cases}$$

which is a two-dimensional step-up potential.

(a) To examine plane wave scattering from such a step, consider a solution of the form

$$\psi(\mathbf{r}; t) = \begin{cases} Ie^{i(\mathbf{k}_1 \cdot \mathbf{r} - \omega t)} + Re^{i(\mathbf{k}_1' \cdot \mathbf{r} - \omega t)} & \text{for } y < 0 \\ Te^{i(\mathbf{k}_2 \cdot \mathbf{r} - \omega t)} & \text{for } y > 0 \end{cases}$$

where the wavevectors $\mathbf{k}_1$, $\mathbf{k}_1'$, $\mathbf{k}_2$ are defined in Fig. 16.22. Match the wavefunction along the $y = 0$ boundary to find a relation between θ_1, θ_1', and θ_2, and compare to Snell's law of refraction.

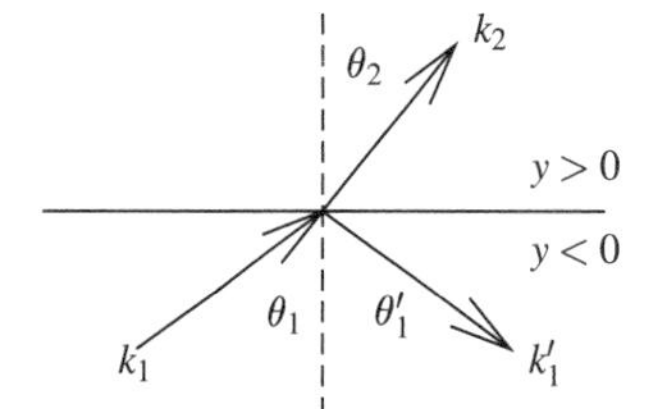

FIGURE 16.22. *Wavenumbers and reflection-refraction angles for two-dimensional step-up potential.*

[16] See, e.g., Roy and Sannigrahi (1979).

(b) Match the derivatives of the wavefunction at the boundary (which direction?) to determine the reflection and transmission probabilities, namely, $|R/I|^2$ and $|T/I|^2$. Show that your results reduce to those in Sec. 12.2 for normal incidence, namely, $\theta_1 = 0$.

(c) For a given angle of incidence, θ, what is the minimum incident energy below which all of the incident particles will be reflected?

P16.3 Find the solutions to the Schrödinger equation for the 2D potential

$$V(x, y) = \begin{cases} 0 & \text{for } x < 0 \text{ and } y < 0 \\ +\infty & \text{otherwise} \end{cases}$$

which is like a "corner reflector." What would you expect for the behavior of a wavepacket incident on such a potential from various angles and for various incident energies?

P16.4 (a) Write down the Schrödinger equation describing a particle moving in a vertical plane subject to a constant downward gravitational force, and show that it is separable.

(b) Use previously obtained results for the free particle and uniformly accelerating wavepackets to write down a wavepacket solution, $\psi_{WP}(x, y; t)$, for this problem.

(c) Evaluate $\langle x \rangle_t$ and $\langle y \rangle_t$ for the wavepacket.

P16.5 Consider a wavefunction in the 2D infinite square box of Sec. 16.1.1 given at $t = 0$ by

$$\psi(x, y; 0) = Nx(L - x)y(L - y)$$

(a) Find N such that $\psi(x, y, 0)$ is normalized.

(b) What is the probability that a measurement of the energy of a particle described by this state would yield the ground state energy of the system at time $t = 0$? What is this probability at later times?

P16.6 Show that the exchange operator $\hat{E}_{x,y}$ in Eqn. (16.18) is Hermitian and that its eigenvalues are ± 1.

P16.7 **Accidental degeneracies?**

(a) Find the energy eigenvalues for a particle in an infinite *rectangular* well with sides of lengths $L_1 \neq L_2$. Show that, in general, there are no degenerate energy levels.

(b) Show that if L_1 and L_2 are commensurate, that is, if $L_1/L_2 = p/q$ is a ratio of integers, that two different levels characterized by pairs of integers (n_1, n_2) and $(pn_2/q, qn_1/p)$ can be degenerate. Show that an example is when $L_1 = 2L_2$ and the pairs $(4, 1)$ and $(2, 2)$ give rise to degenerate energy states. This phenomenon is often called *accidental degeneracy* as it is not due to any obvious symmetry. (Exchange symmetry is not an obviously useful idea for this asymmetric box.)

(c) For the special cases discussed in (b), consider a "bigger" square infinite well of size $L = qL_1 = pL_2$ on a side, and show that the original box "fits into" the lower left-hand corner. Show that the degenerate wavefunctions in the original box, when extended to the larger box, are simply the standard degenerate pairs discussed in the text. This is illustrated in Fig. 16.23 for the explicit example in part (b).

P16.8 Show that there are N different states of the 2D simple harmonic oscillator which have energy $(N+1)\hbar\omega$; i.e., calculate the degeneracy of each level. Use the solution in Cartesian components.

P16.9 Evaluate the expectation values of x, y, $\hat{p}_x$, and $\hat{p}_y$ in any energy eigenstate of the 2D SHO of the form in Eqn. (16.24). Show that the expectation value of $\hat{L}_z$ also vanishes in any such state. How then can we have states of the 2D SHO with definite nonzero values of quantized angular momentum?

P16.10 Investigate possible accidental degeneracies in the energy spectrum of the nonisotropic 2D harmonic oscillator. What happens, for example, when $k_x = 4k_y$? If you find any degeneracies, can you find any similarities with the discussion of P16.7?

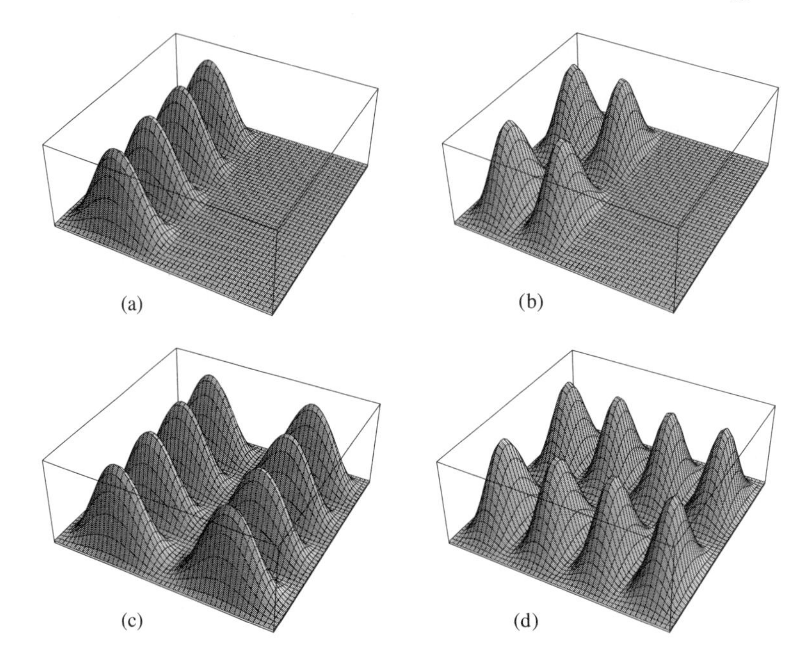

FIGURE 16.23. Accidental degeneracies in two-dimensional rectangular boxes: (a) and (b) correspond to the degenerate $(n_x, n_y) = (4, 1)$ and $(2, 2)$ levels in a box with $L_x = 2L_y$; (c) and (d) correspond to the same levels in the "extended" square box with $L = L_x = 2L_y$ where the levels are "naturally" degenerate.

P16.11 2D harmonic oscillator wavepackets

(a) Using the explicit 1D SHO wavepackets in Sec 10.4, write down the probability distribution for a wavepacket undergoing uniform circular motion in the isotropic 2D harmonic oscillator potential. *Hint:* Choose a wavepacket for the x coordinate with appropriate initial position but vanishing initial momentum, and oppositely for the y coordinate packet.

(b) Calculate $\langle x\rangle_t$, $\langle y\rangle_t$, $\langle \hat{p}_x\rangle_t$, and $\langle \hat{p}_y\rangle_t$ for this state, and show that they behave as expected.

(c) Calculate the expectation value of the angular momentum $\langle \hat{L}_z\rangle$, and show explicitly that it is conserved, i.e., constant in time.

(d) Show that you can write the probability density for a wavepacket representing counterclockwise motion in the form

$$|\psi(r, \theta; t)|^2 = \frac{1}{\pi L^2(t)} \exp\left(-(r^2) - 2rx_0\cos(\theta - \omega t) + x_0^2\right)$$

(e) Repeat parts (a) to (c) for a wavepacket representing a more general elliptical classical path.

(f) Repeat parts (a) to (c) for a wavepacket under the influence of a "nonisotropic" spring of the form in Eqn. (16.27).

P16.12 **(a)** Show that

$$e^{i\alpha\hat{L}_z/\hbar} f(\theta) = f(\theta + \alpha)$$

(b) Calculate $[\hat{L}_z, \theta]$.

P16.13 Using the defining relations Eqn. (16.47) or (16.48), derive Eqns. (16.49), (16.50), and (16.51), and show that

$$\frac{\partial^2}{\partial x^2} + \frac{\partial^2}{\partial y^2} = \frac{\partial^2}{\partial r^2} + \frac{1}{r}\frac{\partial}{\partial r} + \frac{1}{r^2}\frac{\partial^2}{\partial \theta^2}$$

P16.14 Substitute the trial solution Eqn. (16.71) into Bessel's equation, and show that $\alpha = -1/2$ gives the next-to-leading behavior.

P16.15 Use the methods outlined in Sec. 6.2 to derive the classical radial probability distribution for a particle in the circular infinite well. *Hint:* Use the energy relation in the form

$$E = \frac{1}{2}\mu \dot{r}^2 + \frac{L^2}{2\mu r^2}$$

to find an expression relating dr and dt. See Robinett (1996a).

(a) Show that your result can be written in the form

$$P_{CL}(r) = \frac{r}{\sqrt{R^2 - R_{\min}^2}\sqrt{r^2 - R_{\min}^2}}$$

where $R_{\min} = \sqrt{L^2/2\mu E}$ is the distance of closest approach. Discuss the limiting cases of purely radial and purely angular motion.

(b) Calculate $\langle r \rangle$ and Δr, and discuss their behavior in the same limiting cases.

P16.16 Consider the wavefunction in the infinite circular well of radius a given by $\psi(r,\theta) = N(a-r)\sin^2(\theta)$.

(a) Find N such that ψ is properly normalized.

(b) What is the probability that a measurement of L_z in this state will yield $0\hbar$?; $\pm 1\hbar$?; $\pm 2\hbar$? Any other value? You should be able to obtain definite numerical answers for this part. *Hint:* What is the appropriate expansion theorem?

(c) What is the probability that a measurement of the energy finds this particle to be in the ground state of the well? Your answer will be in terms of integrals with Bessel functions that you don't need to evaluate numerically.

P16.17 For the infinite circular well of radius R, the angular wavefunctions, $\Theta_m(\theta)$, ensure that eigenstates with different values of m will be orthogonal. For a given value of m, show that one must have

$$\int_0^R dr\, r J_m(k_{n_1,m} r) J_m(k_{n_2,m} r) = 0 \qquad \text{if } n_1 \neq n_2$$

If you have access to and expertise with an all-purpose computer mathematics package such as *Mathematica®*, confirm this by numerical integration for several cases if you can.

P16.18 **Centrifugal force in the circular infinite well:** Classically, a particle undergoing uniform circular motion would require a force given by

$$F_c(R) = \frac{\mu v^2}{R} = \mu\omega^2 R = \frac{\mu R^2 \omega^2}{R} = \frac{2T_{\text{rot}}}{R}$$

to keep it in motion. Calculate the force exerted by the rapidly spinning ball in the infinite-well problem by considering the change in energy when the wall is slowly moved outward a small amount dR, i.e.,

$$dW = \frac{\hbar^2 a_{(n_r,m)}^2}{2\mu R^2} - \frac{\hbar^2 a_{(n_r,m)}^2}{2\mu (R+dR)^2} = F \cdot dR$$

Show that the force exerted by the wall on the particle is consistent with the classical result.

P16.19 Variational calculation for the circular infinite well:

(a) Make a variational estimate of the ground state energy of the circular infinite well with radius a. Since this state will necessarily have $m = 0$, this amounts to evaluating the energy functional

$$E[\psi] = \frac{\langle\psi|\hat{H}_r|\psi\rangle}{\langle\psi|\psi\rangle}$$

where

$$\hat{H}_r = -\frac{\hbar^2}{2\mu}\left(\frac{d^2}{dr^2} + \frac{1}{r}\frac{d}{dr}\right)$$

Assume a trial wavefunction of the form $\psi(r,\theta) = R(r) = a^\lambda - r^\lambda$ where λ is used as the variational parameter. Compare your answer to the exact ground state energy obtained from Eqn. (16.81) and the table of Bessel function zeroes in Eqn. (16.82).

(b) Because the angular wavefunctions $\Theta_m(\theta)$ form an orthogonal set, one can actually make a rigorous variational estimate for the lowest energy state for every value of m (why?). Recalling the required behavior of the wavefunction near the origin, use a trial wavefunction of the form

$$\psi_{0,m}(r,\theta) = R_{0,m}(r)e^{im\theta} \qquad \text{where} \qquad R_{0,m}(r) = (a^\lambda - r^\lambda)r^m$$

Show that the variational energy is given by

$$E_{\text{var}}(\lambda; m) = \frac{\hbar^2}{2\mu a^2}\frac{(1+m)(1+\lambda+m)(2+\lambda+2m)}{(\lambda+m)}$$

with a minimum value

$$E_{\text{var}}^{\text{min}}(m) = \frac{\hbar^2}{2\mu a^2}\left[\frac{(1+m)(4+2m+(3+m)\sqrt{2+m})}{\sqrt{2+m}}\right]$$

Compare this to your answer in part (a) and to the exact answers for $m = 1, 2, 3, 4$ in Eqn. (16.82).

(c) Sketch the variational radial wavefunctions versus r for increasing values of m, and note that the probability is increasingly peaked near the boundary. Using these approximate wavefunctions, find the value of r at which the probability, that is, $r|R_{0,m}(r)|^2$, is peaked, and show that it approaches a as $m \to \infty$.

P16.20 Semicircular infinite circular well: Consider a particle of mass μ in a "half" infinite circular potential well defined by

$$V(r,\theta) = \begin{cases} 0 & \text{for } 0 < \theta < \pi \text{ and } r < R \\ \infty & \text{otherwise} \end{cases}$$

(a) Find the allowed energies and wavefunctions in terms of those of the "full" infinite circular well. Discuss the degeneracy of each level (if any).

(b) Show that the angular momentum operator, $\hat{L}_z$, is still Hermitian, and discuss why.

(c) Show that $\hat{L}_z$ no longer commutes with the Hamiltonian, so that $\langle\hat{L}_z\rangle_t$ need no longer be constant in time.

(d) Consider the wavefunction

$$\psi(r,\theta;0) = AJ_0(k_{(0,1)}r)\sin(\theta) + BJ_0(k_{(0,2)}r)\sin(2\theta)$$

Find the wavefunction for later times, and evaluate $\langle\hat{L}_z\rangle_t$; show that it is not constant. Show that any initial wavefunction with only even m or only odd m components will have a constant (and vanishing) angular momentum.

P16.21 2D annular infinite well: Consider 2D potential with circular symmetry corresponding to two infinite walls at $r = b, a$, defined via

$$V(r) = \begin{cases} \infty & \text{for } r < b \\ 0 & \text{for } b < r < a \\ \infty & \text{for } r > a \end{cases}$$

(a) Find the allowed solutions, and derive the condition that determines the energy eigenvalues for each value of m.

(b) Estimate the ground state energy when $\delta = a - b \ll a$, i.e., the two walls are very close together. Is there any similarity in this limit to a long rectangular potential of dimensions $2\pi a \times \delta$?

P16.22 2D harmonic oscillator solutions in polar coordinates:

(a) Confirm that the $L_{n_r}^{(k)}(z)$ given in Eqn. (16.93) actually solve Eqn. (16.88); for the case of $n_r = 0$, show that the solutions are trivial.

(b) Show that the degeneracy of the level with energy $E = N_s\hbar\omega$ is N_s.

(c) Compare the wavefunctions for the states with $E = (N+1)\hbar\omega$ in Cartesian and polar coordinates, and show explicitly that they can be written as linear combinations of each other for the cases $N = 0, 1, 2$.

P16.23 Consider a particle of mass m moving in the two-dimensional "one-quarter" isotropic harmonic oscillator potential

$$V(r, \theta) = \begin{cases} Kr^2/2 & \text{for } 0 < \theta < \pi/2 \\ \infty & \text{otherwise} \end{cases}$$

(a) What are the allowed energy levels and wavefunctions in such a potential? How do they compare to the "full" harmonic oscillator?

P16.24 Jastrow-Laughlin wavefunctions:

(a) Consider a particle in the 2D harmonic oscillator with a wavefunction corresponding to $n_r = 0$, and show that it can be written in the form

$$\begin{aligned} \psi(\mathbf{r}) = \psi_{0,m}(r, \theta) &= N r^{|m|} e^{-r^2/2\rho^2} e^{im\theta} \\ &= N' w^m \exp(-|w|^2/2) = \psi(w) \end{aligned}$$

where w is a *complex coordinate* given by $w \equiv (x + iy)/\rho$. Show that this state has an angular momentum given by $m\hbar$.

(b) Consider a *two-particle state* given by

$$\psi(w_1, w_2) = (w_1 - w_2)^m e^{-|w_1|^2/2 - |w_2|^2/2}$$

where $w_{1,2}$ are the complex coordinates for particles 1, 2. Show that this wavefunction is antisymmetric under the exchange of particles, $1 \leftrightarrow 2$, provided that m is odd. Show that the total angular momentum of this system, determined by the operator

$$\hat{L}_z^{(tot)} = \hat{L}_z^{(1)} + \hat{L}_z^{(2)} = \frac{\hbar}{i}\frac{\partial}{\partial\theta_1} + \frac{\hbar}{i}\frac{\partial}{\partial\theta_2}$$

is $m\hbar$.

(c) Finally, show that the multiparticle wavefunction

$$\psi(w_1, w_2, \ldots, w_N) = \prod_{j>k}^{N} (w_j - w_k)^m e^{-\sum_l |w_l|^2/2}$$

is completely antisymmetric under the exchange of any two particles (if m is odd) and that it is an eigenstate of the total angular momentum operator

$$\hat{L}_z^{(\text{tot})} = \sum_{k=1}^{N} \hat{L}_z^{(k)}$$

Note: These wavefunctions have been extensively used in the study of the so-called *fractional quantum Hall effect* by Laughlin.[17]

P16.25 **(a)** The nodal lines for the circular well are guaranteed to be orthogonal to each other, as in Fig. 16.18(b), for energy eigenstates of the form $\psi(r, \theta) = J_m(k_n r)\cos(m\theta)$ or $J_m(k_n r)\sin(m\theta)$ as the zeroes of $\psi(r, \theta)$ are for fixed values of r and θ, respectively, and they form an orthogonal coordinate system. Does the same hold true for a linear combination of these two degenerate states?

(b) The nodal lines for the square well for the degenerate wavefunctions $u_{n,m}(x, y)$ and $u_{m,n}(x, y)$ will be orthogonal, as in Fig. 16.18(a); how about for a general linear combination of the two? Sketch several possibilities.

P16.26 Write a short computer program that finds the separations between adjacent energy levels, $\Delta E = E_n - E_{n-1}$, for a particle of mass m in a rectangular infinite well. Let the area of the well be L^2 and fixed, but allow the aspect ratio (the ratio of lengths of each side) to vary. You might want to include some of the following steps:

1. Find all of the E_{n_x,n_y} for $n_x, n_y < N_{\text{max}}$ for some value of N_{max}; the E_{n_x,n_y} will not be arranged in increasing order.
2. Use a sorting algorithm (such as a "bubble sort") to put them in proper order; you might want to store these values in a list or array.
3. Find the successive differences, and store them in a sorted array.
4. Bin the ΔE values with some appropriate bin size, and plot dN/N versus ΔE.

Plot your distributions for several cases, including a square well $L_x = L_y = L$, a well with $L = L_x = 2L_y$, and one with sides in an irrational ratio, such as $L = L_x = \pi L_y/2$. Do you see any differences due to the varying amounts of symmetry in these systems?

[17] See Laughlin (1983).

CHAPTER 17

The Schrödinger Equation in Three Dimensions

Most "real-world" applications of quantum mechanics arise in three-dimensional systems where the Schrödinger equation for one particle takes the form

$$\hat{H}\psi(\mathbf{r}) = E\psi(\mathbf{r}) \tag{17.1}$$

where the Hamiltonian

$$\hat{H} = \frac{\hat{\mathbf{p}}^2}{2m} + V(\mathbf{r}) \tag{17.2}$$

is written in terms of the gradient in three dimensions, $\hat{\mathbf{p}} \equiv (\hbar/i)\boldsymbol{\nabla}$. Two-body problems with a Hamiltonian of the form

$$\hat{H} = \frac{\hat{\mathbf{p}}_1{}^2}{2m_1} + \frac{\hat{\mathbf{p}}_2{}^2}{2m_2} + V(\mathbf{r}_1 - \mathbf{r}_2) \tag{17.3}$$

are intrinsically no more difficult as the use of relative and center-of-mass coordinates,

$$\mathbf{r}_{\text{rel}} = \mathbf{r}_1 - \mathbf{r}_2 \tag{17.4}$$

$$\mathbf{R}_{\text{cm}} = \frac{m_1\mathbf{r}_1 + m_2\mathbf{r}_2}{m_1 + m_2} \tag{17.5}$$

just as in one dimension (Sec. 15.3), leads to an effective one-particle Schrödinger equation

$$\left(\frac{\hat{\mathbf{p}}^2}{2\mu} + V(\mathbf{r})\right)\psi(\mathbf{r}) = E\psi(\mathbf{r}) \tag{17.6}$$

for the relative coordinate $\mathbf{r} \equiv \mathbf{r}_{\text{rel}}$ with the reduced mass μ as the mass parameter.

Explicit solutions of Eqn. (17.6) require the use of a specific coordinate system, the choice of which, in turn, usually depends on the form of the potential $V(\mathbf{r})$. Solutions in Cartesian coordinates, $\mathbf{r} = (x, y, z)$, which generalize the results of Chap. 6 and Sec. 16.1 are possible, and were implicitly used in the discussion in Sec. 8.3. For problems with a symmetry about a particular rotation axis, cylindrical coordinates given by $\mathbf{r} = (r, \theta, z)$ are often useful (see Sec. 19.5 for an example), and the two-dimensional polar coordinate results of Sec. 16.3 can be trivially extended. Solutions using less familiar systems, such as parabolic coordinates (see P18.14 and Sec. 20.4), are even sometimes employed. For many 3D problems for which there is no preferred axis, a separation of variables using spherical coordinates is most natural, and that is the subject of the next section.

17.1 SPHERICAL COORDINATES AND ANGULAR MOMENTUM

For many important cases, the potential is spherically symmetric, i.e., $V(\mathbf{r}) = V(r)$, and it is natural to attempt solutions using spherical coordinates, namely,

$$\begin{aligned} x &= r\sin(\theta)\cos(\phi) \\ y &= r\sin(\theta)\sin(\phi) \\ z &= r\cos(\theta) \end{aligned} \tag{17.7}$$

or

$$\begin{aligned} r &= \sqrt{x^2 + y^2 + z^2} \\ \tan(\phi) &= y/x \\ \cos(\theta) &= z/\sqrt{x^2 + y^2 + z^2} \end{aligned} \tag{17.8}$$

as illustrated in Fig. 17.1.

The kinetic energy operator can always be written as

$$\hat{T} = \frac{\hat{\mathbf{p}}^2}{2m} = -\frac{\hbar^2}{2m}\nabla^2 = -\frac{\hbar^2}{2m}\left(\frac{\partial^2}{\partial x^2} + \frac{\partial^2}{\partial y^2} + \frac{\partial^2}{\partial z^2}\right) \tag{17.9}$$

Using the defining relations for spherical coordinates, Eqns. (17.7) and (17.8), one can (somewhat tediously) show that $\hat{T}$ can be written in the form

$$\hat{T} = -\frac{\hbar^2}{2m}\left[\frac{\partial^2}{\partial r^2} + \frac{2}{r}\frac{\partial}{\partial r} + \frac{1}{r^2}\left(\frac{\partial^2}{\partial \theta^2} + \cot(\theta)\frac{\partial}{\partial \theta} + \frac{1}{\sin^2(\theta)}\frac{\partial^2}{\partial \phi^2}\right)\right] \tag{17.10}$$

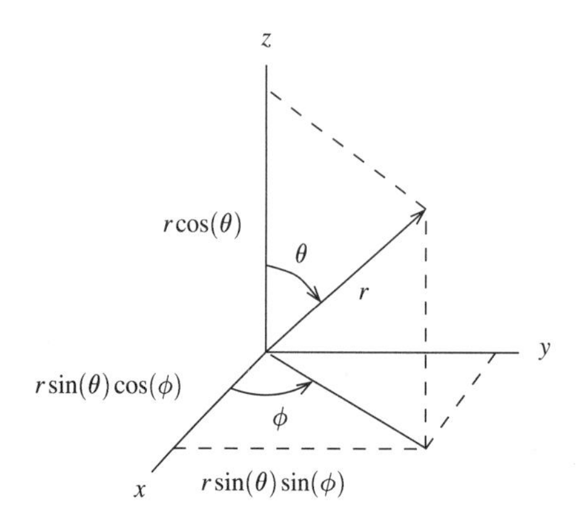

FIGURE 17.1. *Definitions of spherical coordinates.*

which, not surprisingly, is seen to measure the "wiggliness" of the wavefunction in the r, θ, and ϕ directions. Motivated by the results of Chap. 16, we expect an intimate connection between the θ and ϕ derivative terms, the rotational kinetic energy, and the angular momentum, now in three dimensions.

The vector angular momentum operator,

$$\hat{\mathbf{L}} = \mathbf{r} \times \hat{\mathbf{p}} \tag{17.11}$$

has the Cartesian components

$$\hat{L}_x = y\hat{p}_z - z\hat{p}_y \tag{17.12}$$
$$\hat{L}_y = z\hat{p}_x - x\hat{p}_z \tag{17.13}$$
$$\hat{L}_z = x\hat{p}_y - y\hat{p}_x \tag{17.14}$$

which are all obviously individually Hermitian. A change to spherical coordinates shows that

$$\hat{L}_z = \frac{\hbar}{i}\frac{\partial}{\partial\phi} \tag{17.15}$$

which is similar to Eqn. (16.51) in 2D. One should note, however, the change in notation as the azimuthal angle in three dimensions is conventionally labeled ϕ. The other components have a slightly more complicated form, namely,

$$\hat{L}_x = \frac{\hbar}{i}\left(-\sin(\phi)\frac{\partial}{\partial\theta} - \cot(\theta)\cos(\phi)\frac{\partial}{\partial\phi}\right) \tag{17.16}$$

and

$$\hat{L}_y = \frac{\hbar}{i}\left(\cos(\phi)\frac{\partial}{\partial\theta} - \cot(\theta)\sin(\phi)\frac{\partial}{\partial\phi}\right) \tag{17.17}$$

which together give information on the polar angle θ.

For an object with different moments of inertia about each of the three axes, the classical rotational kinetic energy is

$$T_{\text{rot}} = \frac{L_x^2}{2I_x} + \frac{L_y^2}{2I_y} + \frac{L_z^2}{2I_z} \tag{17.18}$$

which can be written as

$$T_{\text{rot}} = \frac{\mathbf{L}^2}{2I} \tag{17.19}$$

if $I_x = I_y = I_z = I$. For a single point particle in a spherically symmetric potential, one has $I = mr^2$, while for a pair of point masses the moment of inertia about the center of mass is $I = \mu r^2$. Motivated by this, we can easily show (P17.2) that

$$\hat{\mathbf{L}}^2 = \hat{L}_x^2 + \hat{L}_y^2 + \hat{L}_z^2 = -\hbar^2\left(\frac{\partial^2}{\partial\theta^2} + \cot(\theta)\frac{\partial}{\partial\theta} + \frac{1}{\sin^2(\theta)}\frac{\partial}{\partial\phi^2}\right) \tag{17.20}$$

which only acts on the angular degrees of freedom. This shows that the kinetic energy operator Eqn. (17.9) can be written in the short-hand form

$$\hat{T} = \hat{T}_r + \frac{\hat{\mathbf{L}}^2}{2\mu r^2} \tag{17.21}$$

where the *radial kinetic energy operator* is

$$\hat{T}_r = -\frac{\hbar^2}{2\mu}\left(\frac{\partial^2}{\partial r^2} + \frac{2}{r}\frac{\partial}{\partial r}\right) \tag{17.22}$$

For a spherically symmetric potential, $V(r)$, the Hamiltonian is then clearly separable, and if we write a solution of the form $\psi(r, \theta, \phi) = R(r)Y(\theta, \phi)$, we obtain

$$-\frac{2\mu}{\hbar^2}\frac{r^2}{R(r)}\left(\hat{T}_r + V(r) - E\right)R(r) = \frac{1}{\hbar^2}\frac{1}{Y(\theta, \phi)}\left[\hat{\mathbf{L}}^2 Y(\theta, \phi)\right] = l(l+1) \tag{17.23}$$

where we have written the separation constant as $l(l+1)$ in anticipation of the result that the $Y(\theta, \phi)$ are the eigenfunctions of the square of the angular momentum operator given by

$$\hat{\mathbf{L}}^2 Y(\theta, \phi) = l(l+1)\hbar^2 Y(\theta, \phi) \tag{17.24}$$

These eigenfunctions can be studied once and for all for any problem with spherical symmetry, and their properties will be discussed extensively in the next section.

For a given value of $l(l+1)$, the corresponding dynamical equation that determines the energy eigenvalues is the *radial Schrödinger equation,*

$$-\frac{\hbar^2}{2\mu}\left(\frac{d^2R(r)}{dr^2} + \frac{2}{r}\frac{dR(r)}{dr}\right) + \left(V(r) + \frac{l(l+1)\hbar^2}{2\mu r^2}\right)R(r) = ER(r) \tag{17.25}$$

The normalization condition

$$1 = \int d\mathbf{r}|\psi(\mathbf{r})|^2 = \int_0^\infty r^2\,dr \int d\Omega\,|\psi(r, \theta, \phi)|^2 \tag{17.26}$$

can then be enforced by demanding that

$$1 = \int_0^{2\pi} d\phi \int_0^{\pi} \sin(\theta)\,d\theta\,|Y(\theta, \phi)|^2 \tag{17.27}$$

and

$$1 = \int_0^\infty r^2\,dr\,|R(r)|^2 \tag{17.28}$$

separately. In three dimensions, an interesting simplification occurs if we write

$$R(r) = \frac{u(r)}{r} \tag{17.29}$$

in that the radial equation, Eqn. (17.25), becomes

$$-\frac{\hbar^2}{2\mu}\frac{d^2u(r)}{dr^2} + \left(V(r) + \frac{l(l+1)\hbar^2}{2\mu r^2}\right)u(r) = Eu(r) \tag{17.30}$$

which is of the form of a standard one-dimensional Schrödinger equation with the inclusion of the centrifugal term $l(l+1)\hbar^2/2\mu r^2$. The normalization condition is also similar, as we now require that

$$\int_0^\infty |u(r)|^2\,dr = 1 \tag{17.31}$$

Because $R(r)$ should be well behaved at the origin, we must assume that

$$\lim_{r \to 0} u(r) = 0 \tag{17.32}$$

Taken together, this form is reminiscent of the "half" potential well problems considered in Sec. 9.3, described by some potential $V(x)$ for $x > 0$, but with an infinite wall at the origin. In that case, we could choose the odd solutions of the "full well" problem as they satisfied the Schrödinger equation for $x > 0$ and the boundary condition at the wall.

In a central potential, the classical angular momentum vector, all of its components as well as its magnitude, will be conserved quantities and can, in principle, be known with arbitrary precision; we wish to understand to what extent this holds true in quantum systems. We first note that $\hat{\mathbf{L}}^2$ commutes with the Hamiltonian since

$$[\hat{H}, \hat{\mathbf{L}}^2] = \left[\hat{T}_r + V(r) + \frac{1}{2\mu r^2}\hat{\mathbf{L}}^2, \hat{\mathbf{L}}^2\right] = 0 \tag{17.33}$$

since $\hat{\mathbf{L}}^2$ acts only on angular variables and certainly commutes with itself. This implies that:

- The solutions of the Schrödinger equation for a rotationally invariant potential can have definite values of both the energy E and total angular momentum $l(l+1)\hbar^2$.
- The total angular momentum will be a constant in time.

While we may thus be able to know the "length" of the observable corresponding to $\hat{\mathbf{L}}^2$ precisely, simultaneous exact measurements of the individual components are not possible, as we have

$$\begin{aligned}
[\hat{L}_x, \hat{L}_y] &= [y\hat{p}_z - z\hat{p}_y, z\hat{p}_x - x\hat{p}_z] \\
&= [y\hat{p}_z, z\hat{p}_x] - [y\hat{p}_z, x\hat{p}_z] - [z\hat{p}_y, z\hat{p}_x] + [z\hat{p}_y, x\hat{p}_z] \\
&= y[\hat{p}_z, z]\hat{p}_x + x[z, \hat{p}_z]\hat{p}_y \\
&= \frac{\hbar}{i}\left(y\hat{p}_x - x\hat{p}_y\right) \\
&= i\hbar\hat{L}_z \neq 0
\end{aligned} \tag{17.34}$$

One similarly finds that

$$[\hat{L}_y, \hat{L}_z] = i\hbar\hat{L}_x \qquad \text{and} \qquad [\hat{L}_z, \hat{L}_x] = i\hbar\hat{L}_y \tag{17.35}$$

both of which have the same form as Eqn. (17.34) with cyclic permutations of x, y, z.

On the other hand, one does find that

$$\begin{aligned}
[\hat{L}_x, \hat{\mathbf{L}}^2] &= [\hat{L}_x, \hat{L}_x^2 + \hat{L}_y^2 + \hat{L}_z^2] \\
&= \hat{L}_y[\hat{L}_x, \hat{L}_y] + [\hat{L}_x, \hat{L}_y]\hat{L}_y \\
&\quad + \hat{L}_z[\hat{L}_x, \hat{L}_z] + [\hat{L}_x, \hat{L}_z]\hat{L}_z \\
&= i\hbar\left(\hat{L}_y\hat{L}_z + \hat{L}_x\hat{L}_y - \hat{L}_z\hat{L}_y - \hat{L}_y\hat{L}_x\right) \\
&= 0
\end{aligned} \tag{17.36}$$

as well as

$$[\hat{L}_y, \hat{\mathbf{L}}^2] = [\hat{L}_z, \hat{\mathbf{L}}^2] = 0 \tag{17.37}$$

so that one can measure the pairs $(L_x, \mathbf{L}^2)$ *or* $(L_y, \mathbf{L}^2)$ *or* $(L_z, \mathbf{L}^2)$ with arbitrary precision. Since each component operator also commutes with the Hamiltonian, we find that:

- The maximally large set of commuting operators for a general spherically symmetric potential will correspond to precisely determined values of the energy, the total angular momentum squared, and one component of **L**. The conventional choice is to look for simultaneous eigenfunctions of $\hat{H}$, $\hat{\mathbf{L}}^2$, and $\hat{L}_z$.

Because the eigenfunctions of $\hat{\mathbf{L}}^2$ and $\hat{L}_z$ are important for any spherically symmetric problem, we devote the next section to the examination of their properties.

17.2 EIGENFUNCTIONS OF ANGULAR MOMENTUM

17.2.1 Methods of Derivation

The derivation of the eigenfunctions of the angular momentum operators is a standard one in many books on quantum mechanics. We will briefly discuss several of the usual methods of analysis of the properties of the $Y(\theta, \phi)$ before turning to attempts to visualize them, discussing their classical limit, and their application in important physical systems. We begin with a differential-equation-based approach, and then discuss the usefulness of operator methods to this problem.

The eigenvalue problem for $\hat{\mathbf{L}}^2$ can be written as

$$\begin{aligned}\hat{\mathbf{L}}^2 Y(\theta, \phi) &= -\hbar^2 \left(\frac{\partial^2 Y}{\partial \theta^2} + \cot(\theta) \frac{\partial Y}{\partial \theta} + \frac{1}{\sin^2(\theta)} \frac{\partial^2 Y}{\partial \phi^2} \right) \\ &= l(l+1)\hbar^2 Y(\theta, \phi) \end{aligned} \tag{17.38}$$

The corresponding problem for $\hat{L}_z$ has already been discussed in the last chapter, in a different notation, where we found that

$$\Phi_m(\phi) = \frac{1}{\sqrt{2\pi}} e^{im\phi} \tag{17.39}$$

satisfies

$$\hat{L}_z \Phi_m(\phi) = \frac{\hbar}{i} \frac{\partial}{\partial \phi} \Phi_m(\phi) = m\hbar \Phi_m(\phi) \tag{17.40}$$

for integral values of m. We are thus led to write

$$Y_{l,m}(\theta, \phi) = \Theta_{l,m}(\theta) \Phi_m(\phi) \tag{17.41}$$

so that Eqn. (17.38) can be seen to be separable as

$$\begin{aligned}-\frac{\sin^2(\theta)}{\Theta_{l,m}(\theta)} \left[\frac{d^2 \Theta_{l,m}}{d\theta^2} + \cot(\theta) \frac{d\Theta_{l,m}}{d\theta} + l(l+1)\Theta_{l,m} \right] &= \frac{1}{\Phi_m(\phi)} \frac{d^2 \Phi_m}{d\phi^2} \\ &= -m^2 \end{aligned} \tag{17.42}$$

The new information on the total angular momentum eigenvalues, $l(l+1)$, is contained in the equation for $\Theta_{l,m}(\theta)$, namely,

$$\sin^2(\theta) \frac{d^2 \Theta_{l,m}(\theta)}{d\theta^2} + \sin(\theta)\cos(\theta) \frac{d\Theta_{l,m}(\theta)}{d\theta} + \left(l(l+1)\sin^2(\theta) - m^2 \right) \Theta_{l,m}(\theta) = 0 \tag{17.43}$$

This equation can be turned into one of the "handbook" variety by noting that with the substitution, $z = \cos(\theta)$, Eqn. (17.43) becomes

$$(1 - z^2)\frac{d^2\Theta_{l,m}(z)}{dz^2} - 2z\frac{d\Theta_{l,m}(z)}{dz} + \left(l(l+1) - \frac{m^2}{z^2}\right)\Theta_{l,m}(z) = 0 \tag{17.44}$$

which was studied by Legendre. We will not proceed further along these lines except to quote the result that the well-known solutions to this equation are polynomials in z, labeled $P_l^m(z)$, called the *associated Legendre polynomials*[1] which require l to be a nonnegative integer. We will determine and tabulate more of their properties using other methods below.

A possibly more profitable and interesting approach is to use ladder operators, factorization techniques, and a more formal notation. For example, we can denote the angular wavefunctions in quantum state notation via

$$Y_{l,m}(\theta, \phi) \Longrightarrow |Y_{l,m}\rangle \tag{17.45}$$

so that, for example,

$$\hat{\mathbf{L}}^2|Y_{l,m}\rangle = l(l+1)\hbar^2|Y_{l,m}\rangle \tag{17.46}$$

and the normalization condition on the angular wavefunctions reads

$$1 = \int d\Omega |Y_{l,m}(\theta, \phi)|^2 = \langle Y_{l,m}|Y_{l,m}\rangle \tag{17.47}$$

In this language, we can argue that

$$\begin{aligned}\langle Y_{l,m}|\hat{\mathbf{L}}^2|Y_{l,m}\rangle &= \langle Y_{l,m}|\hat{L}_x^2 + \hat{L}_y^2 + \hat{L}_z^2|Y_{l,m}\rangle \\ &= \langle \hat{L}_x Y_{l,m}|\hat{L}_x Y_{l,m}\rangle + \langle \hat{L}_y Y_{l,m}|\hat{L}_y Y_{l,m}\rangle + \langle \hat{L}_z Y_{l,m}|\hat{L}_z Y_{l,m}\rangle \\ &\geq 0\end{aligned} \tag{17.48}$$

where we have used the fact that the $\hat{L}_{x,y,z}$ are Hermitian, and that the norm of any vector is nonnegative. This immediately implies, of course, that $l(l+1) \geq 0$, which we take to mean that $l \geq 0$ as well.

A pair of operators that will have the effect of raising and lowering operators can be defined via

$$\hat{L}_+ = \hat{L}_x + i\hat{L}_y \qquad \text{and} \qquad \hat{L}_- = \hat{L}_x - i\hat{L}_y \tag{17.49}$$

which are not Hermitian, but which do satisfy $\hat{L}_+^\dagger = \hat{L}_-$ and $\hat{L}_-^\dagger = \hat{L}_+$. A product relation is

$$\begin{aligned}\hat{L}_+\hat{L}_- &= \left(\hat{L}_x + i\hat{L}_y\right)\left(\hat{L}_x - i\hat{L}_y\right) \\ &= \hat{L}_x^2 + \hat{L}_y^2 - i\hbar[\hat{L}_x, \hat{L}_y] \\ &= \hat{L}_x^2 + \hat{L}_y^2 + \hbar\hat{L}_z\end{aligned} \tag{17.50}$$

which gives

$$\hat{\mathbf{L}}^2 = \hat{L}_+\hat{L}_- + \hat{L}_z^2 - \hbar\hat{L}_z \tag{17.51}$$

with a similar derivation giving

$$\hat{\mathbf{L}}^2 = \hat{L}_-\hat{L}_+ + \hat{L}_z^2 + \hbar\hat{L}_z \tag{17.52}$$

[1] For details, see the mathematical handbook by Abramowitz and Stegun (1963) or App. C.6.

Equally simple manipulations imply that

$$[\hat{L}_+, \hat{L}_-] = 2\hbar\hat{L}_z \tag{17.53}$$

$$[\hat{L}_+, \hat{L}_z] = -\hbar\hat{L}_+ \tag{17.54}$$

$$[\hat{L}_-, \hat{L}_z] = \hbar\hat{L}_- \tag{17.55}$$

and

$$[\hat{L}_+, \hat{\mathbf{L}}^2] = [\hat{L}_-, \hat{\mathbf{L}}^2] = 0 \tag{17.56}$$

The ladder operators, $\hat{L}_\pm$, have no effect on the l quantum number when acting on a $|Y_{l,m}\rangle$ state since, because of Eqn. (17.56),

$$\hat{\mathbf{L}}^2\left(\hat{L}_+|Y_{l,m}\rangle\right) = \hat{L}_+\left(\hat{\mathbf{L}}^2|Y_{l,m}\rangle\right) = l(l+1)\hbar^2\left(\hat{L}_+|Y_{l,m}\rangle\right) \tag{17.57}$$

and similarly for $\hat{L}_-$. On the other hand, using Eqn. (17.54), we have

$$\hat{L}_z\left(\hat{L}_+|Y_{l,m}\rangle\right) = \left(\hat{L}_+\hat{L}_z + \hbar\hat{L}_z\right)|Y_{l,m}\rangle = (m+1)\hbar\left(\hat{L}_+|Y_{l,m}\rangle\right) \tag{17.58}$$

so that the application of the raising operator $\hat{L}_+$ to the state $|Y_{l,m}\rangle$ leaves the l value unchanged but increases the m value by one unit of $\hbar$. This implies that

$$\hat{L}_+|Y_{l,m}\rangle \propto |Y_{l,m+1}\rangle \tag{17.59}$$

with the similar relation for the lowering operator

$$\hat{L}_-|Y_{l,m}\rangle \propto |Y_{l,m-1}\rangle \tag{17.60}$$

For a fixed value of l, the raising and lowering operators move one up and down the ladder in m in unit steps of $\hbar$ as shown in Fig. 17.2. The normalized versions of these last two relations can be shown (P17.3) to be

$$\hat{L}_+|Y_{l,m}\rangle = \sqrt{l(l+1) - m(m+1)}\hbar|Y_{l,m+1}\rangle \tag{17.61}$$

$$\hat{L}_-|Y_{l,m}\rangle = \sqrt{l(l+1) - m(m-1)}\hbar|Y_{l,m-1}\rangle \tag{17.62}$$

More constraints can be obtained by using the fact that

$$\langle Y_{l,m}|\hat{L}_+\hat{L}_-|Y_{l,m}\rangle = \langle\hat{L}_- Y_{l,m}|\hat{L}_- Y_{l,m}\rangle \geq 0 \tag{17.63}$$

which gives

$$\langle Y_{l,m}|\hat{\mathbf{L}}^2 - \hat{L}_z^2 - \hbar\hat{L}_z|Y_{l,m}\rangle = \hbar^2\left(l(l+1) - m^2 - m\right) \geq 0 \tag{17.64}$$

so that $l(l+1) \geq m(m+1)$—a similar restriction, namely, that $l(l+1) \geq m(m-1)$ can be derived by considering the expectation value of $\hat{L}_-\hat{L}_+$. These bounds are illustrated in

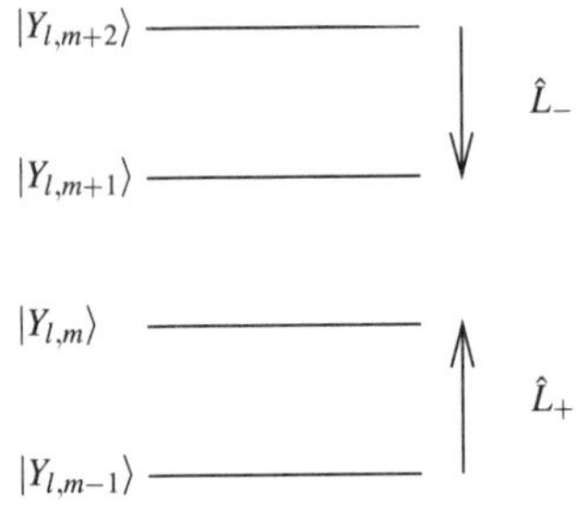

$|Y_{l,m+2}\rangle$

$\hat{L}_-$

$|Y_{l,m+1}\rangle$

$|Y_{l,m}\rangle$

$\hat{L}_+$

$|Y_{l,m-1}\rangle$

FIGURE 17.2. *Effect of the raising and lowering operators for the spherical harmonics,* $|Y_{l,m}|\rangle$.

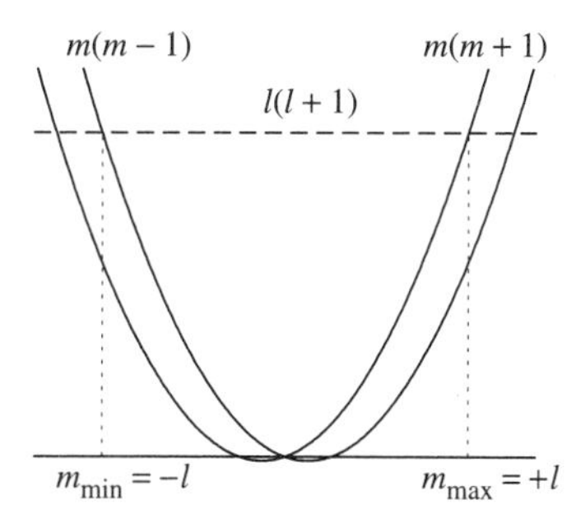

FIGURE 17.3. Bounds on maximum and minimum values of m for a given l.

Fig. 17.3, from which it is clear that one obtains $-l \geq m \geq l$ for a given l value. We thus know that there is a state with a maximal value of m for each l, say m_+, and clearly the raising operator must annihilate that state, namely,

$$\hat{L}_+|Y_{l,m_+}\rangle = 0 \tag{17.65}$$

In this case, the inequality of Eqn. (17.64) is saturated, and we have

$$l(l+1) = m_+(m_+ + 1) \qquad \text{or} \qquad m_+ = +l \tag{17.66}$$

and $\hat{L}_-$ acting on the state of minimal m implies that $m_- = -l$. We are finding that for a fixed value of (integral) l the allowed values of m are given by

$$m = -l, -(l-1), \ldots, +(l-1), +l \tag{17.67}$$

We see that

- For a given value of l, there are $(2l+1)$ values of m given by Eqn. (17.67)

The quantized values of the magnitude of $\mathbf{L}^2$ and L_z are often presented in a vector diagram, as shown in Fig. 17.4, which will be useful in discussing the classical limit.

The raising and lowering operators are also useful in constructing explicit solutions for the angular momentum eigenfunctions. Given a value of l, for the maximal value of $m_+ = l$, we have

$$Y_{l,l}(\theta, \phi) \propto \Theta_{l,l}(\theta)e^{il\phi} \tag{17.68}$$

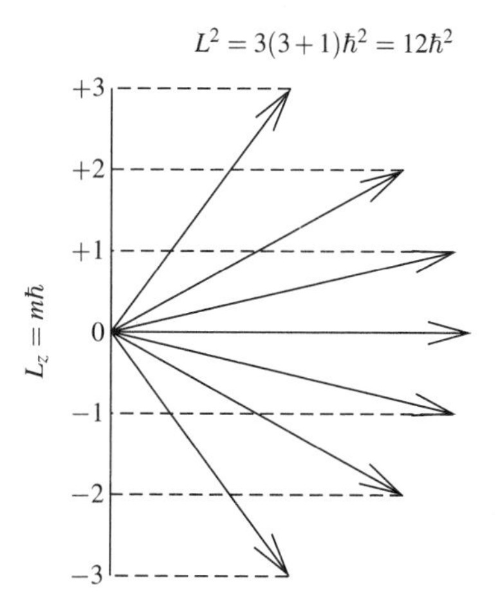

FIGURE 17.4. Vector diagram for quantized angular momentum.

which, by Eqn. (17.65), must satisfy

$$0 = \hat{L}_+ Y_{l,l}(\theta,\phi) = \hbar e^{i\phi}\left(\frac{\partial}{\partial\theta} + i\cot(\theta)\frac{\partial}{\partial\phi}\right)\left(\Theta_{l,l}(\theta)e^{il\phi}\right)$$

$$= \hbar e^{i\phi}e^{il\phi}\left(\frac{d\Theta_{l,l}}{d\theta} - l\cot(\theta)\Theta_{l,l}(\theta)\right) \tag{17.69}$$

This can be immediately integrated to give

$$\Theta_{l,l}(\theta) \propto \sin^l(\theta) \tag{17.70}$$

In the notation of the associated Legendre polynomials, this implies that $P_l^l(z) \propto (1-z^2)^{l/2}$. This special case is useful for two reasons, one of which is that this form will be the appropriate one for the classical description of planar orbits. Perhaps more importantly, the other members of the "family" for a given l value which have lower values of m are easily obtained by repeated applications of the lowering operator, $\hat{L}_-$, for example,

$$Y_{l,l-1}(\theta,\phi) \propto \hat{L}_- Y_{l,l}(\theta,\phi)$$

$$= \hbar e^{-i\phi}\left(\frac{\partial}{\partial\theta} - i\cot(\theta)\frac{\partial}{\partial\phi}\right)\left[\sin^l(\theta)e^{il\phi}\right] \tag{17.71}$$

and so forth. By repeated applications of these methods, explicit representations for the $Y_{l,m}(\theta,\phi)$ can be constructed for any values of l, m. We will restrict ourselves to simply listing many of their most useful properties.

- The $Y_{l,m}(\theta,\phi)$ are collectively called *spherical harmonics* (by analogy with the eigenfunctions of the infinite well which are, in turn, similar to the various "harmonics" of a vibrating string).
- The properly normalized spherical harmonics are given by

$$Y_{l,m}(\theta,\phi) = (-1)^m\left[\frac{2l+1}{4\pi}\frac{(l-m)!}{(l+m)!}\right]^{1/2} P_l^m(\cos(\theta))e^{im\phi} \tag{17.72}$$

  at least for values of $m \geq 0$. For negative values of m, these results are extended via the relation

$$Y_{l,-m}(\theta,\phi) = (-1)^m Y_{l,m}^*(\theta,\phi) \tag{17.73}$$

  Recall that the phase of an eigenfunction is arbitrary, so that the factors of -1 here are a convention.
- The associated Legendre polynomials appearing in Eqn. (17.72) can be constructed using the relation

$$P_l^m(z) = (-1)^{l+m}\frac{(l+m)!}{(l-m)!}\frac{1}{2^l l!}(1-z^2)^{-m/2}\left(\frac{d}{dz}\right)^{l-m}\left(1-z^2\right)^l \tag{17.74}$$

- For use in various problems, it is useful to have the simplest examples for $l = 0, 1, 2$:

$$Y_{0,0} = \frac{1}{\sqrt{4\pi}} \tag{17.75}$$

$$Y_{1,1} = -\sqrt{\frac{3}{8\pi}}e^{i\phi}\sin(\theta) \tag{17.76}$$

$$Y_{1,0} = \sqrt{\frac{3}{4\pi}}\cos(\theta) \tag{17.77}$$

$$Y_{2,2} = \sqrt{\frac{15}{32\pi}}e^{i2\phi}\sin^2(\theta) \tag{17.78}$$

$$Y_{2,1} = -\sqrt{\frac{15}{8\pi}}e^{i\phi}\sin(\theta)\cos(\theta) \tag{17.79}$$

$$Y_{2,0} = \sqrt{\frac{5}{16\pi}}\left(3\cos^2(\theta) - 1\right) \tag{17.80}$$

- From general principles, we know that the (properly normalized) $Y_{l,m}$ form an orthonormal set, namely, that

$$\langle Y_{l',m'}|Y_{l,m}\rangle = \int d\Omega\, Y^*_{l',m'}(\theta,\phi)\, Y_{l,m}(\theta,\phi) = \delta_{l,l'}\delta_{m,m'} \tag{17.81}$$

- The following average values are sometimes useful, namely,

$$\langle \sin^2(\theta)\rangle_{l,m} = \frac{2(l^2 + l - 1 + m^2)}{(2l-1)(2l+3)} \tag{17.82}$$

$$\langle \cos^2(\theta)\rangle_{l,m} = \frac{2(l^2 + l - 1/2 - m^2)}{(2l-1)(2l+3)} \tag{17.83}$$

where

$$\langle f(\theta,\phi)\rangle_{l,m} = \int d\Omega\, f(\theta,\phi)|Y_{l,m}(\theta,\phi)|^2 \tag{17.84}$$

- The spherical harmonics, i.e., the set of all $Y_{l,m}(\theta,\phi)$, comprise a complete set of functions over the two-dimensional angular space described by θ, ϕ. If one has a well-behaved angular function, $f(\theta,\phi)$, which itself is properly normalized, i.e., $\int d\Omega |f(\theta,\phi)|^2 = 1$, then one can write

$$f(\theta,\phi) = \sum_{l=0}^{\infty}\sum_{m=-l}^{m=+l} a_{l,m}Y_{l,m}(\theta,\phi) \tag{17.85}$$

Using Eqn. (17.81), the expansion coefficients can be obtained via

$$a_{l,m} = \int d\Omega\, Y^*_{l,m}(\theta,\phi)\, f(\theta,\phi) \tag{17.86}$$

and will satisfy

$$1 = \sum_{l=0}^{\infty}\sum_{m=-l}^{m=+l} |a_{l,m}|^2 \tag{17.87}$$

The $a_{l,m}$ have the usual probabilistic interpretation, specifically that $|a_{l,m}|^2$ is the probability that a measurement of $\mathbf{L}^2$ *and* L_z will yield the values $l(l+1)\hbar^2$ *and* $m\hbar$, respectively.

- In one-dimensional problems, the notion of parity was an important one; in that case, this was accomplished by the parity operator whose effect was to let $x \longrightarrow -x$. In three dimensions, the parity operator acts via

$$\hat{P}f(\mathbf{r}) = \hat{P}f(x, y, z) = f(-x, -y, -z) = f(-\mathbf{r}) \qquad (17.88)$$

at least in Cartesian coordinates. From Fig. 17.1, we can see that in spherical coordinates the corresponding effect is

$$\psi(\mathbf{r}) \longrightarrow \psi(-\mathbf{r}) \Longrightarrow \psi(r, \theta, \phi) \longrightarrow \psi(r, \pi - \theta, \pi + \phi) \qquad (17.89)$$

From direct examination of the defining relations of the $Y_{l,m}$, we find that

$$Y_{l,m}(\theta + \pi, \phi) = (-1)^{l+2|m|} Y_{l,m}(\theta, \phi) = (-1)^l Y_{l,m}(\theta, \phi) \qquad (17.90)$$

(since $2|m|$ is integral), so that the parity is given by $(-1)^l$; this is consistent with the fact that the eigenvalues of $\hat{P}$ are ± 1. For two-particle systems where $\mathbf{r} = \mathbf{r}_{\text{rel}} = \mathbf{r}_1 - \mathbf{r}_2$, the effect of particle interchange is to give $\mathbf{r}_1 \leftrightarrow \mathbf{r}_2$ so that $\mathbf{r} \longrightarrow -\mathbf{r}$. Knowledge of the parity of a two-particle system is then necessary for understanding the properties of the system under exchange, and is of fundamental importance for systems of indistinguishable particles.

Example 17.1 Let us consider a rigid rotator whose Hamiltonian is given by $\hat{H} = \hat{\mathbf{L}}^2/2I$ and whose (angular) wavefunction is given by

$$\psi(\theta, \phi) = N\left[Y_{0,0}(\theta, \phi) + (1 + 3i)Y_{1,-1}(\theta, \phi) + 2Y_{2,-1}(\theta, \phi) + Y_{2,0}(\theta, \phi)\right] \qquad (17.91)$$

The normalization constant N is obtained by invoking Eqn. (17.87) to find

$$1 = \sum_{l=0}^{\infty} \sum_{m=-l}^{m=+l} |a_{l,m}|^2 = N^2(1 + (1 + 9) + 4 + 1) \longrightarrow N = \frac{1}{4} \qquad (17.92)$$

We then find the following probabilities,

$$\begin{aligned}
\text{Prob}(l = 0) &= \frac{1}{16} \\
\text{Prob}(m = 0) &= \frac{1 + 1}{16} = \frac{1}{8} \\
\text{Prob}(L_z = -\hbar) &= \frac{10 + 4}{16} = \frac{7}{8} \\
\text{Prob}(\mathbf{L}^2 = 6\hbar^2) &= \frac{4 + 1}{16} = \frac{5}{16} \\
\text{Prob}(E = 2\hbar^2/2I) &= \frac{10}{16} = \frac{5}{8}
\end{aligned} \qquad (17.93)$$

the wavefunction at later times is given by

$$\begin{aligned}
\psi(\theta, \phi; t) = N\Big[&Y_{0,0}(\theta, \phi)e^{-iE_0t/\hbar} + (1 + 3i)Y_{1,-1}(\theta, \phi)e^{-iE_1t/\hbar} \\
&+ (2Y_{2,-1}(\theta, \phi) + Y_{2,0}(\theta, \phi))e^{-iE_2t/\hbar}\Big]
\end{aligned} \qquad (17.94)$$

where $E_l = l(l + 1)\hbar^2/2I$, and we have the expectation values

$$\begin{aligned}
\langle \hat{E} \rangle_t &= \left(\frac{1}{16}\right)\frac{0(0 + 1)\hbar^2}{2I} + \left(\frac{10}{16}\right)\frac{1(1 + 1)\hbar^2}{2I} + \left(\frac{4 + 1}{16}\right)\frac{2(2 + 1)\hbar^2}{2I} = \frac{50}{32}\hbar^2 I \\
\langle \hat{P} \rangle_t &= \frac{1}{16}\left(1 - 10 + 4 + 1\right) = -\frac{2}{5}
\end{aligned} \qquad (17.95)$$

for the energy and parity, respectively.

17.2.2 Visualization and Applications

The angular dependence of quantum wavefunctions, determined by the spherical harmonics, can be visualized in 3D plots where the "radius" vector at a given value of (θ, ϕ) is given by the magnitude of $|Y_{l,m}(\theta, \phi)|^2$; for example, the spherically symmetric $|Y_{0,0}(\theta, \phi)|^2$ is shown in Fig. 17.5. For larger values of l, we choose instead to plot

$$\left[Re(Y_{l,m}(\theta, \phi))\right]^2 \propto \left[P_l^m(\cos(\theta))\right]^2 \cos^2(m\phi) \tag{17.96}$$

For $m_{\max} = +l$, the ϕ term has the maximum wiggliness in the azimuthal direction. As m is decreased from this value in unit values, one node is "removed" from the ϕ direction and added to the θ direction as the Legendre polynomials acquire more and more nodes. This behavior is illustrated in Fig. 17.6 for the case of $l = 3$ and $m = 3, 2, 1, 0$. It is also consistent with the "sharing" of rotational kinetic energy given by

$$T_\phi = \left\langle \frac{\hat{L}_z^2}{2I} \right\rangle_{l,m} = \frac{\hbar^2}{2I} m^2 \tag{17.97}$$

and

$$T_\theta = \left\langle \frac{\hat{L}_x^2 + \hat{L}_y^2}{2I} \right\rangle_{l,m} = \frac{\hbar^2}{2I}\left(l(l+1) - m^2\right) \tag{17.98}$$

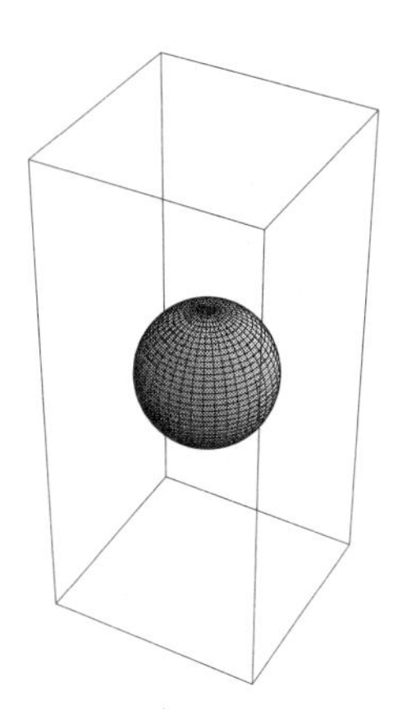

FIGURE 17.5. Visualization of the spherical harmonics, $Y_{0,0}(\theta, \phi)$.

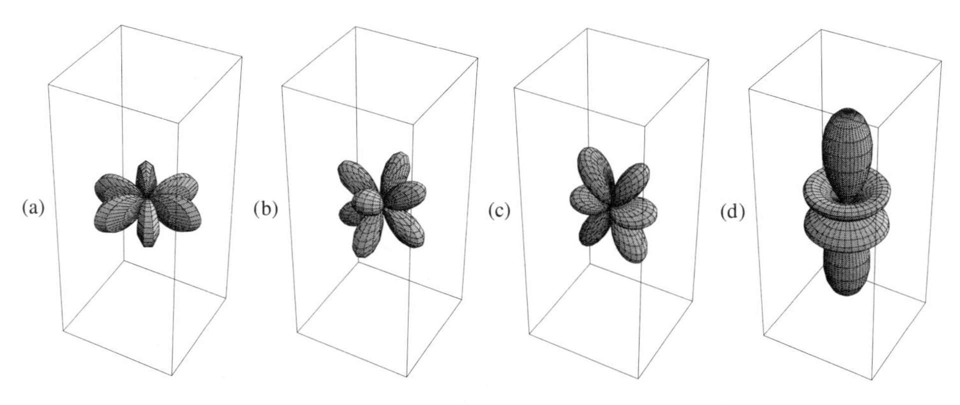

FIGURE 17.6. Visualization of the spherical harmonics, $Re[Y_{3,m}(\theta, \phi)]$ for (a) $m = 3$, (b) $m = 2$, (c) $m = 1$, and (d) $m = 0$.

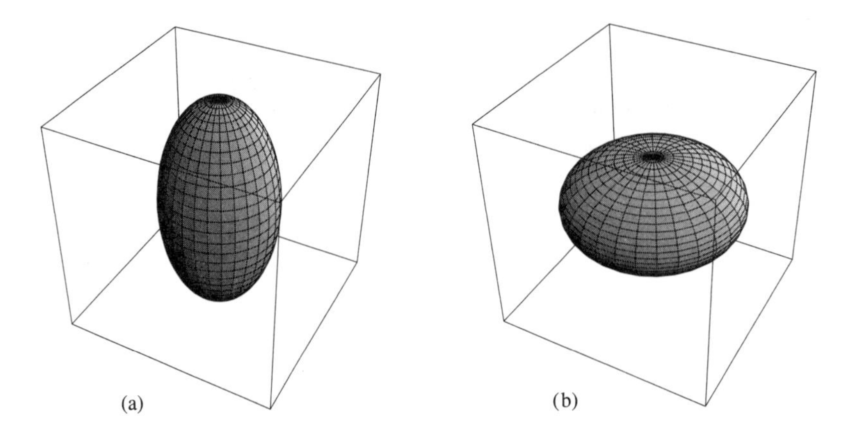

FIGURE 17.7. *(a) Prolate and (b) oblate spheroids with quadrupole moments that are positive and negative, respectively.*

The connection of the spherical harmonics to the description of "shapes" is familiar in *multipole expansions* in classical physics. Many textbooks on electromagnetism[2] show that the electric potential due to an arbitrary charge distribution at a large distance, R, from the origin has a systematic expansion of the form[3]

$$\phi(R) = \frac{1}{4\pi\epsilon_0}\left[\frac{1}{R}\int d\mathbf{r}\,\rho(\mathbf{r}) + \frac{1}{R^2}\int d\mathbf{r}\,z\,\rho(\mathbf{r}) + \frac{1}{R^3}\int d\mathbf{r}\,(3z^2 - r^2)\,\rho(\mathbf{r}) + \cdots\right] \tag{17.99}$$

where $\rho(\mathbf{r})$ is the charge density; similar expressions are useful for the long-range gravitational potential of a mass distribution. The integral in the first term simply counts the net charge and gives the *monopole term,* the second term gives the *electric dipole moment,* while the third describes the *quadrupole moment.* For a quantum mechanical distribution of charge described by a wavefunction of the form $\psi(\mathbf{r}) = R(r)Y_{l,m}(\theta, \phi)$, the charge density is $\rho(\mathbf{r}) = q|\psi(\mathbf{r})|^2$, and the quadrupole moment can be written in the form

$$Q = \int d\mathbf{r}\,(3z^2 - r^2)\,\rho(\mathbf{r}) \tag{17.100}$$

$$= q\langle R(r)|r^2|R(r)\rangle\langle Y_{l,m}|(3\cos^2(\theta) - 1)|Y_{l,m}\rangle \tag{17.101}$$

which is related to an average of $Y_{2,0}(\theta, \phi) \propto P_2^0(\cos(\theta))$. The correlations are

$$Q > 0 \longleftrightarrow \langle z^2\rangle > \langle x^2\rangle, \langle y^2\rangle \longleftrightarrow \text{"prolate"}$$

$$Q < 0 \longleftrightarrow \langle z^2\rangle < \langle x^2\rangle, \langle y^2\rangle \longleftrightarrow \text{"oblate"}$$

which is illustrated in Fig. 17.7 for two spheroidal shapes.

17.2.3 Classical Limit of Rotational Motion

One of the consequences of the conservation of angular momentum for the classical motion of point particles in a spherically symmetry potential is that their orbits must be planar. Because $\mathbf{L} = \mathbf{r}(t) \times \mathbf{p}(t)$, the plane containing $\mathbf{r}(t)$ must always be perpendicular to the

---

[2]See, e.g., Reitz, Milford, and Christy (1993).

[3]This form assumes a specific set of axes for simplicity.

vector **L**, which is fixed in time. If we arbitrarily call this the z direction, we also have $\mathbf{L} = L_z\hat{\mathbf{z}}$.

We know that quantum mechanically we can have precisely known eigenvalues for both the operators $\hat{\mathbf{L}}^2$ and $\hat{L}_z$ simultaneously. If, in addition, we had $\mathbf{L}^2 = L_z^2$ for the eigenvalues, we then know that the values corresponding to both $\hat{L}_x$ and $\hat{L}_y$ would have to vanish identically, which is inconsistent with Eqn. (17.35). We see this effect in Fig. 17.4 where the maximal projection of **L** along the z axis will make an angle given by

$$\cos(\alpha) \equiv \frac{L_z}{|\mathbf{L}|} = \frac{l}{\sqrt{l(l+1)}} = \frac{1}{\sqrt{1+1/l}} \tag{17.102}$$

since $m_{\text{max}} = +l$. The angle α, which measures the degree to which the total angular momentum vector can be made perpendicular to the classical plane of rotation, can thus be seen to become arbitrarily small for a macroscopic system since $\alpha \to 0$ as $l \to \infty$. An estimate of the rate at which this happens can be made by expanding the terms in Eqn. (17.102) in this limit to obtain

$$1 - \frac{1}{2}\alpha^2 + \cdots = 1 - \frac{1}{2l} + \cdots \tag{17.103}$$

or $\alpha \propto 1/\sqrt{l}$.

This effect is also reflected in the behavior of the probability density for the angular variables given by the $|Y_{l,l}(\theta, \phi)|^2$ in this limit. The squares of these spherical harmonics are plotted in Fig. 17.8 versus θ for increasingly large values of l: the tendency for the probability to be more and more concentrated in the plane corresponding to the classical orbit, i.e., for $\theta = \pi/2$, is clear. One can show (P17.6) that the quantum probability distribution in this limit is given by

$$P_{ll}(\theta) = |Y_{l,l}(\theta, \phi)|^2 \propto \sqrt{l}\sin^{2l}(\theta) \tag{17.104}$$

which exhibits the behavior shown in Fig. 17.8. The width of this distribution decreases with increasingly l, and one can estimate (in a somewhat "rough-and-ready" fashion) the range in θ over which the probability of finding the particle is spread. Calling this range

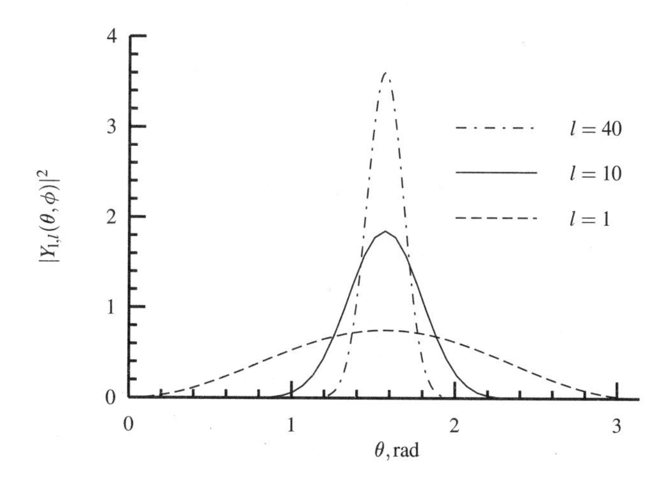

FIGURE 17.8. $|Y_{l,l}(\theta, \phi)|^2$ *versus* θ *for increasingly large values of* l *illustrating the approach to the classical limit. To see that the orbits become increasingly planar in this limit, turn your head 90°.*

δ, we write

$$\sin^{2l}\left(\frac{\pi}{2} \pm \delta\right) = \cos^{2l}(\pm\delta)$$
$$= \left(1 - \frac{1}{2}\delta^2 + \cdots\right)^{2l}$$
$$= 1 - \delta^2 l + \cdots \qquad (17.105)$$

so that $\delta \propto 1\sqrt{l}$, which is similar to the result for the angle α. This is another nice example of the correspondence principle and shows the classical limit of planar orbits in a striking way.

17.3 DIATOMIC MOLECULES

17.3.1 Rigid Rotators

Even though two atoms are known to bind via an interaction of the general form in Fig. 10.2, as a first approximation in our discussion of rotational states of molecules, we can consider a diatomic molecule to consist simply of two atoms with reduced mass μ with a constant separation r_0, i.e., joined by the famous "massless, inextensible rod" of introductory mechanics problems. In this approximation, the Hamiltonian is just that of a *rigid rotator*

$$\hat{H} = \frac{\hat{\mathbf{L}}^2}{2I} \qquad (17.106)$$

where $I = \mu r_0^2$ and the rotational energies are then given by

$$E_l = \frac{\hbar^2 l(l+1)}{2\mu r_0^2} = l(l+1)E_0 \qquad (17.107)$$

Transitions between states due to the emission or absorption of a photon are possible provided they satisfy certain *selection rules* (to be discussed below). In this case, the initial and final values of the angular momentum quantum number must change by one, i.e., $\Delta l = \pm 1$; this implies, for example, that the photon energies measured in an absorption experiment will satisfy

$$E_\gamma^{(l)} = E_l - E_{l-1} = 2lE_0 \qquad (17.108)$$

for a purely rigid rotator. For diatomic molecules with $r_0 \sim 1$ to 2 Å, $\mu \sim 1$ to 20 amu, and $l \sim 1$ to 10, these energies are in the range $E_\gamma \sim 10^{-3} - 10^{-4}$ eV. This corresponds to the wavelengths of the order $\lambda \sim 1$ mm to 1 cm, i.e., in the far infrared to microwave regions.

This simple model implies that the ratio $E_\gamma^{(l)}/2l$ should be constant with increasing l for a given series of rotational lines. Data for H-Cl are plotted in Fig. 17.9 in this format (with the vertical scale greatly magnified!) to show that there is a systematic deviation (downwards) for larger values of l. This effect can be understood semiquantitatively as being due to the "stretch" of the diatomic molecule in response to the increasing angular momentum. If we approximate the interatomic potential near equilibrium (r_0) by a harmonic oscillator of the form $V(r) = K(r - r_0)^2/2$, the classical force equation implies that

$$ma_c = \frac{L^2}{\mu r^3} = K(r - r_0) = F_c \qquad (17.109)$$

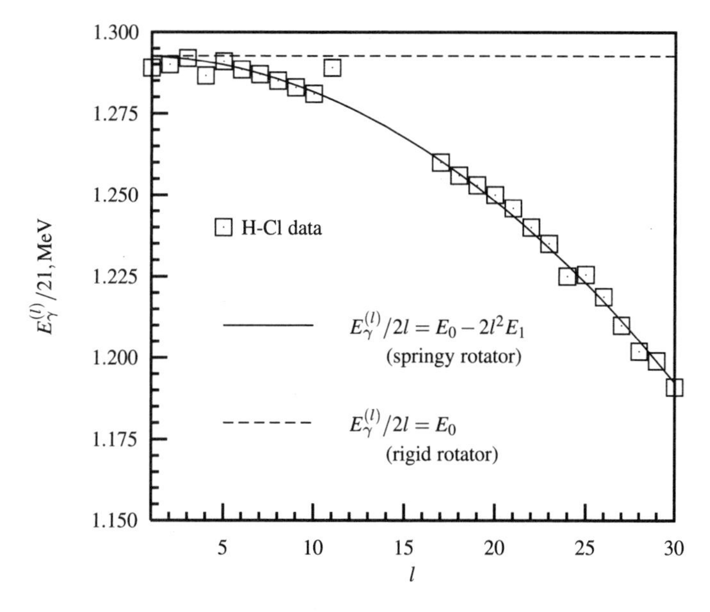

FIGURE 17.9. *Photon energies for $\Delta l = \pm 1$ transitions for a diatomic molecule (H-Cl). The data are plotted to indicate the pattern of energy levels for a "springy rotator."*

where we have written the centripetal acceleration in terms of L. If we call the deviation from equilibrium (the stretch) δ, we find

$$r - r_0 = \delta \approx \frac{L^2}{\mu r_0^3 K} \tag{17.110}$$

The energy now is given by

$$\begin{aligned} E &= \frac{L^2}{2\mu r^2} + \frac{1}{2}K(r - r_0)^2 \\ &= \frac{L^2}{2\mu(r_0 + \delta)^2} + \frac{1}{2}K\delta^2 \\ &\approx \frac{L^2}{2\mu r_0^2} - \frac{L^4}{2\mu^2 r_0^6 K} + \mathcal{O}(L^6) + \cdots \end{aligned} \tag{17.111}$$

Using the quantized values of $\mathbf{L}^2$, this gives the expression

$$E_l = l(l+1)E_0 - [l(l+1)]^2 E_1 \qquad \text{where} \qquad E_1 \equiv \frac{\hbar^4}{2\mu^2 r_0^6 K} \tag{17.112}$$

for the spectrum of a "springy rotator"; systems with "stiffer springs," i.e., ones with larger values of K will have their spectrum changed less. The corresponding photon energies satisfying the $|\Delta l| = 1$ selection rule would then have energies

$$E_\gamma^{(l)} = 2lE_0 - 4l^3 E_1 \qquad \text{or} \qquad \frac{E_\gamma^{(l)}}{2l} = E_0 - 2l^2 E_1 \tag{17.113}$$

and the solid curve in Fig. 17.9 indicates a fit to the data of this form. The fitted values are

$$E_0 \sim 1.927 \times 10^{-3}\,\text{eV} \qquad \text{and} \qquad E_1 \sim 5.6 \times 10^{-5}\text{eV} \tag{17.114}$$

17.3.2 Molecular Energy Levels

To understand the rich structure of quantized energy levels available to diatomic molecules, we consider how one might calculate the energy spectrum from first principles, indicating what role the electronic and nuclear motions play, and how vibrational and rotational states are connected to electronic excitations.

For a diatomic molecule consisting of two atoms with nuclear charges Z_1, Z_2, there will be $Z_1 + Z_2$ electrons, and the complete multibody Hamiltonian can be formally written as

$$\begin{aligned}\hat{H} &= \frac{\hat{\mathbf{P}}_1^2}{2M_1} + \frac{\hat{\mathbf{P}}_2^2}{2M_2} + \sum_{i=1}^{Z_1+Z_2} \frac{\hat{\mathbf{p}}_i^2}{2m_e} \\ &- \sum_{i=1}^{Z_1+Z_2} \frac{KZ_1e^2}{|\mathbf{R}_1 - \mathbf{r}_i|} - \sum_{i=1}^{Z_1+Z_2} \frac{KZ_2e^2}{|\mathbf{R}_2 - \mathbf{r}_i|} \\ &+ \frac{KZ_1Z_2e^2}{|\mathbf{R}_1 - \mathbf{R}_2|} + \sum_{i \neq j=1}^{Z_1+Z_2} \frac{Ke^2}{|r_i - r_j|}\end{aligned} \tag{17.115}$$

where $\mathbf{R}_{1,2}$ and $\mathbf{r}_i$ ($M_{1,2}$ and m_e) are the nuclear and electronic coordinates (masses), respectively. The mutual interactions are simply the Coulomb attractions or repulsions between all combinations of nuclei and electrons.

Because the nuclei are so much heavier than the electrons, the electrons can "respond" to changes in the nuclear positions rapidly, and it makes sense to initially approximate the nuclei as being fixed with some arbitrary separation $R = |\mathbf{R}_1 - \mathbf{R}_2|$. The electronic configuration, for the ground state or any excited state, can then (in principle, this is after all, an imagined calculation) be determined by solving the multiparticle Schrödinger equation. The "effective" Hamiltonian for the two nuclei can now be written in the form

$$\hat{H}_N = \frac{\hat{\mathbf{P}}_1^2}{2M_1} + \frac{\hat{\mathbf{P}}_1^2}{2M_1} + V(R) \tag{17.116}$$

where $V(R)$ now includes not only the Coulomb repulsion of the nuclei but also the effective potential due to the electron configuration. The terms in Eqn. (17.115) involving the electron coordinates are averaged over.

This procedure can be repeated for different values of R and can be used to "map out" the interatomic potential through which the nuclei interact. The schematic potential is shown in Fig. 17.10, both for a ground state electronic configuration, and for an electronic excited state. The two potentials are separated by roughly 1–10 eV, i.e., an atomic energy difference.

The vibrational energy levels considered in Sec. 10.3 are then due to the nuclear motion in each attractive well, and have energy splittings typically in the range 10^{-1} to 10^{-2} eV and are superimposed on each $V(R)$ curve as shown. The "spring constants" inferred from such data are of the order $K_{\text{eff}} \sim \text{eV/Å}^2$ and are consistent with this picture, as Å-size changes in the nuclear separation will change the energy of the electronic configuration by eV's. Finally, the vibrational states are much more closely spaced with splittings of the order 10^{-3} to 10^{-4} eV.

A useful mnenomic device for understanding this hierarchy of energy splittings is as follows. Typical electronic energies are determined by e, $\hbar$, m_e via

$$E_e \sim \frac{e^2}{a_0} \sim \frac{m_e e^4}{\hbar^2} \qquad \text{where} \qquad a_0 \sim \frac{\hbar^2}{m_e e^2} \tag{17.117}$$

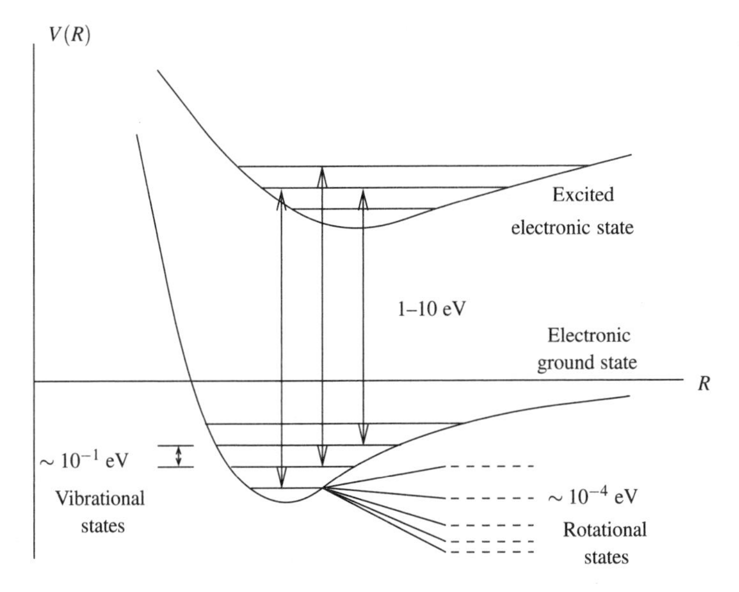

FIGURE 17.10. *Schematic plot of the effective interatomic potential in the electronic ground state and in an excited electronic state. Levels indicating vibrational and rotational states are also indicated, along with their relative energy splittings.*

The spring constants for molecular vibrational motion then scale as

$$K_{\text{eff}} \sim \frac{E_e}{a_0^2} \sim \frac{m_e^3 e^8}{\hbar^6} \tag{17.118}$$

so that the quantized energies go as

$$E_{\text{vib}} \sim \hbar\sqrt{\frac{K_{\text{eff}}}{M}} \sim \left(\frac{m_e e^4}{\hbar^2}\right)\sqrt{\frac{m_e}{M}} \tag{17.119}$$

where M is a nuclear mass. Finally, rotational states vary roughly as

$$E_{\text{rot}} \sim \frac{\mathbf{L}^2}{2\mu r_0^2} \sim \frac{\hbar^2}{M a_0^2} \sim \left(\frac{m_e e^4}{\hbar^2}\right)\frac{m_e}{M} \tag{17.120}$$

so that the electronic, rotational, and vibrational energies are roughly in the ratio

$$E_e/E_{\text{vib}}/E_{\text{rot}} : 1 \bigg/ \sqrt{\frac{m_e}{M}} \bigg/ \frac{m_e}{M} \tag{17.121}$$

Transitions involving changes in the electronic configuration will have typical energies in the 1 to 10 eV region corresponding to visible wavelengths. Because of the many rotational and vibrational states available in both the initial and final states, each electronic transition will actually correspond to many possible lines, and *band spectra* are seen instead of sharp lines as in atomic spectra. Changes in the vibrational state within a given electronic configuration will still allow for many different rotational states, so that fine structure can be resolved. The energy levels of a "pure" rotator-vibrator system are given by

$$E_{n,l} = \hbar\omega(n + 1/2) + l(l+1)E_0 \tag{17.122}$$

so that

$$E_\gamma^{(n,l)} = \hbar\omega + 2lE_0 \tag{17.123}$$

if we use the selection rule $\Delta n = \pm 1$ for vibrational state transitions, which we discuss next.

17.3.3 Selection Rules

Because of their importance in determining the observed patterns of emission and absorption in atomic and molecular spectroscopy (and beyond), we present here a "bare-bones" discussion of the physical principles underlying the selection rules[4] that we have freely used in our discussions of electromagnetic transitions between vibrational and rotational energy levels.

The description of the emission or absorption of a photon as a charged particle (or distribution of charges) changes its quantum state requires knowledge of the coupling of a charge to an external electromagnetic field, as discussed in Sec. 19.2.2. The interaction term relevant for our discussion can be written in the form

$$\frac{1}{m}\hat{\mathbf{p}} \cdot \mathbf{A} = \frac{1}{m}\hat{\mathbf{p}} \cdot \boldsymbol{\epsilon} e^{-i\mathbf{k}\cdot\mathbf{r}} \tag{17.124}$$

where $\mathbf{A}$ is the so-called *vector potential.* In this case, it is written in terms of the polarization vector of the photon, $\boldsymbol{\epsilon}$, and its plane wavefunction where the wavenumber satisfies $k = 2\pi/\lambda = E_\gamma/\hbar c$.

The amplitude that, when squared, describes the probability for a radiative transition between initial and final states, ψ_i and ψ_f, is given by

$$\text{Amp}(\psi_i \to \psi_f + \gamma) \propto \frac{1}{m}\left\langle \psi_f \left| \hat{\mathbf{p}} \cdot \boldsymbol{\epsilon} e^{-i\mathbf{k}\cdot\mathbf{r}} \right| \psi_i \right\rangle \tag{17.125}$$

If the bound state wavefunctions are localized on some length scale R, the argument of the exponential function will be at most of order kR over the region of integration where the overlap integral "gets its support," i.e., is nonvanishing. For many atomic, molecular, and nuclear systems, this factor satisfies $kR \ll 1$, and the exponential can be approximated by unity. For example, in radiative transitions between atomic energy levels, one has

$$kR \sim \frac{E_\gamma a_0}{\hbar c} \sim \frac{(2\,\text{eV})(1\,\text{Å})}{1973\,\text{eV Å}} \sim 10^{-3} \tag{17.126}$$

We will see that this approximation corresponds to considering only electric dipole radiation (and not higher multipoles).

In this limit, we can write the required matrix element in the form

$$\frac{1}{m}\boldsymbol{\epsilon} \cdot \langle \psi_f | \hat{\mathbf{p}} | \psi_i \rangle = \boldsymbol{\epsilon} \cdot \frac{d}{dt}\langle \psi_f | \mathbf{r} | \psi_i \rangle \tag{17.127}$$

which is seen to be essentially the time rate of change of the *dipole matrix element* between initial and final states. This expression can be simplified by evaluating the time derivative using the relation

[4]For more complete discussions, see, e.g., Gasiorowicz (1974).

$$\begin{aligned}\frac{d}{dt}\langle\psi_f|\mathbf{r}|\psi_i\rangle &= \frac{i}{\hbar}\langle\psi_f|[\hat{H},\mathbf{r}]|\psi_i\rangle \\ &= \frac{i(E_f - E_i)}{\hbar}\langle\psi_f|\mathbf{r}|\psi_i\rangle\end{aligned} \tag{17.128}$$

This gives the important result that the amplitude governing electric dipole radiation between two quantum states is proportional to the matrix element

$$\text{Amp}(\psi_i \to \psi_f + \gamma) \propto \boldsymbol{\epsilon}\cdot\langle\psi_f|\mathbf{r}|\psi_i\rangle \tag{17.129}$$

and this dependence is the basis of many selection rules.

For example, for vibrational states described by eigenfunctions of the simple harmonic oscillator (considered here in one dimension, for simplicity), we find that the matrix elements satisfy

$$\langle\psi_{n_f}|x|\psi_{n_i}\rangle = 0 \qquad \text{unless} \qquad \Delta n = n_f - n_i = \pm 1 \tag{17.130}$$

using the results of Eqn. (10.47).

In the case of the rigid rotator, the components of the vector $\mathbf{r} = (x, y, z)$ can be written in terms of the spherical harmonics $Y_{l=1,m}(\theta,\phi)$ with $m = 0, \pm 1$, so that we need to examine the structure of the matrix elements

$$\langle Y_{l_f,m_f}|Y_{1,m}|Y_{l_i,m_i}\rangle \tag{17.131}$$

The azimuthal integrations over ϕ will have the general form

$$\int_0^{2\pi} d\phi\, e^{i(m_i+m-m_f)} \tag{17.132}$$

which will vanish unless

$$\Delta m = m_f - m_i = m = +1, 0, -1 \tag{17.133}$$

The selection rules for l are determined by the integration over the polar angle θ. Arguments similar to those leading to the addition of angular momentum rules in Sec. 17.5 imply that the product wavefunction $Y_{1,m}Y_{l_i,m_i}$ can be written in terms of individual spherical harmonics given by the addition rule $l_i + 1 \longrightarrow l_i - 1, l_i, l_i + 1$. Because of the orthogonality of the spherical harmonics, the only overlap integrals that are then nonvanishing are those for which $l_f - l_i = +1, 0, -1$. However, when $l_f = l_i$, it is easy to see that Eqn. (17.131) vanishes by parity arguments since

$$\langle Y_{l,m}|x, y, z|Y_{l,m'}\rangle = 0 \tag{17.134}$$

This implies the selection rule

$$\Delta l = l_f - l_i = \pm 1 \tag{17.135}$$

and it is sometimes said that "quantum numbers change by one unit in dipole transitions."

Several other comments complete our discussion:

- The spectra of molecules consisting of identical atoms form a special case that can be examined on both physical and more formal grounds.
  - *Physical argument:* For diatomic molecules consisting of unlike atoms, such as C-O or H-Cl, there can be a permanent dipole moment that can then radiate when "shaken" (in transitions between vibrational states) or "spun" (in transitions between rotational states). For molecules of identical atoms, such as C_2 or O_2, the

symmetry between the two atoms implies that no dipole moment can exist (after all, which way would it point?), and so electric dipole radiation is forbidden.

- *Formal argument:* Such systems of indistinguishable particles must have wavefunctions that have the correct symmetry under exchange, i.e.,

$$\psi(\mathbf{r}_1, \mathbf{r}_2) = \pm\psi(\mathbf{r}_2, \mathbf{r}_1) \tag{17.136}$$

which implies that the wavefunction for the relative coordinate, $\mathbf{r} = \mathbf{r}_1 - \mathbf{r}_2$ must satisfy

$$\psi(\mathbf{r}) = \pm\psi(-\mathbf{r}) \tag{17.137}$$

i.e., have a definite parity. The eigenfunctions of the harmonic oscillator, for example, have parity given by $P = (-1)^n$, so that the allowed wavefunctions for indistinguishable particles must have either n odd or even, and allowed states are separated by two units of n. Since the selection rule Eqn. (17.130) implies that $\Delta n = \pm 1$, electric dipole transitions are forbidden.

- Under circumstances where electric dipole transitions are not allowed, the next term in the expansion of $e^{-i\mathbf{k}\cdot\mathbf{r}}$ is required. One must then consider the matrix elements of higher powers of $\mathbf{r}$ that can then connect states with $|\Delta n|$ or $|\Delta l|$ greater than one. Such transitions, corresponding to quadrupole, octopole, etc., terms will have much smaller amplitudes due to the extra powers of kR. They are not "forbidden" but may well have unobservably small intensities in a given experiment.

17.4 SPIN AND ANGULAR MOMENTUM

To derive a quantum description of rotational motion, we have so far followed a path quite similar to that for quantizing classical oscillatory motion. Certain classical variables were generalized to quantum operators that yielded eigenvalue problems; the spherical harmonic functions, $Y_{l,m}(\theta, \phi)$, were obtained as the solutions corresponding to the quantized values of orbital angular momentum and its z component.

If we wish to associate the spin degree of freedom with a half-integral value of angular momentum, a representation in terms of spherical harmonics is clearly inappropriate. To see this, if we formally attempt to use $l = 1/2$ and $m = \pm 1/2$, we find that

$$Y_{l,m} = Y_{1/2,\pm 1/2}(\theta, \phi) \propto \sqrt{\sin(\theta)}e^{\pm i\phi/2} \tag{17.138}$$

This identification does not satisfy any of the functional relations for the spherical harmonics since, for example,

$$\hat{L}_- Y_{1/2,+1/2} \propto \sqrt{\sin(\theta)}e^{\pm i\phi/2} \propto \frac{\cos(\theta)}{\sqrt{\sin(\theta)}}e^{-i\phi/2} \tag{17.139}$$

which is not, in turn, proportional to $Y_{1/2,-1/2}$ or even well-behaved at $\theta = 0$.

It is perhaps not surprising that this formulation for quantized values of angular momentum does not extend to intrinsic angular momentum or spin, as the derivation leading to the $Y_{l,m}$ was based on generalizing classical orbital motion. Clearly, another representation of the angular momentum operators and their eigenfunctions is required, which is more abstract.

As an example, consider a 3×3 matrix representation of the eigenfunctions corresponding to $l = 1$; this formulation can be easily generalized to any value of l using $(2l + 1) \times (2l + 1)$ matrices. We can explicitly write

$$\mathbf{L}_x = \frac{\hbar}{2}\begin{pmatrix} 0 & \sqrt{2} & 0 \\ \sqrt{2} & 0 & \sqrt{2} \\ 0 & \sqrt{2} & 0 \end{pmatrix} \tag{17.140}$$

$$\mathbf{L}_y = \frac{\hbar}{2i}\begin{pmatrix} 0 & \sqrt{2} & 0 \\ -\sqrt{2} & 0 & \sqrt{2} \\ 0 & -\sqrt{2} & 0 \end{pmatrix} \tag{17.141}$$

and

$$\mathbf{L}_z = \hbar\begin{pmatrix} 1 & 0 & 0 \\ 0 & 0 & 0 \\ 0 & 0 & -1 \end{pmatrix} \tag{17.142}$$

The $2l + 1$ values of L_z are located along the diagonal, while the $\sqrt{2} = \sqrt{l(l+1)}$ is appropriate for $l = 1$. One can easily check (P17.11) that the usual commutation relations are obeyed as

$$[\mathbf{L}_x, \mathbf{L}_y] = i\hbar\mathbf{L}_z \tag{17.143}$$

and that

$$\mathbf{L}^2 = \mathbf{L}_x^2 + \mathbf{L}_y^2 + \mathbf{L}_z^2 = \hbar^2\begin{pmatrix} 2 & 0 & 0 \\ 0 & 2 & 0 \\ 0 & 0 & 2 \end{pmatrix} \tag{17.144}$$

The simultaneous eigenfunctions of $\mathbf{L}^2$ and $\mathbf{L}_z$, with eigenvalues $l(l+1)\hbar^2 = 2\hbar^2$ and $L_z = +1, 0, -1$, respectively, are given by

$$v^{(+1)} = \begin{pmatrix} 1 \\ 0 \\ 0 \end{pmatrix} \qquad v^{(0)} = \begin{pmatrix} 0 \\ 1 \\ 0 \end{pmatrix} \qquad v^{(-1)} = \begin{pmatrix} 0 \\ 0 \\ 1 \end{pmatrix} \tag{17.145}$$

A general $l = 1$ "wavefunction" is then written in the form

$$\begin{pmatrix} \alpha \\ \beta \\ \gamma \end{pmatrix} = \alpha v^{(+1)} + \beta v^{(0)} + \gamma v^{(-1)} \tag{17.146}$$

where $|\alpha|^2 + |\beta|^2 + |\gamma|^2 = 1$ is the normalization condition. Raising and lowering operators are easily generalized as well, and one has, for example,

$$\mathbf{L}_+ = \mathbf{L}_x + i\mathbf{L}_y = \sqrt{2}\hbar\begin{pmatrix} 0 & 1 & 0 \\ 0 & 0 & 1 \\ 0 & 0 & 0 \end{pmatrix} \tag{17.147}$$

One can then check explicit relations such as

$$\mathbf{L}_+ v^{(-1)} = \sqrt{2}\hbar\begin{pmatrix} 0 \\ 1 \\ 0 \end{pmatrix} = \sqrt{2}\hbar v^{(0)} \tag{17.148}$$

so that the normalizations of Eqn. (17.61) and (17.62) are maintained.

Example 17.2 It is an instructive exercise to find the eigenvalues and eigenfunctions of the operator

$$\mathbf{L}_\phi = \cos(\phi)\mathbf{L}_x + \sin(\phi)\mathbf{L}_y = \frac{\hbar}{\sqrt{2}}\begin{pmatrix} 0 & e^{-i\phi} & 0 \\ e^{i\phi} & 0 & e^{-i\phi} \\ 0 & e^{i\phi} & 0 \end{pmatrix} \tag{17.149}$$

Writing the eigenvalue problem as

$$\mathbf{L}_\phi \begin{pmatrix} \alpha \\ \beta \\ \gamma \end{pmatrix} = L_\phi \begin{pmatrix} \alpha \\ \beta \\ \gamma \end{pmatrix} \tag{17.150}$$

we find the determinant condition on the eigenvalues L_ϕ,

$$\det\begin{pmatrix} L_\phi & -\hbar e^{-i\phi}/\sqrt{2} & 0 \\ -\hbar e^{i\phi}/\sqrt{2} & L_\phi & -\hbar e^{-i\phi}/\sqrt{2} \\ 0 & -\hbar e^{i\phi}/\sqrt{2} & L_\phi \end{pmatrix} = 0 \tag{17.151}$$

This implies that $L_\phi^3 - \hbar^2 L_\phi = 0$ or $L_\phi = +\hbar, 0, -\hbar$ as expected; the eigenfunction corresponding to $L_\phi = +\hbar$ is easily found to be

$$v_\phi^{(+1)} = \begin{pmatrix} e^{-i\phi}/\sqrt{2} \\ 1 \\ e^{i\phi}/\sqrt{2} \end{pmatrix} \tag{17.152}$$

with similar results for $v_\phi^{(0)}, v_\phi^{(-1)}$.

A relabeling $\phi \to \phi + 2\pi$ should not have any effect on the results, as the two labels for the same angle are physically equivalent. It is easy to check that $|v_{\phi+2\pi}|^2 = |v_\phi|^2$ so that the probability densities are invariant as they should be. We note that the stronger condition, $v_{\phi+2\pi} = v_\phi$, also holds, so that the wavefunction returns to the same phase on one rotation. While such expectations might seem obvious for any classical system, the corresponding results for the nonclassical spin degree of freedom will be different.

This matrix representation of quantized angular momenta can now be generalized to treat spin-1/2 particles. Specifically, we define

$$\mathbf{S}_x = \frac{\hbar}{2}\begin{pmatrix} 0 & 1 \\ 1 & 0 \end{pmatrix} \tag{17.153}$$

$$\mathbf{S}_y = \frac{\hbar}{2i}\begin{pmatrix} 0 & 1 \\ -1 & 0 \end{pmatrix} \tag{17.154}$$

and

$$\mathbf{S}_x = \frac{\hbar}{2}\begin{pmatrix} 1 & 0 \\ 0 & -1 \end{pmatrix} \tag{17.155}$$

which satisfy the standard commutation relations

$$[\mathbf{S}_x, \mathbf{S}_y] = i\hbar S_z \tag{17.156}$$

and give

$$\mathbf{S}^2 = \hbar^2\begin{pmatrix} 3/4 & 0 \\ 0 & 3/4 \end{pmatrix} = \left[\frac{1}{2}\left(\frac{1}{2}+1\right)\hbar^2\right]1 \tag{17.157}$$

which is appropriate for $S = 1/2$. The eigenvectors of $\hat{\mathbf{S}}^2$ and $\mathbf{S}_z$ are just the spinors introduced in Sec. 16.4:

$$\chi_+ = \begin{pmatrix} 1 \\ 0 \end{pmatrix} \quad \text{and} \quad \chi_- = \begin{pmatrix} 0 \\ 1 \end{pmatrix} \tag{17.158}$$

with eigenvalues $S_z = +1/2$ and $-1/2$, respectively; a general spinor is written as

$$\chi = \alpha^{(+)}\chi_+ + \alpha^{(-)}\chi_- = \begin{pmatrix} \alpha^{(+)} \\ \alpha^{(-)} \end{pmatrix} \tag{17.159}$$

We can repeat the exercise in Ex. 17.2 to examine the eigenvalues and eigenfunctions of the "rotated" spin matrix operator

$$\mathbf{S}_\phi = \cos(\phi)\mathbf{S}_x + \sin(\phi)\mathbf{S}_y = \frac{\hbar}{2}\begin{pmatrix} 0 & e^{-i\phi} \\ e^{i\phi} & 0 \end{pmatrix} \tag{17.160}$$

The eigenvalues of $\mathbf{S}_\phi$ are found to be $+\hbar/2$, $-\hbar/2$ as expected, with corresponding eigenvectors

$$v^{(+1/2)} = \frac{1}{\sqrt{2}}\begin{pmatrix} e^{-i\phi/2} \\ e^{i\phi/2} \end{pmatrix} \qquad v^{(-1/2)} = \frac{1}{\sqrt{2}}\begin{pmatrix} e^{-i\phi/2} \\ -e^{i\phi/2} \end{pmatrix} \tag{17.161}$$

Both spinors satisfy the relation

$$|v^{(\pm 1/2)}_{\phi+2\pi}|^2 = |v^{(\pm 1/2)}_{\phi}|^2 \tag{17.162}$$

so the probability density is unchanged by the relabeling of angle as it should be. The spinor wavefunctions or amplitudes, however, do change sign since

$$v^{(\pm 1/2)}_{\phi+2\pi} = -v^{(\pm 1/2)}_{\phi} \tag{17.163}$$

and a rotation by 4π is required to return them to their original phase. This phase change under rotations is typical of fermionic spin-1/2 wavefunctions and has no classical analog; as with any phase behavior, it is best tested in interference experiments as discussed in Sec. 19.7.1.

Before proceeding, we note that it is often useful to write the spin matrices in dimensionless form by defining

$$\mathbf{S} = \frac{\hbar}{2}\boldsymbol{\sigma} \tag{17.164}$$

where

$$\sigma_x = \begin{pmatrix} 0 & 1 \\ 1 & 0 \end{pmatrix} \quad \sigma_x = \begin{pmatrix} 0 & -i \\ i & 0 \end{pmatrix} \quad \sigma_z = \begin{pmatrix} 1 & 0 \\ 0 & -1 \end{pmatrix} \tag{17.165}$$

which satisfy

$$[\sigma_i, \sigma_j] = 2i\epsilon_{ijk}\sigma_k \tag{17.166}$$

where $i, j, k = x, y, z$ in any permutation and $\epsilon_{i,j,k}$ is the totally antisymmetric symbol (App. D.1). The σ are called the *Pauli matrices,* and they are Hermitian (that is, $\mathbf{M}^\dagger = (\mathbf{M}^*)^T = \mathbf{M}$). If the behavior of the spinor wavefunction under rotations is decidedly "unclassical," the dynamics of a charged particle with spin-1/2 does have some classical analogs, namely, their interactions with an external magnetic field.

We will see in Sec. 19.5 that the Hamiltonian for a charged particle in a uniform magnetic field will have a term of the form

$$\hat{H} = -\left(\frac{q}{2m}\right)\hat{\mathbf{L}} \cdot \mathbf{B} = -\hat{\mathbf{M}} \cdot \mathbf{B} \tag{17.167}$$

where we have defined a magnetic moment operator via

$$\hat{\mathbf{M}} = \frac{q}{2m}\hat{\mathbf{L}} \tag{17.168}$$

Equation (17.167) is just the analog of the classical energy of a magnetic dipole in an external **B** field. The relation Eqn. (17.168) between the magnetic moment and angular momentum is consistent with that of a classically rotating-point particle.

Particles with intrinsic spin often have microscopic magnetic moments given by the similar relation

$$\mathbf{M} = g\frac{q}{2m}\mathbf{S} \tag{17.169}$$

where q, m are the charge and mass, respectively, and g is a dimensionless number called the *gyromagnetic ratio*. As the name implies, g measures the ratio of the magnetic moment to the (intrinsic) angular momentum vector. For a classical rotating particle or a classical spinning object for which the mass and charge densities are proportional, one can show that $g = 1$ (P17.12). In contrast, the g value for a pointlike, charged spin-1/2 particle can be derived using the machinery of relativistic quantum mechanics[5] (via the so-called *Dirac equation*), and one finds $g = 2$. Small, calculable corrections to this value arise due to effects from quantum field theory; for the electron these have the form

$$g_e = 2 + \frac{\alpha}{\pi} - 0.65696\left(\frac{\alpha}{\pi}\right)^2 + \cdots \approx 2.0023193048 \tag{17.170}$$

and we will take $g_e = 2$ for simplicity. The magnetic moment vector for such a pointlike spin-1/2 particle can then be written in the form

$$\mathbf{M} = \frac{g}{2}\left(\frac{q\hbar}{2m}\right)\boldsymbol{\sigma} = \mu\boldsymbol{\sigma} \tag{17.171}$$

where the *magneton* is defined via $\mu = q\hbar/2m$.

Based on the prediction of the Dirac theory, for the neutron and proton we would expect to have

$$\mu_p = \frac{e\hbar}{2m_p} \equiv \mu_N \qquad \mu_n = 0 \tag{17.172}$$

since the charge of the neutron vanishes. Instead, one finds

$$\mu_p \approx 2.79\mu_N \qquad \mu_n \approx -1.91\mu_N \tag{17.173}$$

or, more precisely,

$$\frac{g_p}{2} = 2.79284739(6) \qquad \text{and} \qquad \frac{g_n}{2} = -1.9130427(5) \tag{17.174}$$

[5]See, e.g., Griffiths (1987).

These large deviations from pointlike structure can be understood if the neutron and proton are bound states of more fundamental constituents, and are best explained in terms of the *quark model* (P17.18).

The magnetic moments of such particles provide a "handle" with which to manipulate their spin orientation via their interactions with externally applied magnetic fields. The interaction Hamiltonian is given by

$$\hat{H} = -\mathbf{M} \cdot \mathbf{B} = -\frac{g\mu}{2}\boldsymbol{\sigma} \cdot \mathbf{B} \tag{17.175}$$

For a uniform field along the z direction, the dynamical equation of motion for the spinor wavefunction is simply the Schrödinger equation, which in this case has the form

$$i\hbar\frac{\partial}{\partial t}\chi(t) = \hat{H}\chi(t) = -\frac{g\mu}{2}B\sigma_z\chi(t) \tag{17.176}$$

or

$$\begin{pmatrix}\dot{\alpha}^{(+)}(t)\\ \dot{\alpha}^{(-)}(t)\end{pmatrix} = i\left(\frac{g\mu B}{2\hbar}\right)\begin{pmatrix}\alpha^{(+)}(t)\\ -\alpha^{(-)}(t)\end{pmatrix} \tag{17.177}$$

which has the solutions

$$\alpha^{(\pm)}(t) = \alpha^{(\pm)}(0)e^{\pm i\omega t} \tag{17.178}$$

where $\omega = g\mu B/2\hbar$. These solutions correspond to energies

$$E_{\pm} = \hbar\omega_{\pm} = \pm\frac{g\mu B}{2} \tag{17.179}$$

where the spin is parallel $(-)$ or antiparallel $(+)$ to the field direction.

Example 17.3 Spin Precession in a Uniform Field Suppose that initially one has a spinor wavefunction that is an eigenvector of $\mathbf{S}_x$ with eigenvalue $+\hbar/2$, i.e., one knows that the spin is "pointing" along the $+x$ direction; this corresponds to

$$\chi(0) = \frac{1}{\sqrt{2}}\begin{pmatrix}1\\1\end{pmatrix} \tag{17.180}$$

If a uniform magnetic field in the z direction is applied, the time development of this spinor is given by Eqn. (17.178) so that

$$\chi(t) = \frac{1}{\sqrt{2}}\begin{pmatrix}e^{i\omega t}\\ e^{-i\omega t}\end{pmatrix} \tag{17.181}$$

At later times, the expectation value of the spin along the x axis is given by

$$\begin{aligned}\langle \mathbf{S}_x\rangle_t &= \langle \chi(t)|\mathbf{S}_x|\chi(t)\rangle \\ &= \left[\frac{1}{\sqrt{2}}(e^{-i\omega t}, e^{i\omega t})\right]\left[\frac{\hbar}{2}\begin{pmatrix}0 & 1\\ 1 & 0\end{pmatrix}\right]\left[\frac{1}{\sqrt{2}}\begin{pmatrix}e^{i\omega t}\\ e^{-i\omega t}\end{pmatrix}\right] \\ &= \frac{\hbar}{2}\cos(2\omega t)\end{aligned} \tag{17.182}$$

while $\langle \mathbf{S}_y\rangle_t = -\hbar\sin(2\omega t)/2$. We see that:

- The magnetic moment precesses around the z axis at an angular velocity (2ω), which is twice the rate at which the phase of the spinor wavefunction changes (ω).

This precession can also be derived (P17.13) in a more formal way, which makes its classical analog more obvious. This phenomenon is the basis for one experimental test of the spinor nature of the spin-1/2 fermion wavefunction (Sec. 19.7.1).

17.5 ADDITION OF ANGULAR MOMENTUM

For classical particles that do not interact with each other, the total value of some observable quantities, such as energy, momentum, and the like, is often obtained by simply summing the values for each particle. This can occur in quantum mechanical systems as well, where "noninteracting" often implies that the multiparticle wavefunction can be written in a factorized, product form.

For example, for two noninteracting energy eigenstates, one might have

$$\psi(x_1, x_2; t) = \left[\psi_{E_1}(x_1)e^{-iE_1t/\hbar}\right]\left[\psi_{E_2}(x_2)e^{-iE_2t/\hbar}\right] \tag{17.183}$$

which is also an energy eigenstate since

$$\hat{E}\psi(x_1, x_2; t) = i\hbar\frac{\partial}{\partial t}\psi(x_1, x_2; t) = (E_1 + E_2)\psi(x_1, x_2; t) \tag{17.184}$$

Free-particle momentum eigenstates also behave in this manner since

$$\hat{\mathbf{P}}_{\text{tot}}\left[e^{i\mathbf{p}_1\cdot\mathbf{r}_1/\hbar}e^{i\mathbf{p}_2\cdot\mathbf{r}_2/\hbar}\right] = (\mathbf{p}_1 + \mathbf{p}_2)\left[e^{i\mathbf{p}_1\cdot\mathbf{r}_1/\hbar}e^{i\mathbf{p}_2\cdot\mathbf{r}_2/\hbar}\right] \tag{17.185}$$

since $\hat{\mathbf{P}}_{\text{tot}} = (\hbar/i)(\boldsymbol{\nabla}_1 + \boldsymbol{\nabla}_2)$.

For two particles that are described by eigenstates of $\hat{\mathbf{L}}_1^2$, $\hat{\mathbf{L}}_2^2$, and $\hat{L}_{1,z}$, $\hat{L}_{2,z}$, respectively, the situation is more complex. The operators corresponding to the total angular momentum squared and its corresponding z component are given by

$$\hat{\mathbf{J}}^2 = \left(\hat{\mathbf{L}}_1 + \hat{\mathbf{L}}_2\right)^2 = \hat{\mathbf{L}}_1^2 + 2\hat{\mathbf{L}}_1\cdot\hat{\mathbf{L}}_2 + \hat{\mathbf{L}}_2^2 \tag{17.186}$$

$$\hat{J}_z = \hat{L}_{1,z} + \hat{L}_{2,z} \tag{17.187}$$

and we wish to find the corresponding eigenstates, $\psi_{(J,J_z)}$, which satisfy

$$\hat{\mathbf{J}}^2\psi_{(J,J_z)} = J(J+1)\hbar^2\psi_{(J,J_z)} \quad \text{and} \quad \hat{J}_z\psi_{(J,J_z)} = J_z\hbar\psi_{(J,J_z)} \tag{17.188}$$

We especially wish to know if simple products of the respective spherical harmonics will also be eigenstates of these operators. We will thus consider fixed values of l_1 and l_2 with all of their respective values of m_1, m_2.

A product wavefunction of the form $Y_{(l_1,m_1)}(\theta_1, \phi_1)Y_{(l_2,m_2)}(\theta_2, \phi_2)$ is clearly an eigenstate of $\hat{J}_z$ since

$$\begin{aligned}\hat{J}_zY_{l_1,m_1}Y_{l_2,m_2} &= \left(\hat{L}_{1,z} + \hat{L}_{2,z}\right)Y_{l_1,m_1}Y_{l_2,m_2} \\ &= (m_1 + m_2)\hbar Y_{l_1,m_1}Y_{l_2,m_2}\end{aligned} \tag{17.189}$$

or $J_z = (m_1 + m_2)\hbar$. The same is *not* true for $\hat{\mathbf{J}}^2$ because of the "offending cross-term" in $\hat{\mathbf{J}}^2$, namely,

$$\hat{\mathbf{L}}_1\cdot\hat{\mathbf{L}}_2 = \hat{L}_{1,x}\hat{L}_{2,x} + \hat{L}_{1,y}\hat{L}_{2,y} + \hat{L}_{1,z}\hat{L}_{2,z} \tag{17.190}$$

since individual spherical harmonics cannot be simultaneous eigenfunctions of all three components of $\hat{\mathbf{L}}$. More concretely, the operator $\hat{\mathbf{L}}_1 \cdot \hat{\mathbf{L}}_2$ acting on a specific $Y_{l_1,m_1} Y_{l_2,m_2}$ will introduce new values of m_1, m_2. This can be seen most easily by noting that

$$\begin{aligned} \hat{\mathbf{J}}^2 &= \hat{\mathbf{L}}_1^2 + 2\hat{\mathbf{L}}_1 \cdot \hat{\mathbf{L}}_2 + \hat{\mathbf{L}}_2^2 \\ &= \hat{\mathbf{L}}_1^2 + \hat{\mathbf{L}}_2^2 + 2\hat{L}_{1,z}\hat{L}_{2,z} + \hat{L}_{1,+}\hat{L}_{2,-} + \hat{L}_{1,-}\hat{L}_{2,+} \end{aligned} \tag{17.191}$$

and we recall that the action of the raising and lowering operators for each label can be obtained via

$$\begin{aligned} \hat{L}_+ Y_{l,m} &= \sqrt{l(l+1) - m(m+1)}\hbar Y_{l,m+1} \\ &= \sqrt{(l+m+1)(l-m)}\hbar Y_{l,m+1} \end{aligned} \tag{17.192}$$

and

$$\hat{L}_- Y_{l,m} = \sqrt{(l-m+1)(l+m)}\hbar Y_{l,m-1} \tag{17.193}$$

We can, however, hope to find special linear combinations of spherical harmonics that do satisfy Eqn. (17.188), namely,

$$\psi_{(J,J_z)} = \sum_{m_1=-l_1}^{m_1=+l_1} \sum_{m_2=-l_2}^{m_2=+l_2} C(J, J_z; l_1, m_1, l_2, m_2) Y_{l_1,m_1} Y_{l_2,m_2} \tag{17.194}$$

Such an expansion is called a *Clebsch-Gordan series,* and the $C(J, J_z, l_1, m_1, l_2, m_2)$ are referred to as the *Clebsch-Gordan coefficients.* The allowed values of J and J_z can then be determined by looking for eigenstates of this form.

We will proceed by simply stating the result for the possible values of J that are allowed in such an expansion, providing some prima facie evidence to justify the answer, and then discuss the general procedure for its proof.

One finds that for a fixed value of l_1 and l_2:

- The allowed values for the total angular momentum quantum number, J, are in the range

$$J = |l_1 - l_2|, |l_1 + l_2 - 1|, \ldots, l_1 + l_2 - 1, l_1 + l_2 \tag{17.195}$$

  or, equivalently, J occurs in integral steps in the range

$$J_{\text{min}} = |l_1 - l_2| \leq J \leq (l_1 + l_2) = J_{\text{max}} \tag{17.196}$$

These limiting values are consistent with a classical picture of vector addition as when $\mathbf{L}_1$ and $\mathbf{L}_2$ are aligned or parallel (antialigned or antiparallel), their magnitude would be $L_1 + L_2$ ($|L_1 - L_2|$). In addition, the enumeration of distinct quantum states is consistent in the two pictures. For fixed values of l_1 and l_2, there are a total of $(2l_1 + 1)(2l_2 + 1)$ product wavefunctions of the form $Y_{l_1,m_1} Y_{l_2,m_2}$. For the eigenfunctions of total angular momentum, $\psi_{(J,J_z)}$, the equivalent counting is given by (P17.15):

$$\sum_{J=|l_1-l_2|}^{J=l_1+l_2} (2J+1) = (2l_1+1)(2l_2+1) \tag{17.197}$$

as well. The values of m_1, m_2, and J_z are then obviously determined via $J_z = m_1 + m_2$, which implies that

$$C(J, J_z; l_1, m_1, l_2, m_2) = 0 \qquad \text{if} \qquad m_1 + m_2 \neq J_z \tag{17.198}$$

This observation also suggests that we analyze the Clebsch-Gordan series by looking systematically at $\psi_{(J,J_z)}$ for specific values of J_z.

Consider, for example, the unique product wavefunction with the maximal value of J_z, namely, $J_{\max} = l_1 + l_2$; it is given by

$$\psi_{(J,J_z)} = Y_{l_1,l_1}(\theta_1, \phi_1) Y_{l_2,l_2}(\theta_2, \phi_2) \tag{17.199}$$

which is obviously an eigenfunction of $\hat{J}_z = \hat{L}_{1,z} + \hat{L}_{2,z}$, with $J_z = (l_1 + l_2)\hbar$. It is also an eigenfunction of $\hat{\mathbf{J}}^2$ since, using Eqn. (17.192) and the fact that the $\hat{L}_{1,+}$ and $\hat{L}_{2,+}$ operators annihilate both spherical harmonics, we find

$$\begin{aligned} \hat{\mathbf{J}}^2 \psi_{(J,J_z)} &= \left[l_1(l_1+1) + l_2(l_2+1) + 2l_1 l_2\right] \hbar^2 \psi_{(J,J_z)} \\ &= [(l_1+l_2)(l_1+l_2+1)]\hbar^2 \psi_{(J,J_z)} \\ &= J(J+1)\hbar^2 \psi_{(J,J_z)} \end{aligned} \tag{17.200}$$

or $J = l_1 + l_2$. This is, of course, the only possible value for $J_z = l_1 + l_2$.

For the next lowest value of J_z, there are two possible terms that can contribute to an eigenfunction of $\hat{J}_z$ with $J_z = J_{\max} - 1 = (l_1 + l_2 - 1)$, so we write

$$\psi_{(J,J_z)} = \alpha Y_{l_1,l_1-1}(\theta_1, \phi_1) Y_{l_2,l_2}(\theta_2, \phi_2) + \beta Y_{l_1,l_1}(\theta_1, \phi_1) Y_{l_2,l_2-1}(\theta_2, \phi_2) \tag{17.201}$$

In this case we have

$$\begin{aligned} &\hat{\mathbf{J}}^2 \psi_{J,(J_z)} = \\ &\left[\alpha(l_1(l_1+1) + l_2(l_2+1) + 2(l_1-1)l_2 + \beta(\sqrt{4l_1 l_2})\right] \hbar^2 Y_{l_1,l_1-1} Y_{l_2,l_2} \\ &+ \left[\alpha(\sqrt{4l_1 l_2}) + \beta(l_1(l_1+1) + l_2(l_2+1) + 2(l_2-1)l_1)\right] \hbar^2 Y_{l_1,l_1} Y_{l_2,l_2-1} \\ &= J(J+1)\hbar^2 \left[\alpha Y_{l_1,l_1-1} Y_{l_2,l_2} + \beta Y_{l_1,l_1} Y_{l_2,l_2-1}\right] \\ &= J(J+1)\hbar^2 \psi_{(J,J_z)} \end{aligned} \tag{17.202}$$

Equating coefficients, we find a set of coupled equations for α and β that can be written in the form

$$\mathbf{M}\begin{pmatrix}\alpha \\ \beta\end{pmatrix} = J(J+1)\begin{pmatrix}\alpha \\ \beta\end{pmatrix} \tag{17.203}$$

where

$$\mathbf{M} = \begin{pmatrix} l_1(l_1+l_2+1) + l_2(l_1+l_2-1) & 2\sqrt{l_1 l_2} \\ 2\sqrt{l_1 l_2} & l_1(l_1+l_2-1) + l_2(l_1+l_2+1) \end{pmatrix} \tag{17.204}$$

This matrix eigenvalue problem Eqn. (17.203) will have a solution, provided that

$$\det\left|\mathbf{M} - J(J+1)1\right| = 0 \tag{17.205}$$

or equivalently

$$\begin{aligned} [l_1(l_1+l_2+1) l_2(l_1+l_2-1) - J(J+1)] \cdot [l_1(l_1+l_2-1) + l_2(l_1+l_2+1) - J(J+1)] & \\ = 4l_1 l_2 & \qquad (17.206) \\ = (2l_1)(2l_2) & \qquad \text{Choice 1} \\ = (-2l_2)(-2l_1) & \qquad \text{Choice 2} \end{aligned}$$

where we have written the right-hand side in two (hopefully) suggestive ways, as the two solutions to Eqn. (17.206) correspond to the two choices. The second of these yields

$$\begin{aligned} l_1(l_1 + l_2 + 1) + l_2(l_1 + l_2 - 1) + 2l_2 &= J(J + 1) \\ l_1(l_1 + l_2 - 1) + l_2(l_1 + l_2 + 1) + 2l_1 &= J(J + 1) \end{aligned} \tag{17.207}$$

which both give

$$(l_1 + l_2)(l_1 + l_2 + 1) = J(J + 1) \Longrightarrow J = l_1 + l_2 \tag{17.208}$$

The first choice in Eqn. (17.206) corresponds to

$$\begin{aligned} l_1(l_1 + l_2 + 1) + l_2(l_1 + l_2 - 1) - 2l_1 &= J(J + 1) \\ l_1(l_1 + l_2 - 1) + l_2(l_1 + l_2 + 1) - 2l_2 &= J(J + 1) \end{aligned} \tag{17.209}$$

which implies

$$(l_1 + l_2)(l_1 + l_2 - 1) = J(J + 1) \Longrightarrow J = l_1 + l_2 - 1 \tag{17.210}$$

These two possibilities for J are the only ones consistent with $J_z = J_{\max} - 1 = l_1 + l_2 - 1$. The values of α and β can be obtained in either case, and the normalized wavefunctions can be written as

$$\psi_{(l_1+l_2, l_1+l_2-1)} = \sqrt{\frac{l_1}{l_1 + l_2}} Y_{l_1,l_1-1} Y_{l_2,l_2} + \sqrt{\frac{l_2}{l_1 + l_2}} Y_{l_1,l_1} Y_{l_2,l_2-1} \tag{17.211}$$

$$\psi_{(l_1+l_2-1, l_1+l_2-1)} = \sqrt{\frac{l_2}{l_1 + l_2}} Y_{l_1,l_1-1} Y_{l_2,l_2} - \sqrt{\frac{l_1}{l_1 + l_2}} Y_{l_1,l_1} Y_{l_2,l_2-1} \tag{17.212}$$

which are obviously mutually orthogonal.

One can proceed in a similar fashion and construct all of the (J, J_z) eigenstates in this manner, but it is also possible to derive the same states more "mechanically" by repeated use of the "total" lowering operator $\hat{J}_- = \hat{L}_{1,-} + \hat{L}_{2,-}$.

To accomplish this, we note first that one must formally have

$$\hat{J}_- \psi_{(l_1+l_2, l_1+l_2)} = \sqrt{2(l_1 + l_2)}\hbar\, \psi_{(l_1+l_2, l_1+l_2-1)} \tag{17.213}$$

since $\hat{\mathbf{J}}$ satisfies all of the usual properties of an angular momentum operator. We then also have, more explicitly,

$$\begin{aligned} \hat{J}_- \psi_{(l_1+l_2, l_1+l_2)} &= \left(\hat{L}_{1,-} + \hat{L}_{2,-}\right) Y_{l_1,l_1} Y_{l_2,l_2} \\ &= \sqrt{2l_1} Y_{l_1,l_1-1} Y_{l_2,l_2} + \sqrt{2l_2} Y_{l_1,l_1} Y_{l_2,l_2-1} \end{aligned} \tag{17.214}$$

which together with (17.213) give the form in Eqn. (17.211). The remaining combination Eqn. (17.212) can be constructed by finding the orthogonal linear combination of states.

This procedure is the basis for a systematic method of constructing the entire set of eigenfunctions of both $\hat{\mathbf{J}}^2$ and $\hat{J}_z$, namely:

- Start with the (unique) state of largest J_z as in Eqn. (17.199), which gives $\psi_{(J_{\max}, J_{\max})}$
- Use $\hat{J}_-$ on this wavefunction as in Eqns. (17.213) and (17.214) to obtain $\psi_{J_{\max}, J_{\max}-1}$; the state orthogonal to this is necessarily $\psi_{(J_{\max}-1, J_{\max}-1)}$.
- Operate on *both* these states with $\hat{J}_-$ to obtain $\psi_{(J_{\max}, J_{\max}-2)}$ and $\psi_{(J_{\max}-1, J_{\max}-2)}$ respectively; the state orthogonal to both of these must, in turn, be $\psi_{(J_{\max}-2, J_{\max}-2)}$.
- Repeat until all the $\psi_{(J, J_z)}$ states are constructed; along the way, $\hat{J}_-$ will eventually begin to annihilate states it acts on until only the state $\psi_{(J_{\max}, -J_{\max})}$ is left.

This procedure is illustrated below.

$$
\begin{array}{ccccc}
\psi_{(J_{\max},J_{\max})} & & & & \\
\downarrow \text{ via } \hat{J}_- & & & & \\
\psi_{(J_{\max},J_{\max}-1)} & \xrightarrow{\text{orthogonality}} & \psi_{(J_{\max}-1,J_{\max}-1)} & & \\
\downarrow \text{ via } \hat{J}_- & & \downarrow \text{ via } \hat{J}_- & & \\
\psi_{(J_{\max},J_{\max}-2)} & & \psi_{(J_{\max}-1,J_{\max}-2)} & \xrightarrow{\text{orthogonality}} & \psi_{(J_{\max}-2,J_{\max}-2)}
\end{array}
$$

and so forth.

Example 17.4 Two-electron Spin Wavefunctions We illustrate the utility of this procedure by deriving the spin wavefunctions of a two-electron system. If we label the spinors of the two spin-1/2 particles as $\chi_{\pm}^{(1)}$ and $\chi_{\pm}^{(2)}$, respectively, we are interested in eigenfunctions of the total spin angular momentum operator squared, $\hat{\mathbf{S}}^2 = (\hat{\mathbf{S}}_1 + \hat{\mathbf{S}}_2)^2$. Using the spin addition rules, we know that the total spin quantum number, S, will be given by

$$S_1 + S_2 = 1/2 + 1/2 \longrightarrow S = 0, 1 \tag{17.215}$$

with a total of four states; these will be labeled by (S, S_z) as $(1, +1)$, $(1, 0)$, $(1, -1)$, and $(0, 0)$.

We obviously have

$$\chi_{(1,+1)} = \chi_+^{(1)} \chi_+^{(2)} \tag{17.216}$$

as the state of highest S_z. Acting on this with the "total" lowering operator $\hat{S}_- = \hat{S}_{1,-} + \hat{S}_{2,-}$ requires the effect of the individual raising and lowering operators on single particle spinors

$$\hat{S}_- \chi_+ = \hbar \chi_- \qquad \text{and} \qquad \hat{S}_+ \chi_- = \hbar \chi_+ \tag{17.217}$$

which are consistent with Eqns. (17.192) and (17.193).

We then have

$$
\begin{aligned}
\sqrt{2}\hbar \chi_{(1,0)} &= \hat{S}_- \chi_{(1,+1)} \\
&= \left(\hat{S}_{1,-} + \hat{S}_{2,-}\right) \chi_+^{(1)} \chi_+^{(2)} \\
&= \hbar \left(\chi_+^{(1)} \chi_-^{(2)} + \chi_-^{(1)} \chi_+^{(2)}\right)
\end{aligned} \tag{17.218}
$$

or

$$\chi_{(1,0)} = \frac{1}{\sqrt{2}} \left(\chi_+^{(1)} \chi_-^{(2)} + \chi_-^{(1)} \chi_+^{(2)}\right) \tag{17.219}$$

The orthogonal combination is

$$\chi_{(0,0)} = \frac{1}{\sqrt{2}} \left(\chi_+^{(1)} \chi_-^{(2)} - \chi_-^{(1)} \chi_+^{(2)}\right) \tag{17.220}$$

Acting on $\psi_{(0,0)}$ with $\hat{S}_-$ once gives

$$
\begin{aligned}
\hat{S}_- \psi_{(0,0)} &= \left(\hat{S}_{1,-} + \hat{S}_{2,-}\right) \frac{1}{\sqrt{2}} \left(\chi_+^{(1)} \chi_-^{(2)} - \chi_-^{(1)} \chi_+^{(2)}\right) \\
&= 0
\end{aligned} \tag{17.221}
$$

as expected since $\hat{S}_- \chi_{(0,0)} \propto \chi_{(0,-1)}$, which doesn't exist. The last state is obtained using

$$
\begin{aligned}
\sqrt{2}\hbar\chi_{(1,-1)} &= \hat{S}_{-}\chi_{(1,0)} \\
&= \left(\hat{S}_{1,-} + \hat{S}_{2,-}\right)\frac{1}{\sqrt{2}}\left(\chi_{+}^{(1)}\chi_{-}^{(2)} + \chi_{-}^{(1)}\chi_{+}^{(2)}\right) \\
&= \frac{\hbar}{\sqrt{2}}2\chi_{-}^{(1)}\chi_{-}^{(2)}
\end{aligned} \tag{17.222}
$$

or $\chi_{(1,-1)} = \chi_{-}^{(1)}\chi_{-}^{(2)}$, also as expected.

We note that the Clebsch-Gordan coefficients in these spinor wavefunctions have exactly the same form as given in Eqns. (17.211) and (17.212), with $l_1 = l_2 = 1/2$. This formulation in terms of raising and lowering operators treats integral and half-integral angular momenta in the same manner. Similar results can be obtained for the addition of orbital and intrinsic angular momenta, $\mathbf{L} + \mathbf{S}$ (P17.16), which are useful in the study of single-electron atoms. The addition of more than two spins or orbital angular momenta requires only repeated use of the Clebsch-Gordan series for two angular momenta. The allowed values of J and the corresponding wavefunctions corresponding to a sum such as $L_1 + L_2 + L_3 + \cdots + L_N$ can be obtained by "adding" the first two, then "adding" the third to all possible results of the first step, and so on. This is illustrated in P17.17 and in the next example.

Example 17.5 Deuteron Wavefunction The deuteron is a bound state of two spin-1/2 particles, the neutron and proton. The ground state is known to have total angular momentum $J = 1$ and even parity, and we wish to construct the most general wavefunction that satisfies these requirements; for definiteness, we will consider $\psi_{(J,J_z)} = \psi_{(1,+1)}$. This problem requires the angular momentum or spinor wavefunctions corresponding to the angular momentum addition problem

$$
S_1 + S_2 + L = \frac{1}{2} + \frac{1}{2} + l \tag{17.223}
$$

The spin degrees of freedom can be combined as in Ex. 17.4 to give $S_{\text{tot}} = 0, 1$ with corresponding wavefunctions $\chi_{(0,0)}$ and $\chi_{(1,\pm1)}$, $\chi_{(1,0)}$.

For $l = 0$, the total angular momentum J comes from the spin degrees of freedom alone, and we have

$$
\psi_{(1,+1)}^{s} = R_s(r)Y_{(0,0)}(\theta,\phi)\chi_{(1,+1)} = R_s(r)Y_{(0,0)}(\theta,\phi)\chi_{+}^{(p)}\chi_{+}^{(n)} \tag{17.224}
$$

where $R_s(r)$ is a normalized $l = 0$ radial wavefunction.

When $l = 1$, there are two possible combinations that give $J = 1$. If $S_{\text{tot}} = 0$, then we have

$$
\psi_{(1,+1)}^{p,A} = R_p(r)Y_{1,+1}(\theta,\phi)\frac{1}{\sqrt{2}}\left(\chi_{+}^{(p)}\chi_{-}^{(n)} - \chi_{-}^{(p)}\chi_{+}^{(n)}\right) \tag{17.225}
$$

For the case of $S_{\text{tot}} + l = 1 + 1$, we can use the Clebsch-Gordan coefficients derived in Eqn. (17.212) to show that

$$
\begin{aligned}
\psi_{(1,+1)}^{p,B} &= R_p(r)\left[\frac{1}{\sqrt{2}}Y_{1,+1}(\theta,\phi)\chi_{(1,0)} - \frac{1}{\sqrt{2}}Y_{1,0}(\theta,\phi)\chi_{(1,+1)}\right] \\
&= R_p(r)\left[\frac{1}{2}Y_{1,+1}(\theta,\phi)(\chi_{+}^{(p)}\chi_{-}^{(n)} + \chi_{-}^{(p)}\chi_{+}^{(n)}) - \frac{1}{\sqrt{2}}Y_{1,0}(\theta,\phi)\chi_{+}^{(p)}\chi_{+}^{(n)}\right]
\end{aligned} \tag{17.226}
$$

We recall, however, that the parity of the spherical harmonics is given by $(-1)^l$, so that these states have odd parity and thus cannot contribute to the even parity deuteron ground state wavefunction.

The final possibility occurs when $l = 2$. In this case, $S_{tot} = 0$ cannot combine with $l = 2$ to give $J = 1$, so we require the Clebsch-Gordan series for the $(1, +1)$ state occurring in the angular momentum sum $1 + 2$. The necessary expressions are given in P17.19, and we find that

$$\psi^d_{(1,+1)} = R_d(r)\left[\sqrt{\frac{3}{5}}Y_{2,+2}(\theta,\phi)\chi^{(p)}_-\chi^{(n)}_- \right.$$
$$-\sqrt{\frac{3}{10}}Y_{2,+1}(\theta,\phi)\frac{1}{\sqrt{2}}(\chi^{(p)}_+\chi^{(n)}_- + \chi^{(p)}_-\chi^{(n)}_+)$$
$$\left. +\sqrt{\frac{1}{10}}Y_{2,0}(\theta,\phi)\chi^{(p)}_+\chi^{(n)}_+\right] \tag{17.227}$$

For $l \geq 3$, it is not possible to construct a $J = 1$ wavefunction.

The observed magnetic moment of the deuteron system provides information on the structure of the ground state wavefunction. The total magnetic moment of this composite system is due to the intrinsic magnetic moments of the neutron and proton, as well as the orbital angular momentum of the (charged) proton, but not the (neutral) neutron. The magnetic moment operator (along the z axis) can then be written as

$$M_z = \mu_N\left(\frac{1}{2}\frac{L_z}{\hbar} + \frac{g_p}{2}\sigma^{(p)}_z + \frac{g_n}{2}\sigma^{(n)}_z\right) \tag{17.228}$$

In the center-of-mass system, the angular momentum of the proton is half that of the total system; this accounts for the factor of 1/2 multiplying L_z.

In a pure S-wave state, the magnetic moment is given by

$$\langle\psi_{(1,+1)}|M_z|\psi_{1,+1}\rangle = \left[\frac{g_p + g_n}{2}\right]\mu_N = (0.8798047(7))\mu_N \tag{17.229}$$

which is only slightly different from the observed value of

$$M^{(\text{exp})}_z = 0.8574376(4)\mu_N \tag{17.230}$$

The small discrepancy is still much larger than the measured errors and can be partly explained by assuming that the ground state wavefunction has a small admixture of the d state in Eqn. (17.227), i.e.,

$$\psi^{\text{mix}}_{(1,+1)} = a_s\psi^s_{(1,+1)} + a_d\psi^d_{(1,+1)} \tag{17.231}$$

where $a_s^2 + a_d^2 = 1$. Using the explicit form of Eqn. (17.227), one can show (P17.20) that

$$\langle\psi^{\text{mix}}_{(1,+1)}|M_z|\psi^{\text{mix}}_{1,+1}\rangle/\mu_N = a_s^2\left[\frac{g_p + g_n}{2}\right] + a_d^2\left[\frac{3 - (g_p + g_n)}{4}\right] \tag{17.232}$$

The experimental value can then be explained by assuming $a_d^2 \approx 0.04$ or a roughly 4% mixture of d wave in the ground state. This calculation ignores possibly important spin-orbit interactions, relativistic effects, and meson exchanges that can also affect the magnetic moment. Information from other deuteron observables, however, including its quadrupole moment (P17.21), and from scattering experiments confirm the approximately 4% d state admixture.[6]

---

[6]The understanding of magnetic moments of nuclei in terms of nuclear models is discussed in many texts on nuclear physics, e.g., Krane (1988).

17.6 FREE PARTICLE IN SPHERICAL COORDINATES

We close this chapter with a brief discussion of the solutions of the free-particle Schrödinger equation in three dimensions in the language of spherical coordinates. They will be used extensively in our discussion of quantum scattering in Chap. 20.

The radial equation for a free particle simply reads

$$-\frac{\hbar^2}{2\mu}\left(\frac{d^2R(r)}{dr^2}+\frac{2}{r}\frac{dR(r)}{dr}\right)+\frac{l(l+1)\hbar^2}{2\mu r^2}R(r) = ER(r) \tag{17.233}$$

and the obvious change of coordinates $z = kr$, where $\hbar k = \sqrt{2mE}$ gives the dimensionless equation

$$\frac{d^2R(z)}{dz^2}+\frac{2}{z}\frac{dR(z)}{dz}+\left(1-\frac{l(l+1)}{z^2}\right)R(z) \tag{17.234}$$

There is an obvious similarity to the corresponding radial equation in two dimensions in Eqn. (16.68), and we use our experience there to simplify Eqn. (17.234). In the 2D case, the solutions for large z were of the form $\sin(z)/\sqrt{z}$, $\cos(z)/\sqrt{z}$; the factors of $\sqrt{z}$ (when squared) were understood to compensate exactly the "measure" factor $(r\,dr)$ to give a constant amplitude wavefunction, consistent with a free particle. In the 3D case, we would then expect a $1/z$ dependence at large z since $d\mathbf{r} \propto r^2\,dr$. Motivated by this connection, we look for solutions of the form $R(z) = F(z)/\sqrt{z}$ and find that the radial equation has the form

$$\frac{d^2F(z)}{dz^2}+\frac{1}{z}\frac{dF(z)}{dz}+\left(1-\frac{(l+1/2)^2}{z^2}\right)F(z) \tag{17.235}$$

This is exactly of the form of the standard (or cylindrical) Bessel differential equation (App. C.4), but for half-integral values of $m = l + 1/2$. The corresponding solutions are called *spherical Bessel functions*, and conventionally written as

$$j_l(z) = \sqrt{\frac{\pi}{2z}}J_{l+1/2}(z) \qquad \text{and} \qquad n_l(z) = \sqrt{\frac{\pi}{2z}}N_{l+1/2}(z) \tag{17.236}$$

where the $j_l(z)$ ($n_l(z)$) are called the *regular (irregular) solutions* since they are well behaved (diverge) near the origin.

The spherical Bessel functions are actually simpler in form than their two-dimensional relatives, as they can be written in closed form, namely,

$$j_l(z) = (-z)^l\left(\frac{1}{z}\frac{d}{dz}\right)^l\left(\frac{\sin(z)}{z}\right) \tag{17.237}$$

and

$$n_l(z) = -(-z)^l\left(\frac{1}{z}\frac{d}{dz}\right)^l\left(\frac{\cos(z)}{z}\right) \tag{17.238}$$

The first few functions have the explicit form given by:

$$j_0(z) = \frac{\sin(z)}{z} \qquad n_0(z) = -\frac{\cos(z)}{z} \tag{17.239}$$

$$j_1(z) = \frac{\sin(z)}{z^2} - \frac{\cos(z)}{z} \qquad n_1(z) = -\frac{\cos(z)}{z^2} - \frac{\sin(z)}{z} \tag{17.240}$$

and

$$j_2(z) = \left(\frac{3}{z^3} - \frac{1}{z}\right)\sin(z) - \frac{3}{z^2}\cos(z) \tag{17.241}$$

$$n_2(z) = -\left(\frac{3}{z^2} - \frac{1}{z}\right)\cos(z) - \frac{3}{z^2}\sin(z) \tag{17.242}$$

From general arguments (P17.22), we know that the wavefunctions must go as z^{+l} or $z^{-(l+1)}$ near the origin, and with the standard normalization one has

$$j_l(z) \sim \frac{z^l}{1 \cdot 3 \cdot 5 \cdots (2l+1)} = \frac{z^l}{(2l+1)!!} \tag{17.243}$$

$$n_l(z) \sim -\frac{1 \cdot 3 \cdot 5 \cdots (2l-1)}{z^{l+1}} = \frac{(2l-1)!!}{z^{l+1}} \tag{17.244}$$

for $z \ll l$, where $(2l+1)!! \equiv (2l+1) \cdot (2l-1) \cdots 5 \cdot 3 \cdot 1$. The asymptotic behavior for $z \gg l$ is given by

$$j_l(z) \sim \frac{1}{z}\sin\left(z - \frac{l\pi}{2}\right) \qquad n_l(z) \sim -\frac{1}{z}\cos\left(z - \frac{l\pi}{2}\right) \tag{17.245}$$

17.7 QUESTIONS AND PROBLEMS

Q17.1 A classical point particle at the end of a proverbial "massless stick pivoted at one end without friction" will rotate with constant angular velocity with a well-defined angular coordinate given by $\phi(t) = \omega t + \phi_0$. How would you construct a quantum wavepacket that gives this in the classical limit?

Q17.2 Products of momentum eigenfunctions, as in Eqn. (17.185), are also eigenfunctions of $\hat{P}^2_{\text{tot}} = (\hat{p}_1 + \hat{p}_2)^2$. Explain why the cross-terms in $\hat{P}^2$ do not "mess things up" in the way that the cross terms in $\hat{\mathbf{J}}^2 = \left(\hat{\mathbf{L}}_1 + \hat{\mathbf{L}}_2\right)^2$ do.

Q17.3 In the early days of nuclear physics, the decay of the neutron was a puzzle, as the only particles directly observed in the process were consistent with the process $n \rightarrow pe^-$ where the electron was observed with a continuous spectrum of energies. Why were people worried about the lack of conservation of angular momentum and energy? On the basis of these issues, Wolfgang Pauli predicted the existence of the massless (but difficult to observe) neutrino, so that the process is really $n \rightarrow pe^- \overline{\nu}_e$. How did his idea "rescue" the fundamental conservations laws?

P17.1 Show that the angular momentum operators $\hat{L}_x$ and $\hat{L}_y$ in spherical coordinates [Eqns. (17.16) and (17.17)] are Hermitian by confirming that, for example,

$$\langle \hat{L}_x \rangle = \int_0^{\pi} \sin(\theta) d\theta \int_0^{2\pi} d\phi \, f^*(\theta, \phi) \, \hat{L}_x \, f(\theta, \phi)$$

is real. What assumptions do you have to make about the arbitrary $f(\theta, \phi)$ you used?

P17.2 **(a)** Starting with Eqn. (17.9) and using the chain rule, show that the kinetic energy operator in spherical coordinates is given by Eqn. (17.10).

(b) Using the expressions for $\hat{L}_{x,y,z}$ in spherical coordinates, show that the total angular momentum squared in spherical coordinates is given by Eqn. (17.20).

P17.3 Derive Eqns. (17.61) and (17.62) by defining

$$\hat{L}_{\pm} |Y_{l,m}\rangle = C_{l,m}^{(\pm)} |Y_{l,m\pm 1}\rangle$$

using relations such as

$$\langle Y_{l,m} | \hat{L}_+ \hat{L}_- | Y_{l,m} \rangle = \langle \hat{L}_- Y_{l,m} | \hat{L}_- Y_{l,m} \rangle$$

and Eqns. (17.51) and (17.52).

P17.4 **(a)** Evaluate the expectation value of $\hat{L}_{x,y}$ and $\hat{L}^2_{x,y}$ in a state described by a spherical harmonic; e.g., calculate

$$\langle Y_{l,m} | \hat{L}_{x,y} | Y_{l,m} \rangle$$

and so forth.

(b) Assume that you have a general angular wavefunction of the form $\psi = \sum_{l,m} a_{l,m} |Y_{l,m}\rangle$. Express the expectation values of $\hat{L}_x$ and $\hat{L}_y$ in terms of the $a_{l,m}$. *Hint:* Use the definitions of raising and lowering operators and Eqns. (17.61) and (17.62).

P17.5 **(a)** Calculate the following commutators:

$$[\hat{L}_y, x] \qquad [\hat{L}_y, y] \qquad [\hat{L}_y, z]$$

and

$$[\hat{L}_y, \hat{p}_x] \qquad [\hat{L}_y, \hat{p}_y] \qquad [\hat{L}_y, \hat{p}_z]$$

(b) Using these results, show that if $|\psi\rangle$ is an eigenstate of $\hat{L}_y$, then $\langle \psi | \hat{p}_x | \psi \rangle$ and $\langle \psi | \hat{p}_z | \psi \rangle$ both vanish. Similarly show that $\langle \psi | x | \psi \rangle$ and $\langle \psi | z | \psi \rangle$ vanish.

P17.6 Using Eqn. (17.72), show that the angular probability density corresponding to $P_{l,l}(\theta) = |Y_{l,l}(\theta, \phi)|^2$ satisfies

$$P_{l,l}(\theta) \propto \sqrt{l} \sin^{2l}(\theta)$$

Hint: Use Stirling's approximate formula (App. C.9) for the factorial function, namely,

$$x! \sim e^{-x} x^x \sqrt{2\pi x} \left(1 + \mathcal{O}\left(\frac{1}{x} \right) \right)$$

which is useful for $x \gg 1$.

P17.7 An angular wavefunction is given by

$$f(\theta, \phi) = N \left[1 + \sin(\phi) \sin(\theta) \right]$$

(a) Find N such that $f(\theta, \phi)$ is properly normalized.

(b) Find the expansion of $f(\theta, \phi)$ in spherical harmonics.

(c) For a particle described by $f(\theta, \phi)$, what are $\langle \hat{L}_z \rangle$, ΔL_z, $\langle \mathbf{L}^2 \rangle$?

(d) What are the probabilities of finding $L_z = -\hbar$? $L_z = +2\hbar$?

P17.8 Show that the expectation value of θ in the state $Y_{l,l}(\theta, \phi)$ is identically equal to $\pi/2$, independent of l. If you have the ability to perform numerical integrations (for example, by using a multipurpose mathematical language such as *Mathematica*®), evaluate

$$(\Delta\theta)^2 = \left\langle \left(\theta - \frac{\pi}{2}\right)^2 \right\rangle = \int d\Omega |Y_{l,l}(\theta,\phi)|^2 \left(\theta - \frac{\pi}{2}\right)^2$$

and show that for large l it approaches $\Delta\theta = 1/\sqrt{2l}$.

P17.9 **(a)** Show that the quadrupole moment of a particle of charge e in a classical planar circular orbit of radius R is given by $Q = -eR^2$.

(b) For the same particle described by the spherical harmonic $Y_{l,l}(\theta,\phi)$, show that

$$Q = -e\left(\frac{2l}{2l+3}\right)\langle r^2 \rangle$$

and comment on the approach to the classical limit.

P17.10 **Nuclear rotational levels:** Quantized rotational energy levels are observed in heavy nuclei. For nuclei with even numbers of both neutrons and protons (so-called *even-even nuclei*), the nuclear shape is symmetric with respect to a reflection in the origin, and this restricts the allowed quantum numbers for rotation just as for molecules with identical nuclei.

(a) Since the nuclear orientation, described by (θ,ϕ), is only defined up to a rotation by 180°, show that the fact that the rotational wavefunction should be single valued implies that only *even* values of the angular momentum are allowed, i.e., $J = 0, 2, 4, \ldots$ and so forth.

(b) Data[7] for one such system (^{160}Dy) is given here:

J	E, KeV
0	0
2	86.8
4	283.8
6	581.2
8	967.2

Show that the data are approximately described by the formula

$$E_J = \frac{\hbar^2}{2I} J(J+1)$$

(c) Estimate the numerical value of the nuclear moment of inertia, I, and compare it to the classical value for a rigid, rotating sphere of uniform density, namely,

$$I = \frac{2}{5} MR^2$$

where $M = 160$ amu and $R(A) \approx (1.2 - 1.4)A^{1/3}$ F. Comment on your result.

(d) What would the classical angular frequency, ω, of the $J = 8$ state be?

(e) Try to fit the data to the form of a "spring rotator" in Eqn. (17.112) by plotting $E/J(J+1)$ versus $J(J+1)$.

P17.11 **(a)** Verify that the 3×3 matrix representation of the angular momentum operators in Eqns. (17.140) to (17.142) satisfy the commutation relation $[\mathrm{L}_x, \mathrm{L}_y] = i\hbar \mathrm{L}_z$.

(b) Show that the eigenvalues of L_ϕ in Ex. 17.2 are $L_\phi = +\hbar, 0, -\hbar$.

[7] From Johnson, Ryde, and Hjorthh (1972).

P17.12 **(a)** The magnetic moment of a planar current loop is given by $|\mathbf{M}| = IA$, where I is the current and A is the area enclosed. For a charged particle undergoing classical uniform circular motion, show that $|\mathbf{M}| = qrv/2$. Since the orbital angular momentum is $|\mathbf{L}| = mvr$, show that the classical gyromagnetic ratio is given by Eqn. (17.168) with $g = 1$.

(b) A particle moving in an inverse-square law force field can also have elliptical orbits. Calculate the gyromagnetic ratio for any such orbit, and show that $g = 1$. *Hint:* Use the classical relations discussed in Sec. 18.2. The area of an ellipse is $A = \pi ab$, where a, b are the semimajor and semiminor axes.

(c) Show that you obtain the same result for g for a charged disk with uniform charge and mass density rotating about its center, as well as for a uniform charged sphere. *Hint:* Consider the disk as lots of little loops, and the sphere as lots of little disks.

P17.13 **Precession of a magnetic dipole in a uniform field:**

(a) Consider a magnetic moment given by

$$\mathbf{M} = \frac{q}{2m}\mathbf{L}$$

described by the Hamiltonian $\hat{H} = -\mathbf{M}\cdot\mathbf{B}$, with $\mathbf{B}$ a uniform field. Use the quantum mechanical evolution equation for expectation values

$$\frac{d}{dt}\langle\hat{\mathbf{L}}\rangle_t = \frac{i}{\hbar}\langle[\hat{H}, \hat{\mathbf{L}}]\rangle_t$$

to show that

$$\frac{d}{dt}\langle\hat{\mathbf{L}}\rangle = \left(\frac{q}{2m}\right)\langle\hat{\mathbf{L}}\rangle_t \times \mathbf{B} = \langle\mathbf{M}\rangle_t \times \mathbf{B}$$

and show that this reproduces the result of Ex. 17.3. Discuss the connection with the classical equation of motion for the precession of a magnetic dipole.

(b) Consider the time development of $\langle\hat{\mathbf{L}}\cdot\hat{\mathbf{L}}\rangle_t$ in a similar manner, and show that it is a constant. *Hint:* Generalize the result in P13.14 for the expectation values of products of operators.

P17.14 **Conservation of probability for precessing spinors:**

(a) Calculate the Hamiltonian, $\hat{H}$, for the interaction of a spin-1/2 particle in a (possibly time-dependent) magnetic field with arbitrary direction,

$$\mathbf{B}(t) = (B_x(t), B_y(t), B_z(t))$$

and show that $\hat{H}$ is Hermitian.

(b) Use the Schrödinger equation,

$$i\hbar\frac{\partial}{\partial t}\chi(t) = \hat{H}\chi(t)$$

and its complex conjugate equation to show that $|\chi(t)|^2 = |\alpha_+(t)|^2 + |\alpha_-(t)|^2$ is constant in time so that probability is conserved.

P17.15 Explicitly confirm the summation in Eqn. (17.197) using the results of App. B.2.

P17.16 Consider the angular momentum "addition" problem of a spin $S = 1/2$ particle combining with orbital angular momentum L. Using the methods in Sec. 17.5, show explicitly that the allowed values of J are $l + 1/2$ and $l - 1/2$. Do this by writing

$$\psi_{(J,J_z)} = \alpha Y_{l,m}(\theta,\phi)\chi_+ + \beta Y_{l,m+1}(\theta,\phi)\chi_-$$

and finding the values for J that give

$$\hat{\mathbf{J}}^2\psi_{(J,J_z)} = \left(\hat{\mathbf{L}} + \hat{\mathbf{S}}\right)^2\psi_{(J,J_z)} = J(J+1)\hbar^2\,\psi_{(J,J_z)}$$

Also find the Clebsch-Gordan coefficients for both solutions, and show, for example, that

$$\psi_{(l+1/2,m+1/2)} = \sqrt{\frac{l+m+1}{2l+1}}Y_{(l,m)}\chi_+ + \sqrt{\frac{l-m}{2l+1}}Y_{(l,m+1)}\chi_-$$

and find an expression for $\psi_{(l-1/2,m+1/2)}$.

P17.17 Addition of three spins: Consider the product wavefunction of three spin-1/2 particles, i.e.,

$$\chi_{(S,S_z)} = \sum_{\pm} C\left[S, S_z; \frac{1}{2}, \pm\frac{1}{2}; \frac{1}{2}, \pm\frac{1}{2}; \frac{1}{2}, \pm\frac{1}{2}\right]\chi_\pm^{(1)}\chi_\pm^{(2)}\chi_\pm^{(3)}$$

where the summation is over all of the $2\times2\times2 = 8$ possible values of S_z. This corresponds to the problem of "adding" three spins. Use the spin addition rules to show that

$$\frac{1}{2}+\frac{1}{2}+\frac{1}{2} \longrightarrow (0,1)+\frac{1}{2} \longrightarrow \left.\frac{1}{2}\right|_A \quad \text{or} \quad \left.\frac{1}{2}\right|_B \quad \text{or} \quad \frac{3}{2}$$

where $1/2|_{A,B}$ are two distinct spin $-1/2$ states. Find the product wavefunctions corresponding to the possible values of S. Do this in two steps by "adding" the first two spins to find the possible $S = 0, 1$ wavefunctions, and then adding the third spin-1/2. Verify that the wavefunctions are mutually orthogonal and that the action of the raising and lowering operators $\mathbf{S}_{\pm,\text{tot}} = \mathbf{S}_{\pm,1} + \mathbf{S}_{\pm,2} + \mathbf{S}_{\pm,3}$ moves one up and down in S_z appropriately in a state of given S.

P17.18 Neutron and proton spin in the quark model: an exercise with spinors: The nonpointlike values of the gyromagnetic ratios for the neutron (n) and proton (p) is one piece of evidence that they are composite objects. A simple version of the *quark model* states that they can each be thought of as bound states of more fundamental, pointlike, spin-1/2 objects called *quarks;* the *up-quark* with $Q_u = +2/3$ (in terms of the basic charge e) and the *down-quark* with $Q_d = -1/3$ with a common mass m_q. The proton is taken to be a $p = (u_1u_2d)$ bound state, where $u_{1,2}$ are two different u-type quarks, while $N = (ud_1d_2)$. This model can be used to understand the values of the n, p magnetic moments as follows. The magnetic moment operator for the composite system can be written as

$$M_{z,\text{tot}} = \sum_{i=u,d}\left(\frac{Q_i e\hbar}{2m_q}\right)\sigma_z^{(i)}$$

The spin-dependent part of the proton wavefunction can be written as

$$\chi_+^{(p)} = \frac{1}{\sqrt{6}}\left(\chi_+^{(u_1)}\chi_-^{(u_2)}\chi_+^{(d)} + \chi_-^{(u_1)}\chi_+^{(u_2)}\chi_+^{(d)} - 2\chi_+^{(u_1)}\chi_+^{(u_2)}\chi_-^{(d)}\right)$$

Evaluate the expectation value of the magnetic moment of the proton by showing that

$$\langle\chi_+^{(p)}|M_{z,\text{tot}}|\chi_+^{(p)}\rangle = \left(\frac{e\hbar}{2m_q}\right)\langle\chi_+^{(p)}|Q_u\sigma_z^{(u_1)} + Q_u\sigma_z^{(u_2)} + Q_d\sigma_z^{(d)}|\chi_+^{(p)}\rangle$$

$$= \frac{e\hbar}{2m_q}\frac{1}{3}(4Q_u - Q_d)$$

The corresponding result for the neutron is obtained by letting $u \leftrightarrow d$. Show that this implies that

$$\frac{\mu_n}{\mu_p} = -\frac{2}{3}$$

and compare this to the experimental values

$$\frac{\mu_n}{\mu_p} = \frac{-1.913}{2.793} = -0.685$$

P17.19 Using the methods outlined in Sec. 17.5, calculate the Clebsch-Gordan coefficients for $J_z = J_{max} - 2 = l_1 + l_2 - 2$.

(a) For $J_z = l_1 + l_2$ and $J_z = l_1 + l_2 - 1$, act on Eqns. (17.211) and (17.212) with the lowering operator.

(b) For $J = l_1 + l_2 - 2$, use orthogonality, and show that your result can be written in the form

$$\psi_{(l_1+l_2-2,l_1+l_2-2)} = \alpha Y_{l_1,l_1-2}Y_{l_2,l_2} + \beta Y_{l_1,l_1-1}Y_{l_2,l_2-1} + \gamma Y_{l_1,l_1}Y_{l_2,l_2-2}$$

where

$$\alpha = \sqrt{\frac{l_2(2l_2+1)}{(l_1+l_2-1)(2l_1+2l_2-1)}}$$

$$\beta = -\sqrt{\frac{(2l_1-1)(2l_2-1)}{(l_1+l_2-1)(2l_1+2l_2-1)}}$$

$$\gamma = \sqrt{\frac{l_1(2l_1+1)}{(l_1+l_2-1)(2l_1+2l_2-1)}}$$

P17.20 **(a)** Using the wavefunction in Eqn. (17.231), show that the matrix element of the magnetic moment operator, Eqn. (17.228), gives the result in Eqn. (17.232).

(b) Show that $a_d^2 \approx 0.04$ reproduces the observed magnetic moment.

P17.21 **Quadrupole moment of the deuteron:** The quadrupole moment of the deuteron can be defined via

$$eQ = \frac{1}{4}e\int d\mathbf{r}|\psi(\mathbf{r})|^2(3z^2 - r^2)$$

where the "extra" factor of 1/4 arises from the fact that only the proton contributes to the charge distribution. Write this as

$$Q = \frac{1}{4}\int d\mathbf{r}\, r^2|\psi^{\text{mix}}(\mathbf{r})|^2(3\cos^2(\theta) - 1)$$

and use the explicit deuteron wavefunction in Eqn. (17.231) to show that

$$Q = a_s a_d \frac{\sqrt{2}}{10}\langle r^2\rangle_{sd} - a_d^2\frac{1}{20}\langle r^2\rangle_{dd}$$

where

$$\langle r^2\rangle_{sd} = \langle R_s|r^2|R_d\rangle \qquad \text{and} \qquad \langle r^2\rangle_{dd} = \langle R_d|r^2|R_d\rangle$$

Explain how this result could be used to check the magnitude and *sign* of a_d.

P17.22 Find the short-distance behavior of the 3D radial equation, Eqn. (17.233), by assuming a solution of the form $R(r) = r^\alpha G(r)$, and show that $\alpha = +l$ and $-(l+1)$.

P17.23 Obtain a variational estimate of the ground state energy of a particle of mass m in a spherical infinite well of radius R, and compare it to the exact answer. *Hint:* Use a trial solution of the form $R(r;a) = N(R^a - r^a)$.

P17.24 Solve for the energy eigenvalues and wavefunctions for a particle of mass m in a three-dimensional square box of length L on each side. Generalize to a rectangular parallelepiped of sides L_1, L_2, L_3. Discuss the degeneracies, accidental or otherwise, for various cases.

P17.25 **The harmonic oscillator in 3D:** The isotropic harmonic oscillator in three dimensions is defined by the potential

$$V(\mathbf{r}) = \frac{1}{2}m\omega^2\mathbf{r}^2 = \frac{1}{2}m\omega^2(x^2 + y^2 + z^2)$$

(a) Solve the energy eigenvalues and eigenfunctions using Cartesian coordinates. Show that the energies can be written as

$$E_{n_x,n_y,n_z} = \left(n_x + n_y + n_z + \frac{3}{2}\right)\hbar\omega = \left(N + \frac{3}{2}\right)\hbar\omega = E_N$$

Show that the number of states with energy E_N (i.e., the degeneracy) is given by $(N+1)(N+2)/2$.

(b) Solve this problem using spherical coordinates, making use of the same kind of variable changes used in the 1D and 2D oscillators. Show that the same energy spectrum is obtained, which can be written in the form

$$E_{n_r,l} = \left(2n_r + l + \frac{3}{2}\right)\hbar\omega$$

where n_r counts the number of radial nodes. Compare the degeneracy of each level in this notation to that found in part (a). Show that the wavefunctions can be written in the form

$$\psi(r,\theta,\phi) \propto y^l e^{-y^2/2} L_{n_r}^{l+1/2}(y^2) Y_{l,m}(\theta,\phi)$$

where $r = \rho y$ and $\rho^2 = \hbar/\mu\omega$. Show that the wavefunctions for the first few lowest-lying levels in the two solution schemes (Cartesian versus spherical) are linear combinations of each other.

CHAPTER 18

The Hydrogen Atom

One of the first great triumphs of classical mechanics was the analysis of the motion of two bodies (earth-sun, moon-earth, or, perhaps, even earth-apple) moving under the influence of the inverse-square law of gravitation using the then-new tool of calculus. The use of sophisticated computational methods later allowed for the precise calculation of the trajectories of a collection of particles, the solar system. The predictions obtained became so accurate that observed deviations from theory could confidently be used as a signal of new physics phenomena; examples include the prediction of new planets[1] or the need for general relativistic effects as inferred by Einstein.[2]

The situation is somewhat similar in quantum mechanics. The study of the hydrogen atom system, consisting of an electron and proton interacting (to first approximation) only via their Coulomb attraction, was one of the first successes of both the early quantum ideas of Bohr, and later of the complete machinery of wave and matrix mechanics. Extensions to include other effects, such as relativity and spin, as well as the description of multielectron atoms and the resulting understanding of the periodic table, constitute one of the triumphs of theoretical physics in this century.

In this chapter, we focus on some of the simplest aspects of hydrogen and hydrogen-like atoms, with a brief discussion of some of the complications arising from more complex atomic systems, using helium and lithium as examples.

18.1 HYDROGEN ATOM WAVEFUNCTIONS AND ENERGIES

We begin by considering hydrogen-like atoms consisting of a single electron and a point-like, positively charged nucleus of mass M and charge $+Ze$. This includes not only hydrogen itself (where $Z = 1$) but also other atomic states such as singly ionized helium,

[1]Adams and Leverrier independently deduced the mass and orbital parameters of Neptune using Newtonian celestial mechanics.
[2]See, e.g., Weinberg (1972).

He^+ ($Z = 2$), doubly ionized lithium, Li^{++} ($Z = 3$), and so forth. The Hamiltonian for the two-particle system is

$$\hat{H} = \frac{\hat{\mathbf{p}}_e^2}{2m_e} + \frac{\hat{\mathbf{p}}_N^2}{2M} - \frac{KZe^2}{|\mathbf{r}_e - \mathbf{r}_N|} \tag{18.1}$$

The usual change to center of mass and relative coordinates produces the Hamiltonian

$$\hat{H} = \frac{\hat{\mathbf{p}}^2}{2\mu} - \frac{KZe^2}{r} \tag{18.2}$$

where $\mathbf{r} = \mathbf{r}_e - \mathbf{r}_N$ is the relative coordinate, $\mathbf{p}$ is its corresponding momentum operator, and $\mu = m_e M/(m_e + M)$ is the reduced mass.

Just as with the harmonic oscillator, the energy spectrum and wavefunctions for the hydrogen atom problem can be derived using operator methods,[3] but we will follow a straightforward differential equation approach. A solution in spherical coordinates of the standard form, $\psi(\mathbf{r}) = R(r)Y_{l,m}(\theta, \phi)$, then gives the radial equation

$$\frac{d^2R(r)}{dr^2} + \frac{2}{r}\frac{dR(r)}{dr} + \frac{2\mu}{\hbar^2}\left(E + \frac{KZe^2}{r} - \frac{l(l+1)\hbar^2}{2\mu r^2}\right)R(r) = 0 \tag{18.3}$$

where we write $E = -|E|$ when looking for negative energy bound states. The method of solution is then a standard one for which we only present the "highlights." A suitable change of variables,

$$\rho = \left(\frac{8\mu|E|}{\hbar^2}\right)^{1/2} r \tag{18.4}$$

reduces Eqn. (18.3) to dimensionless form

$$\frac{d^2R}{d\rho^2} + \frac{2}{r}\frac{dR}{d\rho} + \left(\frac{\sigma}{\rho} - \frac{l(l+1)}{\rho^2} - \frac{1}{4}\right)R = 0 \tag{18.5}$$

The parameter σ is given by

$$\sigma = \frac{KZe^2}{\hbar^2}\left(\frac{\mu}{2|E|}\right)^{1/2} = Z\alpha\left(\frac{\mu c^2}{2|E|}\right)^{1/2} \tag{18.6}$$

where, for convenience, we use the *fine structure constant,* $\alpha \equiv Ke^2/\hbar c$, as introduced in Ex. 1.1. We note that this change of variable depends on the value of the quantized energy, $|E_n|$, and so will vary with quantum number. The behavior of $R(\rho)$ for both large and small ρ is easily extracted (P18.1), giving $R(\rho) \longrightarrow e^{-\rho/2}$ and $R(\rho) \longrightarrow \rho^l$ in those two limits. These can be incorporated by writing

$$R(\rho) = G(\rho)\rho^l e^{-\rho/2} \tag{18.7}$$

to obtain

$$\frac{d^2G(\rho)}{d\rho^2} + \frac{dG(\rho)}{d\rho}\left(\frac{2l+2}{\rho} - 1\right) + G(\rho)\left(\frac{\sigma - l - 1}{\rho}\right) = 0 \tag{18.8}$$

A series solution to Eqn. (18.8) of the form $G(\rho) = \sum_{s=0}^{\infty} a^s \rho_s$ yields the recursion relation

$$\frac{a_{s+1}}{a_s} = \frac{s - (\sigma - l - 1)}{(s+1)(s+2l+1)} \longrightarrow \frac{1}{s} \Longrightarrow G(\rho) \longrightarrow e^{\rho} \tag{18.9}$$

[3]See, e.g., Ohanian (1990) for such a derivation at the level of this text.

unless the series terminates. The integral value of s for which the solutions of Eqn. (18.8) are then well behaved, s_{max}, counts the number of radial nodes, so we find

$$s_{max} \equiv n_r = \sigma - l - 1 \tag{18.10}$$

or

$$\sigma = n_r + l + 1 \equiv n \tag{18.11}$$

where n is called the *principal quantum number;* it satisfies $n = 1, 2, \ldots$ and $n \geq l + 1$.

From Eqn. (18.6), we see that the energies are quantized in terms of n since

$$E_n = -\frac{1}{2}\mu c^2 (Z\alpha)^2 \frac{1}{n^2} \approx -Z^2 \frac{13.6\,\text{eV}}{n^2} \tag{18.12}$$

when $M >> m_e$. This form is useful because:

- It shows how the Balmer formula for the spectral lines for hydrogen-like atoms arises. The wavelengths of the photons emitted in transitions from one state to another are given by

$$\hbar c \frac{2\pi}{\lambda} = \hbar\omega = E_\gamma = E_m - E_n \tag{18.13}$$

or

$$\frac{1}{\lambda} = R_\infty \left(1 + \frac{m_e}{M}\right)^{-1} Z^2 \left(\frac{1}{n^2} - \frac{1}{m^2}\right) \tag{18.14}$$

This expression gives the so-called *Rydberg constant*

$$R_\infty = \frac{m_e c \alpha^2}{4\pi\hbar} \approx 1.097 \times 10^7\,\text{m}^{-1} \approx \frac{1}{911}\,\text{Å}^{-1} \tag{18.15}$$

in terms of fundamental constants of nature; Eqn. (18.14) also explicitly exhibits the dependence on the nuclear charge and mass.

- When written as

$$E_n = -\frac{1}{2}\mu \left(\frac{Z\alpha c}{n}\right)^2 = -\frac{1}{2}\mu v^2 \tag{18.16}$$

it shows that the magnitude of the typical velocity is given by $v \sim c(Z\alpha/n)$ which, in turn, sets the scale for when relativistic effects become important.

The energy spectrum is illustrated in Fig. 18.1 in such a way as to emphasize that the energy levels do not depend on the angular momentum quantum number l (as, for example, in Fig. 16.10) but only on n.

The radial wavefunctions are written in terms of the variable ρ, which we can express in the form

$$\rho = \left(\frac{8\mu|E|}{\hbar^2}\right)^{1/2} r = \left(\frac{2Z}{na_0}\right) r \tag{18.17}$$

where we have defined the *Bohr radius* via

$$a_0 = \frac{\hbar^2}{m_e K e^2} = \frac{\hbar}{m_e c \alpha} \approx 0.529\,\text{Å} \tag{18.18}$$

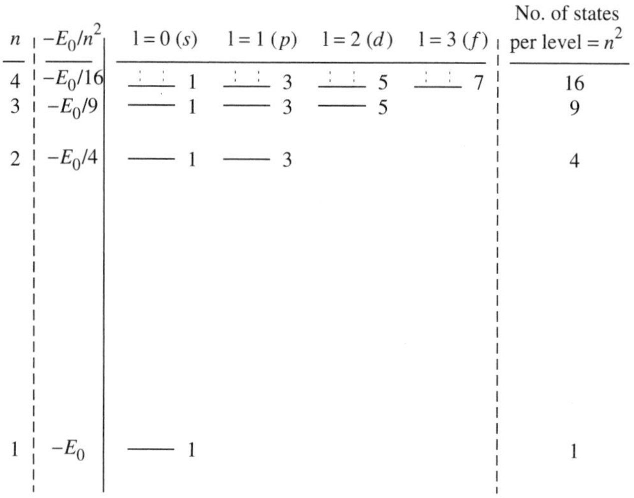

No. of states
n
$-E_0/n^2$
l = 0 (s)
l = 1 (p)
l = 2 (d)
l = 3 (f)
per level = n^2
4
$-E_0/16$
1
3
5
7
16
3
$-E_0/9$
1
3
5
9
2
$-E_0/4$
1
3
4
1
$-E_0$
1
1

FIGURE 18.1. *Part of the energy spectrum for hydrogen-like atoms; the energies correspond to* $-E_0 = -\mu c^2(Z\alpha)^2/2$. *Standard spectroscopic notation is used for the various angular momentum states, and the degeneracies of each level are also shown.*

which is defined for an infinitely heavy nucleus ($\mu \rightarrow m_e$); this length sets the scale for many atomic physics problems. Finally, this implies that the quantized energies can also be written in the form

$$E_n = -\frac{1}{2}\frac{Ke^2}{a_0}\frac{Z^2}{n^2} \tag{18.19}$$

which is sometimes useful, as it emphasizes the electrostatic origin of the binding energy. Note that the expressions for the energies in Eqns. (18.12) and (18.19), and the radial variable (18.17), depend only on the combination Z/n.

l = 0
n = 1, 2, 3
$|rR_{10}(r)|^2$
$|rR_{20}(r)|^2$
$|rR_{30}(r)|^2$
l = 1
n = 2, 3
$|rR_{21}(r)|^2$
$|rR_{31}(r)|^2$
l = 2
n = 3
$|rR_{32}(r)|^2$
0
5
10
15
20
r/a_0

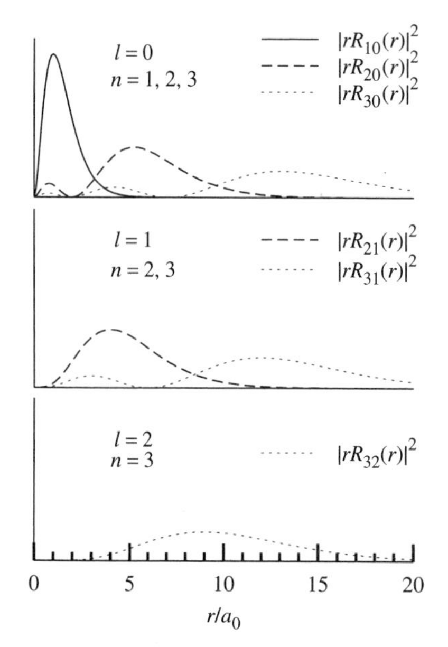

FIGURE 18.2. *Radial wavefunctions for the hydrogen atom where* $r^2|R_{n,l}(r)|^2$ *versus* r/a_0 *is plotted;* $n = 1, 2, 3$ *states are shown by solid, dashed, and dotted lines, respectively.*

Comparison of Eqn. (18.8) with standard forms (as in App. C.7) shows that the solutions $G(\rho)$ can be expressed in terms of Laguerre polynomials; this gives for the radial wavefunction

$$R(r) = R_{n,l}(\rho) \propto \rho^l L_{n-l-1}^{2l+1}(\rho) e^{-\rho/2} \tag{18.20}$$

The properly normalized radial wavefunctions for the lowest-lying states (along with the traditional spectroscopic label) are:

$$1s: \quad R_{1,0}(r) = 2\left(\frac{Z}{a_0}\right)^{3/2} e^{-Zr/a_0} \tag{18.21}$$

$$2s: \quad R_{2,0}(r) = 2\left(\frac{Z}{2a_0}\right)^{3/2} \left[1 - \left(\frac{Zr}{2a_0}\right)\right] e^{-Zr/2a_0} \tag{18.22}$$

$$2p: \quad R_{2,1}(r) = \frac{1}{\sqrt{3}}\left(\frac{Z}{a_0}\right)^{3/2} \frac{Zr}{a_0} e^{-Zr/2a_0} \tag{18.23}$$

$$3s: \quad R_{3,0}(r) = 2\left(\frac{Z}{3a_0}\right)^{3/2} \left[1 - 2\left(\frac{Zr}{3a_0}\right) + 2\left(\frac{Zr}{3a_0}\right)^2\right] e^{-Zr/3a_0} \tag{18.24}$$

$$3p: \quad R_{3,1}(r) = \frac{4\sqrt{2}}{9}\left(\frac{Z}{3a_0}\right)^{3/2} \left(\frac{Zr}{a_0}\right)\left[1 - \frac{1}{2}\left(\frac{Zr}{3a_0}\right)\right] e^{-Zr/3a_0} \tag{18.25}$$

$$3d: \quad R_{3,2}(r) = \frac{2\sqrt{2}}{27\sqrt{5}}\left(\frac{Z}{3a_0}\right)^{3/2} \left(\frac{Zr}{a_0}\right)^2 e^{-Zr/3a_0} \tag{18.26}$$

We plot these in Fig. 18.2. The general form for the special case $R_{n,n-1}(r)$ is easy to derive (P18.3) and is given by

$$R_{n,n-1}(r) = \left[\left(\frac{2Z}{na_0}\right)^{2n+1} \frac{1}{(2n)!}\right]^{1/2} r^{n-1} e^{-Zr/na_0} \tag{18.27}$$

It will be useful to have at hand expressions for various expectation values for the radial variable,

$$\langle r^k \rangle = \int d\mathbf{r}\, r^k |\psi(\mathbf{r})|^2 = \int_0^\infty r^{k+2} |R_{n,l}(r)|^2 \, dr \tag{18.28}$$

which we collect here without proof:

$$\langle r \rangle = \frac{1}{2}\left(\frac{a_0}{Z}\right)\left[3n^2 - l(l+1)\right] \tag{18.29}$$

$$\langle r^2 \rangle = \frac{1}{2}\left(\frac{a_0 n}{Z}\right)^2 \left[5n^2 + 1 - 3l(l+1)\right] \tag{18.30}$$

$$\left\langle \frac{1}{r} \right\rangle = \frac{1}{n^2}\left(\frac{Z}{a_0}\right) \tag{18.31}$$

$$\left\langle \frac{1}{r^2} \right\rangle = \left(\frac{Z}{a_0}\right)^2 \left[\frac{1}{n^3(l+1/2)}\right] \tag{18.32}$$

$$\left\langle \frac{1}{r^3} \right\rangle = \left(\frac{Z}{a_0} \right)^3 \left[\frac{1}{n^3 l(l+1)(l+1/2)(l+1)} \right] \tag{18.33}$$

$$\left\langle \frac{1}{r^4} \right\rangle = \left(\frac{Z}{a_0} \right)^4 \left[\frac{3n^2 - l(l+1)}{2n^5(l-1/2)l(l+1/2)(l+1)(l+3/2)} \right] \tag{18.34}$$

These solutions form a complete and orthonormal set of states in terms of which any (properly behaved) 3D wavefunction can be expanded, namely,

$$\psi(r, \theta, \phi) = \sum_{n=1}^{\infty} \sum_{l=0}^{n-1} \sum_{m=-l}^{+l} a_{n,l,m} R_{n,l}(r) Y_{l,m}(\theta, \phi) \tag{18.35}$$

as well as

$$\begin{aligned} \langle \psi_{n,l,m} \mid \psi_{n',l',m'} \rangle &= \left(\int_0^{\infty} r^2 R^*_{n,l}(r) R_{n',l'}(r)\, dr \right) \left(\int d\Omega\, Y^*_{l,m}(\theta, \phi) Y_{l',m'}(\theta, \phi) \right) \\ &= \delta_{n,n'}\, \delta_{l,l'}\, \delta_{m,m'} \end{aligned} \tag{18.36}$$

18.2 THE CLASSICAL LIMIT OF THE QUANTUM KEPLER PROBLEM

The motion of a particle under the influence of an inverse-square force law is one of the best studied of all classical mechanics problems.[4] We briefly review here some aspects of the solutions in the context of the hydrogen atom problem, using $V(r) = Ke^2/r$; the scaling with Z for ions is easily obtained.

The trajectories for bound state motion (i.e., for $E < 0$) consist of elliptical paths as in Fig. 18.3 with the center of mass at one focus (labeled F); circular orbits are a special case. (The case of unbound motion where $E > 0$, which is of relevance to scattering problems, gives rise to hyperbolic orbits that can be studied in a similar fashion.) The semimajor and semiminor axes, a and b, are given in terms of the energy and angular momentum via

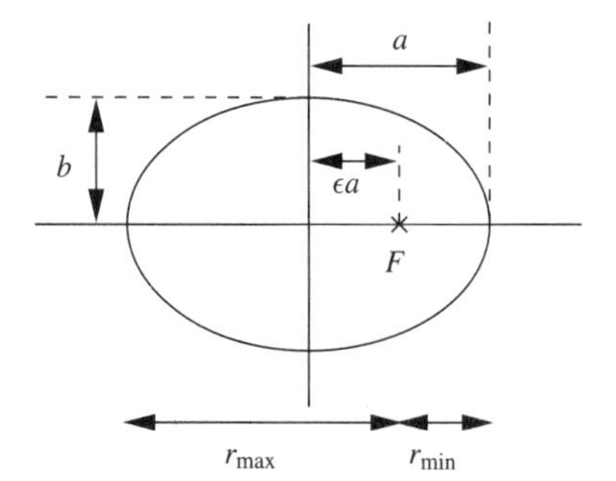

FIGURE 18.3. *Definition of parameters of elliptical orbits.*

[4]For a concise treatment, see Barger and Olsson (1995).

$$|E| = \frac{Ke^2}{2a} = \frac{\mathbf{L}^2}{2\mu b^2} \tag{18.37}$$

where μ is the reduced mass. They are also related by

$$b = a\sqrt{1-\epsilon^2} \qquad \text{or} \qquad \epsilon^2 = 1 - \frac{b^2}{a^2} = 1 - \frac{2\mathbf{L}^2|E|}{\mu(Ke^2)^2} \tag{18.38}$$

where ϵ is the eccentricity of the elliptical orbit; $\epsilon = 0$ corresponds to circular orbits, while $\epsilon \to 1$ gives purely radial motion with no angular momentum.

The equation determining the elliptical orbit can be written in the form

$$r(\theta) = \frac{\alpha}{1 + \epsilon\cos(\theta)} \qquad \text{where} \qquad \alpha = \frac{\mathbf{L}^2}{\mu K e^2} = a(1-\epsilon^2) \tag{18.39}$$

Using this expression, one finds that the distances of closest and furthest approach[5] are given by

$$r_{\text{min}} = a(1-\epsilon) \qquad \text{and} \qquad r_{\text{max}} = a(1+\epsilon) \tag{18.40}$$

The period of the orbit, τ, is given in terms of the semimajor axis via *Kepler's third law,* namely,

$$\tau^2 = \left(\frac{4\pi^2\mu}{Ke^2}\right)a^3 \tag{18.41}$$

We can now use the corresponding quantum results, namely,

$$E = E_n = -\frac{Ke^2}{2a_0}\frac{1}{n^2} \qquad \text{and} \qquad \mathbf{L}^2 = l(l+1)\hbar^2 \tag{18.42}$$

to see that $a = a_0 n^2$ is the semimajor axis. The facts that (1) the classical energy for the Kepler problem depends only on the value of the semimajor axis a (via Eqn. (18.37)), while (2) the quantum energy depends only on n, are obviously related.

The "effective potential" in which the electron moves is then

$$V_{\mathit{eff}}(r) = -\frac{Ke^2}{r} + \frac{l(l+1)\hbar^2}{2\mu r} \tag{18.43}$$

is shown in Fig.18.4 for (n, l) equal to (20, 19), (20, 15), and (20, 0) to illustrate the classical turning points of the motion.

We can use Eqns. (18.37) and (18.38) to show that

$$\epsilon^2 = 1 - \frac{l(l+1)}{n^2} \tag{18.44}$$

Classical circular motion corresponds to no nodes in the radial wavefunction ($n_r = 0$), which gives $l = n - 1$ or

$$\epsilon^2 = \frac{1}{n} \tag{18.45}$$

so that truly circular orbits are achieved only for $n \to \infty$. The corresponding quantum wavefunctions are given by Eqn. (18.27) as $R_{n,n-1} \propto r^{n-1}e^{-r/na_0}$, with corresponding probability density for r given by

$$P(r) \propto r^2|R_{n,n-1}(r)|^2 = r^{2n}e^{-2r/na_0} \tag{18.46}$$

[5]In this context, these classical radii might be called *peri-nucleus* and *apo-nucleus,* respectively, by analogy with *perigee* and *apogee* for the corresponding quantities for earth orbit.

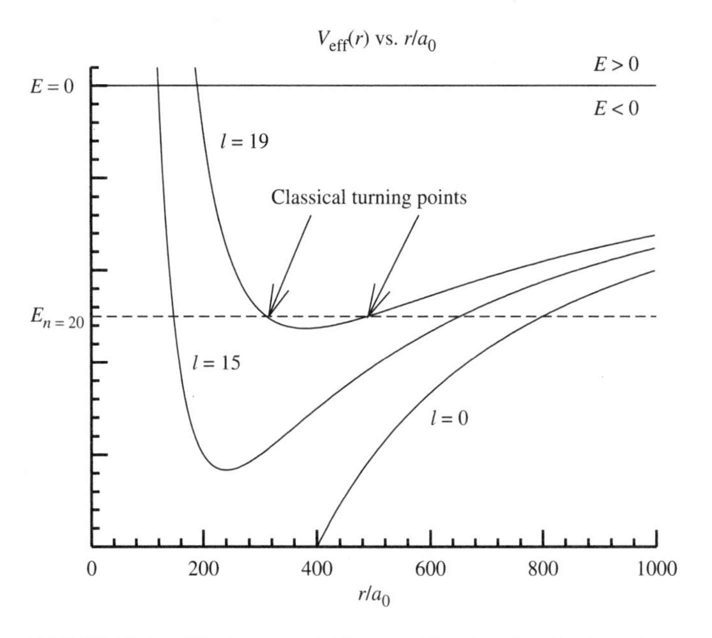

FIGURE 18.4. *Effective potential for $n = 20$ and various l values; the dashed horizontal line shows the $n = 20$ energy level.*

The maximum of this distribution is given by

$$\frac{dP(r)}{dr} = 2nr^{2n-1}e^{-r/na_0} - \frac{2}{na_0}r^{2n}e^{-r/na_0} = 0 \tag{18.47}$$

or $r = a_0 n^2$; this is just the classical radius for circular motion for this case in which $\epsilon \to 0$ and $r_{max} = r_{min}$.

The classical probability distributions can also be derived by standard methods and compared with the quantum results for any value of n and l. To do this, we first write

$$E = \frac{1}{2}\mu v^2 - \frac{Ke^2}{r} + \frac{\mathbf{L}^2}{2\mu r^2} \tag{18.48}$$

in the form

$$v = \frac{dr}{dt} = \sqrt{\frac{2}{\mu}\left(E + \frac{Ke^2}{r} - \frac{\mathbf{L}^2}{2\mu r^2}\right)} \tag{18.49}$$

or substituting the quantum values

$$dt = \sqrt{\frac{a_0 n^2 \mu}{Ke^2}}\, r\, dr/\sqrt{2ra_0n^2 - l(l+1)n^2a_0^2 - r^2} \tag{18.50}$$

With the period given by Eqn. (18.41), we use

$$d\text{Prob} \equiv P_{CL}(r)\, dr = \frac{dt}{\tau/2} \tag{18.51}$$

to find

$$P_{CL}^{(n,l)}(r) = \frac{1}{\pi a_n n^2}\frac{r}{\sqrt{2ra_0n^2 - l(l+1)n^2a_0 - r^2}} \tag{18.52}$$

which is defined only in the interval $(r_{\min}, r_{\max})$ with

$$r_{\min,\max} = a_0 n^2 \left(1 \mp \sqrt{1 - \frac{l(l+1)}{n^2}}\right) \tag{18.53}$$

These classical distributions can be compared with the quantum wavefunctions (including the measure!)

$$P_{qm}^{(n,l)} = r^2 |R_{n,l}(r)|^2 \tag{18.54}$$

and we show results in Fig. 18.5 for the same set of (n, l) values in Fig. 18.4.

Finally, we note that wavepacket solutions for the hydrogen atom in circular[6] or elliptical[7] orbits can be constructed which spread in time, as expected from earlier one- and two-dimensional examples. Just as with the infinite well, however, Coulomb wavepackets can also exhibit so-called *revivals* or *reformations* in which the wavepacket returns to a well-defined localized state. Localized electronic wavepackets can be excited in Rydberg atoms using picosecond laser pulses,[8] and revival phenomena have been observed; such experiments provide a unique laboratory for the study of the connection between quantum and classical dynamics.

Many attractive potentials admit approximately elliptical orbits, but the inverse-square force law is known to be special in that the orientation of the ellipse (as measured,

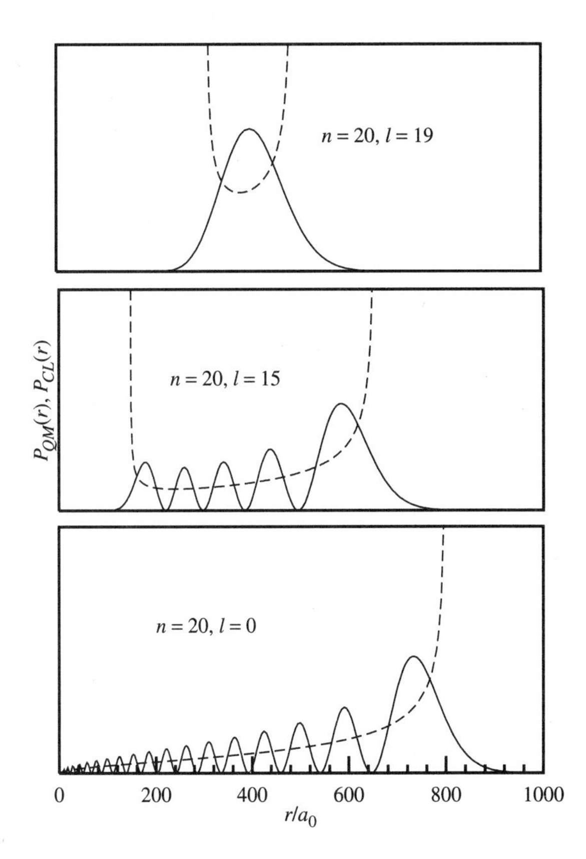

FIGURE 18.5. *The classical (dashed) and quantum (solid) probability distributions for the radial coordinate;* $P_{CL}(r)$ *and* $P_{QM}(r)$ *versus r are shown for the same (n, l) values and over the same range in* r/a_0 *as in Fig. 18.4.*

---

[6] See, e.g., Brown (1973) for an early example.

[7] See, e.g., Nauenberg (1972).

[8] See the review by Alber and Zoller (1991).

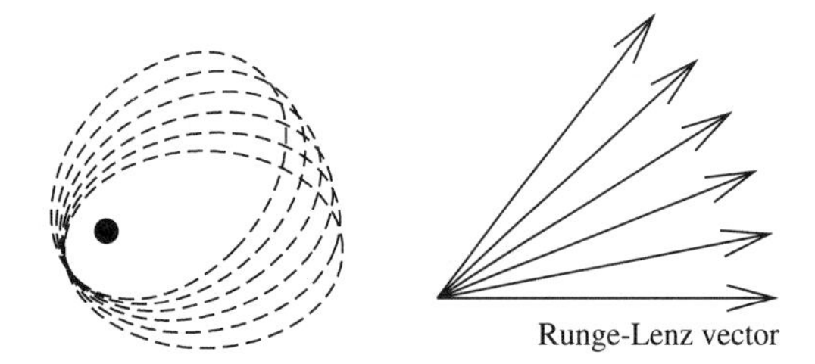

FIGURE 18.6. Precessing elliptical orbit in a non-$1/r^2$ force field and the corresponding behavior of the Runge-Lenz vector.

say, by the direction of the semimajor axis) stays fixed with time; that is, the ellipse does not precess as shown in Fig. 18.6. This is related to the fact that while for any central potential the angular momentum vector **L** is conserved, there is an additional conserved vector, unique to the $1/r$ potential. This quantity is called the *Lenz-Runge vector* and can be written as

$$\mathbf{R} = \frac{\mathbf{r}}{r} - \left(\frac{1}{mKe^2}\right)\mathbf{p} \times \mathbf{L} \tag{18.55}$$

It can be shown (P18.17) that classically the vector **R**

- Points along the semimajor axis of the ellipse as in Fig. 18.7,
- Is fixed in time (i.e., conserved)
- Has magnitude equal to the eccentricity squared, i.e.,

$$\mathbf{R}^2 = 1 - \frac{2\mathbf{L}^2|E|}{\mu(Ke^2)^2} = \epsilon^2 \tag{18.56}$$

This last relation implies an extra degeneracy in the problem due to this simple relation between three conserved quantities. Even relatively small deviations from the $1/r^2$ force law of classical gravity, as due, for example, to the corrections from general relativity,[9] will "break the symmetry," allow **R** to vary in time, and cause a precession of the elliptical orbit. The precession of the orbit of Mercury is due to non-$1/r^2$ effects arising from corrections to the Newtonian potential given by general relativity.

The corresponding quantum vector operator, given by appropriately replacing **p** and **L** by their operator counterparts, i.e.,

$$\hat{\mathbf{R}} = \frac{\mathbf{r}}{r} - \left(\frac{1}{2mKe^2}\right)\left[\hat{\mathbf{p}} \times \hat{\mathbf{L}} - \hat{\mathbf{L}} \times \hat{\mathbf{p}}\right] \tag{18.57}$$

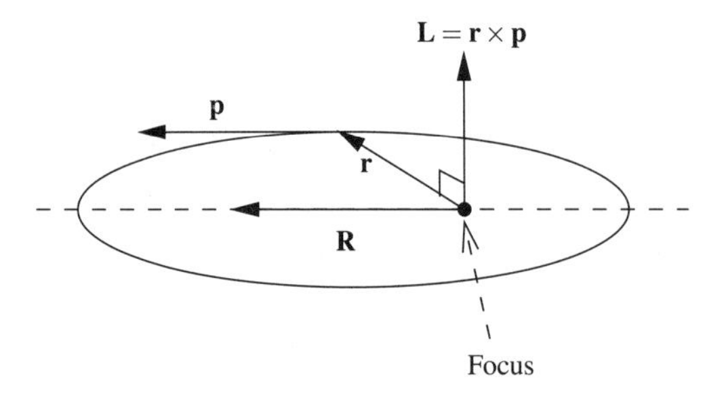

FIGURE 18.7. Relationship of the Runge-Lenz and angular momentum vectors for a classical orbit in a $1/r$ potential.

[9]See Marion and Thornton (1988).

can be shown (P18.18) to commute with the Hamiltonian and therefore to be conserved in the quantum version of the Kepler problem. (Note the "antisymmetrized" quantum version of the classical expression involving the vector cross-product.)

This has profound consequences as the remarkable simplicity of the hydrogen atom energy spectrum is, in part, due to this additional conserved quantity. The conservation of angular momentum only guarantees that the energy eigenvalues of all of the $2l+1$ different m values for a given value of l will be degenerate in any purely central potential. The conservation of $\mathbf{R}$ can be shown to imply that the states with $l = 0, 1, \ldots, n-1$ for a given value of the principal quantum number n will also be degenerate; this gives the total of

$$\sum_{l=0}^{n-1}(2l + 1) = n^2 \tag{18.58}$$

states with same energy. Any additional physical effects that alter the $1/r$ potential, such as relativistic effects (this time special relativity), will then cause splittings of the energy levels and give rise to a less "symmetric" spectrum.

Example 18.1 Level Splittings in Non-Coulombic Potentials As a pedagogical example,[10] consider the slightly modified Coulomb potential given by

$$V(r) = -\frac{ZKe^2}{r}\left(1 + b\frac{a_0}{r}\right) \tag{18.59}$$

where b is dimensionless. The Schrödinger equation for this potential can be solved using the same techniques as above (P18.13), and one finds that the energy levels are now

$$E_{n,l} = -\frac{1}{2}\mu c^2(Z\alpha^2)^2 \frac{1}{[n + \sqrt{(l+1/2)^2 - 2b} - (l+1/2)]^2} \tag{18.60}$$

The energy levels now depend on both n and l, with the larger angular momentum states affected less by the non-Coulombic interaction. (Can you provide an explanation of this last fact?)

18.3 OTHER "HYDROGENIC" ATOMS

There are a number of two-particle, bound state systems, many of whose properties can be understood rather simply on the basis of straightforward extensions of the hydrogen atom. In this section we catalog some of the properties of these "exotic atoms" and describe what we learn from them about the particles and forces involved.

18.3.1 Rydberg Atoms

An atom in which one electron is highly excited (but not ionized), and hence far from the remaining "ionic core," exhibits hydrogenic-like behavior for sufficiently large values of the principal quantum number, n. This is plausible since for sufficiently large distances from the inner electrons, the excited electron will be influenced only by the net charge of the core and not by its detailed structure; the total charge (nuclear charge plus core electrons) will be $+e$, just as for hydrogen. Such highly excited states are often called *Rydberg atoms*.[11]

[10] More realistic effects that give rise to level splittings will be considered in P18.11 and P18.12.

[11] For reviews, see Stebbings (1983) or Gallagher (1994).

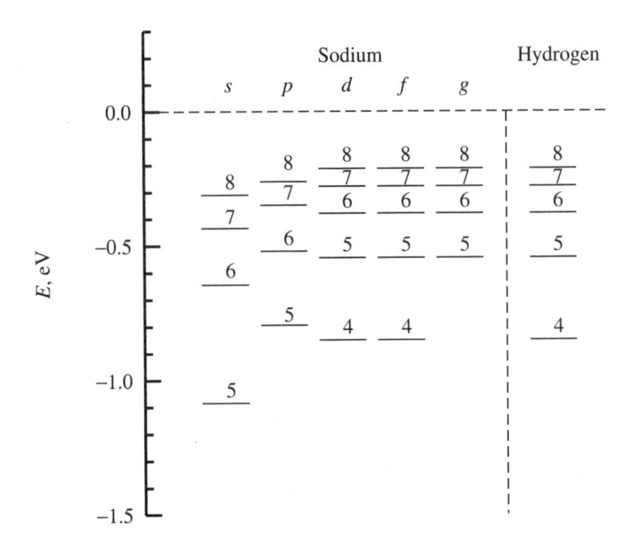

FIGURE 18.8. Energy levels for sodium compared to hydrogen.

The alkali metals—the elements in Column IA of the periodic table, lithium (Li), sodium (Na), potassium (K), cesium (Cs), etc.—are often used for studies of Rydberg states, as they consist of a single unpaired electron outside a closed (noble gas atom) electron shell and therefore most readily exhibit hydrogen-like behavior. They are the "work horses" of atomic physics research and applications.[12] The incredible frequency resolution of modern tunable lasers makes it possible to populate individual Rydberg states with values of principal quantum number $n > 200$.

The energy levels of such an alkali metal are illustrated in Fig. 18.8, where the spectrum of sodium is compared with that for hydrogen. The states with the highest orbital angular momentum l show the closest agreement for a given n, as they represent circular orbits that have the largest angular momentum barrier near the origin. Such orbits are kept furthest away from the core region where they would experience the largest deviations from the $1/r$ potential and are called *nonpenetrating*. States with lower values of l correspond classically to elliptical orbits of large eccentricity that can make excursions into the core region, giving rise to nonhydrogenic behavior; they are called *penetrating orbits*. The classical elliptical orbits for the angular momentum states allowed for $n = 6$ are shown in Fig. 18.9, where the size of the "ionic core" is roughly indicated by the star.

The spectrum of such Rydberg states is characterized by an *effective quantum number*, n^*, or the related notion of a *quantum defect*, δ_l, defined via

$$E_{n,l} = -\frac{E_0}{(n^*)^2} = -\frac{E_0}{(n-\delta_l)^2} \tag{18.61}$$

where $E_0 = \mu c^2\alpha^2/2 \approx 13.6$ eV. The degeneracy of the spectrum for the pure Coulomb potential is lifted by the additional effects of the ionic core, and δ_l measures the splitting due to these interactions. In the simple model of Ex. 18.1, for example, the quantum defect due to the addition of a small b/r^2 potential is

$$\delta_l = -\left[\sqrt{(l+1/2)^2 - 2b} - (l+1/2)\right] \approx \frac{b}{l} \qquad \text{for } l >> 1 \tag{18.62}$$

[12]Recall that the fundamental unit of time, the second, is now defined as being "equal to 9,192,631,770 periods of the radiation corresponding to the transition between the two hyperfine levels of the ground state of ^{133}Cs."

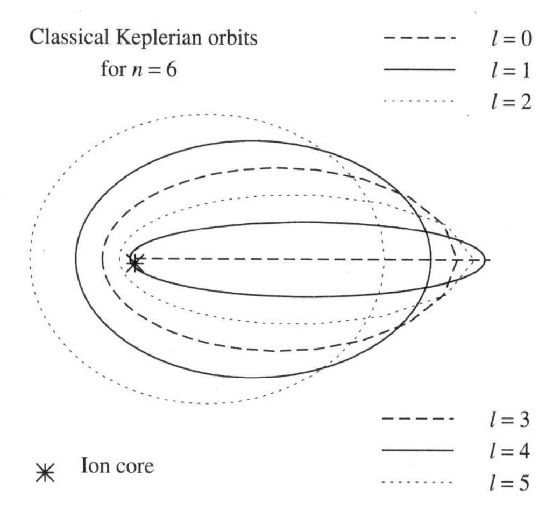

FIGURE 18.9. Classical elliptical orbits corresponding to the $n = 6$ levels of sodium for allowed values of $l = 0, 1, 2, 3, 4, 5$. Orbits with large values of l are less "penetrating," i.e., have less of an overlap with the ion core.

For small values of the quantum defect, one can write

$$E_{n,l} = -\frac{E_0}{n^2(1 - \delta_l/n)^2} \approx -\frac{E_0}{n^2} - \frac{2E_0\delta_l}{n^3} + \cdots \tag{18.63}$$

and the fact that the energy shift goes as $1/n^3$ can be understood as being due to the fact that the probability of the valence electron being near the core scales is n^{-3}.

Rydberg atoms are extremely loosely bound and are characterized by their large sizes. Because $\langle r \rangle \sim a_0 n^2$, such atoms can approach almost macroscopic sizes; for $n = 100$, the diameter of a Rydberg atom is $\sim 1\ \mu$m, roughly the size of a biological cell, and classical arguments are often more appropriate than quantum ones.

Example 18.2 "Sieve" for Rydberg Atoms The large sizes of Rydberg atoms make possible "slit" experiments in which the results do not depend on the quantum mechanical wave properties of the system but, rather, on their classical physical size. In one such experiment,[13] beams of Rydberg atoms were allowed to drift toward an array of rectangular, micrometer size slits in gold foil; the average slit size was $2\ \mu\text{m} \times 10\ \mu\text{m}$. If one assumes that the Rydberg atoms have an effective classical radius given by ka_0n^2, where k is a dimensionless constant, one can argue from Fig. 18.10 that if the center of the atom is more than $d = l/2 - ka_0n^2$ from the center of the slit, the atom will not pass through it (being ionized instead on contact with the foil). The transmission probability, T, is then determined solely by geometry, and is

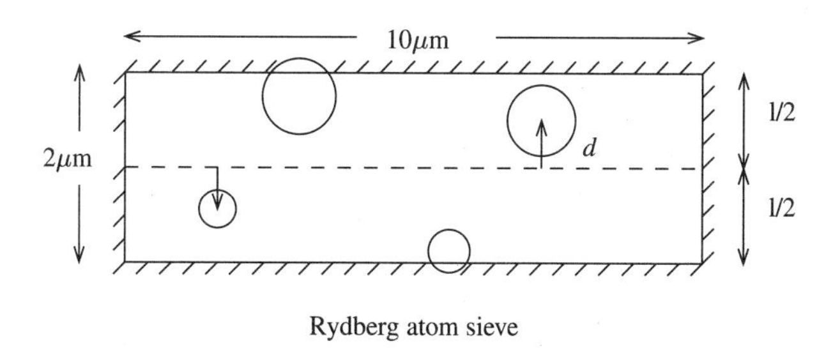

FIGURE 18.10. Geometry of the Rydberg "sieve" in Ex. 18.2.

[13]See Fabre et al. (1983) for details.

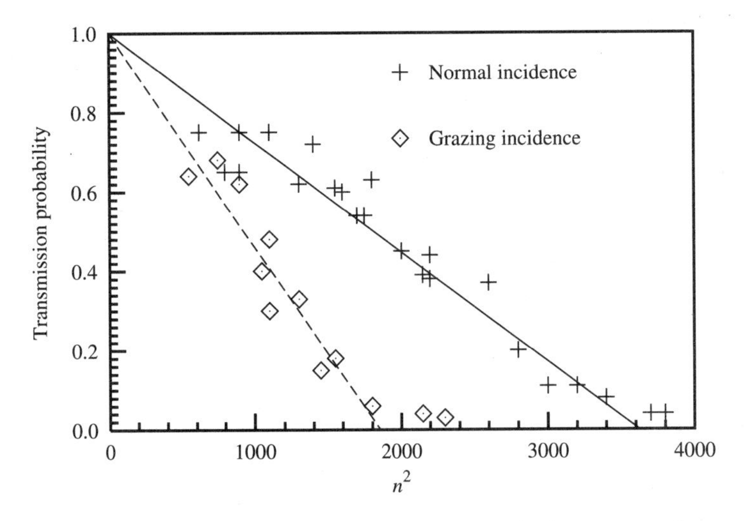

FIGURE 18.11. *Transmission probability versus principal quantum number squared, n^2, for the Rydberg "sieve"; the data are taken from Fabre et al. (1983).*

given by

$$T = \frac{d}{l/2} = 1 - \frac{2ka_0}{l}n^2 \tag{18.64}$$

which predicts a $A - Bn^2$ dependence of the transmission and a "cutoff" size for the atoms. The data are plotted in Fig. 18.11 where the crosses correspond to normal incidence of the atomic beam (beam perpendicular to foil), while the diamonds are for incidence at an angle such that the effective width of the slit, l, is reduced by a factor of 2. The predictions of the simple "hard-sphere" model described above are indicated by the straight lines for the two cases.

18.3.2 Muonic Atoms

Hydrogen-like atoms can be formed in which a *muon* is bound to a nucleus of charge Ze; such states are called *muonic atoms*.[14] Muons are "heavy electrons" in that they have the same electric charge ($-e$) and spin (1/2), but are roughly 207 times heavier ($m_\mu c^2 = 105.6$ MeV instead of $m_e c^2 = 0.511$ MeV). Beams of muons can be slowed in matter, captured by charged nuclei in a highly excited state (i.e., large values of n, l), and undergo successive radiative decays to the ground state with a characteristic time $\tau_{\text{capture}} \sim 10^{-13} - 10^{-14}$ sec; the classical period of rotation once in the ground state is roughly $\tau_{\text{orbit}} \sim 10^{-18}/Z$ sec (P18.21). Even though muons are unstable (decaying because of the weak interaction via $\mu \to e + \nu_\mu + \overline{\nu}_e$), their lifetime, $\tau_\mu \approx 2 \times 10^{-6}$ sec, is long compared to the other characteristic time scales in the problem.[15] Because it is not identical to the other electrons present in the atom, there is no exclusion principle constraint on the muon, and it can exist in any energy level. The short lifetime and the lack of suitably intense muon beams guarantees that atoms with two more muons for which the Pauli principle would be important cannot be realized in the lab.

[14]For a review, see Hüfner (1977).

[15]If one scales the muon lifetime to the average human lifespan, the capture time to the ground state (orbital period) would be roughly 10–100 sec (1 msec).

The dominating feature of such *muonic atoms* is the extremely small size of the Bohr orbits, especially the ground state, which can be obtained from a simple scaling of the results for hydrogen—Eqn. (18.18). The muonic Bohr radius is roughly

$$a_0^{(\mu)} = a_0^{(e)} \frac{m_e}{m_\mu} \approx 0.0025\,\text{Å} = 250\,\text{F} \tag{18.65}$$

which is roughly halfway between atomic (Å) and nuclear (F) length scales when measured logarithmically. Atoms with all their electrons replaced by muons would be truly Lilliputian[16] and would have a very different chemistry (P18.22). Such scaling arguments clearly fail for muonic atoms formed with nuclei of sufficiently large Z. For example, for Bi^{209}, which has $Z = 83$ ($A = 209$), the ground state radius obtained from scaling H-atom results would be roughly $\langle r \rangle \sim 250\,F/83 \sim 3$ F, while the nuclear radius is approximately $R_N \sim 7$ F; the simplifying assumption of a pointlike charged nucleus is strongly violated.

Such arguments do imply that the muon spends much more time "inside" the nucleus than does the electron; the hydrogen wavefunctions in Eqns. (18.21) to (18.26), along with the appropriate "measure," imply that the probability of finding either particle inside a radius $R << a_0^{(e,\mu)}/Z$ scales roughly as

$$\text{Prob}_{n,l}(r < R) \propto \left(\frac{ZR}{n a_0^{(e,\mu)}} \right)^{2l+3} \tag{18.66}$$

The effect of the finite size of the nucleus on the observable energy levels arises from the fact that the nuclear charge is "spread out" and thus leads to a modification of the Coulomb potential at short distances. If, for example, the nuclear charge is modeled as a uniform, spherically symmetric charge density of radius R, the electrostatic potential energy takes the form (P18.23)

$$V(r) = \begin{cases} -(ZKe^2/2R^3)(3R^2 - r^2) & \text{for } r < R \\ -ZKe^2/r & \text{for } r > R \end{cases} \tag{18.67}$$

as shown in Fig. 18.12. If ZR/na_0 is not too large, the effects of the modification can be estimated by perturbation theory (P18.24) using

$$V(r) = -\frac{ZKe^2}{r} + \Delta V(r) \tag{18.68}$$

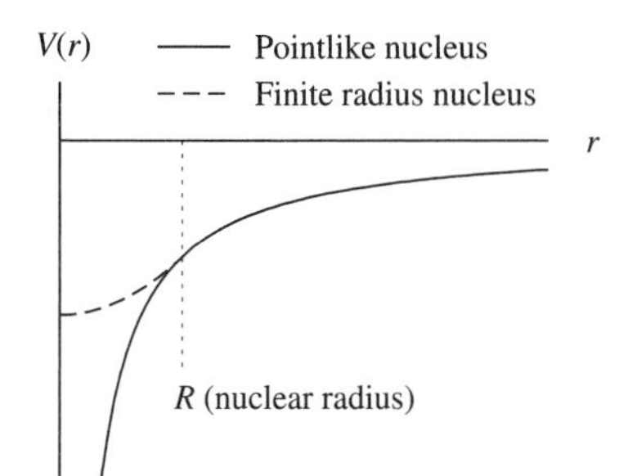

FIGURE 18.12. *Electrostatic potential for pointlike nucleus, i.e., Coulomb's law, (solid) and finite radius nucleus (dashed) in Eqn. (18.67).*

[16]Recall that the size of the inhabitants (and other flora and fauna) described by Jonathan Swift (1726) in *Gulliver's Travels* was consistently 12 times smaller than "normal."

where

$$\Delta V(r) = +\frac{ZKe^2}{r} - \frac{ZKe^2}{2R^3}(3R^2 - r^2) \quad \text{for } r < R \tag{18.69}$$

and vanishing elsewhere. If the energy level shifts are too large, one can solve the radial equation using Eqn. (18.67) directly (P18.25). For this simple model, the nuclear radius R is the only variable, and some of the first systematic measurements of the variation of nuclear size with atomic number were carried out using data from muonic transitions.

18.3.3 Pionic and Kaonic Atoms

Heavier charged particles, like the pion ($m_\pi c^2 = 139.6$ MeV, $\tau(\pi^\pm) = 2.6 \times 10^{-8}$ sec) and the kaon ($m_K c^2 = 493$ MeV, $\tau(K^\pm) = 1.2 \times 10^{-8}$ sec), can form atomic systems with nuclei characterized by even smaller radii. Their decay lifetimes, while shorter than for the muon, are still longer than their production or orbit times. Both particles have intrinsic spin $J = 0$, and the appropriate relativistic wave equation in a Coulomb potential in this case can be solved exactly (P18.11). In addition, many of the complicating features in the spectrum due to electron spin (discussed in more detail in Sec. 19.5.2) are absent, and the energy spectrum is actually simpler than for hydrogen.

The most important new feature in this case, however, is that the π and K interact with the nucleus not only via their electromagnetic interaction (Coulomb attraction) but also via the so-called *strong nuclear interaction.* One implication of this interaction is that bound pions and kaons in low-lying states with a large overlap with the nucleus can be absorbed; this leads to another typical time scale that has important implications for the observed spectrum of such states.

Example 18.3 Lifetime of Kaonic Atoms Due to Nuclear Absorption We can use the scaling behavior of Eqn. (18.66) to make a "back-of-the-envelope" estimate of the lifetime of a kaon in states of varying quantum numbers.[17] For definiteness, we assume a particular variant for the strong kaon-nucleus potential of the form

$$V_{KN}(r) = \begin{cases} -V_0 - iV_{ABS} & \text{for } r < R \\ 0 & \text{for } r > R \end{cases} \tag{18.70}$$

where R is the nuclear radius. The real part of the potential, V_0, describes the attraction due to the strong nuclear force, while the imaginary part (recall Sec. 5.1 and P5.3) is responsible for absorption. A perturbation theory estimate of the shift in energy can be made by using the estimate of the time spent inside the nucleus in Eqn. (18.66), giving

$$\Delta E_0 - i\Delta E_{ABS} = \Delta E = -(V_0 + iV_{ABS})\left(\frac{ZR}{na_0^{(K)}}\right)^{2l+3} \tag{18.71}$$

The lifetime due to the imaginary part of this shift is then given by Eqn. (5.20) as

$$\tau_{ABS} = \frac{\hbar}{2\Delta E_{ABS}} = \frac{\hbar}{2V_{ABS}}\left(\frac{na_0^{(K)}}{ZR}\right)^{2l+3} \tag{18.72}$$

The longest-lived states for a given n will have $l = n - 1$, i.e., the most nearly circular orbits with the smallest penetration into the nucleus. To estimate the magnitude of this time scale, we can use values for a typical medium-light nucleus, namely, $V_{ABS} \sim 80$ MeV, $Z = 6$,

[17] A more sophisticated analysis at the level of this text making use of WKB tunneling ideas is given in Kaufmann (1977).

$R = 2.9$ F, and the kaon Bohr radius obtained from scaling $a_0^{(K)} \sim 55$ F; these combine to give

$$\tau_{ABS} \sim (4 \times 10^{-22}\ \text{sec})(3.2n)^{2n+1} \tag{18.73}$$

in the state $(n, l = n - 1)$, and one finds that only for the state $(n, l) = (6, 5)$ and higher states is $\tau_{\text{decay}}/\tau_{ABS} > 1$. The essentially instantaneous absorption of the kaons in the lowest-lying states has the phenomenological implication that photons from radiative transitions of the type $(n, l) \to (n - 1, l - 1)$ are only observed for relatively large values of the principal quantum number; lower-lying states are quickly "gobbled up" and don't appear in the spectrum.

18.3.4 Positronium

One of the "cleanest" of two-body systems interacting via electromagnetic forces is the bound state of the electron (e^-) and its own antiparticle, the *positron* (e^+); such a system is called *positronium*.[18] There are several novel aspects of this system:

- If the light-heavy, electron-nucleus system bears some resemblance to the classical planet-star solar system, positronium is more like a binary star system. The mass of the positron is identical to that of the electron, so the system is characterized by a reduced mass $\mu = m_e/2$ with a corresponding scaling of energy levels and radii; for example, the Bohr radius is $a_0^{(P)} = 2a_0 \approx 1.06$ Å.
- While the positron itself is stable, the fate of this matter-antimatter bound state system is determined by its eventual self-annihilation. The process responsible for the annihilation decay of an s-wave (i.e., $l = 0$) state with total spin $S = 0$ via a decay into two photons can be visualized using *Feynman diagrams* as in Fig. 18.13. The symmetric combination of the two amplitudes corresponding to this process must be taken as the final state in a system of two indistinguishable bosons (since $J_\gamma = 1$).

Example 18.4 Positron Annihilation In order to understand better the decay process for positronium, we consider first the annihilation rate of a free electron-positron pair. If we consider a beam of positrons of velocity $v = \beta c$ incident on a target containing a number density of (stationary) electrons given by n_e, the annihilation rate is given by

$$\lambda = \sigma \cdot v \cdot n_e \tag{18.74}$$

where $\sigma(e^+e^- \to \gamma\gamma)$ is the cross section for the process. In this case, it is given by

$$\sigma_{\text{free}} = \sigma(e^+e^- \to \gamma\gamma) = \frac{\pi r_0^2}{\beta} \tag{18.75}$$

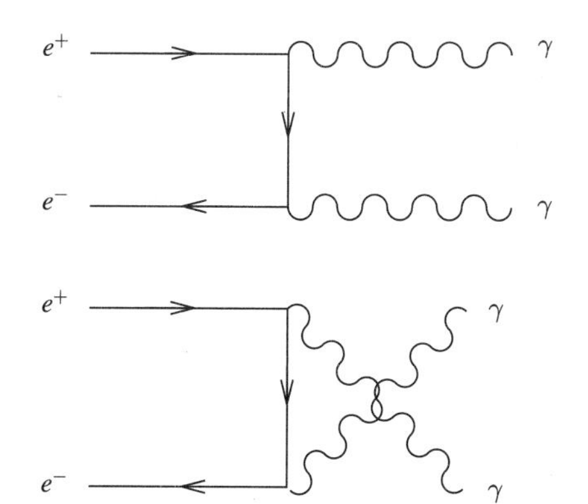

FIGURE 18.13. Feynman diagrams contributing to electron-positron annihilation via $e^+e^- \to \gamma\gamma$.

[18] For a review of the bound state structure of positronium, see Bethe and Salpeter (1957).

where r_0 is the *classical electron radius*[19] given by

$$r_0 = \frac{Ke^2}{m_e c^2} \tag{18.76}$$

and $\beta = v/c$ is the (dimensionless) speed. This process is of some importance in the study of condensed-matter systems (especially metals) where the measurement of this interaction rate provides direct information on the local electron density, n_e, encountered by the positron.[20]

The decay lifetime of a *bound* $^1S_0$ positronium state via two-photon decay can be obtained from Eqn. (18.74) by using the following facts:

- The lifetime, τ, and decay rate are related via $\lambda = 1/\tau$.
- The cross section for the bound state decay is

$$\sigma(^1S_0(e^+e^-) \rightarrow \gamma\gamma) = 4\sigma_{\text{free}} = 4\frac{\pi r_0^2}{\beta} \tag{18.77}$$

  The additional factor of 4 is easily explained by spin-counting arguments. Only the $S = 0$ combination contributes to two-photon decays, and in the free annihilation we have to average over all possible spin combinations (one $S = 0$ and three $S = 1$ states) of the e^+e^- pair and only 1/4 of the time are the free e^+e^- pair in the appropriate spin state. The spin-singlet bound state is, by definition, always in a pure $S = 0$ state.
- The electron density "seen" by the positron in the ground state system is given by

$$n_e = |\psi_{100}(\mathbf{r} = 0)|^2 \tag{18.78}$$

  The presence of the wavefunction squared of the e^+e^- system, evaluated at vanishing separation $\mathbf{r} = \mathbf{r}_{e^+} - \mathbf{r}_{e^-} = 0$, is not surprising as it measures the probability for the particle-antiparticle pair to be "on top of each other"; it also has the appropriate dimensions, giving roughly one "target electron" in a volume corresponding to one Bohr radius.

These expressions can be combined (P18.26) to show that

$$\frac{1}{\tau_{\text{decay}}} = \frac{1}{2}\frac{m_e c^2 \alpha^5}{\hbar} \tag{18.79}$$

or $\tau_{\text{decay}} = 1.25 \times 10^{-10}$ sec for the $n = 1$ state; this is much longer than the classical period.

18.3.5 Quarkonia

The quark model identifies many of the strongly interacting particles produced in high-energy collisions as bound quark-antiquark states or *quarkonia*.[21] While the equations governing the $q\overline{q}$ interaction are well known, they are technically very difficult to solve, even numerically. In certain cases, however, it is appropriate to describe their mutual force by an effective potential of the form

$$V_{q\overline{q}}(r) = -\frac{4}{3}\frac{\alpha_s \hbar c}{r} + kr \tag{18.80}$$

and make use of nonrelativistic quantum mechanics. The linear term corresponds to a long-range confining potential that is responsible for the fact that free, isolated quarks (with their

---

[19] This radius is such that a classical distribution of charge would have electrostatic self-energy of the order of its rest mass.

[20] See Hautojärvi and Vehanen (1979).

[21] See Griffiths (1987) for a discussion of the phenomenology of $q\overline{q}$ states, and Quigg and Rosner (1979) for applications of quantum mechanics to quarkonia.

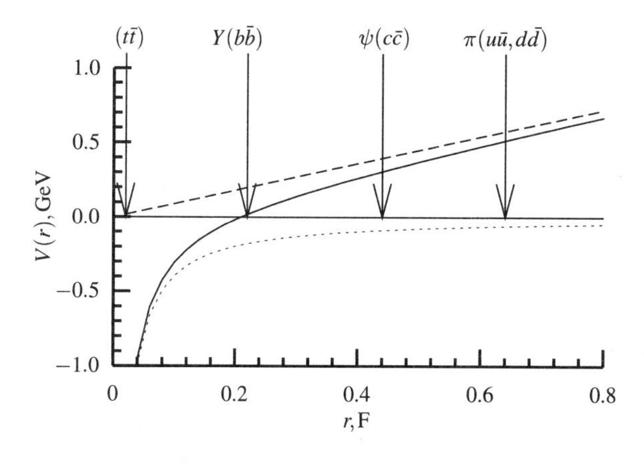

FIGURE 18.14. Quark-antiquark potential, $V_{q\bar{q}}(r)$, versus r. The solid curve is the total linear plus (color) Coulomb form, while the dashed (dotted) curve is the separate linear confining ("color Coulomb") piece. The RMS radii for the ground states (calculated from potential models) of $c\bar{c}$, $b\bar{b}$, and $t\bar{t}$ states are shown; for the pion, the experimental value of the charge radius is used.

exotic fractional charges) have not been observed. The first term, which may be called the *color Coulomb* term, is of the same form as the electrostatic potential

$$V_{\text{Coul}}(r) = -\frac{\alpha\hbar c}{r} \tag{18.81}$$

but the corresponding "color fine structure constant" is given by $\alpha_s \approx 0.1 - 0.2$ instead of $\alpha = 1/137$.

Nature has been generous in providing six different varieties (or "flavors") of quarks (with their corresponding antiquarks) of different masses which can form $q\bar{q}$ bound states; their labels[22] and approximate values for their effective masses are given by

$$\begin{array}{lll} u(0.3\ \text{GeV}/c^2) & d(0.3\ \text{GeV}/c^2) & s(0.5\ \text{GeV}/c^2) \\ c(1.5\ \text{GeV}/c^2) & b(5.0\ \text{GeV}/c^2) & t(\sim 180\ \text{GeV}/c^2) \end{array}$$

We plot the $q\bar{q}$ potential in Fig. 18.14, and indicate the location of the RMS radii of the lowest energy bound states for the three heaviest "flavors" (for which a treatment using nonrelativistic quantum mechanics make sense). Not surprisingly, we see that the heavier quarks have increasingly smaller "color Bohr radii," similar to the case of muonic and kaonic constituents in the Coulomb potential of the nucleus; the heavier the particle, the deeper a given state lies in the potential well.

Example 18.5 Feynman-Hellmann Theorem We can formalize this result that larger constituent masses will have a specified quantum state at lower energy in a given potential by making use of a general result due to Feynman and Hellmann. Suppose that the Hamiltonian of a system depends on a parameter λ in some well-defined manner, $\hat{H} = \hat{H}(\lambda)$, with energy eigenvalues satisfying $E(\lambda)\psi = \hat{H}(\lambda)\psi$. The dependence of the energy eigenvalue of a specific state on λ is then simply given by (P18.28)

$$\frac{\partial E(\lambda)}{\partial\lambda} = \left\langle \frac{\partial \hat{H}(\lambda)}{\partial\lambda} \right\rangle \tag{18.82}$$

This simple result has been utilized extensively in molecular physics, but, in our case, we wish to examine the dependence on the reduced mass of the system. We thus write

[22]The labels are shorthand for the names up (u), down (d), strange (s), charm (c), bottom (b) and top (t), respectively; the somewhat whimsical nomenclature has some historical and classificatory significance. Evidence leading to the discovery of the top quark by two experimental groups is described in Abe et al. (1995).

$$\frac{\partial \hat{H}}{\partial \mu} = -\frac{1}{\mu}\left(-\frac{\hbar^2}{2\mu}\nabla^2\right) = -\frac{1}{\mu}\left(\hat{H} - V(r)\right) \tag{18.83}$$

which gives

$$\frac{\partial E}{\partial \mu} = -\frac{1}{\mu}\langle \hat{H} - V(r)\rangle = -\frac{1}{\mu}\left(E - \langle V\rangle\right) < 0 \tag{18.84}$$

so that $E(\mu)$ decreases as μ increases.

18.4 MULTIELECTRON ATOMS

The continued study of the hydrogen atom can provide a model system in which one can most simply discuss relativistic effects (P18.12), the interactions of the orbital motion and spin degree of freedom of the electron (Sec. 19.6.2), and the effects of external electric (Sec. 19.4.2) and magnetic (Sec. 19.6.1) fields.

If, however, we wish to gain more insight into the structure of matter, we should discuss the structure of multielectron atoms. In the next two sections we focus on the ground state properties of the simplest two- (helium-like) and three-electron (lithium-like) atoms (and ions) as they elucidate many of the general properties of atomic structure that, in turn, results in the periodic table of the elements.

As we focus on the ground states of such systems, we will find it useful to collect in Table 18.1 a compilation of the ionization potentials of neutral and partially ionized atoms. We show values corresponding to the first five elements (i.e., values of the nuclear charge Z ranging from 1 to 5) from which we extract many of the experimental values quoted below.

The traditional notation of spectroscopy is used, so that I indicates the neutral atom, II the singly ionized atom, and so forth. For example, the energy required to extract a single electron from neutral lithium is $E_{\mathrm{I}}(\mathrm{Li}) = 5.39172\,\mathrm{eV}$; the second electron removed from singly ionized Li^+ 'costs' $E_{\mathrm{II}}(\mathrm{Li}) = 75.64108\,\mathrm{eV}$ while it requires $E_{\mathrm{III}}(\mathrm{Li}) = 122.4529\,\mathrm{eV}$ to remove the last electron from the hydrogen-like Li^{++}. Taken together, these imply that the total electronic binding energy of lithium is

$$-(E_{\mathrm{I}}(\mathrm{Li}) + E_{\mathrm{II}}(\mathrm{Li}) + E_{\mathrm{III}}(\mathrm{Li})) = -203.4862\,\mathrm{eV} \tag{18.85}$$

We can already make good use of Table 18.1 by noting that the measured ionization potentials of the hydrogen-like atoms H, He^+, Li^{++}, etc., are given by the uppermost value

TABLE 18.1 *The Ionization Potentials (in* eV*) for Neutral and Partially Ionized Atoms*

Z	Element	I	II	III	IV	V
1	H	13.59844				
2	He	24.58741	54.41778			
3	Li	5.39172	75.64018	122.45429		
4	Be	9.32263	18.21116	153.89661	217.71865	
5	B	8.29803	25.15484	37.93064	259.3721	340.22580

in each column; they are seen to be in the ratio

$$\begin{array}{ccccccccc} E_{\rm I}({\rm H}) & : & E_{\rm II}({\rm He}) & : & E_{\rm III}({\rm Li}) & : & E_{\rm IV}({\rm B}) & : & E_{\rm V}({\rm Be}) \\ 13.59844 & : & 54.41778 & : & 122.4529 & : & 217.71865 & : & 340.22580 \\ 1 & : & 4.001766 & : & 9.004923 & : & 16.01056 & : & 25.01947 \end{array} \quad (18.86)$$

and to compare well with the Z^2 scaling predicted by Eqn. (18.1).

18.4.1 Helium-like Atoms

In contrast to the elegant, nonprecessing, elliptical orbits found in the classical two-body problem evolving under an inverse-square law force, the classical three-body problem corresponding to the helium atom exhibits extremely complex motion, even displaying chaotic behavior. Many of the semiclassical methods (such as Bohr-Sommerfeld quantization) used in the early days of quantum mechanics are built on knowledge of the classical trajectories, and such chaotic solutions prevented the broad application of these techniques to the helium atom and other systems.[23] The situation circa 1922 is described[24] by the following comments:

> Numerous attempts have been made to construct quantum theory models of the normal helium. In Bohr's own model... the two electrons revolve about the nucleus at extremities of a diameter.... In no case is the agreement satisfactory, so that apparently none of these models can be correct if the Sommerfeld quantization conditions are accepted.... We must bear in mind that the extreme chemical stability of helium indicates that the arrangement of the two electrons is particularly simple and symmetrical, for an electron revolving in an orbit outside that of its mate would presumably be a valence electron.

This difficulty is also apparent in the lack of closed-form solutions of the Schrödinger equation. The most accurate studies of the helium atom have utilized variational methods instead, and we will follow that approach in what follows.

The spatial coordinates of the two electrons in a helium-like atom are shown in Fig. 18.15 as measured from the nucleus. Because of the large nuclear mass, we will

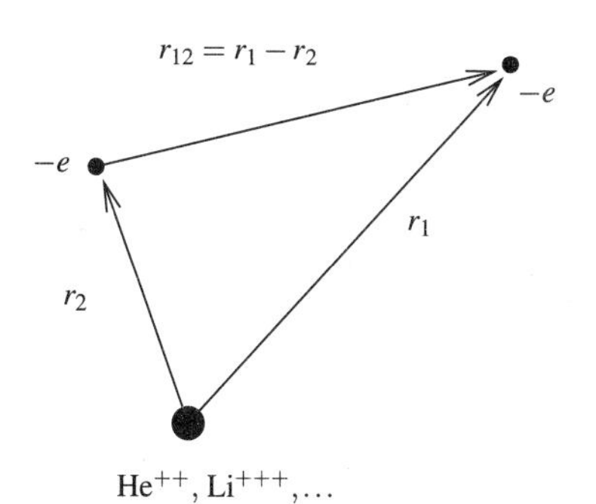

FIGURE 18.15. Coordinates used for two-electron atoms. For neutral helium, one has $He = He^{++} + 2e^-$, while for singly ionized lithium, one has $Li^+ = Li^{+++} + 2e^-$, and so forth.

---

[23]Einstein evidently understood the connection between what is now called *chaos* and the problems with the quantum theory of the helium atom. The semiclassical quantization of the helium atom using periodic orbit trajectories has, in fact, only been accomplished relatively recently; for a discussion, see Heller and Tomsovic (1993) and Gutzwiller (1990).

[24]Comments by Van Vleck (1922) in an article entitled "The Dilemma of the Helium Atom."

make the approximation of taking the nucleus to be at rest. The Hamiltonian for the two-electron system is then

$$\hat{H} = \frac{\hat{\mathbf{p}}_1^2}{2m_e} + \frac{\hat{\mathbf{p}}_2^2}{2m_e} - \frac{ZKe^2}{r_1} - \frac{ZKe^2}{r_2} + \frac{Ke^2}{|\mathbf{r}_1 - \mathbf{r}_2|} \tag{18.87}$$

where Z is the nuclear charge. The neutral helium atom corresponds to $Z = 2$, but one can also consider the family of singly ionized atoms or ions, namely, Li^+, B^{++}, Be^{+++}; we will only present numerical values explicitly for helium.

The binding energy of helium can be read from Table 18.1 to be

$$-(E_{\mathrm{I}}(H) + E_{\mathrm{II}}(H)) = -(24.58741 + 54.41778)\,\mathrm{eV} \approx -79.0\,\mathrm{eV} = E_{\mathrm{exp}} \tag{18.88}$$

If, for the moment, we ignore the effect of the electron-electron repulsion, the last term in Eqn. (18.87), the Schrödinger equation is separable. The ground state position-space wavefunction is simply the product of hydrogen-like ground state solutions, but we also require an antisymmetric spin wavefunction to satisfy the spin statistics theorem for the two indistinguishable electrons; the result is

$$\psi(\mathbf{r}_1, \mathbf{r}_2) = \psi_{100}(\mathbf{r}_1)\psi_{100}(\mathbf{r}_2)\chi_{(0,0)} \tag{18.89}$$

where

$$\chi_{(0,0)} = \frac{1}{\sqrt{2}}\left(\chi_\uparrow^{(1)}\chi_\downarrow^{(2)} - \chi_\downarrow^{(1)}\chi_\uparrow^{(2)}\right) \tag{18.90}$$

The corresponding energy

$$E_0 = 2E_H = 2\left(-\frac{Ke^2}{2a_0}\right)Z^2 \tag{18.91}$$

or

$$E_0 = 2(-13.6\,\mathrm{eV})4 = -108.8\,\mathrm{eV} \tag{18.92}$$

for helium, which is obviously too low.

We can estimate the effect of the electron-electron repulsion, using perturbation theory, by evaluating

$$\begin{aligned} \Delta E_{ee} &= \left\langle \psi(\mathbf{r}_1, \mathbf{r}_2) \left| \frac{Ke^2}{|\mathbf{r}_1 - \mathbf{r}_2|} \right| \psi(\mathbf{r}_1, \mathbf{r}_2) \right\rangle \\ &= \int r_1^2 dr_1 d\Omega_1 \int r_2^2 dr_2 d\Omega_2 |\psi_{100}(\mathbf{r}_1)\psi_{100}(\mathbf{r}_2)|^2 \frac{Ke^2}{|\mathbf{r}_1 - \mathbf{r}_2|} \end{aligned} \tag{18.93}$$

The physical meaning of this expression can be made clearer by recalling that the local charge density of each electron is given by

$$\rho(r_{1,2}) = -e|\psi(\mathbf{r}_{1,2})|^2 \tag{18.94}$$

so that Eqn. (18.93) reduces to

$$\Delta E_{ee} = K\int d\mathbf{r}_1 d\mathbf{r}_2 \frac{\rho(\mathbf{r}_1)\rho(\mathbf{r}_2)}{|\mathbf{r}_1 - \mathbf{r}_2|} = \int \frac{K dq_1 dq_2}{|\mathbf{r}_1 - \mathbf{r}_2|} \tag{18.95}$$

which is just the total Coulomb energy due to the two-electron charge distributions.

To evaluate this expression, we first recall that

$$|\mathbf{r}_1 - \mathbf{r}_2| = \sqrt{r_1^2 + r_2^2 - 2r_1 r_2 \cos(\theta_{1,2})} \tag{18.96}$$

where $\theta_{1,2}$ is the angle between $\mathbf{r}_{12}$. We can use the arbitrariness in the definition of the coordinate system for this (or any other) problem to choose to define the z axis such that it "points along" one of the vectors, say, $\mathbf{r}_2$, so that $\theta_{12} = \theta_2$. In this case we have

$$\begin{aligned}
\int d\Omega_2 \frac{1}{|\mathbf{r}_1 - \mathbf{r}_2|} &= \int_0^{2\pi} d\phi_2 \int_0^{\pi} \sin(\theta_2)\, d\theta_2 \left(r_1^2 + r_2^2 - 2r_1 r_2 \cos(\theta_2)\right)^{-1/2} \\
&= 2\pi \left. \frac{\left(r_1^2 + r_2^2 - 2r_1 r_2 \cos(\theta_2)\right)^{1/2}}{r_1 r_2} \right|_0^{\pi} \\
&= \frac{2\pi}{r_1 r_2} \left(\sqrt{(r_1 + r_2)^2} - \sqrt{(r_1 - r_2)^2} \right) \\
&= \frac{2\pi}{r_1 r_2} \left[(r_1 + r_2) - |r_1 - r_2| \right] \\
&= \begin{cases} 4\pi / r_1 & \text{for } r_1 > r_2 \\ 4\pi / r_2 & \text{for } r_1 < r_2 \end{cases}
\end{aligned} \tag{18.97}$$

The radial integrals can then be performed giving

$$\begin{aligned}
\Delta E_{ee} &= \frac{16 K e^2 Z^6}{a_0^6} \int_0^{\infty} r_1^2 dr_2 e^{-Zr_1/a_0} \\
&\quad \times \left(\int_0^{r_1} dr_2 r_2^2 \frac{1}{r_1} e^{-Zr_2/a_0} + \int_{r_1}^{\infty} dr_2 r_2^2 \frac{1}{r_2} e^{-Zr_2/a_0} \right) \\
&= \frac{5}{8} \frac{K e^2}{a_0} Z
\end{aligned} \tag{18.98}$$

For helium, this yields the perturbation theory estimate

$$E_{\text{pert}} = E_0 + \Delta E_{ee} = -108.8 + 34.0 = -74.8\,\text{eV} \tag{18.99}$$

which is much closer to the observed value.

While this value takes into account the interaction of the electrons, it makes no allowance for the fact that one electron will effectively "shield" the other from the nuclear charge some fraction of the time. This effect can be taken into account in a variational trial wavefunction of the same form as Eqn. (18.89), but with the physical nuclear charge, Z, replaced by an "effective charge," Z^*, which is taken as the variational parameter; we thus take

$$\psi_T(\mathbf{r}_1, \mathbf{r}_2) = \psi_{100}(\mathbf{r}_1; Z^*) \psi_{100}(\mathbf{r}_2; Z^*) \tag{18.100}$$

where we have suppressed the spinor wavefunctions that play no role in the energy minimization. The energy functional,

$$E[\psi_T] = \langle \psi_T | \hat{H} | \psi_T \rangle \tag{18.101}$$

is most easily evaluated by rewriting the Hamiltonian in a clever way, namely,

$$\begin{aligned}
\hat{H} &= \left(\frac{\hat{\mathbf{p}}_1^2}{2m_e} - \frac{Z^* K e^2}{r_1} \right) + \left(\frac{\hat{\mathbf{p}}_2^2}{2m_e} - \frac{Z^* K e^2}{r_2} \right) \\
&\quad + \frac{(Z^* - Z) K e^2}{r_1} + \frac{(Z^* - Z) K e^2}{r_2} + \frac{K e^2}{|\mathbf{r}_1 - \mathbf{r}_2|}
\end{aligned} \tag{18.102}$$

The terms in Eqn. (18.101) can be evaluated in turn. For example, one has

$$\left\langle \psi_T \left| \frac{\hat{\mathbf{p}}_{1,2}^2}{2m_e} - \frac{Z^* K e^2}{r_{1,2}} \right| \psi_T \right\rangle = -\frac{Ke^2}{2a_0}(Z^*)^2 \tag{18.103}$$

since $\psi_{100}(\mathbf{r}; Z^*)$ is the ground state wavefunction corresponding to the effective charge Z^*. The expectation value of $\langle 1/r \rangle$ in Eqn. (18.31) implies that

$$\left\langle \psi_T \left| \frac{(Z^* - Z)Ke^2}{r_{1,2}} \right| \psi_T \right\rangle = (Z^* - Z)\frac{Ke^2}{a_0} Z^* \tag{18.104}$$

while Eqn. (18.98) generalizes to

$$\Delta E_{ee}(Z^*) = \left\langle \psi_T \left| \frac{Ke^2}{|\mathbf{r}_1 - \mathbf{r}_2|} \right| \psi_T \right\rangle = \frac{5}{8}\frac{Ke^2}{a_0} Z^* \tag{18.105}$$

The variational estimate of the energy is then

$$E[\psi_T] = E_{\text{var}}(Z^*) = -\frac{Ke^2}{a_0}\left(4Z^*(Z - 5/16) - 2(Z^*)^2\right) \tag{18.106}$$

which can be minimized to yield

$$\frac{dE_{\text{var}}(Z^*)}{dZ^*} = 0 \Longrightarrow Z^* = Z - \frac{5}{16} \tag{18.107}$$

In this picture, each electron is "inside the orbit" of the other (and hence shields the nuclear charge) something like 1/3 of the time. The variational minimum energy can then be written in the form akin to Eqn. (18.91):

$$E_{\text{var}}^{(\text{min})} = -2\left(\frac{Ke^2}{2a_0}\right)\left(Z - \frac{5}{16}\right)^2 \tag{18.108}$$

which gives

$$E_{\text{var}} = -77.45\,\text{eV} \tag{18.109}$$

for helium, which is lower than the perturbation theory result and hence closer to the experimental value. We collect the various estimates for He in Table 18.2. The reader is encouraged to fill in the remaining entries using the data in Table 18.1 and Eqns. (18.91), (18.98), and (18.108), and compare the entries with the experimental values for other states.

TABLE 18.2 *Experimental Values and Theoretical Estimates of the Ground State Energies of Various Helium-like Atoms*

Z	Element	E_{exp}(eV)	$E_0 = 2E_H$(eV)	ΔE_{ee}(eV)	E_{pert}(eV)	E_{var}(eV)
2	He	−79.0	−108.8	+34.0	−74.79	−77.45
3	Li^+					
4	Be^{++}					
5	B^{+++}					

One can systematically improve on this simple variational estimate[25] by including more variational parameters to model angular and radial correlations, and by including the effects of nuclear motion and relativity.[26]

For excited states in which the two electrons can be in different energy levels, and hence have different spatial wavefunctions, the role of spin becomes more important. For the ground state, only the antisymmetric spin state corresponding to $S = 0$ is allowed, but for other states, the total wavefunction can have the proper symmetry under exchange with an $S = 1$ spin wavefunction. Such states are called *para-helium* ($S = 0$ or spin-singlet) and *ortho-helium* ($S = 1$ or spin triplet), respectively. For example, if one electron is excited to the (nlm) hydrogen-like state, we can write the combinations

$$S = 0: \frac{1}{\sqrt{2}}\left(\psi_{100}(\mathbf{r}_1)\psi_{nlm}(\mathbf{r}_2) + \psi_{100}(\mathbf{r}_2)\psi_{nlm}(\mathbf{r}_1)\right)\chi_{(0,0)} \tag{18.110}$$

$$S = 1: \frac{1}{\sqrt{2}}\left(\psi_{100}(\mathbf{r}_1)\psi_{nlm}(\mathbf{r}_2) - \psi_{100}(\mathbf{r}_2)\psi_{nlm}(\mathbf{r}_1)\right)\chi_{(1,S_z)} \tag{18.111}$$

where the χ_{1,S_z} are given in Ex. 17.4.

The perturbation theory estimate of the energies of these two states can be seen to be of the form

$$E_{\text{pert}} = (E_1 + E_n) + J_{nl} \pm K_{nl} \tag{18.112}$$

where $+, -$ corresponds to $S = 0, S = 1$. The Coulomb energy, J_{nl}, has the same form as Eqn. (18.93), with one ψ_{100} replaced by ψ_{nlm}, but with the same physical interpretation. The so-called *exchange energy* is given by

$$K_{nl} = \int d\mathbf{r}_1 \int d\mathbf{r}_2 \psi_{100}(\mathbf{r}_1)\psi_{nlm}(\mathbf{r}_2)\frac{Ke^2}{|\mathbf{r}_1 - \mathbf{r}_2|}\psi_{100}(\mathbf{r}_2)\psi_{nlm}(\mathbf{r}_1) \tag{18.113}$$

and has no classical analog as it arises from the symmetry constraints on the quantum wavefunction due to indistinguishability. While the integral defining K_{nl} can be evaluated quite generally, we only require the fact that it is positive.

This implies that the spin singlet ($S = 0$) states are shifted up in energy relative to the spin triplet ($S = 1$) levels. The shift arises because the spatially symmetric (antisymmetric) wavefunction of the $S = 0$ ($S = 1$) state implies that the electrons are closer together (further apart) yielding more (less) Coulomb repulsion. The level structure of helium is shown in Fig. 18.16 illustrating the effect. The dotted lines indicate the value of hydrogen energy levels as measured from the first ionization energy, $E_{\text{II}}(\text{He})$.

18.4.2 Lithium-like Atoms

The next most complicated systems are three-electron atoms such as Li, Be^+, B^{++}, and so forth, and we can fairly easily extend a variational analysis to such states. The Hamiltonian is the generalization of Eqn. (18.87), namely,

$$\hat{H} = \sum_{i=1}^{3}\left(\frac{\hat{\mathbf{p}}_i^2}{2m_e} + \frac{ZKe^2}{r_i}\right) + \sum_{i>j=1}^{3}\frac{Ke^2}{|\mathbf{r}_i - \mathbf{r}_j|} \tag{18.114}$$

---

[25]This approach was first used by Hylleraas (1928).

[26]A nice discussion at the level of this text appears in Park (1992).

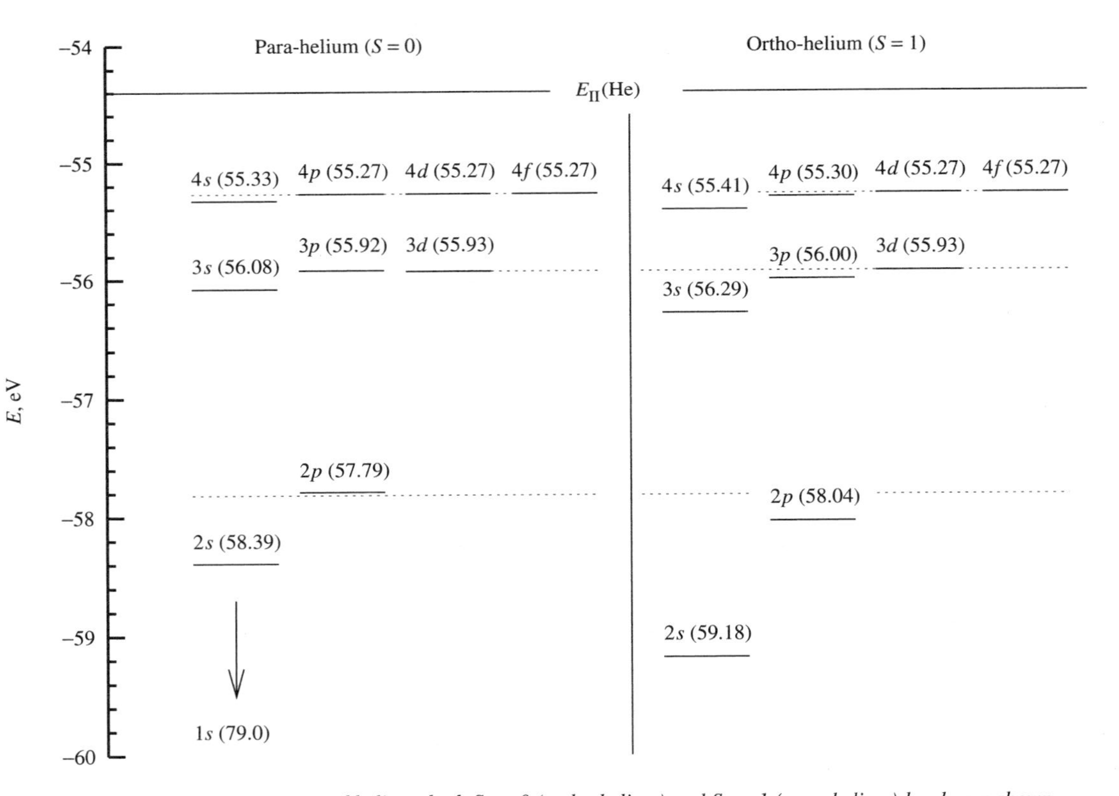

FIGURE 18.16. *Energy spectrum of helium; both $S = 0$ (ortho-helium) and $S = 1$ (para-helium) levels are shown (solid). The energy of singly ionized He corresponding to $E_{II}(He) = -2^2 E_I(H) \approx -54.4\,eV$ is shown for reference. The states are compared to the energy levels for hydrogen as measured from $E_{II}(He)$ (dotted lines).*

TABLE 18.3 One- and Two-Parameter Variational Estimates of the Ground State Energies of Lithium-like Atoms

Z	Element	E_{exp}(eV)	$Z_1 = Z_2$	E_{var}(eV)	Z_1	Z_2	E_{var}(eV)
3	Li	−203.5	2.537	−196.5	2.67	1.37	−200.8
4	Be^+	−389.9	3.537	−382.0	3.67	2.44	−386.2
5	B^{++}	−637.2	4.537	−628.9	4.67	3.45	−632.8

The ground state configuration will consist of two electrons in the $1s$ state with opposite spins and one electron (either spin-up or spin-down) in one of the first excited states; for simplicity, we consider it in the $2s$ state. A properly normalized and antisymmetrized variational wavefunction that describes this configuration is given by the Slater determinant

$$\psi_T(\mathbf{r}_1, \mathbf{r}_2, \mathbf{r}_3; Z_1, Z_2) = \frac{1}{\sqrt{3!}} \begin{vmatrix} \psi_1(\mathbf{r}_1; Z_1)\chi_\uparrow^1 & \psi_1(\mathbf{r}_1; Z_1)\chi_\uparrow^1 & \psi_2(\mathbf{r}_1; Z_2)\chi_\uparrow^1 \\ \psi_1(\mathbf{r}_2; Z_1)\chi_\uparrow^2 & \psi_1(\mathbf{r}_2; Z_1)\chi_\uparrow^2 & \psi_2(\mathbf{r}_2; Z_2)\chi_\uparrow^2 \\ \psi_1(\mathbf{r}_3; Z_1)\chi_\uparrow^3 & \psi_1(\mathbf{r}_3; Z_1)\chi_\uparrow^3 & \psi_2(\mathbf{r}_3; Z_2)\chi_\uparrow^3 \end{vmatrix}$$

$$= \frac{1}{\sqrt{6}}\left(\psi_1(\mathbf{r}_1; Z_1)\psi_1(\mathbf{r}_2; Z_1)\psi_2(\mathbf{r}_3; Z_2)\chi_\uparrow^1\chi_\downarrow^2\chi_\uparrow^3 + \cdots\right) \qquad (18.115)$$

where $\psi_{1,2}(\mathbf{r}; Z_{1,2})$ are the $n = 1, 2$ s-state hydrogen wavefunctions in Eqns. 18.21 and 18.22. We allow for the possibility of different effective charges, as we expect the electron in the outer orbital will be shielded more effectively from the nucleus by the inner-shell electrons.

The variational energy can be evaluated in a straightforward (if tedious) manner using the tricks described above to obtain

$$E[\psi] = E(Z_1, Z_2) = \frac{Ke^2}{a_0}\left(-Z_1^2 - \frac{Z_2^2}{8} + 2(Z_1 - Z) + \frac{(Z_2 - Z)Z_2}{4} + \frac{5}{8}Z_1 + f(Z_1, Z_2)\right) \qquad (18.116)$$

where

$$f(Z_1, Z_2) = \frac{2Z_1Z_2}{(2Z_1 + Z_2)^5}\left(8Z_1^4 + 20Z_1^3Z_2 + 12Z_1^2Z_2^2 + 10Z_1Z_2^3 + Z_2^4\right) \qquad (18.117)$$

We can easily minimize Eqn. (18.116) analytically for the case when $Z_1 = Z_2 = Z^*$ (P18.32) or numerically when both Z_1, Z_2 are allowed to "float," and we show the results in Table 18.3 for several lithium-like atoms for both cases; the experimental values are evaluated using the data in Table 18.1.

Several comments can be made:

- The two-parameter variational energy estimate is lower, and hence closer to the ground state energy, than the one-parameter fit; this is guaranteed since the latter is simply a special case of the former.
- The effective charges in the $1s$ and $2s$ states are different in just the way expected above. The values of Z_1 for the two inner-shell electrons are almost identical to those preferred by the helium-like atoms; namely, $Z^* = Z - 5/16$. This effect can be

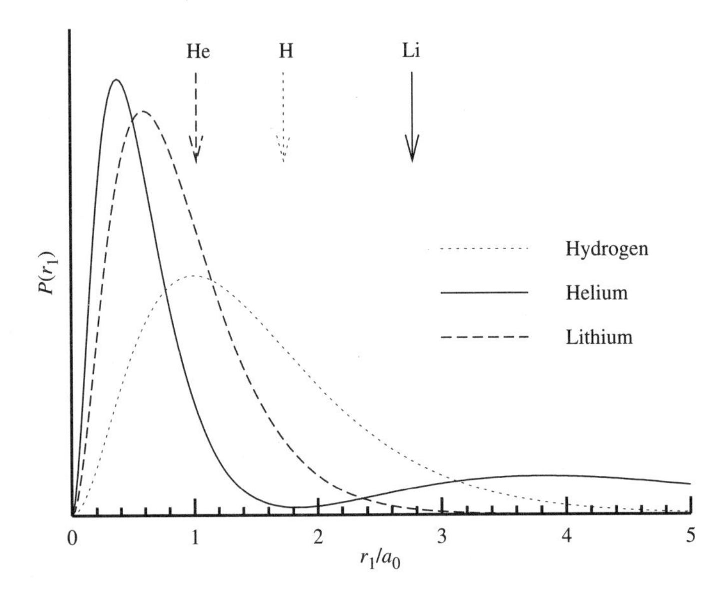

FIGURE 18.17. *Single-electron radial probability density ($P(r_l)$) for hydrogen (dotted), helium (dashed), and lithium (solid). The exact solution in Eqn. (18.21) is used for hydrogen, while the variational wavefunctions of Eqns. (18.100) and (18.115) are used for helium and lithium. The arrows indicate the location of the RMS radius.*

understood on the basis of a Gauss' law argument; the inner electrons should "feel" no effect of a spherically symmetric distribution of charge "outside" themselves.

- The values of the effective charges of the outermost electron are roughly consistent with the observed "first" ionization potentials in Table 18.1. The energy required to ionize the first electron from Li, Be, and B are given by $E_{\rm I} = 5.39$ eV, 18.21 eV, and 37.93 eV, respectively. If we estimate the effective charge seen by the $2s$ electron via

$$E_{\rm I} = -\frac{13.6\,{\rm eV}}{2^2}(Z^*)^2 \tag{18.118}$$

we find $Z^* = 1.26, 2.31, 3.34$. These values are to be compared to values of Z_2 from the variational estimate in the second-to-last row of Table 18.3.

Finally, we can obtain some intuition about the spatial distribution of charge in the simplest atoms considered so far by plotting the radial probability density, $P(r) = r^2|\psi(r)|^2$, for finding a single electron in each system. We use the exact result for hydrogen and the optimized variational wavefunctions for helium and lithium. In the latter cases, we ask for the probability of finding a specific electron, say 1, at some radius. These distributions are obtained by integrating Eqns. (18.100) and (18.115) over the radial coordinates of the unobserved electrons in each case. The results are shown in Fig. 18.17 where the value of the RMS radius, $\sqrt{\langle r^2 \rangle}$, is also indicated as a measure of the size of the atomic system.

18.4.3 The Periodic Table

These results serve to motivate some final comments on the structure of the periodic table arising from the continued "filling" of energy levels with more electrons.

- The inner-core electrons in a heavier atom give rise to a non-Coulombic potential that lifts the degeneracy among l values for a given n; as discussed in the context of Rydberg atoms, the ($l = 0$) s states are most "penetrating," see the largest effective charge, and hence are lower in energy than the corresponding ($l = 1$) p levels, and are "filled in" first. The electron configuration of the first 10 elements (in standard spectroscopic notation) reflects this fact with the assignments

Z	Element	Configuration
1	H	$1s^1$
2	He	$1s^2$
3	Li	$1s^2 2s^1$
4	Be	$1s^2 2s^2$
5	B	$1s^2 2s^2 2p^1$
6	C	$1s^2 2s^2 2p^2$
7	N	$1s^2 2s^2 2p^3$
8	O	$1s^2 2s^2 2p^4$
9	F	$1s^2 2s^2 2p^5$
10	Ne	$1s^2 2s^2 2p^6$

where the first (numerical) label corresponds to the n value, the second to the angular momentum, and the exponent counts the number of electrons in each l state.

- The elements corresponding to completely filled levels are said to have "closed electron shells." The large energy gap to the next excited state implies that the elements with $Z = 2$ (helium) and $Z = 10$ (neon) will be chemically inert, and these correspond to the first two of the *noble gases*.[27]
- There is a relatively large energy difference between the $3p$ and $3d$ levels, so that $Z = 18$ also corresponds to an inert gas; the configurations up to the next noble gas are

11	Na	$1s^1 2s^2 2p^6 3s^1$
12	Mg	$1s^2 2s^2 2p^6 3s^2$
13	Al	$1s^2 2s^2 2p^6 3s^2 3p^1$
14	Si	$1s^2 2s^2 2p^6 3s^2 3p^2$
15	P	$1s^2 2s^2 2p^6 3s^2 3p^3$
16	S	$1s^2 2s^2 2p^6 3s^2 3p^4$
17	Cl	$1s^2 2s^2 2p^6 3s^2 3p^5$
18	Ar	$1s^2 2s^2 2p^6 3s^2 3p^6$

- The level splittings become so great for the $n = 3$ levels that the $4s$ level actually lies lower than the $3d$ level, so that potassium ($Z = 19$) has the structure $1s^2 2s^2 2p^6 3s^2 3p^6 4s^1$.
- The pattern of ionization energies in Fig. 1.5(a) reflects the partial and complete filling of energy levels; more nearly closed shells are harder to ionize. (Atomic radii, calculated in much the same way as for Figure 18.17, show similar evidence for such closed-shell structure; this is illustrated in Figure 18.18.) The *polar-*

[27] As its name implies, helium was first discovered, not via its "terrestrial" chemical properties but rather by spectroscopic measurements of the sun's chromosphere. Evidence for argon was obtained by comparing the density of atmospheric nitrogen with that prepared by other means.

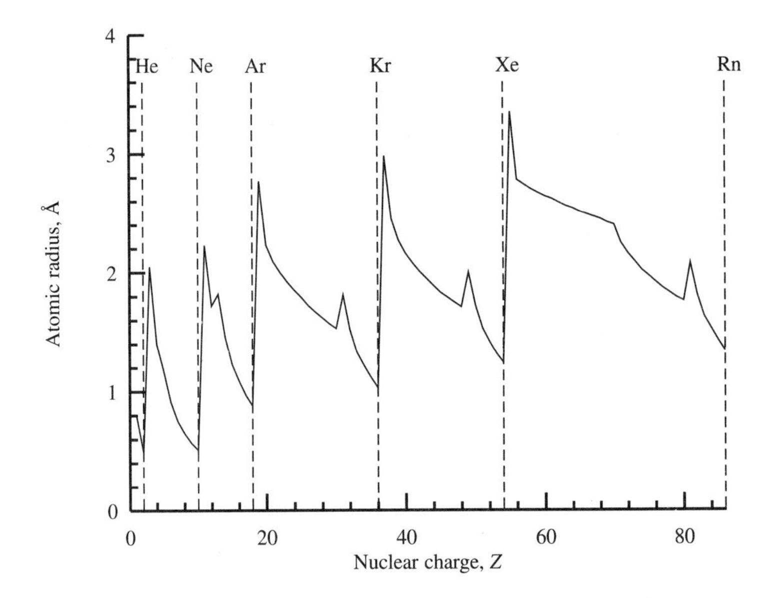

FIGURE 18.18. *(Calculated) atomic radii for elements in the periodic table; the correlation with the closed electron shells corresponding to noble gases is clear.*

izability[28] of each element is a measure of its tendency to interact with an external electric field, distorting in order to minimize the field energy, and shows the opposite correlation with nuclear charge; more nearly closed shells have a decreased polarizability. To see the strong anticorrelation between these two atomic properties, we plot their values for each element in Fig. 18.19, partly for future reference.

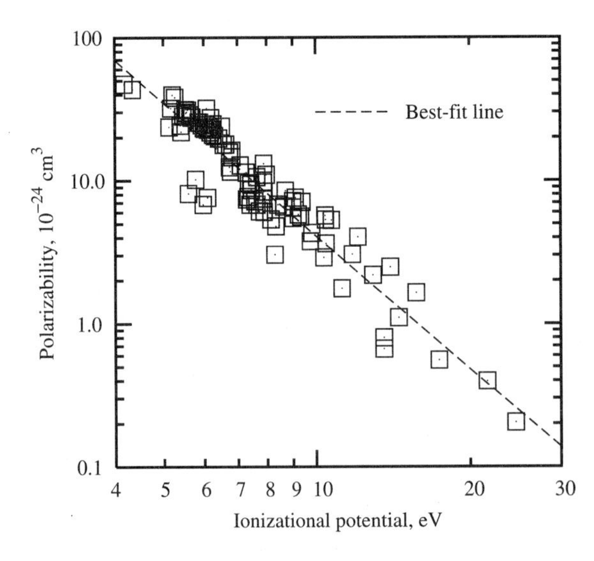

FIGURE 18.19. *Ionization potential (in eV) plotted against polarizability for elements in the periodic table; the straight line corresponds to a "best straight-line fit" through the data points.*

[28]See Sec. 19.4.

18.5 QUESTIONS AND PROBLEMS

Q18.1 If the volume occupied by a hydrogen atom in its ground state were scaled to be the size of your fist, how big a volume would the Rydberg state of hydrogen with the largest value of n observed (roughly $n \approx 350$) be in these scaled units? How big would a muonic atom with $Z = 50$ in its ground state be in these fist units?

Q18.2 In Bohr's original paper deriving the Balmer law for the spectrum of hydrogen, Bohr speculates on why emission from states with large values of n had not been observed in laboratory experiments, while values up to $n \approx 30$ (at that time) had been detected in astrophysical spectra. Discuss why this might be so by considering two additional facts: (1) the hydrogen gas in astrophysical situations is far less dense than in laboratory experiments, but there are large regions of it; (2) the radius of the Bohr orbits increases like n^2, so that the geometrical cross-section scales like n^4.

Q18.3 What would the energy spectrum of hydrogen-like atoms look like in two dimensions? What would be the structure of the periodic table in 2D? For one attempt at this "what-if" question, see Asturias and Aragón (1985).

Q18.4 What property of the muon is responsible for the interest shown in *muon-catalyzed fusion* (μCF)?[29] What one property of the muon would you change if you wanted to make μCF more probable? *Hint:* Recall the discussion in Sec. 12.5.4. of the role of the Coulomb barrier in inhibiting fusion reactions.

Q18.5 The most precise values for the masses of the μ^-, π^-, and K^- have, at one time or another, all been obtained using spectroscopic data from "exotic atoms." How do you think this was accomplished?

Q18.6 The linear piece of the quark-antiquark potential implies that the $q\bar{q}$ pair feel a constant force at large separations. Estimate the magnitude of this force from Fig. 18.14, first in GeV/F and then in tons!

P18.1 Show that the small and large ρ behavior of the hydrogen atom wavefunctions are given by Eqn. (18.7).

P18.2 Radio astronomers searching for radiation generated when electrons and protons recombine have obtained data[30] showing Balmer lines such as in Fig. 18.20. The units on both axes have been suppressed (since they're "strange" astronomy units), so this is a qualitative question. The $n\alpha$ ($n\beta$) labels correspond to transitions between hydrogenic energy levels of the type $(n + 1) \to n$ $((n + 2) \to n)$ for the various elements listed. Show that the three 109 lines and the one 137 line are consistent with the Balmer formula in Eqn. (18.14), when reduced mass effects are included. *Hint:* What is being plotted on the horizontal axis? Evaluate the wavelengths of the lines being described and discuss why astronomers detected them.

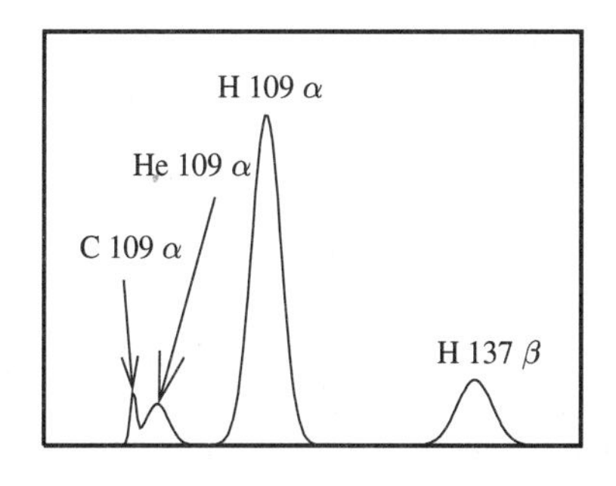

FIGURE 18.20. Schematic picture of spectral lines for highly excited Rydberg atoms as seen by radio astronomers.

[29] See Jones (1989).
[30] Adapted from Kleppner (1986).

P18.3 Show explicitly that the "quasi-circular" radial wavefunctions $R_{n,n-1}(r)$ in Eqn. (18.27) satisfy the radial equation and are properly normalized.

P18.4 For the "quasi-circular" wavefunctions with $l = n - 1$, evaluate $\Delta r^2 = \langle r^2 \rangle - \langle r \rangle^2$. Show that $(\Delta r)^2 / \langle r^2 \rangle \approx 1/2n$.

P18.5 Prove the relation, valid only for S states,

$$|\psi_{n,0,0}(0)|^2 = \frac{\mu}{2\pi\hbar^2} \left\langle \psi_{n,0,0} \left| \frac{dV(r)}{dr} \right| \psi_{n,0,0} \right\rangle$$

for any spherically symmetric potential. Use it to then evaluate the S state wavefunction at the origin squared for any value of n, and compare to the explicit cases listed in Eqns. (18.21), (18.22), and (18.24).

P18.6 **(a)** Use the variational method to estimate the ground state energy of the hydrogen atom by using the trial wavefunction

$$\psi(r; a) = e^{-r/a}$$

where a is the variational parameter. You should, of course, get the correct answer since this is the correct functional form.

(b) Repeat, but use a trial wavefunction of the form

$$\psi(r; b) = e^{-r^2/2b^2}$$

and find the fractional error made in the energy. Show (numerically) that the overlap of the true ground state wavefunction and the variational estimate is roughly

$$\langle \psi_{(1,0,0)}(r) | \psi(r, b_{\min}) \rangle \approx 0.9782$$

(c) Repeat part (b), but use the trial wavefunction

$$\psi(r; c) = \frac{1}{(r^2 + c^2)^2}$$

P18.7 **Virial theorem:**

(a) Generalize the proof of the virial theorem in P13.12 to three dimensions, and confirm it for the case of the Coulomb potential by showing that

$$\langle \hat{T} \rangle_{n,l} = \left\langle \frac{\hat{\mathbf{p}}^2}{2m} \right\rangle_{n,l} = -\frac{1}{2} \left\langle \frac{Ke^2}{r} \right\rangle_{n,l} = \frac{1}{2} \langle \mathbf{r} \cdot \nabla V(r) \rangle_{n,l}$$

(b) Use these results to determine the fraction of energy stored in radial kinetic, rotational kinetic, and potential energy in a hydrogen atom energy eigenstate characterized by (n, l).

P18.8 **Tritium decays:** The nucleus consisting of two neutrons and a single proton (hence with $Z = +1$) is called the *triton*, t (by analogy with the deuteron), and the system consisting of a single electron and a triton is called a *tritium atom*. The triton is unstable against radioactive decays via the process $t \rightarrow {}^3\text{He} + e^- + \overline{\nu}_e$ where ${}^3\text{He}$ consists of one neutron and two protons (i.e., $Z = +2$). On an atomic time scale, the decay and the ejection of the $e^- + \overline{\nu}_e$ happen almost instantaneously. After such a decay, the bound electron suddenly finds itself in a Coulomb field with twice the nuclear charge.

(a) What is the probability that an electron that was originally in the ground state will remain in the ground state of the new system? In the $n = 2s$ state?

(b) What is the probability of being in an $l = 1$ state after the decay? *Hint:* This is just an "expansion in eigenstates problem" similar to Ex. 7.3, but in the context of a real physical system.

P18.9 **Momentum space wavefunctions for hydrogen:** Calculate the momentum space wavefunctions for the three lowest-lying s-wave states of hydrogen, $\psi_{n,l,m}(r)$, i.e., for $n = 1, 2, 3$ and $l = m = 0$. The appropriate three-dimensional Fourier transform is given by

$$\phi_{n,l,m}(\mathbf{p}) = \frac{1}{\sqrt{(2\pi\hbar)^3}} \int d\mathbf{r} e^{-i\mathbf{p}\cdot\mathbf{r}/\hbar} \psi_{n,l,m}(\mathbf{r})$$

Hint: Since there is no preferred direction for s states, write $\mathbf{p}\cdot\mathbf{r} = pr\cos(\theta)$, and let θ be measured along the z axis so that

$$\phi(\mathbf{p}) = \frac{1}{\sqrt{(2\pi\hbar)^2}} \int_0^\infty r^2 dr \psi(r) \int_0^{2\pi} d\phi \int_0^\pi \sin(\theta) e^{-ipr\cos(\theta)/\hbar}$$

For example, for the ground state, show that

$$|\phi_{1,0,0}(\mathbf{p})|^2 = \frac{8}{\pi^2} \frac{p_0^5}{(p^2 + p_0^2)^4}$$

P18.10 For Compton scattering from unbound electrons, the relation between the wavelength of the scattering X rays and the incident ones is

$$\lambda' - \lambda = \frac{2\pi\hbar}{m_e c}(1 - \cos(\theta))$$

where θ is the angle of the scattered photon from the incident direction. If the scattering instead takes place from electrons bound in a hydrogen atom, show that there will be a spread in the scattered wavelengths of the order $\delta\lambda'/\lambda \sim \alpha$ due to the momentum spread of the target electrons. *Hint:* Use the result of the previous problem.

P18.11 **Relativistic wave equation:** The Klein-Gordon equation (first discussed in Sec. 3.1) can be used to solve the hydrogen atom problem in a relativistically correct way; it is only appropriate for spin-0 particles, and thus does not apply directly to electrons and the "real" hydrogen atom. It is useful for pionic and kaonic atoms, and does give some idea of the form of the relativistic corrections.

(a) Assuming a solution of the form $\phi(\mathbf{r}) = R(r)Y_{l,m}(\theta,\phi)$, show that the Klein-Gordan wave equation for the Coulomb potential, namely,

$$\left(E + \frac{ZKe^2}{r}\right)^2 \phi(\mathbf{r}) = \left(\hat{\mathbf{p}}^2 c^2 + (mc^2)^2\right)\phi(\mathbf{r})$$

gives the radial equation

$$\frac{d^2R(r)}{dr^2} + \frac{2}{r}\frac{dR(r)}{dr} + \left(\frac{B}{r} - \frac{C}{r^2} - |A|\right)R(r) = 0$$

where

$$|A| = \frac{|E^2 - (mc^2)^2|}{(\hbar c)^2} \qquad B = \frac{2EZKe^2}{(\hbar c)^2} \qquad C = l(l+1) - (Z\alpha)^2$$

(b) Attempt a solution of the form $R(r) = r^\gamma e^{-r/\beta} G(r)$, and show that $\beta = 1/|A|$ and

$$\gamma = \sqrt{(l + 1/2)^2 - (Z\alpha)^2} - 1/2$$

and obtain the differential equation for $G(r)$.

(c) Show that the condition for the power series solution for $G(r)$ to terminate is

$$\frac{B}{2\sqrt{|A|}} = (n_r + l + 1) + \gamma - l = n + \gamma - l$$

where n_r and n are the usual quantum numbers.

(d) Show that the quantized energies now depend on *both n and l* (in contrast to the nonrelativistic case) and are given by

$$E_{n,l} = mc^2\left(1 + \frac{(Z\alpha)^2}{(n + \sqrt{(l+1/2)^2 - (Z\alpha)^2} - (l+1/2))^2}\right)^{-1/2}$$

(e) Expand this result in powers of $Z\alpha$, and show that

$$E_{n,l} = mc^2\left[1 - \frac{(Z\alpha)^2}{2n^2}\left(1 + \frac{(Z\alpha)^2}{n}\left(\frac{1}{l+1/2} - \frac{3}{4n}\right) + \cdots\right)\right]$$

(f) Compare the $\mathcal{O}(Z\alpha)^2$ terms to the energy levels in Eqn. (18.12), and the $\mathcal{O}(Z\alpha)^4$ terms to the relativistic corrections derived in the next problem.

P18.12 **Relativistic corrections in hydrogen:** In Sec. 1.2, we showed that the expansion of the relativistically correct kinetic energy is given by

$$E = \sqrt{\mathbf{p}^2c^2 + (mc^2)^2} \approx mc^2 + \frac{\mathbf{p}^2}{2m} - \frac{1}{8}\frac{(\mathbf{p}^2)^2}{m^3c^2} + \cdots \tag{18.119}$$

Evaluate the effect of the $\mathcal{O}((\mathbf{p}^2)^2)$ term (now expressed as an operator) on the hydrogen atom wavefunctions using perturbation theory. Show that the first-order energy shift for the state $\psi_{n,l,m}$ is given by

$$\Delta E_{n,l}^{\mathrm{REL}} = -\frac{1}{2}mc^2(Z\alpha)^4\frac{1}{n^3}\left(\frac{2}{2l+1} - \frac{3}{4n}\right)$$

Hint: Write $\hat{\mathbf{p}}^2$ in terms of the unperturbed Hamiltonian using

$$\hat{\mathbf{p}}^2 = 2m\left(\hat{H} + \frac{ZKe^2}{r}\right)$$

and use the average values of $\langle r^n \rangle$.

P18.13 Solve the Schrödinger equation for the potential

$$V(r) = -\frac{ZKe^2}{r}\left(1 + b\frac{a_0}{r}\right)$$

and show that the energy levels are given by

$$E_{n,l} = -\frac{1}{2}\mu c^2(Z\alpha)^2\frac{1}{[n + \sqrt{(l+1/2)^2 - 2b} - (l+1/2)]^2}$$

where $n = n_r + l + 1$ has its usual meaning. *Hint:* Use the method outlined in P18.11.

P18.14 **Hydrogen atom in parabolic coordinates:**

(a) Show that the Hamiltonian for the hydrogen atom problem is also separable in *parabolic coordinates*,[31] namely,

$$\begin{aligned} q_1 &= r - z = r(1 - \cos(\theta)) \\ q_2 &= r + z = r(1 + \cos(\theta)) \\ \phi &= \phi \end{aligned}$$

where (r, θ, ϕ) are the usual spherical coordinates. *Hint:* If you use the expression

$$\nabla^2 = \frac{4}{(q_1 + q_2)}\left[\frac{\partial}{\partial q_1}\left(q_1\frac{\partial}{\partial q_1}\right) + \frac{\partial}{\partial q_2}\left(q_2\frac{\partial}{\partial q_2}\right)\right] + \frac{1}{q_1q_2}\frac{\partial^2}{\partial \phi^2}$$

you should prove it first.

[31] See, e.g., Schiff (1955).

(b) Try a solution of the form

$$\psi(q_1, q_2, \phi) = Q_1(q_1)Q_2(q_2)e^{im\phi}$$

obtain the equations for $Q_{1,2}$, and show how the quantized eigenvalues arise. *Note:* This form of solution is useful when the z axis is singled out for some reason, as with the Stark effect in Sec. 19.4.2 or in scattering problems as in Chap. 20. It also provides more evidence for the remarkable symmetry properties of this problem.

P18.15 Classical probability distributions: Show explicitly that the form of the classical probability distribution in Eqn. (18.52) is properly normalized, i.e., show that

$$1 = \int_{r_{\min}}^{r_{\max}} P_{\text{CL}}^{(n,l)}(r)\, dr = \frac{1}{\pi a_0 n^2} \int_{r_{\min}}^{r_{\max}} dr \frac{r dr}{\sqrt{2ra_0 n^2 - l(l+1)n^2 a_0^2 - r^2}}$$

Evaluate the *classical* expectation values $\langle r \rangle$, $\langle r^2 \rangle$, and $\langle 1/r \rangle$, and compare them to the quantum results in Eqns. (18.29), (18.30), and (18.31).

P18.16 Bohr-Sommerfeld quantization of the hydrogen atom: A semiclassical picture of the hydrogen atom considers only planar orbits with variables r, θ and their corresponding momenta.

(a) Show that the WKB-type quantization condition on the angular variable gives

$$\int_{\theta_{\min}}^{\theta_{\max}} L d\theta = n_\theta h = 2\pi n_\theta \hbar \qquad \text{or} \qquad L = n_\theta \hbar$$

(b) The corresponding condition for the radial coordinate is

$$\int_{r_{\min}}^{r_{\max}} p_r\, dr = \int_{r_{\min}}^{r_{\max}} \sqrt{2\mu} \sqrt{E - \frac{L^2}{2\mu r^2} + \frac{ZKe^2}{r}}\, dr = \pi n_r \hbar$$

Use the results of App. B.1 to evaluate the integrals, and show that the resulting energy quantization condition is

$$E = -\frac{1}{2}\mu c^2 \frac{(Z\alpha)^2}{(n_r + n_\theta)^2}$$

P18.17 Classical Lenz-Runge vector:

(a) Using the definition of Eqn. (18.55) and the equations of motion, show that $\mathbf{R}$ is constant in time, namely, that $\dot{\mathbf{R}} = 0$.

(b) Derive Eqn. (18.56) by squaring Eqn. (18.55).

(c) Show that the expression for $\mathbf{r} \cdot \mathbf{R} = r|\mathbf{R}|\cos(\theta)$ gives the equation for an ellipse in Eqn. (18.39).

P18.18 Quantum Lenz-Runge vector:

(a) Using the definition in Eqn. (18.57), show that $\hat{\mathbf{R}}$ is conserved by proving that it commutes with the Hamiltonian

$$\hat{H} = \frac{\hat{\mathbf{p}}^2}{2\mu} - \frac{Ke^2}{r}$$

(b) Show that $\hat{\mathbf{R}}$ is Hermitian.

(c) Show that the vector operator defined via

$$\hat{\mathbf{S}} = \sqrt{\frac{\mu(Ke^2)^2}{2|E|}}\hat{\mathbf{R}}$$

satisfies

$$[\hat{S}_x, \hat{L}_y] = i\hbar \hat{S}_z \qquad \text{and} \qquad [\hat{S}_x, \hat{S}_y] = i\hbar \hat{L}_z$$

and all the usual permutations. (This requires a reasonable amount of algebra!)

(d) Define the generalized angular momentum operators

$$\hat{M}_{ij} = \left(\frac{\hbar}{i}\right)\left(x_i \frac{\partial}{\partial x_j} - x_j \frac{\partial}{\partial x_i}\right)$$

for $i, j = 1, 2, 3$. Show that they satisfy the commutation relation

$$\left[\hat{M}_{ij}, \hat{M}_{kl}\right] = i\hbar\left(\hat{M}_{ik}\delta_{jl} + \hat{M}_{jl}\delta_{ik} - \hat{M}_{il}\delta_{jk} - \hat{M}_{jk}\delta_{il}\right)$$

Show that the $\hat{M}_{ij}$ are equivalent to the familiar $\hat{\mathbf{L}}$ with the identifications

$$\hat{M}_{12} = \hat{L}_z \qquad \hat{M}_{23} = \hat{L}_x \qquad \hat{M}_{31} = \hat{L}_y$$

Finally, show that $\hat{\mathbf{S}}$ can be included by extending the index label to 4 and identifying

$$\hat{M}_{14} = \hat{S}_x \qquad \hat{M}_{24} = \hat{S}_y \qquad \hat{M}_{34} = \hat{S}_z$$

Note: If one thinks of the $i, j = 1, 2, 3$ sector as representing angular momentum operators in three dimensions, this exercise shows that there is actually a kind of four-dimensional symmetry to the inverse-square law problem, giving it its enhanced symmetries.

P18.19 **Model for the quantum defect in alkali atoms:** For values of the angular momentum l of the excited electron in a Rydberg atom satisfying $l > l_{\text{core}}$ (where l_{core} is the maximum possible angular momentum of the inner shell electrons), there is little actual penetration of the ionic core. One of the most important interactions then responsible for the quantum defect is the so-called *polarization energy* of the outer electron with the dipole field of the core (which is induced by the electron itself). This potential has the form

$$V_{\text{POL}}(r) = -\frac{1}{2}\frac{\alpha_D (Ke^2)^2}{r^4} \sim -\frac{b}{2}\frac{Ke^2 a_0^3}{r^4}$$

where the *dipolar polarizability*, $\alpha_D = ba_0^3/K$, is written in terms of a dimensionless constant b as suggested by results from Sec. 19.4.2. Evaluate the effect of this potential on hydrogen atom states of arbitrary n, l using perturbation theory, and show that the resulting quantum defect scales as $\delta_l \sim 1/l^5$.

P18.20 **Scattering of Rydberg atoms**[32]**:** The cross sections for collisions (of a certain type) of highly excited $l = 0$ sodium atoms are listed below along with the effective quantum number, n^*:

State	n^*	$\sigma(10^9\ \text{Å}^2)$
16s	15.2	0.78 ± 0.18
18s	17.2	1.25 ± 0.13
20s	19.2	2.16 ± 0.66
23s	22.2	3.8 ± 1.0
25s	24.2	3.8 ± 1.0
27s	26.2	5.8 ± 1.8

and the data can be "fit" via the formula

$$\sigma = (3.3 \pm 0.6) \times 10^4 (n^*)^{3.7\pm0.5}$$

[32]The data and fit to data are taken from Gallagher et al. (1982).

Plot these data along with the fit on log-log paper to see if this is true. Can you explain why the observed dependence on n^* is consistent with $\sigma \propto (n^*)^4$?

P18.21 Estimate the classical period of motion in a hydrogen-like state with quantum numbers n, l for a given reduced mass and nuclear charge Z.

P18.22 **Scaling laws for muon-chemistry:** Suppose that the muon (assumed for the moment to be stable) replaced each electron in all atoms. If we call the ratio of the muon-to-electron mass $R = m_\mu/m_e \approx 200$, by what power of R (if any) would the following quantities scale?

(a) Size of typical atom.
(b) Binding energies.
(c) Density of matter.
(d) Effective "spring constant" in a diatomic molecule.
(e) Typical vibrational energy level in diatomic molecule.
(f) Typical rotational energy level in diatomic molecule.

P18.23 Use Gauss's law to show that the electric field for a uniform spherical distribution of charge of radius R and total charge Ze is given by

$$|\mathbf{E}(r)| = \begin{cases} ZKer/R^3 & \text{for } r < R \\ ZKe/r^2 & \text{for } r > R \end{cases}$$

Use this result to derive Eqn. (18.67).

P18.24 Estimate the shift in the $1s$, $2s$, and $2p$ energy levels due to the modified Coulomb potential in Eqn. (18.69) using perturbation theory. Assume that $ZR/a_0^{(e,\mu)} << 1$. For the ground state, for what value of Z does perturbation theory begin to become unreliable; i.e., when does $E^{(1)}/E^{(0)} \sim 0.1$?

P18.25 With the change of variables discussed in Sec. 17.1, namely, $u(r) = rR(r)$, rewrite the 3D Schrödinger equation for s states in the form shown in Eqn. (17.30).
You may have already written a short computer program to integrate numerically the Schrödinger equation in one dimension to analyze problems in Chap. 11. Use your program to solve numerically for the ground state energy of the muonic atom with $Z = 60$ and $R = 6.3$ F with the "smeared out" Coulomb potential in Eqn. (18.67). Compare your answer to the result obtained by simply "scaling" the hydrogen result using the muon mass and appropriate charge Z, and also to the experimental value of roughly $E_0 \approx -6.88$ MeV. How does your wavefunction (if your program can generate it) differ from the Bohr ground state? *Hint:* If the step size in your program is $\Delta x = \epsilon$, use the starting data $u(0) = 0$ and $u(\Delta x = \epsilon) = \Delta x$.

P18.26 Combine the appropriate expressions in Ex. 18.4 to obtain Eqn. (18.79), and evaluate it numerically.

P18.27 **Top-antitop bound states?**

(a) Estimate the period of a $t\bar{t}$ bound state in a color Coulomb potential in its ground state by using analogies with the hydrogen atom and positronium; recall that $\alpha_s \approx 0.1 - 0.2$ replaces the fine-structure constant α, and don't forget reduced mass effects.

(b) The top quark (and its antiparticle) decay with a lifetime given in "particle physics notation" by

$$\lambda_t = \frac{1}{\tau_t} \approx \frac{G_F m_t^3}{8\pi\sqrt{2}}$$

where $G_F = 1.166 \times 10^{-5}\,\text{GeV}^{-2}$ as in P1.16. As in that problem, add enough powers of $\hbar$ and c to make this equation dimensionally correct, and use it to evaluate τ_t. Do these quarks live long enough to make a recognizable quarkonium state?

P18.28 **Feynman-Hellmann theorem:**

(a) Using the fact that

$$\langle E\rangle(\lambda) = \langle\psi(\lambda)|\hat{H}(\lambda)|\psi(\lambda)\rangle$$

prove Eqn. (18.82).

(b) Check the Feynman-Hellmann theorem explicitly by calculating both sides for the Coulomb potential using the nuclear charge Z as the parameter.

(c) Repeat using the simple harmonic oscillator potential and the spring constant K as the parameter.

(d) Repeat using the standard infinite well and the mass, m, as the parameter.

P18.29 **Helium-like atoms:** Calculate the values in the blank entries in Table 18.2. Use Table 18.1 to calculate the experimental values and Eqn. (18.91), (18.98), and (18.108) to evaluate the zero-order, perturbation theory and variational estimates.

P18.30 **Hydrogen ion:** The hydrogen ion consists of a hydrogen atom with an additional electron; it is a helium-like atom but with $Z = 1$. Estimate its ground state energy using the variational formula in Eqn. (18.108), and compare your result to the minimum energy of a hydrogen atom and a free electron. Does your result imply that the hydrogen ion does not have a bound state?

P18.31 Evaluate the exchange integral in Eqn. (18.113) for the case of $(nlm) = (200)$, and use your result to compare to the ortho-para-helium splitting for the $2s$ state as read from Fig. 18.17.

P18.32 Derive the variational estimate in Eqn. (18.116) for the lithium ground state. Find the value of $Z = Z_1 = Z_2$ that minimizes Eqn. (18.116).

CHAPTER 19

Gravity and Electromagnetism in Quantum Mechanics

In introductory physics courses, we are introduced (usually at some length) to two of the four fundamental forces of nature,[1] namely, gravity and electromagnetism. In this chapter, we discuss some examples of the description of these forces and their effects in quantum mechanical terms.

19.1 CLASSICAL GRAVITY AND QUANTUM MECHANICS

The observation that gravity is the most obvious force in the macroscopic world,[2] while electromagnetism is the dominant interaction at the microscopic level of atoms and molecules, is easily explained by two facts:

1. The gravitational force between an electron and a proton is 40 orders of magnitude smaller than the electrostatic attraction between them (P19.1), independent of their separation.
2. The electric charges of the electron and proton, while of opposite sign, are experimentally remarkably close in magnitude; the best experimental limits[3] imply

[1]The other important interactions, the so-called weak and strong interactions, are of most relevance in the subatomic domain and are discussed in any good text on subatomic (i.e., nuclear or elementary particle) physics; see, e.g, Perkins (1987).
[2]Everyone, after all, has fallen down!
[3]Taken from the *Review of Particle Properties* (1994).

that $|Q_p + Q_e|/e \leq 10^{-21}$. This implies that normal matter, in bulk, is electrically neutral, allowing gravitational interactions to dominate, since its interaction strength (i.e., its mass) increases with the size of the system, while the net charge is roughly zero.

These facts suggest that it might not be possible to find a terrestrial system in which gravity and quantum mechanics simultaneously play an important and fundamental role.[4]

Gravitational effects are occasionally used to novel effect in laboratory atomic systems. One notable example[5] is the "atomic fountain" that "launches" cesium atoms upward in an apparatus used to make precise frequency measurements and uses the weak pull of gravity to increase the interaction time. Other electrically neutral particles, most notably neutrons, have been used in a variety of experiments that test both the "classical mechanics" of subatomic particles and the quantum effects of local gravity on the quantum mechanical wavefunction, and that is the main topic of this section.

Beams of reactor neutrons have been measured[6] to "fall" in the earth's gravitational field, following parabolic trajectories, just like textbook projectiles.[7] This is purely a classical effect and would be a "wavepacket" limit of a particle in a linear potential (as in P16.4).

A much more interesting quantum interference effect was observed using *two* beams of neutrons that were split and then allowed to recombine as in Fig. 19.1. All such interference experiments rely on the wave phenomenon of the addition of different wave amplitudes, differing by a phase; the electron interference patterns in Fig. 1.2 are an example where the phase difference arises from a difference in path length, and we review this effect in Sec. 19.8.

In one version of this experiment,[8] the paths followed by the two neutron beams differ in that they find themselves in regions of different gravitational potential energy. The resulting phase difference is most easily estimated by using WKB-type wavefunctions (as in Sec. 11.4.1), i.e., we use

$$\psi(x_1) \sim \psi(x_2)\exp\left(i\int_{x_1}^{x_2} p(x)dx/\hbar\right) \tag{19.1}$$

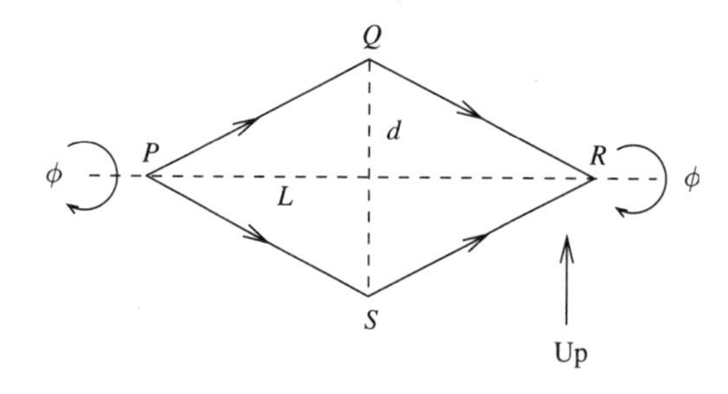

FIGURE 19.1. *Geometry for interference experiment showing interference effects as a result of quantum phase of neutron wavefunction due to gravitational potential.*

[4]Recall, however, the compact astrophysical objects discussed in Sec. 8.4.3.

[5]See, e.g., Gibble and Chu (1993).

[6]See, e.g., Dabbs et al. (1965) who even measure the local acceleration of gravity using neutron fall to be $g_{\text{exp}} = 975.4 \pm 3.1\text{cm/sec}^2$, compared to the known local value of $g = 979.74\text{cm/sec}^2$ at the site of the experiment.

[7]The usual approximation of neglecting air resistance is, in this case, presumably an excellent one!

[8]See Colella and Overhauser (1980) for details and the original references.

for the phase of the neutron wavefunction. If we consider the gravitational potential, $V(y) = mgy$ [where $y = y(x) = (d/L)x$ along the path] as a small perturbation, we have

$$p(x) = \sqrt{2m(E - V(x))}$$
$$= \sqrt{2mE}\left(1 - \frac{V(x)}{E}\right)^{1/2}$$
$$\approx \sqrt{2mE}\left(1 - \frac{1}{2}\frac{V(x)}{E} + \cdots\right) \tag{19.2}$$

The first term will be the same for both upper and lower arms (provided the path lengths are identical), so that the interference comes from the second term and depends on the path followed through the potential. We then have the resulting total phase increase in going along the upper arm PQR given by

$$\phi_{PQR} = -\frac{1}{\hbar}\sqrt{\frac{m}{2E}}2\int_0^L mgy(x)\,dx = \frac{m^2 g}{\hbar p}\frac{dL}{2} \tag{19.3}$$

where we use $E = p^2/2m$. The total difference in phase between the upper (PQR) and lower (PSR) paths can then be written as

$$\Delta\phi = -\frac{m^2 g A \lambda}{2\pi\hbar^2} \tag{19.4}$$

where $A = 2dL$ is the area of the $PQRS$ surface, and we have used the de Broglie relation. Using values that are relevant to the real apparatus, namely, $A \approx (2 - 3\text{cm})^2$, and thermal neutrons ($E \approx k_B T \approx 0.03$ eV), we find that $\Delta\phi \approx 28$; this would correspond to $\Delta\phi/2\pi \approx 5$ fringe shifts in a standard interference experiment. To eliminate systematic effects and make the interference pattern more obvious, the neutron counting rate was measured at point R (as this measured the total recombined amplitude squared) as the device was rotated around the axis PR. As the angle is changed, the effective path difference and relative phase are altered; this resulted in the pattern shown in Fig. 19.2, which illustrates the effect. More sophisticated versions of this experimental technique[9] have even proved

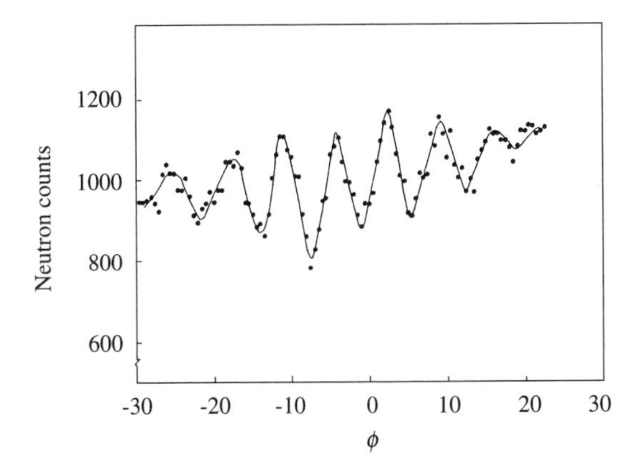

FIGURE 19.2 *Interference pattern from neutron interferometer showing the effect of gravity on the phase of the neutron wavefunction; data are from Colella et al. (1975). (Permission of Roberto Colella.)*

[9]See Staudenmann et al. (1980).

sensitive to the earth's rotation and succeeded in measuring the resulting Coriolis effect on the neutron's path.

19.2 ELECTROMAGNETIC FIELDS

The classical equations governing the motion of a charged particle moving under the influence of external electric and magnetic fields are given by Newton's law of motion with the Lorentz force law, namely,

$$m\ddot{\mathbf{a}}(t) = \mathbf{F} = q\left[\mathbf{E}(\mathbf{r}, t) + \mathbf{v}(t) \times \mathbf{B}(\mathbf{r}, t)\right] \tag{19.5}$$

We will learn below how to incorporate the interactions of electromagnetic (hereafter EM) fields with charged particles into a quantum description of matter, but we first review the classical description of the EM fields themselves.[10]

19.2.1 Classical Electric and Magnetic Fields

One of the major intellectual triumphs of classical physics[11] was the coherent presentation and extension of the then-known laws of electricity and magnetism by Maxwell. In modern language, these can be written in the following form:

$$\nabla \cdot \mathbf{E}(\mathbf{r}, t) = \frac{1}{\epsilon}\rho(\mathbf{r}, t) \qquad \text{Gauss's law} \tag{19.6}$$

$$\nabla \cdot \mathbf{B}(\mathbf{r}, t) = 0 \qquad \text{“No-name” law} \tag{19.7}$$

$$\nabla \times \mathbf{E}(\mathbf{r}, t) = -\frac{\partial}{\partial t}\mathbf{B}(\mathbf{r}, t) \qquad \text{Faraday's law} \tag{19.8}$$

$$\nabla \times \mathbf{B}(\mathbf{r}, t) = \mu\mathbf{J}(\mathbf{r}, t) + \mu\epsilon\frac{\partial}{\partial t}\mathbf{E}(\mathbf{r}, t) \qquad \text{Ampere's law} \tag{19.9}$$

The electric (**E**) and magnetic (**B**) fields are determined by the local charge (ρ) and current (**J**) densities. As a simple example, recall that Gauss's law implies that the electric field arising from a point electric charge in vacuum is given by

$$\mathbf{E}(\mathbf{r}) = \frac{q}{4\pi\epsilon_0}\frac{\hat{\mathbf{r}}}{r^2} = \frac{Kq}{r^2}\hat{\mathbf{r}} \tag{19.10}$$

giving the familiar inverse-square law for the Coulomb force. The corresponding equation for the magnetic field, Eqn. (19.7), implies that there are no point magnetic charges. We also note that Maxwell's equations are often written in an even more compact form using the auxiliary fields $\mathbf{D} = \epsilon\mathbf{E}$ and $\mathbf{B} = \mu\mathbf{H}$, using the fundamental constants of electricity (the electric permittivity, ϵ) and magnetism (the magnetic permeability, μ).

The important concept of the conservation of electric charge is implicitly contained in Maxwell's equations and can be recovered by taking $\nabla\cdot$ (Ampere's law) and using Gauss's law to find

[10]One intriguing aspect of the quantization of the electromagnetic field itself is presented in Sec. 19.9.

[11]Feynman (1964) has said "... there can be little doubt that the most significant event of the 19th century will be judged as Maxwell's discovery of the laws of electrodynamics. The American Civil War will pale into provincial insignificance in comparison with this important scientific event of the same decade."

$$\frac{\partial \rho(\mathbf{r}, t)}{\partial t} + \nabla \cdot \mathbf{J}(\mathbf{r}, t) = 0 \tag{19.11}$$

This so-called *equation of continuity* has the identical form as that arising from the conservation of probability in Sec. 5.1.

One of the most important consequences of Maxwell's equations is the connection with the electromagnetic wave equation. One can, for example, take $\nabla \times$ (Ampere's law), and use Eqns. (19.7) and (19.8), and the vector identity

$$\nabla \times \left(\nabla \times \mathbf{B}\right) = \nabla \left(\nabla \cdot \mathbf{B}\right) - \nabla^2 \mathbf{B} \tag{19.12}$$

to show that

$$\frac{\partial^2 \mathbf{B}(\mathbf{r}, t)}{\partial t^2} = \frac{1}{\epsilon \mu} \nabla^2 \mathbf{B}(\mathbf{r}, t) = c^2 \nabla^2 \mathbf{B}(\mathbf{r}, t) \tag{19.13}$$

in the absence of charges; the electric field $\mathbf{E}(\mathbf{r}, t)$ satisfies an identical wave equation. The derivation of the connection between the three fundamental constants $c = 1/\sqrt{\epsilon_0 \mu_0}$ was one of the main results of Maxwell's contribution. Plane wave solutions of Eqn. (19.13) of the form

$$\mathbf{B}(\mathbf{r}, t) = \mathbf{B}_0 e^{i(\mathbf{k}\cdot\mathbf{r} - \omega t)} \qquad \text{and} \qquad \mathbf{E}(\mathbf{r}, t) = \mathbf{E}_0 e^{i(\mathbf{k}\cdot\mathbf{r} - \omega t)} \tag{19.14}$$

must satisfy $\omega^2 = |\mathbf{k}|^2 c^2$. They must also, however, be consistent with each of the individual Maxwell equations. Faraday's law, for example, implies that

$$\mathbf{k} \times \mathbf{E}_0 = -\omega \mathbf{B}_0 \tag{19.15}$$

while Ampere's law requires that

$$\mathbf{k} \times \mathbf{B}_0 = c^2 \omega \mathbf{E}_0 \tag{19.16}$$

These imply that the electric and magnetic fields in an EM wave must be *transverse,* that is, perpendicular to the direction of propagation determined by $\mathbf{k}$, and must have magnitudes related by $|\mathbf{B}_0| = |\mathbf{E}_0|/c$.

A final important result describing the flow of energy can be obtained by taking $\mathbf{B} \cdot$ Faraday's law $- \mathbf{E} \cdot$ Ampere's law (for simplicity, in the case where there is no current density) where one finds

$$\frac{\partial u(\mathbf{r}, t)}{\partial t} + \nabla \cdot \mathbf{S}(\mathbf{r}, t) = 0 \tag{19.17}$$

The quantity $u(\mathbf{r}, t)$ is given by

$$u(\mathbf{r}, t) = u_E(\mathbf{r}, t) + u_B(\mathbf{r}, t) = \frac{1}{2}\epsilon |\mathbf{E}(\mathbf{r}, t)|^2 + \frac{1}{2\mu} |\mathbf{B}(\mathbf{r}, t)|^2 \tag{19.18}$$

and can be shown to correspond to the local energy density stored in the electric and magnetic fields. The so-called *Poynting vector,* given by

$$\mathbf{S}(\mathbf{r}, t) = \frac{1}{\mu} \mathbf{E}(\mathbf{r}, t) \times \mathbf{B}(\mathbf{r}, t) \tag{19.19}$$

then describes the rate of energy flow per unit time per unit area; $\mathbf{S}$ also has the units of *intensity,* namely, power per unit area. Eqn. (19.17) then expresses the manner in which electromagnetic energy "flows" through a system. Using the relativistic connection between energy and momentum, we also note that $\mathbf{S}/c$ has the units of momentum per unit area per unit time or force per area or pressure and can be used to describe *radiation pressure.* The related quantity, $\mathbf{S}/c^2$, has the units of *momentum density* or momentum per unit volume; we can write $\mathbf{S}/c^2 = d\mathbf{p}/dV$.

Example 19.1 Laser Fields Modern high-power lasers used in atomic physics research can achieve intensities of the order $I = 10^{22}\text{W/m}^2$. The strength of the electric field component corresponding to this intensity is given by

$$|\mathbf{S}| = \frac{1}{\mu_0}|\mathbf{E}||\mathbf{B}| = \frac{1}{\mu_0 c}|\mathbf{E}|^2 \tag{19.20}$$

so that $|\mathbf{E}| \sim 2 \times 10^{12} N/C$; this can also be expressed in somewhat different units as $|\mathbf{E}| \sim 200\text{V/Å}$. A standard value to which to compare this field strength is the magnitude of the electric field felt by the electron in the hydrogen atom, namely, the field arising from a charge e at a distance of roughly $a_0 \sim 0.53\text{Å}$. This "typical" atomic field strength is approximately

$$E_c \equiv \frac{Ke}{a_0^2} = \left(\frac{Ke^2}{\hbar c}\right)\left(\frac{\hbar c}{a_0^2}\right)\frac{1}{e} \approx \left(\frac{1}{137}\right)\left(\frac{1973\text{ eV}\cdot\text{Å}}{(0.53\text{ Å})^2}\right)\frac{1}{e} \approx 51\frac{\text{V}}{\text{Å}} \tag{19.21}$$

Another useful relation arises when one thinks of the laser pulse as an ensemble of photons, each carrying quantized energy $\hbar\omega$ at the speed of light. The Poynting vector can then be written in the form

$$|\mathbf{S}| = n_\gamma \hbar\omega c \tag{19.22}$$

where n_γ is the number density of photons. For a laser of the intensity above, operating at $\lambda = 1000\text{ nm} = 10^4\text{ Å}$, corresponding to a photon energy $\hbar\omega \approx 1$ eV, the relation Eqn. (19.22) implies that there are roughly $2 \times 10^{22}\,\gamma/\text{m}^3$. An atom in such a laser field can find itself immersed in a relatively dense "photon gas."

Example 19.2 Uncertainty Principle Constraints on E and B Fields It is clear from the discussions above that the fact that electromagnetic radiation can carry momentum is not a consequence of any particle-like (i.e., photon) interpretation but arises naturally from classical considerations. We can, however, combine the expression for the momentum density in EM fields,

$$\frac{d\mathbf{p}}{dV} = \frac{1}{\mu_o c^2}\mathbf{E}\times\mathbf{B} \tag{19.23}$$

with the standard Heisenberg uncertainty principle to "motivate" (not prove) an interesting limit to the measurability of electromagnetic field strengths. If we take one component of Eqn. (19.23) and consider the momentum in some small volume, δV, we can write

$$\frac{p_x}{\delta V} \sim \left(\mathbf{E}\times\mathbf{B}\right)_x = \frac{1}{\mu_0 c^2}\left(E_y B_z - E_z B_y\right) \tag{19.24}$$

We can argue that the *uncertainty* in this momentum component will be of the same order as the uncertainties in the values of the field components, so that

$$\left(\delta V \frac{1}{\mu_0 c^2}\Delta E_y \Delta B_z\right)\Delta x \sim \Delta p_x \Delta x \gtrsim \frac{\hbar}{2} \tag{19.25}$$

where the last bound comes from the quantum mechanical uncertainty principle. Taken together, these imply that

$$\Delta E_y \Delta B_z \gtrsim \frac{\hbar}{\mu_0 c^2}\frac{1}{\Delta x \delta V} \tag{19.26}$$

This interesting result can be derived in a somewhat more careful way by examining the ways in which one might attempt to measure simultaneously any two such components of the EM fields.[12]

It suggests that the values of the electromagnetic fields become increasingly uncertain as one attempts to measure them on increasingly smaller distance scales. The values of the **E** and **B** fields, and hence of the energy contained in the electromagnetic field, can be thought of as fluctuating wildly at short distances.

In classical mechanics, one often finds it useful to use the concept of a potential energy function $V(\mathbf{r})$ instead of the Newtonian force $\mathbf{F}(\mathbf{r})$; the connection between the two is $\mathbf{F}(\mathbf{r}) = -\nabla V(\mathbf{r})$. This relation makes it clear that different choices of the "zero of potential" can have no physical meaning as $V(\mathbf{r}) \rightarrow V(\mathbf{r}) + V_0$ yields the same measurable force.

A similar, but more subtle and deep, situation arises in electrodynamics where one can express the (physical) electric and magnetic fields in terms of scalar ($\phi(\mathbf{r}, t)$) and vector ($\mathbf{A}(\mathbf{r}, t)$) potentials via

$$\mathbf{B}(\mathbf{r}, t) = \nabla \times \mathbf{A}(\mathbf{r}, t) \tag{19.27}$$

$$\mathbf{E}(\mathbf{r}, t) = -\nabla\phi(\mathbf{r}, t) - \frac{\partial}{\partial t}\mathbf{A}(\mathbf{r}, t) \tag{19.28}$$

It is easy to show that the fields produced by the potentials ϕ, $\mathbf{A}$ will be identical to those produced by the new potentials ϕ', $\mathbf{A}'$ provided they are related by the transformations

$$\mathbf{A}'(\mathbf{r}, t) = \mathbf{A}(\mathbf{r}, t) + \nabla f(\mathbf{r}, t) \tag{19.29}$$

$$\phi'(\mathbf{r}, t) = \phi(\mathbf{r}, t) - \frac{\partial}{\partial t} f(\mathbf{r}, t) \tag{19.30}$$

where $f(\mathbf{r}, t)$ is an arbitrary scalar function. There are thus an infinite number of different electromagnetic potentials that correspond to a given configuration of measureable fields. Such a change in potentials is called a *gauge transformation,* and will be seen to play an important role in the quantum mechanical treatment of charged particle interactions.

Example 19.3 Scalar and Vector Potentials We will consider below the familiar cases of charged particles acted on by uniform electric and magnetic fields, and we consider the electromagnetic potentials, ϕ and $\mathbf{A}$, that can describe these cases.

A uniform electric field, given by $\mathbf{E}(\mathbf{r}, t) = \mathbf{E}_0$ can be obtained from potentials

$$\phi(\mathbf{r}, t) = -\mathbf{E}_0 \cdot \mathbf{r} \qquad \text{and} \qquad \mathbf{A}(\mathbf{r}, t) = 0 \tag{19.31}$$

which is the standard choice. However, the same field can be described by the potentials

$$\phi'(\mathbf{r}, t) = 0 \qquad \text{and} \qquad \mathbf{A}'(\mathbf{r}, t) = -\mathbf{E}_0 t \tag{19.32}$$

It is easy to see that the transformation relating these two sets of potentials via Eqns. (19.29) and (19.30) is generated by $f(\mathbf{r}, t) = -\mathbf{E}_0 \cdot \mathbf{r}t$.

A uniform magnetic field in the z direction, $\mathbf{B} = B_0\hat{\mathbf{z}} = (0, 0, B_0)$ can be obtained from the potentials

$$\phi(\mathbf{r}, t) = 0 \qquad \text{and} \qquad \mathbf{A}(\mathbf{r}, t) = \frac{B_0}{2}(-y, x, 0) \tag{19.33}$$

[12] See, e.g., Landé (1951).

One of the many other possible choices is

$$\phi'(\mathbf{r}, t) = 0 \qquad \text{and} \qquad \mathbf{A}'(\mathbf{r}, t) = B_0(-y, 0, 0) \tag{19.34}$$

and these two sets can be seen to be gauge equivalent with $f(\mathbf{r}, t) = -B_0 xy/2$.

19.2.2 E and B Fields in Quantum Mechanics

The standard procedure we have adopted to extend the equations of motion of classical mechanics to the quantum Schrödinger equation (in position space) has been through a generalization of the Hamiltonian formulation of classical mechanics. For the free particle in one dimension, this consisted of simply making the identification

$$H_{\text{classical}} = \frac{p^2}{2m} \Longrightarrow \hat{H} = \frac{\hat{p}^2}{2m} \tag{19.35}$$

We show in App. E that the classical Hamiltonian function appropriate for a charged particle (of charge q) acted on by external electric and magnetic fields (now in three dimensions) is given by

$$H_{\text{classical}} = \frac{1}{2m}\big(\mathbf{p} - q\mathbf{A}(\mathbf{r}, t)\big)^2 + q\phi(\mathbf{r}, t) \tag{19.36}$$

This choice for $H_{\text{classical}}$ reproduces the correct equations of motion in Eqn. (19.5).

The corresponding quantum mechanical Hamiltonian is obtained by replacing the momentum variable by its operator counterpart, thereby giving the Schrödinger equation

$$\hat{H}\psi(\mathbf{r}, t) = \hat{E}\psi(\mathbf{r}, t) = i\hbar\frac{\partial}{\partial t}\psi(\mathbf{r}, t) \tag{19.37}$$

where

$$\hat{H} = \frac{1}{2m}\big(\hat{\mathbf{p}} - q\mathbf{A}(\mathbf{r}, t)\big)^2 + q\phi(\mathbf{r}, t) \tag{19.38}$$

One must, of course, be careful of the ordering of any differential operators as

$$\begin{aligned}\left[\hat{\mathbf{p}} - q\mathbf{A}(\mathbf{r}, t)\right]^2 \psi(\mathbf{r}) = &-\hbar^2\nabla^2\psi(\mathbf{r}, t) + iq\hbar\nabla\cdot\big(\mathbf{A}(\mathbf{r}, t)\psi(\mathbf{r}, t)\big) \\ &+ iq\hbar\mathbf{A}(\mathbf{r}, t)\cdot\big(\nabla\psi(\mathbf{r}, t)\big) + q^2[\mathbf{A}(\mathbf{r}, t)\cdot\mathbf{A}(\mathbf{r}, t)]\psi(\mathbf{r}, t)\end{aligned} \tag{19.39}$$

The classical equations of motion depend on the **E** and **B** fields themselves and are obviously invariant under any gauge transformation. The Hamiltonian that now appears in the Schrödinger equation, however, depends on the electromagnetic potentials explicitly and does change its form under such a transformation. Specifically, under the change in potentials given by Eqns. (19.29) and (19.30), the original Hamiltonian, Eqn. (19.38), is replaced by

$$\begin{aligned}\hat{H}' &= \frac{1}{2m}\Big(\hat{\mathbf{p}} - q\mathbf{A}'(\mathbf{r}, t)\Big)^2 + q\phi'(\mathbf{r}, t) \\ &= \frac{1}{2m}\big(\hat{\mathbf{p}} - q\mathbf{A}(\mathbf{r}, t) - q\nabla f(\mathbf{r}, t)\big)^2 + q\phi'(\mathbf{r}, t) - q\frac{\partial}{\partial t}f(\mathbf{r}, t)\end{aligned} \tag{19.40}$$

which does not manifestly imply the same physical solutions. It is not hard to show, however, that if one also simultaneously changes the original wavefunction $\psi(\mathbf{r}, t)$ by a (possibly time- and space-dependent) phase factor, namely,

$$\psi'(\mathbf{r}, t) = \psi(\mathbf{r}, t)e^{iqf(\mathbf{r},t)/\hbar} \tag{19.41}$$

then

$$\hat{H}\psi(\mathbf{r}, t) = \hat{E}\psi(\mathbf{r}, t) = i\hbar\frac{\partial}{\partial t}\psi(\mathbf{r}, t)$$
$$\Updownarrow$$
$$\hat{H}'\psi'(\mathbf{r}, t) = \hat{E}\psi'(\mathbf{r}, t) = i\hbar\frac{\partial}{\partial t}\psi'(\mathbf{r}, t) \tag{19.42}$$

The probability densities corresponding to ψ' and ψ are identical because

$$|\psi'(\mathbf{r}, t)|^2 = |\psi(\mathbf{r}, t)|^2 \tag{19.43}$$

and the gauge transformation makes no change in the observable physics of the system. (Recall P7.5 where a change in the "zero of potential" is discussed, and a similar result is found, namely, a simple change of phase.) The solutions obtained in different gauges may well look quite different but must correspond to the same physical energy eigenvalues and be related via Eqn. (19.41). It is often useful to use this freedom of gauge to choose the ϕ and $\mathbf{A}$ which make the problem most tractable.

In the next two sections, we deal with several special cases corresponding to uniform electric and magnetic fields where a problem can be solved explicitly in separate gauges, and where the connections between the solutions are easily confirmed. These examples also show that different properties of the solution of a given physical problem may be more apparent in one gauge or another.

19.3 CONSTANT ELECTRIC FIELDS

We first consider the action of a uniform electric field, for simplicity in the $+x$ direction, i.e., $\mathbf{E}_0 = E_0\hat{\mathbf{x}}$ on an otherwise free particle of charge q. The classical solutions correspond to free particle motion in the y, z directions and uniform motion in the x direction, namely,

$$\begin{aligned} x(t) &= \frac{qE_0}{2m}t^2 + v_{0x}t + x_0 \\ y(t) &= v_{0y}t + y_0 \\ z(t) &= v_{0z}t + z_0 \end{aligned} \tag{19.44}$$

The corresponding quantum problem can be defined by the Hamiltonian operator

$$\hat{H} = \frac{\hat{\mathbf{p}}^2}{2m} - qE_0x = \left(\frac{\hat{p}_x^2}{2m} - eE_0x\right) + \frac{\hat{p}_y^2}{2m} + \frac{\hat{p}_z^2}{2m} \tag{19.45}$$

The system is clearly separable, and the wavefunction can be written in the form

$$\psi(\mathbf{r}, t) = \psi(x, t)e^{i(p_y y - p_y^2 t/2m)/\hbar}e^{i(p_z z - p_z^2 t/2m)/\hbar} \tag{19.46}$$

The problem of uniform acceleration in one dimension was discussed in Sec. 5.6.2 where a Gaussian wavepacket solution was found with the form

$$\psi(x,t) = \frac{1}{\sqrt{\beta\sqrt{\pi}(1+it/t_0)}} e^{iFt(x-Ft^2/6m)/\hbar} \times \exp\Big(-(x - Ft^2/2m)^2/2\beta^2(1+it/t_0)\Big) \tag{19.47}$$

In Eqn. (19.47), $F = qE_0$, and the initial Gaussian wavepacket is obviously

$$\psi(x,0) = \frac{1}{\sqrt{\beta\sqrt{\pi}}} e^{-x^2/2\beta^2} \tag{19.48}$$

In solving this problem, we have explicitly used the most familiar gauge in which

$$\phi(\mathbf{r},t) = -Ex \qquad \text{and} \qquad \mathbf{A}(\mathbf{r},t) = 0 \tag{19.49}$$

but we can just as well use the gauge equivalent set

$$\phi(\mathbf{r},t) = 0 \qquad \text{and} \qquad \hat{A}(\mathbf{r},t) = -E_0\hat{\mathbf{x}}t \tag{19.50}$$

These two are related by the gauge function $f(x,t) = -E_0xt$.

In the new gauge, the y and z dependence is unchanged, but the 1D Schrödinger equation for the x behavior now reads

$$\begin{aligned}
\hat{H}\psi'(x) &= \left(\hat{H}_0 + t\hat{H}_1 + t^2\hat{H}_2\right)\psi'(x,t) \\
&= \left(\frac{\hat{p}_x^2}{2m} + \frac{qE_0t}{m}\hat{p}_x + \frac{1}{2m}(qE_0t)^2\right)\psi'(x,t) \\
&= i\hbar\frac{\partial}{\partial t}\psi'(x,t)
\end{aligned} \tag{19.51}$$

where the linear terms in $\hat{p}_x$ simplify because the vector potential does not depend on position. As in Sec. 13.5, we can formally solve this problem by integration of the initial value problem and find

$$\psi'(x,t) = e^{-i(\hat{H}_0t + \hat{H}_1t^2/2 + \hat{H}_2t^3/3)/\hbar}\psi'(x,0) \tag{19.52}$$

In this special case, the three component pieces, $\hat{H}_{0,1,2}$ commute with each other so that their order in the exponential is unimportant, and the time development operator can be written in the simpler form

$$\begin{aligned}
\psi'(x,t) &= e^{-i(\hat{H}_2t^3/3)/\hbar}e^{-i(\hat{H}_1t^2/2)/\hbar}e^{-i(\hat{H}_t)/\hbar}\psi'(x,0) \\
&= e^{-iF^2t^3/6m\hbar}e^{-iFt^2\hat{p}_x/2m\hbar}e^{-i\hat{p}_x^2t/2m\hbar}\psi'(x,0)
\end{aligned} \tag{19.53}$$

and we can consider the effect of each operator in turn. The time development operator for the free particle, namely, $\hat{H}_0 = \hat{p}_x^2/2m$, has a simple effect when acting on a Gaussian initial state (Ex. 13.4), i.e.,

$$\psi'(x,t) = e^{-i\hat{H}_0 t/\hbar}\left(\frac{1}{\sqrt{\beta\sqrt{\pi}}}e^{-x^2/2\beta^2}\right)$$

$$= \frac{1}{\sqrt{\beta(1+it/t_0)\sqrt{\pi}}}e^{-x^2/2\beta^2(1+it/t_0)} \tag{19.54}$$

corresponding to spreading, but with no translation of the central value. The effect of the second operator can be made clear by recalling (P13.6) that the momentum operator generates translations in space via

$$e^{ia\hat{p}/\hbar}\psi(x,t) = \psi(x+a,t) \tag{19.55}$$

so that the spreading wavepacket of Eqn. (19.54) is translated via

$$x \longrightarrow x - \frac{qE_0 t^2}{2m} = x - \frac{Ft^2}{2m} \tag{19.56}$$

The final factor, $e^{-i\hat{H}_2 t^3/3\hbar}$, is a simple phase. Taken together, these three operators acting on the initial wavefunction give

$$\psi'(x,t) = \frac{1}{\sqrt{\beta\sqrt{\pi}(1+it/t_0)}}e^{-i(F^2t^3/6m)/\hbar}$$

$$\exp\left(-(x - Ft^2/2m)^2/2\beta^2(1+it/t_0)\right) \tag{19.57}$$

where $F = qE_0$. This explicitly satisfies

$$\psi'(x,t) = e^{-iqf(x,t)/\hbar}\psi(x,t) = e^{-iFxt/\hbar}\psi(x,t) \tag{19.58}$$

as expected from the gauge transformation.

19.4 ATOMS IN ELECTRIC FIELDS: THE STARK EFFECT

One of the simplest ways to probe and manipulate the energy level structure and wavefunction of an atom is to place it in a constant external electric field. We will consider the quantum version of this problem for hydrogen, where the resulting pattern of energy level shifts is called the *Stark effect.* We begin, however, by considering a simplified classical model to gain some intuition.

19.4.1 Classical Case

As a simple model of a (classical) hydrogen atom, consider a positive point charge $+e$, representing the proton, embedded in a uniformly charged spherical cloud of radius a_0 (the

Bohr radius) with total charge $-e$. The resulting charge density is

$$\rho(\mathbf{r}) = \begin{cases} -3e/4\pi a_0^3 & \text{for } r < a_0 \\ 0 & \text{for } r > a_0 \end{cases} \tag{19.59}$$

Gauss's law can then be used to derive (as in P18.23) the corresponding electric field, giving

$$\mathbf{E}(\mathbf{r}) = \begin{cases} -Ke\mathbf{r}/a_0^3 & \text{for } r < a_0 \\ -Ke\hat{\mathbf{r}}/r^2 & \text{for } r > a_0 \end{cases} \tag{19.60}$$

with corresponding potential energy

$$V(\mathbf{r}) = \begin{cases} -Ke^2/2a_0(3 - r^2/a_0^2) & \text{for } r < a_0 \\ -Ke^2/r & \text{for } r > a_0 \end{cases} \tag{19.61}$$

If an external electric field $\mathbf{E}_0$ (say, in the z direction) is applied, the electron cloud and "nucleus" will shift in opposite directions by a net amount r_0 until a new equilibrium situation is achieved, as in Fig. 19.3. The new system has the separation of charge characteristic of an electric dipole and the "atom" has been *polarized* by the external field. If we assume that the spherical shape of the electron cloud is unchanged for sufficiently small external fields, the net displacement can be determined by balancing the forces on the proton giving

$$0 = F_{\text{external}} + F_{\text{cloud}} = eE_0 - \frac{Ke^2 r_0}{a_0^3} \tag{19.62}$$

or $r_0 = E_0 a_0^3/Ke$; the induced *dipole moment* of the system is $p \equiv er_0 = E_0 a_0^3/K$. The new situation can also be visualized in terms of the potential energy function in Fig. 19.4 where the new minimum is apparent. It is easy to show that the new system has a potential energy lower by an amount

$$\Delta V = \Delta E = -\frac{E_0^2 a_0^3}{2K} \tag{19.63}$$

The interaction energy of an electric dipole in an external field is given by $-\mathbf{p} \cdot \mathbf{E}_0$, so an atom with a permanent dipole moment would have an energy shift linear in the applied field. For the situation above, there is only an induced dipole moment so that energy shift is

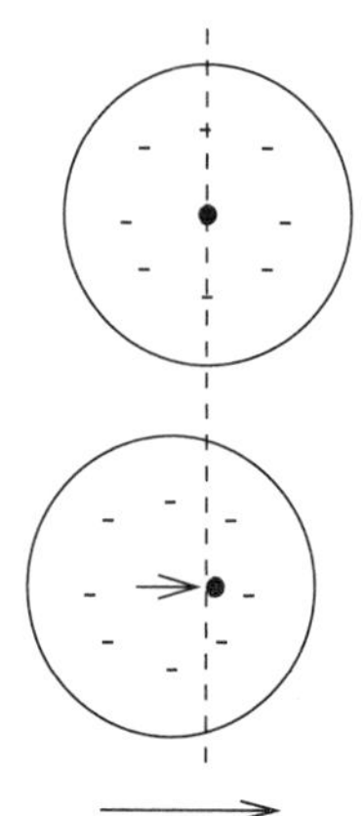

FIGURE 19.3. *Classical picture of polarization of a charge distribution by an external electric field.*

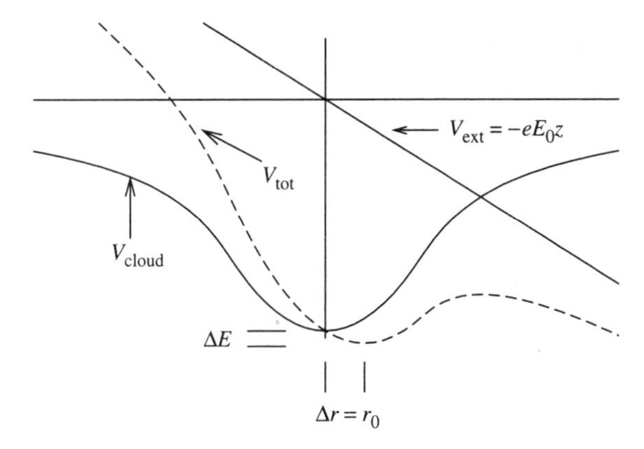

FIGURE 19.4. Energy shift due to polarization.

necessarily quadratic in E_0. The *electric dipole polarizability* (labeled α) is often defined from the observed energy shift ΔE via

$$\Delta E = -\frac{1}{2}\alpha E_0^2 \tag{19.64}$$

The simple model above then implies that $\alpha \propto a_0^3/K$ with a numerical coefficient that varies with the assumed charge distribution. We note that this energy shift has the dimensions given by product of the energy density in the external field ($u_E = \epsilon|\mathbf{E}|^2/2$) times a typical atomic or molecular volume $V \sim a_0^3$. This result also gives the form for the interaction potential of a charged particle with a neutral atom or molecule, the so-called *polarization potential,* via

$$V_{\text{POL}}(r) \sim -\frac{1}{2}\alpha\left(\frac{Ke}{r^2}\right)^2 \propto (\alpha K)\frac{Ke^2}{r^4} \propto \frac{Ke^2a_0^3}{r^4} \tag{19.65}$$

Note that the introduction of the external (linear) potential $V_{\text{ext}} = -eE_0z$ makes the potential "turn over" for sufficiently large values of $|z|$. The existence of a potential minimum even for arbitrarily large values of E_0 guarantees that there will always be a classical bound state; quantum effects, such as tunneling and zero point energy, will make the problem in real atoms more interesting.

19.4.2 Quantum Stark Effect

We now turn to the quantum description of a hydrogen atom in an external field. Being somewhat careful initially, we note that the potential felt by the two charged particles (proton and electron) is

$$V(\mathbf{r}_e, \mathbf{r}_p) = -eE_0z_p + eE_0z_e = eE_0(z_e - z_p) = eE_0z \tag{19.66}$$

where $z = z_{\text{rel}} = z_e - z_p$ is the relative coordinate. The complete Hamiltonian is now

$$\hat{H} = \frac{\hat{\mathbf{p}}^2}{2\mu} - \frac{Ke^2}{r} + eE_0z \tag{19.67}$$

For sufficiently weak fields, we can use perturbation theory[13] to treat the additional term

[13]It has been said (Condon and Shortley (1951)) that the treatment of the Stark effect in hydrogen was the first application of perturbation theory in quantum mechanics.

representing the external field; we will consider its effects on both the ground state and first excited states as examples of second-order and degenerate state perturbation theory, respectively.

For the ground state, $\psi_{100}(\mathbf{r})$, the first-order shift in energy is given by

$$E_1^{(1)} = \langle\psi_{100}|eE_0 z|\psi_{100}\rangle = eE_0 \int d\mathbf{r}\, |\psi_{100}(\mathbf{r})|^2 z = 0 \tag{19.68}$$

from familiar parity arguments. This is also consistent with the classical arguments in which we expect no energy shift linear in the applied field. We must then consider the second-order shift given by

$$E_1^{(2)} = \sum_{n=2}^{\infty}\sum_{l=0}^{n-1}\sum_{m=-l}^{+l} \frac{|\langle\psi_{nlm}|eE_0 z|\psi_{nlm}\rangle|^2}{E_1^{(0)} - E_n^{(0)}} \tag{19.69}$$

Writing $z = r\cos(\theta)$, we require the matrix elements

$$\langle\psi_{nlm}|z|\psi_{100}\rangle = \int d\mathbf{r}\left[R_{n,l}(r)Y_{l,m}^*(\theta,\phi)\right][r\cos(\theta)]\left[R_{1,0}(r)Y_{0,0}(\theta,\phi)\right] \tag{19.70}$$

The angular integration can be easily performed, since

$$Y_{0,0} = \frac{1}{\sqrt{4\pi}} \qquad \text{while} \qquad \cos(\theta) = \sqrt{\frac{4\pi}{3}}\,Y_{1,0}(\theta,\phi) \tag{19.71}$$

which contributes a factor of

$$\int d\Omega\, Y_{l,m}^*(\theta,\phi)\frac{1}{\sqrt{3}}Y_{1,0}(\theta,\phi) = \frac{1}{\sqrt{3}}\delta_{l,1}\delta_{m,0} \tag{19.72}$$

when we use the orthonormality properties of the spherical harmonics. This result is in the form of a selection rule. It also simplifies the overlap integrals required for the radial piece of Eqn. (19.70), as we now require

$$\int_0^\infty dr r^2 R_{n,l}(r)\, r\, R_{1,0}(r)\,\delta_{1,0}\,\delta_{m,0} = \int_0^\infty dr\, r^3 R_{n,1}(r)R_{1,0}(r) \tag{19.73}$$

We will not derive these integrals but simply quote the results,[14] namely,

$$|\langle\psi_{nlm}|z|\psi_{100}\rangle|^2 = \frac{1}{3}\left(\frac{2^8 n^7 (n-1)^{2n-5}}{(n+1)^{2n+5}}\right)a_0^2 \equiv g(n)a_0^2 \tag{19.74}$$

Since $E_n^{(0)} = -Ke^2/2a_0n^2$, we can write the second-order shift in the ground state energy as

$$E_1^{(2)} = (eE_0)^2 \sum_{n=2}^{\infty} \frac{g(n)a_0^2}{[-Ke^2/2a_0(1-1/n^2)]} = -2G\frac{E_0^2 a_0^3}{K} \tag{19.75}$$

[14] See Bethe and Salpeter (1957) for a complete discussion.

where

$$G \equiv \sum_{n=2}^{\infty} \frac{g(n)}{1 - 1/n^2} = \sum_{n=2}^{\infty} \frac{n^2 g(n)}{n^2 - 1} \approx 1.125 \tag{19.76}$$

where the summation can be done numerically (P19.9) or even in closed form.[15] This confirms the classical expectation for the general form of the energy shift, Eqn. (19.63), and gives the polarizability for hydrogen as $\alpha = 4Ga_0^3/K$.

For the first excited state of hydrogen, corresponding to $n = 2$, we have four states that are degenerate in energy in the absence of an external field,

$$n = 2, l = 0 \qquad \psi_{200} = \psi_A \tag{19.77}$$

and

$$n = 2, l = 1, m = \begin{Bmatrix} +1 & \psi_{2,1,1} \equiv \psi_B \\ 0 & \psi_{2,1,0} \equiv \psi_C \\ -1 & \psi_{2,1,-1} \equiv \psi_D \end{Bmatrix} \tag{19.78}$$

Using the formalism of degenerate state perturbation theory (Sec. 11.6.2), we then are required, in principle, to diagonalize the 4×4 matrix corresponding to the equation

$$\begin{pmatrix} E_2^{(0)} + \langle V \rangle_{AA} & \langle V \rangle_{AB} & \langle V \rangle_{AC} & \langle V \rangle_{AD} \\ \langle V \rangle_{BA} & E_2^{(0)} + \langle V \rangle_{BB} & \langle V \rangle_{BC} & \langle V \rangle_{BD} \\ \langle V \rangle_{CA} & \langle V \rangle_{CB} & E_2^{(0)} + \langle V \rangle_{CC} & \langle V \rangle_{CD} \\ \langle V \rangle_{DA} & \langle V \rangle_{DB} & \langle V \rangle_{DC} & E_2^{(0)} + \langle V \rangle_{DD} \end{pmatrix} \begin{pmatrix} \psi_A \\ \psi_B \\ \psi_C \\ \psi_D \end{pmatrix}$$

$$= E_2^{(1)} \begin{pmatrix} \psi_A \\ \psi_B \\ \psi_C \\ \psi_D \end{pmatrix} \tag{19.79}$$

where $V = eE_0 z$. The problem simplifies considerably, for the matrix elements connecting $\psi_B = \psi_{2,1,1}$ to the other states through this interaction vanish, since

$$\langle \psi_B | z | \psi_{A,C,D} \rangle = 0 \qquad \text{because they have different } m \text{ values} \tag{19.80}$$

while

$$\langle \psi_B | z | \psi_B \rangle = 0 \qquad \text{because of parity} \tag{19.81}$$

The same is true of the matrix elements for $\psi_D = \psi_{2,1,-1}$; these two states effectively decouple from the problem, and their energy levels are unchanged. That leaves the 2×2 subspace corresponding to $\psi_{A,C}$. The matrix elements connecting these two also simplify as

$$\langle \psi_A | z | \psi_A \rangle = 0 = \langle \psi_C | z | \psi_C \rangle \tag{19.82}$$

by parity while

$$\langle \psi_A | z | \psi_C \rangle = -3a_0 \tag{19.83}$$

---

[15] See Borowitz (1967).

by direct calculation (P19.8). So, in the $\psi_{A,C}$ subspace, we have the determinant

$$\det\begin{pmatrix} E_2^{(0)} - E_2^{(1)} & -3eE_0 a_0 \\ -3eE_0 a_0 & E_2^{(0)} - E_2^{(1)} \end{pmatrix} = 0 \tag{19.84}$$

or

$$E_2^{(1)} = E_2^{(0)} \pm 3eE_0 a_0 \tag{19.85}$$

and the energies of the $\psi_{2,0,0}$ and $\psi_{2,1,0}$ states are split as shown in Fig. 19.5. The two solutions labeled by $\pm$ correspond to the (normalized) wavefunctions

$$\psi^{(+)} = \frac{1}{\sqrt{2}}\left(\psi_{2,0,0} - \psi_{2,1,0}\right) \tag{19.86}$$

$$\psi^{(-)} = \frac{1}{\sqrt{2}}\left(\psi_{2,0,0} + \psi_{2,1,0}\right) \tag{19.87}$$

To visualize this result, in Fig. 19.5 we plot $V(z)|\psi^{(\pm)}(0,0,z)|^2$ versus z to show the distribution of potential energy in the two cases; the $+$ $(-)$ combinations are obviously shifted up (down) in energy as expected.

It is important to stress that the presence of an energy shift that is linear or first order in E_0 depends critically on the presence of degenerate energy levels. Linear combinations of the $\psi_{2,1,0}$ and $\psi_{2,0,0}$ states are required to produce the dipole moments that can interact via $E = -\mathbf{p}\cdot\mathcal{E}$, and these states must be degenerate (or nearly so) for this mixing to occur.

To exemplify these remarks further, in Fig. 19.6 we show part of the energy spectrum of lithium Rydberg atoms as a function of the strength of the applied field. As discussed in Sec. 18.3.1, the states with the lowest values of l are shifted down in energy due to their interactions with the inner-electron core. In this case, the $15s$ $(l = 0)$ state is lowered in energy much more than the $15p$ $(l = 1)$ state, while the remaining ones with $l = 2, 3, \ldots, 14$ are still almost degenerate. We note the following distinctive features:

- For small field strengths (less than roughly 2500 V/cm), the energy shift due to the external field for the $15s$ state (as well as the $16s$ state) is quadratic instead of linear due to it's "isolation" from any nearby degenerate states. The polarizability of these states can be estimated from the data (P19.16)
- For quite small fields, the $15p$ state also has a quadratic energy shift, but as the Stark level shifts of the $n = 15$ states become comparable to the initial splitting of the $15p$ state from the rest, it "joins in" and contributes to the linear Stark effect pattern. States do not have to be exactly degenerate to require use of degenerate perturbation theory[16] as discussed in Sec. 11.6.2.
- The remaining $n = 15$ states show the standard Stark effect of a linear energy shift with a large number of splittings due to the large degeneracy.
- For large enough fields, the perturbation theory predictions become unreliable, and the pattern of energy levels becomes highly complex. Note however, the many avoided level crossings that are characteristic of "level repulsion" discussed in Sec. 11.6.

We briefly mention two other effects that arise in the quantum description of atomic energy levels in an external electric field:

[16]For a nice discussion of degeneracy effects and dipole moments, see Gasiorowicz (1968).

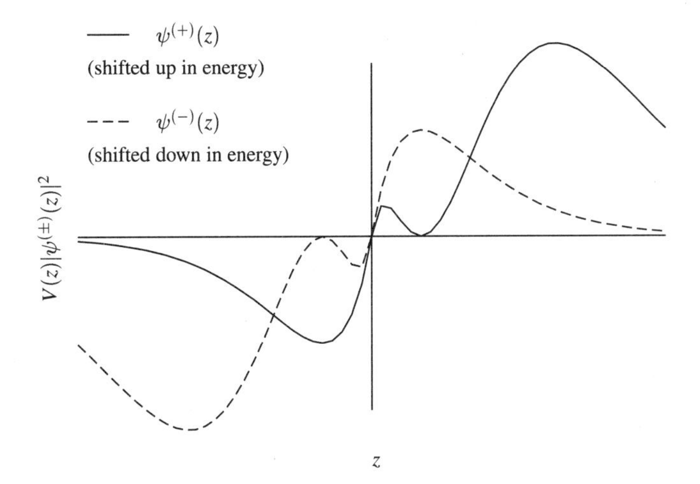

FIGURE 19.5. *$V(z) = mgz$ times $|\psi^{\pm}(0, 0, z)|^2$ for the linear combination states shifted up (+) and down (−) in energy in the linear Stark effect.*

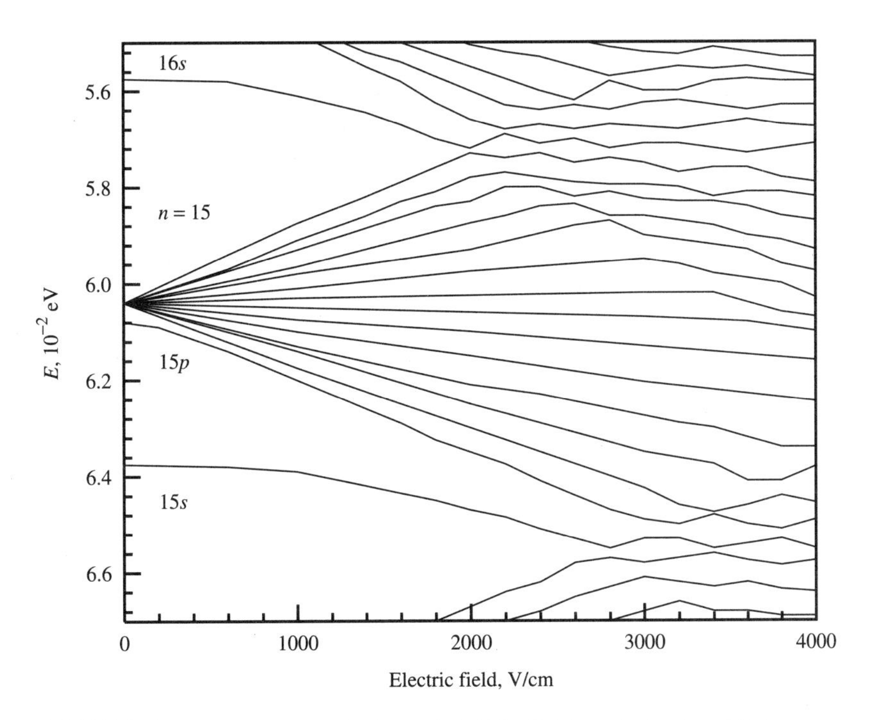

FIGURE 19.6. *Energy versus applied electric field for highly excited states of lithium ($m = 0$ states) illustrating the first-order (linear in E) and second-order (quadratic in E) Stark effects. Data taken from Zimmerman et al. (1979). (Permission of Daniel Kleppner.)*

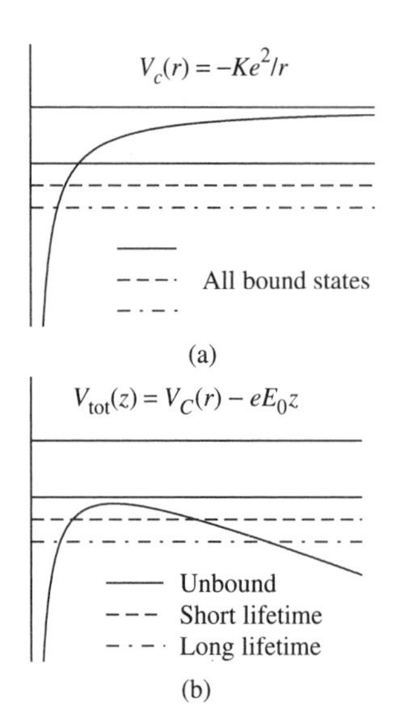

FIGURE 19.7 Potential energy due to Coulomb plus external electric fields showing the possibility of field ionization and tunneling.

- The combined Coulomb plus external field potential felt by an electron (illustrated in Fig. 19.7) implies that there are no absolutely stable bound states in such a potential, as there is always a possibility of quantum tunneling. One can derive estimates (P19.17) of the tunneling probability, which confirm that the lifetimes in modest fields are much longer than the age of the universe, consistent with intuition. If, however, the external field is increased enough, then the potential "turns over," allowing states to become unbound, leading to so-called *field ionization.* Simple estimates of the field necessary to unbind the nth state (P19.18) seem to be confirmed by experiment.
- The emission of photons via radiative transitions from one state to another forms the basis of atomic and molecular spectroscopy, which, in turn, provides the evidence for the quantized energy levels calculated in quantum mechanics. The related (inverse) process of single-photon absorption is conceptually the same; electrons absorb photons of energy $\hbar\omega = |E_n - E_{n'}|$ in transitions from one level to another or are ionized from the nth level if $E_\gamma = \hbar\omega > |E_n|$. Sufficiently intense beams of photons of energy *less than* $|E_n|$ can still ionize atoms[17] via the process of *multiphoton ionization (*MPI*)*, which is illustrated schematically in Fig. 19.8. The absorption of

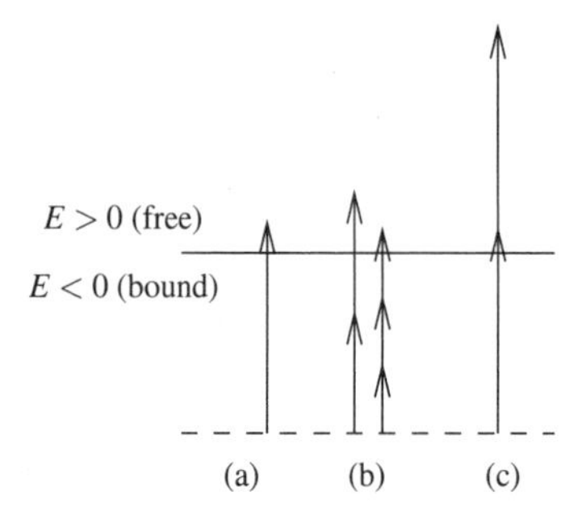

FIGURE 19.8 Schematic diagram showing photon absorption leading to (a) single photon ionization, (b) multiphoton ionization (MPI), and (c) above-threshold ionization (ATI).

[17]For a nice review of multiphoton processes in atoms, see Delone and Krainov (1994).

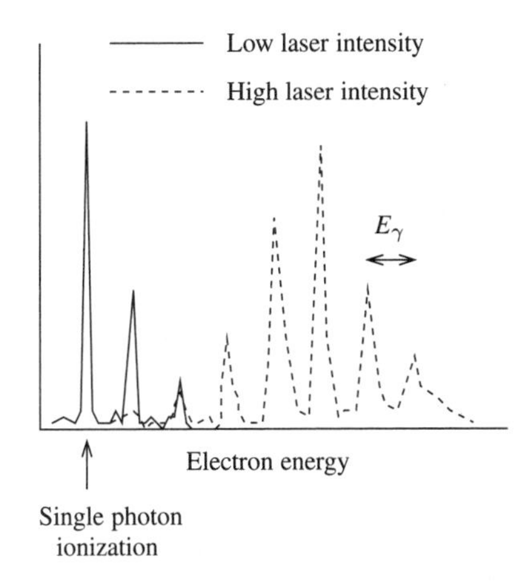

FIGURE 19.9. Number of emitted photoelectrons versus their kinetic energy for two laser intensities. For low laser intensities (solid), single photon ionization dominates, while for larger laser intensities (dashed), the absorption of more than one photon becomes appreciable. The distance between peaks is $E_\gamma = \hbar\omega$, indicating that the kinetic energies are given by $mv^2/2 = nE_\gamma - E_0$.

the "first" photon would not normally allow the electron to be freed, but if the number density of photons is high enough so that there is a chance of a second (or third and so forth) interaction, the ionization can occur as a multistep process. A similar process is called *above-threshold ionization* (ATI), where the final state energy of the electron is measured to be

$$E_e = -|E_{\text{binding}}| + n\hbar\omega \tag{19.88}$$

with $n > 1$ implying that more than one photon absorption has been used to "kick" the electron into the continuum of free-particle states. Clearly, the probability for a k-step ionization process scales with the intensity, I, as I^k. The effect is illustrated schematically in Fig. 19.9 where the same laser beam is used in both (a) and (b)—meaning that the frequency ω or energy of each photon E_γ is unchanged—but the intensity is increased near the critical value for the onset of the multiphoton process.

19.5 CONSTANT MAGNETIC FIELDS

We next consider the problem of a charged particle (charge q) in a uniform magnetic field in the $+z$ direction. The classical solutions consist of helical trajectories, that is, uniform translational motion in the direction parallel to the field $\mathbf{B}_0$ and uniform rotational motion in the plane perpendicular to $\mathbf{B}_0$. For a positive charge q, the circular motion is in the clockwise direction when viewed from above ($z > 0$), as in Fig. 19.10.

To discuss the quantum version, we choose the gauge where

$$\mathbf{A}(\mathbf{r}, t) = \frac{B_0}{2}(-y, x, 0) \tag{19.89}$$

and consider other choices in the problems. The Hamiltonian from Eqn. (19.38) can be written as

$$\hat{H} = \frac{\hat{\mathbf{p}}^2}{2\mu} - \frac{q}{2\mu}(\hat{\mathbf{p}} \cdot \mathbf{A} + \mathbf{A} \cdot \hat{\mathbf{p}}) + \frac{1}{2\mu}\left(\frac{qB_0}{2}\right)^2 (x^2 + y^2) \tag{19.90}$$

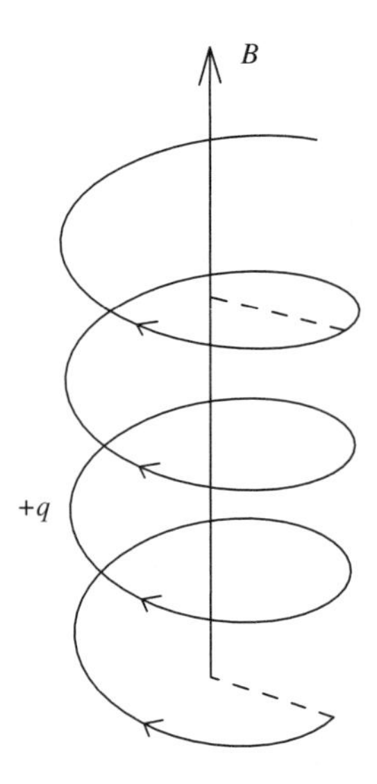

FIGURE 19.10. *Classical helical motion of a positively charged particle in a uniform magnetic field.*

The middle terms can be combined and written in the form

$$-\frac{q}{\mu}\frac{\hbar}{i}\frac{B_0}{2}\left(-\frac{\partial}{\partial y}x + \frac{\partial}{\partial x}y\right) = -\frac{qB_0}{2\mu}\hat{L}_z \tag{19.91}$$

This form is clearly a generalization of the classical result

$$\hat{H} = -\mathbf{M}\cdot\mathbf{B} \tag{19.92}$$

where the magnetic moment is given by

$$\mathbf{M} = g\left(\frac{q}{2\mu}\right)\mathbf{L} \tag{19.93}$$

with the classical value of $g = 1$, as found in P17.12 for a rotating particle.

We can define the so-called *Larmor frequency*

$$\omega_L = \frac{qB_0}{2\mu} = \frac{\omega_c}{2} \tag{19.94}$$

which is half the *cyclotron frequency* that corresponds to the classical frequency of circular motion determined via the Lorentz force law, namely,

$$F = \mu a_c \longrightarrow qvB_0 = \mu\frac{v^2}{r} = \mu\omega_c^2 r \qquad \text{or} \qquad \omega_c = \frac{qB_0}{\mu} \tag{19.95}$$

We then write the Hamiltonian in the form

$$\hat{H} = \frac{\hat{\mathbf{p}}^2}{2\mu} - \omega_L\hat{L}_z + \frac{1}{2}\mu\omega_L^2(x^2 + y^2) \tag{19.96}$$

The corresponding Schrödinger equation is most conveniently solved in cylindrical coordinates, where we assume a solution of the form

$$\psi(\mathbf{r}) = \psi(r, \theta, z) = R(r)e^{im\theta}e^{ik_z z} \tag{19.97}$$

This solution is an eigenfunction of both the $\hat{L}_z$ and z kinetic energy terms that together add terms to the total energy given by $-\hbar\omega_L$ and $\hbar^2 k_z^2/2\mu$, respectively. The remaining two-dimensional problem is essentially the planar harmonic oscillator discussed extensively in Sec. 16.3.3. The resulting solutions are given by

$$\psi(r, \theta, z) \propto r^{|m|} L_{n_r}^{|m|}(r/\rho) e^{im\theta} e^{ik_z z} \tag{19.98}$$

where $\rho^2 = \hbar\mu\omega_L$ with energy eigenvalues

$$E_{n_r,m} = \hbar\omega_L\left(2n_r + |m| - m + 1\right) + \frac{\hbar^2 k_z^2}{2\mu} \tag{19.99}$$

The equally spaced harmonic-oscillator-like energy states are often called *Landau levels.*

As discussed for the 2D oscillator, the classical limit of uniform circular motion in the plane corresponds to the minimum number of radial nodes ($n_r = 0$). In that case, large values of $|m|$ of either sign could give macroscopic energies corresponding to motion in either clockwise or counterclockwise directions. Here, Eqn. (19.99) implies that only $m = -|m|$, i.e., $L_z = -|m|\hbar$ or clockwise motion gives that limit, as expected for the motion of a positive charge.

Using the wavefunction solutions in Eqn. (19.98) for $n_r = 0$ and large $|m|$, one can show that the probability density is strongly peaked at a (planar) radius coordinate given by

$$r^2 = \frac{|m|\hbar}{\mu\omega_L} = \frac{|L_z|}{\mu\omega_L} \tag{19.100}$$

To compare this to the classical case, we can use arguments similar to those in Ex. 1.1 and Sec. 16.3.3, and the Lorentz force law, to find

$$\frac{\mu v^2}{r} = \mu a_c = F = qvB_0 \tag{19.101}$$

The relationship between velocity and momentum is generalized in the presence of a magnetic vector potential to be

$$\mu\mathbf{v} = \mathbf{p} - q\mathbf{A} \tag{19.102}$$

so that

$$\mu\left(\mathbf{r}\times\mathbf{v}\right)_z = \left(\mathbf{r}\times\mathbf{p}\right)_z - q\left(\mathbf{r}\times\mathbf{A}\right)_z \tag{19.103}$$

or

$$-\mu rv = -|L_z| - \frac{1}{2}qB_0 r^2 \tag{19.104}$$

Eqns. (19.101) and (19.104) can then be combined to give

$$r^2 = \frac{2|L_z|}{qB_0} = \frac{|L_z|}{\mu\omega_L} \tag{19.105}$$

which is consistent with the large quantum number limit.

The energies of the system now form a continuous spectrum because of the motion in the z direction. Even for a fixed value of k_z, however, there is still an infinite degeneracy in the quantized energy levels corresponding to the fact that $|m| - m = 0$ for all values of positive m. This degeneracy has its origin in the arbitrariness in the initial conditions for the planar orbits; the x and y coordinates of the center of the circular orbit are not specified even if the total energy and k_z are. This is strictly true only for a region of uniform field that is of infinite extent. If the field is confined in the x and y directions by a box with sides L, not all circular orbits will "fit into" the box. In this more realistic case, the degeneracy of each level[18] can be shown to be

$$\frac{qB_0L^2}{\pi\hbar} \tag{19.106}$$

which scales with the area, L^2, as expected.

19.6 ATOMS IN MAGNETIC FIELDS

It has been said that "magnetism is inseparable from quantum mechanics"[19] since it can be shown that systems interacting via purely classical mechanics in statistical equilibrium can exhibit no magnetic moments, even in response to externally applied magnetic fields. Of the many possible manifestations of magnetic effects in quantum mechanical systems, in this section we consider only three:

1. The problem of a one-electron atom subject to an external **B** field (the Zeeman effect).
2. The interactions of the electron spin in an atom to the "internal" magnetic field caused by the orbital motion of the atomic constituents themselves (the so-called *spin-orbit coupling*).
3. The magnetic couplings of two spin magnetic moments in an atom, giving rise to so-called *hyperfine splittings* in atoms and contributing to spin-spin level shifts in other systems.

19.6.1 The Zeeman Effect: External B Fields

To study the effect of a constant external magnetic field (oriented in the z direction for definiteness) on a one-electron atom, we consider the Hamiltonian for such a system (in a standard gauge) given by

$$\begin{aligned}\hat{H} &= \left(\frac{\hat{\mathbf{p}}^2}{2m_e} - \frac{KZe^2}{r}\right) + \frac{eB}{2m_e}\hat{L}_z + \frac{e^2B^2}{8m_e}(x^2+y^2) \\ &= \hat{H}_{\text{Coul}} + \hat{H}_1 + \hat{H}_2 \end{aligned} \tag{19.107}$$

This is strictly valid only for a spinless electron, as we have ignored the coupling of its intrinsic angular momentum to the external field; this complication is discussed in P19.23.

For sufficiently small applied fields, we can initially neglect the term quadratic in **B**, the so-called *diamagnetic term.* In this case we see that the eigenfunctions of the standard Coulomb problem remain solutions since the spherical harmonics are also eigenfunctions of $\hat{L}_z$. The shift in energy due to the external field is then given by

---

[18]For a careful discussion of the boundary conditions that give rise to this estimate, see Peierls (1955).

[19]See Kittel (1971). He goes on to say "and were the value of $\hbar$ to go to zero, the loss to the science of magnetism is one of the catastrophes that would overwhelm the universe."

$$E_B^{(1)} Y_{l,m} = \hat{H}_1 Y_{l,m} = \left[\left(\frac{e\hbar}{2m_e} B \right) m \right] Y_{l,m} \tag{19.108}$$

or

$$E_B^{(1)} = \left[\frac{e\hbar}{2m_e} B \right] m = (\mu_e B)\, m = m\hbar\omega_L \tag{19.109}$$

We have written this in two complementary forms:

- One form implicitly uses the *Bohr magneton,* $\mu_e = e\hbar/2m_e$, whose numerical value[20] is $\mu_e = 5.788 \times 10^{-5}$eV/T; this emphasizes its identification as being the interaction energy of the magnetic moment due to orbital motion with the external field.
- The second form is written in terms reminiscent of the quantized Landau energy level spacing for a charged particle discussed in Sec. 19.5.

The introduction of $\hat{H}_1$ reduces the (unexpectedly large) symmetry of the pure Coulomb problem by picking out a specific axis, and the energy spectrum is correspondingly "disrupted"; L_z, however, remains a constant of the motion, and the problem is still exactly soluble. The inclusion of $\hat{H}_2$ completely destroys the symmetry, but for small fields the effect of $\hat{H}_2$ can then be estimated via perturbation theory using the standard solutions for hydrogen-like atoms. Using the angular and radial overlap integrals in Eqn. (17.82) and Eqn. (18.30), and the fact that

$$x^2 + y^2 = r^2 \sin^2(\theta) \tag{19.110}$$

we find that

$$E_B^{(2)} \approx \frac{e^2 B^2 a_0^2}{8m_e} F(n, l, m) \tag{19.111}$$

where

$$F(n, l, m) = \left[n^2 (5n^2 + 1 - 3l(l+1)) \frac{(l^2 + l - 1 - m^2)}{(2l - 1)(2l + 3)} \right] \tag{19.112}$$

when evaluated in the state $R_{n,l,m} = R_{n,l}(r) Y_{l,m}(\theta, \phi)$ and where a_0 is the Bohr radius. The prefactor in Eqn. (19.111) can be written in the form

$$\frac{e^2 B^2 a_0^2}{8m_e} = \frac{1}{4} \frac{(\hbar\omega_L)^2}{|E_1|} \tag{19.113}$$

where E_n are the Bohr energy levels for the Coulomb problem. To see how this scales with n, we can choose for definiteness $l = n - 1$ and $m = 0$ and find that

$$E_B^{(2)} \approx \frac{e^2 B^2 a_0^2}{16m_e} n^4 \tag{19.114}$$

In the limit when the effects of $\hat{H}_2$ can no longer be considered as a small perturbation, other calculational tools (such as variational methods or diagonalization of matrices)

[20]In this section, as elsewhere, the appropriate MKSA unit of magnetic field is the Tesla (T), which is related to the Gauss (G) via 1 T = 10^4 G.

must be employed to estimate the energy eigenvalues using the full Hamiltonian of Eqn. (19.107).

Considering only $E_B^{(1)}$, we see that for a given value of l, the $2l + 1$ degenerate levels corresponding to different values of m are split by equal amounts, and the pattern of energy level shifts is shown in Fig. 19.11. Despite the seemingly large number of possible new transitions, the selection rule for the magnetic quantum number for dipole radiation (see Sec. 17.3.3), namely, $\Delta m = +1, 0, -1$, and the uniform splitting imply that the absorption or emission line corresponding to the transition in the $B = 0$ case (E_γ) is only split into three distinct lines ($E_\gamma + \mu_e B \Delta m$). This pattern of shifts in spectral lines is called the *(linear) ordinary Zeeman effect,* and it is observed for small field strengths in atoms in which the total electronic spin is zero. The more usual case where one must also consider the electron spin exhibits more structure and is called the *anomalous Zeeman effect.*

To better understand the relative magnitudes of the energy splittings induced by $\hat{H}_{1,2}$, we first note that the difference between adjacent energy levels in hydrogen goes as

$$\Delta E_n \sim 13.6 \text{ eV}\left(\frac{1}{(n-1)^2} - \frac{1}{n^2}\right) \sim 27 \text{ eV}\frac{1}{n^3} \tag{19.115}$$

The maximum splitting due to $\hat{H}_1$ between the uppermost ($m = +l$) and lowermost ($m = -l$) scales roughly as

$$\Delta E_B^{(1)} \sim \left(5.8 \times 10^{-5} \frac{\text{eV}}{\text{T}}\right) B(2n) \tag{19.116}$$

since $n \geq l$. This implies that

$$\frac{\Delta E_B^{(1)}}{\Delta E_n} \sim \left(\frac{B}{2.3 \times 10^5 \text{ T}}\right) n^4 \tag{19.117}$$

Typical values of the magnetic field range from $B \sim 0.5 \text{ G} = 5 \times 10^{-5}$ T (corresponding to the earth's intrinsic field) or less (if shielding is provided) to $B \sim 1$ to 6 T, which is now standardly available in laboratories. For low-lying states ($n = 1, 2, \ldots$), we see that the (linear) Zeeman splittings are always much smaller than the differences between energy levels. However, Eqn. (19.117) implies that the two become comparable in a *1* Tesla field for $n \sim 22$. We can also write the ratio in Eqn. (19.117) in more symbolic fashion in the form

$$\frac{\Delta E_B^{(1)}}{\Delta E_n} = \left(\frac{B}{B_c}\right) n^4 \quad \text{with} \quad B_c \equiv \frac{\hbar}{e a_0^2} = \frac{\Phi_B}{\pi a_0^2} = 2.3 \times 10^5 \text{ T} \tag{19.118}$$

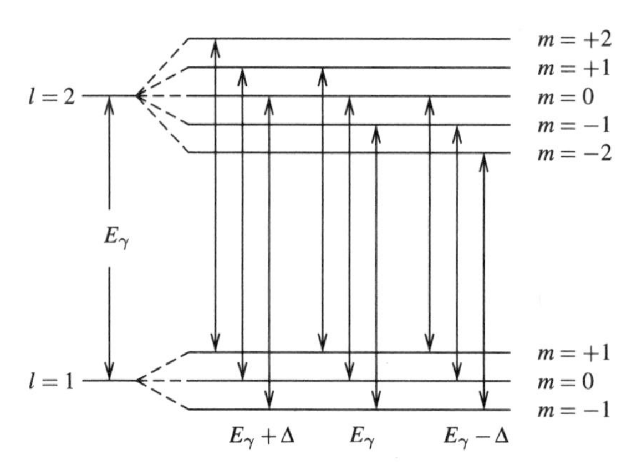

FIGURE 19.11. *Energy level splittings for P (l = 2) and D (l = 3) states illustrating the linear Zeeman effect.*

where $\Phi_B = \pi\hbar/e$ is written in terms of fundamental constants and has the units of a magnetic flux.

For such large fields (B) or principal quantum numbers (n), we must consider the effect of $\hat{H}_2$, and we can use the perturbation theory estimate of Eqn. (19.113) to write

$$\frac{\Delta E_B^{(2)}}{\Delta E_n} \approx \left(\frac{B}{4B_c}\right)^2 n^7 \tag{19.119}$$

This condition implies that the *quadratic Zeeman effect* (with energy shifts varying as B^2) will dominate over the more familiar linear Zeeman effect for highly excited Rydberg states (n large) or for sufficiently large magnetic field strengths.[21]

As an example of the effect of the diamagnetic term in the Hamiltonian, $\hat{H}_2$, in the large field limit, we show in Fig. 19.12 the spectra of highly excited ($n = 27$ to 29) sodium Rydberg states in magnetic fields up to $B = 6$ T. The experiment was performed by selectively choosing specific m states so that the linear Zeeman effect is absent. The theoretical predictions (light lines), obtained by matrix diagonalization of the Hamiltonian, are shown in Fig. 19.12 along with the experimental excitation curves. The energy (note the units!) is plotted versus B^2 (note the horizontal scale!) to show the quadratic (parabolic) dependence on B; note also the avoided level crossings.

19.6.2 Spin-orbit Splittings: Internal B Fields

Even in the absence of externally applied magnetic fields, the charged particles (and their associated magnetic moments) in an atom are still subject to magnetic forces due to the "motion" of the other charges. To understand the physical origin of this effect, we consider first the classical problem of a charged particle in uniform circular motion, as in Fig. 19.13. The circulating charge is equivalent to a current loop, and the magnetic field at the center

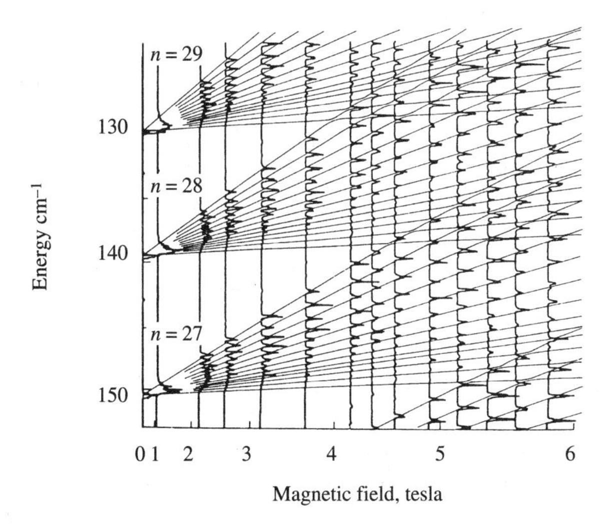

FIGURE 19.12. Energy versus applied magnetic field squared (note the horizontal scale!) illustrating the effect of diamagnetic H_2 term in the Hamiltonian. Data are from Zimmerman et al. (1978) from measurements of lithium atoms. (Permission of Daniel Kleppner.)

---

[21] It is estimated that when $B > 100$ T, one has $E_B^{(2)} > E_B^{(1)}$. Such large values of the magnetic field can be found at the surfaces of collapsed astrophysical objects such as white dwarf or neutron stars, and quadratic Zeeman effects due to them have likely been seen in spectral lines from such objects.

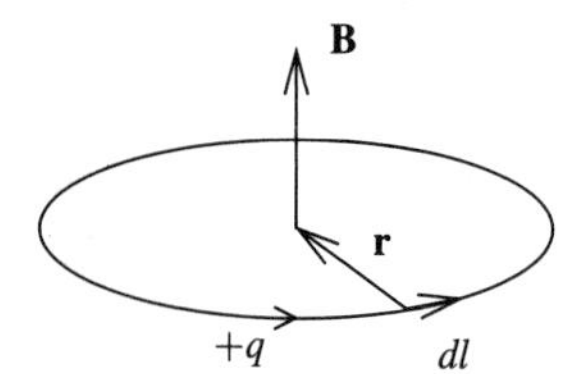

FIGURE 19.13. Classical picture of the magnetic field at center of current loop.

of the loop can be calculated classically using the Biot-Savart law. The small contribution from one "part of the circuit" can be integrated to obtain the familiar result

$$d\mathbf{B} = \frac{\mu_0}{4\pi}\frac{I d\mathbf{l} \times \mathbf{r}}{r^3} = \frac{\mu_0}{4\pi}\frac{dq\mathbf{v} \times \mathbf{r}}{r^3} \Longrightarrow \mathbf{B}_{\text{cent}} = \frac{\mu_0}{4\pi}\frac{qv}{r^2}\hat{\mathbf{k}} \tag{19.120}$$

If we

- Associate the circulating charge $q = Ze$ with the nuclear charge (remember that the electron and nucleus are in orbit around each other so the electron "sees" a positively charged current loop)
- Recall that the fundamental constant of magnetism can be rewritten using $c^2 = 1/\mu_0\epsilon_0$
- Use the classical relation for the angular momentum $|\mathbf{L}| = rp = rmv$

we can estimate that an atomic electron "sees" an effective magnetic field of the order

$$\mathbf{B}_{\text{cent}} = \frac{\mu_0}{4\pi}\frac{Ze}{mr^3}\mathbf{L} = \frac{Ze^2}{4\pi\epsilon_0}\frac{1}{mc^2r^3}\mathbf{L} \tag{19.121}$$

We can make an initial estimate of the magnitude of this field by letting $|\mathbf{L}| \sim \hbar$ and $r \sim a_0$ and find that $B_{\text{cent}} \sim 12$ T; this is much larger than typical external fields, and can cause sizable level splittings.

The magnetic moment associated with the electron, namely,

$$\mathbf{M} = \frac{gq}{2m}\mathbf{S} = -\frac{e}{m_e}\mathbf{S} \tag{19.122}$$

will then have an interaction energy in the atomic field given by

$$E = -\mathbf{M} \cdot \mathbf{B} = \frac{ZKe^2}{m^2c^2}\frac{1}{r^3}\mathbf{L} \cdot \mathbf{S} \tag{19.123}$$

where $K = 1/4\pi\epsilon_0$, as usual.

This classical estimate of the *spin-orbit (SO) interaction* (so called because it exhibits the coupling between the electron spin and the orbital motion) neglects other important relativistic effects (note the explicit factors of c that appear), and the corresponding quantum mechanical expression is most convincingly derived by a nonrelativistic reduction of the Dirac equation for the electron. When this is done, the precise expression for the spin-orbit coupling term differs by only a factor of 1/2 from our simple estimate, so that

$$\hat{H}_{\text{SO}} = \frac{ZKe^2}{2m^2c^2}\frac{1}{r^3}\hat{\mathbf{L}} \cdot \hat{\mathbf{S}} \tag{19.124}$$

To evaluate the effect of this interaction on the spectrum of a one-electron atom, we first need to calculate the effect of the $\mathbf{L} \cdot \mathbf{S}$ term. The total angular momentum due to orbital and spin contributions will be given by

$$l + 1/2 \Longrightarrow j = l + 1/2, l - 1/2 \tag{19.125}$$

with wavefunctions discussed in P17.16. The more formal relation among the corresponding operators

$$\mathbf{J} = \mathbf{L} + \mathbf{S} \tag{19.126}$$

can be manipulated to yield

$$\mathbf{J}^2 = (\mathbf{L} + \mathbf{S})^2 \Longrightarrow \mathbf{L} \cdot \mathbf{S} = \frac{1}{2}\left(\mathbf{J}^2 - \mathbf{L}^2 - \mathbf{S}^2\right) \tag{19.127}$$

Recalling that S is fixed, we then have

$$\mathbf{L} \cdot \mathbf{S} = \frac{1}{2}\left(j(j+1) - l(l+1) - \frac{3}{4}\right)\hbar^2 \tag{19.128}$$

or

$$\mathbf{L} \cdot \mathbf{S} = \begin{cases} l/2 & \text{for } j = l + 1/2 \\ -(l+1)/2 & \text{for } j = l - 1/2 \end{cases} \tag{19.129}$$

which implies that states with $j = l + 1/2(l - 1/2)$ are shifted up (down) in energy. For states with $l > 0$, the original $2(2l + 1)$ degenerate levels are split into two distinct sets of states with $2(l + 1/2) + 1 = 2l + 2$ and $2(l - 1/2) + 1 = 2l$ levels, respectively. For $l = 0$, only $j = 1/2$ is allowed and there is no splitting. This pattern of energy level shifts is sometimes called *Lande's interval rule.*

The fact that $\hat{H}_{SO}$ is relativistic in origin suggests that its effects may be small for low Z recall Eqn. (18.16) and allows us to use perturbation theory to estimate its size. We require the expression

$$\left\langle \frac{1}{r^3} \right\rangle_{n,l} = \left(\frac{Z^3}{a_0^3}\right) \frac{2}{n^3 l(2l+1)(l+1)} \tag{19.130}$$

The splittings due to this spin-orbit coupling can then be written as

$$\Delta E_{n,l}^{SO} = \frac{Ke^2}{2m_e^2c^2} \frac{Z^4\hbar^2}{a_0^3} \frac{1}{n^3 l(2l+1)(l+1)} \begin{Bmatrix} l & \text{for } j = l + 1/2 \\ -(l+1) & \text{for } j = l - 1/2 \end{Bmatrix} \tag{19.131}$$

We note that this expression is not obviously well defined for $l = 0$, being a ratio of the form 0/0. A careful derivation using the Dirac equation shows that the s-state energy levels are, in fact, shifted by just the amount predicted by Eqn. (19.131) by simply canceling the factors of l for the $j = l + 1/2$ case.[22]

[22]This result may seem counterintuitive as we would expect no spin-orbit contribution for s states for which there is no angular momentum. In fact, a more careful evaluation of the effect of Eqn. (19.124) for the case of $l = 0$, not based on perturbation theory, shows that $\langle 1/r^3 \rangle$ is indeed finite, so that it gives no effect on s states. There is, however, another term that arises in the nonrelativistic reduction of the Dirac equation (which has no simple classical correspondence) which is only nonvanishing for s states, and which gives exactly the effect as Eqn. (19.131), which we then take as *operationally* correct for all values of l. For a thorough discussion, see Condon and Shortley (1951).

Using the expression for the unperturbed energy levels, we can rewrite Eqn. (19.131) in the more compact form

$$\frac{\Delta E^{SO}_{n,l}}{|E_n|} = (Z\alpha)^2 \frac{1}{nl(2l+1)(l+1)} \begin{Bmatrix} l & \text{for } j = l + 1/2 \\ -(l+1) & \text{for } j = l - 1/2 \end{Bmatrix} \tag{19.132}$$

Several comments can be made:

- The factor of $(Z\alpha)^2$ in Eqn. (19.132) implies that spin-orbit splittings are typically $Z^2 \times 10^{-5}$ smaller than the unperturbed energy levels; this justifies our use of perturbation theory.
- At this same level of precision, we must also include relativistic corrections to the nonrelativistic approximation for the kinetic energy (as in Sec. 1.4), which we standardly use. We can use the results of P18.12 in the form

$$\frac{\Delta E^{\text{rel}}_{n,l}}{|E_n|} = -\frac{(Z\alpha)^2}{n}\left[\frac{2}{2l+1} - \frac{3}{4n}\right] \tag{19.133}$$

  and taken together with Eqn. (19.132), these two relativistic corrections successfully reproduce the results of the Dirac equation analysis up to order $(Z\alpha)^4$, namely,

$$\frac{\Delta E^{\text{rel}}_{n,l} + \Delta E^{SO}_{n,l}}{|E_n|} = -\frac{(Z\alpha)^2}{n}\left[\frac{2}{2j+1} - \frac{3}{4n}\right] \tag{19.134}$$

  for both $j = l \pm 1/2$. The spin splittings due to these combined relativistic effects are shown schematically in Fig. 19.14 for adjacent S and P states. The spectroscopic notation nL_j is used to distinguish states of different principle quantum number (n), orbital angular momentum ($L = S, P, D, F, \ldots$ for $l = 0, 1, 2, 3, \ldots$) and total angular momentum $j = l \pm 1/2$.
- The splitting of the $(n+1)P \to nS$ transition line into two distinct lines is due to the energy difference between the $P_{3/2,1/2}$ energy levels, which in turn arises solely from the spin-orbit coupling, since the relativistic correction is the same for fixed l, n. This splitting is responsible for the distinctive pattern of *doublets* in the line spectra of alkali atoms; the pair of closely spaced lines called the *sodium*

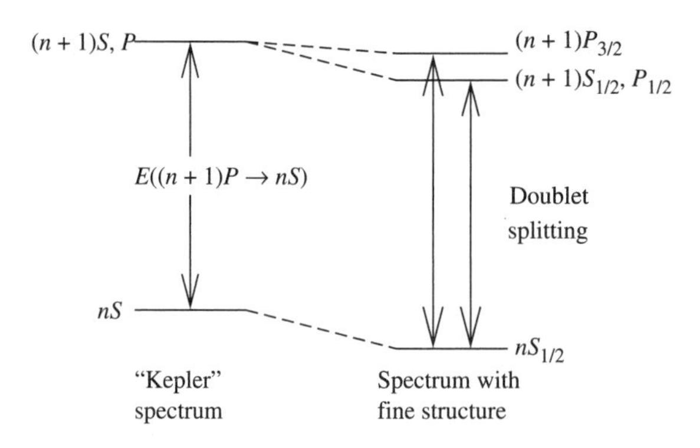

FIGURE 19.14. *Splittings of energy levels due to fine structure (spin-orbit plus relativistic kinetic energy effects) on S and P states. The "doublet" splitting relevant for $(n+1)P \to nS$ transitions is due to the spin-orbit coupling only.*

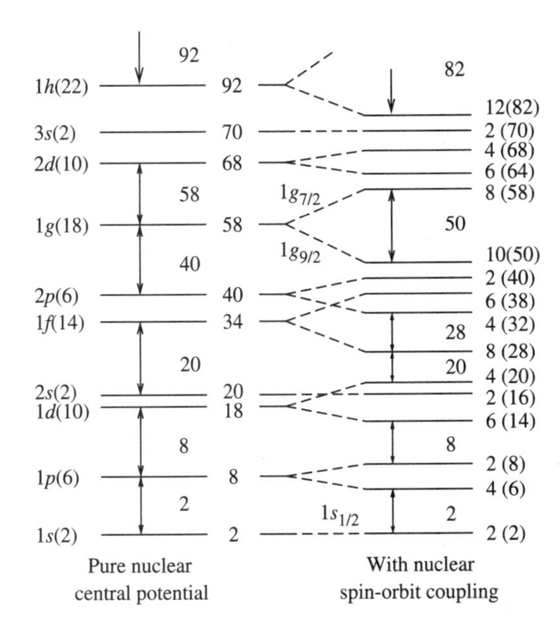

FIGURE 19.15. *Energy level scheme for typical central nuclear potential (left) and including spin-orbit interactions (right). The inclusion of the spin-orbit couplings reproduces the observed "magic numbers."*

doublet[23] is perhaps the most famous example, and corresponds to the transition of the outermost electron from the $4P \to 3S$ level.

Spin-orbit coupling terms are also present in nuclear systems where they play an important role in understanding the observed pattern of the so-called *magic numbers* implied by nuclear shell structure. Just as with atomic systems, large energy gaps between relatively closely spaced sets of quantized energy levels in nuclei can give rise to especially stable configurations as the available states are "filled" with neutrons and protons; many observable properties are correlated with nuclei whose Z or N is equal to $2, 8, 20, 28, 50, 82, 126$, as in Fig. 1.5(b).

Calculations for various models described by purely central potentials, $V(\mathbf{r}) = V(r)$, as in Fig. 19.15, give levels consistent with the first few such magic numbers, but fail to reproduce the observed pattern for heavier nuclei. If one includes a strong spin-orbit coupling[24] of the form $V_{SO}(r)\mathbf{L} \cdot \mathbf{S}$, with the appropriate sign for $V_{SO}(r)$, the level structure is changed dramatically for larger angular momentum states and nicely reproduces observed nuclear shell structure. For example, the lowest-lying $l = 4$ state, here labeled $1g(18)$, can accommodate $2(2 \cdot 4 + 1) = 18$ spin-1/2 neutrons or protons. The total angular momentum of a single nucleon in such a state is $J = L + S = 4 + 1/2 = 7/2$ (8 states) and $J = 9/2$ (10 states), and the fine-structure interaction splits these two combinations into different shells.

19.6.3 Hyperfine Splittings: Magnetic Dipole-Dipole Interactions

The magnetic moment of an atomic electron can also interact with the magnetic dipole field associated with the nuclear magnetic moment. In the context of hydrogen (and other

[23] This fact is the basis for the only suggestion for a quantum mechanics experiment in this book, namely, looking at the emission spectrum of sodium (using a diffraction grating) by appropriately igniting some table salt, i.e., sodium chloride; see Crawford (1968) for more concrete suggestions.

[24] This observation was made independently by Maria Goeppart Mayer and H. D. Jensen, for which they shared the Nobel Prize in 1963.

atomic systems), the resulting spin-spin interactions give rise to small level shifts called hyperfine splittings, and are also present in other "hydrogenic" bound state systems.

The magnetic field from a *point* dipole, $\mathbf{M}$, is given by

$$\mathbf{B}(\mathbf{r}) = \frac{\mu_0}{4\pi}\left(\frac{3(\mathbf{M}\cdot\mathbf{r})\mathbf{r}}{r^5} - \frac{\mathbf{M}}{r^3} + \frac{8\pi}{3}\mathbf{M}\delta(\mathbf{r})\right) \tag{19.135}$$

The first two terms are the familiar result derived in most standard texts on electricity and magnetism as the dipole field of a distant current loop; the third term is more subtle[25] and arises when one considers point dipoles. The corresponding classical energy of one dipole in the field of another is then simply

$$\begin{aligned} E &= -\mathbf{M}\cdot\mathbf{B} = \hat{H}_{\text{dip}} \\ &= -\frac{\mu_0}{4\pi}\left(\frac{3(\mathbf{M}_N\cdot\mathbf{r})(\mathbf{M}_e\cdot\mathbf{r})}{r^5} - \frac{\mathbf{M}_N\cdot\mathbf{M}_e}{r^3} + \frac{8\pi}{3}\mathbf{M}_N\cdot\mathbf{M}_e\,\delta(\mathbf{r})\right) \end{aligned} \tag{19.136}$$

where we have specialized to nuclear (N) and electronic (e) moments given by

$$\mathbf{M}_N = g_N\frac{Ze}{2M}\mathbf{S}_N \qquad \mathbf{M}_e = -g_e\frac{e}{2m_e}\mathbf{S}_e \approx -\frac{e}{m_e}\mathbf{S}_e \tag{19.137}$$

where M, Z, g_N are the nuclear mass, charge, and gyromagnetic ratio, respectively. The Hamiltonian describing this dipole-dipole interaction is given by this form where $\mathbf{S}_{e,N}$ are associated with spin operators.

The effect of this spin-spin interaction on the ground state of a hydrogen-like atom can be estimated using perturbation theory via $\Delta E_{1,0}^{(h.f.s)} = \langle\psi_{1,0,0}|\hat{H}_{\text{dip}}|\psi_{1,0,0}\rangle$, which we can evaluate using several observations:

1. The expectation value of the "standard" dipole term in Eqn. (19.136) vanishes when evaluated in the spherically symmetric ground state.
2. The "point dipole" term then contributes a factor of

$$\langle\psi_{1,0,0}|\delta(\mathbf{r})|\psi_{1,0,0}\rangle = |\psi_{1,0,0}(0)|^2 = \frac{1}{\pi}\left(\frac{Z}{a_0}\right)^3 \tag{19.138}$$

3. The magnetic permeability can be replaced in favor of $\mu_0 = 1/(c^2\epsilon_0)$.
4. The product of spin operators can be performed using the standard trick, namely,

$$\mathbf{S}_e\cdot\mathbf{S}_N = \frac{1}{2}\left(\mathbf{S}^2 - \mathbf{S}_e^2 - \mathbf{S}_N^2\right) \tag{19.139}$$

where $\mathbf{S} = \mathbf{S}_e + \mathbf{S}_N$ in general. For the case of hydrogen when the total spin can be $S = 1/2 + 1/2 = 0, 1$, we have

$$\mathbf{S}_e\cdot\mathbf{S}_p = \hbar^2 \qquad \begin{cases} 1(1+1) - 3/4 - 3/4 = 1/2 & \text{for } S = 1 \\ 0 - 3/4 - 3/4 = -3/2 & \text{for } S = 0 \end{cases} \tag{19.140}$$

These can be combined to find the spin-dependent shifts to the ground state levels of hydrogen, namely,

$$\Delta E_{1,0}^{(h.f.s)} = \frac{Ke^2}{2m_e^2c^2}\frac{Z^4\hbar^2}{a_0^3}\left[g_p\frac{4}{3}\frac{m_e}{m_p}\right] \qquad \begin{cases} +1 & \text{for } S = 1 \\ -3 & \text{for } S = 0 \end{cases} \tag{19.141}$$

[25]For a nice discussion, see Griffiths (1981).

which we have put in a form that can be compared more readily to the expression for fine-structure splitting in Eqn. (19.131). We note that:

- The $S = 0$ or ${}^1S_0$ state is split down relative to the $S = 1$ or ${}^3S_1$ state (where the notation ${}^SL_J$ is used).
- This hyperfine splitting (hence h.f.s.) is suppressed relative to the fine-structure effects by the factor in square brackets, namely, $m_e/M \lesssim 1/1800$ and motivates its name.
- The photon energy corresponding to transitions from the $S = 1$ to $S = 0$ states is roughly $E_\gamma = 5.9 \times 10^{-6}$ eV, which corresponds to a frequency and wavelength of roughly $f \approx 1420$ Hz and $\lambda \approx 21$ cm, respectively. Interstellar hydrogen atoms can undergo collisional excitations that populate the (slightly) excited $S = 1$ state from the ground state ($S = 0$), and the resulting 21-cm radio emission line is extensively used by astronomers to map the concentrations of hydrogen gas. For example, Doppler-shifted line profiles of the 21-cm line spectrum have been used to map out the spiral-arm structure of our galaxy (as in Fig. 1.8). This physical feature of the most basic atom in nature has been described[26] as "a unique, objective standard frequency, which must be known to every observer in the universe," and has given rise to suggestions[27] that it be used (or at least monitored) for interstellar communications.

The spin-spin level splittings due to magnetic dipole-dipole interactions are present in other systems as well. In the e^+e^- positronium system, for example, the effect is present with two important changes:

- Because both particles have the same mass, the hyperfine splitting is not suppressed relative to spin-orbit, fine-structure effects, as the m_e/M_p term in Eqn. (19.141) is now of order unity.
- An additional physical mechanism contributes to the effective interaction in Eqn. (19.136) due to the possibility of "annihilation type" interactions, as shown in Fig. 19.16.

As mentioned in Sec. 18.3.5, quark-antiquark bound states of many different masses exist which can be used to probe the mass dependence of the $q - \overline{q}$ system. There are

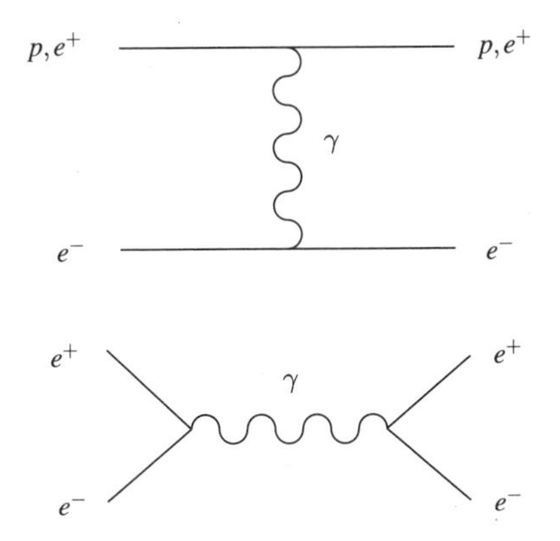

FIGURE 19.16. *Feynman diagrams leading to hyperfine splittings for e^+e^- bound states.*

[26]See Cocconi and Morrison (1959).

[27]For a guide to the scientific literature on search for extraterrestrial civilizations, see Kuiper (1989).

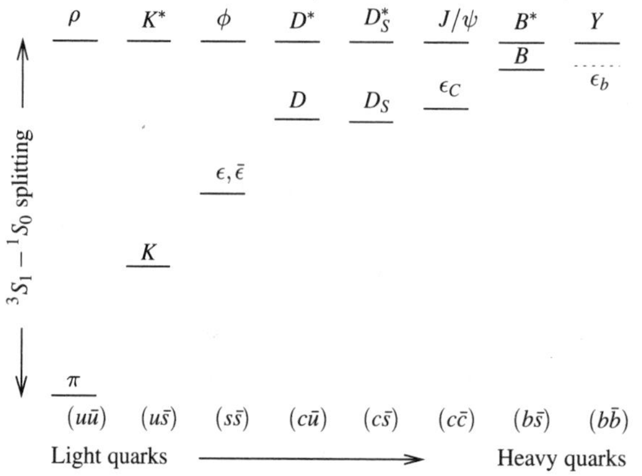

ρ K^* ϕ D^* D_S^* J/ψ B^* Y

B ϵ_b ϵ_C D D_S

$\epsilon, \bar{\epsilon}$

K

π

$^3S_1 - {}^1S_0$ splitting

$(u\bar{u})$ $(u\bar{s})$ $(s\bar{s})$ $(c\bar{u})$ $(c\bar{s})$ $(c\bar{c})$ $(b\bar{s})$ $(b\bar{b})$

Light quarks → Heavy quarks

FIGURE 19.17. Energy differences between $^3S_1$ and $^1S_0$ $q_1\overline{q}_2$ bound states for different quark varieties of increasing masses. "Color dipole-dipole" interactions lead to hyperfine splittings proportional to $1/m_1m_2$.

spin-dependent couplings between quarks which are the "color Coulomb" (see Sec. 18.3.4) analog of the magnetic dipole-dipole interaction leading to hyperfine splittings; they can be written in the form

$$\hat{H}_{\text{color-dipole}} \propto \mathbf{M}_1 \cdot \mathbf{M}_2 \propto \left(\frac{\mathbf{S}_1}{m_1}\right) \cdot \left(\frac{\mathbf{S}_2}{m_2}\right) \tag{19.142}$$

so that the $^3S_1 - {}^1S_0$ splitting for quark-antiquark states should scale as

$$\Delta E(^3S_1 - {}^1S_0) \propto \frac{1}{m_1 m_2} \tag{19.143}$$

We plot in Fig. 19.17 the splitting between the lowest lying $^3S_1$ and $^1S_0$ states for a given pair of quark types. The mass of the heaviest quark in the system increases from left to right, while the spin-splittings decrease, as suggested by the $1/m_1m_2$ factor in Eqn. (19.143).

19.7 SPINS IN MAGNETIC FIELDS

19.7.1 Measuring the Spinor Nature of the Neutron Wavefunction

We have focused on the interaction of charged particles with **E** and **B** fields. Electrically neutral particles can have nontrivial electromagnetic interactions via the coupling of a magnetic dipole moment with external fields. This was discussed in Sec. 17.4 in connection with the spinor wavefunction of spin-1/2 particles, where it was shown that such an interaction can provide a "handle" on the precession of the spin vector. This fact was the basis for the observation[28] that the phase behavior of the neutron spinor wavefunction, with its predicted phase change of -1 on rotation by 2π, could be tested in such systems.

The angle through which a magnetic moment precesses in a field in a small time dt is given by Eqn. (17.179) as

$$d\theta = \omega_L t = \frac{\mu B}{\hbar} dt \tag{19.144}$$

This can be integrated over any path over which the neutron moves with constant speed v to yield

[28]First made by Bernstein (1967) and Aharonov and Susskind (1967).

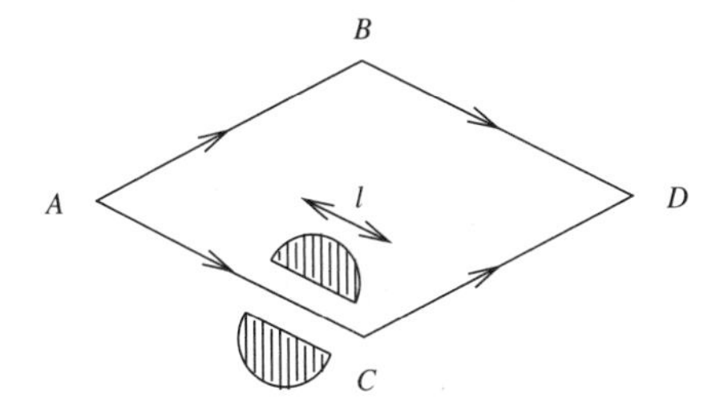

FIGURE 19.18. Schematic representation of experiment designed to test the phase change on rotation of the neutron's spinor wavefunction.

$$\theta = \int d\theta = \frac{\mu_n}{\hbar}\int B\,dt = \frac{g\mu}{\hbar}\frac{1}{v}\int_{\text{path}} \mathbf{B}\cdot d\mathbf{l} \tag{19.145}$$

and v is the classical speed. Recall that the neutron's magnetic moment is roughly $\mu_n = -1.93\,\mu_N$, where $\mu_N = e\hbar/2m_n$.

After having its magnetic moment precess through the angle θ, the spinor wavefunction becomes

$$\begin{pmatrix}\alpha^+(\theta)\\ \alpha^-(\theta)\end{pmatrix} = \begin{pmatrix}e^{i\theta/2} & 0\\ 0 & e^{-i\theta/2}\end{pmatrix}\begin{pmatrix}\alpha^+(0)\\ \alpha^-(0)\end{pmatrix} = \begin{pmatrix}\alpha^+(0)e^{i\theta/2}\\ \alpha^-(0)e^{-i\theta/2}\end{pmatrix} \tag{19.146}$$

Now consider the experiment described schematically by Fig. 19.18 in which a beam of neutrons is split between two paths, one which traverses a magnetic field, and one in a field-free region. If the initial neutron beam is unpolarized (equal amounts of "up" and "down"), then the neutrons along the path ABD experience no change in phase and

$$\psi_{ABD} = \psi_A = \frac{1}{\sqrt{2}}\begin{pmatrix}1\\1\end{pmatrix} \tag{19.147}$$

The pieces of the neutron spin wavefunction which do traverse the magnetic field pick up phases given by

$$\psi_{ACD}(\theta) = \begin{pmatrix}e^{i\theta/2} & 0\\ 0 & e^{-i\theta/2}\end{pmatrix}\psi_A = \frac{1}{\sqrt{2}}\begin{pmatrix}e^{i\theta/2}\\ e^{-i\theta/2}\end{pmatrix} \tag{19.148}$$

When the beams are recombined, the total wavefunction is given by

$$\psi_{\text{tot}}(\theta) = \psi_{ABD}(\theta) + \psi_{ACD}(\theta) = \frac{1}{\sqrt{2}}\begin{pmatrix}1 + e^{i\theta/2}\\ 1 + e^{-i\theta/2}\end{pmatrix} \tag{19.149}$$

so that the probability density is proportional to

$$|\psi_{\text{tot}}(\theta)|^2 = \frac{1}{2}\left(\left|1 + e^{i\theta/2}\right|^2 + \left|1 + e^{i\theta/2}\right|^2\right) \propto 1 + \cos\left(\frac{\theta}{2}\right) \tag{19.150}$$

The counting rates at either detector C_1 or C_2 measure the beam intensity and hence this probability. If one scales the intensity obtained with various amounts of "magnetic path length" $\int_{\text{path}} \mathbf{B}\cdot d\mathbf{l}$ to that with the field turned off, one should then have

$$\frac{I(\theta)}{I(0)} = \frac{|\psi_{\text{tot}}(\theta)|^2}{|\psi_{\text{tot}}(0)|^2} = \frac{1 + \cos(\theta/2)}{2} \tag{19.151}$$

This formula exhibits the typical spin-1/2 phase change on rotation by 2π; constructive interference corresponding to a return to the same phase is only seen after a rotation of the magnetic moment by an angle of $\theta = 4\pi$.

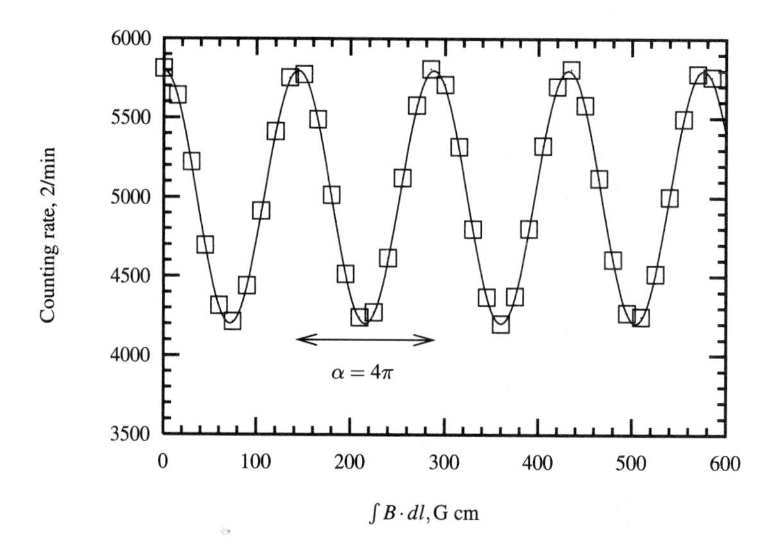

FIGURE 19.19. Measured neutron intensity versus magnetic path length illustrating the "sign flip" for a spin-1/2 wavefunction rotated through 2π. [Data from Rauch et al. (1975).]

Two sets of experiments using neutron interferometers (as in Sec. 19.1) were performed[29] soon after the predictions were made, and results from one of them are shown in Fig. 19.19. Numerical values for this data set are analyzed in P19.28.

19.7.2 Spin Resonance

The steady precession of a magnetic moment around the direction of a static field can exhibit striking resonance effects if a time-varying magnetic field is applied at right angles to the original field. This can be understood in two relatively simple ways:

- We illustrate in Fig. 19.20 a large, uniform field in the z direction and the precession of the magnetic moment **M** (and hence spin direction) around it. The presence of a small, rotating $\mathbf{B}_A$ field will, in general, have little effect as the torque it induces on **M** will average out to zero if it rotates at a different rate (or even different direction) than does **M**. If, however, the magnetic field "stays in phase" with the moment by

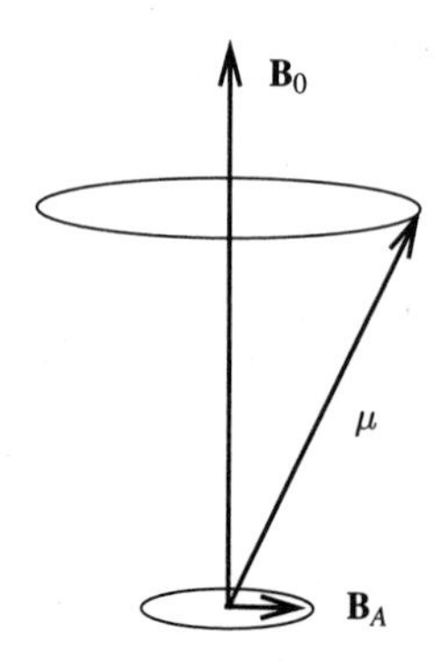

FIGURE 19.20. Magnetic moment precessing around static field B_0 plus additional, rotating magnetic field B_A.

[29] See Werner et al. (1975) and Rauch et al. (1975).

rotating at a rate $\omega = \omega_{\text{precess}} = g\mu B/\hbar$, its torque can act continuously and induce dramatic changes in the rotational motion.

- We have seen (Sec. 17.4) that in a uniform field the energy levels of the spin-1/2 particle are given by

$$E_{\pm} = \pm\hbar\omega_{\text{phase}} = \pm\frac{g\mu B}{2} \tag{19.152}$$

which implies a splitting between the parallel and antiparallel configurations of spins given by $\Delta E = g\mu B$. This implies that electromagnetic radiation of energy $E_\gamma = \hbar\omega = \Delta E$ or angular frequencies $\omega = g\mu B/\hbar$ can preferentially be absorbed.

In either viewpoint, the spin system will exhibit resonance behavior when subjected to external magnetic fields when $\omega = \omega_{\text{precess}}$, and we will analyze the dynamical equations of motion for the spinor wavefunction in a somewhat formal way in order to derive the exact form of the resonance response.

We assume a large, static field ($\mathbf{B}_0$) in the z direction and a rotating field of arbitrary magnitude ($\mathbf{B}_A$) at right angles, given by

$$\mathbf{B}_0 = (0, 0, B_0) \tag{19.153}$$

$$\mathbf{B}_A = (B_A\cos(\omega t), -B_A\sin(\omega t), 0) \tag{19.154}$$

The spin Hamiltonian is then given by

$$\hat{H} = -\frac{g\mu}{2}\mathbf{B}\cdot\boldsymbol{\sigma} = -\frac{g\mu}{\hbar}\left(B_A^{(x)}\sigma_x + B_A^{(y)}\sigma_y + B_A^{(z)}\sigma_z\right)$$

$$= -\frac{g\mu}{2}\begin{pmatrix} B_0 & B_A e^{i\omega t} \\ B_A e^{-i\omega t} & -B_0 \end{pmatrix} \tag{19.155}$$

The corresponding Schrödinger equation for the spinor coordinates is

$$i\hbar\frac{\partial}{\partial t}\begin{pmatrix}\alpha_+(t)\\ \alpha_-(t)\end{pmatrix} = \hat{H}\begin{pmatrix}\alpha_+(t)\\ \alpha_-(t)\end{pmatrix} \tag{19.156}$$

which can be written in the form

$$\begin{pmatrix}\dot{\alpha}_+(t)\\ \dot{\alpha}_-(t)\end{pmatrix} = \frac{i}{2}\begin{pmatrix}\omega_0 & \omega_A e^{i\omega t}\\ \omega_A e^{-i\omega t} & -\omega_0\end{pmatrix}\begin{pmatrix}\alpha_+(t)\\ \alpha_-(t)\end{pmatrix} \tag{19.157}$$

where we have defined two new frequencies,

$$\omega_0 = \frac{g\mu B_0}{\hbar} \quad \text{and} \quad \omega_A = \frac{g\mu B_A}{\hbar} \tag{19.158}$$

in addition to the frequency ω describing the time rate of change of the external field. These then give the coupled differential equations

$$2\dot{\alpha}_+(t) = i\left(\omega_0\alpha_+(t) + \omega_A e^{i\omega t}\alpha_-(t)\right) \tag{19.159}$$

$$2\dot{\alpha}_-(t) = i\left(\omega_A e^{-i\omega t}\alpha_+(t) - \omega_0\alpha_-(t)\right) \tag{19.160}$$

We assume that the spinor is initially in the "up" direction so that the

$$\begin{pmatrix} \alpha_+(0) \\ \alpha_-(0) \end{pmatrix} = \begin{pmatrix} 1 \\ 0 \end{pmatrix} \tag{19.161}$$

and attempt to find the solutions of Eqn. (19.157) subject to these initial conditions.

Motivated by the time dependence of the spinor wavefunction when $B_A = 0$, we attempt a solution of the form

$$\alpha_+(t) = A_+ e^{i\omega_+ t} \qquad \alpha_-(t) = A_- e^{i\omega_- t} \tag{19.162}$$

so that Eqns. (19.159) and (19.160) become

$$2\omega_+ A_+ = \omega_0 A_+ + \omega_A A_- e^{i(\omega + \omega_- - \omega_+)t} \tag{19.163}$$

$$2\omega_- A_- = \omega_A A_+ e^{-i(\omega + \omega_+ - \omega_-)t} - \omega_0 A_- \tag{19.164}$$

The time dependence can be eliminated if we choose

$$\omega_+ = \omega_- + \omega \tag{19.165}$$

which gives now the coupled *algebraic* equations in matrix form

$$\begin{pmatrix} (2\omega_+ - \omega_0) & -\omega_A \\ -\omega_A & (2\omega_- + \omega_0) \end{pmatrix} \begin{pmatrix} A_+ \\ A_- \end{pmatrix} = 0 \tag{19.166}$$

For a consistent solution, the determinant of coefficients must vanish, and if we combine this condition with Eqn. (19.165), we find that

$$\begin{aligned} \omega_- &= -\frac{\omega}{2} \pm \frac{1}{2}\sqrt{(\omega - \omega_0)^2 + \omega_A^2} \\ &= -\frac{\omega}{2} \pm \Delta\omega \end{aligned} \tag{19.167}$$

so that $\omega_+ = +\omega/2 \pm \Delta\omega$.

As with any set of coupled second-order differential equations, the result for $\alpha_-(t)$ will consist of a linear combination of the two independent solutions, namely,

$$\alpha_-(t) = Ae^{-i\omega t/2} e^{i\Delta\omega t} + Be^{-i\omega t/2} e^{-i\Delta\omega t} \tag{19.168}$$

The initial condition $\alpha_-(0) = 0$ implies that $B = -A$, while its derivative is given by $\dot{\alpha}_-(0) = 2iA\Delta\omega$. Substituting these values into Eqn. (19.160) and using the other initial condition on the spinor upper component, namely, $\alpha_+(0) = 1$, implies that

$$2iA = \frac{i\omega_A}{2\Delta\omega} \tag{19.169}$$

Our "big result" is that the total time dependence is given by

$$\alpha_-(t) = \frac{i\omega_A}{2\Delta\omega} \sin(\Delta\omega t) e^{-i\omega t/2} \tag{19.170}$$

This spinor amplitude gives information on the probability that the spin will have "flipped" to a state antiparallel to $\mathbf{B}_0$ at later times since

$$\text{Prob(spin down)}(t) = |\alpha_-(t)|^2 = \left[\frac{\omega_A^2}{\omega_A^2 + (\omega - \omega_0)^2}\right] \sin^2(\Delta\omega t) \qquad (19.171)$$

Observations on this result include:

- The prefactor in Eqn. (19.171), namely,

$$\frac{\omega_A^2}{\omega_A^2 + (\omega - \omega_0)^2} \qquad (19.172)$$

  exhibits the *Lorentzian line shape* discussed in P5.4, which is typical of resonance phenomena, and we will illustrate its form for various values of ω_A in Fig. 19.21. The spin flip amplitude has its maximum value when

$$\omega = \omega_0 = \frac{g\mu B_0}{\hbar} = \omega_{\text{precess}} \qquad (19.173)$$

  as expected, but we also note that the "sharpness" of the resonance is proportional to ω_A. In order to measure the resonant frequency as precisely as possible (which is the hallmark of the method), we want to make B_A small, which often means that the sample must be shielded from stray fields not under the control of the experimenter (the earth's field, for example, or stray RF signals).
- When applied to an unpaired electron spin, this method is called *electron spin resonance* or ESR, and for typical external field strengths of $B_0 \approx 0.3T$, the resonant frequencies, wavelengths, and photon energies are

$$f \approx 10\text{ GHz} \qquad \lambda \approx 3\text{ cm} \qquad E_\gamma = \Delta E \approx 4 \times 10^{-5}\text{ eV} \qquad (19.174)$$

  so that ESR usually requires microwave techniques. When applied to the nuclear magnetic moments of free or unpaired protons (or other nuclei with nonvanishing spins), the technique is called *nuclear magnetic resonance* or NMR. The corresponding resonance parameters for protons in a 1 T field are

$$f \approx 40\text{ MHz} \qquad \lambda \approx 7\text{ m} \qquad E_\gamma = \Delta E \approx 2 \times 10^{-7}\text{ eV} \qquad (19.175)$$

  which shows that radio frequencies (RF) are required for the time-dependent fields.

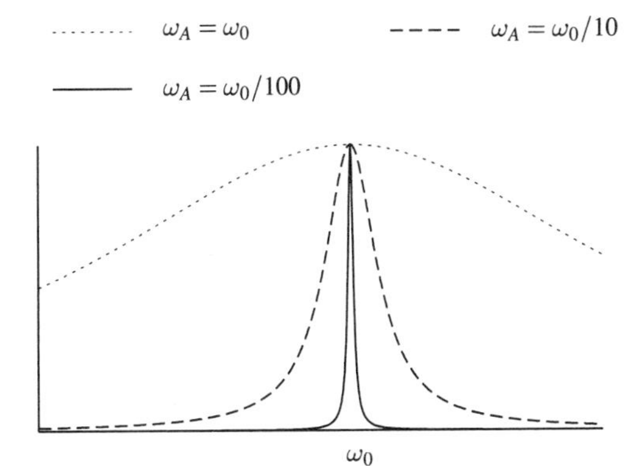

FIGURE 19.21. Probability of a spin flip versus frequency of applied field ω showing Lorentzian line shape for spin resonance. Different values of ω_A are shown, indicating the effect of the magnitude of B_A on the "sharpness" of the resonance peak.

- Even with large field strengths B_0, the energy "gain" from being parallel to the field is much smaller than ordinary thermal fluctuations, so that there is only a small excess of spin-up (N_+) versus spin-down (N_-) states in a typical sample. For example, the population ratio in the two states is given by their respective Boltzmann factors, namely,

$$\frac{N_+}{N_-} = \frac{e^{-E_+/kT}}{e^{-E_-/kT}} = e^{-\Delta E/kT} \tag{19.176}$$

where k is Boltzmann's constant and T is the temperature. At room temperatures, where $kT \approx 1/40$ eV, this implies that

$$\left(\frac{N_+ - N_-}{N_+ + N_-}\right) \sim \frac{1}{2}\left(1 - e^{-\Delta E/kT}\right) \sim \frac{\Delta E}{2kT} \approx 4 \times 10^{-6} \tag{19.177}$$

for the proton values above.

- ESR and NMR have been extensively applied in the fields of nuclear and solid-state physics as well as in chemistry. The precision with which frequency measurements can be made can then be translated into very accurate determinations of nuclear magnetic moments. Nuclei with known moments can then be used to probe the electronic environment in solids or determine the structure of complex molecules. The technique is perhaps best known, however, when it is applied to probing the protons in the human body (recall Fig. 1.7) where it is known as *magnetic resonance imaging* or MRI.[30]

19.8 THE AHARONOV-BOHM EFFECT

We have so far considered the electromagnetic interactions of charged particles and magnetic dipoles in regions where there is a nonvanishing **B** field. We now describe a strikingly "quantum" phenomenon in which the phase of the wavefunction of a charged particle is changed even in a field-free region. The phase can be calculated in terms of the vector potential **A** and shows that the potentials themselves can play an important role.

We have stressed that a given configuration of electric and magnetic fields can be derived from an arbitrarily large number of different electromagnetic potentials, ϕ and **A**. This fact even carries over to the case where there are no magnetic fields present. The restriction

$$\nabla \times \mathbf{A}(\mathbf{r}, t) = \mathbf{B}(\mathbf{r}, t) = 0 \tag{19.178}$$

can still be satisfied by the gradient of any scalar function, namely, $\mathbf{A}(\mathbf{r}, t) = \nabla f(\mathbf{r}, t)$; this fact is also consistent with the gauge transformation Eqns. (19.29) and (19.30) starting with a vanishing **A**. This relation can be inverted to give

$$f(\mathbf{r}, t) = \int^{\mathbf{r}} d\mathbf{r}' \cdot \mathbf{A}(\mathbf{r}', t) \tag{19.179}$$

[30]For a review of the physical principles and clinical applications of MRI, see Partain et al. (1988).

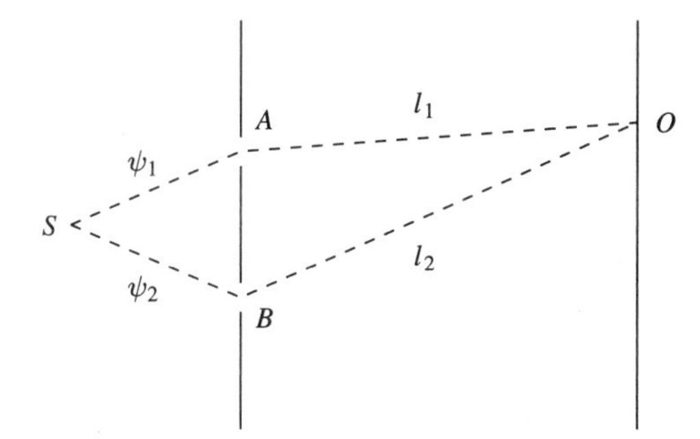

FIGURE 19.22. Path length geometry for classical wave interference.

so that the phase factor connecting a free-particle wavefunction to its gauge equivalent partner can be written as

$$\psi'(\mathbf{r}, t) = e^{iqf(\mathbf{r},t)/\hbar}\left[\psi_{\text{free}}(\mathbf{r}, t)\right] = \exp\left(iq\int^{\mathbf{r}} d\mathbf{r}' \cdot \mathbf{A}(\mathbf{r}', t)/\hbar\right)\psi_{\text{free}}(\mathbf{r}, t) \quad (19.180)$$

The phase factor is seemingly arbitrary as it depends on the choice of gauge, and is unmeasurable by itself as $|\psi'|^2 = |\psi|^2$.

We have already been reminded in Sec. 9.1 and 19.7.1 that interference experiments are necessary to probe the relative phases of wavefunctions, so, before proceeding further, we recall the wave physics behind electron (or any wave) interference experiments, as illustrated in Fig. 19.22. The wavefunctions from the source, S, travel different path lengths and acquire different phase factors. When the electron beams are recombined, the wave amplitude can be expressed as

$$\psi^O_{\text{tot}} = \psi^O_1 + \psi^O_2 = \psi^S_1 e^{ikl_1} + \psi^S_2 e^{ikl_2} = \left(\psi^S_1 + \psi^S_2 e^{i\theta}\right)e^{ikl_1} \quad (19.181)$$

where

$$\theta = k(l_2 - l_1) = k\Delta l = \frac{2\pi}{\lambda}\Delta l \quad (19.182)$$

This implies the familiar result that a difference in path length equal to an integral number of wavelengths will give constructive interference.

Aharonov and Bohm[31] suggested a conceptually simple version of such an interference experiment to probe the quantum phase induced by the vector potential, illustrated in Fig. 19.23. A small region of magnetic field is present (in the circular area shown), but the particles classically move in field-free regions along the paths SAO and SBO. The geometrical path lengths along SAO and SBO are identical, but there can still be a phase difference in the two paths due to a variation in the vector potential traversed, despite the fact that neither particle is in a magnetic field. Similarly to Eqn. (19.181), we have

$$\psi^O_{\text{tot}} = \psi^O_1 + \psi^O_2 = \psi^S_1 e^{i\theta_1} + \psi^S_2 e^{i\theta_2} = \left(\psi^S_1 + \psi^S_2 e^{i(\theta_2-\theta_1)}\right)e^{i\theta_1} \quad (19.183)$$

[31] See Aharonov and Bohm (1957).

where

$$\theta_1 = \frac{q}{\hbar}\int_{SAO}^{\mathbf{r}} d\mathbf{r}' \cdot \mathbf{A}(\mathbf{r}', t) \qquad \text{and} \qquad \theta_2 = \frac{q}{\hbar}\int_{SBO}^{\mathbf{r}} d\mathbf{r}' \cdot \mathbf{A}(\mathbf{r}', t) \tag{19.184}$$

The phase difference that can give rise to interference effects can be written as

$$\begin{aligned}
\Delta\theta &= \theta_2 - \theta_1 \\
&= \frac{q}{\hbar}\left(\int_{SAO}^{\mathbf{r}} d\mathbf{r}' \cdot \mathbf{A}(\mathbf{r}', t) - \int_{SBO}^{\mathbf{r}} d\mathbf{r}' \cdot \mathbf{A}(\mathbf{r}', t)\right) \\
&= \frac{q}{\hbar}\oint_{SAOBS} d\mathbf{r}' \cdot \mathbf{A}(\mathbf{r}', t)
\end{aligned} \tag{19.185}$$

The line integral of **A** around the closed path *SAOBS* can be rewritten using Stoke's theorem in the form of an area integral:

$$\begin{aligned}
\Delta\theta &= \frac{q}{\hbar}\int_{\text{area}} \nabla' \times \mathbf{A}(\mathbf{r}', t) \cdot d\mathbf{S} \\
&= \frac{q}{\hbar}\int_{\text{area}} \mathbf{B} \cdot d\mathbf{S} \\
&= \frac{q}{\hbar}\Phi_B
\end{aligned} \tag{19.186}$$

where Φ_B is the magnetic flux enclosed by the path. This result is striking as it says that

- The phase of the wavefunction does depend on the (gauge-dependent) vector potential **A** in a nontrivial way, even in a region where the physical magnetic field vanishes.
- That phase dependence gives rise to observable interference effects.
- The quantity that actually determines the interference pattern, however, is the enclosed magnetic flux, which is a perfectly gauge-invariant quantity.

The experimental verification of this prediction was first performed[32] by the use of a micron size iron "whisker" containing the field.

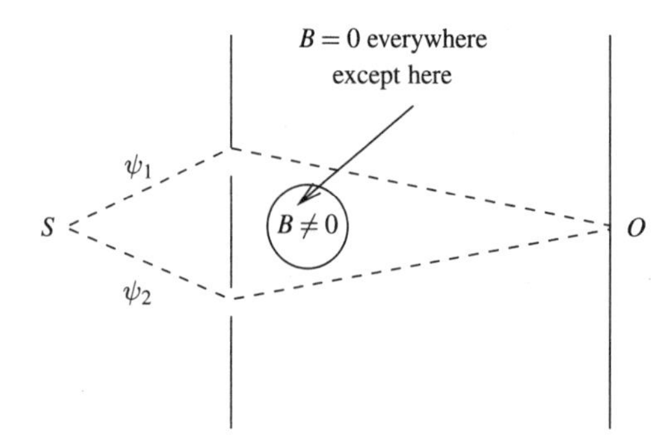

FIGURE 19.23. *Schematic representation of Aharonov-Bohm experiments.*

[32]See Chambers (1960).

19.9 QUANTUM ASPECTS OF THE ELECTROMAGNETIC FIELD: THE CASIMIR EFFECT

So far, we have focused on the quantum behavior of particles acting under the influence of classical electromagnetic fields; even the Aharonov-Bohm effect relied on the effects of the classical vector potential **A** on the quantum wavefunction of the charged particle. We conclude this chapter with a discussion of one remarkable aspect that arises when one applies quantum ideas to the electromagnetic field itself.

In Sec. 10.1, we used energy methods to argue that the electromagnetic field could be thought of as a set of harmonic oscillators, each with frequency $\omega_{\mathbf{k}} = |\mathbf{k}|c$. We have now seen at least two important examples of how the zero point energy of a quantum mechanical system can have not only observable but also dramatic consequences: the minimum vibrational energy of a diatomic molecule and the zero point energy contribution to the total energy of the electrons in a solid such as a metal or white dwarf star. We now describe how the zero point energy of the EM field oscillators gives rise to a remarkable new quantum effect.

Using our results for the harmonic oscillator energy spectrum, we know that the smallest possible total energy (labeled here by the subscript "vac" for *vacuum*) of the electromagnetic field is obtained when

$$E_{\text{vac}} = \sum_{\mathbf{k},\lambda} \frac{1}{2}\hbar\omega_{\mathbf{k}} = \frac{1}{2}\hbar c \sum_{\mathbf{k},\lambda} \sqrt{k_x^2 + k_y^2 + k_z^2} \tag{19.187}$$

where the sum is over wavevectors **k** and polarizations λ, which are available to the system. The summation is actually a continuous integral, as all values of **k** are allowed.

The sum is manifestly divergent, as it contains contributions from an infinite number of nonnegative terms. This is one of the first indications of the troublesome infinities that occur in quantum field theory. In our case, we will use the particle in the infinite well system yet again as a model system to see what physically interesting quantities might be calculable, despite the obvious divergences present.

In that system, the introduction of the walls, by their imposition of boundary conditions on the wavefunction, not only made the allowed values of wavenumber k, and hence momentum p, discrete but also introduced a lowest allowed value of energy, which was not present in the free-particle case. The change in energy between the free and confined case also necessarily led to a repulsive force on the walls (P6.17), tending to expand the system; this force could even be understood using classical arguments as well.

By analogy, the introduction of, say, conducting surfaces into the vacuum will also impose boundary conditions and change the ground state quantum energy of the EM field. The *difference* in ground state energies between the no-conductor (vac) and conductor present (wall) cases and its dependence on the geometry of the conducting surfaces (the "infinite walls") turns out to be a physically meaningful quantity, and will give rise to forces on the surfaces that are measureable in macroscopic experiments. This observation was first made by Casimir,[33] and we will follow his analysis and use the simplest geometry possible, namely, two large conducting plates of side L separated by a finite distance

[33] For a historical and critical review of the Casimir effect at our level, see Elizalde and Romeo (1991).

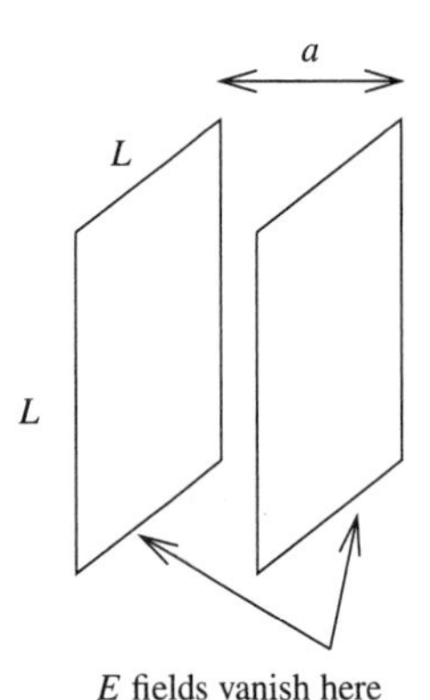

FIGURE 19.24. *Parallel-plate geometry for Casimir effect. The presence of the conducting plates imposes boundary conditions on the E field, which changes the zero point energy of the quantized EM field.*

$a \ll L$, as in Fig. 19.24, and examine the zero point energy *inside* the region both with and without the plates present.

The boundary conditions with the plates present imply that the values of k_z will be quantized via $k_z = n\pi/a$ where $n = 1, 2, \ldots$, and so forth. Recalling that there are two transverse polarization states for each allowed mode, we find that

$$\begin{aligned} E_{\text{walls}} &= \frac{\hbar c}{2} \sum \sqrt{k_x^2 + k_y^2 + k_z^2} \\ &= \frac{\hbar c}{2} \int_{-\infty}^{+\infty} \frac{L dk_x}{2\pi} \int_{-\infty}^{+\infty} \frac{L dk_y}{2\pi} \left[\sqrt{k_x^2 + k_y^2} + 2 \sum_{n=1}^{\infty} \left(k_x^2 + k_y^2 + \frac{n^2\pi^2}{a^2} \right)^{1/2} \right] \end{aligned} \tag{19.188}$$

where we have integrated (summed) over the continuous (discrete) values of k_x, k_y (k_z), and included only one polarization state for the special case of $k_z = 0$. Once again, this is formally an infinite quantity, but we should compare it to the zero point energy *not* in the presence of the conducting walls, namely,

$$E_{\text{vac}} = \frac{\hbar c}{2} \int_{-\infty}^{+\infty} \frac{L dk_x}{2\pi} \int_{-\infty}^{+\infty} \frac{L dk_y}{2\pi} \int_{-\infty}^{+\infty} \frac{a dk_z}{2\pi} \sqrt{k_x^2 + k_y^2 + k_z^2} \tag{19.189}$$

The *difference* of these two quantities, divided by L^2 gives the net energy per unit surface area, $\mathcal{E}$, which will be directly related to the force between the plates.

If we write $k_x^2 + k_y^2 = k^2$ so that $dk_x dk_y = 2\pi k dk$, and relabel the continuous k_z integral in Eqn. (19.189) using the natural change of variables $k_z = n\pi/a$, we find that

$$\begin{aligned} \mathcal{E} &\equiv \frac{E_{\text{walls}} - E_{\text{vac}}}{L^2} \\ &= \frac{\hbar c}{2\pi} \int_0^{\infty} k\,dk \left(\frac{k}{2} + \sum_{n=1}^{\infty} \sqrt{k^2 + n^2\pi^2/a^2} - \int_0^{\infty} dn \sqrt{k^2 + n^2\pi^2/a^2} \right) \end{aligned} \tag{19.190}$$

Changing variables once more to $z = ak/\pi$, we find that $\mathcal{E}$ can be written in the form

$$\mathcal{E} = \frac{\hbar c}{2\pi} \left(\frac{\pi}{a} \right)^3 \left[\frac{1}{2} E(0) + \sum_{n=1}^{\infty} E(n) - \int_0^{\infty} dn\, E(n) \right] \tag{19.191}$$

where we have defined

$$E(n) \equiv \int_0^{\infty} dz\, z \sqrt{z^2 + n^2} \tag{19.192}$$

Such differences of infinite series and integrals can be evaluated using the so-called *Euler-Maclaurin formula* (discussed in App. B.2), namely,

$$\frac{1}{2}E(0) + \sum_{n=1}^{\infty} E(N) - \int_0^{\infty} dn\, E(n) = -\frac{1}{6 \cdot 2!}E'(0) + \frac{1}{30 \cdot 4!}E'''(0) + \cdots \quad (19.193)$$

provided the individual sums and integrals are properly convergent. In our case that is simply not the case, so let us examine in more detail the assumptions that we have made.

We have to acknowledge that a description of conducting plates that impose *perfect* boundary conditions (i.e., on which the electric field must vanish identically) is certainly incorrect for large enough values of k. The limit of large k corresponds to small values of λ, and when λ is small compared to the typical atomic or molecular size, a classical description in terms of continuous, perfect conductors must fail. In that limit, the walls will be "permeable" to the fields, the discreteness of k vectors will disappear, and the infinite sum will approach the infinite integral, yielding a cancellation, and hence a finite result, for Eqn. (19.190). To account for this more physical behavior for large values of k, we can introduce a convergence factor in the integrals defining $E(n)$ to ensure that both values are finite, i.e.,

$$E(n) \longrightarrow \int_0^{\infty} dz\, z\sqrt{z^2 + n^2}\, f(z) \quad (19.194)$$

where we assume that $f(z)$ satisfies the requirements that

$$f(x) = \begin{cases} 1 & \text{for } z \ll z_{\text{max}} \\ 0 & \text{for } z_{\text{max}} \ll z \end{cases} \quad (19.195)$$

The desired physical result is to model the *cancellation* of the sum/integral terms for large k, but this convergence factor mimics this by making *both* terms finite. Such a procedure will be justified a posteriori provided that the final result we obtain does not depend in any meaningful way on the form of the cutoff function.[34] As a tractable example, we can use

$$f(x) = \begin{cases} 1 & \text{for } z \le \Lambda \\ 0 & \text{for } z > \Lambda \end{cases} \quad (19.196)$$

in which case, one has

$$E(n) = \frac{1}{3}[(\Lambda^2 + n^2)^{3/2} - n^3]. \quad (19.197)$$

Substitution into Eqn. (19.193) then gives

$$E'(0) = 0 \qquad \text{and} \qquad E'''(0) = -6 \quad (19.198)$$

and all higher derivatives vanishing, all *independent* of the cutoff Λ. This implies that

$$\mathcal{E} = -\frac{\pi^3}{720}\frac{\hbar c}{a^3} \quad (19.199)$$

which, most importantly, implies that there is a nonvanishing (attractive!) force per unit area between the two plates, $\mathcal{F}$, given by

[34] For a more thorough discussion of this point, see, e.g., Itzykson and Zuber (1980).

$$\mathcal{F} = -\frac{\pi^3}{240}\frac{\hbar c}{a^4} \tag{19.200}$$

which was first derived by Casimir. This quantum effect was experimentally verified by Sparnaay[35] in an ingenious series of experiments that probed both the magnitude and the characteristic $1/a^4$ dependence. We note that

- While the dependence of the result on the parameters of the problem is easily derivable from dimensional analysis (P19.31), the crucial *sign* of the effect is not as easily understood. In fact, it is highly geometry-dependent, as the similar result for the cubical cavity (P19.32) leads to a repulsive force.
- Our simplified analysis has ignored the contributions outside the infinite walls but a more thorough study[36] shows that the results above are, in fact, valid.

19.10 QUESTIONS AND PROBLEMS

Q19.1 In the discussion of the Stark effect, we found that only states with $m = 0$ contributed to the second-order perturbation theory result. Discuss why this should be so, concentrating on the relationship between the perturbation eE_0z and $\hat{L}_z$.

Q19.2 Suppose one wished to calculate the Stark shifts for a hydrogen-like ion. How would the form of the perturbation change, and what additional approximations, if any, would one have to make? By what factor would the polarizability for the ground state of such ions change, i.e., how would Eqn. (19.75) scale with Z? How would you attempt to evaluate the polarizability of the helium atom?

Q19.3 Evaluate the "generalized slope" of the best-fit line in log-log plot in Fig. 18.19, and discuss the relationship between polarizability and ionization potential that it implies.

Q19.4 One is often told that a microwave oven works along these lines:

- A time-dependent external electric field interacts with the permanent dipole moment of the water molecules in the food causing them to rotate.
- The resulting rotational kinetic energy is transferred to neighboring molecules, resulting in the desired increase in overall temperature.

For an oven operating at $f \approx 2.5\,\text{GHz}$, evaluate the energy of a single microwave photon, and compare it to the minimum energy necessary to excite a typical rotational state as discussed in Sec. 17.3. Based on your answer, decide whether the microwave oven is a classical or a quantum device, and then explain briefly how you think it works.

Q19.5 Can you see any connection between the infinite vacuum energy of the EM field in Eqn. (19.187) and the uncertainty principle relation of Eqn. (19.26)?

Q19.6 What kind of spectroscopic evidence would show that sunspots are correlated with regions of high magnetic-field activity?

Q19.7 It was mentioned that the 21-cm hyperfine line of hydrogen might be useful as a "universal" standard of frequency. The Pioneer-10 and -11 spacecraft carried engraved plaques that were designed to be the first material artifacts of mankind designed to escape the solar system carrying a message (as opposed, for example, to electromagnetic transmissions). These "cosmic greeting cards" specified all distances and times in terms of this

[35] See E. Sparnaay (1958).

[36] See Itzykson and Zuber (1980) for references.

frequency. What do you think the message consisted of? What would you have included in such a message? What fundamental concepts in physics would you think important and possible to communicate to another intelligent civilization?

Q19.8 How would one use NMR phenomena as the basis for an imaging technique as in MRI? How does one know where in the body the radio frequency photons from the external field were absorbed?

P19.1 **(a)** Calculate the ratio of the gravitational force to the electrostatic force between a proton and an electron.

(b) A rough limit on any possible difference between the magnitudes of the electron and proton charges can be inferred from cosmology; we will use the notation $|Q_p + Q_e| = \Delta e$ for any such difference. The electrostatic repulsion between two (supposedly neutral) hydrogen atoms could, in principle, overwhelm their gravitational attraction if Δe were too large; such a repulsion could make the universe expand at a rate much larger than observed. Show that demanding that any such repulsive force be less than 10% of the mutual gravitational attraction of two hydrogen atoms implies that $\Delta e/e \lesssim 10^{-18}$.

P19.2 **Neutron wave optics:**[37]

(a) What is the wavelength λ and wavenumber k for thermal neutrons, i.e., ones for which $E = p^2/2m = k_B T/2$, with $T = 300$ K?

(b) The behavior of neutrons in matter can be characterized by an effective *index of refraction* given by

$$n = \sqrt{1 - \frac{4\pi N b}{k^2}}$$

where N is the number of nuclei per unit volume, and b is the so-called *scattering length.* Estimate $1 - n$ for thermal neutrons if $b \approx 5\,\text{F}$ and $N \approx 10^{29}\,\text{m}^{-3}$.

(c) Using (b), estimate the "glancing angle" at which thermal neutrons incident on matter will experience total internal reflection. Can you imagine how this effect is used to provide a "clean" neutron beam far from a "dirty" reactor? *Hint:* How do fiber optic cables work?

(d) Neutrons are projected horizontally from a height H with velocity v and follow parabolic trajectories. Show that the critical height, H_c, at which all of the neutrons "skid" along the surface because of internal reflection is given by

$$H_c = \frac{2\pi m_n^2}{g\hbar^2} N b \tag{19.201}$$

This effect is the basis for the *neutron-gravity refractometer.*

P19.3 **(a)** How many photons per second enter your eye from a 100-W lightbulb 1 meter away; estimate the size of your iris in such a situation.

(b) What is the number density of photons (n_γ/m^3) 1 kilometer away from a 50,000 W AM radio station operating at $f = 1000$ kHz; assume that the antenna radiates its energy uniformly (which is not a very good approximation).

(c) What is the radiation pressure (in N/m^2) due to sunlight at the earth's surface; the solar luminosity is roughly 4×10^{26} W. Compare it to atmospheric pressure. Why do you "feel" the sunlight's energy but not its momentum?

(d) Solar radiation pressure is responsible for "sweeping" the solar system clean of dust particles below a certain size. Estimate the radius of dust particles for which the radiation force overcomes the gravitational attraction of the sun. Is this pressure

[37] For a review of applications of neutron scattering to solid-state physics, see Dobbrzynski and Blinowski (1994).

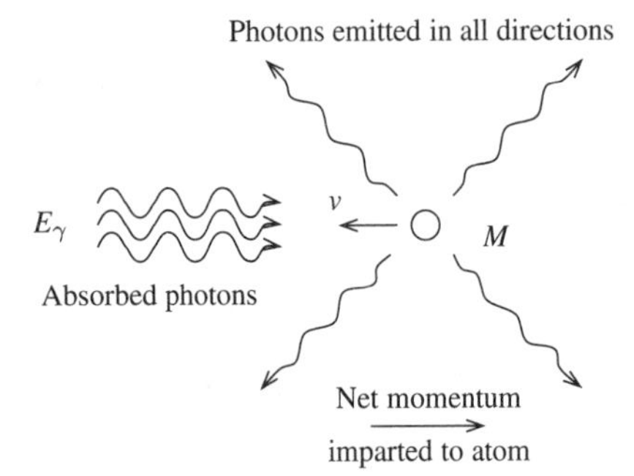

FIGURE 19.25. Schematic diagram of absorption of laser light by atom leading to net force.

large enough to "push" anything else around, even in space? [See Clarke (1972) and Wright (1992) for some ideas on this subject.]

P19.4 **"Stopping atoms with laser light":**[38] Laser light can be used to exert a force on atoms, as shown in the schematic diagram in Fig. 19.25. Photons that are absorbed to excite the atom transfer a net momentum in the $+z$ direction, while the photons emitted in the subsequent decay are emitted isotropically so that there is a net momentum transfer.

(a) Recalling the Doppler effect, show that the photon E_γ energy required to excite the atomic transition of energy E_A is $E_\gamma = (1 - v/c)E_A$.

(b) Show that each absorption results, on average, in a reduction in speed given by $\Delta v = \hbar k/M$, where M is the atom mass.

(c) For a sodium atom ($M = 23$ amu) with initial speed $v_0 \sim 600$ m/s, how many such absorptions, N, are required to bring it to rest if $\lambda = 5890$ Å.

(d) What is the approximate spread in longitudinal (i.e., perpendicular to the laser beam) speed after this many absorptions? *Hint:* This is a statistical process, so estimate the error in N in part (c).

(e) If the time between photon absorptions is limited by the radiative lifetime of the atomic state to be $\Delta t \sim 10^{-8}$ sec, estimate the average force on the atom using $F = \Delta p/\Delta t$, and compare it to the force of gravity.

P19.5 **Doppler cooling of atoms,**[39] **or "optical molasses":** Imagine an atom as shown in Fig. 19.26 which is now irradiated by two equal intensity lasers with frequencies just below a resonant frequency of the atom, namely, $E_\gamma \lesssim E_A$. The probability of absorption of each beam has the Lorentzian form in P5.4, and is given by

$$\frac{1}{(E_\gamma - E_a)^2 + (\hbar/2\tau)^2} = \frac{1}{(\Delta E)^2 + (\hbar/2\tau)^2}$$

where τ is the lifetime of the stated excited at E_A.

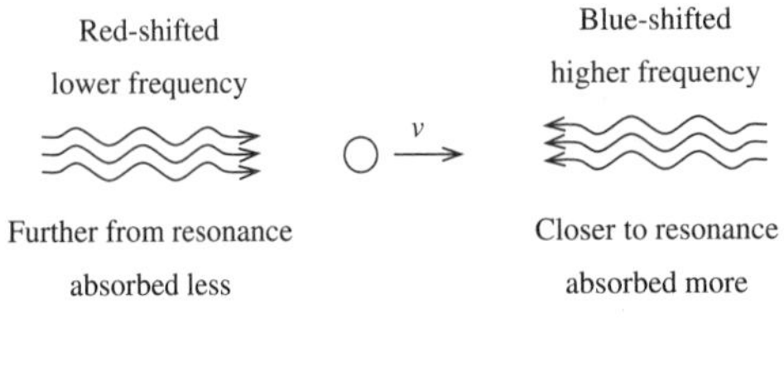

Red-shifted
lower frequency
v
Blue-shifted
higher frequency
Further from resonance
absorbed less
Closer to resonance
absorbed more

FIGURE 19.26. Geometry of opposing laser beams leading to "optical molasses" or velocity-dependent damping force giving atom trapping.

[38]This is the title of a paper by Prodan et al. (1985), which discusses this effect.

[39]For a discussion, see Cohen-Tannoudji (1990).

(a) Show that if the atom moves to the right, say, with speed v, that there is a net force on the atom proportional to

$$F_{\text{net}} \propto -\frac{|\Delta E|}{(\Delta E)^2 + (\hbar/2\tau)^2}\left(\frac{v}{c}\right) \propto -v$$

(b) For what value of ΔE is this force maximized?

(c) A linear restoring force proportional to $-x$ is like a spring, but one proportional to $-v$ is like a viscous damping force, hence the name "optical molasses." Solve the classical equation of motion for such an object with initial conditions $x(0) = 0$ and $v(0) = v_0$.

P19.6 Show that the vector potential for a uniform field in an arbitrary direction, $\mathbf{B}_0$, can be written as $\mathbf{A} = -\mathbf{r} \times \mathbf{B}_0/2$.

P19.7 Show that the combined effects of a gauge transformation on the Hamiltonian via Eqn. (19.40) and on the wavefunction via Eqn. (19.41) reproduces the original Schrödinger equation.

P19.8 For a system described by the Hamiltonian

$$\hat{H} = \frac{1}{2\mu}\left(\mathbf{p} - q\mathbf{A}\right)^2$$

show that the probability current that satisfies the equation of continuity,

$$\frac{\partial}{\partial t}\left(\psi^*(\mathbf{r}, t)\psi(\mathbf{r}, t)\right) + \nabla \cdot \mathbf{J}(\mathbf{r}, t) = 0$$

is given by

$$\mathbf{J} = \frac{\hbar}{2\mu i}\left(\psi^* \nabla \psi - \psi \nabla \psi^* - \frac{2iq}{\hbar}\mathbf{A}|\psi|^2\right)$$

P19.9 Write a short computer program to evaluate the series in Eqn. (19.76), and comment on the convergence.

P19.10 Evaluate the matrix element in Eqn. (19.83).

P19.11 In Sec. 19.4.2, we considered a uniform electric field in a specific direction, $V(\mathbf{r}) = e\mathcal{E}z$. Suppose the field is in an arbitrary direction, namely, $V(\mathbf{r}) = e\boldsymbol{\mathcal{E}} \cdot \mathbf{r}$, where $\boldsymbol{\mathcal{E}} = (\mathcal{E}_x, \mathcal{E}_y, \mathcal{E}_z)$.

(a) Why must the results for the first- and second-order shifts for the ground state be the same as in Eqns. (19.68) and (19.75)?

(b) Show explicitly that this is so. Use the relations

$$x = -\sqrt{\frac{2\pi}{3}}\, r\left(Y_{1,1}(\theta, \phi) - Y_{1,-1}(\theta, \phi)\right) \qquad y = i\sqrt{\frac{2\pi}{3}}\, r\left(Y_{1,1}(\theta, \phi) + Y_{1,-1}(\theta, \phi)\right)$$

(c) What happens to the shifts in the first excited states? Is the pattern of splitting the same? Are the linear combinations that split the same?

(d) Explicitly work out the splitting of the first excited state for a uniform field of the form $V(\mathbf{r}) = e\mathcal{E}x$.

P19.12 **Variational method for Stark effect:** Use the variational method to estimate the ground state energy for the Stark Hamiltonian $\hat{H} = \hat{H}_{\text{Coul}} + e\mathcal{E}z$ using the trial wavefunction

$$\psi(r) = \cos(\theta)\psi_{1,0,0}(\mathbf{r}) + \sin(\theta)\psi_{2,1,0}(\mathbf{r})$$

with θ as the variational parameter. Evaluate the energy shift due to the electric field by calculating $\Delta E = E(\mathcal{E} \neq 0) - E(\mathcal{E} = 0)$; show that it is proportional to $\mathcal{E}^2$; and evaluate α using Eqn. (19.64).

P19.13 Work out the pattern of level splittings for the Stark effect for the $n = 3$ case of hydrogen. You will have to consider the nine possible states corresponding to *s, p, d* states.

P19.14 **Stark effect for a 3D harmonic oscillator:**

(a) Consider a 3D isotropic oscillator with potential

$$V(x, y, z) = \frac{1}{2}K(x^2 + y^2 + z^2)$$

with an external field given by $V(z) = e\mathcal{E}z$. Show that the energy eigenvalues and wavefunctions can be obtained exactly by generalizing P10.6.

(b) Use your results to obtain the polarization, α, for the ground state.

(c) What happens to the first and second excited state? Is there a linear Stark effect for any set of levels? Give a reason for your answer.

(d) If the oscillator is nonisotropic, so that

$$V(x, y, z) = \frac{1}{2}(K_x x^2 + K_y y^2 + K_z z^2)$$

show that the problem is still soluble exactly. Show that the energy shift due to the external field in an arbitrary direction, $V(\mathbf{r}) = e\mathcal{E} \cdot \mathbf{r}$ can be written in the form

$$\Delta E = -\frac{1}{2}\mathcal{E} \cdot \alpha \cdot \mathcal{E}$$

where α is now a *polarization tensor.*

P19.15 Consider a particle of mass m and charge e in a 3D cubical box of side L. Estimate the dipole polarizability of the particle in the ground state. Is there a first-order Stark effect for the first excited state (which is triply degenerate)?

P19.16 Use the data in Fig. 19.6 to estimate the polarizability of the (nondegenerate) 15*s* state. Compare the value you obtain with the ground state value given by $\alpha = 4Ga_0^3/K$, and explain any differences.

P19.17 Use the tunneling formula for field emission in Sec. 12.5.1 as a very rough guide to estimate the lifetime of the electron in a hydrogen atom placed in an external electric field $\mathcal{E}$. (This amounts to neglecting the effect of the Coulomb attraction.) Show that this gives

$$\tau_{\text{tunnel}} \sim \tau_H \exp\left(\frac{4\sqrt{2}}{3}\sqrt{\frac{m_e W}{\hbar^2}}\frac{W}{e\mathcal{E}}\right)$$

where τ_H is a characteristic atomic time scale. Obtain a numerical estimate of the lifetime by using $W \sim E_1 \sim 13.6$ eV, $\mathcal{E} = 2000$ V/m, and τ_H as the classical orbital period.

P19.18 An electron in an external electric field has the potential energy function

$$V(\mathbf{r}) = -\frac{Ke^2}{r} - e\mathcal{E}z$$

Use this to argue that a Rydberg atom in a state of effective quantum number n^* will be ionized by an external field of strength

$$\mathcal{E} = \frac{E_c}{16(n^*)^4}$$

where E_c is the "typical" atomic field in Eqn. (19.21).

P19.19 **(a)** Imagine that the values on the next page have been obtained in an experiment looking for multiphoton ionization of hydrogen atoms from their ground state; they give the yield of emitted electrons versus the laser intensity, I.

$I(TW/\text{cm}^2)$	**Electrons** (10^{10})
4	3
7	18
20	300
60	8000

From the "data" (which is not from any particular experiment), estimate the value of k in a power-law fit to the data of the form

$$\text{Electron yield} \propto (\text{laser power})^k$$

(b) Estimate how many photons are required for multiphoton ionization of hydrogen if the photon wavelength used was $\lambda = 2480$ Å, and compare the results to your answer in part (a).

P19.20 **Uniform magnetic field in different gauges:** Consider the problem of a charged particle in a uniform field in the $+z$ direction described by the vector potential $\mathbf{A} = (0, xB_0, 0)$.

(a) Show that the Hamiltonian can be written in the form

$$\hat{H} = \frac{\hat{p}_x^2}{2\mu} + \frac{(qB_0)^2}{2\mu}x^2 + \frac{\hat{p}_y^2}{2\mu} - \frac{qB_0}{\mu}\hat{p}_y + \frac{\hat{p}_z^2}{2\mu}$$

(b) Try a solution of the form

$$\psi(x, y, z) = X(x)e^{i(k_y y + k_z z)}$$

and show that the equation for $X(x)$ reduces to that for a shifted harmonic oscillator as in P10.6.

(c) Find the quantized energy eigenvalues, and show that they give the same spectrum as Eqn. (19.99) with the same degeneracy.

(d) The wavefunctions corresponding to this solution, and those for the case where $\mathbf{A} = B_0/2(-y, x, 0)$, should be related by a simple phase as in Eqn. (19.41). Is it easy to see this relationship?

P19.21 **Velocity selectors:** The motion of a charged particle in combined (but constant) electric $\mathbf{E}_0$ and magnetic $\mathbf{B}_0$ fields can be quite complicated in general. For a particle with initial velocity given by

$$\mathbf{v}_{\text{cross}} = \frac{\mathbf{E}_0 \times \mathbf{B}_0}{|\mathbf{B}_0|^2}$$

the electric and magnetic forces cancel, and the particle moves in a straight-line trajectory. This is the basis for a classical "velocity selector." Solve for the energy eigenvalue spectrum and wavefunctions for the quantum version of this problem for uniform electric and magnetic fields in the $-x$ and $+z$ directions, respectively; use the potentials

$$\phi(\mathbf{r}) = E_0 x \qquad \text{and} \qquad \mathbf{A}(\mathbf{r}) = (0, xB_0, 0)$$

Discuss to what extent any "velocity selection effect" is still present in the quantum system.

P19.22 **Hydrogen atom in a different gauge:** One almost always solves the hydrogen atom problem by (implicitly) using the gauge

$$\phi(\mathbf{r}, t) = \frac{Ke}{r} \qquad \text{and} \qquad \mathbf{A}(\mathbf{r}, t) = 0$$

to describe the Coulomb field of the proton.

(a) Show that this configuration can also be derived by the potentials

$$\phi(\mathbf{r}, t) = 0 \qquad \text{and} \qquad \mathbf{A}(\mathbf{r}, t) = -\frac{K e \mathbf{r}}{r^2} t$$

and find the gauge function $f(\mathbf{r}, t)$ that connects it to the choice in (a).

(b) Set up the corresponding Hamiltonian, $\hat{H}'$, and show explicitly that $\psi'_{nlm} = \exp(iqf(\mathbf{r}, t)/\hbar)\psi_{nlm}$ solves the Schrödinger equation in the new gauge.

(c) Discuss how the semiclassical arguments in Ex. 1.1 are changed in this case; recall that when there is a vector potential, one has $m\mathbf{v} = \mathbf{p} - q\mathbf{A}$.

P19.23 **The "anomalous" Zeeman effect:** If we include the effect of electron spin and its coupling to the external magnetic field in the Zeeman effect, we need to consider the Hamiltonian

$$\hat{H}_1 = \frac{eB}{2m_e}\left(L_z + 2S_z\right)$$

if we assume that $g_e \approx 2.0$ for the electron. Evaluate the expectation value of this term in the appropriate coupled states in P17.16 for both $j = l + 1/2$ and $j = l - 1/2$, and show that your result can be expressed in the form

$$\Delta E_B^{(1)} = \frac{e\hbar B}{m_e} \frac{m_j}{(2l+1)} \begin{cases} l+1 & \text{for } j = l + 1/2 \\ l & \text{for } j = l - 1/2 \end{cases}$$

P19.24 **Diamagnetism:** The quadratic term in the Hamiltonian for a uniform magnetic field in the standard gauge is given by

$$\frac{e^2 B^2}{8m}(x^2 + y^2)$$

(a) Using perturbation theory, show that the corresponding change in energy is

$$\Delta E = \frac{e^2 B^2}{12m}\langle r^2 \rangle$$

for spherically symmetric distributions.

(b) Using the definition, $\Delta E = -\boldsymbol{\mu} \cdot \mathbf{B}$, show that the induced magnetic dipole is

$$\mu = -\frac{e^2 \langle r^2 \rangle}{6m} B$$

An induced magnetic moment opposite to the applied field is termed *diamagnetic*.

(c) Can you derive this result classically by looking at the magnetic moment induced by the charges rotating in the uniform field?

P19.25 **Islands of isomerism:** Using the nuclear shell energy level diagram in Fig. 19.15, explain why there are many long-lived excited states (so-called *isomers*) for odd-A nuclei for values of either *N* or *Z just below* the observed magic numbers. *Hint:* The rate for radiative decays between two energy levels is proportional to $(kR)^{2l+1}$ where l is the change in angular momentum between the initial and final state; $E_i - E_f = \Delta E = E_\gamma = \hbar kc$ is the photon energy in the transition.

P19.26 Use the results of P17.25 to show that the "magic numbers" for the 3D harmonic oscillator potential are 2, 8, 20, 40, 70, 112, 168, ... and so forth. *Hint:* Use the degeneracy of the spectrum to find the *total* number of spin-1/2 particles that can be accommodated upon the closing of each shell. Since the harmonic oscillator has such a simple spectrum, the shell structure is obvious.

P19.27 **(a)** Use the expression in Eqn. (19.141) to find the numerical value of $\Delta E, f$, and λ for the energy, frequency, and wavelength of the photon emitted in the ${}^3S_1 \rightarrow {}^3S_0$ hyperfine transition in hydrogen.

(b) Generalize your result to find the corresponding quantity for deuterium. You will need to know that (1) the deuteron magnetic moment is $\mu_D = 0.8798\mu_N$, (2) $M_D \approx 2m_p$, and (3) the deuteron spin is $S = 1$.

P19.28 Use the data in Fig. 19.19 to show that the angle θ through which the neutron magnetic moment must precess to change its phase by -1 is $352 \pm 19°$. Use the following numerical values:

(a) The observed "period" in the magnetic path length is $\int_{\text{path}} \mathbf{B} \cdot d\mathbf{l} = 1.44 \pm 8\, G \cdot cm$.

(b) The gyromagnetic ratio of the neutron is $g_n = -1.93$.

(c) The neutron magnetic moment is given by $g_n e\hbar/2m_n$.

(d) The wavelength of the neutrons used was 1.82 ± 0.01 Å. This combined with the mass of the neutron 1.67×10^{-27} kg gives the neutron velocity v.

P19.29 The data in Fig. 1.6 show a resonant absorption curve for Li^7 nuclei where the frequency of the external oscillating field, f_{ext}, is held fixed but the magnitude of the static field is increased. Use the values in that figure to estimate the g factor for this nucleus.

P19.30 **Flux quantization:** Consider an electron moving in the magnetic field geometry of the Aharonov-Bohm effect in Sec. 19.8. Show that on making one complete circuit around the localized magnetic field region that the electron wavefunction acquires a phase given by

$$e^{ie\Phi/\hbar}$$

Use the fact that the electron wavefunction must be single-valued at any point to show that this implies that magnetic flux must be quantized in, i.e.,

$$\Phi_n = \left(\frac{2\pi\hbar}{e}\right) n \qquad \text{where } n = 0, 1, 2, \ldots$$

P19.31 Use dimensional analysis to construct a combination of $\hbar$, c, and a that forms a quantity with the units of force per unit area.

P19.32 **The Casimir effect in a conducting cavity:** The Casimir result for the difference in energies between vacuum and a situation with conducting walls can be generalized to other geometries besides parallel plates. It can be shown[40] that the shift in energy densities due to a rectangular conducting cavity of dimensions L_1, L_2, L_3 is given by

$$\Delta u(L_1, L_2, L_3) = -\frac{\hbar c}{16\pi^2}\left[\sum_{m_{1,2,3}=-\infty}^{+\infty}{}' \frac{1}{(m_1^2 L_1^2 + m_2^2 L_2^2 + m_3^2 L_3^2)^2} - \frac{\pi^3}{3}\frac{1}{L_1 L_2 L_3}\left(\frac{1}{L_1} + \frac{1}{L_2} + \frac{1}{L_2}\right)\right]$$

The total shift in energy is given by $\Delta E = L_1 L_2 L_3 \Delta u$. The summation, $\sum'$, is over all the integers, but does not include the point $(m_1, m_2, m_3) = (0, 0, 0)$. The corresponding pressures or force per unit areas are given by

$$\mathcal{F}_i(L_1, L_2, L_3) = -\frac{\partial}{\partial L_i}\left[L_i \Delta u(L_1, L_2, L_3)\right]$$

(a) Show that in the limit that $L_2 = L_3 \equiv L \ggg L_1 \equiv a$, one recovers the Casimir result for large parallel plates, namely, that

$$\mathcal{E}(a) = -\frac{\hbar c}{a^3}\frac{\pi^2}{720}$$

where $\mathcal{E}(a) = a\Delta u(L, L, a)$. You might need the summation

[40] See Luksoz (1971).

$$\zeta(4) = \sum_{m=1}^{\infty} \frac{1}{m^4} = \frac{\pi^4}{90}$$

(b) For the case of a cubical cavity, i.e., $L_1 = L_2 = L_3 = L$, show that

$$\Delta u(L) = \frac{\hbar c}{16\pi^2 L^4}\left(\sum{}' \frac{1}{(m_1^2 + m_2^2 + m_3^2)^2} - \pi^3\right)$$

Write a short computer program to evaluate the multidimensional summation numerically, and show that

$$\sum{}' \frac{1}{(m_1^2 + m_2^2 + m_3^2)^2} \approx 16.5323\ldots$$

Use this to show that the change in energy is roughly

$$\Delta E = L^3 \Delta u(L) \approx 0.0916\left(\frac{\hbar c}{L}\right)$$

which leads to a *repulsive* force, i.e., the vacuum fluctuations of the EM field tend to make the cube expand.

Note: Such calculations have a historical interest motivated by one model for the electron put forward by Casimir. He presumed that the electron was a thin shell of charge (which would not be stable against its own Coulomb repulsion) balanced by the (hopefully) attractive Casimir force arising from the "conducting sphere." Dimensional analysis would then predict that the total pressure for a spherical region of radius a would have the form

$$\mathcal{F}_{\text{total}} = \mathcal{F}_{\text{Coulomb}} + \mathcal{F}_{\text{Casimir}} = C_1 \frac{Ke^2}{a^4} - C_2 \frac{\hbar c}{a^4}$$

where $C_{1,2}$ are dimensionless constants that could, in principle, be calculated. This pressure could be made to vanish, independent of the radius a, provided that the Casimir force had the appropriate (attractive) sign. Such a cancellation would imply an exciting geometrical relation between the fundamental constants of EM, quantum mechanics, and relativity ($e, \epsilon_0, \hbar, c$), namely,

$$\alpha = \frac{Ke^2}{\hbar c} = \frac{C_2}{C_1}$$

The change in energy for a spherical conducting region was calculated (numerically) with the result[41]

$$\Delta E \approx 0.09\left(\frac{\hbar c}{2a}\right)$$

which is approximately equal to that for a cubical shell with $L = 2a$. The crucial sign difference therefore ruled out Casimir's simple model. Such calculations do show, however, some of the subtleties inherent in quantum field theory. It is not entirely inappropriate that the last problem in this chapter is an example of something that *didn't* work. The exploration of the quantum world in many diverse areas of research is ongoing, and many interesting questions remain to be answered or are even yet to be asked.

[41] See Boyer (1969).

CHAPTER 20

Scattering in Three Dimensions

While much of our knowledge of the microscopic world has been gleaned from bound state problems (e.g., the spectroscopy of atoms, molecules, nuclei, etc.), scattering experiments have also provided a wealth of information on the nature of the constituents of matter and their interactions on atomic scales and below.[1] Because much of the formalism of scattering is similar in the classical and quantum formulations, we begin by discussing some generalities shared by the two descriptions.

A typical scattering experiment has the layout already shown schematically in Fig. 12.1. A beam of incident scatterers with a given flux or intensity (number of particles per unit area per unit time) impinges on some target, described here by a scattering potential; this flux can be written as

$$J_{\rm inc} = \frac{dN_{\rm inc}}{dA\, dt} \tag{20.1}$$

Classically, individual scatterers are deflected along well-defined trajectories, while in the quantum case, the "scattered" wavefunction gives probabilistic information on the likelihood of finding a particle scattered through a given angle. In either case, the number of particles per unit time which are detected in a small region of solid angle, $d\Omega$, located at a given angular deflection specified by (θ, ϕ), can be counted; this can be written

$$\frac{dN_{\rm sc}}{d\Omega\, dt} \tag{20.2}$$

The *differential cross section* for scattering is defined by the ratio of these two quantities via

$$\frac{d\sigma}{d\Omega}(\theta, \phi) = \left(\frac{dN_{\rm sc}}{d\Omega\, dt}\right) \Big/ \left(\frac{dN_{\rm inc}}{dA\, dt}\right) \tag{20.3}$$

[1]The sentence you just read, for example, was "detected" by performing a scattering experiment involving optical photons and a system of biological analog detectors.

From this definition, it is clear that $d\sigma/d\Omega$ has the dimensions of an area. The *total cross section* corresponds to scatterings through any angle and is given by

$$\sigma = \int d\Omega \, \frac{d\sigma}{d\Omega} = \int_0^{2\pi} d\phi \int_0^{\pi} \sin(\theta)\, d\theta \frac{d\sigma}{d\Omega}(\theta, \phi) \tag{20.4}$$

20.1 CLASSICAL TRAJECTORIES AND CROSS SECTIONS

For a particle obeying classical mechanics, the trajectory for the unbound motion corresponding to a scattering event is deterministically predictable given the interaction potential and the initial conditions. The path of any scatterer in the incident beam can be followed, in principle, and its angular deflection determined as precisely as required.

For a central force, any given trajectory will be planar (why?), and we illustrate two paths corresponding to slightly different initial conditions in Fig. 20.1. As the z axis is often defined to be the "beam line," we show events in the $z - y$ plane, with the center of the (here repulsive) scattering potential at the origin. The equations of motion for any incident particle can be integrated (perhaps numerically, if necessary) to find the trajectory via

$$m\ddot{\mathbf{r}}(t) = \mathbf{F}(r) \tag{20.5}$$

where we use the initial conditions

$$\begin{aligned} y(t = -\infty) &= b \qquad & v_\infty \equiv \dot{y}(t = -\infty) &= \sqrt{2mE} \\ z(t = -\infty) &= -\infty \qquad & \dot{z}(t = -\infty) &= 0 \end{aligned} \tag{20.6}$$

where we have used the fact that the initial kinetic energy is given by $E = mv_\infty^2/2$. In this language, b is called the *impact parameter*, and would be the distance of closest approach to the scattering center in the absence of any force.

In Fig. 20.1, two trajectories, differing by a small amount db, and scattering into angles separated by $d\theta$ are shown. The number of particles scattered per unit time into the angular region $(\theta, \theta + d\theta)$ with any value of ϕ can be written as

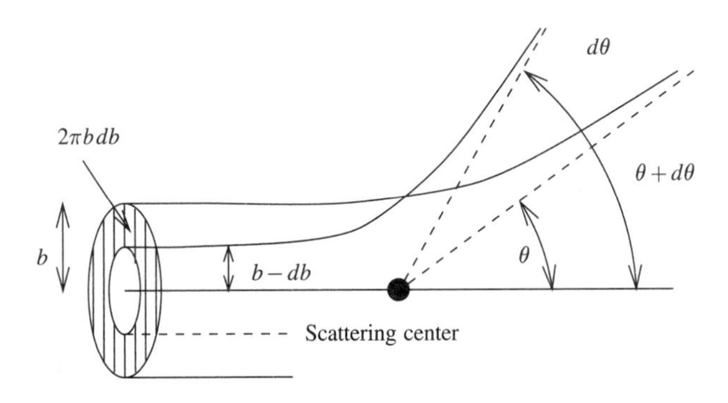

FIGURE 20.1. *Scattering trajectories corresponding to different impact parameters b give different scattering angles θ. All of the particles in the beam in the hatched area of $d\sigma = 2\pi b\, db$ are scattered into the angular region $(\theta, \theta + d\theta)$.*

$$\frac{dN_{sc}}{dt} = d\sigma \frac{dN_{inc}}{dA\,dt} \tag{20.7}$$

where

$$d\sigma = 2\pi b\,db \tag{20.8}$$

which we can also write as

$$\frac{d\sigma}{d\theta} = 2\pi b(\theta)\left|\frac{db(\theta)}{d\theta}\right| \tag{20.9}$$

The absolute value is required, since θ increases (larger angle scattering) as $b(\theta)$ decreases.

Since we have "integrated" over all ϕ values, in this case $d\Omega = 2\pi \sin(\theta)\,d\theta$, and using the definition of Eqn. (20.3), we find that

$$\frac{d\sigma}{d\Omega} = \left[\frac{1}{2\pi \sin(\theta)}\right] 2\pi b(\theta)\left|\frac{db(\theta)}{d\theta}\right| = \frac{b(\theta)}{\sin(\theta)}\left|\frac{db(\theta)}{d\theta}\right| \tag{20.10}$$

Knowledge of $b(\theta)$, obtained directly from Newton's laws or other methods, is then sufficient to calculate the scattering cross section. At this stage, the relation of one initial parameter, namely, $b = y(t=-\infty)$, on the scattering process, i.e., $\theta(b)$, is evident. The influence of $v_\infty = \dot{y}(t=-\infty)$ or, equivalently, the energy E is less explicit, but arises through the equations of motion to give $\theta(b)$ as well.

Example 20.1 "Specular" Scattering from a Hard Sphere A simple case that can be treated using only geometrical methods involves the scattering of small, light particles from a heavy, impenetrable sphere of radius R. The classical trajectory is shown in Fig. 20.2 where the incident particles experience equal angle ($\alpha = \alpha_{inc} = \alpha_{ref}$) scattering relative to the tangent plane at the intersection point; hence the name *specular* or mirrorlike. Instead of using an exclusively geometrical approach, however, we will determine $\theta(b)$ using a more "functional" approach that can be generalized to other problems (P20.1, P20.2).

From the diagram, it is clear that the scattering angle satisfies $\theta = 2\phi$, while $\tan(\phi) = |dy/dz|$, evaluated at $y = b$, where b is the impact parameter. The spherical scattering surface is determined by $x^2 + y^2 + z^2 = R^2$, so that in the $y - z$ plane we have

$$\tan\left(\frac{\theta}{2}\right) = \tan(\phi) = \left|\frac{dy}{dz}\right| = \left|\frac{z}{y}\right| = \frac{\sqrt{R^2 - b^2}}{b} \tag{20.11}$$

Solving for $b(\theta)$ gives

$$b(\theta) = R\cos\left(\frac{\theta}{2}\right) \qquad \text{so that} \qquad \left|\frac{db}{d\theta}\right| = \frac{R}{2}\sin\left(\frac{\theta}{2}\right) \tag{20.12}$$

Taken together and using the fact that $\sin(\theta) = 2\sin(\theta/2)\cos(\theta/2)$, these imply that

$$\frac{d\sigma}{d\theta} = \pi R^2 \sin\left(\frac{\theta}{2}\right)\cos\left(\frac{\theta}{2}\right) = \frac{\pi R^2}{2}\sin(\theta) \tag{20.13}$$

or equivalently

$$\frac{d\sigma}{d\Omega} = \frac{R^2}{4} \tag{20.14}$$

The total cross section is

$$\sigma = \int d\Omega \frac{d\sigma}{d\Omega} = \frac{R^2}{4}\int d\Omega = \pi R^2 \tag{20.15}$$

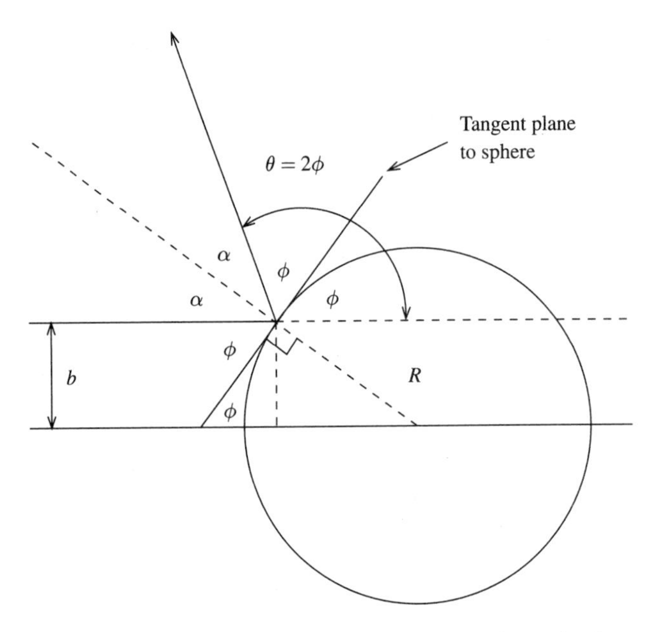

FIGURE 20.2. *Geometry for "specular" scattering from a hard sphere.*

which is just the effective cross-sectional area presented to the scatters by the sphere; this is just the so-called *geometrical cross section* of the object. A cross section that is independent of polar angle θ, as in Eqn. (20.14), is called *isotropic,* and is consistent with the simple spherical shape of the scatterer in this example.

A less-trivial case is evident when one considers scattering from ellipsoids, as illustrated in Fig. 20.3. The axis along the z direction is multiplied by a dimensionless factor f, while the transverse (x, y) dimensions are unchanged (so as to keep the geometrical cross section presented to the incident beam the same). The surface of the ellipsoid is now determined by the relation $x^2 + y^2 + z^2/f^2 = R^2$, and one can show (P20.1) that

$$b(\theta) = R\frac{1}{\sqrt{1 + f^2\tan^2(\theta/2)}} \tag{20.16}$$

corresponding to the differential cross section

$$\frac{d\sigma}{d\Omega} = \frac{R^2}{4}\left[\frac{2f}{1 + f^2 + (1 - f^2)\cos(\theta)}\right]^2 \tag{20.17}$$

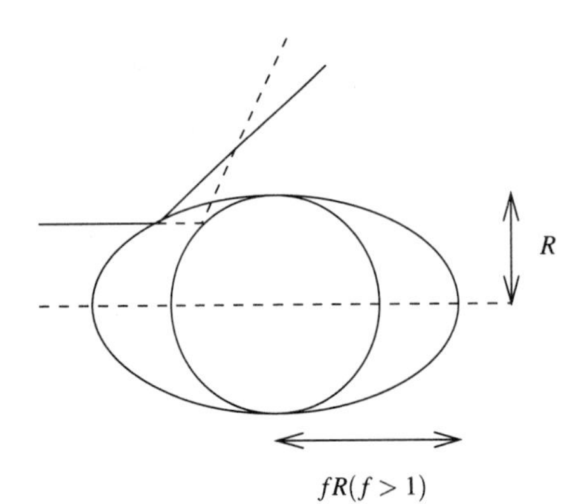

FIGURE 20.3. *Ellipsoidal hard scatterers.*

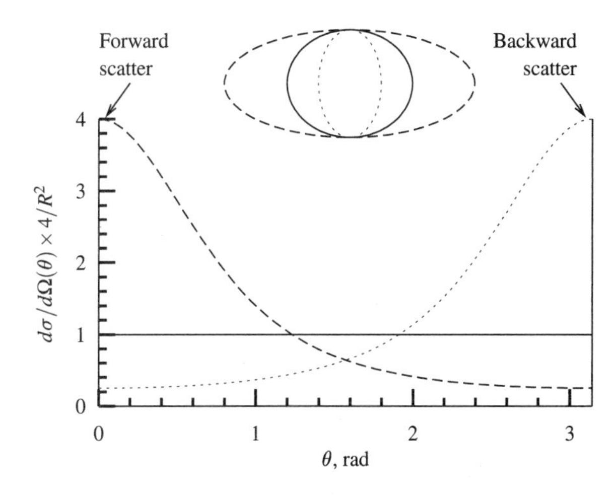

FIGURE 20.4. Angular dependence of differential cross section for various ellipsoidal shapes. Spherical or $f = 1$ (solid), and ellipsoidal shapes with $f > 1$ (dashed) and $f < 1$ (dotted) are illustrated.

which still satisfies $\sigma = \int d\Omega(d\sigma/d\Omega) = \pi R^2$ as intended. The resulting angular dependence is shown in Fig. 20.4 for various values of f. We note that $f > 1$ ($f < 1$) corresponds to more forward (backward) scattering consistent with more glancing (large angle) collisions than with a sphere.

While this is a very artificial example (there is no energy dependence, for example), it does illustrate how the angular dependence of the scattering cross section can give information on the form of the scattering potential, in this case the "shape" of the scatters.

For scattering from nontrivial central forces, the trajectory can be obtained from the equations of motion or by using energy methods. The latter case is often more direct as we do not require detailed knowledge of the time dependence of the coordinates, i.e., $r(t)$, $\theta(t)$, but rather the path in space, i.e., $r(\theta)$. For example, one can rewrite

$$E = \frac{1}{2}m\dot{r}^2 + \frac{L^2}{2mr^2} + V(r) \tag{20.18}$$

in the form

$$\sqrt{\frac{2}{m}\left(E - \frac{L^2}{2mr^2} - V(r)\right)} = \frac{dr}{dt} = \frac{dr}{d\theta}\frac{d\theta}{dt} = \frac{dr}{d\theta}\dot{\theta} \tag{20.19}$$

The angular momentum in the process can be written as $L = mr^2\dot{\theta}$, and one can write

$$d\theta = \left(\frac{L}{r^2\sqrt{2m(E - L^2/2mr^2 - V(r))}}\right)dr \tag{20.20}$$

so that the angle through which the particle moves as it travels between two radial distances is

$$\Delta\theta = \int_{r_1}^{r_2} dr \frac{L}{r^2\sqrt{2m(E - L^2/2mr^2 - V(r))}} \tag{20.21}$$

For a scattering process, we have the situation pictured in Fig. 20.5. The angle Θ is the angular deflection experienced by the particle as it moves from infinity to the distance of closest approach, r_{min}. Since the scattering event is symmetrical about this point, the net scattering angle is easily seen to satisfy

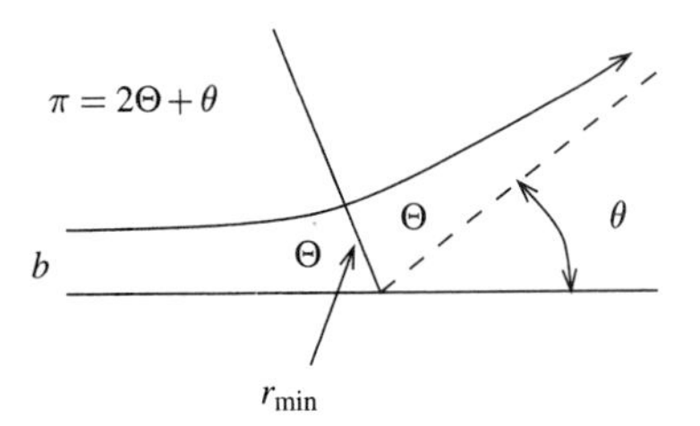

FIGURE 20.5. Scattering trajectories in a central potential showing distance of closest approach.

$$2\Theta + \theta = \pi \qquad \text{or} \qquad \Theta = \frac{\pi}{2} - \frac{\theta}{2} \tag{20.22}$$

The angular momentum can be determined by the impact parameter from the relation

$$L = mv_\infty b \tag{20.23}$$

which can be written in terms of the total energy since $E = mv_\infty^2/2$ giving

$$L = b\sqrt{2mE} \tag{20.24}$$

Using Eqn. (20.20), one has

$$\Theta = \int_{r_{\min}}^{\infty} dr \frac{b}{r^2\sqrt{1 - b^2/r^2 - V(r)/E}} \tag{20.25}$$

This makes it clear that the scattering trajectories are determined by the energy E, the angular momentum L (via b), and the form of the potential. If this can be evaluated explicitly, one obtains $b(\Theta) = b((\pi - \theta)/2)$, and one can then evaluate the differential cross section using Eqn. (20.10). The integral can be evaluated in closed form for only a few special cases, and it is lucky that the inverse-square law force is one of them.

Example 20.2 Classical Coulomb Scattering The potential corresponding to the scattering of charged particles via Coulomb's law can be written in the general form

$$V(r) = \frac{A}{r} \tag{20.26}$$

where the constant $A = \pm Z_1 Z_2 K e^2$ allows for the electrostatic interaction of like sign and opposite sign point charges of arbitrary magnitudes. The necessary integral in Eqn. (20.25) has the form

$$\Theta = \int_{r_{\min}}^{\infty} dr \frac{b}{r\sqrt{r^2 - (A/E)r - b^2}} \tag{20.27}$$

where $r_{\min}$ is determined by the vanishing of the denominator, namely,

$$r_{\min} = \sqrt{b^2 + (A/2E)^2} + (A/2E) \tag{20.28}$$

We note that for large energies ($E \to \infty$) or vanishing charges ($A \to 0$), one has $r_{\min} \to b$ as expected. The integral can be evaluated using the results of App. B.2, and using trigonometric identities, one finds

$$\cos(\Theta) = \frac{a}{\sqrt{a^2 + b^2}} \tag{20.29}$$

where $a = A/2E$. This corresponds to

$$b(\theta) = a\cot\left(\frac{\theta}{2}\right) \qquad \text{and} \qquad \frac{db(\theta)}{d\theta} = -\frac{a}{2\sin^2(\theta/2)} \tag{20.30}$$

One then finds that

$$\frac{d\sigma}{d\theta} = 2\pi b(\theta)\left|\frac{db(\theta)}{d\theta}\right| = \pi a^2 \frac{\cos(\theta/2)}{\sin^3(\theta/2)} \tag{20.31}$$

or

$$\frac{d\sigma}{d\Omega} = \frac{a^2}{4}\frac{1}{\sin^4(\theta/2)} = \left[\frac{KZ_1Z_2e^2}{4E}\right]^2 \frac{1}{\sin^4(\theta)} \tag{20.32}$$

This result is called the *Rutherford formula,* and exhibits the famous $1/\sin^4(\theta/2)$ dependence that is characteristic of Coulomb scattering from a pointlike charge.

If one attempts to evaluate the total cross section using Eqn. (20.4), one obtains an infinite result. This divergence is due to the essentially infinite range of the $1/r$ potential, and the "infinity" in the integral comes from scatterings as $\theta \to 0$. From Eqn. (20.30), we can see that this corresponds to arbitrarily large-impact parameters where the Coulomb potential can just "tickle" the scatterer but still contribute to the total cross section. This infinity is not, however, realized in nature due to the overall electrical neutrality of matter. If, for example, we scatter electrons from charged nuclei, then for sufficiently large values of b, the beam electrons are outside the atomic electron cloud, the system appears electrically neutral, and Coulomb scattering from the nucleus is shielded.

Given the special nature of the $1/r$ potential, it may not be surprising that the classical formula can be evaluated exactly; what is more unexpected is that the Rutherford formula also emerges from the quantum treatment as well.

20.2 PROBABILITY, STATISTICS, AND CROSS SECTIONS

The quantum description of scattering phenomena discussed in the next section will be seen to be as different from the classical notion of following trajectories as it was for bound state problems. Nevertheless, the operational definition of cross section in Eqn. (20.3) implies that the same "counting" processes will be utilized in any scattering experiment, and in this section we briefly discuss the statistical nature of the measurement of a scattering cross section.

For purposes of illustration, we imagine scattering from the spherical ($f = 1$) and ellipsoidal (here with $f = 2$) shapes in Ex. 20.1. Incident particles with random impact parameters are generated (via computer), and allowed to scatter from the objects, and the differential cross sections are "measured" in the following way:

- For a given (random) impact parameter, the scattering angle is determined via Eqn. (20.16).
- The number of particles, dN_θ, scattered into angular bins of "width" $d\theta = \pi/18 = 10°$ centered at $\theta = 5°, 15°, \ldots$ are counted; the statistical error in each "bin" is taken to be $\sqrt{dN_\theta}$. The total number of particles that were scattered is N_{tot}. (After all, some of the particles simply miss the target and are not counted as having scattered. In a real scattering experiment, it is sometimes said that these unscattered particles simply "go down the beam pipe.")
- The differential cross section at a given value of θ is scaled to the total cross section and estimated using

$$\frac{d\sigma(\theta)}{d\Omega} \Big/ \sigma \approx \frac{(dN_\theta \pm \sqrt{dN_\theta})}{2\pi \sin(\theta)\, d\theta} \Big/ N_{\text{tot}} \tag{20.33}$$

since $d\Omega = 2\pi \sin(\theta)\, d\theta$, as we have counted the values for all values of ϕ in the range $(0, 2\pi)$. All of the relevant values for one such data set are shown explicitly in Table 20.1 as an example.

The resulting "experimental" cross sections are plotted along with the "theoretical" predictions [given by Eqn. (20.17)] for data sets with two different values of N_{tot} in Figs. 20.6(a) and (b). The agreement between "theory" and "experiment" is consistent with the statistical errors, and it seems that we have a "good fit" in both cases. We can clearly distinguish between the two values of f, even for the smaller data set.

Suppose instead that we had been given a set of such data points and been asked to find the shape of the scatterers that had produced it, namely, the value of f. We are now asked to match some experimental data points to a theoretical curve, and to extract the parameters that give the best fit. While there are many methods for quantifying the "goodness of fit" between theory and experiment, we discuss one here that is conceptually simple: a *least squares fit* that makes use of the so-called χ^2 or *chi-squared distribution.*

Assume that we have N_{bin} experimental data points and their errors, labeled y_i^{exp} and Δy_i, respectively; the corresponding theoretical predictions are called y_i^{th}. A useful measure of the deviation of the observed points from their expected values which takes into account the random error in the measurement is the so-called χ^2 given by

$$\chi^2 = \sum_{i=1}^{N_{\text{bin}}} \frac{(y_i^{\text{exp}} - y_i^{\text{th}})^2}{(\Delta y_i)^2} \tag{20.34}$$

This expression measures the deviation of each data point from the prediction in units of the observed error of that measurement; for a Gaussian-type distribution, we expect

TABLE 20.1 *Computer-Generated Data for "Specular" Scattering from an Ellipsoid*

θ interval (deg)	dN_θ	$(d\sigma/d\Omega)/\sigma\,\vert_{\text{exp}}$	$(d\sigma/d\Omega)/\sigma\,\vert_{th}$	θ	$\vert$exp − th$\vert$/err
(0, 10)	229	0.3071 ± 0.0203	0.3141	5	0.38
(10, 20)	652	0.2944 ± 0.0115	0.2881	15	0.55
(20, 30)	822	0.2273 ± 0.0079	0.2447	25	2.19
(30, 40)	1000	0.2038 ± 0.0064	0.1970	35	1.06
(40, 50)	901	0.1489 ± 0.0050	0.1536	45	0.95
(50, 60)	850	0.1213 ± 0.0042	0.1184	55	0.69
(60, 70)	695	0.0896 ± 0.0034	0.0941	65	0.52
(70, 80)	628	0.0756 ± 0.0030	0.0714	75	1.52
(80, 90)	480	0.0563 ± 0.0026	0.0567	85	0.15
(90, 100)	385	0.0452 ± 0.0023	0.0456	95	0.36
(100, 110)	294	0.0356 ± 0.0021	0.0382	105	1.24
(110, 120)	271	0.0350 ± 0.0021	0.0324	115	1.19
(120, 130)	203	0.0290 ± 0.0020	0.0282	125	0.38
(130, 140)	148	0.0246 ± 0.0020	0.0251	135	0.32
(140, 150)	112	0.0228 ± 0.0022	0.0229	145	0.03
(150, 160)	78	0.0216 ± 0.0024	0.0214	155	0.08
(160, 170)	42	0.0190 ± 0.0029	0.0204	165	0.45
(170, 180)	12	0.0161 ± 0.0046	0.0199	175	0.83

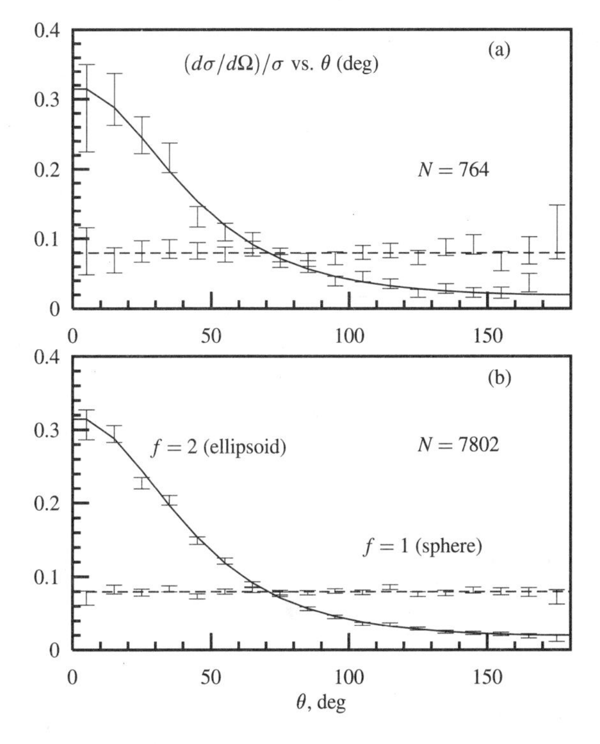

FIGURE 20.6. *Computer-generated "experimental" data for scattering from ellipsoidal shapes with $f = 1, 2$; (a) and (b) correspond to $N = 764$ and $N = 7802$ "hits," respectively.*

measurements (the y_i^{exp}) to fall within 1σ (2σ) [i.e., 1 (2) units of Δy_i] of their expected value (the y_i^{th}) something like 68% (90%) of the time. Thus, even if our theory is "perfect" and matches the physical situation exactly, we expect each term in Eqn. (20.34) to be of order unity; this is illustrated in Table 20.1 where $|y_i^{\text{exp}} - y_i^{\text{th}}|/\Delta y_i$ is shown for each bin. We thus expect that χ^2 for N experimental measurement bins should scale roughly as N. This is why $\chi_N^2 = \chi^2/N$ or the "chi-squared per degree of freedom" is often calculated as a measure of the goodness of fit. Values of χ^2/N much larger than 1 indicate a bad fit to data, while values much less than unity are also suspect as giving too good a fit. The values of χ^2/N for various values of N (and information on their probability distribution) are given in Fig. 20.7.

The notion of χ^2 can then be used to quantify the goodness of fit and to find the values in a theoretical parameterization that give the closest agreement with data. As an example, assume that the theoretical formula that gives the y_i^{th} depends on a single parameter, f, so that $y_i^{\text{th}} = y_i^{\text{th}}(f)$. For any given value of f, we can calculate the resulting $\chi^2(f)$/N and determine, either analytically or numerically, the value of f that minimizes the χ^2 error. The extension to many parameters, i.e. $y_i^{\text{th}} = y_i^{\text{th}}(f_1, f_2, \ldots, f_{N_{\text{fit}}})$, is conceptually the same, except that we have to minimize the resulting $\chi^2(f_1, f_2, \ldots)$ in a multidimensional parameter space. Given a function characterized by enough parameters (call the number N_{fit}), we can, of course, reproduce almost any set of data points,[2] and the χ^2 will usually decrease (artificially) as we increase N_{fit} because of this. To take this effect into account, one usually uses $N = N_{\text{eff}} = N_{\text{bin}} - N_{\text{fit}}$ as the effective number of degrees of freedom in evaluating χ^2/N.

[2]Imagine, for example, fitting a straight line through two data points, a parabola through three, and so forth.

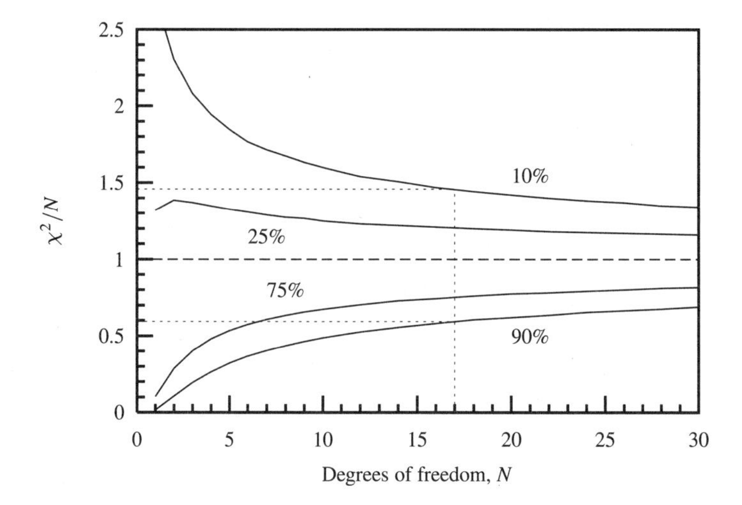

FIGURE 20.7. *Values of χ^2/N versus the number of degrees of freedom N for various confidence levels. For example, for $N = 17$, we would expect to find $\chi^2/N \gtrsim 0.59$ ($\gtrsim 1.46$) 90% (10%) of the time. Numerical values taken from the* CRC Handbook of Mathematics *(1995).*

We can apply this method to the data in Fig. 20.6 and find the χ^2/N_{eff} for various values of f for the different data sets we plot in Fig. 20.8. Because we fit only for one parameter, f, we use $N_{\text{eff}} = 18 - 1 = 17$ as the effective number of data points. We see that the data are centered at the appropriate values of f in each case as expected. From Fig. 20.7, we estimate that the likelihood of having a χ^2/N_{eff} larger than roughly 1.5 or smaller than about 0.6 is approximately 10% in each case, so we take the upper value as our measure of when the fit becomes unacceptable; this value is shown as a solid horizontal

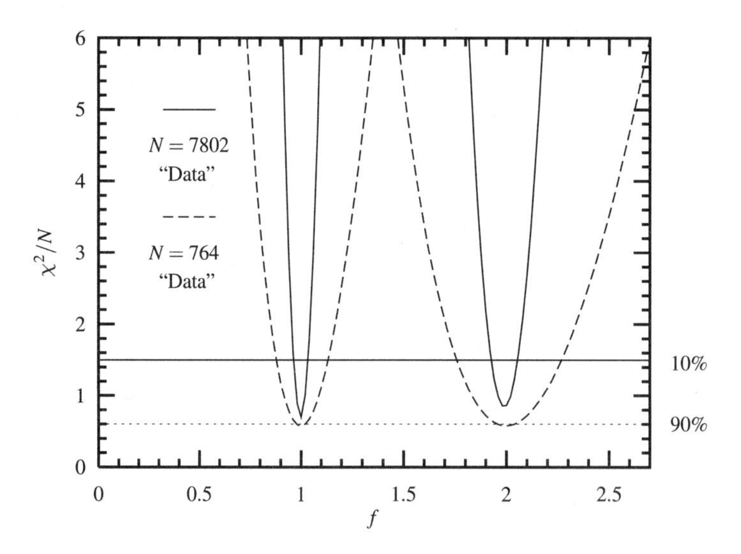

FIGURE 20.8. *Plot of $\chi^2(f)/N$ for the data in Fig. 20.6, showing the minimum at the appropriate value of f. For the larger data sets, the spread in f values is smaller, indicating a more precise value of f extracted from "experiment."*

line, and its intersection with the $\chi^2(f)/N_{\rm eff}$ curves gives a measure of the expected error in our extraction of f in either case.

Figure 20.8 also illustrates the correlation between the amount of data used in the fit and our confidence in the value of f. The data sets with larger numbers of values do not have an appreciably smaller $\chi^2/N_{\rm eff}$ (since that value should always be roughly unity); in this case the χ^2/N gets closer to unity as the size of the data set grows. On the other hand, the *spread* in f values around the minimum value does decrease as we take more data, consistent with our prejudice that we should be able to find a more precise value of f if we do more experiments. For example, from Fig. 20.8, we find $f \approx 2.0 \pm 0.25$ for $N_{\rm tot} = 764$, and $f \approx 2.00 \pm 0.06$ for $N_{\rm tot} = 7802$.

20.3 QUANTUM SCATTERING

20.3.1 Cross Section and Flux

While it is possible to describe quantum scattering in three dimensions using wavepackets, we will instead use an approach similar to that discussed extensively in 1D, namely, the scattering of plane wave states. In three dimensions, the incident plane wave impinges on the scattering potential as in Fig. 12.1, giving rise to a spherical wavefront. Plane wave solutions representing incident particles of momentum $\mathbf{p}$ will have the form

$$\psi_{\rm inc}(\mathbf{r}) = \frac{1}{L^{3/2}} e^{i\mathbf{p}\cdot\mathbf{r}/\hbar} = \frac{1}{L^{3/2}} e^{i\mathbf{k}\cdot\mathbf{r}} \tag{20.35}$$

where we include the (arbitrary) factor $1/L^{3/2}$ to denote the dimensions of a three-dimensional wavefunction. We will often assume that the beam of scatterers is incident along the z direction so that $\mathbf{k}\cdot\mathbf{r} = kz = kr\cos(\theta)$.

The scattered wave will have the form

$$\psi_{sc}(\mathbf{r}) = \frac{1}{L^{3/2}} \left[f(\hat{\mathbf{n}} = \hat{\mathbf{r}}) \frac{e^{ikr}}{r} \right] = \frac{1}{L^{3/2}} f(\theta, \phi) \frac{e^{ikr}}{r} \tag{20.36}$$

where $\hat{\mathbf{n}} = \hat{\mathbf{r}}$ denotes a unit vector in the direction of the detector located at angular position (θ, ϕ). The quantity $f(\theta, \phi)$ has the dimensions of length and is called the *scattering amplitude;* we will see that it determines the scattering cross section. The total wavefunction has the form

$$\psi(\mathbf{r}) = \psi_{\rm inc}(\mathbf{r}) + \psi_{sc}(\mathbf{r}) = \frac{1}{L^{3/2}} \left(e^{i\mathbf{k}\cdot\mathbf{r}} + f(\theta, \phi) \frac{e^{ikr}}{r} \right) \tag{20.37}$$

The flux of incident and scattered particles will be related to the probability flux for this wavefunction, namely,

$$\mathbf{j}(\mathbf{r}) = \frac{\hbar}{2mi} \left(\psi^*(\mathbf{r}) \nabla \psi(\mathbf{r}) - \psi(\mathbf{r}) \nabla \psi^*(\mathbf{r}) \right) \tag{20.38}$$

where we can evaluate the two terms in

$$\nabla\psi(\mathbf{r}) = \nabla\psi_{\rm inc}(\mathbf{r}) + \nabla\psi_{sc}(\mathbf{r}) \tag{20.39}$$

separately. We clearly have

$$\nabla\psi_{\text{inc}}(\mathbf{r}) = \frac{1}{L^{3/2}}\left[i\mathbf{k}e^{i\mathbf{k}\cdot\mathbf{r}}\right] \tag{20.40}$$

while using the expression for the gradient operator in spherical coordinates we find

$$\begin{aligned}\nabla\psi_{sc}(\mathbf{r}) = \frac{1}{L^{3/2}}\Bigg[&\hat{\mathbf{r}}f(\theta,\phi)\left\{ik\frac{e^{ikr}}{r} - \frac{e^{ikr}}{r^2}\right\}\\ &+ \hat{\theta}\,\frac{1}{r}\frac{\partial f(\theta,\phi)}{\partial\theta}\frac{e^{ikr}}{r} + \hat{\phi}\frac{1}{r}\frac{\partial f(\theta,\phi)}{\partial\phi}\frac{e^{ikr}}{r}\Bigg]\end{aligned} \tag{20.41}$$

where we have used a standard expression for the gradient operator in terms of spherical coordinates when acting on $\psi_{sc}(\mathbf{r})$.

That part of the flux from Eqn. (20.38) which contains only $\psi_{\text{inc}}(\mathbf{r})$ has the familiar form

$$\mathbf{j}_{\text{inc}} = \frac{1}{L^3}\frac{\hbar\mathbf{k}}{m} = \frac{1}{L^3}\mathbf{v} \tag{20.42}$$

along the initial direction specified by $\mathbf{k}$ or the velocity $\mathbf{v}$. This quantity has the dimensions of number of particles per unit area per unit time, and we identify it with Eqn. (20.1). The flux corresponding to $\psi_{sc}(\mathbf{r})$ has the form

$$\mathbf{j}_{sc} = \hat{\mathbf{r}}\frac{1}{L^3}\left(\frac{p}{m}\right)\frac{|f(\theta,\phi)|}{r^2} + \mathcal{O}\left(\frac{1}{r^3}\right) \tag{20.43}$$

where we will be able to eventually neglect the terms of order $1/r^3$ or higher. The number of particles scattered into a small area of detector per unit time is given by

$$\begin{aligned}\mathbf{j}_{sc}\cdot d\mathbf{A} &= \left[\hat{\mathbf{r}}\frac{1}{L^3}\left(\frac{p}{m}\right)\frac{|f(\theta,\phi)|}{r^2}\right]\cdot\left[\hat{\mathbf{r}}\,r^2\,d\Omega\right]\\ &= \frac{1}{L^3}\left(\frac{p}{m}\right)|f(\theta,\phi)|^2\,d\Omega\end{aligned} \tag{20.44}$$

where r and $d\Omega$ are the distance to and the angle subtended by the detector, respectively. This implies that

$$\frac{dN_{sc}}{d\Omega\,dt} = \frac{1}{L^3}\frac{p}{m}|f(\theta,\phi)|^2 \tag{20.45}$$

and we see that any terms that drop off as $1/r^3$ or faster will not contribute to a measurement at macroscopic distances r; this justifies the neglect of such terms in Eqn. (20.43). There are also cross-terms in Eqn. (20.38) which arise due to the "interference" between ψ_{inc} and ψ_{sc}. They all contain factors of the form

$$e^{\pm ikr(1-\cos(\theta))} \tag{20.46}$$

and we note that for a detector far away from the scattering center, the factor in the exponential satisfies $kr >> 1$. This implies that the phase of these contributions oscillates very rapidly, and the positive and negative, real and imaginary, parts average to zero.[3]

[3]This is not true, of course, for $\theta \to 0$, but then this contribution is not counted as having been scattered anyway.

Finally, using the definition of cross section in Eqn. (20.3) and Eqns. (20.42) and (20.45), we find that the differential cross section is simply given by the scattering amplitude via

$$\frac{d\sigma(\theta,\phi)}{d\Omega} = |f(\theta,\phi)|^2 \tag{20.47}$$

The problem of determining the cross section in quantum mechanics then reduces to finding $f(\theta,\phi)$ for a given incident energy and scattering potential.

20.3.2 Wave Equation for Scattering and the Born Approximation

The information on scattering probabilities encoded in $f(\theta,\phi)$ is presumably contained in the large r behavior of the Schrödinger wavefunction for unbound states. One approach to the evaluation of the scattering amplitude involves rewriting the Schrödinger equation in a form that "builds in" as much of the initial conditions of a scattering problem as possible. We can trivially rewrite

$$\left(-\frac{\hbar^2}{2m}\nabla^2 + V(\mathbf{r})\right)\psi(\mathbf{r}) = E\psi(\mathbf{r}) \tag{20.48}$$

in the form

$$\left(\nabla^2 + k^2\right)\psi(\mathbf{r}) = \left(\frac{2m}{\hbar^2}\right)V(\mathbf{r})\,\psi(\mathbf{r}) \tag{20.49}$$

where $k^2 = 2mE$, and which is more reminiscent of the classical wave equation. Using standard techniques for the study of such differential equations, we can rewrite Eqn. (20.49) as an *integral equation* in the form

$$\psi(\mathbf{r}) = e^{i\mathbf{k}\cdot\mathbf{r}} - \frac{m}{2\pi\hbar^2}\int \frac{e^{ik|\mathbf{r}-\mathbf{r}'|}}{|\mathbf{r}-\mathbf{r}'|}\,V(\mathbf{r}')\,\psi(\mathbf{r}')\,d\mathbf{r}' \tag{20.50}$$

which is called the *Lippman-Schwinger* equation.

We will not describe the manipulations by which Eqn. (20.50) can be derived, but will rather content ourselves in showing that it is equivalent to Eqn. (20.49). This can be accomplished by acting on both sides of Eqn. (20.50) with the "wave operator" $\nabla^2 + k^2$ and reproducing Eqn. (20.49). The first term in Eqn. (20.50) clearly satisfies

$$\left(\nabla^2 + k^2\right)e^{i\mathbf{k}\cdot\mathbf{r}} = (-\mathbf{k}^2 + k^2)e^{i\mathbf{k}\cdot\mathbf{r}} = 0 \tag{20.51}$$

The wave operator acts on the second term inside the integral sign and requires evaluation of

$$\left(\nabla^2 + k^2\right)\left(\frac{e^{ik|\mathbf{r}-\mathbf{r}'|}}{|\mathbf{r}-\mathbf{r}'|}\right) \tag{20.52}$$

This, in turn, means we need to calculate

$$\nabla^2\left(\frac{e^{ikr}}{r}\right) = -k^2\left(\frac{e^{ikr}}{r}\right) + e^{ikr}\nabla^2\left(\frac{1}{r}\right) \tag{20.53}$$

A standard way to derive the action of the gradient squared on the function $1/r$ is to appeal to a familiar result from electrostatics. The electric field of a point charge (located at the origin for definiteness) will certainly satisfy Gauss's law in the form

$$\nabla \cdot \mathbf{E}(\mathbf{r}) = -\frac{1}{\epsilon_0}\rho(\mathbf{r}) = -\frac{1}{\epsilon_0}q\delta(\mathbf{r}) \tag{20.54}$$

where the three-dimensional delta function can be understood to mean

$$\delta(\mathbf{r}) = \delta(x)\delta(y)\delta(z) \tag{20.55}$$

and this form is consistent with the charge density of truly pointlike charge. The Coulomb field can be written in the form

$$\frac{q}{4\pi\epsilon_0}\frac{\hat{\mathbf{r}}}{r^2} = \mathbf{E}(\mathbf{r}) = -\nabla\phi(\mathbf{r}) = -\nabla\left(\frac{q}{4\pi\epsilon_0}\frac{1}{r}\right) \tag{20.56}$$

which, when combined with Eqn. (20.54), yields

$$\nabla^2\left(\frac{1}{r}\right) = -4\pi\delta(\mathbf{r}) \tag{20.57}$$

We then find that

$$\begin{aligned}\left(\nabla^2 + k^2\right)\left(\frac{e^{ik|\mathbf{r}-\mathbf{r}'|}}{|\mathbf{r}-\mathbf{r}'|}\right) &= -4\pi e^{ik|\mathbf{r}-\mathbf{r}'|}\delta(\mathbf{r}-\mathbf{r}') \\ &= -4\pi\delta(\mathbf{r}-\mathbf{r}')\end{aligned} \tag{20.58}$$

This implies that

$$\begin{aligned}\left(\nabla^2 + k^2\right)\psi(\mathbf{r}) &= -\frac{m}{2\pi\hbar^2}(-4\pi)\int \delta(\mathbf{r}-\mathbf{r}')\,V(\mathbf{r}')\,\psi(\mathbf{r}')\,d\mathbf{r}' \\ &= \left(\frac{2m}{\hbar^2}\right)V(\mathbf{r})\psi(\mathbf{r})\end{aligned} \tag{20.59}$$

as hoped.

The usefulness of Eqn. (20.50) in extracting the scattering amplitude can be seen by expanding the exponential in the limit $|\mathbf{r}| >> |\mathbf{r}'|$, i.e., at an observation point far from the region where the scattering potential is acting. One has

$$\frac{e^{ik|\mathbf{r}-\mathbf{r}'|}}{|\mathbf{r}-\mathbf{r}'|} \approx \frac{e^{ikr}}{r}e^{-ik\hat{\mathbf{r}}\cdot\mathbf{r}'} = \frac{e^{ikr}}{r}e^{-i\mathbf{k}'\cdot\mathbf{r}'} \tag{20.60}$$

where $\mathbf{k}' \equiv k\hat{\mathbf{r}}$ is the wavevector in the direction of the detected scattered particle. We thus have

$$\psi(\mathbf{r}) \stackrel{r\text{ large}}{\longrightarrow} e^{i\mathbf{k}\cdot\mathbf{r}} - \left[\frac{m}{2\pi\hbar^2}\int e^{-i\mathbf{k}'\cdot\mathbf{r}'}\,V(\mathbf{r}')\psi(\mathbf{r}')\,d\mathbf{r}'\right]\frac{e^{ikr}}{r} \tag{20.61}$$

which is exactly of the form Eqn. (20.37) with the scattering amplitude given by

$$f(\theta,\phi) = -\frac{m}{2\pi\hbar^2}\int e^{-i\mathbf{k}'\cdot\mathbf{r}}\,V(\mathbf{r})\,\psi(\mathbf{r})\,d\mathbf{r} \tag{20.62}$$

where we have dropped the prime label on the integration variable for convenience; the dependence on the scattering angles, (θ, ϕ), is contained in the scattered wavevector, $\mathbf{k}' = k\hat{\mathbf{r}}$.

The evaluation of the scattering amplitude via Eqn. (20.62) still requires knowledge of the exact unbound wavefunction, $\psi(\mathbf{r})$, for a given energy in the field of the potential. A formal solution of Eqn. (20.50) which lends itself to a systematic approximation method can be obtained by iteration as follows:

- The incident plane wave will have the solution in the absence of the scattering potential, so label

$$\psi^{(0)}(\mathbf{r}) = e^{i\mathbf{k}\cdot\mathbf{r}} \tag{20.63}$$

  as the zeroth-order solution.
- Substitute this into the integral on the right-hand side of Eqn. (20.50). Writing

$$G(\mathbf{r}, \mathbf{r}') = -\frac{m}{2\pi\hbar^2}\frac{e^{ik|\mathbf{r}-\mathbf{r}'|}}{|\mathbf{r}-\mathbf{r}'|} \tag{20.64}$$

  for simplicity, this yields the "next best guess"

$$\psi^{(1)}(\mathbf{r}) = \psi^{(0)}(\mathbf{r}) + \int G(\mathbf{r}, \mathbf{r}')\psi^{(0)}(\mathbf{r}')\,V(\mathbf{r}')\,d\mathbf{r}' \tag{20.65}$$

- Iterate by using $\psi^{(1)}$ to obtain $\psi^{(2)}$ and so forth; the solution can then be written as an infinite series in the form

$$\begin{aligned}\psi(\mathbf{r}) = {} & \psi^{(0)}(\mathbf{r}) + \int d\mathbf{r}'\,G(\mathbf{r}, \mathbf{r}')\,V(\mathbf{r}')\,\psi^{(0)}(\mathbf{r}') \\ & + \int d\mathbf{r}' \int d\mathbf{r}''\,G(\mathbf{r}, \mathbf{r}')\,V(\mathbf{r}')\,G(\mathbf{r}', \mathbf{r}'')\,V(\mathbf{r}'')\,\psi^{(0)}(\mathbf{r}'') \\ & + \cdots\end{aligned} \tag{20.66}$$

If the scattering potential is sufficiently weak,[4] then the first nontrivial iteration is used; this is called the *Born approximation* and gives the scattering amplitude as

$$\begin{aligned}f_B(\theta, \phi) &= -\frac{m}{2\pi\hbar^2}\int e^{-i\mathbf{k}'\cdot\mathbf{r}}V(\mathbf{r})e^{i\mathbf{k}\cdot\mathbf{r}}\,d\mathbf{r} \\ &= -\frac{m}{2\pi\hbar^2}\int e^{-i\mathbf{q}\cdot\mathbf{r}}V(\mathbf{r})\,d\mathbf{r}\end{aligned} \tag{20.67}$$

The vector $\mathbf{q} \equiv \mathbf{k}' - \mathbf{k}$ is related to the momentum change of the particle as a result of the collision since $\mathbf{q} = (\mathbf{p}' - \mathbf{p})/\hbar$. To this order of approximation, the scattering amplitude is simply the (three-dimensional) Fourier transform of the potential, $V(\mathbf{r})$, with respect to $\mathbf{q}$. For central potentials for which there is no interesting ϕ dependence, it is useful to write $q = |\mathbf{q}|$ in terms of the scattering angle using the geometry in Fig. 20.9. We find

$$\begin{aligned}\mathbf{q}^2 = (\mathbf{k}' - \mathbf{k})^2 &= 2k^2 - 2\mathbf{k}'\cdot\mathbf{k} \\ &= 2k^2(1 - \cos(\theta)) \\ &= 4k^2\sin^2(\theta/2)\end{aligned} \tag{20.68}$$

[4]See Saxon (1968) for a discussion of the conditions under which this is a valid assumption.

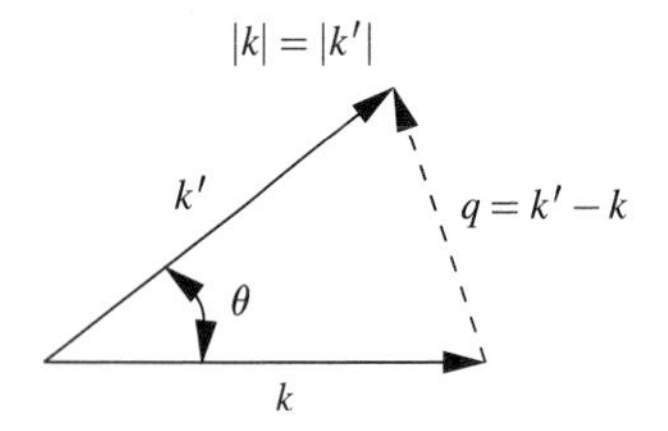

FIGURE 20.9. Incident (k) and scattered (k') wavevectors showing momentum transfer, $\Delta p = \hbar q = \hbar(k' - k)$.

or

$$q = 2k\sin(\theta/2) = q_{\max}\sin(\theta/2) \tag{20.69}$$

with $q_{\max} = 2k = 2\sqrt{2mE}/\hbar$; the angular dependence of the scattering angle depends only on q, so we often write $f(\theta, \phi) = f(q)$. In this case, we can choose the polar axis to lie along the z direction so that $\mathbf{q}\cdot\mathbf{r} = qr\cos(\theta)$ and

$$\begin{aligned}
\int e^{-i\mathbf{q}\cdot\mathbf{r}}\, V(r)\, d\mathbf{r} &= \int_0^\infty dr\, r^2\, V(r) \int_0^{2\pi} d\phi \int_0^\pi d\theta\, \sin(\theta) e^{-iqr\cos(\theta)} \\
&= \frac{2\pi}{iq}\int_0^\infty dr\, rV(r)\left(e^{iqr} - e^{-iqr}\right) \\
&= \frac{4\pi}{q}\int_0^\infty dr\, r\sin(qr)\, V(r)
\end{aligned} \tag{20.70}$$

or

$$f_B(q) = -\frac{2m}{\hbar^2 q}\int_0^\infty dr\, r\, \sin(qr)\, V(r) \tag{20.71}$$

which is sometimes useful.

Example 20.3 Scattering from a Finite Well As an example, let us calculate the scattering cross section for the finite attractive well in three dimensions, defined via

$$V(r) = \begin{cases} -V_0 & \text{for } r < a \\ 0 & \text{for } r > a \end{cases} \tag{20.72}$$

using the Born approximation. Besides illustrating several rather general aspects of scattering theory, this potential has some relevance to the (short-range) nucleon-nucleon potential as noted in Sec. 9.4.2.

The scattering amplitude from Eqn. (20.70) is given by

$$\begin{aligned}
f_B(q) &= \frac{2mV_0}{\hbar^2 q}\int_0^a r\sin(qr)\, dr \\
&= \frac{2mV_0 a^3}{\hbar^2}\left(\frac{\sin(qa)}{(qa)^3} - \frac{\cos(qa)}{(qa)^2}\right) \\
&= \frac{2mV_0 a^3}{3\hbar^2}\left(3\frac{j_1(z)}{z}\right) \\
&= \left(\frac{2mV_0 a^3}{3\hbar^2}\right) G(z)
\end{aligned} \tag{20.73}$$

where $z = qa$ and $f_B(q)$ can be seen to have the appropriate dimensions. We have made use of the fact that the result can be written in terms of the spherical Bessel function since

$$j_1(z) = \frac{\sin(z)}{z^2} - \frac{\cos(z)}{z} \tag{20.74}$$

and written $G(z)$ in such a way that $G(0) = 1$ for convenience.

The differential cross section is simply given by

$$\frac{d\sigma}{d\Omega}(\theta) = |f_B(q)|^2 = \left(\frac{2mV_0a^3}{3\hbar^2}\right)^2 |G(z)|^2 \tag{20.75}$$

where $z = aq_{max}\sin(\theta/2)$. We exhibit the angular dependence of the scattering in Fig. 20.10 for several values of $z_{max} = aq_{max} = 2a\sqrt{2mE}/\hbar$ corresponding to low and high energies; z_{max} is an appropriate figure of merit as its definition implies that

$$E = \frac{z_{max}^2\hbar^2}{8ma^2} \tag{20.76}$$

and it measures E relative to a typical bound state energy scale. The plot in Fig. 20.10 shows some typical behavior of the differential cross section for scattering from a finite range potential, and we make some general comments based on this one example.

- *Low energies:* For low energies, the scattering is uniform in angle, implying an isotropic cross section. This result can be seen more generally from Eqn. (20.67) when we note that low energies imply that $\mathbf{q} \cdot \mathbf{r} \to 0$, so that

$$\begin{aligned} f_B(q) &\approx -\frac{m}{2\pi\hbar^2}\int V(r)\,d\mathbf{r} \\ &\approx -\frac{m}{2\pi\hbar^2} V(0)\,d^3r \end{aligned} \tag{20.77}$$

  where d^3r is roughly the volume over which the potential is nonvanishing; in our example this gives the exact result, namely,

$$f_B(q) = \frac{m}{2\pi\hbar^2}V_0\left(\frac{4\pi}{3}a^3\right) = \frac{2mV_0a^3}{3\hbar^2} \tag{20.78}$$

  More generally, $f_b(q)$ is independent of angle for low energies.
- *High energies:* For high energies, the cross section obviously becomes more and more peaked in the forward ($\theta \approx 0$) direction, and the total cross section—given, recall, by

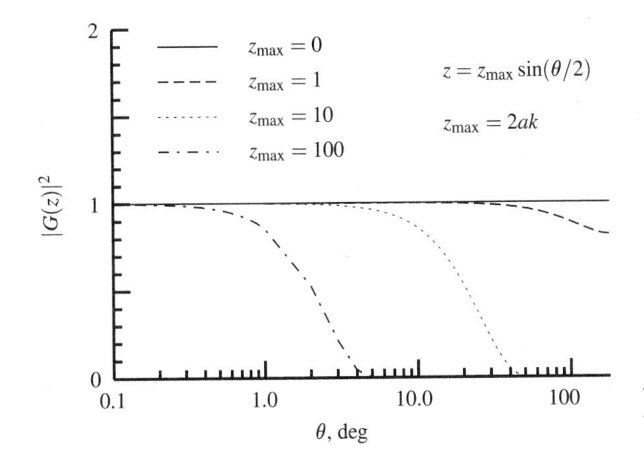

FIGURE 20.10. *Plot of differential cross section (via $|G(z)|^2$) versus angle for scattering from finite well of radius **a** for various values of ka showing forward diffractive peak.*

the "area" under the $d\sigma/d\Omega$ versus θ curve weighted by $2\pi\sin(\theta)$—becomes smaller. Both of these effects can also be understood rather generally from Eqn. (20.67).

When the argument of the exponential phase, $2ka\sin(\theta/2)$, becomes large, the phase factor oscillates rapidly, and the contribution to the total integral from that region becomes small due to cancellations; a rough cutoff for when this happens is when

$$2ka\sin\left(\frac{\theta}{2}\right) \longrightarrow ka\,\theta \approx 1 \tag{20.79}$$

As the energy increases, so that $ka >> 1$, this occurs for smaller and smaller angles, and the differential cross section becomes nonnegligible only for angles satisfying

$$\theta_{\max} \lesssim \frac{1}{ka} \tag{20.80}$$

This effect is reminiscent of classical wave diffraction, and the forward peak is often called the *main diffractive peak* for this reason. The location of the first zero of $G(z)$, determined by $j_1(z) = 0$, can be used to specify the first diffraction minimum; one finds $2ka\sin(\theta/2) \approx 4.49$. Not coincidentally, the condition for the diffraction pattern for circular apertures in optics (familiar from problems involving the design of optical instruments) is quite similar to this relation (P20.6).

The decrease of the total cross section with energy can now be understood by integrating $d\sigma/d\Omega$; one has roughly

$$\begin{aligned}\sigma = \int d\Omega\frac{d\sigma}{d\Omega} &\approx 2\pi\int_0^{1/ka} |f(\theta)|^2\sin(\theta)\,d\theta \\ &\approx 2\pi|f(0)|^2\int_0^{1/ka} \theta\,d\theta \\ &\approx \frac{\pi|f(0)|^2}{a^2k^2} \\ &\propto \frac{1}{E}\end{aligned} \tag{20.81}$$

20.4 ELECTROMAGNETIC SCATTERING

Because of its importance, we consider the scattering of a charged particle via the Coulomb potential separately. Much of the symmetry of the $1/r^2$ force carries over from bound state problems to scattering, yielding simple, closed-form solutions for the cross section for an inverse-square law force. Deviations from that result can then signal the presence of new physics, as did the splitting of degenerate energy levels in the hydrogen atom spectrum.

The long range of the $1/r$ potential means that many of the standard formulations of scattering theory are not directly applicable.[5] We will be content with quoting some standard results using the Born approximation.

In order to calculate the scattering amplitude for the Coulomb potential, it is necessary to introduce a convergence factor and write

$$V_C(r) = \frac{Ke^2}{r} \longrightarrow V_C(r;\mu) = \frac{Ke^2}{r}e^{-\mu r} \tag{20.82}$$

[5] For a very thorough discussion and further references, see Schiff (1955) and Sakurai (1994).

for the interaction of two like-sign charges $\pm e$. This is like the Yukawa potential considered in P3.9, with a finite range of roughly $R \sim 1/\mu$; we will let $\mu \to 0$ at the end of the calculation. The Born approximation then gives

$$
\begin{aligned}
f_B(\theta) = f_B(q) &= -\left(\frac{2m}{\hbar^2}\right)\frac{Ke^2}{q}\int_0^\infty dr\, e^{-\mu r}\sin(qr) \\
&= -\left(\frac{2mKe^2}{\hbar^2}\right)\frac{1}{\mu^2+q^2} \\
&\longrightarrow -\left(\frac{2mKe^2}{\hbar^2}\right)\frac{1}{q^2} \qquad \text{as } \mu \to 0
\end{aligned} \tag{20.83}
$$

The differential cross section can then be written as

$$
\frac{d\sigma}{d\Omega} = \left(\frac{2mKe^2}{\hbar^2 q^2}\right)^2 = \left(\frac{Ke^2}{4E}\right)^2 \frac{1}{\sin^4(\theta/2)} \tag{20.84}
$$

using the definitions of $q = 2k\sin(\theta/2)$ and $\hbar k = \sqrt{2mE}$. Somewhat amazingly, this is the same *Rutherford cross section* result obtained using purely classical techniques, and some comments can be made:

- The fact that this cross section does not contain any explicit factors of $\hbar$ when expressed in terms of the observable energy E is special to the case of Coulomb scattering, and is also suggestive of the unique correspondence to the classical result.
- The result is easily generalized to the scattering of charges $Z_1 e$ by a potential of charge $Z_2 e$ by simply letting $e^2 \to Z_1 Z_2 e^2$, and has been verified for many systems. Data for scattering of oxygen nuclei ($Z_1 = 8$) on gold nuclei ($Z_2 = 79$) are shown in Fig. 20.11, confirming the characteristic $1/\sin^4(\theta/2)$ dependence.
- The exact expression for the Coulomb scattering amplitude can be derived by solving the Schrödinger equation in parabolic coordinates,[6] with the result

$$
f_C(\theta) = f_B(\theta)e^{-i\eta\ln(\sin^2(\theta/2))+2i\beta} \tag{20.85}
$$

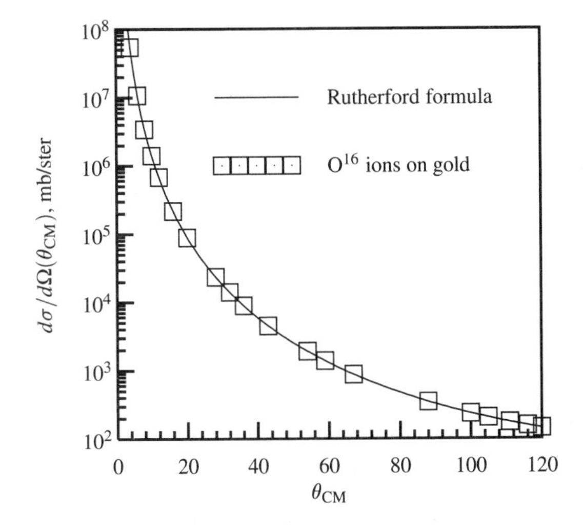

FIGURE 20.11. *Differential cross section for Coulomb scattering of oxygen on gold nuclei illustrating Rutherford formula; data taken from Bromley, Kuehner, and Almqvist (1961).*

[6]See, e.g., Schiff (1968).

where

$$\eta \equiv \frac{mZ_1Z_2Ke^2}{\hbar^2 k} = \frac{Z_1Z_2Ke^2}{\hbar v} \tag{20.86}$$

where v is the velocity of the particle. The additional phase factor is given in terms of η via

$$e^{2i\beta} = \frac{\Gamma(1+i\eta)}{\Gamma(1-i\eta)} \tag{20.87}$$

where $\Gamma(x)$ is the generalized factorial function discussed in App. C.9. The η factor does depend explicitly on $\hbar$, and so is a true quantum effect, but it does not seem to have any observable effects as $|f_C(\theta)|^2 = |f_B(\theta)|^2$. Interference effects, such as occur in the scattering of identical particles, can show the effects of such phases, and we will discuss this in Sec. 20.6.2.

- It is interesting to note that the same exact result can be used to give the Coulomb wavefunction in the other limit, namely, $r \approx 0$. One obtains

$$|\psi_C(r \to 0)|^2 \propto \frac{2\eta\pi}{v(e^{2\eta\pi}-1)} \tag{20.88}$$

For very slow particles, corresponding to $\eta >> 1$, we find different behavior for the attractive ($\eta < 0$) and repulsive ($\eta > 0$) case, namely

$$|\psi_C(0)|^2 \approx \begin{cases} \frac{2\pi|\eta|}{v} & \text{for } \eta < 0 \\ \frac{2\pi\eta}{v} e^{-2\pi\eta} & \text{for } \eta > 0 \end{cases} \tag{20.89}$$

The exponential suppression for like-sign charges ($\eta > 0$) is simply the Gamow tunneling factor derived in Sec. 12.5.3.

Another case of interest is the scattering of a point charge (here of magnitude e) from a distribution of charge that is not pointlike (but with total charge still equal to e). A classic example is the scattering of electrons from charged nuclei where the fact that the nuclear charge is "spread out" over a finite size becomes apparent. This situation can be handled by recalling that the potential of a point charge in the field of a nontrivial charge density can be written in the form

$$V(r) = Ke^2 \int \frac{\rho(\mathbf{r}')d\mathbf{r}'}{|\mathbf{r}-\mathbf{r}'|} \tag{20.90}$$

where we have written the scattering charge density as $e\rho(\mathbf{r})$; $\rho(\mathbf{r})$ is normalized such that

$$\int d\mathbf{r}'\rho(\mathbf{r}') = 1 \tag{20.91}$$

or Z for an object with larger charge ($Q = Ze$). A point charge corresponds to $\rho(\mathbf{r}) = \delta(\mathbf{r})$, and Eqn. (20.90) reproduces Coulomb's law in that limit.

If we substitute this into the Born amplitude, we require the Fourier transform

$$\int d\mathbf{r}\, e^{-i\mathbf{q}\cdot\mathbf{r}}\left[\int d\mathbf{r}' \frac{\rho(\mathbf{r}')}{|\mathbf{r}-\mathbf{r}'|}\right] = \left[\int d\mathbf{r}'\, \rho(\mathbf{r}')e^{-i\mathbf{q}\cdot\mathbf{r}'}\right]\left[\int d\mathbf{r}\, \frac{e^{-i\mathbf{q}\cdot(\mathbf{r}-\mathbf{r}')}}{|\mathbf{r}-\mathbf{r}'|}\right]$$

$$\equiv F(\mathbf{q})\left[\int d\mathbf{s}\, \frac{1}{s} e^{-i\mathbf{q}\cdot\mathbf{s}}\right] \tag{20.92}$$

where we have written $\mathbf{s} = \mathbf{r} - \mathbf{r}'$ for convenience. The last term in Eqn. (20.92) is simply the Fourier transform of the Coulomb potential, which, using regularization tricks, we have evaluated above. The first term, defined via

$$F(\mathbf{q}) = \int d\mathbf{r}\, \rho(\mathbf{r}) e^{-i\mathbf{q}\cdot\mathbf{r}} \tag{20.93}$$

is the Fourier transform of the (dimensionless) charge density, and it describes the modification to the Rutherford cross section due to the "smearing out" of the charge; $F(\mathbf{q})$ is called the *form factor*. The two terms seem to "factorize" from each other, so that one can write

$$\frac{d\sigma}{d\Omega} = \left(\frac{d\sigma}{d\Omega}\right)_R |F(\mathbf{q})|^2 \tag{20.94}$$

so that $|F(\mathbf{q})|$ can be measured as the ratio of the observed cross section to the Rutherford prediction $(d\sigma/d\Omega)_R)$, as one varies q. Several comments can be made:

- The charge density is normalized such that

$$F(0) = \int d\mathbf{r}\, \rho(\mathbf{r}) = 1 \tag{20.95}$$

  The physical interpretation of this result is that small values of q correspond to very long wavelengths, and such probes "see" only the total charge of the scatterer and cannot resolve its structure; small q can also arise if $\theta \approx 0$ and no scattering occurs.
- We can expand $F(\mathbf{q})$ for small q and find

$$\begin{aligned} F(\mathbf{q}) &= \int d\mathbf{r}\, \rho(\mathbf{r}) e^{-i\mathbf{q}\cdot\mathbf{r}} \\ &= \int \rho(\mathbf{r}) d\mathbf{r} - i \sum_j q_j \int x_j\, \rho(\mathbf{r}) - \frac{1}{2} \sum_{j,k} q_j q_k \int x_j x_k\, \rho(\mathbf{r}) d\mathbf{r} + \cdots \\ &= 1 - i \sum_i q_j \langle x_j \rangle - \frac{1}{2} \sum_{j,k} q_j q_k \langle x_j x_k \rangle + \cdots \end{aligned} \tag{20.96}$$

  where

$$\langle f(\mathbf{r}) \rangle = \int f(\mathbf{r})\, \rho(\mathbf{r})\, d\mathbf{r} \tag{20.97}$$

  For a spherically symmetric charge distribution with $\rho(\mathbf{r}) = \rho(r)$, the average values satisfy

$$\langle x_j \rangle = 0 \qquad \text{and} \qquad \langle x_j x_k \rangle = \frac{1}{3} \langle r^2 \rangle \delta_{j,k} \tag{20.98}$$

  since $\langle x^2 \rangle = \langle y^2 \rangle = \langle z^2 \rangle$. This implies that

$$F(\mathbf{q}) = F(q) = 1 - \frac{1}{6} q^2 \langle r^2 \rangle + \cdots \tag{20.99}$$

  and knowledge of the form factor for small but finite q values gives information on the spatial extent of the charge distribution, its "charge radius," $\langle r^2 \rangle$.

Example 20.4 Nuclear Sizes Much of the information on the size of nuclei has been obtained using the experimental determination of form factors from electron-nucleus scattering experiments. As a simple model of the charge distribution of a nucleus, consider the density given by

$$\rho(\mathbf{r}) = \rho(r) = \begin{cases} 3/4\pi R^3 & \text{for } r < R \\ 0 & \text{for } r > R \end{cases} \tag{20.100}$$

that is, a constant nuclear charge density up to sharply defined nuclear radius, R. The form factor for this distribution is given by the same Fourier transform considered in Ex. 20.3, and we find that

$$F(q) = \frac{3j_1(kR)}{qR} = 3\left(\frac{\sin(qR)}{(qR)^3} - \frac{\cos(qR)}{(qR)^2}\right) \tag{20.101}$$

The differential cross section for electron scattering on oxygen (O^{16}) nuclei for two different incident energies is shown in Fig. 20.12, where a pronounced dip is seen at roughly 44° for the 420 MeV data. The dashed curve is the Rutherford cross section multiplied by the form factor in Eqn. (20.101) with an appropriately chosen value of R. The simple model accounts for many of the gross features of the actual data; the assumption of a "sharp" nuclear edge is responsible for the bad fit near the first diffraction minimum; more realistic nuclear models with a smoother nuclear surface work better and give a less deep diffraction "dip."

To make quantitative contact with the data, we note that electrons of these energies are ultrarelativistic, implying that their energies satisfy $E = pc = \hbar kc$. Using the value of the first zero of $j_1(z)$ mentioned above, we find that

$$qR = 4.49 \implies R = \frac{4.49\,\hbar c}{2E\sin(\theta/2)} \approx 2.8\text{ F} \tag{20.102}$$

A common parameterization of nuclear sizes which gives a good fit to many nuclei is given by $R(A) = (1.2\text{ F})A^{1/3}$, where A is the atomic mass of the nucleus; here, $A = 16$, which gives $R \approx 3$ F. The data for lower energy exhibit the same diffraction minimum, but at larger angles consistent with Eqn. (20.102).

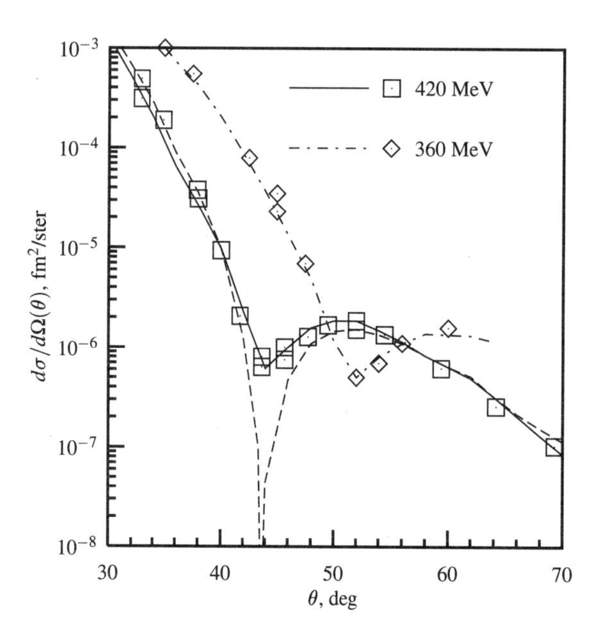

FIGURE 20.12. Differential cross section for electron scattering on oxygen nuclei. Note that the first "diffractive dip" moves outward in angle as the energy is decreased. The dashed curve corresponds to a nuclear charge distribution with "sharp edges," while the solid curves are for a more realistic parameterization, which is smoother. Adapted from Ehrenberg et al. (1959).

We can also use form factor ideas to understand the shielding of the nuclear charge by its atomic electron cloud, and how this effect gives a total cross section for scattering via neutral atoms, which does not diverge as the simple Rutherford formula suggests. For example, we might write the charge density for a neutral hydrogen atom in its ground state as

$$e\rho_H(r) = +e\,\delta(\mathbf{r}) - e|\psi_{1,0,0}(\mathbf{r})|^2 \tag{20.103}$$

where $\psi_{1,0,0}(\mathbf{r})$ is the electron wavefunction; this corresponds to a positive pointlike nucleus and "spread-out" negative electron cloud. Using

$$|\psi_{1,0,0}(\mathbf{r})|^2 = \frac{1}{\pi a_0^3} e^{-2r/a_0} \tag{20.104}$$

we find that the form factor is

$$F(q) = 1 - \frac{1}{(1 + (qa_0/2)^2)^2} \tag{20.105}$$

so that

$$\begin{aligned} |F(q \to 0)|^2 &\approx \left|1 - \left(1 - 2\frac{q^2 a_0^2}{4} + \cdots\right)\right|^2 \\ &\approx \frac{q^4 a_0^4}{4} \end{aligned} \tag{20.106}$$

When multiplied by the Rutherford cross section in Eqn. (20.94), this eliminates the $1/\sin^4(\theta/2)$ divergence in the cross section.

One final comment can be made concerning the scattering of point charges (such as electrons) by other subatomic particles via electromagnetic forces. Besides its charge that gives rise to Rutherford scattering via the Coulomb interaction, a proton, for example, has an intrinsic magnetic moment. An incident electron can then also experience magnetic scattering, calculable classically, for example, via the Lorentz force $\mathbf{F} = -e\mathbf{v} \times \mathbf{B}$. The form of the magnetic potential for a point magnetic dipole,

$$\mathbf{A}(\mathbf{r}) = \frac{\mu_0}{4\pi}\frac{\mathbf{m} \times \mathbf{r}}{r^3} \sim \frac{1}{r^2} \tag{20.107}$$

shows that it is shorter range ($1/r^2$) than the Coulomb interaction ($1/r$), so that electrostatic interactions will dominate at long distances. On the other hand, this also implies that the magnetic force can come to dominate the scattering if r is small enough. Using Eqn. (20.28), we note that the classical distance of closest approach for an inverse law force can be written

$$r_{\min} = \left(\frac{A}{2E}\right)\left[\frac{1}{\sin(\theta/2)} + 1\right] \tag{20.108}$$

which implies that magnetic scattering can become important for sufficiently large energies and/or scattering angles. This effect is clearly observed in elastic electron-proton scattering where the magnetic contribution comes to dominate the cross section for $q^2 \gtrsim 2\ \mathrm{GeV}^2$. It is even more important in electron-neutron scattering at all energies as it is the dominant effect due to the vanishing charge of the neutron.

20.5 PARTIAL WAVE EXPANSIONS

The methods used so far to calculate scattering amplitudes have the flavor of perturbation theory, and in this section we present an alternative method of evaluating $f(\theta)$, which is based more directly on matching of solutions of the Schrödinger equation with the form of the incident and scattered wave of Eqn. (20.37). It also makes connection somewhat more directly to the notion of classical scattering trajectories being determined by energy and angular momentum values.

We first recall that the solutions of the free-particle radial in spherical coordinates,

$$-\frac{\hbar^2}{2m}\left(\frac{d^2R(r)}{dr^2}+\frac{2}{r}\frac{dR(r)}{dr}\right)+\frac{l(l+1)}{2mr^2}R(r) = ER(r) \tag{20.109}$$

which are well behaved at the origin have the form

$$R_{E,l}(r) = j_l(kr) \longrightarrow \frac{\sin(kr - l\pi/2)}{kr} \tag{20.110}$$

where $k = \sqrt{2mE}/\hbar$. Far from the interaction point, where the potential is negligible, $V(r) \approx 0$, the scattered wavefunction must also satisfy Eqn. (20.109), but with the more general solution

$$R_{sc}(r) = \alpha_l j_l(kr) + \beta_l n_l(kr) \tag{20.111}$$

since one is far from the origin where the $n_l(kr)$ are poorly behaved. We can then write the α_l, β_l in a suggestive form and examine the large r limit and find

$$\begin{aligned} R_{sc}(r) &= a_l\big(\cos(\delta_l)j_l(lr) - \sin(\delta_l)n_l(kr)\big) \\ &\longrightarrow \frac{a_l\big(\cos(\delta_l)\sin(kr - l\pi/2) + \sin(\delta_l)\cos(kr - l\pi/2)\big)}{kr} \\ &= a_l\left[\frac{\sin(kr - l\pi/2 + \delta_l)}{kr}\right] \end{aligned} \tag{20.112}$$

This result is similar to the one-dimensional scattering examples considered in Sec. 12.4, where the reflected and transmitted waves differed in amplitude and phase from the incident plane wave. For this reason, δ_l is called the *phase shift of the l-th partial wave.* Attractive (repulsive) potentials imply that $\delta_l > 0$ ($\delta_l < 0$) corresponds to the wave being "pulled in" ("pushed out") by the scattering center, resulting in a phase delay (advance). The phase shifts can, in principle, be extracted from the exact radial wavefunctions, if known, for each l value by examining their large r behavior, which is guaranteed to be of the form in Eqn. (20.111); from that form, one has

$$-\frac{\beta_l}{\alpha_l} = \frac{\sin(\delta_l)}{\cos(\delta_l)} = \tan(\delta_l) \tag{20.113}$$

The complete solution of the scattering wavefunction can then be written for large r in the form

$$\begin{aligned} \psi(\mathbf{r}) &\longrightarrow \sum_{l=0}^{\infty}\sum_{m=-l}^{+l} a_l \frac{\sin(kr - l\pi/2 + \delta_l)}{kr} Y_{l,m}(\theta,\phi) \\ &\longrightarrow \sum_{l=0}^{\infty} a_l \frac{\sin(kr - l\pi/2 + \delta_l)}{kr} P_l(\cos(\theta)) \end{aligned} \tag{20.114}$$

for a central potential, which implies no ϕ and hence no m dependence. We then wish to match this form onto the standard scattering solution for a plane wave incident along the z axis:

$$\psi = e^{ikz} + f(\theta)\frac{e^{ikr}}{r} \tag{20.115}$$

to determine the scattering amplitude.

To accomplish this, we use the relation

$$e^{ikz} = e^{ikr\cos(\theta)} = \sum_{l=0}^{\infty}(2l+1)\, i^l\, j_l(kr)\, P_l(\cos(\theta)) \tag{20.116}$$

which we quote, but do not prove. The strategy is then to:

- Equate Eqn. (20.114) with Eqn. (20.115) using the identity Eqn. (20.116).
- Use the trigonometric relation

$$\sin(z) = \frac{e^{iz} - e^{-iz}}{2i} \tag{20.117}$$

  to rewrite the sine functions in terms of complex exponentials.
- Equate the coefficients of the linearly independent terms proportional to e^{ikr} and e^{-ikr}.

The coefficients of the e^{-ikr} terms can be seen to imply that

$$a_l = (2l+1)i^l e^{i\delta_l} \tag{20.118}$$

while the e^{ikr} terms impose a relation on the scattering amplitude, namely,

$$f(\theta) = \frac{2i}{k}\sum_{l=0}^{\infty}\left[a_l e^{i\delta_l} - (2l+1)i^l\right] e^{-il\pi/2}\, P_l(\cos(\theta)) \tag{20.119}$$

Using Eqns. (20.118) and (20.119) and the fact that $e^{-il\pi/2} = (1/i)^l$, we find the very useful result

$$\begin{aligned} f(\theta) &= \frac{1}{2ik}\sum_{l=0}^{\infty} e^{-il\pi/2}(2l+1)i^l[e^{2i\delta_l} - 1]\, P_l(\cos(\theta)) \\ &= \frac{1}{k}\sum_{l=0}^{\infty}(2l+1)e^{i\delta_l}\sin(\delta_l)\, P_l(\cos(\theta)) \end{aligned} \tag{20.120}$$

The total cross section is then given by

$$\sigma = \int d\Omega \frac{d\sigma}{d\Omega} = \int d\Omega |f(\theta)|^2 = \frac{4\pi}{k^2}\sum_{l=0}^{\infty}(2l+1)\sin^2(\delta_l) \tag{20.121}$$

where we use the orthonormality properties of the Legendre polynomials, namely,

$$\int d\Omega\, P_l(\cos(\theta))\, P_{l'}(\cos(\theta)) = \frac{4\pi}{(2l+1)}\delta_{l,l'} \tag{20.122}$$

Example 20.5 Hard-sphere Scattering One of the simplest scattering problems in classical mechanics is that of scattering from an impenetrable sphere considered in Ex. 20.1. We consider here the same problem in quantum mechanics using the method of partial waves; the appropriate central potential is given by

$$V(r) = \begin{cases} 0 & \text{for } r > a \\ +\infty & \text{for } r < a \end{cases} \tag{20.123}$$

The solution for $r < a$ must vanish, and it must also match onto the most general free-particle solution for a given partial wave for $r > a$, namely,

$$R_l(r) = \alpha_l j_l(kr) + \beta_l n_l(kr) \tag{20.124}$$

which implies that

$$\frac{\beta_l}{\alpha_l} = -\frac{j_l(ka)}{n_l(ka)} \tag{20.125}$$

Using Eqn. (20.113), we find that

$$\tan(\delta_l) = \frac{j_l(ka)}{n_l(ka)} \qquad \text{or} \qquad \sin^2(\delta_l) = \frac{j_l^2(ka)}{j_l^2(ka) + n_l^2(ka)} \tag{20.126}$$

Using the expansion in Sec. 17.6 for the spherical Bessel functions, we find that for $ka << 1$

$$\tan(\delta_l) \approx \frac{(ka)^{(2l+1)}}{(2l+1)(1 \cdot 3 \cdots (2l-1))^2} \tag{20.127}$$

so that for low enough energies, only the $l = 0$ wave will contribute appreciably to the scattering. The angular momentum barrier effectively keeps the scatterers apart, and only s-wave scattering is important. This is also consistent with the conclusions drawn from the Born approximation, namely, that low-energy scattering from finite potentials has an isotropic angular distribution. Conversely, if the effective range of some interaction is roughly R, then the contribution to the effective potential due to rotational motion will equal the incident energy when

$$\frac{l(l+1)\hbar^2}{2mR^2} \approx E \qquad \text{or} \qquad l_{\text{max}} \approx \frac{\sqrt{2mE}}{\hbar} R \approx kR \tag{20.128}$$

implying that angular momentum values up to the order of l_{max} will contribute to the scattering amplitude, consistent with Eqn. (20.127).

In this example, when $ka << 1$, we have $\sin(\delta_0) \approx \delta_0 \approx ka$, which implies that the total cross section is

$$\sigma = \frac{4\pi}{k^2}\sin^2(\delta_0) \approx \frac{4\pi k^2 a^2}{k^2} = 4\pi a^2 \tag{20.129}$$

Based on our experience with the Rutherford cross section, we might have expected that any cross section that does not contain an explicit factor of $\hbar$ would necessarily correspond to the classical result, but we find here a cross section that is 4 times larger than the geometrical area presented to the scatterers. To understand better the approach to the classical limit, we consider now the limit where $ka >> 1$ as well; in that case, one finds

$$\tan(\delta_l) \longrightarrow -\frac{\sin(kr - l\pi/2)}{\cos(kr - l\pi/2)} \tag{20.130}$$

so that $\delta_l \longrightarrow -(kr - l\pi/2) >> 1$ in magnitude, so that many partial waves contribute to the total cross section. We can estimate the summation in Eqn. (20.121) in the following heuristic manner:

- Each contributing factor of $\sin^2(\delta_l)$ with a large argument will average to 1/2, as that is the value of $\sin^2(x)$ when averaged over many cycles.

- The summation will be cut off at a value of $l_{max} \approx ka$, and we approximate

$$\sum_{l=0}^{l_{max}}(2l+1) \approx \int_0^{l_{max}}(2l+1)dl \approx l_{max}^2 \approx (ka)^2 \tag{20.131}$$

- The total cross section would then be given by

$$\sigma = \frac{4\pi}{k^2}\left(\frac{1}{2}\right)(ka)^2 = 2\pi a^2 \tag{20.132}$$

While this is not a proof, a careful numerical evaluation of the exact cross section given by

$$\sigma = \frac{4\pi}{k}\sum_{l=0}^{\infty}(2l+1)\frac{j_l^2(ka)}{(j_l^2(ka)+n_l^2(ka))} \tag{20.133}$$

confirms the semiquantitative argument, and the results are shown in Fig. 20.13. Even at high energy, the total cross section is still twice the classical value. This result is perhaps more surprising than Eqn. (20.129), since when $ka >> 1$, we have $2\pi a >> \lambda$, and we could, in principle, make wavepackets of dimensions much smaller than the scattering object, which should follow classical trajectories as in Sec. 20.1.

The answer to where the "extra" πa^2 comes from can be seen much more clearly by examination of the differential cross section, $d\sigma(\theta)/d\Omega$ versus θ, for increasingly large values of ka, as shown in Fig. 20.14. In units of a^2, the classical differential cross section is given by Eqn. (20.14), yielding the isotropic distribution

$$\left[\frac{d\sigma}{d\Omega}(\theta)\right]_{CL}\frac{1}{a^2} = \frac{1}{4} \tag{20.134}$$

and for $ka << 1$, we obtain 4 times that result, as noted above. On the other hand, for $ka >> 1$, the cross section shows an obvious forward (diffractive) peaking consistent with $\theta_{max} \sim 1/ka$, as mentioned above, with a long "tail" that approaches the "flat" value given by Eqn. (20.134). Fig. 20.14 shows that the extra cross section is all contained in the forward diffraction peak. For macroscopic systems, the value of θ_{max} is so small as to make this contribution unobservable (P20.11), and the remaining cross section is consistent with the purely classical result.

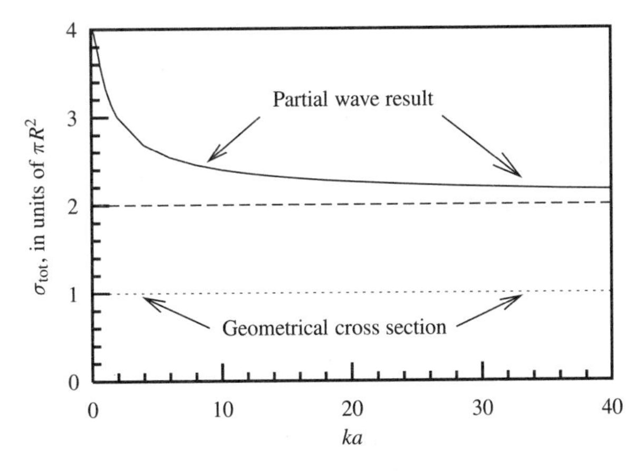

FIGURE 20.13. *Energy dependence (via ka) of total cross section for scattering from a hard sphere using partial wave expansions. Note that even for large energies $\sigma_{TOT} = 2\pi R^2$ instead of the expected classical geometrical cross section, $\sigma_{CL} = \pi R^2$.*

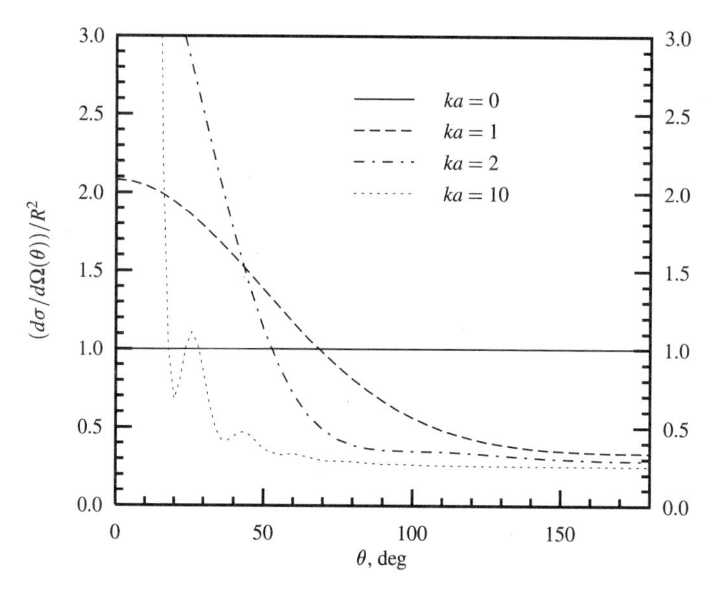

FIGURE 20.14. *Differential cross section for scattering from a hard sphere using a partial wave expansion. For increasing energies (or ka), the "extra" πR^2 in the total cross section is all in the very forward direction and unobservable.*

20.6 SCATTERING OF PARTICLES

Thus far, we have considered only so-called *potential scattering* where particles interact with fixed external sources of potential, $V(r)$. This can only be an idealization, as such processes do not conserve momentum, as illustrated in Fig. 20.9; the momentum transfer $\Delta\mathbf{p} = \hbar\mathbf{q}$ to the particle comes from "nowhere." It is more realistic to consider the collisions of two bodies that scatter under the influence of their mutual interaction potential, $V(\mathbf{r}_1 - \mathbf{r}_2)$. The formalism developed in Sec. 15.3.2 for separating two-body problems using relative and center-of-mass coordinates is again directly applicable in this situation. The equivalent one-particle Schrödinger equation for the relative coordinate, now with reduced mass μ, will give the information on the scattering process. All of the results we have derived for scattering can then be taken over directly, provided the center-of-mass coordinate is fixed. Many of the complications of analyzing scattering experiments will arise for simple geometrical reasons due to the motion of the center of mass of the scattering system.

In the next two sections, we describe (1) how the description of the scattering process depends on the motion of the center of mass of the two-body system and (2) how the spin statistics theorem affects the scattering of indistinguishable particles.

20.6.1 Frames of Reference

For a two-body problem, the center-of-mass coordinate is given by

$$\mathbf{R}_{\mathrm{cm}} = \frac{m_1\mathbf{r}_1 + m_2\mathbf{r}_2}{m_1 + m_2} \tag{20.135}$$

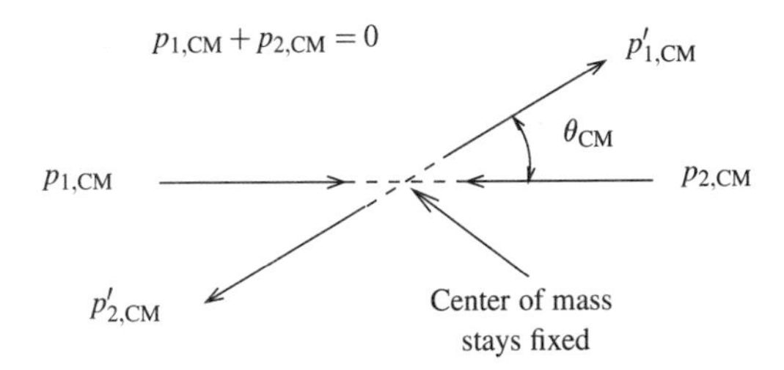

FIGURE 20.15. Center-of-mass frame of reference for scattering.

which also gives a relation involving the momenta, namely,

$$\mathbf{P}_{\mathrm{cm}} = M\mathbf{V}_{\mathrm{cm}} = m_1\mathbf{v}_1 + m_2\mathbf{v}_2 = \mathbf{p}_1 + \mathbf{p}_2 \tag{20.136}$$

Let us first analyze scattering in a frame of reference in which $\mathbf{R}_{\mathrm{cm}}$ is fixed so that the total momentum vanishes as in Fig. 20.15. In this frame, the particles approach each other with equal and opposite momenta (but not necessarily velocities). This is called the *center-of-mass frame,* hereafter abbreviated as CM, and is the one in which our scattering formalism (Born approximation, partial-wave analysis, etc.) applies directly. We will carefully distinguish between the center-of-mass variables (cm) defined above and the center-of-mass frame of reference (CM).

Using conservation of energy for an elastic collision and conservation of momentum in the special case where $\mathbf{P}_{\mathrm{cm}} = 0$, one can show that that the *magnitude* of $\mathbf{p}_{1,\mathrm{CM}}$ (and hence $\mathbf{p}_{2,\mathrm{CM}}$) cannot change, but only be scattered through an angle θ_{CM}, as shown. This also implies that the speed of each particle is unchanged in the collision, for example

$$|\mathbf{v}'_{1,\mathrm{CM}}| = |\mathbf{v}_{1,\mathrm{CM}}| \tag{20.137}$$

where we use the subscript CM to remind ourselves that this is generally only true in this special frame of reference. This effect can be easily visualized in one dimension where conservation of kinetic energy and momentum read

$$\frac{p_1^2}{2m_1} + \frac{p_2^2}{2m_2} = E_{\mathrm{CM}} \qquad \text{and} \qquad p_1 + p_2 = 0 \tag{20.138}$$

which corresponds to the intersection of a line and an ellipse, as shown in Fig. 20.16. The two solutions correspond to a change in direction of $p_{1,\mathrm{CM}}$ and $p_{2,\mathrm{CM}}$, but a constant magnitude for each.

Actual scattering experiments are more often realized by colliding beams of particles with stationary targets, as shown in Fig. 20.17. This arrangement is conventionally called the *laboratory frame of reference*. This notation is something of a misnomer (or at least an anachronism), as many modern experiments do utilize colliding beams of particles of

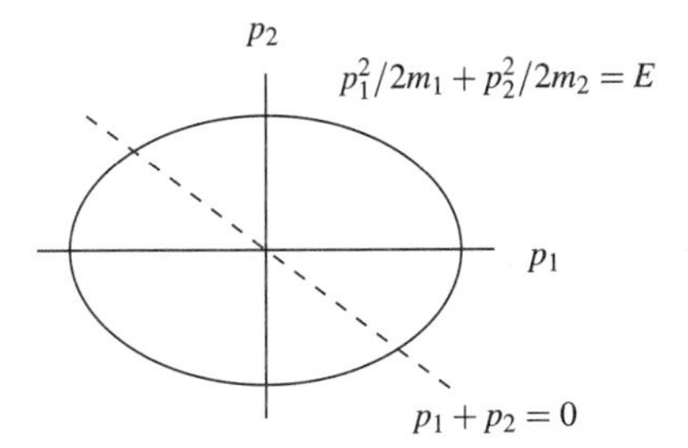

FIGURE 20.16. Conservation of energy and momentum for a two-body collision in one dimension.

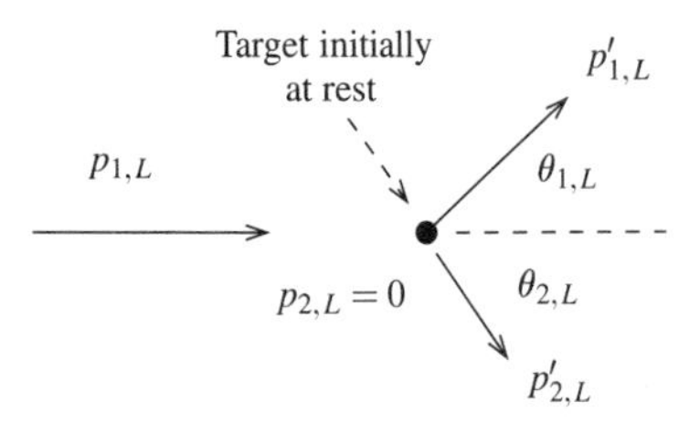

FIGURE 20.17. *Laboratory frame of reference with target at rest.*

equal and opposite momenta[7] where the center-of-mass system is realized directly in the laboratory; nevertheless, we will use the standard notation. In this arrangement, the total momentum of the system is nonvanishing and given by

$$\mathbf{P}_L = \mathbf{P}_{\text{cm}} = m_1\mathbf{v}_{1,L} + m_2\mathbf{v}_{2,L} = m_1\mathbf{v}_{1,L} \neq 0 \tag{20.139}$$

where we use the subscript L to indicate quantities measured in the lab frame of reference.

The velocities, angles, and energies measured by an observer in the lab frame will not correspond directly to those necessary to analyze the experiment in the CM frame, and our task is to be able to "translate" from one frame to the other. To "view" or analyze this system in the center of mass, we would have to "run alongside" the lab system (as in Fig. 20.18) at a velocity $\mathbf{v}_0$ such that $\mathbf{v}_{1,2}$ are seen to have their values changed so as to satisfy

$$\mathbf{P}_{\text{cm}} = m_1(\mathbf{v}_{1,L} - \mathbf{v}_0) + m_2(-\mathbf{v}_0) = 0 \tag{20.140}$$

which implies that

$$\mathbf{v}_0 = \left(\frac{m_1}{m_1 + m_2}\right)\mathbf{v}_{1,L} \tag{20.141}$$

The velocities and momenta of each particle before the collision in the CM frame are then given by

$$\mathbf{v}_{1,\text{CM}} = \mathbf{v}_{1,L} - \mathbf{v}_0 = \left(\frac{m_2}{m_1 + m_2}\right)\mathbf{v}_{1,L} \tag{20.142}$$

$$\mathbf{v}_{2,\text{CM}} = -\mathbf{v}_0 = -\left(\frac{m_1}{m_1 + m_2}\right)\mathbf{v}_{1,L} \tag{20.143}$$

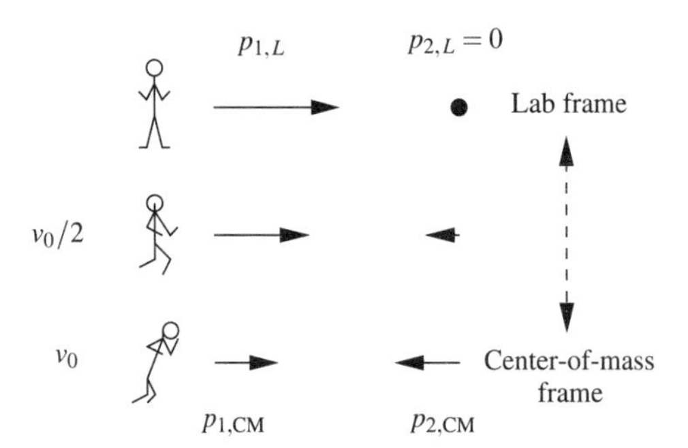

FIGURE 20.18. *Transformation from lab frame to center-of-mass frame.*

[7] This can occur, for example, at facilities that make use of the same arrangements of magnetic fields to accelerate both particles and their antiparticles in opposite directions simultaneously. Since the particle-antiparticle pair have the same mass but opposite charge, the acceleration process gives them equal but opposite momenta; examples include electron-positron and proton-antiproton colliders.

and

$$\mathbf{p}_{1,\mathrm{CM}} = m_1 \mathbf{v}_{1,\mathrm{CM}} = \mu \mathbf{v}_{1,L} = -\mathbf{p}_{2,\mathrm{CM}} \tag{20.144}$$

The kinetic energy available in the CM system for the collision can be written in the form

$$\begin{aligned} E_{\mathrm{CM}} &= \frac{1}{2} m_1 \left(\mathbf{v}_{1,\mathrm{CM}}\right)^2 + \frac{1}{2} m_2 \left(\mathbf{v}_{2,\mathrm{CM}}\right)^2 \\ &= \frac{1}{2}\left(\frac{m_1 m_2}{m_1 + m_2}\right)\left(\mathbf{v}_{1,L}\right)^2 \\ &= \left(\frac{\mu}{m_1}\right)\left[\frac{1}{2} m_1 \left(\mathbf{v}_{1,L}\right)^2\right] \\ &= \frac{\mu}{m_1} E_{1,L} \end{aligned} \tag{20.145}$$

This reminds us that the kinetic energy, among many other observables, is a frame-dependent quantity. We note the two limits:

$$E_{\mathrm{CM}} = \left(\frac{m_2}{m_1 + m_2}\right) E_{1,L} \approx \begin{cases} E_{1,L} & \text{for } m_2 >> m_1 \\ E_{1,L}/2 & \text{for } m_1 = m_2 \end{cases} \tag{20.146}$$

To relate the angles measured in the lab frame to those required for the analysis in the CM frame, we note that the velocity vectors of the scattered particle (1) in the two frames are related by

$$\mathbf{v}'_{1,\mathrm{CM}} = \mathbf{v}'_{1,L} - \mathbf{v}_0 \qquad \text{or} \qquad \mathbf{v}'_{1,\mathrm{CM}} + \mathbf{v}_0 = \mathbf{v}'_{1,L} \tag{20.147}$$

The corresponding relations among the components parallel and perpendicular to the initial direction can be written as

$$v_{1,\mathrm{CM}} \cos(\theta_{\mathrm{CM}}) + v_0 = v'_{1,L} \cos(\theta_L) \tag{20.148}$$

$$v_{1,\mathrm{CM}} \sin(\theta_{\mathrm{CM}}) = v'_{1,L} \sin(\theta_L) \tag{20.149}$$

where we have used Eqn. (20.137).

Taking the ratio of Eqns. (20.149) to (20.148), we find that

$$\tan(\theta_L) = \frac{\sin(\theta_{\mathrm{CM}})}{\cos(\theta_{\mathrm{CM}}) + \gamma} \tag{20.150}$$

where

$$\frac{v_0}{v_{1,\mathrm{CM}}} = \frac{m_1}{m_2} \equiv \gamma \tag{20.151}$$

This relation between the angles can also be usefully written in the forms

$$\cos(\theta_L) = \frac{\cos(\theta_{\mathrm{CM}}) + \gamma}{\sqrt{1 + \gamma^2 + 2\gamma \cos(\theta_{\mathrm{CM}})}} \tag{20.152}$$

$$\sin(\theta_L) = \frac{\sin(\theta_{\mathrm{CM}})}{\sqrt{1 + \gamma^2 + 2\gamma \cos(\theta_{\mathrm{CM}})}} \tag{20.153}$$

We illustrate the relation between the lab and CM angles in Fig. 20.19 for several values of γ. Several limits of this relation should be noted:

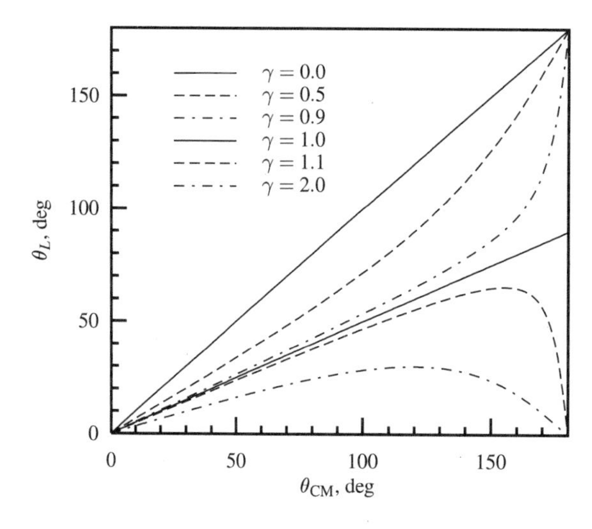

FIGURE 20.19. *Lab (θ_L) versus center-of-mass (θ_{CM}) angle for various values of γ.*

- When the target mass is much heavier than the projectile, we have $m_2 >> m_1$, which implies that $\gamma << 1$ and $\theta_L \approx \theta_{CM}$. This limit corresponds to potential scattering where the heavy target can act to conserve momentum.
- The special case of $m_2 = m_1$ implies that

$$\tan(\theta_L) = \frac{\sin(\theta_{CM})}{\cos(\theta_{CM}) + 1} = \tan\left(\frac{\theta_{CM}}{2}\right) \tag{20.154}$$

  so that $\theta_{CM} = 2\theta_L$; this is important for the case of scattering of identical particles.
- For any value of $m_1 > m_2$, the incident projectile can no longer scatter backwards in the lab frame; even for head-on collisions, the heavier incident particle now proceeds forward. Even more interestingly, there are now *two* center-of-mass angles (see Fig. 20.20) that correspond to a given lab angle; it is interesting to ponder how one could disentangle the two contributions.
- When the projectile is much heavier than the target, i.e., $m_2 << m_1$, Eqns. (20.150) and (20.152) imply that there is a maximum angle through which m_1 can be scattered in the lab frame, which goes as $\theta_L \rightarrow 1/\gamma$ for $\gamma >> 1$. There is not much scattering of a moving truck from a stationary pebble.

The *total scattering cross sections* in the two frames must be equal as σ measures the total probability of particles being scattered in any direction, which should be independent of the frame of reference. The differential cross sections, however, will be frame-dependent because they are defined using different angles. We can only say that the total number of particles scattered into solid angle $d\Omega_L$ at angle (θ_L, ϕ_L) (call it dN_L) must be the same as scattered into $d\Omega_{CM}$ at (θ_{CM}, ϕ_{CM}) (labeled dN_{CM}). We thus have

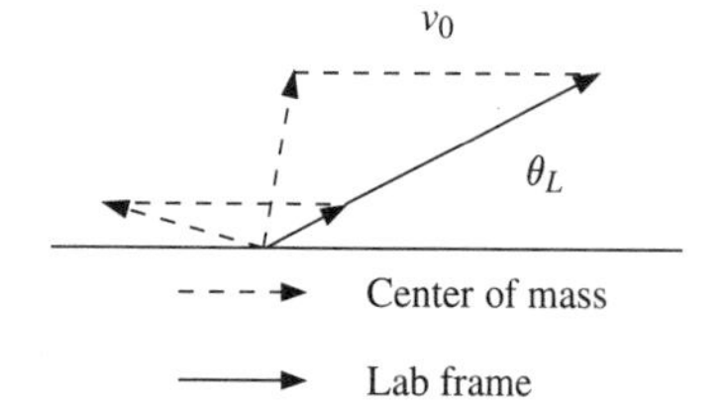

FIGURE 20.20. *Two velocity vectors (dashed arrows) with different θ_{CM} which give velocity vectors in the lab frame (solid arrows) corresponding to the same θ_L, only possible when $m_1 > m_2$.*

$$dN_L \propto \left(\frac{d\sigma}{d\Omega}\right)_L \sin(\theta_L)\, d\theta_L\, d\phi_L = \left(\frac{d\sigma}{d\Omega}\right)_{\text{CM}} \sin(\theta_{\text{CM}})\, d\theta_{\text{CM}}\, d\phi_{\text{CM}} \propto dN_{\text{CM}} \tag{20.155}$$

so that

$$\left(\frac{d\sigma}{d\Omega}\right)_L = \left(\frac{d\sigma}{d\Omega}\right)_{\text{CM}} \left[\frac{\sin(\theta_{\text{CM}})}{\sin(\theta_L)}\right]\left(\frac{d\theta_{\text{CM}}}{d\theta_L}\right) \tag{20.156}$$

The differential relation between $d\theta_L$ and $d\theta_{\text{CM}}$ can be derived by differentiating both sides of Eqn. (20.150), giving

$$\frac{1}{\cos^2(\theta_L)}\, d\theta_L = \frac{1+\gamma\cos(\theta_{\text{CM}})}{(\cos(\theta_{\text{CM}})+\gamma)^2}\, d\theta_{\text{CM}} \tag{20.157}$$

Then, using Eqn. (20.152), we find that

$$\left(\frac{d\sigma}{d\Omega}\right)_L = \frac{\left(1+\gamma^2+2\gamma\cos(\theta_{\text{CM}})\right)^{3/2}}{(1+\gamma\cos(\theta_{\text{CM}}))} \left(\frac{d\sigma}{d\Omega}\right)_{\text{CM}} \tag{20.158}$$

For the special case of equal-mass particles ($\gamma = 1$), we find the useful relation

$$\left(\frac{d\sigma}{d\Omega}\right)_{\text{CM}} = \frac{1}{4\cos(\theta_L)} \left(\frac{d\sigma}{d\Omega}\right)_L \tag{20.159}$$

20.6.2 Identical Particle Effects

We have seen that the bound state wavefunctions of systems of indistinguishable particles have to satisfy strong symmetry constraints imposed by the spin-statistics theorem. The cross sections describing the scattering of identical particles will have to satisfy similar constraints as they are derived from the large-distance behavior of similar unbound wavefunctions.

To see how indistinguishability effects can play a role in the measurement of a scattering cross section, we show in Fig. 20.21 two classical trajectories describing the scattering of identical mass particles in the center of mass. The two configurations for which the scattering angles are related via $(\theta, \phi) \rightarrow (\pi - \theta, \pi + \phi)$ give rise to the same "signal" in the detectors. In a classical description of scattering where we could, in principle, follow the individual trajectories, we would simply add the two contributing scattering probabilities,

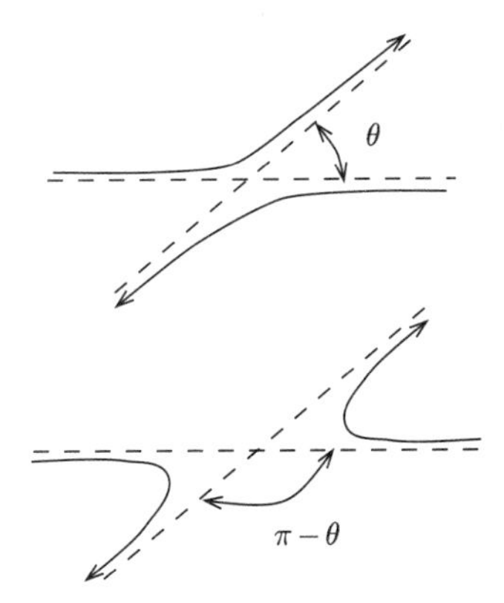

FIGURE 20.21. Indistinguishable scattering geometries.

i.e., cross sections, to obtain

$$\left(\frac{d\sigma}{d\Omega}\right)_{\text{CL}} = \frac{d\sigma(\theta)}{d\Omega} + \frac{d\sigma(\pi - \theta)}{d\Omega} \tag{20.160}$$

In the quantum mechanical case, just as with any wave phenomena, we must combine the scattering amplitudes first and then square to obtain the differential cross section. The relevant amplitudes, namely, $f(\theta)$ and $f(\pi - \theta)$, correspond to the position-space wavefunctions of the two-particle system, and information on the spin degrees of freedom must also be included so as to obtain a total wavefunction with the appropriate symmetry under exchange, just as with the bound state case. Systems of bosons (fermions) must have a total scattering amplitude that is symmetric (antisymmetric) under the interchange of the two particle labels.

The simplest case corresponds to spinless bosons with $J = 0$. Since there are no spin wavefunctions, the scattering amplitude is simply the symmetric combination that gives

$$\begin{aligned}\left(\frac{d\sigma}{d\Omega}\right)_{J=0} &= \left|f(\theta) + f(\pi - \theta)\right|^2 \\ &= \frac{d\sigma(\theta)}{d\Omega} + \frac{d\sigma(\pi - \theta)}{d\Omega} + 2Re[f(\theta)f^*(\pi - \theta)]\end{aligned} \tag{20.161}$$

This form already imposes interesting constraints on the scattering of identical spinless particles as it implies that only *even* partial wave amplitudes will contribute to such scattering; this arises since $\cos(\pi - \theta) = -\cos(\theta)$ and the Legendre polynomials satisfy $P_l(-y) = (-1)^l P_l(y)$.

For spin-1/2 particles, the spinor wavefunctions (Ex. 17.4) describe a total spin of $S = 0$ or 1, which are, respectively, antisymmetric and symmetric under exchange. The corresponding "spatial" wavefunctions must then be the combinations that are symmetric and antisymmetric under $\theta \to \pi - \theta$ in order for the total wavefunction to satisfy the spin statistics for fermions. In addition, we recall two other facts about the "spin counting": (1) There are a total of $2S + 1 = 3$ $S = 1$ states that can contribute and only a single $S = 0$ state, and (2) in an experiment involving unpolarized particles, all $3 + 1 = 4$ combinations of spins are equally likely, and we should average these possibilities. All of these effects taken together give the cross section

$$\begin{aligned}\left(\frac{d\sigma}{d\Omega}\right)_{J=1/2} &= \frac{3}{4}\left|f(\theta) - f(\pi - \theta)\right|^2 + \frac{1}{4}\left|f(\theta) + f(\pi - \theta)\right|^2 \\ &= \frac{d\sigma(\theta)}{d\Omega} + \frac{d\sigma(\pi - \theta)}{d\Omega} - Re[f(\theta)f^*(\pi - \theta)]\end{aligned} \tag{20.162}$$

The Clebsch-Gordan coefficients for the spin sum $1 + 1$ (P17.19) can be used to show that the spinor wavefunctions for the net spin $J = 2, 1, 0$ are symmetric, antisymmetric, and symmetric, respectively, so that performing the spin counting as above, we find for the scattering of identical $J = 1$ particles,

$$\begin{aligned}\left(\frac{d\sigma}{d\Omega}\right)_{J=1} &= \frac{5 + 1}{9}\left|f(\theta) + f(\pi - \theta)\right|^2 + \frac{3}{9}\left|f(\theta) - f(\pi - \theta)\right|^2 \\ &= \frac{d\sigma(\theta)}{d\Omega} + \frac{d\sigma(\pi - \theta)}{d\Omega} - \frac{2}{3}Re[f(\theta)f^*(\pi - \theta)]\end{aligned} \tag{20.163}$$

These examples can be further generalized (P20.11) to show that the cross section for the scattering of identical spin J particles (whether bosons or fermions) can be written in the form

$$\left(\frac{d\sigma}{d\Omega}\right)_J = \frac{d\sigma(\theta)}{d\Omega} + \frac{d\sigma(\pi-\theta)}{d\Omega} + 2\frac{(-1)^{2J}}{(2J+1)} Re[f(\theta)f^*(\pi-\theta)] \quad (20.164)$$

For scattering via the Coulomb potential, the form of the cross-terms can be calculated exactly using Eqn. (20.85), giving

$$\begin{aligned} Re[f_C(\theta)f_C^*(\pi-\theta)] &\propto \frac{1}{\sin^2(\theta/2)\cos^2(\theta/2)} Re\left[e^{-i\eta\ln(\sin^2(\theta/2))}e^{+i\eta\ln(\cos^2(\theta/2))}\right] \\ &= \frac{1}{\sin^2(\theta/2)\cos^2(\theta/2)} Re[e^{-i\eta\ln(\tan^2(\theta/2))}] \\ &= \frac{1}{\sin^2(\theta/2)\cos^2(\theta/2)} \cos(\eta\ln(\tan^2(\theta/2))) \end{aligned} \quad (20.165)$$

where

$$\eta = \frac{Z_1 Z_2 K e^2 \mu}{\hbar^2 k} = Z_1 Z_2 \frac{K e^2}{\hbar c}\sqrt{\frac{\mu c^2}{2E_{\text{cm}}}} \quad (20.166)$$

This form explicitly shows how the classical limit of Eqn. (20.160) is reached: If we formally allow $\hbar \to 0$, then $\eta \to \infty$, and the interference term oscillates so rapidly that any measurement that accepts particles in a finite range of θ (as any real detector must do) will effectively sample the cross-term over many cycles of its argument, and hence average to zero. We also see that the Born approximation for the scattering amplitude misses this essential physics, since in that case we have

$$Re[f_B(\theta)f_B^*(\pi-\theta)] \propto \frac{1}{\sin^2(\theta/2)\cos^2(\theta/2)} \quad (20.167)$$

corresponding to $\eta = 0$

Example 20.6 Coulomb Scattering of Identical Spin-Zero Particles We examine to what extent the effects of indistinguishability can be observed by discussing an experiment involving the scattering of spinless carbon nuclei; specifically, we consider C^{12} for which $Z = 6$ and $\mu = M/2 = 6$ amu. The cross section for electromagnetic scattering of such particles is given by Eqn. (20.161) with the the cross-term in Eqn. (20.165), giving the so-called *Mott cross section*

$$\frac{d\sigma}{d\Omega} = \left(\frac{Z^2 K e^2}{4E_{\text{CM}}}\right)^2 \left[\frac{1}{\sin^4(\theta/2)} + \frac{1}{\cos^4(\theta/2)} + \frac{2\cos(\eta\ln(\tan^2(\theta/2)))}{\sin^2(\theta/2)\cos^2(\theta)}\right] \quad (20.168)$$

where

$$\eta = 0.158\, Z^2 \sqrt{\frac{\mu(\text{amu})}{E_{\text{CM}}(\text{ MeV})}} \quad (20.169)$$

We show in Fig. 20.22 the experimental data for two different energies; the upper curves correspond to $E_{\text{CM}} = 3.0$ MeV, while the lower ones are for $E_{\text{CM}} = 12.5$ MeV. The solid

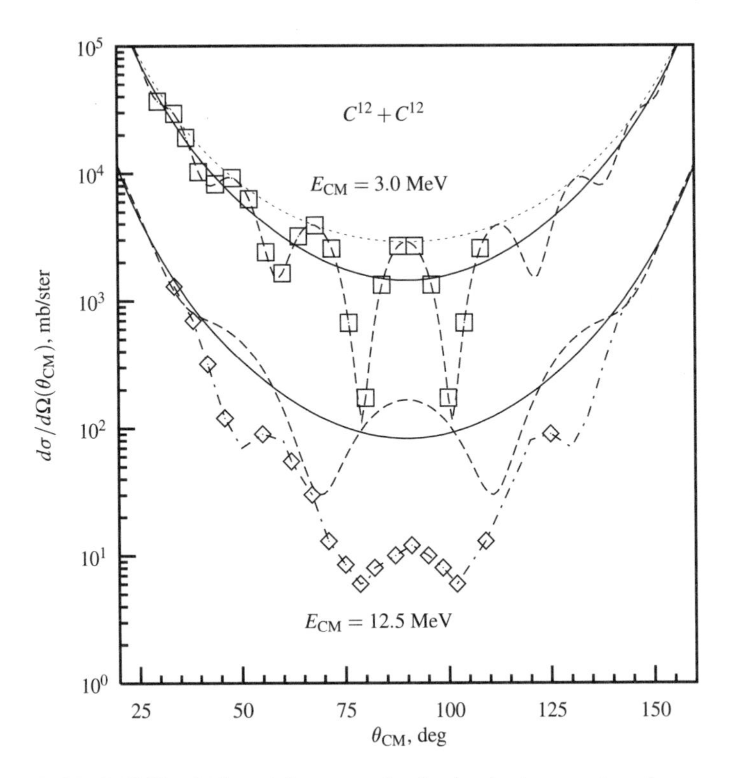

FIGURE 20.22. *Differential cross section for the elastic scattering of spinless $C^{12}+C^{12}$ nuclei versus center-of-mass angle θ_{CM} for two values of E_{CM}. The upper set of curves correspond to $E_{CM} = 3.0$ MeV, while the lower ones are for $E_{CM} = 12.5$ MeV. In each case, the solid lines are the Rutherford prediction, Eqn.(20.160), while the dashed line is the Mott prediction, Eqn.(20.168), using the exact Coulomb scattering amplitude. For the top set, the dotted line shows the Mott prediction, but using the approximate Born amplitudes, which miss the η dependence. For energies below the Coulomb barrier, pure electromagnetic scattering dominates, and the Mott prediction agrees very well with experiment. For higher energies, nuclear interactions also contribute, and the dot-dash line for the lower set shows a fit including both Coulomb and nuclear scattering.*

curves in both cases show the prediction for the "classical" result of Eqn. (20.160), while the dashed lines show the Mott prediction of Eqn. (20.168), which includes the interference effects. The dotted curve for the lower-energy data indicates the Mott result, but with $\eta = 0$ corresponding to the Born amplitude.

The 3.0 MeV data clearly agree with the Mott cross section, and gives clear evidence for the interference effects due to quantum mechanics *and* the exact Coulomb scattering amplitudes. The higher-energy data (fitted with the dot-dash curve) do not agree with purely electromagnetic scattering, as described by quantum mechanics, because of the need to include the strong nuclear force; we briefly discuss these effects below.

For two charged objects of finite radii R_1 and R_2, the electrostatic potential energy just when they "touch" is given

$$E_{\rm cm}^{\rm top} \approx \frac{Z_1 Z_2 K e^2}{R_{\rm sep}} \tag{20.170}$$

where $R_{\rm sep} \approx (R_1 + R_2)$ and $E_{\rm cm}^{\rm top}$ refers to the "top" of the Coulomb barrier. Using $R = (1.2\ {\rm F})A^{1/3}$ for each nucleus, this implies that $E_{\rm cm}^{\rm top} \approx 9.4$ MeV for this system. At the lower energy, the two nuclei are kept sufficiently far apart by their Coulomb repulsion that they never "feel" the finite range nuclear force. For higher energies, nuclear interactions come into play, and scattering amplitudes due to this force must be incorporated in addition to the purely electromagnetic scattering. We can even roughly estimate the number of partial waves that must be taken into account to fit the data by using energy conservation to write

$$E_{\text{cm}} = E_{\text{Coulomb}} + E_{\text{rot}} \tag{20.171}$$

$$12.5 \text{ MeV} = 9.4 \text{ MeV} + \frac{l(l+1)\hbar^2}{2\mu R_{\text{sep}}^2} \tag{20.172}$$

which implies that $l(l+1) \approx 31.6$ or $l_{\text{max}} \approx 5.1$. Since only *even* partial waves contribute in this case (why?), the experimental analysis actually required $l_{\text{max}} = 6$ to obtain agreement with the data.

20.7 QUESTIONS AND PROBLEMS

Q20.1 What would the picture corresponding to Fig. 18.7 for the angular momentum and Lenz-Runge vectors look like for the case of unbound, hyperbolic orbits found in the Coulomb scattering problem?

Q20.2 In Fig. 20.6(a) and (b), why do the points at the extreme values of angle (smallest and largest θ) have the largest error bars?

Q20.3 If we think of a total cross section as an effective area presented to scatterers, discuss why relativistic "boosts," i.e., changes in coordinate systems moving at constant speed relative to one another, should have no effect on σ_{tot}. *Hint:* How do length contractions work in relativity?

Q20.4 If the incident particles are heavier than the target particles (initially at rest) in a scattering experiment, scatterings from two distinct center-of-mass angles give contributions to the same lab angle. How do you tell one scatter from the other (or can you)?

Q20.5 Can you use the methods outlined in this chapter (e.g., the Born approximation or partial wave expansions) to discuss quantum scattering in two or one dimensions? [For references, see Eberly (1965) (1D) and Lapidus (1982a) or Adhikari (1986) (2D).]

P20.1 **(a)** Use the method outlined in Ex. 20.1 to find the hard-scattering cross section for the ellipsoidal shape discussed there. Specifically, show that

$$\tan\left(\frac{\theta}{2}\right) = \left|\frac{dy}{dz}\right| = \frac{\sqrt{R^2 - b^2}}{fb}$$

and thereby derive Eqn. (20.16).

(b) Show that this gives the cross section in Eqn. (20.17), and integrate it explicitly to see that $\sigma = \pi R^2$.

P20.2 Find the differential cross section for classical hard scattering from a paraboloid shape; in this case, the surface is given by the relation

$$z = a\left(\frac{x^2 + y^2}{R^2} - 1\right)$$

for $x^2 + y^2 \leq R^2$.

(a) Show that there is a minimum angle below which there is no scattering, given by

$$\tan\left(\frac{\theta_{\text{min}}}{2}\right) = \frac{R}{2a}$$

(b) Show that the cross section can be written in the form

$$\frac{d\sigma}{d\Omega} = \frac{1}{16}\frac{R^4}{a^2}\frac{1}{\sin^4(\theta/2)}$$

(c) Confirm that the total cross section is still the geometrical one given by $\sigma = \pi R^2$.
(d) Sketch $(d\sigma(\theta)/d\Omega)/\sigma$ versus θ for $a = 0.5R,\ R,\ 5R$, and discuss your results.
(e) Does the $1/\sin^4(\theta/2)$ dependence have anything to do with Rutherford scattering?

P20.3 Calculate the classical scattering cross section from an impenetrable sphere of radius a by using Eqn. (20.25), and show that your result agrees with the result in Ex. 20.1. The appropriate potential for this case is

$$V(r) = \begin{cases} +\infty & \text{for } r < a \\ 0 & \text{for } r > a \end{cases}$$

P20.4 The data in Table 20.2 were obtained from a computer "scattering experiment" of the type described in Ex. 20.1 and Sec. 20.2. They correspond to an ellipsoid of some unknown value of f, and tabulate the number of scattered particles detected in angular bins of "width" 10°.
(a) Use the "data" to evaluate $(d\sigma/d\Omega)/\sigma$ versus θ, including error bars.
(b) Use the theoretical formula in Eqn. (20.17), and evaluate χ^2/N_{eff} as a function of f; you will have to do this numerically.
(c) Estimate the value of f that gives the best fit, and the likely uncertainty in f using the criteria discussed in the text.

TABLE 20.2 *Number of Scattered Particles, N_{sc}, in 10° Bins from the "Specular" Scattering from an Ellipsoid with $f \neq 1$. (What value of f is most consistent with these "data"?)*

θ	N_{sc}	θ	N_{sc}	θ	N_{sc}
(0, 10)	30	(60, 70)	383	(120, 130)	588
(10, 20)	90	(70, 80)	462	(130, 140)	470
(20, 30)	144	(80, 90)	529	(140, 150)	454
(30, 40)	203	(90, 100)	528	(150, 160)	354
(40, 50)	295	(100, 110)	566	(160, 170)	213
(50, 60)	314	(110, 120)	567	(170, 180)	71

P20.5 **Two-dimensional scattering:** Use the methods and ideas outlined in Sec. 20.1 to discuss "specular" scattering in two dimensions. First define and then calculate the analogs of the differential and total cross sections for scattering from a "hard ellipse" of arbitrary parameter f; discuss your results.

P20.6 **Straight line fits:** Suppose you have data of the form $(x_i, y_i \pm \Delta_i)$ with $i = 1, 2, \ldots, N$ and where the x_i are known exactly and the corresponding experimental values have experimental errors as shown. It is desired to find the best straight-line fit through such data of the form $y^{th} = y(x) = ax + b$. Show how to do this by minimization of χ^2 by considering

$$\chi^2(a, b) = \sum_{i=1}^{N} \frac{(ax_i + b - y_i^{\exp})^2}{\Delta_i^2}$$

and finding the values of a, b that minimize this expression. Show that your answers can be written in the form

$$a = \left(\frac{[1][xy] - [x][y]}{[x^2][1] - [x]^2}\right) \qquad b = \left(\frac{[x^2][y] - [x][xy]}{[x^2][1] - [x]^2}\right)$$

where the "average" value [ ] is defined via

$$[f(x)] = \sum_{i=1}^{N} \frac{f(x_i)}{\Delta_i^2}$$

so that, for example,

$$[1] = \sum_{i=1}^{N} \frac{1}{\Delta_i^2} \qquad [x] = \sum_{i=1}^{N} \frac{x_i}{\Delta_i^2}$$

and so forth.

P20.7 The condition for the first diffraction minimum for a circular aperture is quoted in many introductory physics books and is given by $\sin(\theta) \approx 1.22\lambda/D$. Compare this to the corresponding angle for scattering from a finite well, discussed in Ex. 20.3.

P20.8 Using the Born approximation, calculate the differential and total cross sections for the two potentials

$$V_1(r) = V_0\left(\frac{a}{r}\right)e^{-r/a} \qquad V_2(r) = V_0 e^{-r^2/a^2}$$

Show that your answers can be written in the form

$$\sigma_1 = 4\pi\left(\frac{2mV_0a^3}{\hbar^2}\right)^2 \frac{1}{(1+4a^2k^2)}$$

$$\sigma_2 = 2\left(\frac{mV_0\pi a^3}{2\hbar^2}\right)^2 \frac{(1-e^{-2a^2k^2})}{(ka)^2}$$

Discuss whether the general conclusions about scattering in the low- and high-energy limits discussed in Ex. 20.3 are satisfied. Do the differential cross sections show any interesting structure (dips or zeroes) as a function of θ because of diffraction effects? Why or why not?

P20.9 The form factor of the proton obtained from electron scattering is fit rather well over the region $(\hbar q)^2 \approx 0 - 2\ \text{GeV}^2$ with a *dipole formula*, namely,

$$F(q^2) = \frac{1}{(1+(\hbar q/Mc)^2)^2}$$

where $Mc^2 \approx 0.84\ \text{GeV} = 840\ \text{MeV}$.

(a) Show that the charge radius of the proton is roughly $\sqrt{\langle r^2 \rangle} \approx 0.88$ F.

(b) To what distribution of charge, $\rho(\mathbf{r})$, does this correspond?

Note: The form factors for the corresponding magnetic scattering for both proton and neutron have similar forms,[8] implying that the charge and "magnetism" for both particles are distributed in the same way.

P20.10 **Partial waves for the finite well:** Consider particles of mass m scattered from a three-dimensional finite well given by

$$V(r) = \begin{cases} -V_0 & \text{for } r < a \\ 0 & \text{for } r > a \end{cases}$$

and shown in Fig. 20.23. For low-enough energies, we need only consider s-wave scattering, so, using the results of Sec. 17.6, write a solution of the form

[8] See, e.g., Perkins (1987).

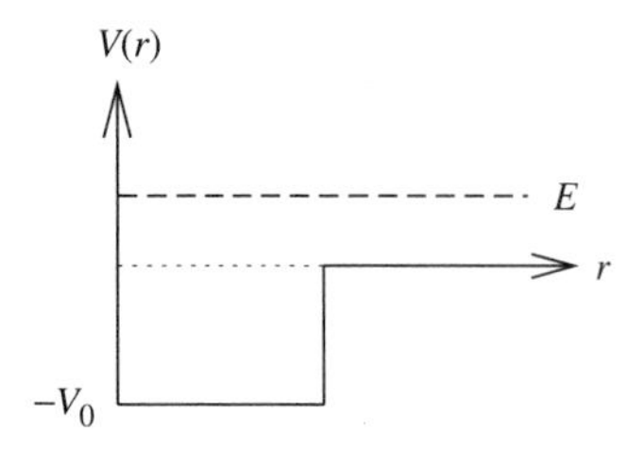

FIGURE 20.23. *Finite square well for scattering problems.*

$$u(r) = rR(r) = \begin{cases} A\sin(\kappa r) & \text{for } r < a \\ B\sin(kr) + C\cos(kr) & \text{for } r > a \end{cases}$$

where $\hbar^2 k^2 = 2mE$ and $\hbar^2\kappa^2 = 2m(E + V_0)$.

(a) Impose the appropriate boundary conditions at $r = a$, find the ratio C/B, and show that that the $l = 0$ phase shift is given by

$$\delta_0 \approx (ka)\left(\frac{\tan(\kappa a)}{\kappa a} - 1\right)$$

when $ka << 1$.

(b) Find the total cross section, and show that there are energies for which this approximation to σ becomes large, in fact, formally infinite in this approximation. Return to the derivation and show that at these energies a better approximation is $\sin(\delta_0) \approx 1$ so that

$$\sigma = \frac{4\pi}{k^2}$$

Such energies correspond to *resonant states*.

(c) Show that there are energies for which the cross section (in this approximation) vanishes. Show that these energies correspond to the *transmission* resonances considered in Sec. 12.4.1.

(d) Consider the case of a repulsive well with $-V_0 \longrightarrow +V_0$. Show that the expression above for the total cross section can be written as

$$\sigma = 4\pi a^2\left(\frac{\tanh(\kappa' a)}{\kappa' a} - 1\right)^2$$

and find an expression for κ'. Show that when $V_0 \to +\infty$, one obtains the $l = 0$ partial wave result for the total cross section.

P20.11 If we scatter 1 gram (1g) BBs with a speed of 1 cm/sec from a bowling ball, inside what angles is the "other half" of the quantum cross section, i.e., the forward diffractive peak, contained?

P20.12 Justify the formula in Eqn. (20.164) for the cross section for scattering of indistinguishable spin-J particles.

P20.13 The data[9] in Table 20.3 for the differential cross section for $\alpha - \alpha$ scattering were obtained in the lab frame of reference with $E_L = 0.6$ MeV corresponding to [via Eqn. (20.146)] $E_{\text{cm}} = 0.3$ MeV. The lab scattering angles, θ_L, have been transformed into θ_{cm} using Eqn. (20.154). All cross-section values are in mb/ster (where 1 mb $= 10^{-3}\,b$) and have a statistical error of roughly 1%. The α particles have $J = 0$.

[9]Taken from Heydenburg and Temmer (1956).

TABLE 20.3 Differential Cross Sections (in mb/ster) in the Lab Frame versus Lab Angle for $\alpha-\alpha$ Scattering at $E_L = 0.6$ MeV

		Expt'l	Expt'l	Theory	Theory	Theory
θ_L	θ_{cm}	$(d\sigma/d\Omega)_L$	$(d\sigma/d\Omega)_{cm}$	$(d\sigma/d\Omega)_{J=0}$	$(d\sigma/d\Omega)_{J=1/2}$	$(d\sigma/d\Omega)_{CL}$
15	30	187.4				
20	40	47.6				
25	50	18.2				
30	60	12.2				
35	70	11.4				
40	80	11.4				
45	90	10.8				
50	100	9.35				
55	110	7.90				
60	120	7.01				
65	130	8.26				
70	140	16.6				
75	150	48.6				

(a) Use the transformation from lab to center-of-mass differential cross sections to fill in the fourth column.

(b) Use the Mott scattering formula in Eqn. (20.168) to fill in the fifth column, and compare your results to the experimental data; use $Z = 2$ and $\mu = M_\alpha/2 = 2$ amu.

(c) Also calculate the classical and $J = 1/2$ cross sections in Eqns. (20.160) and (20.162), and complete the last two columns for comparison.

(d) Repeat the Coulomb barrier argument in Ex. 20.6 to see if you would expect purely electromagnetic scattering to dominate for these data.

P20.14 Proton-proton scattering at sufficiently large energies or scattering angles can get contributions not only from the electrostatic potential but also from the strong nuclear force. For moderate energies, we need only consider the $l = 0$ partial wave in a phase-shift analysis so that the scattering amplitude is given by

$$f(\theta) = f_C(\theta) + \frac{1}{k} e^{i\delta_0} \sin(\delta_0)$$

where $f_C(\theta)$ is given by Eqn. (20.85). The resulting cross section is then given by Eqn. (20.162) using this amplitude.

(a) Show that the cross section in this approximation is given by

$$\frac{d\sigma}{d\Omega} = \left(\frac{Z^2 K e^2}{4E_{CM}}\right)^2 \left[F_C(\theta) + F_{INT}(\theta, \delta_0) + F_S(\theta, \delta_0)\right]$$

where

$$F_C(\theta) = \frac{1}{\sin^4(\theta/2)} + \frac{1}{\cos^4(\theta/2)} - \frac{\cos(\eta \ln(\tan^2(\theta/2)))}{\sin^2(\theta/2)\cos(\theta^2)}$$

$$F_{INT}(\theta, \delta_0) = -\frac{2}{\eta} \sin(\delta_0) \left[\frac{\cos(\delta_0 + \eta \log(\sin^2(\theta/2)))}{\sin^2(\theta/2)} + \frac{\cos(\delta_0 + \eta \log(\cos^2(\theta/2)))}{\cos^2(\theta/2)}\right]$$

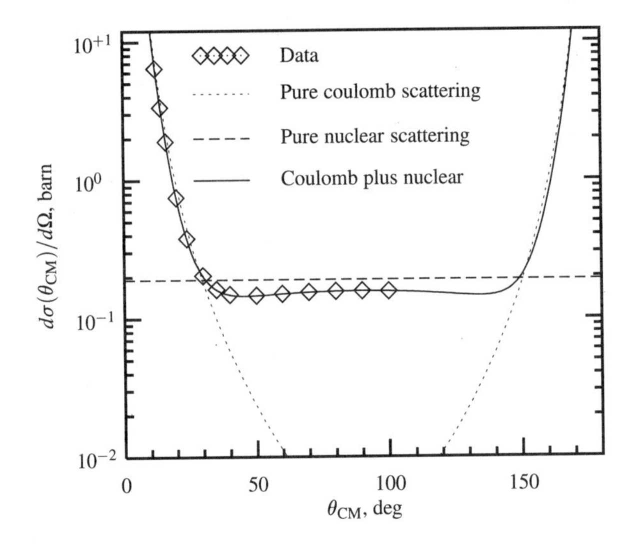

FIGURE 20.24. *Differential cross section versus center of mass angle (θ_{CM}) for proton-proton scattering at $E_{CM} = 1.21$ MeV.*

and

$$F_S(\theta, \delta_0) = \frac{4}{\eta^2} \sin^2(\delta_0)$$

where F_{INT} gives the interference between the Coulomb and nuclear amplitudes. (In this case, it turns out that one can ignore the effect of the β phase term in the Coulomb amplitude in Eqn. (20.85).)

(b) Data[10] on pp scattering at $E_{CM} = 1.21$ MeV is shown in Fig. 20.24, which shows the effect. Use the data to *estimate* the magnitude and sign of δ_0. Based on your answer, is the strong nuclear proton-proton attractive or repulsive?

[10]Taken from Knecht, Dahl, and Messelt (1966).

Complex Numbers and Functions

Because the formalism of quantum mechanics requires the manipulation of complex variables, we review here some of the basic definitions and formulae governing their properties. The imaginary unit, i, is defined[1] via $i = \sqrt{-1}$, and a general *complex number* is given by

$$z = a + ib \tag{A.1}$$

where a, b themselves have purely real values; the values a and b are called the *real* and *imaginary* parts of z, respectively; these are often written in the form

$$a = Re(z) \qquad b = Im(z) \tag{A.2}$$

Complex numbers obey standard algebraic relations. For example, if $z_{1,2} = a_{1,2} + ib_{1,2}$, we have addition and subtraction given by

$$z_1 \pm z_2 = (a_1 \pm a_2) + i(b_1 \pm b_2) \tag{A.3}$$

while multiplication is defined via

$$z_1 z_2 = (a_1 a_2 - b_1 b_2) + i(a_1 b_2 + a_2 b_1) \tag{A.4}$$

More complicated functions can often be evaluated using series expansions. For example, for θ real we can write

$$e^{i\theta} = 1 + (i\theta) + \frac{(i\theta)^2}{2!} + \cdots = \cos(\theta) + i\sin(\theta) \tag{A.5}$$

The *complex conjugate* of a complex number is defined via

$$z^* \equiv a - ib \tag{A.6}$$

that is, by letting $i \to -i$. A useful relation is

$$|z|^2 \equiv zz^* = (a + ib)(a - ib) = a^2 + b^2 \tag{A.7}$$

[1] Engineers often use the notation $j = \sqrt{-1}$.

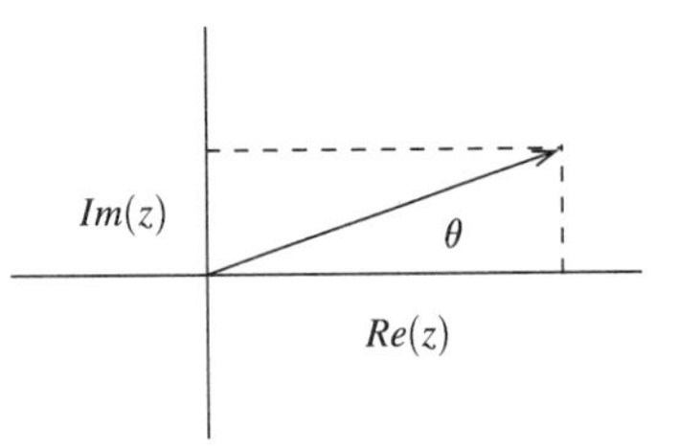

FIGURE A.1 *Representation of a complex number, z, as a vector in the complex plane.*

which defines the *modulus* of a complex number via

$$|z| = \sqrt{a^2 + b^2} \tag{A.8}$$

This quantity is the analog of the absolute value of a real number. We can make use of the identity (A.5) to also write

$$a + ib = z \equiv |z|e^{i\theta} = |z|\cos(\theta) + i|z|\sin(\theta) \tag{A.9}$$

where θ is called the *phase* or *argument* of the complex number z; it is given by

$$\tan(\theta) = \frac{b}{a} = \frac{Im(z)}{Re(z)} \tag{A.10}$$

This form for complex numbers is useful as it shows that

$$|z_1 z_2| = |z_1||z_2| \tag{A.11}$$

Obviously $z^* = |z|e^{-i\theta}$, so complex conjugation "flips the phase" of z but keeps its modulus fixed. A complex number with $|z| = 1$, i.e., of the form $z = e^{i\theta}$, is often said to be "just a phase." A general complex number can be represented as a point (or vector) in the complex plane, as shown in Fig. A.1, and addition and subtraction can be given a vector interpretation.

Some useful formulae (for θ real) are

$$\cos(\theta) = \frac{e^{i\theta} + e^{-i\theta}}{2} \qquad \sin(\theta) = \frac{e^{i\theta} - e^{-i\theta}}{2i} \tag{A.12}$$

which are easily derived by combining Eqn. (A.5) and its complex conjugate. One also has

$$\cos(i\theta) = \cosh(\theta) \qquad \sin(i\theta) = i\sinh(\theta) \tag{A.13}$$

Other familiar trig identities are easily proved using complex notation. For example, one has

$$\begin{aligned} 2\sin\left(\frac{\alpha+\beta}{2}\right)\cos\left(\frac{\alpha-\beta}{2}\right) &= 2\left(\frac{e^{i(\alpha+\beta)/2} - e^{-i(\alpha+\beta)/2}}{2i}\right)\left(\frac{e^{i(\alpha-\beta)/2} + e^{i(\alpha-\beta)/2}}{2}\right) \\ &= \sin(\alpha) + \sin(\beta) \end{aligned} \tag{A.14}$$

One also has

$$\sin(\alpha \pm \beta) = \sin(\alpha)\cos(\beta) \pm \cos(\alpha)\sin(\beta) \tag{A.15}$$

$$\cos(\alpha \pm \beta) = \cos(\alpha)\cos(\beta) \mp \sin(\alpha)\sin(\beta) \tag{A.16}$$

and the related special cases

$$\sin(2\alpha) = 2\sin(\alpha)\cos(\alpha) \tag{A.17}$$

$$\cos(2\alpha) = \cos^2(\alpha) - \sin^2(\alpha) \tag{A.18}$$

or

$$\sin^2(\alpha) = \frac{1}{2}(1 - \cos(2\alpha)) \tag{A.19}$$

$$\cos^2(\alpha) = \frac{1}{2}(1 + \cos(2\alpha)) \tag{A.20}$$

A.1 PROBLEMS

PA.1 Calculate the result of dividing two complex numbers; specifically, if

$$Re(w) + iIm(w) = w = \frac{z_1}{z_2} = \frac{a_1 + ib_1}{a_2 + ib_2}$$

find explicit expressions for $Re(w), Im(w)$.

PA.2 Use the series expansions in App. B.2 to prove Eqn. (A.5).

PA.3 Find a general expression for the modulus of $|w|$ if

$$w = \exp\left(\frac{1}{a + ib}\right)$$

where a, b are real; find a numerical value if $a = 2$ and $b = 1$.

PA.4 Verify the identity

$$|e^{i\alpha} + e^{i\beta}| = \left|2\cos\left(\frac{\alpha + \beta}{2}\right)\right|$$

if α, β are real. Derive a similar identity for $|e^{i\alpha} - e^{i\beta}|$

PA.5 If $z = 2 - 3i$, find $\sqrt{z}$.

APPENDIX B

Integrals, Summations, and Series Expansion

B.1 INTEGRALS

In this section, we collect many of the nontrivial indefinite and definite integrals that may be needed for some of the exercises in the text. Some of them are evaluated using sophisticated methods (such as contour integration), but we are only interested in using this collection as a reference. The reader is urged to consult other mathematical handbooks or especially to make use of symbolic manipulation programs such as *Mathematica*®.[1]

We begin by recalling that the simple rule for the differentiation of product functions

$$\frac{d}{dx}\big(f(x)g(x)\big) = \frac{df(x)}{dx}g(x) + f(x)\frac{dg(x)}{dx} \tag{B.1}$$

is the basis for the *integration by parts* (or IBP) method that we use frequently, namely,

$$\int_a^b dx\frac{df(x)}{dx}g(x) = -\int_a^b dx\, f(x)\frac{dg(x)}{dx} + \big(f(x)g(x)\big)_a^b \tag{B.2}$$

Some standard indefinite integrals follow:

$$\int \frac{dx}{x^2 + a^2} = \frac{1}{a}\tan^{-1}\left(\frac{x}{a}\right) \tag{B.3}$$

$$\int \frac{dx}{a^2 - x^2} = \frac{1}{2a}\log\left(\frac{a+x}{a-x}\right) \tag{B.4}$$

$$\int (\sin(ax))(\sin(bx))\,dx = \frac{\sin(a-b)x}{2(a-b)} - \frac{\sin(a+b)x}{2(a+b)} \qquad (a^2 \neq b^2) \tag{B.5}$$

[1] See Wolfram (1991).

$$\int (\cos(ax))(\cos(bx))\,dx = \frac{\sin(a-b)x}{2(a-b)} + \frac{\sin(a+b)x}{2(a+b)} \qquad (a^2 \neq b^2) \tag{B.6}$$

$$\int (\sin(ax))(\cos(bx))\,dx = -\frac{\cos(a-b)x}{2(a-b)} - \frac{\cos(a+b)x}{2(a+b)} \qquad (a^2 \neq b^2) \tag{B.7}$$

$$\int dx\, x \sin(ax) = \frac{1}{a^2}\sin(ax) - \frac{x}{a}\cos(ax) \tag{B.8}$$

$$\int dx\, x \cos(ax) = \frac{1}{a^2}\cos(ax) + \frac{x}{a}\cos(ax) \tag{B.9}$$

$$\int dx\, x^2 \sin(ax) = \frac{2x}{a^2}\sin(ax) - \frac{a^2x^2-2}{a^3}\cos(ax) \tag{B.10}$$

$$\int dx\, x^2 \cos(ax) = \frac{2x}{a^2}\cos(ax) + \frac{a^2x^2-2}{a^3}\sin(ax) \tag{B.11}$$

$$\int dx\, x^4 \sin(ax) = \frac{4x(a^2x^2-6)}{a^4}\sin(ax) - \frac{(a^4x^4-12a^2x^2+24)}{a^5}\cos(ax) \tag{B.12}$$

$$\int dx\, x^4 \cos(ax) = \frac{4x(a^2x^2-6)}{a^4}\cos(ax) + \frac{(a^4x^4-12a^2x^2+24)}{a^5}\sin(ax) \tag{B.13}$$

$$\int e^{ax}\,dx = \frac{1}{a}e^{ax} \tag{B.14}$$

$$\int xe^{ax}\,dx = \frac{1}{a^2}(ax-1)e^{ax} \tag{B.15}$$

$$\int x^2e^{ax}\,dx = \frac{1}{a^3}(a^2x^2-2ax+2)e^{ax} \tag{B.16}$$

Some definite integrals:

$$\int_{-\infty}^{+\infty} \frac{\sin(x)^2}{x^2}\,dx = \pi \tag{B.17}$$

$$\int_{-\infty}^{+\infty} \frac{\sin(x)^4}{x^2}\,dx = \frac{\pi}{2} \tag{B.18}$$

$$\int_{-\infty}^{+\infty} \frac{(1-\cos(x))}{x^2}\,dx = \pi \tag{B.19}$$

$$\int_{-\infty}^{+\infty} \frac{(1-\cos(x))^2}{x^2}\,dx = \pi \tag{B.20}$$

$$\int_{-\infty}^{+\infty} \frac{\sin(x)\cos(x)}{x}\,dx = \frac{\pi}{2} \tag{B.21}$$

$$\int_{-\infty}^{+\infty} \frac{\sin(x_1 - x)\sin(x_2 - x)}{(x - x_1)(x - x_2)} = \pi \frac{\sin(x_1 - x_2)}{(x_1 - x_2)} \tag{B.22}$$

$$\int_0^{\infty} \frac{\cos(mx)}{x^2 + a^2}\,dx = \frac{\pi}{2|a|} e^{-|ma|} \tag{B.23}$$

$$\int_0^{\infty} \frac{\cos(mx)\cos(nx)}{x^2 + a^2}\,dx = \frac{\pi}{a}\left(e^{-|(m-n)a|} + e^{-|(m+n)a|}\right) \tag{B.24}$$

$$\int_0^{\infty} \frac{\sin(mx)\sin(nx)}{x^2 + a^2}\,dx = \frac{\pi}{a}\left(e^{-|(m-n)a|} - e^{-|(m+n)a|}\right) \tag{B.25}$$

$$\int_0^{\infty} \cos(mx)e^{-ax}\,dx = \frac{a}{a^2 + m^2} \qquad (a > 0) \tag{B.26}$$

$$\int_0^{\infty} \sin(mx)e^{-ax}\,dx = \frac{m}{a^2 + m^2} \qquad (a > 0) \tag{B.27}$$

These integrals make use of the Euler Gamma function discussed in App. C.9.

$$\int_0^{\infty} x^n e^{-x}\,dx = n! = \Gamma(n+1) \tag{B.28}$$

$$\int_0^{\infty} dx\, x^n e^{-(ax)^m} = \frac{1}{ma^{n+1}}\Gamma\left(\frac{n+1}{m}\right) \tag{B.29}$$

$$\int_0^{1} \frac{x^m\,dx}{\sqrt{1 - x^n}} = \frac{\Gamma(1/2)\Gamma((m+1)/n)}{n\Gamma(1/2 + (m+1)/n)} \tag{B.30}$$

Integrals containing Gaussian terms of the form $\exp(-ax^2)$ are of special importance, and we discuss their evaluation in slightly more detail. The standard trick for the evaluation of the basic integral

$$I \equiv \int_{-\infty}^{+\infty} dx\, \exp(-x^2) \tag{B.31}$$

is to consider

$$\begin{aligned} I^2 = I \cdot I &= \left(\int_{-\infty}^{+\infty} dx\, \exp(-x^2)\right) \cdot \left(\int_{-\infty}^{+\infty} dy\, \exp(-y^2)\right) \\ &= \int_{-\infty}^{+\infty}\int_{-\infty}^{+\infty} dx\,dy\, \exp(-x^2 - y^2) \\ &= \int_0^{\infty}\int_0^{2\pi} r\,dr\,d\theta\, \exp(-r^2) \\ &= 2\pi \int_0^{\infty} dr\,r\, \exp(-r^2) \\ &= \pi \end{aligned} \tag{B.32}$$

so that $I = \sqrt{\pi}$.

The more general basic integral is

$$I(a) = \int_{-\infty}^{+\infty} dx \, \exp(-ax^2) = \sqrt{\frac{\pi}{a}} \tag{B.33}$$

and a related one is

$$\begin{aligned} I(a,b) &\equiv \int_{-\infty}^{+\infty} dx \, \exp(-ax^2 - bx) \\ &= \int_{-\infty}^{+\infty} dx \, \exp(-a(x^2 + bx/a + b^2/4a^2 - b^2/4a)) \\ &= \exp(b^2/4a) \int_{-\infty}^{+\infty} dx \, \exp(-a(x + b/a)^2) \\ &= \exp(b^2/4a) \sqrt{\frac{\pi}{a}} \end{aligned} \tag{B.34}$$

where we have used a standard method of completing the square and shifting variables. One can generalize these expressions further by noting that

$$\begin{aligned} J(a,b;n) &= \int_{-\infty}^{+\infty} dx \, x^n \exp(-ax^2 - bx) \\ &= \left(-\frac{\partial}{\partial b}\right)^n I(a,b) \\ &= \left(-\frac{\partial}{\partial b}\right)^n \left[\exp(b^2/4a)\sqrt{\frac{\pi}{a}}\right] \end{aligned} \tag{B.35}$$

Integrals containing ax^2+bx+c arise in the study of the classical limit of the hydrogen atom. If we define $X = ax^2 + bx + c$ and $q = 4ac - b^2$, one has

$$\int \frac{dx}{\sqrt{X}} = -\frac{1}{\sqrt{-a}} \sin^{-1}\left(\frac{2ax + b}{\sqrt{-q}}\right) \qquad (a < 0) \tag{B.36}$$

$$\int \frac{x\,dx}{\sqrt{X}} = \frac{\sqrt{X}}{a} - \frac{b}{2a} \int \frac{dx}{\sqrt{X}} \tag{B.37}$$

$$\int \frac{x^2\,dx}{\sqrt{X}} = \left(\frac{x}{2a} - \frac{3b}{4a^2}\right)\sqrt{X} + \frac{3b^2 - 4ac}{8a^2} \int \frac{dx}{\sqrt{X}} \tag{B.38}$$

$$\int \frac{x^3\,dx}{\sqrt{X}} = \left(\frac{x^2}{3a} - \frac{5bx}{12a^2} + \frac{5b^2}{8a^3} - \frac{2c}{3a^2}\right)\sqrt{X} + \left(\frac{3bc}{4a^2} - \frac{5b^3}{16a^3}\right) \int \frac{dx}{\sqrt{X}} \tag{B.39}$$

$$\int \frac{dx}{x\sqrt{X}} = \frac{1}{\sqrt{-c}} \sin^{-1}\left(\frac{bx + 2c}{|x|\sqrt{-q}}\right) \qquad (c < 0) \tag{B.40}$$

B.2 SUMMATIONS AND SERIES EXPANSIONS

We collect here some useful results that evaluate the summations of certain finite and infinite series.

$$\sum_{k=1}^{k=N} x^k = \frac{1 - x^{N+1}}{1 - x} \tag{B.41}$$

$$\sum_{k=1}^{k=N} k = \frac{N(N+1)}{2} \tag{B.42}$$

$$\sum_{k=1}^{k=N} k^2 = \frac{N(N+1)(2N+1)}{6} \tag{B.43}$$

The *Riemann zeta function* is defined via

$$\zeta(s) = 1 + \frac{1}{2^s} + \frac{1}{3^s} + \cdots = \sum_{n=1}^{\infty} \frac{1}{n^s} \tag{B.44}$$

Some special cases are

$$\zeta(2) = \frac{\pi^2}{6} \quad \zeta(4) = \frac{\pi^4}{90} \quad \zeta(6) = \frac{\pi^6}{945} \quad \zeta(8) = \frac{\pi^8}{9450} \tag{B.45}$$

One can also show that

$$\zeta_{\text{odd}}(s) \equiv \frac{1}{1} + \frac{1}{3^s} + \frac{1}{5^s} + \cdots = \sum_{n=1}^{\infty} \frac{1}{(2n-1)^s} = \left(1 - \frac{1}{2^s}\right)\zeta(s) \tag{B.46}$$

so that

$$\zeta_{\text{odd}}(2) = \frac{\pi^2}{8} \qquad \zeta_{\text{odd}}(4) = \frac{\pi^4}{96} \tag{B.47}$$

One also has:

$$S(x) = \sum_{k=1}^{\infty} \frac{1}{((2k-1)^2 - x^2)} = \frac{\pi}{2x} \tan\left(\frac{\pi x}{2}\right) \tag{B.48}$$

$$\begin{aligned} T(x) &= \sum_{k=1}^{\infty} \frac{1}{((2k-1)^2 - x^2)^2} \\ &= \frac{\pi}{16x^3}\left[\pi x \sec^2\left(\frac{\pi x}{2}\right) - 2\tan\left(\frac{\pi x}{2}\right)\right] \end{aligned} \tag{B.49}$$

The *Taylor series expansion* of a (well-behaved) function $f(x)$ about the point $x = a$ is given by

$$\begin{aligned} f(x) &= f(a) + f'(a)(x-a) + \frac{1}{2!}f''(a)(x-a)^2 + \cdots \\ &= \sum_{n=0}^{\infty} \frac{f^{(n)}(a)}{n!}(x-a)^n \end{aligned} \tag{B.50}$$

where

$$f^{(n)}(a) = \left. \frac{d^n f(x)}{dx^n} \right|_{x=a} \tag{B.51}$$

is the nth derivative of $f(x)$ evaluated at $x = a$. Familiar examples include:

$$\begin{aligned} (1 \pm x)^n &= 1 \pm nx + \frac{n(n-1)}{2!}x^2 \pm \frac{n(n-1)(n-2)}{3!}x^3 + \cdots \\ &= \sum_{k=1}^{\infty} (\pm 1)^k \frac{n!}{(n-k)!k!} x^k \qquad \text{for } |x| < 1 \end{aligned} \tag{B.52}$$

$$e^x = 1 + x + \frac{x^2}{2!} + \frac{x^3}{3!} + \cdots = \sum_{k=1}^{\infty} \frac{x^k}{k!} \qquad \text{for all real } x \tag{B.53}$$

$$\ln(1+x) = x - \frac{x}{2} + \frac{x^3}{3} + \cdots = \sum_{k=1}^{\infty} (-1)^k \frac{x^k}{k} \qquad \text{for } -1 < x \leq +1 \tag{B.54}$$

$$\sin(x) = x - \frac{x^3}{3!} + \frac{x^5}{5!} + \cdots = \sum_{k=1}^{\infty} (-1)^k \frac{x^{2k-1}}{(2k-1)!} \qquad \text{for all real } x \tag{B.55}$$

$$\cos(x) = 1 - \frac{x^2}{2!} + \frac{x^4}{4!} + \cdots = \sum_{k=0}^{\infty} (-1)^k \frac{x^{2k}}{(2k)!} \qquad \text{for all real } x \tag{B.56}$$

$$\tan(x) = x - \frac{x^3}{3!} + \frac{2x^5}{15} + \cdots \qquad \text{for } |x| < \pi/2 \tag{B.57}$$

A useful tool to investigate the convergence of a series expansion is the *ratio test.* If an infinite summation is defined via

$$S = \sum_{n=0}^{\infty} \rho_n \tag{B.58}$$

the limit of successive ratios is defined via

$$\rho = \lim_{n \to \infty} \frac{\rho_{n+1}}{\rho_n} \tag{B.59}$$

One then knows that the:

- Series **converges** (S is finite) if $\rho < 1$.
- Series **diverges** (S is infinite) if $\rho > 1$.
- Test is inconclusive (the series may either converge or diverge) if $\rho = 1$.

It is often useful to recall the definition of the (one-dimensional) integral as the "area under the curve." The trapezoidal approximation to the area under $f(x)$ in the interval (a, b) is obtained by splitting the interval into N equal pieces of size $h = (b-a)/N$ which gives

$$\int_a^b dx\, f(x) \approx F_N(a,b) \equiv h\left(\frac{1}{2}f(a) + \sum_{n=1}^{N-1} f(a+nh) + \frac{1}{2}f(b)\right) \tag{B.60}$$

This expression can form the basis for the simplest of numerical integration programs if necessary. The *Euler-Maclaurin* describes the difference between these two approximations via

$$\int_a^b dx\,\left[f(x) - F_N(a,b)\right] = -\frac{B_2}{2!}h^2 f'(x)\Big|_a^b - \frac{B_4}{4!}h^4 f'''(x)\Big|_a^b + \cdots \tag{B.61}$$

where the B_n are the Bernoulli numbers, the first few of which are

$$B_0 = 1 \qquad B_2 = \frac{1}{6} \qquad B_4 = -\frac{1}{30} \qquad B_6 = \frac{1}{42} \tag{B.62}$$

B.3 ASSORTED CALCULUS RESULTS

The gradient-squared operator or *Laplacian operator* in rectangular (Cartesian), cylindrical (polar), or spherical coordinates is given by

$$\nabla^2 f(x,y,z) = \frac{\partial^2 f}{\partial x^2} + \frac{\partial^2 f}{\partial y^2} + \frac{\partial^2 f}{\partial z^2} \tag{B.63}$$

$$\nabla^2 f(r,\theta,z) = \frac{1}{r}\frac{\partial}{\partial r}\left(r\frac{\partial f}{\partial r}\right) + \frac{1}{r^2}\frac{\partial^2 f}{\partial \theta^2} + \frac{\partial^2 f}{\partial z^2} \tag{B.64}$$

$$\nabla^2 f(r,\theta,\phi) = \frac{1}{r^2}\frac{\partial}{\partial r}\left(r^2\frac{\partial f}{\partial r}\right) + \frac{1}{r^2 \sin(\theta)}\frac{\partial}{\partial \theta}\left(\sin(\theta)\frac{\partial f}{\partial \theta}\right) + \frac{1}{r^2 \sin^2(\theta)}\frac{\partial^2 f}{\partial \phi^2} \tag{B.65}$$

If one changes variables in a multidimensional integral, one must also apply the appropriate transformation in the "infinitesimal measure." If one changes variables via

$$x, y \Longrightarrow u(x,y), v(x,y) \tag{B.66}$$

then one has the relation

$$du\,dv = J(w,v;x,y)\,dx\,dy = \det\begin{pmatrix} \partial u/\partial x & \partial u/\partial y \\ \partial v/\partial x & \partial v/\partial y \end{pmatrix} dx\,dy \tag{B.67}$$

with similar extensions to more dimensions; the function $J(u,v;x,y)$ is called the *Jacobian* of the transformation.

B.4 PLOTTING

The functional relationship between two variables is often best exemplified or analyzed (or even discovered in the first place!) by plotting the "data" in an appropriate manner. In this section, we briefly recall some of the basics of plotting techniques. Because linear relations are easiest to visualize, many standard tricks rely on graphing data in such a way as to yield a straight line.

For variables connected by an exponential relation, one has

$$\text{exponential:} \quad y = ae^{bx} \quad \Longrightarrow \quad \ln(y) = \ln(a) + bx$$

which suggests that one plot $\ln(y)$ versus x; this gives a so-called *semilog plot.* A straight-line fit on such a plot implies an exponential relation, and the "generalized slope" is given by

$$b = \frac{(\ln(y_2) - \ln(y_1))}{(x_2 - x_1)} = \frac{\ln(y_2/y_1)}{(x_2 - x_1)}$$

The value of a, or $\ln(a)$, plays the role of an "intercept" and can be extracted from any point on the line once b is known. If the point with $x = 0$ is included, then $y(0) = a$ is the most obvious choice.

For power law relations of the form

$$\text{power law:} \quad y = cx^d \Longrightarrow \ln(y) = \ln(c) + d\ln(x)$$

it is best to graph $\ln(y)$ versus $\ln(x)$ giving a *log-log plot* where the "generalized slope" is now

$$d = \frac{(\ln(y_2) - \ln(y_1))}{(\ln(x_2) - \ln(x_1))} = \frac{\ln(y_2/y_1)}{\ln(x_2/x_1)}$$

B.5 PROBLEMS

PB.1 Derive any of the integrals in Eqns. (B.5) to (B.7) by using complex exponentials.

PB.2 Derive any of the integrals in Eqns. (B.8) to (B.11) using integration-by-parts techniques.

PB.3 Derive the integral in Eqn. (B.8) by differentiating both sides of the relation

$$\int \cos(ax)\,dx = \frac{1}{a}\sin(ax)$$

with respect to a.

PB.4 Evaluate $J(a, b; n)$ in Eqn. (B.35) for $n = 1, 2$. Show that for even n, $J(a, b; n)$ can also be evaluated by differentiating with respect to a, and calculate it in this manner for $n = 2$.

PB.5 Derive Eqn. (B.46).

PB.6 At very low temperatures, the heat capacity (at constant volume) of metals is expected to be given by an expression of the form

$$C_V = \gamma T + AT^3$$

Given values for T and $C_V(T)$, what would be the best way to plot the data to confirm such a relation and to most easily extract γ and A?

Special Functions

In this section, we collect some of the well-known properties of many of the special functions considered in this text. Many are quoted without proof, while some have been discussed in a more physical context throughout the book.

C.1 TRIGONOMETRIC AND EXPONENTIAL FUNCTIONS

Although they are presumably familiar to all students, we briefly discuss the properties of the trigonometric and exponential functions. Since many of the special functions found in mathematical physics arise as solutions to similar differential equations, it is useful to recall here that:

- The differential equation

$$\frac{d^2 f(x)}{dx^2} = -k^2 f(x) \tag{C.1}$$

  has the (conventionally normalized) trig function solutions $f(x) = \sin(kx), \cos(kx)$ while
- the equation

$$\frac{d^2 f(x)}{dx^2} = +\kappa^2 f(x) \tag{C.2}$$

  has exponential solutions $f(x) = e^{\kappa x}, e^{-\kappa x}$.

The intuitive physical connections of these solutions with the oscillatory motion of a particle near a potential energy minimum, case (C.1), versus the "runaway" (or damped) motion of a particle moved away from an unstable potential maximum, case (C.2), can often be generalized to other differential equations to help understand the physical origin of the behavior of the solutions.

In each case, we obtain two, linearly independent solutions, $f_1(x)$, $f_2(x)$. The most general solution of the differential is then obtained by taking a linear combination $a_1 f_1(x) + a_2 f_2(x)$ and using the boundary conditions (in quantum mechanics) or initial conditions (in classical mechanics) to determine the arbitrary coefficients.

C.2 AIRY FUNCTIONS

The Airy differential equation is

$$\frac{d^2 f(x)}{dx^2} = x f(x) \tag{C.3}$$

This has been discussed at several points in the text (Sec. 5.6.2, P5.21, and Sec. 10.7) where various limiting cases have been derived. Here we note the following:

- This problem is related to the quantum version of a particle moving under the influence of a uniform force.
- Using Eqn. (C.2) as a model, for $x > 0$, we expect exponentially damped or growing solutions; these should be consistent with the tunneling wavefunctions of Sec. 9.2.2.
- For $x < 0$, we expect oscillatory solutions, with the period of oscillation *decreasing* for increasing $|x|$ as the "effective wavenumber" grows like $k^2 \sim |y|$; the physical connection to accelerated motion was discussed in Sec. 5.6.2.

The two linearly independent solutions are labeled $Ai(x)$ and $Bi(x)$ and are shown in Fig. C.1. If we introduce the natural variable, $\zeta = 2x^{3/2}/3$, these solutions can be expanded

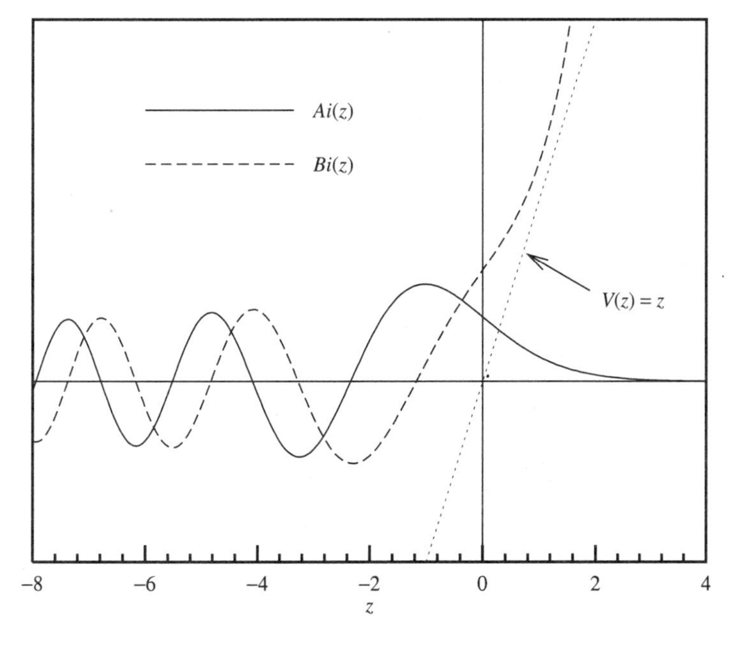

FIGURE C.1. Solutions of the Airy differential equation, Eqn. (C.3).

for large values of x as follows:

$$Ai(x) \longrightarrow \frac{1}{2}\frac{1}{\sqrt{\pi\sqrt{x}}}e^{-\zeta}\left[1 - \frac{c_1}{\zeta} + \cdots\right] \tag{C.4}$$

$$Ai(-x) \longrightarrow \frac{1}{\sqrt{\pi\sqrt{x}}}\left[\sin\left(\zeta + \frac{\pi}{4}\right) - \cos\left(\zeta + \frac{\pi}{4}\right)\frac{c_1}{\zeta} + \cdots\right] \tag{C.5}$$

$$Bi(x) \longrightarrow \frac{1}{\sqrt{\pi\sqrt{x}}}e^{\zeta}\left[1 + \frac{c_1}{\zeta} + \cdots\right] \tag{C.6}$$

$$Bi(-x) \longrightarrow \frac{1}{\sqrt{\pi\sqrt{x}}}\left[\cos\left(\zeta + \frac{\pi}{4}\right) + \sin\left(\zeta + \frac{\pi}{4}\right)\frac{c_1}{\zeta} + \cdots\right] \tag{C.7}$$

where $c_1 = 5/72$.

C.3 HERMITE POLYNOMIALS

The differential equation

$$\frac{d^2 h_n(z)}{dz^2} - 2z\frac{dh_n(z)}{dz} + 2nh_n(z) = 0 \tag{C.8}$$

is called *Hermite's equation* and has the solutions given by *Rodriges's formula*

$$h_n(z) = (-1)^n e^{z^2}\frac{d^n}{dz^n}\left(e^{-z^2}\right) \tag{C.9}$$

which are polynomials of degree n. The solutions are defined over the interval $(-\infty, +\infty)$ and satisfy the normalization condition

$$\int_{-\infty}^{+\infty}\left[h_n(z)\right]^2 e^{-z^2}\,dz = 2^n n!\sqrt{\pi} \tag{C.10}$$

C.4 CYLINDRICAL BESSEL FUNCTIONS

The free-particle Schrödinger and wave equation in two dimensions (three dimensions), when written in polar (cylindrical) coordinates, leads to the equation

$$\frac{d^2 R_m(z)}{dz^2} + \frac{1}{z}\frac{dR_m(z)}{dz} + \left(1 - \frac{m^2}{z^2}\right)R_m(z) = 0 \tag{C.11}$$

where we consider integral values of m. The solutions are generically called *cylindrical Bessel functions,* and for each $|m|$, the two linearly independent solutions are labeled $J_{|m|}(z)$ (Bessel functions of the first kind) or $Y_{|m|}(z)$ (Neumann or Bessel functions of the second kind). Their limiting behavior and properties are discussed and displayed graphically in Sec. 16.3.1.

C.5 SPHERICAL BESSEL FUNCTIONS

The free-particle Schrödinger equation in 3D written in spherical coordinates yields another version of Bessel's equation, namely,

$$\frac{d^2R_l(z)}{dz^2} + \frac{2}{z}\frac{dR_l(z)}{dz} + \left(1 - \frac{l(l+1)}{z^2}\right)R_l(z) = 0 \tag{C.12}$$

with l an integer. Its solutions are the *spherical Bessel functions,* $j_l(z)$ and $n_l(z)$, which can be written in a standard form

$$j_l(z) = \sqrt{\frac{\pi}{2z}}J_{l+1/2}(z) \qquad n_l(z) = \sqrt{\frac{\pi}{2z}}Y_{l+1/2}(z) \tag{C.13}$$

and are discussed in Sec. 17.6.

C.6 LEGENDRE POLYNOMIALS

Legendre's equation is written in the form

$$(1 - z^2)\frac{d^2\Theta_{l,m}(z)}{dz^2} - 2z\frac{d\Theta_{l,m}(z)}{dz} + \left(l(l+1) - \frac{m^2}{z^2}\right)\Theta_{l,m}(z) = 0 \tag{C.14}$$

The solutions are the *associated Legendre polynomials* given by

$$P_l^m(z) = (-1)^m\frac{(1-z^2)^{m/2}}{2^l\, l!}\left(\frac{d}{dz}\right)^{l+m}\left(z^2 - 1\right)^l \tag{C.15}$$

for $m > 0$ and extended to negative m via

$$P_l^{-m}(z) = (-1)^m\frac{(l-m)!}{(l+m)!}\,P_l^m(z) \tag{C.16}$$

They are defined over the interval $(-1, 1)$, and the normalization is such that

$$\begin{aligned}\int_{-1}^{+1} dz\, P_l^m(z)P_{l'}^m(z) &= \int_0^{\pi} \sin(\theta)\, d\theta\, P_l^m(\cos(\theta))P_{l'}^m(\cos(\theta)) \\ &= \frac{2}{2l+1}\frac{(l+m)!}{(l-m)!}\delta_{l,l'}\end{aligned} \tag{C.17}$$

The special case of $m = 0$ gives the *Legendre polynomials,* which are defined via

$$P_l(z) \equiv P_l^{m=0}(z) \tag{C.18}$$

C.7 GENERALIZED LAGUERRE POLYNOMIALS

The differential equation

$$\frac{d^2G(z)}{dz^2} + \left(\frac{\alpha - 1}{z} + 1\right)\frac{dG(z)}{dz} + nG(z) = 0 \tag{C.19}$$

is called *Laguerre's equation* and has polynomial solutions labeled as

$$G(z) = L_n^{(\alpha)}(z) \tag{C.20}$$

which can be generated using *Rodrigues's formula*

$$L_n^{(\alpha)}(z) = \frac{e^z}{n!\, z^\alpha}\left(\frac{d}{dz}\right)^n \left[z^{n+\alpha}e^{-z}\right] \tag{C.21}$$

The solutions are defined over the interval $(0, +\infty)$ and satisfy the normalization condition

$$\int_0^{+\infty} dz\, z^\alpha e^{-z}\left[L_n^{(\alpha)}(z)\right]^2 = \frac{\Gamma(n+\alpha+1)}{n!} \tag{C.22}$$

C.8 THE DIRAC δ FUNCTION

The Dirac δ function was introduced and discussed extensively in Sec. 2.4; here we only list some additional properties and identities. We recall that

$$\int_a^b dx\, f(x)\delta(x-c) = \begin{cases} f(c) & \text{if } a < c < b \\ 0 & \text{otherwise} \end{cases} \tag{C.23}$$

that is, the value of $f(x = c)$ is picked out from the integrand or not, depending on whether the singularity is contained in the region of integration or not. One can also derive (or justify) the following results:

$$\delta(ax) = \frac{1}{|a|}\delta(x) \tag{C.24}$$

$$\begin{aligned} \delta(x^2 - a^2) &= \delta[(x-a)(x+a)] \\ &= \frac{1}{|x+a|}\delta(x-a) + \frac{1}{|x-a|}\delta(x+a) \\ &= \frac{1}{2|a|}\big(\delta(x-a) + \delta(x+a)\big) \end{aligned} \tag{C.25}$$

which is a special case of the more general relation

$$\delta[f(x)] = \sum_i \frac{\delta(x - x_i)}{|df/dx|_{x=x_i}} \tag{C.26}$$

where the sum is over all possible roots of $f(x_i) = 0$. Finally, recall that the *step*- or *Heaviside-function* is defined via

$$\Theta(x-a) = \begin{cases} 0 & \text{for } x < a \\ 1 & \text{for } x > a \end{cases} \tag{C.27}$$

and is given by

$$\Theta'(x-a) = \delta(x-a) \tag{C.28}$$

One can show that $\delta(x)$ can be obtained by taking the limit of the family of functions

$$\delta_\epsilon(x) = \frac{1}{\epsilon\pi}\frac{\sin^2(\epsilon x)}{x^2} \tag{C.29}$$

as $\epsilon \to 0$.

C.9 THE EULER GAMMA FUNCTION

Using integration by parts techniques, it is easy to derive Eqn. (B.28), namely,

$$\int_0^\infty dx\, x^n e^{-x} = n(n-1)(n-2)\cdots 3\cdot 2\cdot 1 \equiv n! \tag{C.30}$$

where $n!$ is read as "n-factorial." The integral can be generalized to noninteger values of n, and the *Gamma function* is defined in this way via

$$\int_0^\infty dx\, x^{n-1}\, e^{-x} \equiv \Gamma(n) \qquad \text{for } n \neq 0, -1, -2, -3, \ldots \tag{C.31}$$

For positive integers, it reduces to the factorial function

$$\Gamma(n) = (n-1)! \qquad \text{for integral } n > 0 \tag{C.32}$$

and also satisfies

$$\Gamma(n+1) = n\Gamma(n) \tag{C.33}$$

$$\Gamma(n)\Gamma(1-n) = \frac{\pi}{\sin(n\pi)} \tag{C.34}$$

Other special values are

$$\Gamma\left(\frac{1}{2}\right) = 2\int_0^\infty e^{-t^2}\, dt = \sqrt{\pi} \tag{C.35}$$

which can be combined with Eqn. (C.34) (for nonnegative integral n) to give

$$\Gamma\left(n+\frac{1}{2}\right) = \frac{1\cdot 3\cdot 5\cdots(2n-1)}{2^n} = \frac{(2n-1)!!}{2^n}\sqrt{\pi} \tag{C.36}$$

which implicitly defines the double-factorial function.

Finally, we note that *Stirling's formula* can be used to estimate the value of the factorial function for large argument, namely,

$$\Gamma(n+1) = n! \sim \sqrt{2\pi n}\left(\frac{n}{e}\right)^n \left(1 + \frac{1}{12n} + \frac{1}{288n^2} + \cdots\right) \tag{C.37}$$

C.10 PROBLEMS

PC.1 Show that the solutions to the Airy differential equation can be written in terms of cylindrical Bessel functions [satisfying Eqn. (C.11) of fractional ($n = \pm 1/3$) order]. For example, a standard result is that

$$Ai(-x) = \frac{1}{3}\sqrt{x}\left[J_{1/3}(y) + J_{-1/3}(y)\right]$$

where $y = 2x^{3/2}/3$.

PC.2 Estimate the value of 20! using Stirling's formula, and compare to the exact value.

APPENDIX D

Vectors, Matrices, and Group Theory

D.1 VECTORS AND MATRICES

We collect here some of the most basic definitions and properties of real, finite-dimensional vectors and matrices. We intentionally ignore all of the subtleties regarding the precise definitions of vectors and tensors and supply only the "bare necessities." Many comments on the generalization of these ideas to complex vectors and infinite dimensional spaces are given in the text.

We will take vectors to be ordered N-tuples of numbers, e.g.,

$$\mathbf{x} = (x_1, x_2, \ldots, x_N) \qquad \mathbf{y} = (y_1, y_2, \ldots, y_N) \tag{D.1}$$

along with a *dot-* or *inner-product* of the form

$$\mathbf{x} \cdot \mathbf{y} = \sum_{i=1}^{N} x_i y_i = x_1 y_1 + x_2 y_2 + \cdots + x_N y_N \tag{D.2}$$

The *norm* of the vector is taken to be

$$|x| = \sqrt{\mathbf{x} \cdot \mathbf{x}} \tag{D.3}$$

A *matrix* will be defined to be a square $N \times N$ array of the form

$$\mathsf{M} = \begin{pmatrix} M_{11} & M_{12} & \cdots & M_{1N} \\ M_{21} & M_{22} & \cdots & M_{2N} \\ \vdots & \vdots & \ddots & \vdots \\ M_{N1} & M_{N2} & \cdots & M_{NN} \end{pmatrix} \tag{D.4}$$

The *unit matrix* is given by

$$\mathbf{1} = \begin{pmatrix} 1 & 0 & \cdots & 0 \\ 0 & 1 & \cdots & 0 \\ \vdots & \vdots & \ddots & 0 \\ 0 & 0 & \cdots & 1 \end{pmatrix} \tag{D.5}$$

Multiplication of a vector by a matrix *on the left* (as with operators) gives a vector, i.e.,

$$\mathbf{x}' = \mathsf{M} \cdot \mathbf{x} \tag{D.6}$$

In component form one can write

$$(x')_i = (\mathsf{M} \cdot \mathbf{x})_i = \sum_j M_{ij} x_j \tag{D.7}$$

or more explicitly

$$\begin{aligned} \mathsf{M} \cdot \mathbf{x} &= \begin{pmatrix} M_{11} & M_{12} & \cdots & M_{1N} \\ M_{21} & M_{22} & \cdots & M_{2N} \\ \vdots & \vdots & \ddots & \vdots \\ M_{N1} & M_{N2} & \cdots & M_{NN} \end{pmatrix} \cdot \begin{pmatrix} x_1 \\ x_2 \\ \vdots \\ x_N \end{pmatrix} \\ &= \begin{pmatrix} M_{11}x_1 + M_{12}x_2 + \cdots + M_{1N}x_N \\ M_{21}x_1 + M_{22}x_2 + \cdots + M_{2N}x_N \\ \vdots \\ M_{N1}x_1 + M_{N2}x_2 + \cdots + M_{NN}x_N \end{pmatrix} \end{aligned} \tag{D.8}$$

The product of two matrices is again a matrix, with the component definition

$$(\mathsf{M} \cdot \mathsf{N})_{ik} = \sum_{j=1}^{N} M_{ij} N_{jk} \tag{D.9}$$

or

the ikth element of $\mathsf{M} \cdot \mathsf{N}$
$\Updownarrow$
(the ith row of M) dotted into (the kth column of N)

The *transpose* of the matrix M, labeled M^T, is obtained by "reflecting" all of its elements along the diagonal, i.e.,

$$(\mathsf{M}^T)_{ij} = (\mathsf{M})_{ji} = M_{ji} \tag{D.10}$$

so that

$$\begin{pmatrix} a & b & c \\ d & e & f \\ g & h & i \end{pmatrix}^T = \begin{pmatrix} a & d & g \\ b & e & h \\ c & f & i \end{pmatrix} \tag{D.11}$$

The generalization of this to complex matrices is the *adjoint* or *Hermitian conjugate* defined via

$$\mathbf{M}^\dagger = (\mathbf{M}^T)^* = (\mathbf{M}^*)^T \tag{D.12}$$

which "flips" the matrix elements and takes their complex conjugate.

The equivalent of the expectation value of an operator in a quantum state is given by

$$\langle x|M|x\rangle \sim \mathbf{x}\cdot\mathbf{M}\cdot\mathbf{x} = \sum_{j=1}^{N}\sum_{k=1}^{N} x_j M_{jk} x_k \tag{D.13}$$

A matrix transformation of the form Eqn. (D.6) generally changes the norm of the vector since

$$\begin{aligned}
\mathbf{x}'\cdot\mathbf{x}' = \sum_i x_i' x_i' &= \sum_{i=1}^{N}\left(\sum_{j=1}^{N} M_{ij}x_j\right)\left(\sum_{k=1}^{N} M_{ik}x_k\right) \\
&= \sum_{j,k=1}^{N} x_j \left[\sum_i (M^T)_{jk} M_{ik}\right] x_k \qquad \text{(D.14)} \\
&= \sum_{j,k=1}^{N} x_j P_{jk} x_k \\
&\neq \sum_j x_j x_j = \mathbf{x}\cdot\mathbf{x} \qquad \text{(D.15)}
\end{aligned}$$

unless one has

$$\sum_i (M^T)_{ji} M_{ik} = P_{jk} = \delta_{j,k} \qquad \text{or} \qquad \mathbf{M}^T\cdot\mathbf{M} = \mathbf{P} = \mathbf{1} \tag{D.16}$$

Matrices satisfying Eqn. (D.16) are said to be *orthogonal.*

Finally, the *determinant* of a matrix is a number formed from the elements of the matrix via

$$\det(\mathbf{M}) = \sum_{i_1,i_2,\ldots,i_N=1}^{N} \epsilon_{(i_1,i_2,\ldots,i_N)} M_{1,i_1} M_{2,i_2} \cdots M_{N,i_N} \tag{D.17}$$

The *totally antisymmetric symbol,*[1] $\epsilon_{(i_1,i_2,\ldots,i_n)}$, is defined via

$$\epsilon_{(i_1,i_2,\ldots,i_N)} = \begin{cases} +1 & \text{if } (i_1, i_2, \ldots, i_N) \text{ is an} \\ & \text{even permutation of } (1, 2, \ldots, N) \\ -1 & \text{if it is an odd permutation} \\ 0 & \text{otherwise} \end{cases} \tag{D.18}$$

It vanishes if any two of its indices are the same and is antisymmetric under the interchange of any pair of indices. Each term in the determinant then consists of a product of one element from each row, with appropriate signs. For example,

$$\det(\mathbf{A}) = \det\begin{pmatrix} a_{11} & a_{12} \\ a_{21} & a_{22} \end{pmatrix} = a_{11}a_{22} - a_{12}a_{21} \tag{D.19}$$

[1]It is also called the *Levi-Civita symbol.*

and

$$\det(\mathbf{B}) = \det\begin{pmatrix} b_{11} & b_{12} & b_{13} \\ b_{21} & b_{22} & b_{23} \\ b_{31} & b_{32} & b_{33} \end{pmatrix}$$
$$= b_{11}b_{22}b_{33} + b_{12}b_{23}b_{31} + b_{13}b_{32}b_{21}$$
$$- b_{13}b_{31}b_{22} - b_{11}b_{23}b_{32} - b_{12}b_{21}b_{33} \quad \text{(D.20)}$$

One important property of determinants is that the interchange of any two rows (or columns) gives the same value, but with an additional factor of (-1); this follows from the definition in Eqn. (D.17) and the antisymmetry of the ϵ symbol.

Equations of the form

$$\mathbf{M} \cdot \mathbf{v}_\lambda = \lambda \mathbf{v}_\lambda \quad \text{(D.21)}$$

are called *eigenvalue problems,* and λ is the eigenvalue and $\mathbf{v}_\lambda$ the corresponding eigenvector. We can also write this as

$$\left(\mathbf{M} - \lambda \mathbf{1}\right) \cdot \mathbf{v} = 0 \quad \text{(D.22)}$$

or in matrix form as

$$\begin{pmatrix} M_{11} - \lambda & M_{12} & \cdots & M_{1N} \\ M_{21} & M_{22} - \lambda & \cdots & M_{2N} \\ \vdots & \vdots & \ddots & \vdots \\ M_{N1} & M_{N2} & \cdots & M_{NN} - \lambda \end{pmatrix} = 0 \quad \text{(D.23)}$$

This is equivalent to a set of N linear equations in N unknowns, and the condition for a solution to exist is that

$$\det\left(\mathbf{M} - \lambda \mathbf{1}\right) = 0 \quad \text{(D.24)}$$

and this condition determines the allowed eigenvalues λ. Real matrices for which

$$\mathbf{M}^T = \mathbf{M} \quad \text{(D.25)}$$

are called *symmetric,* while complex matrices for which

$$\mathbf{M}^\dagger = \mathbf{M} \quad \text{(D.26)}$$

are called *Hermitian,* and both have the properties:

- The eigenvalues of $\mathbf{M}$ are real.
- The eigenvectors of $\mathbf{M}$ corresponding to different eigenvalues are orthogonal.

Example D1 Eigenvalues and eigenvectors of a simple matrix

The eigenvalues of the matrix

$$\mathbf{M} = \begin{pmatrix} 23 & -36 \\ -36 & 2 \end{pmatrix} \quad \text{(D.27)}$$

are determined by the condition

$$\det\begin{pmatrix} 23 - \lambda & -36 \\ -36 & 2 - \lambda \end{pmatrix} = \lambda^2 - 25\lambda - 1250 = 0 \quad \text{(D.28)}$$

or $\lambda_1 = 50$ and $\lambda_2 = -25$. The eigenvector corresponding to λ_1 can be found by insisting that

$$\begin{pmatrix} 23-50 & -36 \\ -36 & 2-50 \end{pmatrix} \begin{pmatrix} a \\ b \end{pmatrix} = 0 \tag{D.29}$$

or

$$\mathbf{v}_1 = \begin{pmatrix} 4/5 \\ -3/5 \end{pmatrix} \tag{D.30}$$

when normalized so that $\mathbf{v}_1 \cdot \mathbf{v}_1 = 1$; one similarly has

$$\mathbf{v}_2 = \begin{pmatrix} 3/5 \\ 4/5 \end{pmatrix} \tag{D.31}$$

and we confirm that $\mathbf{v}_1 \cdot \mathbf{v}_2 = 0$.

Finally, some useful identities involving the scalar and cross-products of vectors are

$$\mathbf{A} \times (\mathbf{B} \times \mathbf{C}) = \mathbf{B}(\mathbf{C} \cdot \mathbf{A}) - \mathbf{C}(\mathbf{A} \cdot \mathbf{B}) \tag{D.32}$$

$$\mathbf{A} \cdot (\mathbf{B} \times \mathbf{C}) = \mathbf{B} \cdot (\mathbf{C} \times \mathbf{A}) = \mathbf{C} \cdot (\mathbf{A} \times \mathbf{B}) \tag{D.33}$$

$$(\mathbf{A} \times \mathbf{B}) \cdot (\mathbf{C} \times \mathbf{D}) = (\mathbf{A} \cdot \mathbf{C})(\mathbf{B} \cdot \mathbf{D}) - (\mathbf{A} \cdot \mathbf{D})(\mathbf{B} \cdot \mathbf{C}) \tag{D.34}$$

D.2 GROUP THEORY

We conclude with a brief definition of a mathematical *group*. A set of elements $G = \{g_1, g_2, \ldots\}$ along with a binary operation (often called *group multiplication*) denoted by $g_1 \cdot g_2$ constitutes a group if it satisfies four conditions:

1. The product of any two group elements is also a group element, i.e., $g_3 = g_1 \cdot g_2$ is from G if g_1, g_2 are also; the group is closed under multiplication.
2. The group multiplication is associative, namely,

$$(g_1 \cdot g_2) \cdot g_3 = g_1 \cdot (g_2 \cdot g_3) \tag{D.35}$$

3. There is a unique group element, labeled I or the identity element, which satisfies

$$I \cdot g_i = g_i \cdot I = g_i \tag{D.36}$$

   for all $g_i \in G$.
4. Every group element, g_i, has a unique inverse element labeled g_i^{-1} that satisfies

$$g_i \cdot g_i^{-1} = g_i^{-1} \cdot g_i = I \tag{D.37}$$

The set of group elements can be finite or infinite. Groups for which the multiplication gives the same answer in either order, i.e., for which $g_i \cdot g_j = g_j \cdot g_i$ for every pair of group elements is called a *commutative* or *Abelian group.*

D.3 PROBLEMS

PD.1 Show that $(\mathbf{A} \cdot \mathbf{B} \cdots \mathbf{Y} \cdot \mathbf{Z})^T = \mathbf{Z}^T \cdot \mathbf{Y}^T \cdots \mathbf{B}^T \cdot \mathbf{A}^T$.

PD.2 Show that the cross-product of two vectors can be written in the form

$$\mathbf{A} \times \mathbf{B} = \det \begin{pmatrix} \hat{\mathbf{i}} & \hat{\mathbf{j}} & \hat{\mathbf{k}} \\ A_x & A_y & A_z \\ B_x & B_y & B_z \end{pmatrix} \tag{D.38}$$

PD.3 Find the eigenvalues and eigenvectors of the Hermitian matrix

$$\mathsf{M} = \begin{pmatrix} 4 & 3 + 2i \\ 3 - 2i & -5 \end{pmatrix}$$

and show explicitly that the two eigenvectors are orthogonal. Note that the dot product of two *complex* vectors is defined via $\mathbf{v}_1^* \cdot \mathbf{v}_2$.

Hamiltonian Formulation of Classical Mechanics

In this appendix, we briefly review some aspects of the Hamiltonian formulation of classical mechanics. The Hamiltonian function for a single particle described by a single coordinate, call it x, is a function of x and the so-called *conjugate momentum*, p_x. Examples of such coordinate pairs include x and $p_x = m\dot{x}$ for translational motion, and θ and $p_\theta = L_z = mr^2\dot{\theta}$ for a rotational system. The Hamiltonian is written as $H = H(x, p_x)$, and x and p_x are initially considered as independent variables.

The dynamical equations of motion for $x(t)$ and $p_x(t)$ are *Hamilton's equations,* namely,

$$\frac{dx}{dt} = \dot{x} = \frac{\partial H}{\partial p_x} \tag{E.1}$$

$$-\frac{dp_x}{dt} = -\dot{p}_x = \frac{\partial H}{\partial x} \tag{E.2}$$

To see the equivalence to Newtonian mechanics, note that the Hamiltonian function

$$H(x, p_x) = \frac{p_x^2}{2m} + V(x) \tag{E.3}$$

gives the equations

$$\dot{x} = \frac{p_x}{m} \qquad \text{and} \qquad -\dot{p}_x = \frac{\partial V(x)}{\partial x} \equiv -F(x) \tag{E.4}$$

or

$$m\ddot{x} = \dot{p}_x = F(x) \tag{E.5}$$

The Hamiltonian function for the degrees of freedom of more than one particle (in one dimension) can be written

$$H = H(x_i, p_i) = \sum_i \frac{p_i^2}{2m_i} + \sum_i V_i(x_i) + \sum_{i>j} V_{ij}(x_i - x_j) \tag{E.6}$$

with the corresponding equations

$$\frac{dx_i}{dt} = \dot{x}_i = \frac{\partial H}{\partial p_i} \tag{E.7}$$

$$-\frac{dp_i}{dt} = -\dot{p}_i = \frac{\partial H}{\partial x_i} \tag{E.8}$$

For two functions that depend on the coordinates of a multivariable problem (and possibly the time coordinate explicitly), $g = g(x_i, p_i; t)$ and $h = h(x_i, p_i; t)$, the *Poisson bracket* is defined via

$$[g, h] = \sum_k \left(\frac{\partial g}{\partial x_k} \frac{\partial h}{\partial p_k} - \frac{\partial g}{\partial p_k} \frac{\partial h}{\partial x_k} \right) \tag{E.9}$$

Any such arbitrary function can depend on time either from an explicit t dependence or via the coordinates $x_i(t)$, $p_i(t)$; a convenient way of exhibiting the time development is

$$\begin{aligned} \frac{dg}{dt} &= \frac{\partial g}{\partial t} + \sum_k \left(\frac{\partial g}{\partial x_k} \frac{dx_k}{dt} + \frac{\partial g}{\partial p_k} \frac{dp_k}{dt} \right) \\ &= \frac{\partial g}{\partial t} + \sum_k \left(\frac{\partial g}{\partial x_k} \frac{\partial H}{\partial p_i} - \frac{\partial g}{\partial p_k} \frac{\partial H}{\partial x_i} \right) \\ &= \frac{\partial g}{\partial t} + [g, H] \end{aligned} \tag{E.10}$$

Note the similarity between this classical relation and Eqn. (13.87) for the time rate of change of expectation values of quantum operators.

Example H.1 Using the Poisson bracket formalism, we can show that angular momentum is conserved for a central potential in three dimensions. The Hamiltonian function is given by

$$H = \frac{1}{2m}\left(p_x^2 + p_y^2 + p_z^2\right) + V(r) \tag{E.11}$$

where $r = \sqrt{x^2 + y^2 + z^2}$. We note that the force is given $\mathbf{F} = -\nabla V(r)$, so that

$$\frac{\partial V(r)}{\partial x} = (-F)\frac{x}{r} \tag{E.12}$$

and so forth.

Considering, for definiteness, the z component of angular momentum given by $L_z = xp_y - yp_x$, one can show that

$$\begin{aligned} \frac{dL_z}{dt} &= \frac{\partial L_z}{\partial t} + [L_z, H] \\ &= \left[\frac{\partial L_z}{\partial x} \frac{\partial H}{\partial p_x} - \frac{\partial L_z}{\partial p_x} \frac{\partial H}{\partial x} \right] + \left[\frac{\partial L_z}{\partial y} \frac{\partial H}{\partial p_y} - \frac{\partial L_z}{\partial p_y} \frac{\partial H}{\partial y} \right] + \left[\frac{\partial L_z}{\partial z} \frac{\partial H}{\partial p_z} - \frac{\partial L_z}{\partial p_z} \frac{\partial H}{\partial z} \right] \\ &= \left[(p_y)\left(\frac{p_x}{m}\right) - (-y)\left(-F\frac{x}{r}\right) \right] + \left[(-p_x)\left(\frac{p_y}{m}\right) - (x)\left(-F\frac{y}{r}\right) \right] \\ &= 0 \end{aligned} \tag{E.13}$$

This relation also suggests that the Poisson bracket of two functions is the classical quantity that can be generalized in quantum mechanics to the commutator of two operators:

$$[\hat{g}, \hat{h}] \equiv \hat{g}\hat{h} - \hat{h}\hat{g} \tag{E.14}$$

via

$$[g, h]_{\text{Poisson}} \longrightarrow [\hat{g}, \hat{h}] = i\hbar[g, h]_{\text{Poisson}} \tag{E.15}$$

The Hamiltonian for a charged particle acted on by electromagnetic fields is written in terms of the potentials, $\phi(\mathbf{r}, t)$ and $\mathbf{A}(\mathbf{r}, t)$ via

$$H(\mathbf{r}, \mathbf{p}) = \frac{1}{2m}\left(\mathbf{p} - q\mathbf{A}(\mathbf{r}, t)\right)^2 + q\phi(\mathbf{r}, t) \tag{E.16}$$

To prove this requires one to show that the corresponding Hamilton's equations reproduce Newton's laws with the Lorentz force. The Hamiltonian in Eqn. (E.16) can be written more explicitly as

$$H = \frac{1}{2m}\left(p_x^2 + p_y^2 + p_z^2\right) - \frac{q}{m}\left(p_x A_x + p_y A_y + p_z A_z\right) + \frac{q^2}{2m}\left(A_x^2 + A_y^2 + A_z^2\right) + q\phi \tag{E.17}$$

Hamilton's equations for the x and p_x coordinates in this case become

$$\dot{x} = \frac{\partial H}{\partial p_x} = \frac{p_x}{m} - \frac{q}{m}A_x \qquad \text{or} \qquad m\dot{x} = p_x - qA_x \tag{E.18}$$

and

$$\begin{aligned} -\dot{p}_x &= \frac{\partial H}{\partial x} \\ &= -\frac{q}{m}\left(p_x\frac{\partial A_x}{\partial x} + p_y\frac{\partial A_y}{\partial x} + p_z\frac{\partial A_z}{\partial x}\right) + \frac{q^2}{2m}\left(A_x\frac{\partial A_x}{\partial x} + A_y\frac{\partial A_y}{\partial x} + A_z\frac{\partial A_x}{\partial x}\right) + q\frac{\partial \phi}{\partial x} \end{aligned} \tag{E.19}$$

These can be combined by differentiating Eqn. (E.18) with respect to t, provided one recalls that

$$m\ddot{x} = \dot{p}_x - q\frac{dA_x}{dt} = \dot{p}_x - q\left(\frac{\partial A_x}{\partial t} + \dot{x}\frac{\partial A_x}{\partial x} + \dot{y}\frac{\partial A_x}{\partial y} + \dot{z}\frac{\partial A_x}{\partial z}\right) \tag{E.20}$$

since $\mathbf{A} = \mathbf{A}(x(t), y(t), z(t); t)$. The resulting equation for the x variable is then

$$m\ddot{x} = q\left(-\frac{\partial \phi}{\partial x} - \frac{\partial A_x}{\partial t} + \dot{y}\left[\frac{\partial A_y}{\partial x} - \frac{\partial A_x}{\partial y}\right] + \dot{z}\left[\frac{\partial A_z}{\partial x} - \frac{\partial A_x}{\partial z}\right]\right) \tag{E.21}$$

or

$$m\ddot{x} = q\left(\mathbf{E}_x + \left(\mathbf{v} \times \mathbf{B}\right)_x\right) \tag{E.22}$$

since

$$\mathbf{E} = -\nabla\phi(\mathbf{r}, r) - \frac{\partial}{\partial t}\mathbf{A}(\mathbf{r}, t) \qquad \text{and} \qquad \mathbf{B} = \nabla \times \mathbf{A}(\mathbf{r}, t) \tag{E.23}$$

The Hamiltonian formalism also provides a way to describe classical probability distributions for position and momentum that can be compared to their quantum counterparts (as discussed in Secs. 6.2, 10.4, and 13.5). One starts with the notion of the *classical phase space*. For one particle in one dimension, this is the space of possible values of x and p as in Fig. E.1. For N particles in three dimensions, it is a $6N$-dimensional space corresponding to the possible values of $\mathbf{r}_i$, $\mathbf{p}_i$. Given a set of initial conditions, the solutions obtained

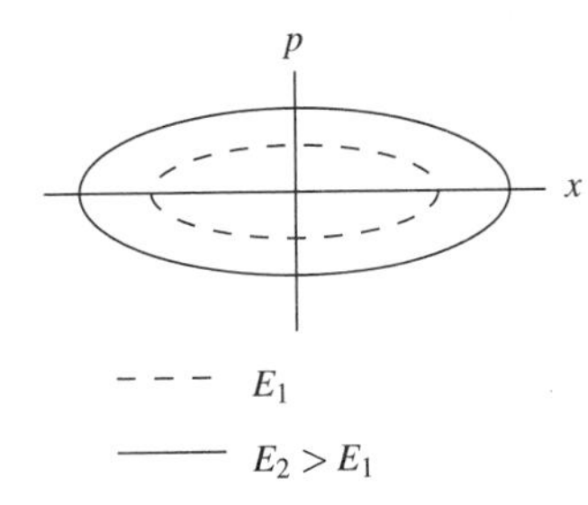

FIGURE E.1 Phase space diagram (plot of allowed values of p and x) for a single particle in a harmonic oscillator potential. The allowed "trajectories" in the parameter space in this case are determined by Eqn. (E.25).

from Newton's (or Hamilton's) equations for $x(t)$ and $p(t)$ trace out a trajectory in the phase space. For example, for a harmonic oscillator with initial conditions $x(0) = A$ and $\dot{x}(0) = 0$, the solutions are obviously

$$x(t) = A\cos(\omega t) \qquad \text{and} \qquad p(t) = -\frac{A\omega}{m}\sin(\omega t) \tag{E.24}$$

which gives the elliptical path in phase space shown in Fig. E.1. The form of this "trajectory" can be determined, even if we specify only the total energy, via the relation

$$E = H = \frac{p^2}{2m} + \frac{1}{2}m\omega^2 x^2 \tag{E.25}$$

The *phase space distribution*, $\rho(x, p)$, for a given value of E can then be written in the form

$$\rho(x, p) = K\delta(E - H(p, x)) \tag{E.26}$$

where the normalization constant is determined by the condition that

$$\int dx \int dp\rho(x, p) = 1 \tag{E.27}$$

The classical probability densities for position (x) or momentum (p) can be derived by integrating over the variable, which is not specified; for example,

$$P_{\text{CL}}(x) = \int dp\rho(x, p) \qquad \text{and} \qquad P_{\text{CL}}(p) = \int dx\rho(x, p) \tag{E.28}$$

As an example, consider the Hamiltonian with a general potential energy function $V(x)$

$$H = \frac{p^2}{2m} + V(x) \tag{E.29}$$

If we write $p_0(x) = \sqrt{E - V(x)}$, the corresponding classical distribution in x will be given by

$$\begin{aligned}
P_{\text{CL}}(x) &= K\int dp\ \delta\left(E - \left(\frac{p^2}{2m} - V(x)\right)\right) \\
&\propto \int dp\ \delta\left(p^2 - 2m(E - V(x))\right) \\
&\equiv \int dp\ \delta\left(p^2 - p_0^2(x)\right) \\
&= \frac{1}{2|p_0(x)|}\int dp\left[\delta(p - p_0(x)) + \delta(p + p_0(x))\right] \\
&\propto \frac{1}{\sqrt{E - V(x)}}
\end{aligned} \tag{E.30}$$

since $p_0^2(x) = 2m(E - V(x))$. This is the same result obtained in Secs. 6.2 and 10.4.2 using more intuitive methods.

E.1 PROBLEMS

PE.1 Write down Hamilton's equations for the angular variable pair θ and $p_\theta = mr^2\dot{\theta}$ where the Hamiltonian is

$$H = \frac{p_\theta^2}{2I} + V(\theta) \tag{E.31}$$

and show how the standard equations for rotational motion arise.

PE.2 Show that the Poisson bracket of the position and momentum coordinates of a multiparticle system satisfy

$$[x_i, x_j] = 0 \qquad [p_i, p_j] = 0 \qquad [x_i, p_j] = \delta_{i,j} \tag{E.32}$$

PE.3 Consider a harmonic oscillator for which the classical Hamiltonian is

$$H = \frac{p_x^2}{2m} + \frac{1}{2}m\omega^2 x^2 \tag{E.33}$$

Use Eqn. (E.11) to show that the function

$$\phi(x, p_x; t) = i\left(\log(A) - \log(x - ip/m\omega)\right) - \omega t \tag{E.34}$$

is actually independent of time for this system. What is the physical significance of this variable?

PE.4 For the classical Kepler problem, defined by the Hamiltonian

$$H = \frac{1}{2m}\mathbf{p}^2 - \frac{k}{r}$$

show that the *Lenz-Runge vector* defined by

$$\mathbf{R} = \frac{\hat{\mathbf{r}}}{r} - \left(\frac{1}{mk}\right)\mathbf{p} \times \mathbf{L}$$

is a constant of the motion, i.e., it is conserved. Do this by showing that $d\mathbf{R}/dt = 0$ using the Poisson bracket formalism.

PE.5 Using the Hamiltonian corresponding to a linear confining potential

$$H = \frac{p^2}{2m} + C|x|$$

and Eqn. (E.28), derive $P_{\mathrm{CL}}(p)$, and show that it agrees with the result obtained in Sec. 10.4.2.

PE.6 What does the classical phase space diagram look like for one particle in the 1D well; i.e., what is the analog of Fig. E.1? For the particle in the potential of PE.5? For an unbound free particle? For a particle subject to a constant force given by $V(x) = -Fx$?

APPENDIX F

Constants

F.1 PHYSICAL CONSTANTS

We collect values of various physical constants used in the text. For dimensionful quantities involved in electricity and magnetism, we consistently use MKSA or SI (*Système International*) units, unless specifically noted. In addition to the usual units of mass (kg), length (m), and time (s), for simplicity, we often express less-familiar dimensionful quantities in terms of force (Newton, N), energy (Joule, J), and charge (Coulomb, C); this sometimes makes the checking of dimensions more straightforward.

Planck's constant	$\hbar$	$= 1.055 \times 10^{-34}$ J · s
		$= 6.528 \times 10^{-16}$ eV · s
	$h = 2\pi\hbar$	$= 6.652 \times 10^{-34}$ J · s
Speed of light	c	$= 2.9972 \times 10^{8}$ m/s
	$\hbar c$	$= 1973$ eV · Å
		$= 197.3$ MeV · F
Electron mass	m_e	$= 9.11 \times 10^{-31}$ kg
	$m_e c^2$	$= 0.511$ MeV
Proton mass	m_p	$= 1.67 \times 10^{-27}$ kg
	$m_p c^2$	$= 938.3$ MeV
Neutron mass	$m_n c^2$	$= 939.6$ MeV
Muon rest mass	$m_\mu c^2$	$= 105.7$ MeV
Fundamental charge	e	$= 1.60 \times 10^{-19}$ C
Electric permittivity	ϵ_0	$= 8.85 \times 10^{-12}$ $C^2/(N \cdot m^2)$
	$K = 1/4\pi\epsilon_0$	$= 8.98 \times 10^{9}$ $N \cdot m^2/C^2$
	Ke^2	$= 2.30 \times 10^{-28}$ J · m
		$= 14.4$ eV · Å
		$= 1.44$ MeV · F
Fine-structure constant	$\alpha = Ke^2\hbar c$	$= 1/137.0$
Permeability constant	μ_0	$= 4\pi \times 10^{-7}$ $N \cdot s^2C^2$
Flux quantum	Φ_B	$= 4.14 \times 10^{-15}$ $T \cdot m^2$
Boltzmann constant	k_B	$= 1.38 \times 10^{-23}$ J/(molecule · K)
Thermal energy at $T = 300$ K	$k_B T$	$= 1/39$ eV
Avogadro constant	N_0	$= 6.02 \times 10^{23}$ molecules/mole
Gas constant	$R = N_0 k_B$	$= 8.31$ J/(mole · K)

Bohr radius	a_0	$= 0.53$ Å
Rydberg constant	R_∞	$= 1.10 \times 10^7\ \text{m}^{-1}$
Rydberg energy	$E_0 = m_e c^2 \alpha^2/2$	$= 13.6$ eV
Gravitational constant	G	$= 6.67 \times 10^{-11}\ \text{N} \cdot \text{m}^2/\text{kg}^2$
Solar mass	$M_\odot$	$= 1.99 \times 10^{30}$ kg
Earth mass	M_e	$= 5.98 \times 10^{24}$ kg
Moon mass	M_m	$= 7.36 \times 10^{22}$ kg
Mean earth-sun distance	AU	$= 1.50 \times 10^{11}$ m

Some useful conversion factors are:

1 Å	$= 10^{-10}$ m
1 F	$= 10^{-15}$ m
1 ly(lightyear)	$= 9.46 \times 10^{15}$ m
1 pc(parsec)	$= 3.09 \times 10^{16}$ m
1 eV	$= 1.60 \times 10^{-19}$ J
1 barn	$= 10^{-28}\ \text{m}^2$
1 G (Gauss)	$= 10^{-4}$ T (Tesla)

Some often used prefixes for powers of ten are

P	peta	10^{15}	m	milli	10^{-3}
T	tera	10^{12}	μ	micro	10^{-6}
G	giga	10^{9}	n	nano	10^{-9}
M	mega	10^{6}	p	pico	10^{-12}
k	kilo	10^{3}	f	femto	10^{-15}

F.2 THE GREEK ALPHABET

Alpha	A	α	Nu	N	ν
Beta	B	β	Xi	Ξ	ξ
Gamma	Γ	γ	Omicron	O	o
Delta	Δ	δ	Pi	Π	π
Epsilon	E	ϵ	Rho	R	ρ
Zeta	Z	ζ	Sigma	Σ	σ
Eta	H	η	Tau	T	τ
Theta	Θ	θ	Upsilon	Υ	υ
Iota	I	ι	Phi	Φ	ϕ
Kappa	K	κ	Chi	X	χ
Lambda	Λ	λ	Psi	Ψ	ψ
Mu	M	μ	Omega	Ω	ω

F.3 GAUSSIAN PROBABILITY DISTRIBUTION

Finding the probability that a variable represented by a Gaussian probability density with mean value μ and standard deviation σ will have a value in some finite region (a, b) requires the evaluation of the "area under the curve," given by

$$\text{Prob}[x \in (a, b)] = \int_a^b dx\, P(x; \mu, \sigma) \tag{F.1}$$

where

$$P(x; \mu, \sigma) = \frac{1}{\sigma\sqrt{2\pi}} e^{-(x-\mu)^2/2\sigma^2} \tag{F.2}$$

Any such problem can be "standardized" in terms of a dimensionless variable by writing

$$z = \frac{x - \mu}{\sigma} \tag{F.3}$$

where z measures the value of x away from the mean μ in units of σ. All of the information required to evaluate such probabilities can be tabulated once and for all in the form of a cumulative probability distribution using this standardized normal random variable by calculating

$$F(z) = \frac{1}{\sqrt{2\pi}} \int_{-\infty}^{z} e^{-t^2/2}\, dt \tag{F.4}$$

This corresponds to the probability of finding the variable t anywhere in the interval $(-\infty, z)$. It is defined such that $F(0) = 0.5$ corresponding to half the probability being on either side of μ. The integral defining $F(z)$ can be evaluated numerically, and values are shown in Table A.1. They can be extended to negative values of z by using $F(-z) = 1 - F(z)$. Finally, the probability of finding the standardized variable in the interval $(z_{\text{min}}, z_{\text{max}})$ is given by

$$\text{Prob}[z \in (z_{\text{min}}, z_{\text{max}})] = F(z_{\text{max}}) - F(z_{\text{min}}) \tag{F.5}$$

TABLE F.1. *Values of the Cumulative Gaussian Probability Distribution Defined by the Integral in Eqn. (F.4)*

z	$F(z)$	z	$F(z)$	z	$F(z)$
0.0	0.5000	1.0	0.8413	2.0	0.9722
0.1	0.5398	1.1	0.8643	2.1	0.9821
0.2	0.5793	1.2	0.8849	2.2	0.9861
0.3	0.6179	1.3	0.9032	2.3	0.9893
0.4	0.6554	1.4	0.9192	2.4	0.9918
0.5	0.6913	1.5	0.9332	2.5	0.9938
0.6	0.7257	1.6	0.9452	2.6	0.9953
0.7	0.7580	1.7	0.9554	2.7	0.9965
0.8	0.7881	1.8	0.9641	2.8	0.9974
0.9	0.8159	1.9	0.9713	2.9	0.9981
1.0	0.8413	2.0	0.9722	3.0	0.9987

Example F.1 Normal Distributions As an example of the use of Table E.1, we can calculate the probability that a measurement of a variable corresponding to a Gaussian distribution with $\mu = 7$ and $\sigma = 2$ will find it in the interval (5.8, 9.4). The corresponding range in the standardized variables are

$$z_{\text{min}} = \frac{5.8 - 7}{2} = -0.6 \qquad z_{\text{max}} = \frac{9.4 - 7}{2} = 1.2 \tag{F.6}$$

The probability in this interval is

$$\begin{aligned} \text{Prob}[z \in (-0.6, 1.2)] &= F(1.2) - F(-0.6) \\ &= 0.8849 - (1.0000 - 0.7257) = 0.6016 \end{aligned} \tag{F.7}$$

or about 60% of the total.

What's Not in This Book

No single textbook can do complete justice to the wide range of topics, the varying levels of mathematical detail, or potential applications of quantum mechanics. We briefly review here some of the topics that have been omitted from this treatment, pointing the reader to some valuable references for further study.

Computer Simulations There now exist several very useful interactive computer packages that allow for the visualization of many aspects of quantum theory; some textbooks now come packaged with software included. Examples are Brandt and Dahmen (1985, 1994); Hiller, Johnston, and Styer (1995); and McMurry (1993).

Applications The use of quantum mechanics covers many disciplines; some relevant sources of more specialized material are:

- Solid-state physics: Kittel (1971) and Ashcroft and Mermin (1976)
- Atomic and molecular physics: Fano and Fano (1972)
- Nuclear physics: Krane (1988)
- Elementary particle physics: Perkins (1987) and Griffiths (1987)
- Quantum chemistry: Lowe (1993)

Foundational Issues in Quantum Theory This general title is meant to include such topics as quantum measurement theory (the famous "collapse of the wavefunction" or the "Schrödinger cat paradox") and the probabilistic interpretation of quantum mechanics. References include Park (1992), Ballentine (1990) and Gribbin (1984), and other sources cited therein.

History of Quantum Mechanics Scientists are often poorly educated when it comes to the history of their own field. The development of quantum mechanics is particularly well documented, and one source of original material is Waerden (1967). It is hard to imagine learning more about the history of 20th-century physics (much of which is directly related to the discovery and subsequent application of quantum theory) than by studying the history of the Nobel Prize awardees and their contributions to the scientific landscape of their times; the physics awardees and their accomplishments are described in some detail (until 1988) by Magill (1989).

Specific Topics We have not included several topics that are covered in many other texts. These include:

- Time-dependent perturbation theory generalizes the results in Chap. 11 for static problems and allows one to calculate transition rates, in decays or scattering problems; it includes the famous "Golden Rule" of Fermi; for discussions, see Gasiorowicz (1974) or Saxon (1968).
- Relativistic quantum mechanics is discussed by Saxon (1968) and especially Baym (1976) at the level of this text; complete discussions of the Dirac or Klein-Gordon equations are usually reserved for more advanced courses, often in graduate school.
- The interaction of radiation with matter (including such topics as spontaneous emission, lasers, and Raman scattering) is discussed in Baym (1976) and Gasiorowicz (1974).

Research News Really readable reviews of the results of recent research can be found in *Physics Today* and *Scientific American.* More detailed reports on many aspects of scientific progress covering many fields can be found in the journals *Nature* and *Science.* One can always hear about the latest breakthroughs from the *New York Times.* And finally, the number of World Wide Web sites related to science (as well as to everything else) is increasing at an exponential rate, so that keyword searches using almost any web browser will find some very useful home pages. The URL

http://www.phys.psu.edu/ROBINETT/robinett.html

is my home page and can be used as a starting point for physics-related searches.

References

Physics, like the arts and humanities, is a human endeavor, and no text should fail to acknowledge the accomplishments of the men and women who discovered, codified, or have helped us understand its content. In most cases, the complete titles for journal articles are included in order to give a better "flavor" of the original paper.

1. Abe, F. et al. (CDF Collaboration) (1995). Observation of top quark production in $p\bar{p}$ collisions with the collider detector at Fermilab, *Phys. Rev. Lett.* 74, 2626; see also, S. Abachi et al. (1995). Observation of the top quark, *Phys. Rev. Lett.* 74, 2632.
2. Abramowitz, M., and I. A. Stegun (1964). *Handbook of Mathematical Functions,* National Bureau of Standards.
3. Adhikari, S. K. (1986). Quantum scattering in two dimensions, *Am. J. Phys.* 54, 362.
4. Aguilera-Navarro, V. C., H. Iwamoto, E. Ley-Koo, and A. H. Zimerman (1981). Quantum bouncer in a closed court, *Am. J. Phys.* 49, 648.
5. Aharonov, Y., and D. Bohm, (1957). Significance of electromagnetic potentials in the quantum theory, *Phys. Rev.* 115, 485.
6. ——, and L. Susskind (1967). Observability of the sign change of spinors under 2π rotations, *Phys. Rev.* 158, 1237.
7. Alber, G., and P. Zoller (1991). Laser excitation of electronic wave packets in Rydberg atoms, *Phys. Rep.* 199, 231.
8. Anderson, J. B. (1975). A random-walk simulation of the Schrödinger equation: H_3^+, *J. Chem. Phys.* 63, 1499.
9. Ashcroft, N., and N. Mermin (1976). *Solid State Physics,* Saunders, Philadelphia.
10. Asturias, F. J., and S. R. Aragón (1985). The hydrogenic atom and the periodic table of the elements in two spatial dimensions, *Am. J. Phys.* 53, 893.
11. Backenstoss, G. (1979). Exotic atoms, in *Progress in Atomic Spectroscopy,* Part B, W. Hanle and H. Kleinpoppen, eds., Plenum, New York.
12. Ballentine, L. E. (1990). *Quantum Mechanics,* Prentice-Hall, Englewood Cliffs, NJ.
13. Barger, V., and M. Olsson (1995). *Classical Mechanics: A Modern Perspective,* 2d ed., McGraw-Hill, New York.
14. ——, and R. J. N. Phillips (1987). *Collider Physics,* Addison-Wesley, Redwood City, CA.
15. Baym, G. (1976). *Lectures on Quantum Mechanics,* Benjamin, Reading, PA.
16. Bernstein, H. J. (1967). Spin precession during interferometry of fermions and the phase factor associated with rotations through 2π radians, *Phys. Rev. Lett.* 18, 1102.
17. Bethe, H. A., and E. E. Salpeter (1957). *Quantum Mechanics of One- and Two-electron Atoms,* Springer-Verlag, Berlin.

18. Blinder, S. M. (1968). Evolution of a Gaussian wavepacket, *Am. J. Phys.* 36, 525.

19. Borowitz, S. (1967). *Fundamentals of Quantum Mechanics: Particles, Waves, and Wave Mechanics,* Benjamin, New York.

20. Boyer, T. H. (1969). Quantum electromagnetic zero-point energy of a conducting spherical shell and the Casimir model for a charged particle, *Phys. Rev.* 174, 1764.

21. Brandt, S., and H. D. Dahmen (1985). *The Picture Book of Quantum Mechanics,* Wiley, New York.

22. ——, and H. D. Dahmen (1994). *Quantum Mechanics on the Personal Computer,* 3d ed., Springer-Verlag, Berlin.

23. Bromley, D. A., J. A. Kuehner, and E. Almqvist (1961). Elastic scattering of identical spin-zero nuclei, *Phys. Rev.* 123, 878.

24. Brown, L. S. (1973). Classical limit of the hydrogen atom, *Am. J. Phys.* 41, 525.

25. Burrows, A. (1990). Neutrinos from supernova explosions, *Ann. Rev. Nucl. Part. Sci.* 40, 181.

26. Butkov, E. (1968). *Mathematical Physics,* Addison-Wesley, Menlo Park, CA.

27. Ceperley, D. (1984). A review of quantum Monte Carlo methods and results for Coulombic systems, in *Monte Carlo Methods in Quantum Problems,* M. H. Kalos, ed., Springer-Verlag, New York.

28. Chambers, R. G. (1960). Shift of an electron interference pattern by enclosed magnetic flux, *Phys. Rev. Lett.* 5, 3.

29. Chapman, M. S. et al. (1995). Optics and interferometry with Na_2 molecules, *Phys. Rev. Lett.* 74, 4783.

30. Chen, J. (1993). *Introduction to Scanning Tunneling Microscopy,* Oxford University Press, New York.

31. Churchill, J. N. (1978). Motion of an electron wave packet in a uniform electric field, *Am. J. Phys.* 46, 537.

32. Cohen-Tannoudji, C. N., and W. D. Phillips (1990). New mechanisms for laser cooling, *Phys. Today,* Sep., 34.

33. Clarke, A. (1972). *The Wind from the Sun,* Harcourt Brace, Orlando, FL.

34. Cocconi, Giuseppe, and P. Morrison (1959). Searching for interstellar communications, *Nature,* 184, 844.

35. Colella, R., A. W. Overhauser, and S. A. Werner (1975). Observation of gravitationally induced quantum interference, *Phys. Rev. Lett.* 34, 1472.

36. ——, and A. W. Overhauser (1980). Neutrons, gravity and quantum mechanics, *Am. Sci.* 68, 70.

37. Condon, E. U., and G. H. Shortley (1951). *The Theory of Atomic Spectra,* Cambridge University Press, Cambridge, MA.

38. Cooper, F., A. Khare, and U. Sukhatme (1995). Supersymmetry and quantum mechanics, *Phys. Rep.* 251, 267.

39. Crawford, F. S. (1968). *Waves,* Vol. 3 of the Berkeley Physics Series, McGraw-Hill, New York.

40. CRC (1995). *CRC Handbook of Chemistry and Physics,* D. Lide, ed., CRC Press, Boca Raton, FL.

41. Dabbs, J., J. Harvey, D. Paya, and F. Horstmann (1965). Gravitational acceleration of free neutrons, *Phys. Rev.* 139, B756.

42. Davis, L., A. S. Goldhaber, and M. M. Nieto (1975). Limit on the photon mass deduced from Pioneer-10 observations of Jupiter's magnetic field, *Phys. Rev. Lett.* 35, 1402.

43. DeLange, O. L., and R. E. Raab (1991). *Operator Methods in Quantum Mechanics,* Clarendon Press, Oxford, England.

44. Delbourgo, R. (1977). On the linear potential hill, *Am. J. Phys.* 45, 1110.

45. Delone, N. B., and V. P. Krainov (1994). *Multiphoton Processes in Atoms,* Springer-Verlag, Berlin.

46. Dobrzynski, L., and K. Blinowski (1994). *Neutrons and Solid State Physics,* Ellis Horwood (Simon & Schuster), New York.

47. Dutt, R., A. Khare, and U. Sukhatme (1988). Supersymmetry, shape invariance, and exactly solvable potentials, *Am. J. Phys.* 56, 163.
48. deLange, O. L., and R. E. Raab (1991). *Operator Methods in Quantum Mechanics,* Clarendon Press, Oxford, England.
49. Eisberg, R., and R. Resnick (assisted by David Caldwell and J. Richard Christman) (1974). *Quantum Physics of Atoms, Molecules, Solids, Nuclei, and Particles,* 2d ed., Wiley, New York.
50. Eberly, J. H. (1965). Quantum scattering theory in one dimension, *Am. J. Phys.* 33, 771.
51. Ehrenberg, H. F., R. Hofstadter, U. Meyer-Berkhour, D. G. Ravenhall, and S. E. Sobotten (1959). High-energy electron scattering and the charge distribution of Carbon-12 and Oxygen-16, *Phys. Rev.* 113, 666.
52. Elizalde, E., and A. Romeo (1991). Essentials of the Casimir effect and its computation, *Am. J. Phys.* 59, 711.
53. Fabre, C., M. Gross, J. M. Raimond, and S. Haroche (1983). Measuring atomic dimensions by transmission of Rydberg atoms through micrometre size slits, *J. Phys. B: At. Mol. Phys.* 16, L671.
54. Fano, U., and L. Fano (1972). *Physics of Atoms and Molecules: An Introduction to the Structure of Matter,* University of Chicago Press, Chicago.
55. Feynman, R. P. (1963). *The Feynman Lectures on Physics,* Addison-Wesley, Reading, MA.
56. ——, and A. R. Hibbs (1965). *Quantum Mechanics and Path Integrals,* McGraw-Hill, New York.
57. Freeman, R. R., and D. Kleppner (1974). Core polarization and quantum defects in high-angular momentum states of alkali atoms, *Phys. Rev.* A14, 1614.
58. French, A. P., and E. F. Taylor (1971). Qualitative plots of bound state wavefunctions, *Am. J. Phys.* 39, 961.
59. Gallagher, T. F., K. A. Safinya, F. Gounand, J. F. Delpech, W. Sandner, and R. Kachru (1982). Resonant Rydberg atom–Rydberg atom collisions, *Phys. Rev.* A25, 1905.
60. —— (1994). *Rydberg Atoms,* Cambridge University Press, Cambridge, MA.
61. Gamow, G. (1946). *Mr. Tompkins in Wonderland: Stories of c, G, and h,* Macmillan, New York. Reprinted 1965 in *Mr. Tompkins in Paperback,* Cambridge University Press, Cambridge, MA.
62. Gasiorowicz, S. (1974). *Quantum Physics,* Wiley, New York.
63. Gibble, K., and S. Chu (1993). Laser-cooled Cs frequency standard and a measurement of the frequency shift due to ultracold collisions, *Phys. Rev. Lett.* 70, 1771.
64. Gleick, J. (1988). *Chaos: Making a New Science,* Penguin, New York.
65. Goldstein J., C. Lebiedzik, and R. W. Robinett (1994). Supersymmetric quantum mechanics: examples with Dirac δ-functions, *Am. J. Phys.* 62, 612.
66. Gribbin, J. R. (1984). *In Search of Schrödinger's Cat: Quantum Physics and Reality,* Bantam, New York.
67. Griffiths, D. (1981). *Introduction to Electrodynamics,* Prentice-Hall, Englewood Cliffs, NJ.
68. —— (1987). *Introduction to Elementary Particles,* Harper & Row, Cambridge, MA.
69. Gutzwiller, M. C. (1990). *Chaos in Classical and Quantum Mechanics,* Springer-Verlag, New York.
70. Hautojärvi, P., and A. Vehanen (1979). Introduction to positron annihilation, in *Positrons in Solids,* P. Hautojärvi, ed., Springer-Verlag, New York.
71. Heller, E. J., and S. Tomsovic (1993). Postmodern quantum mechanics, *Phys. Today,* Jul., 38.
72. Heydenburg, N. P., and G. M. Temmer (1956). Alpha-Alpha scattering at low energies, *Phys. Rev.* 104, 123.
73. Hiller, J. R., I. D. Johnston, and D. F. Styer (1995). *Quantum Mechanics Simulations: The Consortium for Upper-level Physics Software,* Wiley, New York.
74. Hüfner, J., F. Scheck, and C. S. Wu (1977). in Muon Physics, V. Hughes and C. S. Wu, eds. Academic, New York.
75. Itzykson, C., and J.-B. Zuber (1980). *Quantum Field Theory,* McGraw-Hill, New York.

76. Jackson, J. D. (1975). *Classical Electromagnetism,* 2d ed., Wiley, New York.
77. Johnson, A., H. Ryde, and S. A. Hjorth (1972). Nuclear moment of inertia at high rotational frequencies, *Nucl. Phys.* A179, 753.
78. Jones, S. E., J. Rafelski, and H. Monkhorst (1989). *Muon-Catalyzed Fusion,* American Institute of Physics, New York.
79. Kastner, M. A. (1993). Artificial atoms, *Phys. Today,* Jan., 24.
80. Kaufmann, W. B. (1977). Strong interaction effects in hadronic atoms, *Am. J. Phys.* 45, 735.
81. Keith, D. W., C. R. Ekstrom, Q. A. Turchette, and D. E. Pritchard (1991). An interferometer for atoms, *Phys. Rev. Lett.* 66, 2693.
82. Kittel, C. (1971). *Introduction to Solid State Physics,* 4th ed., Wiley, New York.
83. Kleppner, D. (1986). An introduction to Rydberg atoms, in *Atoms in Unusual Situations,* J. P. Briand, ed., Plenum, New York.
84. Knecht, D. J., P. F. Dahl, and S. Messelt (1966). Proton-proton scattering. Revision and analysis of experimental measurements from 1.4 to 3.0 MeV, *Phys. Rev.* 148, 1031.
85. Krane, K. (1988). *Introductory Nuclear Physics,* Wiley, New York.
86. Kuiper, T. B. H. (1989). Resource Letter ETC-1: Extraterrestrial civilizations, *Am. J. Phys.* 57, 12.
87. Landé, A. (1951). *Quantum Mechanics,* Pitman, London.
88. Lapidus, I. R. (1970). One-dimensional model of a diatomic ion, *Am. J. Phys.* 38, 905.
89. ——(1982a). Quantum-mechanical scattering in two dimensions, *Am. J. Phys.* 50, 45.
90. ——(1982b). One-dimensional hydrogen atom in an infinite square well, *Am. J. Phys.* 50, 563.
91. ——(1982c). Resonance scattering from a double δ-function potential, *Am. J. Phys.* 50, 663.
92. ——(1987). Particle in a square well with a δ-function potential, *Am. J. Phys.* 55, 172.
93. Laughlin, R. B. (1983). Anomalous quantum Hall effect: an incompressible quantum fluid with fractionally charged excitations, *Phys. Rev. Lett.* 50, 1395.
94. Lieb, E. (1976). The stability of matter, *Rev. Mod. Phys.* 48, 553.
95. Lieber, M. (1975). Quantum mechanics in momentum space: an illustration, *Am. J. Phys.* 43, 486.
96. Lighthill, M. J. (1959). *Introduction to Fourier Analysis and Generalised Functions,* Cambridge University Press, Cambridge, England.
97. Lowe, J. (1993). *Quantum Chemistry,* 2nd ed., Academic, Boston.
98. Lukosz, W. (1971). Electromagnetic zero-point energy and radiation pressure for a rectangular cavity, *Physica* 56, 109.
99. Lyons, L. (1986). *Statistics for Nuclear and Particle Physics,* Cambridge University Press, Cambridge, England.
100. Magill, F. N. (1989). *The Nobel Prize Winners; Physics,* 3 vols., Salem Press, Pasadena, CA; Vol. 1 (1901–1937), Vol. 2 (1938–1967), Vol. 3 (1968–1988).
101. Mathews, J., and R. L. Walker (1970). *Mathematical Methods of Phsyics,* 2d ed., Benjamin, Menlo Park, CA.
102. Marion, J. B., and S. T. Thornton (1988). *Classical Dynamics of Particles and Systems,* 3d ed., Harcourt Brace, Fort Worth, TX.
103. McDonald, S. W., and A. N. Kaufman (1979). Spectrum and eigenfrequencies for a Hamiltonian with stochastic trajectories, *Phys. Rev. Lett.* 42, 1189.
104. ——, and A. N. Kaufman (1988). Wave chaos in the stadium: statistical properties of short-wave solutions of the Helmholtz equation, *Phys. Rev.* A37, 3067.
105. McMurry, S. M. (1933). *Quantum Mechanics,* Addison-Wesley, Reading, MA.
106. Merli, P. G., G. F. Missiroli, and G. Pozzi (1976). On the statistical aspect of electron interference phenomena, *Am. J. Phys.* 44, 306.
107. Millikan, R. A. (1916). A direct photoelectric determination Of Planck's "h," *Phys. Rev.* 7, 355.
108. ——, and C. F. Eyring (1926). Laws governing the pulling of electrons out of metals by intense electrical fields, *Phys. Rev.* 27, 51.

109. Muller, H. G., P. Agostini, and G. Petite (1992). Multiphoton ionization, in *Atoms in Intense Laser Fields,* M. Gavrila, ed., Academic, Boston.

110. Nauenberg, M. (1989). Quantum wave packets on Kepler elliptic orbits, *Phys. Rev.* A40, 1133.

111. ——, C. Stroud, and J. Yeazell (1994). The classical limit of an atom, *Sci. Am.* 270, Jun., 44.

112. Nussenzveig, H. M. (1992). *Diffraction Effects in Semiclassical Scattering,* Cambridge University Press, New York.

113. Ohanian, H. C. (1990). *Principles of Quantum Mechanics,* Prentice-Hall, Englewood Cliffs, NJ.

114. Park, D. (1992). *Introduction to the Quantum Theory,* McGraw-Hill, New York.

115. Partain, C. L., R. R. Price, J. A. Patton, M. V. Kulkarni, and A. E. James (1988). *Magnetic Resonance Imaging,* Saunders, Philadelphia.

116. Particle Data Group (1994). Review of particle properties, *Phys. Rev.* D50, 1173.

117. Peebles, P. J. E. (1992). *Quantum Mechanics,* Princeton University Press, Princeton, NJ.

118. Perkins, D. H. (1987). *Introduction to High Energy Physics,* 3d ed., Addison-Wesley, Reading, MA.

119. Peierls, R. (1955). *Quantum Theory of Solids,* Oxford, New York.

120. Pippard, A. B. (1978). *The Physics of Vibration, Vol. 1,* Cambridge University Press, Cambridge, MA.

121. ——(1983). *The Physics of Vibration, Vol. 2,* Cambridge University Press, Cambridge.

122. Press, W. H., B. P. Flannery, S. A. Teukolsky, and W. T. Vetterling (1987). *Numerical Recipes: The Art of Scientific Computing,* Cambridge University Press, Cambridge, MA.

123. Prodan, J., A. Migdall, W. D. Phillips, I. So, H. Metcalf, and J. Dalibard (1985). "Stopping atoms with laser light," *Phys. Rev.* 54, 992.

124. Quigg, C., and J. L. Rosner (1979). "Quantum mechanics with applications to quarkonia," *Phys. Rep.* 56, 167.

125. Rabi, I. I., J. R. Zacharias, S. Millman, and P. Kusch (1938). "A new method for measuring nuclear magnetic moment," *Phys. Rev.* 53, 318.

126. Ramberg, E., and G. A. Snow (1990). "Experimental limit on a small violation of the Pauli principle," *Phys. Lett.* B238, 438.

127. Rauch, H., Z. Zeilinger, G. Badurek, A. Wilfing, W. Bauspiess, and U. Bonse (1975). "Verification of coherent spinor rotation of fermions," *Phys. Lett.* 54A, 425.

128. Reed, M.A. (1993). "Quantum dots," *Sci. Am.,* Jan., 118.

129. Reitz, J. R., F. J. Milford, and R. W. Christy (1993). *Foundations of Electromagnetic Theory,* Addison-Wesley, New York.

130. Robinett, R. W. (1995). "Quantum and classical probability distributions for position and momentum," *Am. J. Phys.* 63, 823.

131. ——(1996a). "Visualizing the solutions for the circular infinite well in quantum and classical mechanics," *Am. J. Phys.* 64, 432.

132. ——(1996b, in press). "Quantum mechanical time-development operator for the uniformly accelerated particle," *Am. J. Phys.* (to appear).

133. Roy, C. L., and A. B. Sannigrahi (1979). "Uncertainty relation between angular momentum and angle variable," *Am. J. Phys.* 47, 965.

134. Sakurai, J. J. (1994). *Modern Quantum Mechanics* (Revised edition), Addison-Wesley, Reading.

135. Saxon, D. S. (1968). *Introductory Quantum Mechanics,* McGraw-Hill, New York.

136. Schiff, L. I. (1968). *Quantum Mechanics,* McGraw-Hill, New York.

137. Schrödinger, E. (1943). "The earth's and the sun's permanent magnetic fields in the unitary field theory," *Proc. Roy. Irish Acad.* A49, 135.

138. Schwabl, F. (1990). *Quantum Mechanics,* Springer-Verlag, Berlin.

139. Segre, C. U., and J. D. Sullivan (1976). "Bound-state wave packets," *Am. J. Phys.* 44, 729.

140. Shapiro, S., and S. Teukolsky (1983). *Black Holes, White Dwarfs, and Neutron Stars,* Wiley, New York.

141. Sparnaay, M. J. (1958). "Measurement of attractive forces between flat plates," *Physica* 24, 751.

142. Staelin, D. H., and E. C. Reifenstein III (1968). "Pulsating radio sources near the Crab Nebula," *Science* 162, 1481.

143. Staudenmann, J., S. Werner, R. Colella, and A. W. Overhauser (1980). "Gravity and inertia in quantum mechanics," *Phys. Rev.* A21, 1419.

144. Stebbings, R. F., and F. B. Dunning (1983). *Rydberg States of Atoms and Molecules,* Cambridge University Press, Cambridge.

145. Stroscio, J. A., and D. M. Eigler (1991). "Atomic and molecular manipulation with the scanning tunneling microscope," *Science* 254, 1319.

146. ——, and W. J. Kaiser (editors) (1993). *Scanning Tunneling Microscopy,* Academic Press, Boston.

147. Stöckman, H.-J., and J. Stein (1990). " 'Quantum' chaos in billiards studied by microwave absorption," *Phys. Rev. Lett.* 64, 2215.

148. Swift, J. (1726). *Travels into several remote Nations of the World, in Four Parts, by Lemuel Gulliver,* Printed for Benj. Motte, London. For a modern edition, using a later edition, see *The Annotated Gulliver's Travels* (1980), edited by Isaac Asimov, Clarkson Potter, New York.

149. Tsong, T. T. (1990). *Atom Probe Field Ion Microscopy,* Cambridge University Press, Cambridge.

150. ——, and Müller, A. (1968). *Field ion microscopy,* American Elsevier, New York.

151. Urey, H. C., F. G. Brickwedde, and G. Murphy (1932). "A Hydrogen isotope of mass 2 and its concentration," *Phys. Rev.* 40, 1.

152. Van Vleck, J. H. (1922). "The dilemma of the helium atom," *Phil. Mag.* 44, 842.

153. von Klitzing, K. (1987). "The quantum Hall effect" in *The Physics of the Two-dimensional Electron Gas,* edited by J. T. Devreese and F. M. Peeters, Plenum Press, New York.

154. Walker, F. W., D. Miller, and F. Feiner (1983). "Chart of the Nuclides," 13th Edition, Knolls Atomic Power Laboratory (for the Department of Energy), General Electric Company, San Jose.

155. Waerden, B. L. van der (1967). *Sources of Quantum Mechanics,* North-Holland, Amsterdam.

156. Weinberg, S. (1972). *Gravitation and Cosmology,* Wiley, New York.

157. Werner, S. A., R. Collela, A. W. Overhauser, and C. F. Eagen (1975). "Observation of the phase shift of a neutron due to precession in a magnetic field," *Phys. Rev. Lett.* 35, 1053.

158. Wolfram, S. (1991). *Mathematica: A System for Doing Mathematics by Computer,* Addison-Wesley, Reading, Ma.

159. Wright, J. (1992). *Space Sailing,* Gordon and Breach, Langhorn, Pa.

160. Zewail, A. (1988). "Laser femtochemistry," *Science* 242, 1645.

161. ——(1990). "The birth of molecules," *Sci. Am.* 263, December, 76.

162. Zimmerman, M. L., J. C. Castro, and D. Kleppner (1978). "Diamagnetic structure of Na Rydberg states," *Phys. Rev. Lett.* 40, 1083.

163. ——, M. G. Littman, M. M. Kash, and D. Kleppner (1979). "Stark structure of the Rydberg states of alkali-metal atoms," *Phys. Rev.* A20, 2251.

Index